ENGINEERING DESIGN ASPECTS IN GEOMECHANICS AND GROUND CONTROL

FOR MINING AND OTHER EXCAVATIONS

In honoring my beloved parents

Acknowledgments

Many investigators, authors, university colleagues, mining engineers, support manufacturers, geotechnical and ground control specialists have freely shared their knowledge. To all of them, from whom much of this text is evolved, the author is greatly appreciative. I am especially grateful for permission to reproduce several tables and figures used in the first edition of this book which have appeared elsewhere, including the support of various national and international organizations for some of the researches undertaken. Appropriate acknowledgment is also given individually in the lists of references within the text.

ENGINEERING DESIGN ASPECTS IN GEOMECHANICS AND GROUND CONTROL

FOR MINING AND OTHER EXCAVATIONS

Andy A. Afrouz

Library of Congress Control Number:		2014922661
ISBN:	Hardcover	978-1-5035-2947-2
	Softcover	978-1-5035-2948-9
	eBook	978-1-5035-2949-6

Print information available on the last page.

Rev. date: 04/01/2015

To order additional copies of this book, contact:
Xlibris
1-888-795-4274
www.Xlibris.com
Orders@Xlibris.com
650715

Prologue

This is a second edition of a published book which was previously entitled "Practical Handbook of Rock Mass Classification Systems and Modes of Ground Failure". In the second edition, the author has modified the book to address some of the practical problems facing the Mining and other fields, dealing with the stability of excavated surface and underground. These problems occur during and after a surface or underground excavation proceeds; necessitating adequate Ground Control for the safety of personnel and stability of the whole venture.

The primary purposes of this book is the following:

1 - To discuss and utilize fourteen of the most widely used Rock Mass Classification (RMC) Systems;
2 - To express mathematical interrelationships between the various RMC Systems;
3 - To elaborate on the modes of Ground Failure under static, dynamic, passive and active conditions; and
4 - To provide practical examples and case studies on how the above noted (1), (2) and (3) can be utilized for design of excavations and ground control by its reinforcement or support for any types of surface and underground excavations.

In this undertaking, the original text has been modified as follows:

A - Added an extensive third Chapter to provide more practical examples on the utilization of the initial two chapters of the original book.
B - Changed the original title of the book to more closely reflect its contents. So it is entitled **"Engineering Design aspects in Geomechanics and Ground Control** (for Mining and other Excavations)".
C - Used larger font size, where needed, to provide more legible and comfortable readability; especially in some of the tabulations.
D - Have arranged so that the book becomes less expensive and more accessible to various users worldwide.

This book is for the engineers, researchers, consultants, teachers and students who are engaged in the fields of mining engineering, civil engineering, geological engineering, geomechanics, rock mechanics, ground control, petroleum engineering and those who are active in any kind of surface or underground excavation, drilling and blasting. The interested reader may pursue the subject in greater detail through the appendix and references cited in this book.

Andy A. Afrouz
USA – 2015

Author Biography

Andy A. Afrouz has a PhD from the UK, a post MS and MS from the USA and a BS from the UK. He was a past Professional Engineer from the Province of Ontario, Canada and Australia. He has extensive mining, rock mechanics and ground control experiences, with variety of mine products and at all the professional levels in the USA, UK, Iran, Canada, France, Zambia and Namibia. He has published numerous nationally and internationally peers reviewed researched publications, including 3 books. His interests are mainly in the science, technology, space ventures, human interactions, nature and literature.

Preface

Excavations are of two distinct types. These are:

A - **Service openings-** which include ramps and access to the below surface level activities such as building foundation for buildings below the surface level, driving highways or motorways on the side of a mountain, building bomb shelters, mine accesses, airways, rock haulage drives, shafts, tunnels, underground workshops and crusher chambers. They should have a permanent life life with relatively low maintenance costs.

B - **Production openings-** which have a temporary role in the operational life. They include mine stopes or rooms, stope accesses, service ways and mine production areas. In these cases, it is necessary to assure stability of the excavation only during the productive life of the excavation, which may be as short as a few months.

Appropriate to each of the above noted cases, five main groups of parameters can be considered for control of the excavated ground, safety, reinforcement and / or support purposes. They are:

1 - Rock type and their geomechanical properties.

2 - Primary existence of discontinuities and development of fractured zone around excavations causing stress transformation. This results in a rock mass different from the initial state of the ground before the excavation. Therefore rock mass classification according to its stability

and behavior around the excavation are important factors to account for and be predicted.

3 - Modes of rock mass failure surrounding an excavation, which may differ from the failure criterion of rock samples.

4 - Purpose of the excavation, its useful life and geometry, such as shape and size.

5 - Recognition that the ground control is not necessarily to prevent deformation or closure and failure of the rock mass surrounding the excavation. It is to ensure that large uncontrolled or unacceptable displacements and ultimately failure of the ground do not occur during the anticipated life of the excavation. This may be achieved by attention the the above noted parameters 1 to 4. Bearing accounts for the setting of an acceptable safety, limit of deformation, closure, repair and maintenance costs for the excavations. An appropriate ground reinforcement or method of support can then be selected and designed.

Contents

CHAPTER 1

ROCK MASS CLASSIFICATION (RMC) SYSTEMS

1.1 INTRODUCTION

Qualification and quantification of in situ rock mass are some of the most important aspects of site characterization for input to design of the ground support and reinforcement. These characteristics generally indicate the strength, deformability, and stability of the rock mass. For economical reasons, it is not feasible to measure fully the characteristics of a complex rock mass. This is obvious from many factors influencing the rock mass behavior listed in Table 1.1.[1.1]

Generally rock mass classification yields the following information for ground control application in mining:

1. To classify a particular rock mass into groups, on the basis of similar behavior
2. To provide a basis for determining physical and mechanical behavior of each group
3. To obtain quantitative data for the ground control purposes
4. To achieve a common standard for communication

Table 1.1 Factors Influencing Rock Mass Behavior[1.1]

---

1. Characteristics of Rock

- Rock material structure: lithology, anisotropy, cracks, and pores
- Rock mass structure: discontinuities (joints, faults, bedding planes, etc.), their type, orientation, continuity, roughness, waviness, spacing, and length
- Properties: mechanical, physical, and chemical properties of rock material, discontinuities, and rock mass

2. Specimen and Environmental Conditions

- Moisture content, temperature, and pore-pressure conditions
- Groundwater conditions and chemical environment (weathering)
- Specimen size and shape

3. State of Stress or Strain

- Magnitude of applied stress or strain
- Distribution of stress or strain (uniformly and non-uniformly distributed: tension, compression, bending or torsion; effects of specimen: ends, platens, and machine)

4. Method of Loading

- Type of loading (uniaxial, bi-axial, triaxial, tensile, compressive, or shear components)
- Rate of loading: slow (static) loading — < 10 MPa/s; rapid (dynamic) loading — 10 to 10^5 MPa/s; instantaneous (impact) loading — > 10^5 MPa/s
- Pattern of loading: constant load, gradually increasing monotonically, repetitive (fatigue), pulse, or alternating

---

1.2 CLASSIFICATION OF DEGREE OF WEATHERING

There are five design conditions that define the degree of weathering. These are:

A. MICRO FRESH STATE — determined in the field by means of a 10-power hand lens. This condition exists when there is no oxidation alteration of any of the mineral components in the rock. It is desirable, but not necessary, to make this determination for the ground reinforcement purposes, e.g., cable bolting, except for investigations of rock fragmentation, and concrete aggregate.

A. VISUALLY FRESH STATE — the condition that is not expected to change within the economic limits of the excavation. The mineral components are evaluated with the naked eye. The rock material has a uniform color, usually with shades of grey, green, blue, or black. Such an evaluation is usually acceptable for all excavation and ground reinforcement designs. This class of rocks is representative of maximum unit weight, least relative water absorption, and maximum strength.

B. STAINED STATE — denotes that the rock material is partly or completely discolored due to oxidation. It is weakened from its intact state, but cannot be broken or remolded by means of finger pressure. The mineral components are usually shades of yellow or brown and have a reduced unit weight (expressed as a percentage of the class B unit weight), and a higher absorption of water than the condition B noted above. The strength of rock mass under this condition may or may not vary from that of the class B.

C. PARTIALLY DECOMPOSED STATE — is a condition when the rock can be broken or remolded by means of finger pressure. Its strength is less than the class C. It is weathered and has a higher absorption of water than the class C rocks.

D. COMPLETELY DECOMPOSED STATE — is a state when rocks change to soil. In this case, the soil classification and mechanism is predominant.

1.3 CLASSIFICATION OF DISCONTINUITIES

For geotechnical purpose, a discontinuity may be defined as a boundary or break within the soil or rock mass which marks a change in the mass properties and thereby a change in engineering characteristics. This definition includes features such as lithological boundaries, bedding planes, joints, faults, unconformities, and orientation of micro-fractures. Classification of most common types of discontinuity and their geotechnical significance is noted in Tables 1.2 and 1.3.[1.2] A typical proforma for the collection and recording of discontinuity characteristics is given in Table 1.4.[1.2]

Table 1.2 Classification of Discontinuities Common to All Rock Types[1.2]

Discontinuity type	Physical characteristics	Geotechnical aspects	Comments
Tectonic joints	Persistent fractures resulting from tectonic stresses. Joints often occur as related groups or "sets". Joint systems of conjugate sets may be explained in terms of regional stress field.	Tectonic joints are classified as "shear" or "tensile" according to probable origin. Shear joints are often less rough then tensile joints. Joints may die out laterally resulting in impertinence and high strength.	May only be extrapolated confidently where systematic and where geological origin is understood.
Faults	Fractures along which displacement has occurred. Any scale from millimeters to hundreds of kilometers. Often associated with zones of sheared rock.	Often low shear strength particularly where slickensided or containing gouge. May be associated with high groundwater flow or act as barriers to flow. Deep zones of weathering occur along faults. Recent faults may be seismically active.	Mappable, especially where rocks either side can be matched. Major faults often recognized as photo lineations due to localized erosion.
Sheeting joints	Rough, often widely spaced fractures; parallel to the ground surface; formed under tension as a result of unloading.	May be persistent over tens of meters. Commonly adverse (parallel to slopes). Weathering concentrated along them in otherwise good quality rock.	Readily identified due to individuality and relationship with topography.

Lithological boundaries	Boundaries between different rock types. May be of any angle, shape, and complexity according to geological history.	Often mark distinct changes in engineering properties such as strength, permeability and degree and style of jointing. Commonly form barriers to groundwater flow.	Mappable allowing interpolation and extrapolation providing the geological history is understood.

Table 1.3 Classification of Discontinuities for Particular Rock Types[1.2]

Rock or soil type	Discontinuity type	Physical characteristics	Geotechnical aspects	Comments
Sedimentary	Bedding planes/ bedding plane joints	Parallel to original deposition surface and making a hiatus in deposition. Usually almost horizontal in unfolded rocks.	Often flat and persistent over tens or hundreds of meters. May mark changes in lithology, strength, and permeability. Common close, tight, with considerable cohesion. May become open due to weathering and unloading.	Geologically mappable and, therefore, may be extrapolated providing structure understood. Other sedimentary features such as ripple marks and mud-cracks may aid interpretation and affect shear strength.
	Shale y cleavage	Close parallel discontinuities formed in mudstones during diagnosis and resulting in fissility.		
	Random fissures	Commonly in recent sediments probably due to shrinkage and minor shearing during consolidation. Not extensive but important mass feature.	Controlling influence for strength and permeability for many clays.	Best described in terms of frequency
Igneous	Cooling joints	Systematic sets of hexagonal joints perpendicular to cooling surfaces are common in lavas and sills.	Columnar joints have regular pattern so are easily dealt with. Other joints often widely spaced with variable orientation and nature	Either entirely predictable or fairly random.

		Larger intrusions typified by doming joints and cross joints.		
Metamorphic	Slaty cleavage	Closely spaced, parallel and persistent planar integral discontinuities in fine-grained strong rock.	High cohesion where intact but readily opened to weathering or unloading. Low roughness.	Formed by regional stresses and, therefore, mappable over wide areas.
	Schistosity	Granulate or wavy foliation with parallel alignment of minerals in coarser-grained rocks.	Often foliations coated with minerals such as talc and chlorate giving a low shear strength.	Less mappable than slaty cleavage, but general trends recognizable.

The discontinuity rating is related to the blockiness and is the sum of rating for rock quality designation (RQD) and discontinuity spacing, as shown in Figure 1.1.[1.3] If either type of data is lacking, it can be estimated from Figure 1.2.[1.4]

There are 5 types of joint rating used in the rock mass classification. They are described in the following sections.

1.3.1 Rating Due to Joint Spacing (JS)

The rock mass is classified according to the spacing between joints or the joint sets. To drive the rating from joint spacings (J_s), measured from diamond drilled cores, joints are classified in three groups: minimum spacing, intermediate spacing, and maximum spacing within a range of 0 to 30 degrees plane inclination from the major axis of the core.[1.5] Joints in any one group may represent one or more sets. If more than one set is present, this must be recognized to assess the correct J_s. The true distances obtained are used to arrive at the J_s rating by utilizing Figure 1.3. Ratings of 0 to 30 points for up to three-joint systems can be read off Figure 1.3. Where there is a system of four or more joints, the three closest spaced joints are used. If there is any doubt about the number of joints in a system, a three-joint system should be assumed. At the worst, this will give a conservative rating.

Table 1.4 Proforma for the Collection and Recording of Discontinuity Characteristics[1.2]

GENERAL INFORMATION

Size |_|_|_|_|_|_|_|_| Date (Day Month Year) |_|_|_|_|_|_| Operator |_|_| Discontinuity data sheet No. |_|_| of |_|_|

NATURE AND ORIENTATION OF DISCONTINUITIES

Chainage or No	Type	Dip	Dip direction	Persistence								Remarks

Type	Dip/Dip direction	Persistence	Aperture	Nature of infilling	Consistency of infilling: Soil strength	Consistency of infilling: Rock strength	Roughness	Waviness	Water
0. Fault zone 1. Fault 2. Joint 3. Cleavage 4. Schistosity 5. Shear 6. Fissure 7. Tension crack 8. Foliation 9. Bedding	(Expressed in meters)	(Expressed in meters)	1. Wide (>200 mm) 2. Mod. wide (60–200 mm) 3. Mod. narrow (20–60 mm) 4. Narrow (6–20 mm) 5. Very narrow (2–6 mm) 6. Ext. narrow (<2 mm) 7. Tight	1. Clean 2. Surface staining 3. Noncohesive 4. Cohesive 5. Cemented 6. Calcite 7. Chlorite, talc 8. Others — specify	1. Soft 2. Firm 3. Stiff 4. Hard	5. Weak 6. Mod. strong 7. Strong 8. Very strong	1. Polished 2. Slickensided 3. Smooth 4. Rough 5. Defined ridges 6. Small slope 7. Very rough	Express wavelength and amplitude in meters	1. Dry 2. Seepage 3. Slight flow <0.1 l/sec 4. Mod. flow 0.1–1 l/sec 5. High flow >01 l/sec

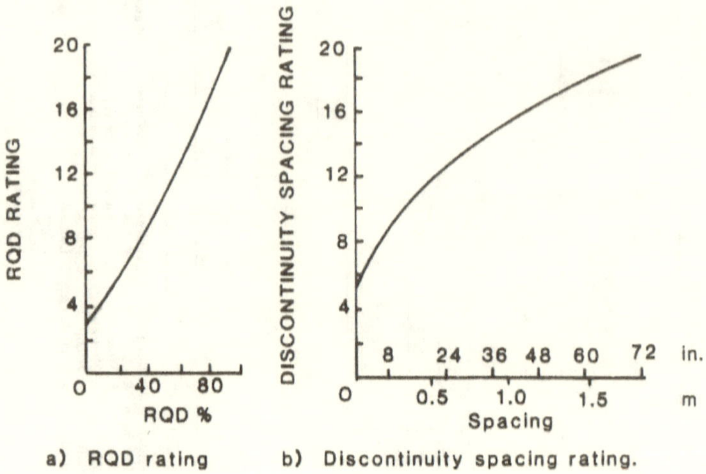

Figure 1.1. Ratings for discontinuity density.[1.3]

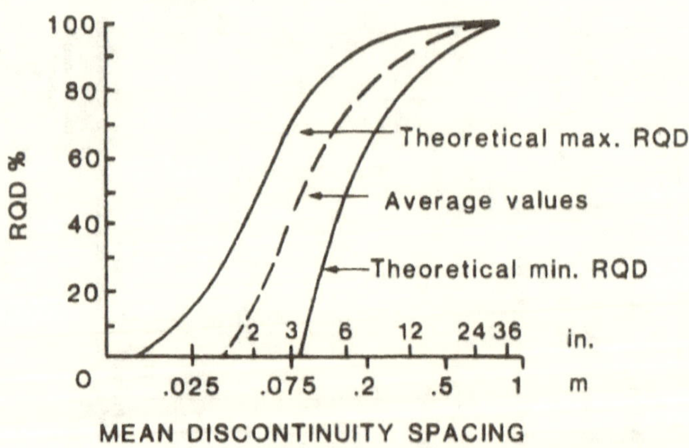

Figure 1.2. Theoretical relationship between RQD and discontinuity spacing.[1.4]

Rock quality classification (RQC) in relation to spacing between discontinuity and its rating, J_s, is given in Table 1.5. The rating of discontinuity condition from 0 to 30 points is given in Table 1.6.

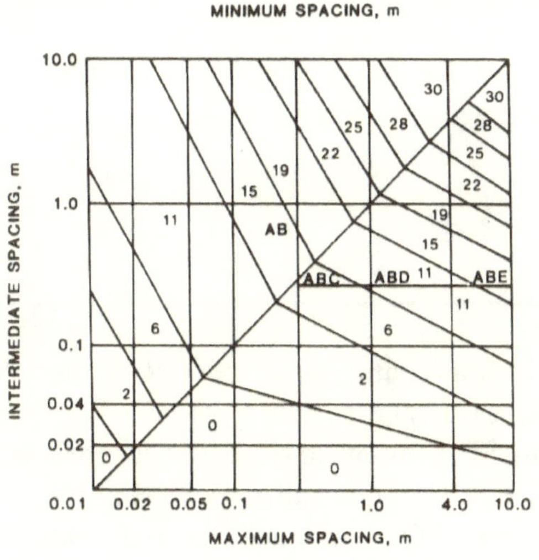

Figure 1.3. Ratings (J_s) for multi-joint systems.[1.5]

Table 1.5 Classification of Rock Quality (RQC) in Relation to the Spacing of Discontinuity

RQC	Discontinuity spacing, m	Rating (J_s)	Rock mass grading
Excellent	>3	30	Solid
Good	1–3	25	Massive
Fair	0.3–1	20	Blocky/seamy
Poor	0.05–0.3	10	Fractured
Very poor	<0.05	5	Crushed and shattered

Modified from References 1.6 and 1.7.

Table 1.6 Rating of Discontinuity Conditions[1.3]

Description of discontinuity					
Wall roughness	VR	R-SR	SR	SM-SK	SM
Wall separation	None	Hairline	Hairline to 2 mm	2–6 mm	6 mm
Joint filling	None	None	Minor clay	Stiff clay gouge	Soft Clay gouge
Wall weathering	F	SL	SO	SO	VS
Rating	30	25	20	10	0

Key: VR = very rough (coarse sandpaper); R = rough (medium or fine sandpaper);
SM = smooth; SR = smooth to slightly rough; SK = slickensided, shiny; F = hard, unweathered, fresh; SL = hard, slightly weathered; SO = softened, strongly weathered; VS = very soft or decomposed.

1.3.2 Rating Due to Joint-Set Number (Jn)

The rock mass is classified according to number of joint sets encountered. It is usually affected by foliation, schistosity, slatey cleavage or bedding, etc. If strongly developed, these parallel joints should be counted as a complete joint set. However, if there are few joints visible, or only occasional breaks appear in the cores due to these features, then it will be more appropriate to count them as random joints when evaluating J_n, as in Table 1.7. For practical purposes, in underground intersections, rating of $3J_n$, and in portals, rating of $2J_n$ should be considered.

1.3.3 Rating Due to Joint Roughness Number (Jr)

The rock mass is classified according to the roughness of the joint surfaces *in situ*. Table 1.8 gives the J_r values.[1.8] For the mean spacing of the relevant joint sets greater than 3 m, a rating of $J_r + 1$ should be used. For planar slickensided joints having lineations, provided the lineations are oriented for minimum strength, $J_r = 0.5$ can be used.

Table 1.7 Rating Due to Joint Set Number (J_n)[1.8]

Rock mass classification	J_n
Massive, no or few joints	0.5–1.0
One joint set	2
One joint set plus random	3
Two joint sets	4
Two joint sets plus random	6
Three joint sets	9
Three joint sets plus random	12
Four or more joint sets, random, heavily jointed, "sugar-cube", etc.	15
Crushed rock, earthlike	20

Note: For intersections use ($3.0 \times J_n$); for portals use ($2.0 \times J_n$).

Table 1.8 Rating Due to Joint Roughness Number (J_r)[1.8]

Rock mass classification	J_r
Discontinuous joints	4
Rough or irregular, undulating	3
Smooth, undulating	2
Slickensided, undulating	1.5
Rough or irregular, planar	1.5
Smooth, planar	1.0
Slickensided, planar	0.5
Zone containing clay minerals thick enough to prevent rock wall contact	1.0
Sandy, gravelly or crushed zone thick enough to prevent rock wall contact	1.0

Note: Add 1.0 if the mean spacing of the relevant joint set is greater than 3 m; $J_r = 0.5$ can be used for planar slickensided joints having lineations, provided the lineations are oriented for minimum strength.

1.3.4 Rating Due to Joint Alteration (Ja)

This is due to existence of any soft infilling such as clay, or alterations between each two-joint surfaces. Table 1.9 presents the J_a values for various rock and joint infilling. Ratio of J_r/J_a is important in determination of shear strength at the joint between two adjacent blocks.[1.8]. This ratio is relevant to the weakest significant joint set or clay filled discontinuity in the given zone. However, if the joint set or discontinuity with the minimum value of J_r/J_a is favorably oriented for stability, then a second less favorably oriented joint set or discontinuity may sometimes be of more significance, and its higher value of J_r/J_a should be used.

Table 1.9 Rating Due to Joint Alteration (J_a)[1.8]

	Rock mass classification	J_a	φ
	Rock Wall Contact		
A.	Tightly healed, hard, nonsoftening, impermeable filling, i.e., quartz or epidote	0.75	–
B.	Unaltered joint, walls, surface staining only	1.0	25–35°
C.	Slightly altered joint walls; non-softening mineral coatings, sandy particles, clay-free disintegrated	2.0	25–30°
	rock, etc.	3.0	20–30°
D.	Silty-, or sandy-clay coating, small clay fraction (non-soft)	4.0	8–16°
E.	Softening or low-friction clay mineral coatings, i.e., kaolin or mica; also chlorate, talc, gypsum, graphite, etc., and small quantities of swelling clays		
	Rock Wall Contact Before 10 cm Shear		
F.	Sandy particles, clay-free, disintegrated rock, etc.	4.0	25–30°
G.	Strongly over-consolidated nonsoftening clay mineral fillings (continuous, but <5 mm thickness)	5.0	16–24°
H.	Medium or low over-consolidation, softening, clay mineral fillings (continuous, but <5 mm thickness)		
J.	Value of J_a depends on percent of swelling clay-size particles in filling and access to water, etc.	8–12	6–12°
	No Rock Wall Contact When Sheared		
K.	Zones or bands of disintegrated or crushed rock and clay	6,8, or 8–12	6–24°
L.	(see G, H, J, for description of clay condition)		
M.		5.0	–
N.	Zones or bands of silty- or sandy-clay, small clay fraction (non-softening)	10–20	6–24°
O.	Thick, continuous zones or bands of clay (see G, H, J for description of clay condition		

1.3.5 Rating Due to Existence of Water in the Joint (Jw)

Ground water can strongly influence rock mass behavior. If an exploratory adit or pilot tunnel is available, measurements of water inflow or joint water pressure may be used to determine the rating increment directly as given in Tables 1.10 and 1.11. The drill core and drilling log can be used to classify rock mass to one of four water conditions. These are: completely dry, moist, water under moderate pressure, or severe water problem.[1.9]

Table 1.10 Roding Due to Joint Water (J_w)[1.8]

	Classification of joint water	J_w	Approx. water pressure (kg/cm^2)
A.	Dry excavations or minor inflow, i.e. < 5 1/min. locally	1.0	<0
B.	Medium inflow or pressure, occasional outwash of joint fillings	0.66	1–2.5
C.	Large inflow or high pressure in competent rock with unfilled joints	0.5	2.5–10
D.	Large inflow or high pressure, considerable outwash of joint fillings	0.33	2.5–10
E.	Exceptionally high inflow or water pressure at blasting, decaying with time	0.2–0.1	>10
F.	Exceptionally high inflow or water pressure continuing without noticeable decay	0.1–0.05	>10

Note: Factors C to F are crude estimates. Increase J_w if drainage measures are installed; special problems caused by ice formation are not considered.

Table 1.11 Increments of Rock Mass Rating Due to Ground Water Condition[1.9]

Inflow per 10m tunnel length (1/min)	OR	Joint water pressure divided by major principal stress	OR	General condition	Rating
None		0		Completely dry	10
				Moist	7
25		0.0–0.2		Water under	4
25–125		0.2–0.5		moderate pressure	
125		0.5		Severe water problems	0

1.4 ROCK QUALITY DESIGNATION (RQD)

This is a quantitative index based on a core recovery procedure, in which the core recovery is determined by incorporation of only those pieces of core that are longer than 100 mm. The standard core size is at least BXM (42 mm diameter) and its length is longer than 2 m, RQD as a percentage can be expressed as follows:[1.10]

RQD % = 100 (Length of core in pieces > or = 100 mm) / Length of run (1.1)

The orientation of the fracture with respect to the borehole is important. If a 3m long BXM borehole is drilled perpendicular to five fracture sets spaced at 65 to 90 mm, with total core length of 0 m longer than 100 mm, the RQD is 0%. If the borehole is drilled at an inclination of 45° to vertical, the spacing between the same fractures will be 91.9 to 127.3 mm (averaging 109.6 mm) corresponding to the RQD of 100 [3 – (2 × 0.0919]/3 = 93.9%. This result is obviously incorrect, the correct procedure is to calculate length of the core pieces longer than 100 mm perpendicular to the fracture planes. It is important to note the following: (1) experience in the determination of RQD of core is necessary before attempting to use the data; (2) where a joint face forms the sidewall of an excavation, the opposite wall should be assessed; (3) weaker bedding planes do not necessarily break when cored; (4) one should not be misled by basting fractures; and (5) shear zones greater than a 2-m width must be classified separately.[1.5] Rock quality classification in terms of RQD is given in Table 1.12.

Table 1.12 Relationship Between RQD and RQC (Rock Quality Condition)

RQD%	RQC	Rating
91–100	Excellent	20
76–90	Good	17
51–75	Fair	13
25–50	Poor	8
1–25	Very poor	3
0–1	Extremely poor	>0.1

Modified from References 1.9 and 1.11.

It is important to note that the RQD determined from the core logs may be different from that obtained *in situ.*

This is due to the influence of factors such as: sample size, core drilling, circulation of water or coolant in coring, drill hole orientation, and excavation techniques. The *in situ* values of RQD were suggested by the author[1.12] to be determined from the following general empirical expression:

$$RQD\% = A^x - B^y \cdot D_v \qquad (1.2)$$

where:

D_v = total number of discontinuities per cubic meter of rock mass, where the plane of discontinuities is not perpendicular to the direction of maximum principal stress.

A, B, x, y = constants relating to the above noted factors in such a way that:

$A^x = 105$ to 120, and $B^y = 2$ to 12.

Determination of *in situ* values of RQD in clay-free rock mass along tunnel was suggested as follows:[1.13]

$$RQD\% = 115 - 3.3\,J_v \qquad (1.3)$$

where J_v = total number of joints per cubic meter of rock mass. For $J_v < 4.5$, RQD = 100.

In practice, for conventional tunnels of up to 10-m width, if the *in situ* RQD > 90% there is no need for support, and for 75 < RQD < 90% bolting will be sufficient.[1.13]

1.5 ROCK MASS QUALITY (Q) SYSTEM

This system was developed at the Norwegian Geotechnical Institute (NGI) from a study of over 200 tunnels. It covers the whole spectrum of rock mass qualities from heavy squeezing ground to sound unjointed rocks, including 13 igneous, 24 metamorphic, 9 sedimentary rock types, and various types of joint fillings.[1.14] The system is also called NGI-classification, or Q-rating system. It is mainly developed to rate rock mass quality in the vicinity of tunnels. The six parameters chosen to describe the rock mass quality (Q) are combined as follows:[1.15]

$$Q = (RQD / J_n)\,(J_r / J_s)\,(J_w / SRF) \qquad (1.4)$$

where:

SRF	=	stress reduction factor (see Table 1.13).
J_w / SRF	=	active stress
J_r / J_s	=	inter-block shear strength ($\simeq \tan \phi$)
RQD / J_n	=	block size

Table 1.13 Stress Reduction Factor (SRF) for Various Ground Conditions[1.15]

	Ground Condition	SRF
	I. Weakness zones intersecting excavation, which may cause loosening of rock mass when tunnel is excavated	
A.	Multiple occurrences of weakness zones containing clay or chemically disintegrated rock, very loose surrounding rock (any depth)	10
B.	Single weakness zones containing clay or chemically disintegrated rock (depth of excavation <50 m)	5
C.	Single weakness zones containing clay or chemically disintegrated rock (depth of excavation >50 m)	2.5
D.	Multiple shear zones in competent rock (clay-free), loose surrounding rock (any depth)	7.5
E.	Single shear zones in competent rock (clay-free) (depth of excavation <50 m)	5.0
F.	Single shear zones in competent rock (clay-free) (depth of excavation >50 m)	2.5
G.	Loose open joints, heavily jointed or "sugar cube" etc. (any depth)	5.0

Note: (1) Reduce these values of SRF by 25–50% if the relevant shear zones only influence but do not intersect the excavation.

II. Competent rock, rock stress problems

		$\sigma c / \sigma_1$	$\sigma t / \sigma_1$	SRF
H.	Low stress, near surface	> 200	> 13	2.5
J.	Medium stress	200 –10	13 – 0.66	1
K.	High stress, very tight structure (usually favorable to stability, may be unfavorable for wall stability)	10 –5	0.66 – 0.33	0.5 – 2.0
L.	Mild rock burst (massive rock)	5 –2.5	0.33 – 0.16	0.5-10.0
M.	Heavy rock burst (massive rock)	< 2.5	< 0.16	10.0 -20.0

Note: (2) For strongly anisotropic virgin stress field (if measured): when $5 \geq \sigma_1/\sigma_2 \geq 10$, reduce σ_c and σ_t to 0.8 σ_c and 0.8 σ_t. When $\sigma_1/\sigma_3 > 10$, reduce σ_c and σ_t to 0.6 σ_c and 0.6 σ_t, wcere: σ_c = unconfined compression strength, and σ_t = tensile strength (point load) and σ_1 and σ_3 are the major and minor principal stresses.
(3) Few case records available where depth of crown below surface is less than span width. Suggest SRF increase from 2.5 to 5 for such cases (see H).

III. Squeezing rock plastic flow of incompetent rock under the influence of high rock pressure

N.	Mild squeezing rock pressure	5-10
O.	Heavy squeezing rock pressure	10-20

IV. Swelling rock chemical swelling activity depending on presence of water

P.	Mild swelling rock pressure	5-10
R.	Heavy swelling rock pressure	10-15

In the rock mass quality system, the Q-values range from 0.001 for exceptionally poor rock masses up to 1000 for exceptionally good rocks.

- When a rock mass contains clay, the factor SRF appropriate to loosening loads should be evaluated (Table 1.13, I). In such cases the strength of intact rock is of little interest.
- When jointing is minimal and clay is completely absent, the strength of the intact rock may become the weakest link, and the stability will then depend on the ratios of σ_c/σ_1 and σ_t/σ_1 in the rock mass (Table 1.13, II).
- A highly anisotropic stress field promotes unfavorable stability conditions and is roughly accounted for as in Table 1.13 (see Note 2).
- In general, the σ_c and σ_t of the rock mass should be evaluated in the direction that is unfavorable to stability. This is especially important where highly anisotropic rock mass is encountered.
- If the *in situ* rock mass is water saturated or will become saturated in future excavations, then the test samples should be saturated.
- For rocks which deteriorate when exposed to the mine air or ground water, a conservative estimate of the SRF should be made.

In the Q-system, the Q-values range from 0.001 for exceptionally poor rock masses up to 1000 for exceptionally good rocks. Table 1.14 describes various ranges encountered.

Results of correlation between RQD and Barton's Q (obtained from five core logs totaling 160 meters in length, driven to the roof, floor, and face of a prospective tunnel in Norway) is summarized in Table 1.15.

1.5.1 Determination of Maximum Unsupported Excavation Span and their Stand-up Time Using Q-System

Investigation into the relationship between existing maximum unsupported excavation span (SPAN) and Q around the excavation standing up for more than 10 years, revealed the following relationship:[1.8]

Table 1.14 Description of Ranges in the Q-System[1.15]

0.001–0.01	Exceptionally poor
0.01–0.1	Extremely poor
0.1–1	Very poor
1–4	Poor
4–10	Fair
10–40	Good
40–100	Very good
100–400	Extremely good
400–1000	Exceptionally good

Table 1.15 Summary of Comparison Between RQD and Q System[1.8]

Rock quality	Best	Medium	Poor
J_n	3	4	9
J_r	2	2	1
J_a	1	2	4
J_w	1	1	0.66
SRF	1	1	2.5
RQD	100	90	70
Q	67	22	0.5

$$SPAN = 2Q^{0.66} = 2(ESR)Q^{0.4} \qquad (1.5)$$

where: ESR = excavation to support ratio, see Table 1.16;
$\propto$ 1/excavation factor of safety.

The maximum design of unsupported span for various rock mass quality and excavation support ratio is given in Figure 1.4. The six parallel lines correspond to various excavation types and support ratios of 0.8 to 5.[1.8]

The closeness with which an unsupported opening can be designed to the envelope of maximum design span will depend on the type and shape of excavation, its use, the Q-value, the degree of safety, and the stand-up time required. If the maximum design span is exceeded, or if some of the above conditional factors are not satisfied, the stand-up time will be less than 10 years, i.e., they will be temporary openings.[1.8] Figure 1.5 illustrates envelopes predicting stand-up time of an excavation in relation to its maximum design span and Q.

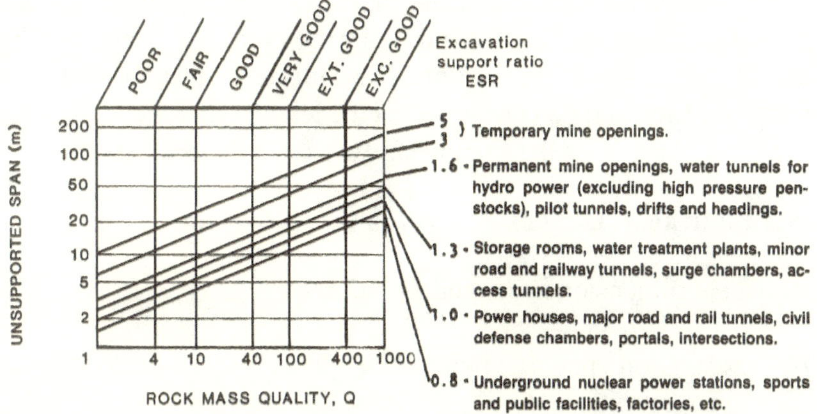

Figure 1.4. Recommended maximum unsupported excavation span for various rock mass quality and excavation support ratios.[1.8]

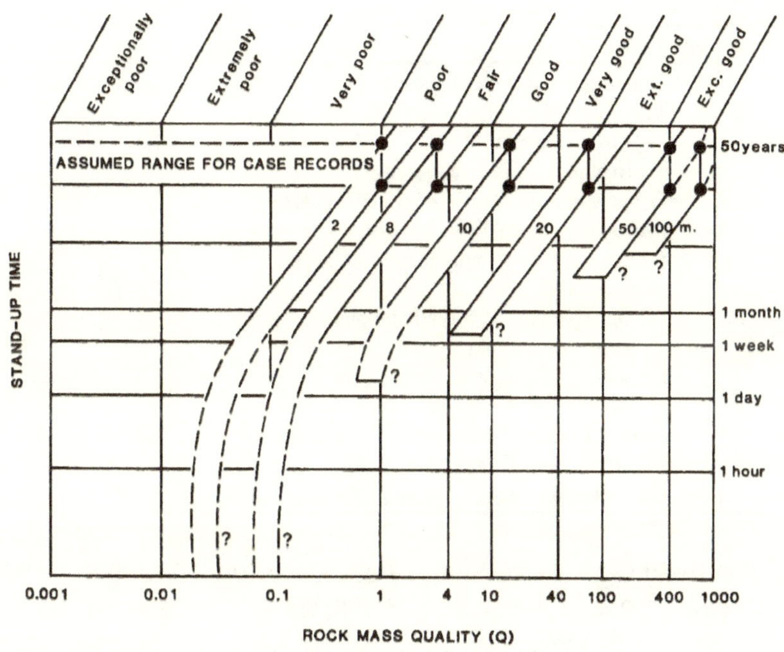

Figure 1.5. Prediction of stand-up time of excavations in relation to the rock mass quality (Q) and the maximum excavation span.[1.8]

The envelopes have been truncated at various time intervals as a concession to the approximate minimum construction periods of the various dimensions of excavation. The equivalent unsupported span at any one time can be considered as the length from the face to the supported zone, or as the span itself, whichever is smaller.

1.5.2 Equivalent Dimension (De) in the Q-System

Equivalent dimension is defined as follows:[1.16]

D_e = excavation span, diameter or height in meter / excavation to support ratio (1.6)

Note: For values of the excavation support ratio (ESR) see Table 1.16.

Table 1.16 The Excavation to Support Ratio (ESR) for Variety of Underground Openings[1.16]

Category	Type of excavation	ESR	No. of cases
A	Temporary mine openings, etc.	3–5	2
B	Vertical shafts		
	Circular section	2.5	1
	Rectangular/square section	2.0	1
C	Permanent mine openings, water tunnels for hydropower (exclude high pressure pen-stocks), pilot tunnels, drifts and headings for large excavations, etc.	1.6	83
D	Storage rooms, water treatment plants, minor road and railway tunnels, surge chambers, access tunnels, etc. (hemispherical caverns?)	1.3	25
E	Power stations, major road and railway tunnels, civil defense chambers, portals, intersections, etc.	1.0	79
F	Underground nuclear power stations, railway stations, sports and public facilities factories, etc.	0.8	2

The method of classifying a rock mass for the Q-system was developed by use of successive reanalysis of case records, until a consistent relationship was obtained between Q, the excavation dimensions, and the support actually utilized.[1.16] The results are given by means of a support chart (Figure 1.6). The box numbering 1 to 38 is used as a reference to the support categories. The support measures that are appropriate to each category are listed in Tables 1.17 to 1.20.

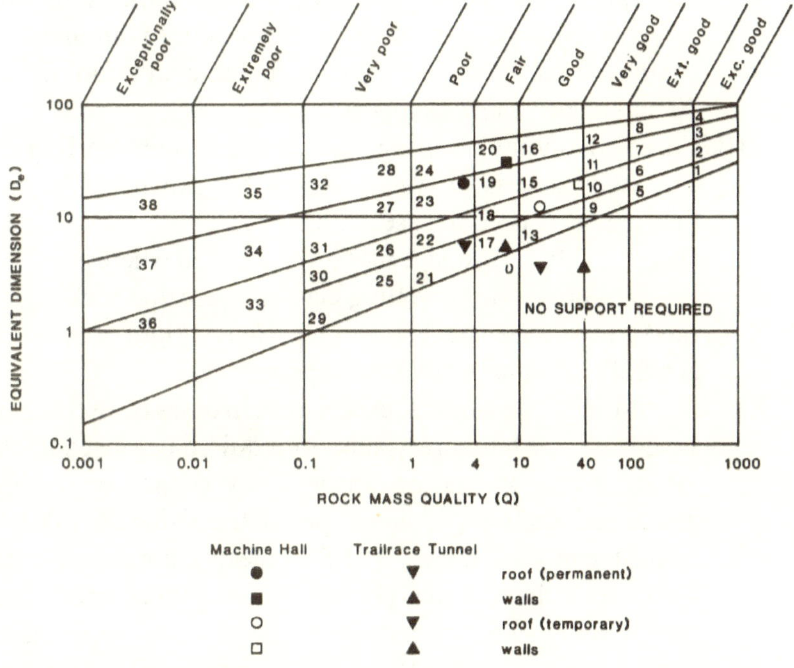

Figure 1.6. Excavation support chart, using Q-system, with box numbering for 38 categories of support.[1.16]

1.5.2.1 Supplementary Notes for Support Tables 1.17 to 1.20[1.16]

I. For cases of heavy rock bursting or "popping", tensioned bolts with enlarged bearing plates are often used, with spacing of about 1 m (occasionally down to 0.8 m). Final support when "popping" activity ceases. (Selmer-Olsen, 1970[1.16])

II. Several bolt lengths often used in same excavation, i.e., 3, 5, and 7 m.

III. Several bolt lengths often used in same excavation, i.e., 2, 3, and 4 m.

IV. Tensioned cable anchors often used to supplement bolt support pressures. Typical spacing 2 to 4 m.

V. Several bolt lengths often used in same excavations, i.e., 6, 8, and 10 m.

VI. Tensioned cable anchors often used to supplement bolt support pressures. Typical spacing 4 to 6 m.

VII. Several of the older generation power stations in this category employ systematic or spot bolting with areas of chain-link mesh, and a free span concrete arch roof (25 to 40 cm) as permanent support.

VIII. Cases involving swelling, for instance montmorillonite clay (with access of water). Room for expansion behind the support is used in cases of heavy swelling. See Selmer-Olsen (1970).[1.16] Drainage measures are used where possible.

IX. Cases not involving swelling clay or squeezing rock.

X. Cases involving squeezing rock. Heavy rigid support is generally used as permanent support.

XI. According to the authors' experience, in cases of swelling or squeezing, the temporary support required before concrete (or shotcrete) arches are formed may consist of bolting (tensioned shell-expansion type) if the value of RQD/J_n is sufficiently high (i.e., > 1.5), possibly combined with shotcrete. If the rock mass is very heavily jointed or crushed (i.e., RQD/J_n <1.5, for example a "sugar cube" shear zone in quartzite), then the temporary support may consist of up to several applications of shotcrete. Systematic bolting (tensioned) may be added after casting the concrete, but it may not be effective when RQD/J_n <1.5, or when a lot of clay is present, unless the bolts are grouted before tensioning. A sufficient length of anchored bolt might also be obtained using quick setting resin anchors in these extremely poor quality rock masses. Serious occurrences of swelling and/or squeezing rock may require that the concrete arches are taken right up to the face, possibly using a shield as temporary shuttering. Temporary support of the working face may also be required in these cases.

XII. For reasons of safety the multiple drift method will often be needed during excavation and supporting of roof arch. Categories 16, 20, 24, 28, 32, and 35 (SPAN/ESR > 15 m only).

XIII. Multiple drift method usually needed during excavation and support of arch, walls and floor in cases of heavy squeezing. Category 38 (SPAN/ESR > 10 m only).

Table 1.17 Support Measures for Rock Masses of "Exceptional", "Extremely Good", "Very Good", and "Good" Quality (Q range: 1000–10)[1.16]

Support category	Conditional factors $\frac{RQD}{J_n}$	$\frac{J_r}{J_a}$	$\frac{SPAN}{ESR}$	Type of support[a]	Supplementary Notes
1[c]	—	—	—	sb(utg)	-
2[c]	—	—	—	sb(utg)	-
3[c]	—	—	—	sb(utg)	-
4[c]	—	—	—	sb(utg)	-
5[c]	—	—	—	sb(utg)	-
6[c]	—	—	—	sb(utg)	-
7[c]	—	—	—	sb(utg)	-
8[c]	—	—	—	sb(utg)	-
9	≥20	—	—	sb(utg)	-
	<20	—	—	B(utg) 2.5–3 m	-
10	≥30	—	—	B(u/g) 2–3 m	-
	<30	—	—	B(utg) 1.5–2 m + clm	-
11[c]	≥30	—	—	B(tg) 2–3 m	-
	<30	—	—	B(tg) 1.5–2 m + clm	-
12[c]	≥30	—	—	B(tg) 2–3 m	-
	<30	—	—	B(tg) 1.5–2 m + clm	-
13	≥10	≥1.5	—	sb(utg)	I
	≥10	<1.5	—	B(utg) 1.5–2 m	I
	<10	≥1.5	—	B(utg) 1.5–2 m	I
	<10	<1.5	—	B(utg) 1.5–2 m + S	I
				2–3 cm	
14	≥10	—	≥15	B(tg) 1.5–2 m + clm	I, II
	<10	—	≥15	B(tg) 1.5–2 m + S(mr)	I, II
	—	—	<15	5–10 cm	I, III
				B(utg) 1.5–2 m +clm	
15	>10	—	—		I, II, IV
	≤10	—	—	B(tg) 1.5–2 m + clm	I, II, IV
				B(tg) 1.5–2 m + S(mr)	
16[c]	>15	—	—	5–10 cm	I, V, VI
(See Note	≤15	—	—		I, V, VI
XII p. 25)				B(tg) 1.5–2 m + clm	
				B(tg) 1.5–2 m + S(mr)	
				10–15 cm	

Note: The type of support to be used in categories 1 to 8 will depend on the blasting technique. Smooth wall blasting and thorough barring-down may remove the need for support. Rough-wall blasting may result in the need for single applications of shotcrete, especially where the excavation height is >25 m. Future case records should differentiate categories 1 to 8.

[a] Key to support tables: sb = spot bolting; B = systematic bolting; (utg) = untensioned, grouted; (tg) = tensioned, (expanding shell type for competent rock masses, grouted post-tensioned in very poor quality rock masses; see Note XI pp. 22 and 25);
[b] Refer to *Supplementary Notes for Support Tables 1.17 to 1.20,* Section 1.5.2.1.
[c] Authors estimates of support. Insufficient case records available for reliable estimation of support requirements.

s = shotcrete; (mr) = mesh reinforced;
clm = chain link mesh;
CCA = cast concrete arch;
(sr) = steel reinforced.

Bolt spacings are given in meters (m). Shotcrete, or cast concrete arch thickness is given in centimeters (cm).

Table 1.18 Support Measures for Rock Masses of "Fair" to "Poor" Quality (Q range: 10–1)[1.16]

Support category	Conditional factors			Type of support[a]	Supplementary Notes
	$\frac{RQD}{J_n}$	$\frac{J_r}{J_a}$	$\frac{SPAN}{ESR}$		
17	>30	—	—	sb(utg)	I
	≥10, ≤30	—	—	B(utg) 1–1.5m	I
	<10	—	≥6 m	B(utg) 1–1.5 m + S 2–3 cm	I
18	<10	—	<6 m	S 2–3 cm	I
	>5	—	≥10 m	B(tg) 1–1.5 m + clm	I, III
	>5	—	<10 m	B(utg) 1–1.5 m + clm	I
	≤5	—	≥10 m	B(tg) 1–1.5 m + S 2–3 cm	I, III
	≤5	—	<10 m	B(utg) 1–1.5 m + S 2–3m	I
19	—	—	≥20 m	B(tg) 1–2 m + S(mr) 10–15 cm	I, II, IV
	—	—	<20 m	B(tg) 1–1.5 m + S(mr) 5–10 cm	I, II
20[c]	—	—	≥35 m	B(tg) 1–2 m + S(mr) 20–25 cm	I, V VI
(See Note XII p. 25)	—	—	<35 m	B(tg) 1–2 m + S(mr) 10–20 cm	I, II, IV
21	≥12.5	≤0.075	—	B(utg) 1 m + S 2–3 cm	I
	<12.5	≤0.75	—	S 2.5–5 cm	I
	—	>0.75	—	B(utg) 1 m	I
22	>10, <30	>1.0	—	B(utg) 1 m + clm	I
	≤10	>1.0	—	S 2.5–7.5 cm	I
	<30	≤1.0	—	B(utg)1 m + S(mr) 2.5–5 cm	I
	≥30	—	—	B(utg) 1 m	I

23	—	—	≥15 m	B(tg) 1–1.5 m + S(mr) 10–15 cm	I, II, IV, VII
	—	—	<15 m	B(utg) 1–1.5 m + S(mr) 5–10 cm	I
24[c]	—	—	≥30 m	B(tg) 1–1.5 m + S(mr) 15–30 cm	I, V, VI
(See Note XII p.25)	—	—	<30 m	B(tg) 1–1.5 m + S(mr) 10–15 cm	I, II, IV

Note: [a] Refer to Table 1.17, footnote [a].
[b] Refer to *Supplementary Notes for Support Tables 1.17 to 1.20,* Section 1.5.2.1.
[c] Authors estimates of support. Insufficient case records available for reliable estimation of support requirements.

Table 1.19 Support Measures for Rock Masses of "Very Poor" Quality (Q range: 1.0–0.1)[1.16]

Support category	Conditional factors RQD / J_n	J_r / J_a	SPAN / ESR	Type of support[a]	Supplementary Notes
25	>10	>0.5	—	B(utg) 1 m + mr or clm	I
	≤10	>0.5	—	B(utg) 1 m + S(mr) 5 m	I
	—	≤0.5	—	B(tg) 1 m + S(mr) 5 cm	I
26	—	—	—	B(tg) 1 m + S(mr) 5–7.5 cm	VIII, X, XI
	—	—	—	B(utg) 1 m + S 2.5–5 cm	I, IX
27	—	—	≥12 m	B(tg) 1 m + S (mr)) 7.5–10 cm	I, IX
	—	—	<12m	B(utg) 1 m + S(mr) 5–7.5 cm	I, IX
	—	—	>12 m	CCA 20–40 cm + B(tg) 1 m	VIII, X, XI
	—	—	<12 m	S(mr) 10–20 cm + B(tg) 1m	VIII, X, XI
28[c]	—	—	≥30 m	B(tg) 1 m + S(mr) 30–40 cm	I, IV, V, IX
(See Note	—	—	≥20, <30 m	B(tg) 1 m + S(mr) 20–30 cm	I, II, IV, IX
XII p.25)	—	—	<20 m	B(tg) 1m + S(mr) 15–20 cm	I, II, IX
	—	—	—	CCA(sr) 30–100 cm + B(tg) 1 m	IV, VIII, X, XI
29[c]	>5	>0.25	—	B(utg) 1 m + S 2–3 cm	—
	≤5	>0.25	—	B(utg) 1 m + S(mr) 5 cm	—
	—	≤0.25	—	B(tg) 1 m + S(mr) 5 cm	—
30	≥5	—	—	B(tg) 1 m + S 2.5–5 cm	IX
	<5	—	—	S(mr) 5–7.5 cm	IX
	—	—	—	B(tg) 1 m + S(mr) 5–7.5 cm	VIII, X, XI
31	>4	—	—	B(tg) 1 m +	IX
	≤4, ≥1.5	—	—	S(mr) 7.5 25 cm	IX
	<1.5	—	—	CCA 20–40 cm + B(tg) 1 m	IX
	—	—	—	CCA(sr) 30–50 cm + B(tg) 1 m	VII, X, XI
32	—	—	≥20 m	B(tg) 1 m + S(mr) 40–60 cm	II, IV, IX
(See Note	—	—	<20 m	B(tg) 1 m + S(mr) 20–40 cm	III, IV, IX
XII p.25)	—	—	—	CCA(sr) 40–120 cm + B(tg) 1m	IV, VIII, X, XI

Note: a Refer to Table 1.17, footnote.
[b] Refer to *Supplementary Notes for Support Tables 1.17 to 1.20,* Section 1.5.2.1.
[c] Authors estimates of support. Insufficient case records available for reliable estimation of support requirements.

Table 1.20 Support Measures for Rock Masses of "Extremely Poor" and "Exceptionally Poor" Quality (Q range: 0.1–0.001)[1.16]

Support category	Conditional factors: RQD / J_n	J_r/J_a	SPAN / ESR	Type of support[a]	Supplementary Notes
33[c]	>2	—	—	B(tg) m + S(mr) 2.5–5 cm	IX
	≤2	—	—	S(mr) 5–10 cm	IX
	—	—	—	S(mr) 7.5–15 cm	VIII, X
34	≥2	≥0.25	—	B(tg) 1 m + S(mr) 5–7.5 cm	IX
	<2	0.25	—	S(mr) 7.5–15 cm	IX
	—	—	—	S(mr) 15–25 cm	IX
	—	—	—	CCA(sr) 20–60 cm + B(tg) 1m	VIII, X, XI
35	—	—	≥15 m	B(tg) 1 m+ S(mr) 30–100 cm	II, IX
	—	—	≥15 m	CCA(sr) 60–200 m +B(tg) 1 m	VIII, X, XI, II
	—	—	<15 m	B(tg) 1 + S(mr) 20–75 cm	IX, III
	—	—	<15 m	CCA(sr) 40–150 cm + B(tg) 1 m	VIII, X, XI, III
36[c]	—	—	—	S(mr) 10–20 cm	IX
	—	—	—	S(mr) 10–20 cm + B(tg) 0.5–1.0 m	VIII, X, XI
37	—	—	—	S(mr) 20–60 cm	IX
	—	—	—	S(mr) 20–60 cm + B(tg) 0.5–1.0 m	VIII, X, XI
38	—	—	≥10 m	CCA(sr) 100–300 cm	IX
(See Note XIII p. 25)	—	—	≥10	CCA(sr) 100–300 cm + B(tg) 1 m	VIII, X, II, XI
	—	—	<10 m	S(mr) 70–200 cm	IX
	—		>10 m	S(mr) 70–200 cm + B(tg) 1 m	VIII, X, III, XI

Note: [a] = Refer to Table 1.17, footnote a.

[b] = Refer to *Supplementary Notes for Support Tables 1.17 to 1.20,* Section 1.5.2.1.

[c] = Authors estimates of support. Insufficient case records available for reliable estimation of support requirements.

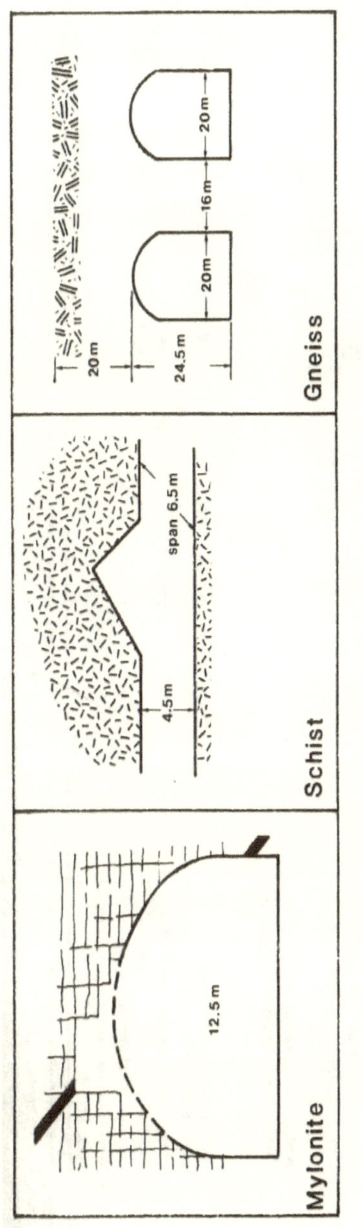

Figure 1.7. Sketches of three case records described in Table 1.2.[1.17]

Table 1.21 Comparison of Support Used and Support Recommended for Three Case Records[1.17]

Case no.	1. Description of Rock Mass 2. Nature of instability 3. Purpose of excavation, location, reference	SPAN (m)	Height (m)	Depth (m)	Support used	$\frac{RQD}{J_n}$	$\frac{J_r}{J_a}$	$\frac{J_w}{SRF}$	Q	ESR	$\frac{SPAN}{ESR}$	Estimate of permanent roof support[a]
						(Code: Tables 1.7–1.10 and 1.12–1.20)						
24	1. 60-m length, including a 1 m wide shear zone in mylonite. Crushed mylonite and nonsoftening clay seams and joint fillings. Intersecting joint set. 2 joint sets plus random, 5–30 cm spacing. Minor water inflows (<31/min). RQD = 60 2. Wedge shaped roof fall 3. Headrace tunnel, Vietas Hydro, N. Sweden; Ref. Cecil 1970.	12.5	6.5	60	Rock bolts, wire mesh, and shotcrete	60 6 $\left(\frac{1C}{2E}\right)$	1.0 6 $\left(\frac{3H}{4K}\right)$	1.0 2.5 $\left(\frac{5A}{6C}\right)$	1.3	1.6	7.8	Category 22; =B 1 m; +S(mr); 2.5–5 cm
48	1. 15-m length, overtrust shear zone in schist, in which there was a 3-cm thick clay (non-softening and graphite seam. Shear zone was 50–100 cm wide and contained smooth, slickensided graphite-coated joint surfaces, 1 joint set, 5–30 cm spacing. Insignificant water inflow. RQD = 10 2. Wedge-shaped roof fall 3. Tailrace tunnel, Bergvattnet, Hydro, N. Sweden; Ref. Cecil 1970	6.5	4.5	50	Rock bolts, wire mesh two shotcrete applications	10 2 $\left(\frac{1A}{2B}\right)$	1.0 10 $\left(\frac{3H}{4O}\right)$	1.0 5 $\left(\frac{5A}{6B}\right)$	0.10	1.6	4.1	Category 31; =B 1 m; +S(mr) 5 cm
77	1. 300-m length, massive gneiss, few joints. Planar, rough-surfaced, unaltered joints. 3-m spacing. Insignificant water inflow. RQD = 100 2. Minor overbreak, no falls or slides 3. Wine and liquor storage rooms. Stockholm; Ref. Cecil 1970	20	24.5	18	50 spot bolts in about 300 m of chamber	100 1.0 $\left(\frac{1E}{2A}\right)$	5 1.0 $\left(\frac{3E}{4B}\right)$	1.0 2.5 $\left(\frac{5A}{6B}\right)$	200	1.3	15.4	Category 0, 5; = None or sb

Note: S = shotcrete, B = systematic bolting, sb = spot bolting, mr = mesh reinforced. Bolt spacing is given in meters. Shotcrete or concrete thickness is given in centimeters.

[a] Right-hand column "Estimate of permanent roof support" is obtained from Tables 1.17 to 1.20.

1.5.3 Effect of Erroneous Evaluation of Rock Mass Quality (Q)

The problem of failing to anticipate unfavorable rock mass parameters — e.g., slickensided joints, swelling clay, high rock pressure, squeezing ground, and large water inflows — may cause individual errors ranging from factors of 1.5 to 2 to a maximum of about 20. If the errors are "unfavorable", two or more large errors out of the six parameters will be virtually certain of causing failure (thereby overestimating Q and underestimating support). However, there is room for several minor errors, especially since both "unfavorable" and "favorable" judgments of the rock mass may be made, thereby balancing effects to some extent. Total errors amounting to a factor of between 2.5 and 4 will be likely to change the support recommendation, since the "width" of most categories is of this order as can be seen from Figure 1.6. Smaller errors than this will only be reflected in slight adjustments to bolt spacing.

One of the most serious errors in engineering judgment is failure to anticipate a clay-filled weakness zone. This may have a "snowball error" effect on Q and therefore results to inadequate support, especially if the clay concerned is of the swelling variety. A hypothetical but realistic example is given below to illustrate this situation.

- Assumed rock mass quality Q_0 = 70/9 x 1.5/3 x 1.0/1.0 = 3.9 (POOR)
  Code to descriptions, Tables 1.7 to 1.10 (1C/2F, 3E/4D, 5A/6J)
- Actual rock mass quality revealed upon excavation Q = 20/9 x 1.0/15 x 0.66/2.5
  = 0.039 (EXT. POOR)
  Code to descriptions, Tables 1.7 to 1.10 (1A/2F, 3H/4R, 5B/6C)
  Apparent safety ratio Q_0/Q is given in Table 1.22.

Table 1.22 Apparent Safety Ratio when Estimating Q[1.16]

Case record no.	ESR	Q_0	Q	Safety ratio (Q_0/Q)
18	1.6	0.37	0.0094	40
19	1.6	0.36	0.028	23
45	1.6	≥14	0.60	≥23
79	1.0	≥ 4	0.05	≥80

In Table 1.22, a safety ratio (Q_0/Q) equal to 100 will be virtually certain of causing failure in the unlikely event that support is not redesigned. The two Q values can be translated into engineering terms by imagining a water tunnel (ESR = 1.6) with both span and height equal to 9 m. The two classifications given above lead to the following estimates for:

a. Permanent roof support ;
b. Permanent wall support ;
c. Temporary roof support ;
d. Temporary wall support.

(The method of estimating b, c, and d is given in the Appendix).

1. a. Category 21 = S(5 cm)
   b. Category 17 = S(2-3 cm); (Note: Q_w(wall) = 3.9 x 2.5)
   c. Category 0 = NONE
   d. Category 0 = NONE; (Temporary support: 1.5 ESR, 5Q)

2. a. Category 34 = CCA(sr) 35 cm + B(tg) 1 m,
   see Supplementary Notes: VIII, XI pp. 20 and 22.
   b. Category 34 = CCA(sr) 35 cm + B(tg) 1 m,
   see Supplementary Notes: VIII, XI pp. 20 and 22.
   (Note: Q_w = 0.039 x 1.0)
   c. Category 30 = B(tg) 1 m + S(mr) 5 cm,
   see Supplementary Notes: VIII, XI pp. 20 and 22.
   d. Category 30 = B(tg) 1 m + S(mr) 5 cm,
   see Supplementary Notes: VIII, XI pp. 20 and 22.

The safety ratio of 100 in the above example is by no means the largest that can occur. For instance if the rock mass was essentially crushed in the weakness zone the safety ratio would exceed 100. However, it is a useful illustration of the "snowball error" that can occur through faulty engineering-geological judgment. All six parameters can be altered unfavorably by an unexpected clay zone.

It should be emphasized that sensitivity analysis of this type can be very informative for the design engineer since there is quite a large store of case records coded in Tables 1.17 to 1.20. The economic consequences of pessimistic assumptions of rock-mass conditions can be compared with those resulting from expected conditions, and the consequences of individual parameter errors can be investigated. It may even be of value to investigate the economic consequences of changing the span of an excavation, if such a choice is available in the design.

1.5.4 Estimation of Support Pressure utilizing Rock Mass Quality (Q)

The support pressure capacity of tensioned or grouted bolts (P) is equal to the yield capacity of one bolt (Y, if adequately anchored) divided by the square of the bolt spacing (s), expressed as follows:

$$P=Y/s^2 \qquad (1.7)$$

where: P = support pressure capacity, kPa
Y = yield capacity of one bolt, kN
s = spacing between bolts, m

If a 10 kN working load is assumed for a 20-mm diameter bolt, the support pressure is as follows:[1.14]

$$P=1/s^2 \qquad (1.8)$$

An empirical equation to relate permanent support pressure and Q is given as follows:[1.14]

$$P_{roof}=66.7\,J_n^{0.5}/J_r \bullet Q^{0.3} \qquad (1.9)$$

$$P_{wall}=66.7\,J_n^{0.5}/J_r \bullet Q_w^{0.3} \qquad (1.10)$$

where: P_{roof} = permanent roof support pressure, kPa
P_{wall} = permanent wall support pressure, kPa
J_n = joint set number
= 0.5 for massive, up to 20 for crushed rock masses
J_r = joint roughness number
= 5 for discontinuity joints, up to 0.5 for crushed zones
Q = rock mass quality
= 0.001 for exceptionally poor, up to 1000 for exceptionally good rock mass
Q_w = wall factor (= 1.0 Q up to 5.0 Q, see Appendix)

Chart for estimating permanent radial support pressure is given as Figure 1.8. The diagonal lines drawn in Figure 1.8, present the J_r values, were plotted from Equations 1.9 and 1.10. Generally, the range of support pressures to be expected lie within the shaded envelope. When Q is higher than about 100, the estimation of support pressure loses its meaning since the excavation is self supporting, with the exception of occasional blocks that require spot bolting.

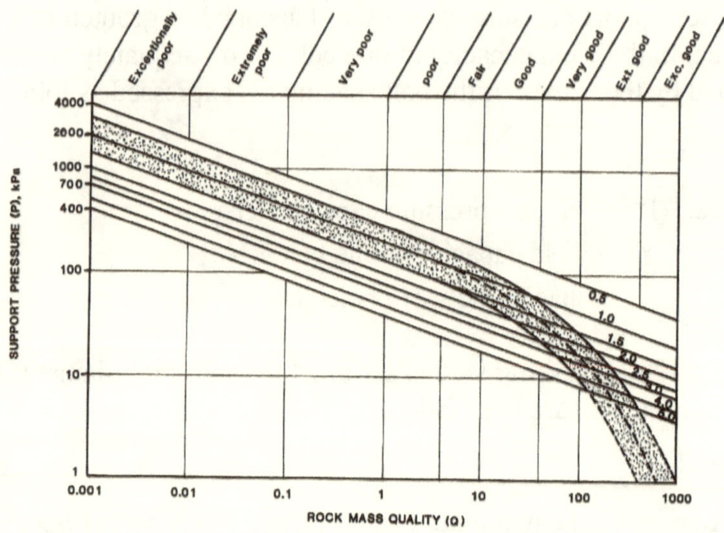

Figure 1.8. Support pressure estimation chart using rock mass quality and joint rating.[1.14]

The proposed relationship between support pressure and Q provides a convenient means for developing classification rules in the static conditions. However, under dynamic conditions, the dynamic stresses resulting from the passage of seismic waves will exceed the static stresses by some factor. An increase of up to 20% was suggested for the case of lined excavations.[1.19]

A modified presentation of support predictions originally proposed by Barton et al.,[1.16] using Q is given in Table 1.23.

Table 1.23 Recommended Support Based Upon NGI Tunneling Quality Index Q[1.11]

Rock mass quality (Q)	Equivalent Dimension $\left(\frac{SPAN}{ESR}\right)$	Block size $\left(\frac{RQD}{J_n}\right)$	Inter-block strength $\left(\frac{J_r}{J_a}\right)$	Approx. support pressure P (MPa)	Spot reinforcement with untensioned grouted dowels	Untensioned grouted dowels on grid spacing indicated	Tensioned rockbolts on grid spacing indicated	Chainlink mesh anchored to bolts and intermediate points	Shotcrete applied directly to rock, thickness indicated	Shotcrete reinforced with weldmesh, thickness indicated	Unreinforced cast concrete arch, thickness indicated	Steel reinforced cast concrete arch, thickness indicated	Notes by Barton, Lien, and Lunde	Notes by Hoek and Brown
1000–400	20–100			<0.001									1	a
400–100	12–88			0.005									1	a
100–40	8.5–19	≥20		0.025									1	a
100–40	8.5–19	≤20		0.025		2.5–3 m								
100–40	14–30	≥30		0.025		2–3 m								
100–40	14–30	<30		0.025		1.5–2 m		X						b
100–40	23–72	≥30		0.025		2–3 m								
100–40	23–72	<30		0.025		1.5–2 m		X						b
40–10	5–14	≥10	≥1.5	0.05	X								2	
40–10	5–14	≥10	<1.5	0.05		1.5–2 m							2	
40–10	5–14	<10	≥1.5	0.05		1.5–2 m							2	
40–10	5–14	<10	<1.5	0.05		1.5–2 m			20–30 mm				2	
40–10	15–23	≥10		0.05			1.5–2 m	X					2, 3	b
40–10	15–23	<10		0.05			1.5–2 m			50–100 mm			2, 3	c
40–10	9–15			0.05		1.5–2 m		X					2, 4	b
40–10	15–40	>10		0.05		1.5–2 m		X					2, 3, 5	b
40–10	15–40	≤10		0.05		1.5–2 m				50–10 mm			2, 3, 5	c
40–10	30–65	>15		0.05		1.5–2 m		X					2, 6, 7, 13	b
40–10	30–65	≤15		0.05		1.5–2 m				100–150 mm			2, 6, 7, 13	c
10–4	3.5–9	≥30		0.10	X								2	a
10–4	3.5–9	≥10 ≤30		0.10		1–1.5 m							2	
4–1	18–30			0.15			1–1.5 m			100–150 mm			2, 3, 5	c
1–0.4	1.5–4.2	>10	>0.6	0.225		1 m		X					2	d
1–0.4	1.5–4.2	≤10	>0.5	0.225		1 m				50 mm			2	c
1–0.4	1.5–4.2		≤0.5	0.225		1 m				50 mm			2	c
1–0.4	3.2–7.5			0.225			1 m			50–75 mm			14, 11, 12	
1–0.4	3.2–7.5			0.225		1 m			25–50 mm				2, 10	
1–0.4	12–18			0.225			1 m			75–100 mm			2, 10	c
1–0.4	6–12			0.225		1 m				50–75 mm			2, 10	c
1–0.4	12–18			0.225			1 m				200–400 mm		14, 11, 12	e

Table 1.23 - Continued.

1–0.4	6–12			0.225			1 m			100–200 mm			14, 11, 12	c
1–0.4	30–38			0.225			1 m			300–400 mm			2, 5, 6, 10, 13	c, f
1–0.4	20–30			0.225			1 m			200–300 mm			2, 3, 5, 10, 13	c
1–0.4	15–20			0.225			1 m			150–200 mm			1, 3, 10, 13	c
1–0.4	15–38			0.225			1 m					300 mm–1 m	5, 9, 10, 12, 13	
0.4–0.1	1–3.1	>5	>0.25	0.3		1 m			20–30 mm					
0.4–0.1	1–3.1	≤5	>0.25	0.3		1 m				50 mm				c
0.4–0.1	1–3.1		≤0.25	0.3			1 m			50 mm				c
0.4–0.1	2.2–6	≥5		0.3			1 m			25–50 mm			10	c
0.4–0.1	2.2–6	<5		0.3						50–75 mm			10	c
0.4–0.1	2.2–6			0.3			1 m			50–75 mm			9, 11, 12	c
0.4–0.1	4–14.5	>4		0.3			1 m			50–125 mm			10	c
10–4	6–9	<10		0.10		1–1.5 m			20–30 mm				2	
10–4	<6	<10		0.10					20–30 mm				2	
10–4	10–15	>5		0.10			1–1.5 m	X					2, 4	a
10–4	7–10	>5		0.10		1–1.5 m		X					2	a
10–4	10–15	≤5		0.10			1–1.5 m		20–30 mm				2, 4	
10–4	7–10	≤5		0.10		1–1.5			20–30 mm				2	
10–4	20–29			0.10			1–2 m			100–150 mm			2, 3, 5	c
10–4	12–20			0.10			1–1.5 m			50–100 mm			2, 3	c
10–4	35–52			0.10			1–2 m			200–250 mm			2, 6, 7, 13	c
10–4	24–35			0.10			1–2 m			100–200 mm			2, 3, 5, 13	c
4–1	2.1–6.5	≥12.5	≤0.75	0.15		1 m			20–30 mm				2	
4–1	2.1–6.5	<12.5	≤0.75	0.15					20–30 mm				2	
4–1	2.1–6.5		>0.75	0.15		1 m							2	
4–1	4.5–11.5	>10	1 < 30	0.15		1 m		X					2	a
4–1	4.5–11.5	≤10	>1	0.15					25–75 mm				2	
4–1	4.5–11.5	<30	≤1	0.15		1 m				25–50 mm			2	c
4–1	4.5–11.5	≥30		0.15		1 m							2	
4–1	15–24			0.15			1–1.5 m			100–150 mm			2, 3, 5, 8	c
4–1	8–15			0.15			1–1.5 m			50–100 mm			2	c
4–1	30–46			0.15			1–1.5 m			150–300 mm			2, 6, 7, 13	c
0.4–0.1	4–14.5	1.5 ≤ 4		0.3						75–250 mm			10	c
0.4–0.1	4–14.5	<1.5		0.3			1 m				200–400 mm		10, 12	e
0.4–0.1	4–14.5			0.3			1 m					300–500 mm	9, 11, 12	
0.4–0.1	20–34			0.3			1 m			400–600 mm			3, 5, 10, 12, 13	f
0.4–0.1	11–20			0.3			1 m			200–400 mm			4, 5, 10, 12, 13	c
0.4–0.1	11–34			0.3			1 m					400 mm–1.2 m	5, 9, 11, 12, 13	
0.1–0.01	1–3.9	≥2		0.6			1 m			25–50 mm			10	c
0.1–0.01	1–3.9	<2		0.6			1 m			50–100 mm			10	c
0.1–0.01	1–3.9			0.6						75–150 mm			9, 11	c

Table 1.23 continued - Recommended support based Upon NGI Tunneling Quality Index Q 1.11

Rock mass quality (Q)	Equivalent Dimension $\left(\frac{SPAN}{ESR}\right)$	Block size $\left(\frac{RQD}{J_n}\right)$	Inter-block strength $\left(\frac{J_r}{J_a}\right)$	Approx. support pressure P (MPa)	Spot reinforcement with untensioned grouted dowels	Untensioned grouted dowels on grid spacing indicated	Tensioned rockbolts on grid spacing indicated	Chainlink mesh anchored to bolts and intermediate points	Shotcrete applied directly to rock, thickness indicated	Shotcrete reinforced with weldmesh, thickness indicated	Unreinforced cast concrete arch, thickness indicated	Steel reinforced cast concrete arch, thickness indicated	Notes by Barton, Lien, and Lunde	Notes by Hoek and Brown
0.1–0.01	2–11	≥2	≥0.25	0.6			1 m			50–75 mm			10	c
0.1–0.01	2–11		<0.25	0.6						150–250 mm			10	c
0.1–0.01	2–11			0.6			1 m					200–600 mm	9, 11, 12	
0.1–0.01	15–28			0.6			1 m			300–1 m			3, 10, 12, 13	c, f
0.1–0.01	15–28			0.6			1 m					600 mm–2 m	3, 9, 11, 12, 13	
0.1–0.01	6.5–15			0.6			1 m			200–750 mm			4, 10, 12, 13	c, f
0.1–0.01	6.5–15			0.6			1 m					400 mm–1.5 m	3, 9, 11, 12, 13	
0.01–0.001	1–2			1.2						100–200 mm			10	c
0.01–0.001	1–2			1.2			0.5–1 m			100–200 mm			9, 11, 12	c
0.01–0.001	1–6.5			1.2						200–600 mm			10	c, f
0.01–0.001	1–6.5			1.2			0.5–1 m			200–600 mm			9, 11, 12	c, f
0.01–0.001	10–20			1.2								1–3 m	10, 14	
0.01–0.001	10–20			1.2			1 m					1–3 m	3, 9, 11, 12, 14	
0.01–0.001	4–10			1.2						700 mm–2 m			10, 14	c, f
0.01–0.001	4–10			1.2			1 m			700 mm–2 m			4, 9, 10, 11, 14	c, f

[a] Supplementary notes by Barton, Lien and Lunde:[1.16]

1. The type of support used in extremely good and exceptionally good rock will depend upon the blasting technique. Smooth wall blasting and thorough barring down may remove the need for support. Rough wall blasting may result in the need for a single application of shotcrete, especially where the excavation height exceeds 25 m.
2. For cases of heavy rock bursting or "popping", tensioned bolts with enlarged bearing plates often used, with spacing about 1 m (occasionally 0.8 m). Final support when "popping" activity ceases.
3. Several bolt lengths often used in same excavation, i.e., 3, 5, and 7 m.
4. Several bolt lengths often used in same excavation, i.e., 2, 3 and 4 m.
5. Tensioned cable anchors often used to supplement bolt support pressures. Typical spacing 2 to 4 m.
6. Several bolt lengths often used in same excavation, i.e., 6, 8, and 10 m.

Table 1.23 continued - Recommended support based Upon NGI Tunneling Quality index Q 1.11.

7. Tensioned cable anchors often used to supplement bolt support pressures. Typical spacing 4 to 6 m.
8. Several older generations power stations in this category employ systematic or spot bolting with areas of chain link mesh, and a free span concrete arch roof (250-400 mm) as permanent support.
9. Cases involving swelling, for instance montmorillonite clay (with access of water). Room for expansion behind the support is used in cases of heavy swelling. Drainage measures are used where possible.
10. Cases not involving swelling day or squeezing rock.
11. Cases involving squeezing rock. Heavy rigid support is generally used as permanent support.
12. According to author's experience (Barton et al.). in cases of swelling or squeezing, the temporary support required before concrete (for shotcrete) arches are formed may consist of bolting (tensioned shell-expansion type) if the value of ROD/J_n is sufficiently high i.e., > 1.5), possibly combined with shotcrete. If the rock mass is very heavily jointed or crushed (i.e., ROD/J_n. < 1.5, for example a "sugar cube" shear zone in quartzite), then the temporary support may consist of up to several applications of shotcrete. Systematic bolting (tensioned) may be added after casting the concrete (or shotcrete) arch to reduce the uneven loading on the concrete, but it may not be effective when ROD/J_n. <1.5, or when a lot of clay is present, unless the bolts are grouted before tensioning. A sufficient length of anchored bolt might also be obtained using quick setting resin anchors in these extremely poor quality rock masses. Serious occurrences of swelling and/or squeezing rock may require that the concrete arches be taken right up to the face, possibly using a shield as temporary shuttering. Temporary support of the working face may also be required in these cases.
13. For reasons of safety the multiple drift method will often be needed during excavation and supporting of roof arch. For Span/ESR > 15 only.
14. Multiple drift method usually needed during excavation and support of arch, walls, and floor in cases of heavy squeezing. For Span/ESR > 20 in exceptionally good rock only.

[b] Supplementary notes by Hoek and Brown:[1 11]

a. In Scandinavia, the use of "Perfobolts" is common. These are perforated hollow tubes which are filled with grout and inserted into drillholes. The grout is extruded to fill the annular space around the tube when a piece of reinforcing rod is pushed into the grout filling the tube. Obviously, there is no way in which these devices can be tensioned although it is common to thread the end of the reinforcing rod and place a normal bearing plate or washer and nut on this end. In North America the use of "Perfobolts" is rare. In mining applications a device known as a "Split set" or "Friction set" (developed by Scott) has become popular. This is a split tube which is forced into a slightly smaller diameter hole than the outer diameter or the tube. The friction between the steel, tube and the rock, particularly when the steel rusts, acts in much the same way as the grout around a reinforcing rod.

Table 1.23 - Continued.

For temporary support these devices are very effective. In Australian mines, untensioned grouted reinforcing is installed by pumping thick grout into drillholes and then simply pushing a piece of threaded reinforcing rod into the grout. The grout is thick enough to remain in an uphole during placing of the rod.

b. Chainlink mesh is sometimes used to catch small pieces of rock which can become loose with time. It should be attached to the rock at intervals of between 1 and 1.5 m and short grouted pins can be used between bolts. Galvanized chainlink mesh should be used where it is intended to be permanent, e.g., in an underground powerhouse.

c. Weldmesh, consisting of steel wires set on a square pattern and welded at each intersection, should be used for the reinforcement of shotcrete since it allows easy access of the shotcrete to the rock. Chainlink mesh should never be used for this purpose since the shotcrete cannot penetrate all the spaces between the wires and air pockets are formed with consequent rusting of the wire. When choosing weldmesh, it is important that the mesh can be handled by one or two men working from the top of a high-lift vehicle and hence the mesh should not be too heavy. Typically, 4.2 mm wires set at 100 mm intervals (designated 100 x 100 x 4.2 weldmesh) are used for reinforcing shotcrete.

d. In poorer quality rock, the use of untensioned grouted dowels as recommended by Barton, Lien, and Lune depends upon immediate installation of these reinforcing elements behind the face. This depends upon integrating the support drilling and installation into the drill-blast-muck cycle and many non-Scandinavian contractors are not prepared 10 consider this system. When it is impossible to ensure that untensioned grouted dowels are going to be installed immediately behind the face, consideration should be given to using tensioned rockbolts which can be grouted at a later stage. This ensures that support is available during the critical excavation stage.

e. Many contractors would consider that a 200 mm thick cast concrete arch is too difficult to construct because there is not enough room between the: shutter and the surrounding rock to permit easy access for pouring concrete and placing vibrators. lbe U.S. Army Corps of Engineers suggests 10 in. (2.S4 mm) as a normal minimum while some contractors prefer 300 mm.

f. Barton, Lien, and Lunde suggest shotcrete thicknesses of up to 2 m. This would require many separate applications and many contractors would regard shotcrete thickness of this magnitude as both impractical and uneconomic, preferring to cast concrete arches instead. A strong argument in favor of shotcrete is that it can be placed very close to the face and hence can be used to provide: early support in poor quality rock masses. Many contractors would argue that a 50 to 100 mm layer is generally sufficient for this purpose, particularly when used in conjunction with tensioned rockbolts as indicated by Barton, Lien, and Lunde, and that the placing of a cast concrete lining at a later stage would be a more effective way to tackle the problem. Obviously, the final choice will depend upon the unit rates for concreting and shotcreting offered by the contractor and. if shotcrete is cheaper upon a practical demonstration by the contractor that he can actually place shotcrete to this thickness.

In North America, the use of concrete or shotcrete linings of up to 2 m thick would be considered unusual and a combination of heavy steel sets and concrete would normally be used to achieve the high support pressures required in very poor ground.

1.5.5 Case Studies

1.5.5.1 Determination of Rock Mass Quality from Surface Exposures

Eight dissimilar rock masses are considered with specifications obtained from Tables 1.7 to 1.14. Their Q-values are determined using equation 1.4 as follows:

Rock type	RQD%	J_n	J_r	J_a	J_w		Q (Eq. 1.4)	Remarks
Granite 1	60	9	3	1	0.66	1	13.3	good
Granite 2	45	15	2	1	0.85	1	5.1	fair
Granite 3	10	20	1	6	0.50	6	0.07	extremely poor
Sandstone	40	10	1	2	0.50	3	0.3	very poor
Limestone	45	9	2	4	0.66	5	0.09	very poor
Mudstone	30	9	1	5	0.66	5	0.09	extremely poor
Quartzite	90	10	3	2	0.85	2	5.7	fair
Black Shale	95	9	4	2	0.85	1	17.9	good

1.5.5.2 Sample Determination

Assume Q was determined to be 5.1. The unsupported excavation factor of safety (FS) is calculated to be 1.5. If the excavation is rectangular in cross section, estimate the SPAN.

***Solution*:** Utilizing Equation 1.6:

ESR = excavation support ratio = 1.5

D_e = equivalent dimension

= 4, from Figure 1.6 and bottom line for maximum unsupported excavation.

Therefore: SPAN = (ESR) (D_e) = (1.5) (4) = 6, m

Other examples are included in the Appendix.

1.6 ROCK QUALITY INDEX (QI)

QI is defined as a percentage ratio of the longitudinal velocity of a seismic wave in a fissured or porous rock mass (V_L^*) to that of the intact rock without fissure or pore (V_L). It is a measure of rock mass quality under dynamic condition and can be expressed as follows:[1.20]

$$QI\% = 100\, V_L / V_L^* \quad (1.11)$$

If the mineral composition is known, V_L^* can be calculated from the following expression:[1.20]

$$1/V = \Sigma_i \,(C_i / V_{L,i}) \quad (1.12)$$

where $V_{L,i}$ is the longitudinal wave velocity in mineral constituent i, which has volume proportion Ci in the rock.

Average velocities of longitudinal waves in rock — forming minerals (V_L) and typical values of V_L for a few rock types are given in Tables 1.24 and 1.25, respectively.[1.20]

Table 1.24 Longitudinal Velocity of Seismic Waves (V_{LI}) in Various Minerals[1.20]

Mineral	$V_{L,i}$ (m/sec)	Mineral	$V_{L,i}$ (m/sec)
Quartz	6050	Calcite	6600
Olivine	8400	Dolomite	7500
Augite	7200	Magnetite	7400
Amphibole	7200	Gypsum	5200
Muscovite	5800	Epidote	7450
Orthoclase	5800	Pyrite	7450
Plagioclase	6250		

Table 1.25 Typical Values of the Longitudinal Velocity of Seismic Wave (V) in Intact Rocks without Fissures or Pores[1.20]

Rock	V (m/sec)
Gabbro	7000
Basalt	6500–7000
Limestone	6000–6500
Dolomite	6500–7000
Sandstone and quartzite	6000
Granitic rocks	5500–6000

Effect of pores on the QI is expressed in the following empirical form:[1.20]

$$QI\% = 100 - 1.6\, n_p\% \quad (1.13)$$

where: $n_p\%$ = percentage porosity of nonfissured rock.

Classification scheme to relate rock mass fissuring, percent porosity and rock quality index is given in Figure 1.9.

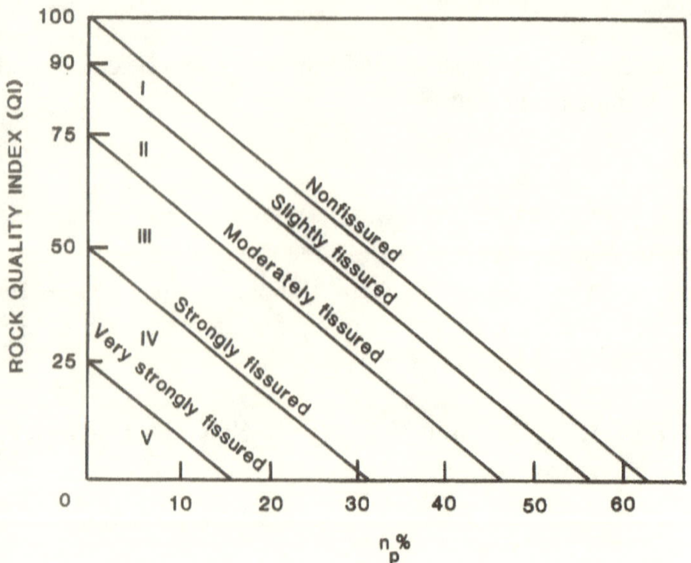

Figure 1.9. Classification scheme for fissuring in rock specimens.[1.20]

1.7 STRENGTH PROPERTIES OF ROCK MASS

This property of rock mass dictates its deformations, failure and ultimately compaction under *in situ* stress conditions. Basically, there are three controlling parameters of rock mass strength: mechanical strength of the ground, the structure, and the stress conditions. There are also numerous interdependent factors affecting the rock mass strength. The most significant of these factors are given as follows.

1.7.1. Specific Gravity (ρ) and Unit Weight (γ)

There are five categories or ranges of apparent specific gravity (ρ) and unit weight (γ). This is given in Table 1.26.

Table 1.26 The Categories of Rock Mass Specific Gravity (ρ), Unit Weight (γ), and Their Standard Symbols[1.21]

Symbol	Definition	Sp. gravity (ρ)	Unit KN /m3	Weight (γ) Lb/ ft3
A	No directional weakness, no oxidation alteration of mineral components, and no weathering	> 2.56	> 26	> 160
B	Nonvisual mineral alignment, uniform color, and no weathering	2.40 –2.56	25 -26	150 -160
C	Visual mineral alignment, partly or completely discolored due to oxidation and weathering, but cannot be remolded by finger pressure	2.23 –2.40	25 - 33	140 -150
D	Non-intersecting planes of weaknesses, remoldable by means of finger pressure to rock and soil	2.02 –2.24	21 -33	130 - 140
E	Intersecting planes of weakness, remolds by means of finger pressure to soil	< 2.08	< 21	< 130

1.7.2 Static Physical and Mechanical Properties

Factors such as density, unconfined compressive strength, point load strength, hardness, Young's Modulus, tensile strength, bending strength, Poisson's ratio, porosity, and degree of saturation are some of the main factors to be considered.

The unconfined compressive strength of rock (σ_c) is interdependent with the degree of weathering. A combination of which was called the strength of rock material, five categories are given in Table 1.27. Various classifications for strength of intact rock is compared in Table 1.28. A major limitation of using intact-rock strength classifications is that they do not provide quantitative data accurate enough for engineering design purposes. They should be utilized conservatively and cautiously and in conjunction with other rock mass properties.

Table 1.27 Modified Unconfined Compressive Strength (σ_c) and Toughness of Intact Rocks in Fine Strength Classifications[a]

Symbol[b]	Rock strength classification	Example of rock types	σ_c lb.f / in2	σ_c kg.f / cm2	σ_c MPa	Toughness (psi)
A	Very high	Basalt, Dolerite, Diabase, Gabro, Quarzite	>30,000	>2,000	>200	> 59
	High	Granite, Gneiss, Marble	15,000–30,000	1,000 – 2,000	100 –200	16 -59
B	Medium	Limestone, Sandstone, Schistoses Shale, Slate	75,000–15,000	500 – 1,000	50 –100	04-06
C	Low	Goal, Schist, Gypsum, Tuff, Siltstone	3,500 – 7,500	250 – 500	25 –50	0.75 - 4
D & E	Very low	Chalk, Rock-salt	<3,500	< 250	< 25	< 0.75

Note: a = Data from References 1.22 & 1.23; b = Refer to Table 1.26.

Some static physical and mechanical properties of rocks obtained by various investigators is given in Tables 1.29 to 1.36 for quick reference. Variation in results of the above noted tables is due to natural variation in the properties of the rock samples tested.

A rock mass quality classification chart is shown in Figure 1.10, which combines factors such as the uniaxial compressive strength of rock with spacing of discontinuities in the rock mass.[1.30] Figure 1.10a is subdivided by use of straight lines. A change in the inclination of these lines gives different weighing from one rock quality index to the other. These subdivisions have been proposed by the British Geological Society.[1.31] Figure 1.10b shows a subdivision that might apply to the quality of rock as a road aggregate. The subdividing lines are vertical, showing that the quality is affected by strength and the fracture spacings in this "application" have no significance. Figure 1.10c shows the situation more likely to obtain in practice. The chart relates to the method of rock excavation.

Table 1.28 Comparison of Various Classifications for Strength (σ_c) of Intact Rock[1.24]

Scale: 05, 07, 1, 2, 3, 4, 5, 6, 7, 8, 10, 20, 30, 40, 50, 70, 100, 200, 300, 400, 700

Classification							
COATES 1964	VERY WEAK	WEAK	STRONG	VERY STRONG			
DEERE & MILLER 1966	VERY LOW STRENGTH	LOW STRENGTH	MEDIUM STRENGTH	HIGH STRENGTH	VERY HIGH STRENGTH		
GEOLOGICAL SOCIETY 1970	VERY WEAK	WEAK	MODERATELY WEAK	MODERATELY STRONG	STRONG	VERY STRONG	EXTREMELY STRONG
BROCH & FRANKLIN 1972 (← SOIL / ROCK →)	EXTREMELY LOW STRENGTH	VERY LOW STRENGTH	LOW STRENGTH	MEDIUM STRENGTH	HIGH STRENGTH	VERY HIGH STRENGTH	EXTREMELY HIGH STRENGTH
JENNINGS 1973	SOIL	VERY SOFT ROCK	SOFT ROCK	HARD ROCK	VERY HARD ROCK	EXTREMELY HARD ROCK	
BIENIAWSKI 1973	SOIL	VERY LOW STRENGTH	LOW STRENGTH	MEDIUM STRENGTH	HIGH STRENGTH	VERY HIGH STRENGTH	
ISRM 1979	VERY LOW	LOW STRENGTH	MODERATE	MEDIUM	HIGH	VERY HIGH	

UNIAXIAL COMPRESSIVE STRENGTH (σ_c) MPo

Note: 1 MP_a = 145 lbf/in.^{2}

Table 1.29 Static Physical and Mechanical Properties of Various Rocks[a]

Rocks	Density, dry (t/m³)	Young's modulus (1000 × kg/cm²)	Poisson's number	Porosity (%)	Compressive strength (kg/cm²)	Tensile strength (kg/cm²)	Bending strength (kg/cm²)
Bartholitic rocks							
Granite-grandiorite	2.5–2.75	300–700	5–8	0.1–2	1200–2800	40–70	100–200
Gabbro	2.92–3.05	600–1000	5–8	2–5	1500–2000	50–80	100–220
Extrusive rocks							
Riolite	2.45–2.60	100–200	5–10	0.4–4	800–1600	50–90	100–220
Dacite	2.50–2.75	80–180	5–11	0.5–5	800–1600	30–80	90–200
Andesite	2.30–2.75	120–350	5–9	0.2–8	400–3200	50–110	130–250
Basalt	2.75–3.00	200–1000	5–7	0.2–1.5	300-4200	60–120	140–260
Diabase	2.90–3.10	300–900	5–8	0.3–0.7	1200–2500	60–130	120–260
Volcanic tuff	1.30–2.20	—	5–10	8–35	50–600	5–45	30–80
Sedimentary rocks							
Sandstone	2.10–2.50	150–170	8–15	1–8	100–1200	15–60	40–160
Limestone (fine)	2.60–2.85	500–800	5–10	0.1–0.8	500–2000	40–70	50–150
Limestone (coarse)	1.55–2.30	—	8	2–16	40–600	10–35	25–70
Limestone (freshwater)	1.55–2.50	—	8–15	1.5–6	400–2000	15–50	30–90
Dolomite	2.20–2.70	200–300	5–12	0.2–4	150–2000	25–60	40–160
Shale-clay	2.45–2.75	—	—	0.2–0.4	—	—	200–300
Metamorphic rocks							
Marble	2.65–2.75	600–900	5–9	0.1–0.5	500–1800	50–80	80–120
Gneiss	2.60–2.78	250–600	5–11	1–5	800–2500	40–70	80–200

Note: [a] = Modified from References 1.1, 1.11, 1.25, and 1.26.

Table 1.30 Typical Static Mechanical Properties of Some Common Rock Types[1.27]

Rock class	Rock type	Unconfined compressive strength σ_c [MPa]	Tensile strength σ_t [MPa]	Modulus of elasticity E [GPa]	Point load index $I_{s(50)}$ [MPa]
Sedimentary rocks	Limestone	50–200	5–20	20–70	0.5–7
	Mudstone	5–15	—	—	0.1–6
	Sandstone	50–150	5–15	15–50	0.2–7
	Siltstone	5–200	2–20	20–50	6–10
	Shale	50–100	2–10	5–30	—
Metamorphic rocks	Gneiss	100–200	5–20	30–70	2–11
	Marble	100–200	5–20	30–70	2–12
	Quartzite	200–400	25–30	50–90	5–15
Igneous rocks	Basalt	100–300	10–15	40–80	9–14
	Gabbro	100–300	10–15	40–100	6–15
	Granite	100–200	5–20	30–70	5–10

Table 1.31 Some Strength Properties of Igneous and Metamorphic Rocks[1.28]

	Relative density (ρ)	Unconfined compressive strength σ_c [MPa]	Point load strength [MPa]	Shore scleroscope hardness	Schmidt hammer hardness	Young's modulus (E) (× 10^3 MPa)
Mount Sorrel Granite	2.68	176.4	11.3	77	54	60.6
Eskdale granite	2.65	198.3	12.0	80	50	56.6
Dalbeattie Granite	2.67	147.8	10.3	74	69	41.1
Markfieldite	2.68	185.2	11.3	78	66	56.2
Granophyre (Cumbria)	2.65	204.7	14.0	85	52	84.3
Andesite (Somerset)	2.79	204.3	14.8	82	67	77.0
Basalt (Derbyshire)	2.91	321.0	16.9	86	61	93.6
Slate[a] (North Wales)	2.67	96.4	7.9	41	42	31.2
Slate[b] (North Wales)		72.3	4.2			
Schist[a] (Aberdeenshire)	2.66	82.7	7.2	47	31	35.5
Schist[b]		71.9	5.7			
Gneiss	2.66	162.0	12.7	68	49	46.0
Hornfels (Cumbria)	2.68	303.1	20.8	79	61	109.3

Note: [a] Tested normal to cleavage or schistocity ; [b] Tested parallel to cleavage or schistocity.

Table 1.32 Some Strength Properties of Arenaceous Sedimentary Rocks[1.28]

	Fell Sandstone (Rothbury)	Chatsworth Grit (Stanton in the Peak)	Bunter Sandstone (Edwinstowe)	Keuper Waterstones (Edwinstowe)	Horton Flags (Helwith Bridge)	Bronllwyn Grit (Llanberis)
Relative density	2.69	2.69	2.68	2.73	2.7	2.71
Dry density (Mg/m$^{3)}$	2.25	2.11	1.87	2.26	2.62	2.63
Porosity	9.8	14.6	25.7	10.1	2.9	1.8
Dry unconfined compressive strength (MPa)	74.1	39.2	11.6	42.0	194.8	195.7
Saturated unconfined compressive strength (MPa)	52.8	24.3	4.8	28.6	179.6	190.7
Point dad strength (MPa)	4.4	2.2	0.7	2.3	10.1	7.4
Scleroscope hardness	42	34	18	28	67	88
Schmidt hardness	37	28	10	21	62	54
Young's modulus (x10^3 MPa)	32.7	25.8	6.4	21.3	67.4	51.1
Permeability (x10^{-9} m/sec)	1740	1960	3500	22.4		

Table 1.33 Some Strength Properties of Carbonate Rocks[1.28]

	Carboniferous limestone (Buxton)	Magnesium limestone (Anstone	Ancaster freestone (Ancaster)	Bath stone (Corsham)	Middle chalk (Hillington)	Upper chalk (North fleet)
Relative density	2.71	2.83	3.70	2.71	2.70	2.69
Dry density (Mg/m^3)	2.58	2.51	2.27	2.30	2.16	1.49
Porosity (%)	2.9	10.4	14.1	15.6	19.8	41.7
Dry unconfined compressive strength (MPa)	106.2	54.6	28.4	15.6	27.2	5.5
Saturated unconfined compressive strength (MPa)	83.9	36.6	16.8	9.3	12.3	1.7
Point load strength (MPa)	3.5	2.7	1.9	0.9	0.4	—
Scleroscope hardness	53	43	38	23	17	6
Schmidt hardness	51	35	30	15	20	9
Young's modulus (x10^3 MPa)	66.9	41.3	19.5	16.1	30.0	4.4
Permeability (x10^{-9} m/s)	0.3	40.9	125.4	160.5	1.4	13.9

Table 1.34 Some Strength Properties of Evaporitic Rocks[1.28]

	Gypsum (Sherburn in Elmet)	Anhydrite (Sandwith	Rock salt (Windsford)	Potash (Loftus)
Relative density	2.36	2.93	2.2	2.05
Dry density (Mg/m^3)	2.19	2.82	2.09	1.98
Porosity (%)	4.6	2.9	4.8	5.1
Unconfined compressive strength (MPa)	27.5	97.5	11.7	25.8
Point load strength (MPa)	2.1	3.7	0.3	0.6
Scleroscope hardness	27	38	12	9
Schmidt hardness	25	40	8	11
Young's modulus (x10^3 MPa)	24.8	63.9	3.8	7.9
Permeability	6.2	0.3	—	—

Table 1.35 Some Strength Properties of Coal Measure Rocks[1.28]

	Mudstone	Siltstone	Shale	Barnsley Hards coal	Deep Duffryn coal
Relative density	2.69	2.67	2.71	1.5	1.2
Dry density (Mg/ m^3)	2.32	2.43	2.35	—	—
Dry unconfined compressive strength (Mpa)	45.5	83.1	20.2	54.0	18.1
Saturated unconfined compressive strength (MPa)	21.3	64.8	—	—	—
Point load strength (MPa)	3.8	6.2	—	4.1	0.9
Scleroscope hardness	32	49	—	—	—
Schmidt hardness	27	39	—	—	—
Young's modulus (x 10^3 MPa)	25	45	5.2	26.5	—

1.7.3 Shear Strength Properties (τ)

The shear strength of a potential failure surface may consist of a single discontinuity plane or a complex path following several discontinuities and involving some fractured intact rock material.

1.7.3.1 Shear strength of rocks with planar unfilled discontinuities

Examples are bedding plane, single fault and joint with no surface roughness or undulations. The shear strength of rocks with planar unconfined discontinuities can be expressed in the following general forms:

$$\tau_p = S_o + (\sigma_n - u)\tan\phi_p \tag{1.14}$$

$$\tau_r = (\sigma n - u)\tan\phi_r \tag{1.15}$$

where:

τ_p = Peak shear strength, MPa ;

τ_r = Residual shear strength, MPa

S_0 = cohesion or cohesive strength, MPa ;

σ_n = Stress normal to the failure plane, MPa ;

u = Groundwater pressure, MPa ;

$\sigma_n - u$ = Effective normal stress, MPa ;

ϕ_p = Peak internal friction angle of failure plane to horizontal, degree ; and

ϕ_r = Residual internal friction angle of failure plane to horizontal, degree

Table 1.36 Some Physical and Mechanical Properties of Shales[1.29]

Physical properties			Probable *in situ* behavior						
	Average range of values								
Laboratory tests and *in situ* observations	Unfavorable	Favorable	High pore pressure	Low bearing capacity	Tendency to rebound	Slope stability problems	Rapid sinking	Rapid erosion	Tunnel support problems
Compressive strength (kPa)	350–2,070	2,070–3,500	x	x					
Modulus of elasticity (MPa)	140–1,400	1,400–14,000		x					x
Cohesive strength (kPa)	35–700	700–>10,500			x	x			x
Angle of internal friction (degrees)	10–20	20–65			x	x			x
Dry density (Mg/m^3)	1.12–1.78	1.78–2.56	x					x(?)	
Potential swell (%)	3–15	1–3			x	x		x	x
Natural moisture content (%)	20–35	5–15	x			x			
Coefficient of permeability (m/sec)	10^{-7}–10^{-12}	$>10^{-7}$	x			x	x		
Predominant clay minerals	Montmorillonite or illite	Kaolinite and chlorite	x			x			
Activity ratio	0.75–>20	0.35–0.75				x			
Wetting and drying cycles	Reduces to grain sizes	Reduces to flakes					x	x	
Spacing of rock defects	Closely spaced	Widely spaced		x		x		x	x
Orientation of rock defects	Adversely oriented	Favorable oriented		x		x			x
State of stress	>Existing overburden load	≈Overburden load			x	x			x

Note: According to S. Irmay (*Isr. J. Technol.*, 6(4), 165–172, 1968), the maximum possible $\phi = 47.4°$. The x relate to the unfavorable range of values.

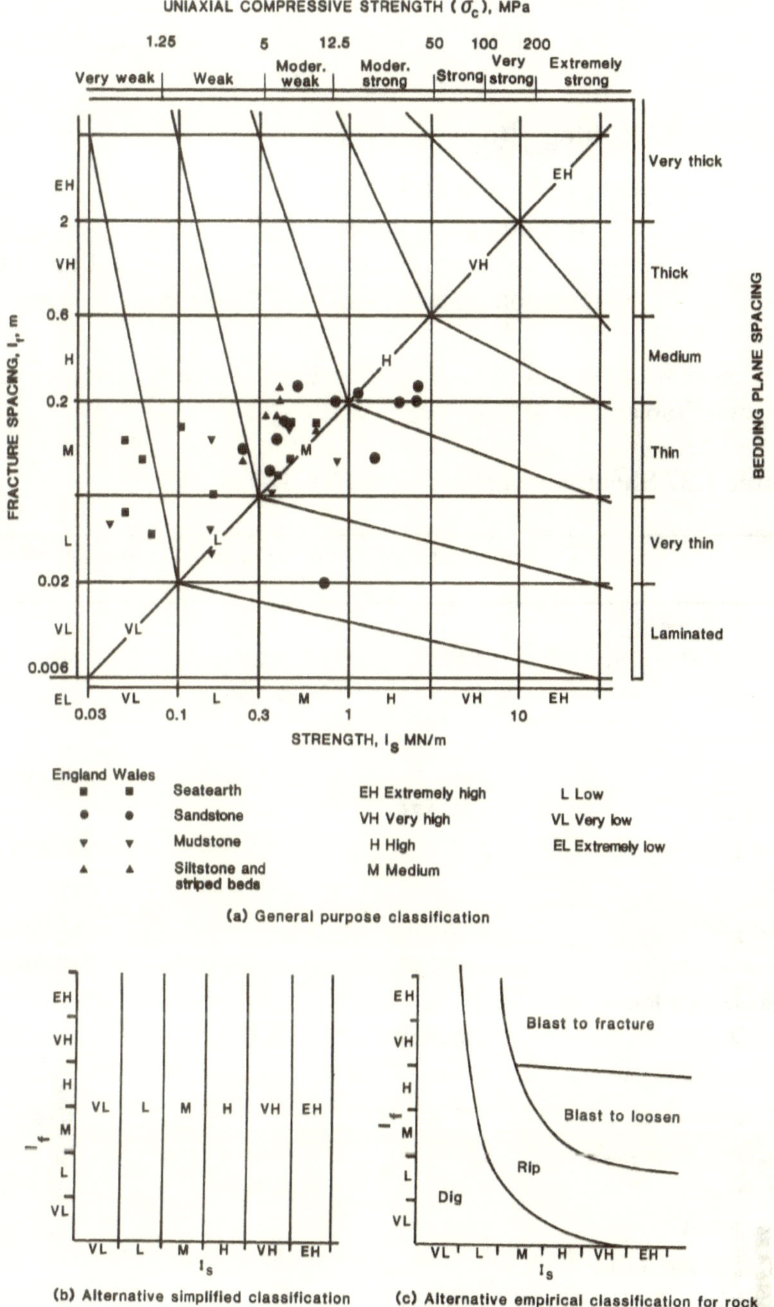

Figure 1.10. Classification of rock mass quality.[1.30]

One of the most recent empirical expressions to describe the shear strength of intact rock (τ) is given in a dimensionless form as follows:[1.32]

$$\tau/\sigma_n = \tan\{50 \log_{10}[(\sigma_1 - \sigma_3)/\sigma_n] + \phi_c\} \qquad (1.16)$$

where: σ_1 = axial stress at failure, MPa
σ_3 = effective confining pressure, MPa
ϕ_c = internal friction angle of the discontinuity filling or the cementation material; ≃25 to 28, degrees

Shear strength and frictional properties of some typical rocks are given in Tables 1.37 to 1.40.

Table 1.37 Shear Strength of Some Rocks[1.1, 1.11, 1.25, 1.26]

	Density, dry (t/m³)	Young's modulus (1000 × kg/cm²)	Shear strength (kg/cm²)
Batholitic Rocks			
Granite-Grandiorite	2.5–2.75	300–700	50–80
Gabbro	2.92–3.05	600–1000	40–85
Extrusive Rocks			
Riolite	2.45–2.60	100–200	40–110
Dacite	2.50–2.75	80–180	30–100
Andesite	2.30–2.75	120–350	50–120
Basalt	2.75–3.00	200–1000	50–130
Diabase	2.90–3.10	300–900	60–100
Volcanic tuff	1.30–2.20	—	10–40
Sedimentary Rocks			
Sandstone	2.10–2.50	150–170	20–60
Limestone (fine)	2.60–2.85	500–800	30–70
Limestone (coarse)	1.55–2.30	—	15–35
Limestone (freshwater)	1.55–2.50	—	20–50
Dolomite	2.20–2.70	200–300	25–70
Shale-clay	2.45–2.72	—	—
Metamorphic Rocks			
Marble	2.65–2.75	600–900	35–80
Gneiss	2.60–2.78	250–600	30–70

Table 1.38 Internal Friction Angle of Some Rocks Within the σ_c Range of 50 to 200 Mpa[1.27]

Rock class	Rock type	Unconfined compressive strength σ_c (MPa)	Angle of friction ϕ (degrees)
Sedimentary rocks	Limestone	50–200	33–40
	Mudstone	5–15	—
	Sandstone	50–150	25–35
	Siltstone	5–200	27–31
	Shale	50–100	27
Metamorphic rocks	Gneiss	100–200	23–23
	Marble	100–200	25–35
	Quartzite	200-400	48
Igneous rocks	Basalt	100–300	31–38
	Gabbro	100–300	—
	Granite	100–200	29–35

Table 1.39 Frictional Properties of Some Rocks[1.25]

Description of type/ material	Unit weight (γ) (Saturated/dry)		Friction angle (ϕ) (degrees)	Cohesion (S_0)	
	lb/ft³	kN/m³		lb/ft²	KPa
Loose sand, uniform grain size	118/90	19/14	28–34[a]		
Dense sand, uniform grain size	130/109	21/17	32–40[a]		
Loose sand, mixed grain size	124/99	20/16	34–40[a]		
Dense sand, mixed grain size	135/116	21/18	38–46[a]		
Gravel, uniform grain size	140/130	22/20	34–37[a]		
Sand and gravel, mixed grain size	120/110	19/17	48–45[a]		
Basalt	140/110	22/17	40–50[a]		
Chalk	80/62	13/10	30–40[a]		

Granite	125/110	20/17	45–50[a]		
Limestone	120/100	19/16	35–40[a]		
Sandstone	110/80	17/13	35–45[a]		
Shale	125/100	20/16	30–35[a]		
Soft bentonite	80/30	13/6	7–13	200–400	10 -20
Very soft organic clay	90/40	14/6	12–16	200–600	10 -30
Soft, slightly organic clay	100/60	16/10	22–27	400–1000	20 -50
Soft glacial clay	110/76	17/12	27–32	600–1500	30 -70
Stiff glacial clay	130/105	20/17	30–32	1500–3000	70 -150
Glacial till, mixed grain size	145/130	23/20	32–35	3000–5000	150 -250
Hard igneous rocks — granite, basalt, porphyry	160–190[b]	25–30	35–45	720,000–1,150,000	35000-55000
Metamorphic rocks — quartzite, gneiss, slate	160–180	25–28	30–40	400,000–800,000	20000-40000
Hard sedimentary rocks — limestone, dolomite, sandstone	150–180	23–28	35–45	200,000–600,000	10000-30000
Soft sedimentary rocks — sandstone, coal, chalk, shale	110–150	17–23	25–35	20,000–4,000,000	1000-20000

Note a = Higher friction angles in cohesion-less materials occur at low confining or normal stresses.
b = For intact rock, the unit weight of the material does not vary significantly between saturated and dry states with the exception of materials such as porous sandstones.

The influence of porosity and water upon the cohesive and frictional properties of rock mass and its discontinuities depends upon the rock type (i.e., its mineralogy), and the nature of the discontinuity filling or cementing material. Therefore, it is important to determine the shear strength and frictional properties of rock mass as exact as *in situ* moisture content and conditions. Figure 1.11 shows the relationship between angle of internal friction at failure (ϕ_f) and porosity for various clay, silt, sand, and bentonite fillings.

Table 1.40 Cohesive Strength (S_0) and Internal Friction Angle (ϕ) for Various Rock Mass Conditions and Rock Types[1.25, 1.26]

Material description	S_0 (lb/ft²)	S_0 (lb/ft²)
Very soft soil	35	170
Soft soil	70	340
Firm soil	180	880
Stiff soil	450	2,200
Very stiff soil	1,600	7,800
Very soft rock	3,500	17,000
Soft rock	11,500	56,00
Hard rock	35,000	170,000
Very hard rock	115,000	560,000
Very, very hard rock	230,000	1,000,000

Approximate friction angles for typical rocks

Rock	Intact rock φ	Joint φ	Residual φ
Andesite	45	31–35	28–30
Basalt	48–50	47	
Chalk		35–41	
Diorite	53–55		
Granite	50–64		31–33
Graywacke	45–50		
Limestone	30–60		33–37
Monzonite	48–65		28–32
Porphyry		40	30–34
Quartzite	64	44	26–34
Sandstone	45–50	27–38	25–34
Schist	26–70		
Shale	45–64	37	27–32
Silstone	50	43	
Slate	45–60		24–34

Other materials	Approximate ϕ values
Clay gouge (remolded)	10–20
Calcite shear zone material	20–27
Shale fault material	14–22
Hard rock breccia	22–30
Compacted hard rock aggregate	40
Hard rock fill	38

1.7.3.2 Shear Strength of Rocks with Smooth Inclined Plane of Unfilled Discontinuity

In this case the discontinuity plane is in an inclined angle (i) to the direction of horizontal shear stress (τ). Therefore, unlike the case in Section 1.7.3.1, the shear and normal stresses acting on the inclined failure surface are not τ and σ_n, but are given as follows:[1.25]

$$\tau_i = \tau \cdot \cos^2 i - \sigma_n \cdot \sin i \cdot \cos i \tag{1.17}$$

$$\sigma_i = \sigma_n \cdot \cos^2 i + \tau \cdot \sin i \cdot \cos i \tag{1.18}$$

where: τ_i = shear stress parallel to the inclined discontinuity plane, MPa

σ_i = stress normal to the inclined discontinuity plane, MPa

If the discontinuity surface is cohesion-less (i.e., $S_0 = 0$), then:

$$\tau_i = \sigma_i \cdot \tan \phi \tag{1.19}$$

Substituting Equations 1.17 and 1.18 into Equation 1.19 gives:

$$\tau = \sigma_n \cdot \tan(\phi + i) \tag{1.20}$$

Equation 1.20 is confirmed in practice as well.[1.33]

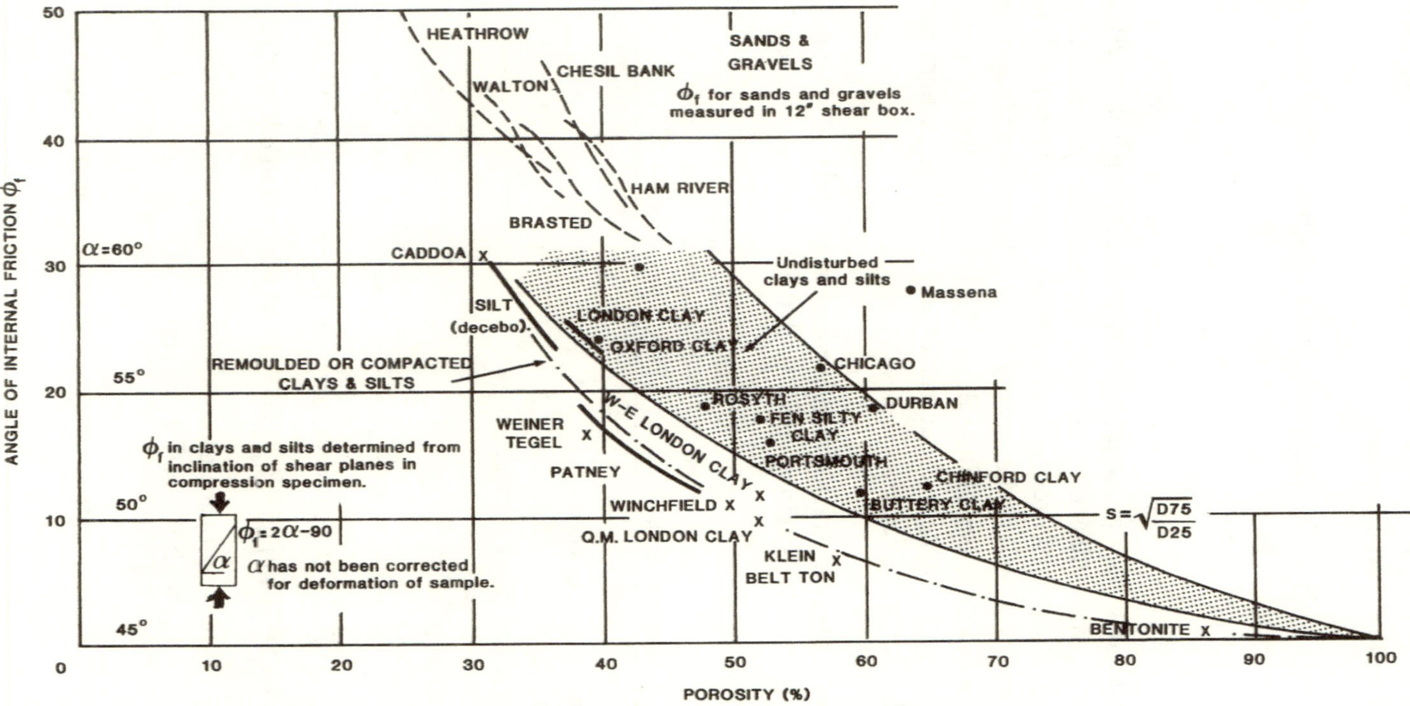

Figure 1.11. Relation between angle of internal friction at failure and porosity for various discontinuity fillings.[1.25]

1.7.3.3 Shear Strength of Rocks with Inclined and Rough Surfaces of Unfilled Discontinuity

Equations 1.17 to 1.20 relate to an inclined discontinuity plane of first order projection angle (i) to the direction of shear stress (τ). In the case of inclined discontinuity plane with small bumps and ripples, the i value is much higher. This later i value is called the second order projections.[1.34] They are effective at very low normal stresses and were measured as (ϕ + i), values of which are given in Table 1.41A.

As the normal stress increases, the second order projection are sheared off and the first order projections take over as the controlling factor. Further increase of the normal stress will cause the first order projections to also shear off, resulting in no further dilation (i.e., i → 0), and eventually the internal friction angle will be the dominant factor at higher stress levels.

Table 1.41 Internal friction and roughness inclination angle (ϕ + I) for various rocks under low normal stresses.

Type of surface	Normal stress σ_n (kg/cm²)	(ϕ + i)
Limestone, slightly rough bedding surfaces	1.57–6.00	71°–77°
Limestone, rough bedding surfaces	3.05–6.80	66°–72°
Shale, closely jointed seam in limestone	0.21	70°–71°
Quartzite, gneiss and amphibolite discontinuity beneath natural slopes	—	80°
Beneath excavated slopes	—	75°
Granite, rough undulating artificial tension fractures	1.5–3.5	69°–72°

The transition from dilation to shearing can be expressed as follows:[1.35]

$$\tau = \sigma_n [(1 - a_s) (v + \tan \phi) + a_s \,.\, \tau r] / [1 - (1 - a_{s)}\ v. \tan \phi] \qquad (1.21)$$

where: a_s = portion of the discontinuity surface which is sheared through projections of intact rock material = $1 - [1 - (\sigma_n / \sigma_j)]^L$

v = dilution rate dv /du at peak shear strength = $[1 - (\sigma_n / \sigma_j)]^K . \tan i$

σ_j = uniaxial compressive strength of the rock material adjacent to the discontinuity

K = constant = 4 for rough discontinuity surfaces

L = constant = 1.5 for rough discontinuity surfaces

τ_r = shear strength of the intact rock material

Simpler methods for driving the shear strength (τ) along discontinuity surfaces are given as follows (Hoek and Brown[1.11]):

$$\tau = A \cdot \sigma_c [(\sigma_n / \sigma_c) T]^n \qquad (1.22)$$

where: A = constant defining the shape of Mohr failure envelope

σ_c = uniaxial compressive strength of individual intact rock pieces within the rock mass, MPa

T = $[m - (m^2 + 4s)^{0.5}] / 2$

m, s = dimensionless constants dependent upon the shape and degree of interlocking of the individual pieces of rock within the mass. The m and s values are very site sensitive and should be chosen accurately (see Table 1.41A).

In general, for disturbed rock masses[1.11]: $m = m_i \cdot \exp[(RMR - 100)/14]$

$$s = \exp[(RMR - 100)/6]$$

m_i = value of m for intact rock

For undisturbed or interlocking rock masses[1.11a]:

$$m = m_i \cdot \exp[(RMR - 100)/28]$$

$$s = \exp[(RMR - 100)/9]$$

From Barton:[1.32] $\tau = \sigma_n \cdot \tan [JRC.\ \log_{10} (JCS / \sigma_n) + \phi_b]$ (1.23)

where: σ_n = effective normal stress, MPa

JRC = roughness coefficient along the discontinuity plane, i.e., joint

JCS = compressive strength of the discontinuity or joint surface, MPa

ϕ_b = basic internal friction angle (see Table 1.42)

Table 1.41A Approximate Relationship Between Rock Mass Quality and Material Constants[1.11A]

Disturbed rock mass m and s Empirical failure criterion $\sigma'_1 = \sigma'_3 + \sqrt{m\sigma_c \cdot \sigma'_3 + s \cdot \sigma_c^2}$ σ'_1 = major principal effective stress σ'_3 = minor principal effective stress σ_c = uniaxial compressive strength of intact rock, and m, s = empirical constants		Undisturbed rock mass m and s values: Carbonate rocks with well developed crystal cleavage (dolomite, limestone, and marble)	Lithified argillaceous rocks [mudstone, siltstone, shale, and slate (normal to cleavage)]	Arenaceous rocks with strong crystals and poorly developed crystal cleavage (sandstone and quartzite)	Fine grained polymineralic igneous crystalline rocks (andesite, dolerite, diabase, and rhyolite)	Coarse grained polymineralic igneous & metamorphic crystalline rocks (amphibolite, gabbro gneiss, granite, norite, quartz-diorite)
Intact rock samples						
Laboratory size specimens free	m	7.00	10.00	15.00	17.00	25.00
from discontinuities	s	1.00	1.00	1.00	1.00	1.00
CSIR rating: RMR = 100	m	7.00	10.00	15.00	17.00	25.00
NGI rating: Q = 500	s	1.00	1.00	1.00	1.00	1.00
Very good quality rock mass						
Tightly interlocking undisturbed rock	m	2.40	3.43	5.14	5.82	8.56
with unweathered joints at 1 to 3 m	s	0.082	0.082	0.082	0.082	0.082
CSIR rating: RMR = 85	m	4.10	5.85	8.78	9.95	14.63
NGI rating: Q = 100	s	0.189	0.189	0.189	0.189	0.189
Good quality rock mass						
Fresh to slightly weathered rock,	m	0.575	0.821	1.231	1.395	2.052
slightly disturbed with joints at 1-3 m	s	0.00293	0.00293	0.00293	0.00293	0.00293
CSIR rating: RMR = 65	m	2.006	2.865	4.298	4.871	7.163
NGI rating: Q = 10	s	0.0205	0.0205	0.0205	0.0205	0.0205
Fair quality rock mass						
Several sets of moderately weathered	m	0.128	0.183	0.275	0.311	0.458
joints spaced at 0.3-1 m	s	0.00009	0.00009	0.00009	0.00009	0.00009
CSIR rating: RMR = 44	m	0.947	1.353	2.030	2.301	3.383
NGI rating: Q = 1	s	0.00198	0.00198	0.00198	0.00198	0.00198
Poor quality rock mass						
Numerous weathered joints at 30-500 mm,	m	0.029	0.041	0.061	0.069	0.102
some gouge; clean compated waste rock	s	0.000003	0.000003	0.000003	0.000003	0.000003
CSIR rating: RMR = 23	m	0.447	0.639	0.959	1.087	1.598
NGI rating: Q = 0.1	s	0.00019	0.00019	0.00019	0.00019	0.00019
Very poor quality rock mass						
Numerous heavily weathered joints	m	0.007	0.010	0.015	0.017	0.025
spaced < 50mm with gouge; waste rock with fines	s	0.0000001	0.0000001	0.0000001	0.0000001	0.0000001
CSIR rating: RMR = 3	m	0.219	0.313	0.469	0.532	0.782
NGI rating: Q = 0.01	s	0.00002	0.00002	0.00002	0.00002	0.00002

Table 1.42 Approximate Values of the Basic Internal Friction Angle (ϕ_b) for Various Dry Rock Types[1.34]

Rock type	ϕ_b (degrees)
Amphibolite	32
Basalt	31–38
Conglomerate	35
Chalk	30
Dolomite	27–31
Gneiss (schistose)	23–29
Granite (fine grain)	29–35
Granite (coarse grain)	31–35
Limestone	33–40
Porphyry	31
Sandstone	25–35
Shale	27
Siltstone	27–31
Slate	25–30

Note: Wet rocks give lower ϕ_b values.

Barton et al.[1.14] have empirically suggested that the ratio of joint roughness (J_r) to the degree of alteration of joint walls or filling materials (J_a) can be a fair approximation to shear strength of the discontinuity (τ) in the following form:

$$\tau = \tan^{-1} (J_r / J_a) \tag{1.24}$$

Table 1.43 gives values of $\tan^{-1} (J_r / J_a)$ for various categories of rock-wall contact described in Tables 1.8 and 1.9.

The shear strength of intact rock material (τ_r) can be determined from one of the following expressions (Fairhurst[1.36]):

$$\tau_r = \sigma_j \{[(1 + n)^{0.5} - 1] / n\} [1 + (n.\sigma) / \sigma j]^{0.5} \tag{1.25}$$

where, n = ratio of uniaxial compressive strength to uniaxial tensile strength of the rock (σ_c / σ_t) $\simeq$ 10, for most hard intact rocks[1.25]

A simpler technique to drive the τ_r value is by using the Mohr -Coulomb's equation:

$$\tau_r = S_{oj} + \sigma \cdot \tan\phi_j \tag{1.26}$$

where:

S_{oj} = cohesion of the joint or discontinuity material, MPa

= $C_h + (C_v - C_h) \cos^2 i$

ϕ_j = internal friction angle of the joint or discontinuity material, degree

C_h = cohesion of discontinuity material whose major principal stress i horizontal, MPa

C_v = cohesion of discontinuity material whose major principal stress is vertical, MPa

Table 1.43 Estimation of Apparent Shear Strength of Discontinuity from Equation 1.24[1.14]

Rock wall contact	J_r	$\tan^{-1}(J_r/J_a)°$, degrees				
		$J_a = 0.75$	**1.0**	**2**	**3**	**4**
Discontinuous joints	4	79	76	63	53	45
Rough, undulating	3	76	72	56	45	37
Smooth, undulating	2	69	63	45	34	27
Slickensided, undulating	1.5	63	56	37	27	21
Rough planar	1.5	63	56	37	27	21
Smooth, planar	1.0	53	45	27	18	14
Slickensided, planar	0.5	34	27	14	9.5	7.1
Rock wall contact when sheared		**$J_a = 4$**	**6**	**8**	**12**	--
Discontinuous joints	4	45	34	27	18	–
Rough, undulating						
Smooth, undulating	2	27	18	14	9.5	--
Slickensided, undulating	1.5	21	14	11	7.1	--
Rough, planar	1.5	21	14	11	7.1	--
Smooth, planar	1.0	14	9.5	7.1	4.7	--
Slickensided, planar	0.5	7	4.7	3.6	2.4	--
		$J_a = 6$	**8**	**12**		
Disintegrated or crushed rock and clay	1.0	9.5	7.1	4.7		
		$J_a = 5$				
Bands of silty- or sandy clay	1.0	11				
		$J_a = 10$	**13**	**20**		
Thick continuous bands of clay	1.0	5.7	4.4	2.9		

1.7.3.4 Shear Strength of Filled Discontinuities

This is a common case encountered in practice. The fill material may provide a stronger cementation to the relatively weaker rock mass. However, the strength properties of fill are generally weaker than those of hard rocks. In the latter case, shear strength of the rock mass is inversely proportional to the fill thickness. A list of shear strength properties for filled joints is given in Table 1.44.

Table 1.44 Shear Strength Properties of Filled Joints[1.37]

Rock type	Description of fill	Peak strength		Residual strength		Tested by
		S_{o}, kg/cm^2	$\phi°$	S_{o}, kg/cm^2	$\phi°$	
Basalt	Clayey basaltic breccias, wide variation from clay to basalt content	2.4	42			Ruiz, Camargo, Midea, and Nieble (1968)
Bentonite	Bentonite seam in chalk	0.15	7.5			Link (1969)
	Thin layers	0.9–1.2	12–17			
	Triaxial tests	0.6–1.0	9–13			Sinclair and Brooker (1967)
Bentonitic shale	Triaxial tests	0–2.7	8.5–29			Sinclair and
	Direct shear tests			0.3	8.5	Brooker (1967)
Clays	Over-consolidated, slips, joints, and minor shears	0–1.8	12–18.5	0–0.03	10.5–16	Skepmton and Petley 1962)
Clay shale	Triaxial tests	0.6	32			Sinclair and Brooker (1967)
Clay shale	Stratification surfaces			0	19–25	Leussink and Muller-Kirchenbauer (1967)
Coal measure rocks	Clay mylonite seams, 1.0 to 2.5-cm thick	0.11–0.13	16	0	11–11.5	Stimpson and Walton (1970
Dolomite	Altered shale bed, approximately 15 cm thick.	0.41	14.5	0.22	17	Pigot and Mackenzie (1964)

Diorite, granodiorite, and porphyry	Clay gouge (2% clay, PI = 17%)	0	26.5			Brawner (1970)
Granite	Clay filled faults	0–1.0	24–45			Rocha (1964)
	Weakened with sandy-loam fault filling	0.5	40			Nose (1964)
	Tectonic shear zone, schistose and broken granites, disintegrated rock and gouge	2.42	42			Evdokimov and Sapcgin (1970
Greywacke	1–2 mm clay in bedding planes			0	21	Drozd (1967)
Limestone	6 mm clay layer			0	13	Krsmanovic et al. (1966)
	1–2 cm clay fillings	1.0	13–14			Krsmanovic and
	<1 mm clay fillings	0.5–2.0	17–21			Popodovic (1966)
Limestone, marl	Interbedded lignite layers	0.8	38			Salas and Uriel
and lignites	Lignite/marl contact	1.0	10			(1964)
Limestone	Marlaceous joints, 2 cm thick	0	25	0	15–24	Bernaix (1969)
Lignite	Layer between lignite and underlying clay	0.14–0.3	15–17.5			Schultze (1957)
Montmorillonite	8 cm seams	3.6	14	0.8	11	Eurenius (1969)
clay	of bentonite (montmorillonite) clay in chalk	0.16–0.2	7.5–11.5			Underwood (1964)
Schists,	10–15-cm thick	0.3–0.8	32			Serafim and
quartzites and	clay filling	6.1–7.4	41			Guerreiro
siliceous schists	Stratification with	3.8	31			(1968)
Slates	thin clay	0.5	33			Coates,
	Stratification with thick clay					McRorie, and Stubbins
	Finely laminated and altered					(1963)
Quartz/kaolin/ pyrolusite	Remolded triaxial tests	0.42–0.9	36–38			

1.8 ROCK MASS FACTOR (j)

This is defined as the ratio of deformity of a rock mass within any readily identifiable lithological and structural component to that of the deformability of the intact rock comprising the component. Consequently, it reflects the effect of discontinuities on the expected performance of intact rock. The value of j depends upon the method of assessing the deformability of the rock mass, and the value beneath an actual foundation will not necessarily be the same as that determined from even a large scale field test, particularly on the more massive rocks. Generally the plate loading test, suitably scaled to the discontinuity spacing, is the best method of determining the field modulus of deformation. An assessment of rock quality and deformability can be made by using water injection tests between packers in drill-holes. High hydraulic conductivity in relation to the fracture frequency means correspondingly low j values and vice versa. In simple joints systems it is theoretically possible to relate joint opening to hydraulic conductivity. According to Hobbs[1.38] the greatest difficulties which occur in jointed rock masses in relation to foundation design are experienced in those where the fracture spacing falls within a range of about 100 to 500 mm, in as much as small variations in fracture spacing and condition result in exceptionally large changes in the j value. Some of the classifications of rock quality in relation discontinuities and the ratio of the compressional wave velocity of the rock mass *in situ* (V_f) to that of the intact rock specimen (V_l) are given in Table 1.45.

Table 1.45 Classification of Rock Mass Factor (j) in Relation to the Incidence of Discontinuities[1.38]

Quality Classification	RQD %	Fracture frequency per meter	Velocity ratio V_f/V_l	Mass (j) factor
Very poor	0–25	Over 15	0–0.2	
Poor	25–50	15–8	0.2–0.4	Less than 2.0
Fair	50–75	8–5	0.4–0.6	0.2–0.5
Good	75–90	5–1	0.6–0.8	0.5–0.8
Excellent	90–100	Less than 1	0.8–1.0	0.8–1.0

1.9 ROCK MASS RATING (RMR)

The RMR, also called the Geomechanics classification system was developed at the South African Council of Scientific and Industrial Research (CSIR) and proposed by Bieniawski.[1.1, 1.4, 1.22, 1.24, 1.39-1.43] It is based on the algebraic sum of six rock mass properties rating, namely:

1. Strength of intact rock material, designated by the uniaxial compressive strength of cored samples (σ_c), or for very low-strength rocks by the point load strength index (see Table 1.46). This is based on the Deere and Miller[1.11, 1.39] classification of intact rock strength.

Table 1.46 Rock Mass Rating Increments for the Strength of Intact Rock Material[1.11, 1.39]

Point index (Mpa)	or	Unconfined compressive strength (MPa	Rating
>8		>200	15
4–8		100–200	12
2–4		50–100	7
1–2		25–50	4
Do not use		10–25	2
Do not use		3–10	1
Do not use		<3	0

2. RQD as described in Section 1.4 and Table 1.12.
3. Spacing of joints or discontinuities, as presented in Section 1.3.1, Figure 1.3, and Table 1.5.
4. Condition and quality of joints or discontinuities, accounting for the separation, persistence or continuity, surface roughness, nature of any fill material present or cementation, and the discontinuity wall conditions (i.e., hard or soft) see Tables 1.6 to 1.8, and 1.47.

Table 1.47 Rock Mass Rating Increments for the Condition and Quality of Joints[1.9]

Description	Rating
Very rough surfaces of limited extent; hard wall rock	25
Slightly rough surfaces; aperture less than 1 mm; hard wall rock	20
Slightly rough surfaces; aperture less than 1 mm; soft wall rock	12
Smooth surfaces, **or** gouge filling 1–5 mm thick, **or** aperture of 1–5 mm; joints extend more than several meters	6
Open joints filled with more than 5 mm of gouge, **or** open more than 5 mm; joints extend more than several meters	0

5. Condition of groundwater, as water presence or inflow and the ratio of joint water pressure to major principal stress, see Tables 1.9 to 1.11.
6. Orientation of joint or discontinuity relative to the excavation. This is evaluated according to Tables 1.48 and 1.49 for tunnels, foundations and slopes.[1.40] When falling or sliding of rock blocks from the roof or walls of an excavation is a possibility, a wedge or block analysis described in Chapter 2 should be used.[1.26] No points are subtracted for very favorable orientations of discontinuities. Up to 12 points are deducted for unfavorable discontinuity orientations in tunnels, up to 25 points for that of foundations, and up to 60 points for the case of slopes.

Table 1.48 The Effect of Joint Strike and Dip in Tunnelling[1.40]

Strike perpendicular to tunnel axis				**Strike parallel to tunnel axis**	
Drive with dip		Drive against dip			Dip 0 - 20°
Dip	Dip	Dip	Dip	Dip	Dip irrespective
45–90°	20–45°	45–90°	20–45°	45–90°	20–45° of strike
Very favorable	Favorable	Fair	Unfavorable	Very unfavorable	Fair Unfavorable

Table 1.49 Adjustment in the Rock Mass Rating (RMR) for Orientation of Discontinuity[1.40]

Assessment of influence of orientation on the work	Rating increment for tunnels	Rating increment for foundations	Rating increments for slopes
Very favorable	0	0	0
Favorable	−2	−2	−5
Fair	−5	−7	−25
Unfavorable	−10	−15	−50
Very unfavorable	−12	−25	−60

It is difficult to apply Tables 1.48 and 1.49 accurately without adequate experience, as a given discontinuity orientation may be favorable or unfavorable depending on the groundwater and condition of joint fill. The orientation of joint sets cannot be determined from normal, routine drilling of rock masses but can be measured from drill core with special tool or procedure.[1.44] Logging of the borehole using a television or camera down-hole will reveal the orientation of discontinuities and absolute orientations will also be obtained from logging shafts and adits.

Summary of the RMR ranges for various class of rock mass is given in Table 1.50, and the complete geomechanical classification of jointed rock masses, or RMR is given in Table 1.51.

Table 1.50 Summary of the Ranges of Rock Mass Rating (RMR) for Various Classifications of Rock Mass[1.42]

Class	Description of rock mass	RMR Sum of rating increments
1	Very good rock	81–100
2	Good rock	61–80
3	Fair rock	41–60
4	Poor rock	21–40
5	Very poor rock	0–20

Table 1.51 Geomechanics Classification of Jointed Rock Masses[1.42]

Parameter	Ranges of values				
	A. Classification parameters and their ratings				
1. Strength of intact rock material:					
i–Point-load strength index (MPa)	>9	5–8	3–4	1–2	For this low range, uniaxial compression test is preferred
ii–Uniaxial compressive strength (MPa)	>201	101–200	51–100	26–50	11–25 4–10 1–3
Rating	15	12	7	4	2 1 0
2. Drill core quality RQD (%)	91–100	76–90	51–75	26–50	<25
Rating	20	17	13	8	3
3. Joint spacing (m)	>4	1.5–3	0.4–1	0.06–0.3	<0.05
Rating	30	25	20	10	5
4. Condition of joints	Very rough surfaces, not continuous, no separation, hard joint wall rock	Slightly rough surfaces, separation <1 mm, hard joint wall rock	Slightly rough surfaces, separation <1 mm, soft joint wall rock	Slickensided surfaces or gouge <5 mm thick or joints open 1-5 mm, continuous joints	Soft gouge >5 mm thick or joints open >5 mm continuous joints
Rating	25	20	12	6	0
5. Groundwater					
Inflow per 10-m tunnel length ($l \cdot min^{-1}$)	None		<25	26–125	>126
Joint water pressure Major principal stress	0		0.1–0.2	0.3–0.5	>0.6
General conditions	Completely dry		Moist only (interstitial water)	Water under moderate pressure	Severe water problem
Rating	10		7	4	0

Continuation of the Table 1.51 - Geomechanics characterization of jointed rock masses.

B. Rating adjustment for joint orientations

Strike and dip orientations of joints:	Very favorable	Favorable	Fair	Unfavorable	Very unfavorable
Ratings for					
i–Tunnels	0	−2	−5	−10	−12
ii–Foundations	0	−2	−7	−15	−25
iii–Slopes	0	−5	−25	−50	−60

C. Rock mass classes determined from total ratings

Ratings	100–81	80–61	60–41	40–21	<20
Class no.	I	II	III	IV	V
Description	Very good rock	Good rock	Fair rock	Poor rock	Very poor rock

D. Meaning of rock classes

Class no.	I	II	III	IV	V
Average stand-up time	10 years for 5-m span	6 months for 4-m span	1 week for 3-m span	5 h for 1.5-m span	10 min for 0.5-m span
Cohesion of the rock mass (kPa)	>301	201–300	151–200	101–150	<100
Friction angle of the rock mass	>46°	41–45°	36–40°	31–35°	<30°

The interpretation of RMR in terms of standing times of underground excavations, its unsupported span, and rock mass strength parameters are given in Part D of Table 1.51 and illustrated Figure 1.12. Time dependent convergence of tunnels (in section D) at 40 < RMR < 71 is shown in Figure 1.13.

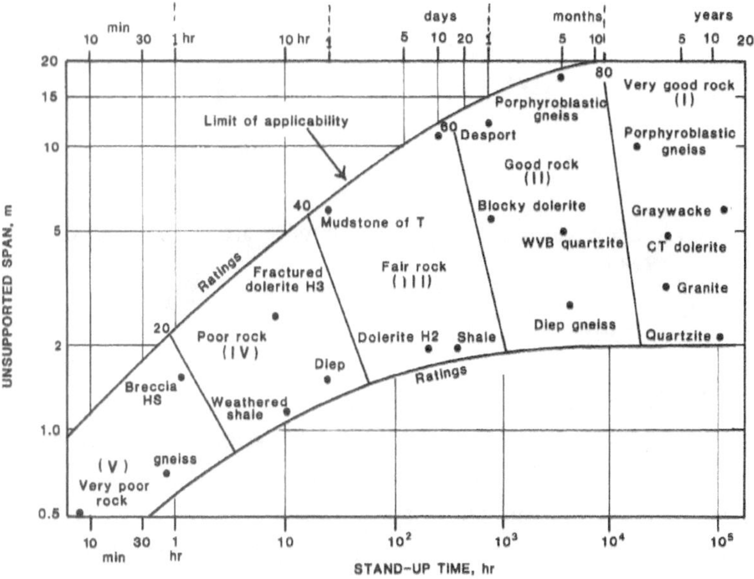

Note: The unsupported span is the length of unsupported section at the face or the width of the tunnel which ever is greater.

Figure 1.12. Geomechanics classification of rock masses or rock mass rating (RMR) applied to predicting tunneling conditions.[1.41]

If a parameter value for calculation of the RMR is not available, it can be estimated and a sensitivity check can then be run on the RMR value for different values chosen.

Proceeding identification of the structural regions, the classification parameters for each structural region are determined from field measurements and recorded on the input data form given in Table 1.52.

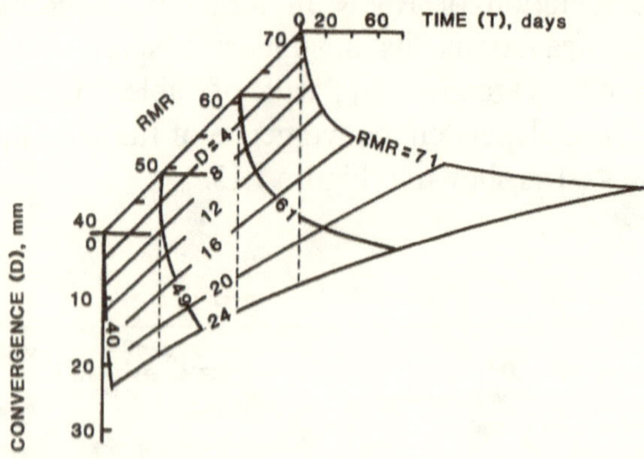

Figure 1.13. Time dependent tunnel convergence at various rock mass rating (RMR).[1.45]

1.9.1 Case Study

A 4 m-wide tunnel is to be driven in a Gneiss having an average uniaxial compressive strength (σ_c) of 130 MPa. The dominant joint set has a spacing of 1.5 m, it dips at 40° against the drive direction, and strikes approximately perpendicular to the long axis of the tunnel. The joints surfaces are slightly rough with hard walls. Coring has revealed an RQD of 45%. Groundwater inflow was observed to be less than 25 l/min/100 m of tunnel length under low moisture. Determine the total RMR for the rock along the tunnel, the stand-up time for unsupported tunnel, the average cohesion, and the internal friction angle of the rock mass.

Solution. Information from the above noted data can be input into the Tables 1.48 and 1.51. The output is given in Table 1.53, with the total RMR being the algebraic sum of all six parameter ratings.

The stand-up time of the unsupported 4-m wide tunnel at RMR of 62 is one month (from Figure 1.12). The average cohesion of the rock mass (S_0) = 205 MPa. The average internal friction angle of the rock mass (ϕ) = 41°. Hence, the tunnel needs to be appropriately reinforced or supported for longer utilization.

Table 1.52 Input Data Form for Determination of the Rock Mass Rating (RMR)[1.41]

Name of project:	
Site of survey:	ROCK TYPE AND ORIGIN
Conducted by:	
Date:	STRUCTURAL REGION

Drill Core Quality RQD [a]		Wall Rock Discontinuities
Excellent quality	90–100σ°	Unweathered ________
Good quality	75–90σ°	Slightly weathered ________
Fair quality	50–75σ°	Moderately weathered ________
Poor quality	25–50σ°	Highly weathered ________
Very poor quality	<25σ°	Residual soil ________

Note: a RQD = Rock Quality Designator

Ground Water		Strength of Intact Rock Material			
Inflow per 10 m of tunnel length	Liters min^{-1}	Designation	Uniaxial compressive strength, MPa	OR	Pore-load strength index, MPa
Or					
Water pressure	Pa	Very high	Over 250		>10
Or		High	100–250		1–10
General conditions (completely dry, damp, wet, dripping, or flowing under low, medium, or high pressure)		Medium high	50–100		2–1
		Moderate	25–50		1–2
		Low	5–25		<1
		Very low	1–5		

Spacing of Discontinuities					
		Set 1	Set 2	Set 3	Set 4
Very wide:	Over 2 m				
Wide:	0.6–2 m				
Moderate:	200–600 mm				
Close:	60–200 mm				
Very close:	<60 mm				

Table 1.52 continued

Strike and Dip Orientations				
Set 1	Strike: _______ (average)	(from_______ to _______)	Dip: _______ (angle)	_______ (direction)
Set 2	Strike: _______	(from_______ to _______)	Dip:_______	_______
Set 3	Strike: _______	(from_______ to _______)	Dip: _______	_______
Set 4	Strike: _______	(from_______ to _______)	Dip: _______	_______

Condition of Discontinuities		Set 1	Set 2	Set 3	Set 4
Persistence (continuity)					
Very low:	<1 m				
Low:	1–3 m				
Medium:	3–10 m				
High:	10–20 m				
Very high:	> 20 m				
Separation (aperture)	<0.1 mm				
Very tight joints:	0.1–0.5 mm				
Tight joints:	0.5–2.5 mm				
Moderately open joints:	2.5–10 mm				
Open joints:	>10 mm				
Very wide aperture					
Roughness (state also if surface are stepped, undulating or planar)					
Very rough surfaces:					
Rough surfaces:					
Slightly rough surfaces:					
Smooth surface:					
Slickensided surfaces:					
Filling (gouge)					
Type:					
Thickness:					
Uniaxial compressive strength, MPa					
Seepage:					

Table 1.52 continued

Major Faults or Folds
Describe major faults and folds specifying their locality, nature, and orientations.

General Remarks and Additional Data
1. For definitions and methods consult ISRM document: *Quantitative Description of Discontinuities in Rock Masses.*
2. The data on this form constitute the minimum required for engineering design. The geologist should, however, supply any further information which he considers relevant.

Note: Refer all directions to the magnetic north.

The following example clarifies determination of Bieniawski's RMR technique.

Table 1.53 Determination of Rock Mass Rating (RMR) for a Tunnel Driven in Gneiss Parameters

	Parameters	Field Observation	RMR
1.	Uniaxial compressive strength of intact rock (σ_c or IRS)	130 MPa	12
2.	Rock quality designation (RQD)	45%	8
3.	Joint spacing	1.5 m	25
4.	Condition of joints	Slightly rough with hard walls	20
5.	Condition of groundwater	<25 l/min/100 m tunnel	7
6.	Strike and dip orientation of joints	Unfavorable (Table 1.48)	–10

Note: Total RMR = 72 – 10 = 62 = Lower end of class II, good rock.

1.10 LAUBSCHER'S ADJUSTED RMR RATING (AR)

On the basis of field observation in chrysolite asbestos mines of Africa, Laubscher[1.5] has modified the Bieniawski's RMR classification in the form given in Tables 1.54, 1.55 and 1.56. This is based on the Bieniawski's six classification parameters divided into subclasses A and B. Also new ranges of ratings for intact rock strength (IRS) are used. The joint spacing and condition of joint parameters are evaluated differently.

The total possible percentage adjustments that can be made to each classification parameter for each of the site factors encountered is shown in Table 1.56.

Table 1.54 Laubscher's Modified Geomechanics Classification Scheme (AR)1.5

Class:	1		2		3		4		5	
Rating:	100–81		80–61		60–41		40–21		20–0	
Description:	Very good		Good		Fair		Poor		Very poor	
Subclasses item:	A	B	A	B	A	B	A	B	A	B
RQD, %	100–91	90–76	75–66	65–56	55–46	45–36	35–26	25–16	15–6	5–0
Rating	20	18	15	13	11	9	7	5	3	0
IRS, MPa	141–136	135–126	125–111	110–96	95–81	80–66	65–51	50–36	35–21	20–6
Rating	10	9	8	7	6	5	4	3	2	1
Joint spacing [a]										
Rating	30									0
Condition of joint [b]	45°				static angle of friction					5°
Rating	30									0
	Inflow per 10-m length, or		0		25 1 min^{-1}		25–125 1 min^{-1}		125 1 min^{-1}	
Groundwater	Joint water pressure		0		0.0–0.2		0.2–0.5		0.5	
	Major principal stress									
Description	Completely dry		Completely dry		Moist only		Moderate pressure		Severe problems	
Rating	10		10		7		4		0	

Note [a] Refer to Figure 1.3 for information ; [b] Refer to Tables 1.55 and 1.56.

1.11 ROCHA'S QUALITY OF ROCK MASS (MR)

This classification is applicable for the tunnel supports. In this classification it was assumed that the pressure on the tunnel support is imposed by a volume of rock that may separate from the rock mass and fall, after the excavation is made. Such rock volume was assumed to have a shape as illustrated in Figure 1.14.

Table 1.55 Assessment of Joint Condition-Adjustments as Combined Percentage of Total Possible Rating of 30[1.5]

Parameter	Description	Percentage adjustment [a]
Joint expression (large scale)	Wavy unidirectional	90–99
	Curved	80–89
	Straight	70–79
Joint expression (small (scale)	Striated	85–99
	Smooth	60–84
	Polished	50–59
Alteration zone	Softer than wall rock	70–99
Joint filling	Coarse hard-sheared	90–99
	Fine hard-sheared	80–89
	Coarse soft-sheared	70–79
	Fine soft-sheared	50–69
	Gouge thickness < irregularities	35–49
	Gouge thickness > irregularities	12–23
	Flowing material > irregularities	0–11

Note: [a] See Table 1.56.

Table 1.56 Total Possible Percentage Reduction to Ratings for Rock Classification Parameters[1.5]

Parameter	RQD	IRS	Joint spacing	Condition of Joints	Total
Weathering	95	96		82	75
Field and induced stresses				120–76	120–76
Changes in stress; strike and dip orientation			70		70
Blasting	93			86	80

In order to quantify the quality of rock mass (MR), four parameters were considered:[1.46] (A) Joint spacing (J_s) ; (B) Joint system or set number (J_n) ; (C) Joint shear strength (τ_j) and (D) Groundwater conditions and pressure (P_w)

Different ratings are allocated to each of the above parameters according to the *in situ* conditions as specified in Table 1.57.

The quality of rock mass is represented by the MR index, varying between zero and 100, and is the total of the four above noted ratings, such that

$$MR = J_s + J_n + \tau_j + P_w \tag{1.27}$$

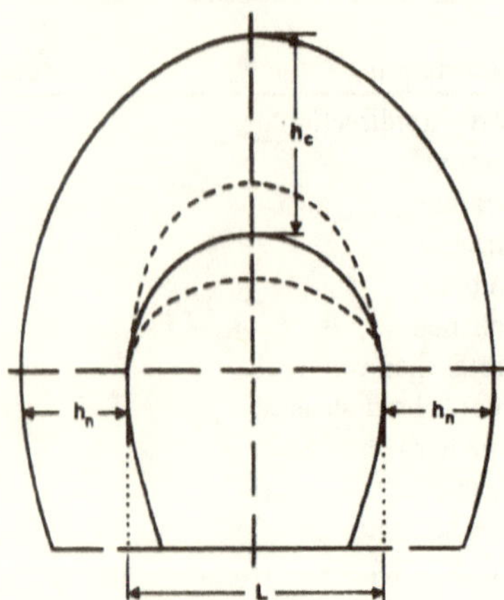

Figure 1.14. Volume of rock that may load a tunnel support.[1.46]

The roof load on the support (h_c) is given by the following equation:

$$h_c = k \cdot L \tag{1.28}$$

where: k = representative quality of the rock mass in fine classifications, given in Table 1.58 and Figure 1.15.

The loads on the support walls (h_n) is given as follows:[1.45]

$$\text{For } MR > 60 \,;\, h_n = 0 \tag{129a}$$

$$\text{For } 50 < MR < 60 \,;\, 0 < h_n < h_c/2 \tag{1.29b}$$

$$\text{For } MR < 50 \,;\, h_n = h_c/2 \tag{1.29c}$$

Table 1.58 Rocha's MR Index, k values and Tunnel Support Considerations[1.46]

Class	MR	k	Support
I	80–100	0–0.05	Sporadic support: e.g., rock-bolting, in accordance with the observed roof conditions
II	60–80	0.05–0.3	Systematic support in the roof
III	50–60	0.3–0.6	Systematic support in the roof. Sporadic support in the walls may also be necessary
IV	30–50	0.6–0.9	Systematic support in both, the roof and walls is necessary
V	0–30	0.9–1	

Table 1.57 Rocha's (MR) Classification Parameters and Their Ratings[1.46]

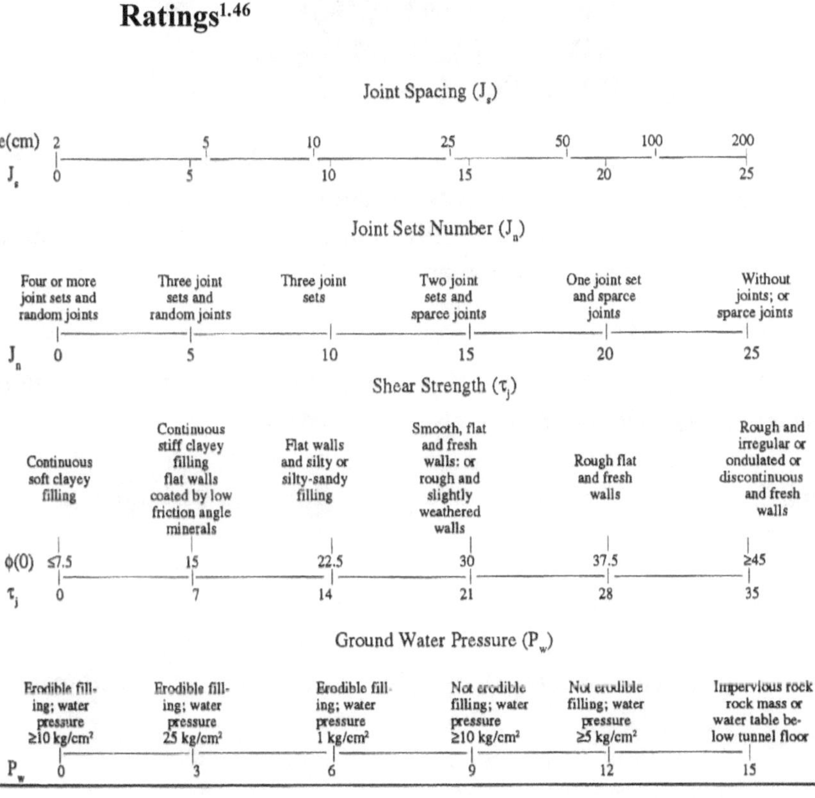

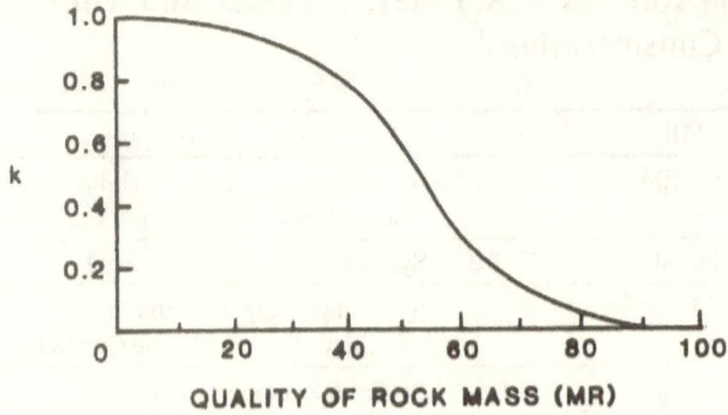

Figure 1.15. Determination of Rocha's k-value from the MR Index.[1.46]

1.12 WICKHAM, TIEDEMANN, AND SKINNER1.47 — ROCK STRUCTURE RATING (RSR)

This classification system is also applicable to the rock masses surrounding tunnels. It involves the sum of three basic parameters called A, B, and C, which are individual evaluations as to the relative effect on the support requirement of various geologic factors:

$$RSR = A + B + C \tag{1.30}$$

Parameter A is a general appraisal of the rock structure through which the tunnel is to be driven, given in Table 1.59.

Parameter B relates to the discontinuity or joint pattern, i.e., joint spacing, dip, and strike relative to the direction of tunnel driven, given in Table 1.60.

Parameter C is given in Table 1.61 and takes into consideration the following factors:

1. The overall quality of rock, indicated by the sum of parameters A and B, (i.e., A + B)
2. Condition of the joint or discontinuity surfaces
3. The anticipated amount of groundwater inflow

The RSR thus obtained reflects the quality or competency of the rock structure with respect to its needs for support, regardless of the tunnel size. The relationship between the RSR and tunnel size is taken into consideration in the determination of respective rib ratios (RR).

For any particular tunnel it is necessary to have a standard datum by which all situations can be compared on a common basis. In the case of steel ribs, predominantly utilized most mining tunnels, the theoretical steel support spacing required for various rib sizes and tunnels diameters are given in Table 1.62.

The relationship between RSR and RR are as follows (see Table 1.63 and Figure 1.16.[1.47])

$$(RSR+30)(RR+80)=8800\,(\text{in ft}) \tag{1.35}$$

Table 1.59 Rock Structure Rating (RSR)— Parameter A, General Area Geology[1.47]

	Basic rock types			
	Hard Medium	Med.	Soft	Decomposed
Igneous	1 2	2	3	4
Metamorphic	1 2	2	3	4
Sedimentary	2 3	3	4	4
	Geological structure			
Rock type	Slightly Faulted or Massive Folded	faulted or folded	Moderately Faulted or Folded	Increasingly Faulted or Folded
Type 1	Max. 30 22	22	15	9
Type 2	27 20	20	13	8
Type 3	24 18	18	12	7
Type 4	19 15	15	10	6

1.12.1 Effect of Rib Ratio (RR) and Tunnel Size on RSR Evaluation

Rib ratios (RR) are determined as follows[1.47]

$$RR = 100\, S_t / S_a \tag{1.31}$$

where: S_t = theoretical spacing for steel rib (datum condition), ft (see Table 1.62)

$= P_r / 331D = P_r \cdot D / P_1$, for circular tunnels (1.32)

P_r = allowable load on the steel support, lb/ft of tunnel width or diameter.

D = tunnel diameter, ft

P_1 = maximum roof load for loose, cohesion-less sand below the groundwater table = $[1.38\,(B + H)]\, B \cdot \gamma$ (1.33)

$= 331\, D^2$ for circular tunnel, and $\gamma = 120$ lb/ft^3 (1.34)

B = width of tunnel, ft

H = height of tunnel, ft

γ = weight per unit volume of the roof rock, lb/ft^3

S_a = actual rib spacing used for a sample tunnel section, ft.

Table 1.60 Rock Structure Rating (RSR) – Parameter B Joint Pattern Relative to the Direction of Drive[1.47]

SPACING IN INCHES / THICKNESS IN INCHES	Strike ⊥ to axis					Strike ∥ to axis		
	Direction of drive					Direction of drive		
	Both	With dip		Against dip		Both		
	Dip of prominent joints					Dip of prominent joints		
	Flat 0–20°	Dipping 20–50°	Vertical 50–90°	Dipping 20–50°	Vertical 50–90°	Flat 0–20°	Dipping 20–50°	Vertical 50–90°
(1) Very closely jointed	9	11	13	10	12	9	9	7
(2) Closely jointed	13	16	19	15	17	14	14	11
(3) Moderately jointed	23	24	28	19	22	23	23	19
(4) Moderate to blocky	30	32	36	25	28	30	28	24
(5) Blocky to massive	36	38	40	33	35	36	34	28
(6) Massive	40	43	Max. 45	37	40	40	38	34

Table 1.61 Rock Structure Rating (RSR) — Parameter C, Joint Condition and Groundwater[1.47]

Anticipated water inflow (gpm/1000 ft. tunnel length)	Sum of parameters A + B					
	13–44			45–75		
	Joint condition [a]					
	Good	Fair	Poor	Good	Fair	Poor
None	22	18	12	max. 25	22	18
Slight (<200 gpm)	19	15	9	23	19	14
Moderate (200–1000 gpm)	15	11	7	21	16	12
Heavy (>1000 gpm)	10	8	6	18	14	10

Note: [a] Good = tight or cemented; Fair = slightly weathered or altered; Poor = severely weathered, altered, or open.

Table 1.62 Theoretical Steel Support Spacing (S_1) of Typical Rib Sizes for Datum Condition (in feet)[1.47]

Rib Size	Tunnel diameter, ft 10'	12'	14'	16'	18'	20'	22'	24'	26'	28'	30'
417.7		1.16									
4H13.0	2.01	1.51	1.16	0.92							
6H15.5	3.19	2.37	1.81	1.42	1.14						
6H20		3.02	2.32	1.82	1.46	1.20					
6H25			2.86	2.25	1.81	1.48	1.23	1.04			
8WF31				3.24	2.61	2.14	1.78	1.51	1.29	1.11	
8WF40					3.37	2.76	2.30	1.95	1.67	1.44	1.25
8WF48						3.34	2.78	2.35	2.01	1.74	1.51
10WF49								2.59	2.22	1.91	1.67
12WF53										2.19	1.91
12WF65										---	2.35

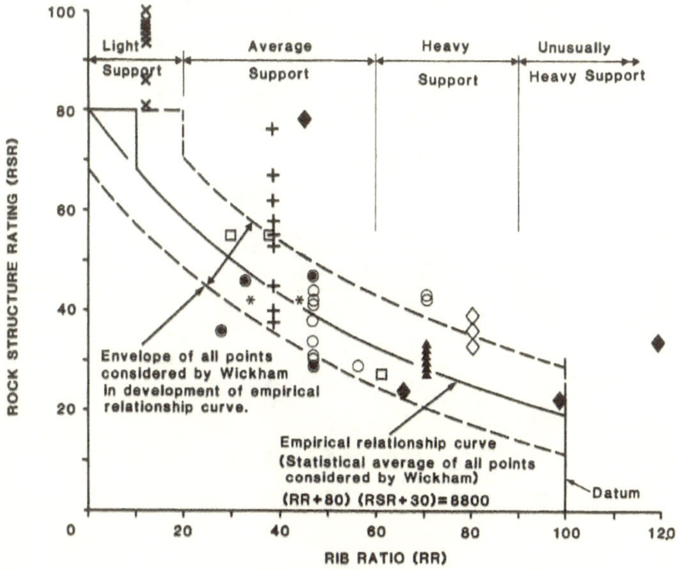

Figure 1.16. Correlation of the rock structure rating (RSR) to rib ratio (RR).[1.47]

Table 1.63 Relationship Between Rock Structure Rating (RSR) and Rib Ratio (RR)

	Heavy support ←		→	← Medium support			→	← Light support			--->
RSR	19	22	28	38	43	50	54	63	68	74	80
RR	100	90	75	50	40	30	25	15	10	5	0

$$RSR = \{8800/[(302\,W_r/D) + 80]\} - 30, \text{ in feet} \quad (1.36)$$

where, W_r = unit rock load, see Table 1.64 = $P_r/1000\,S_{al}$

= $D \cdot RR/302 = (D/302)\,[8800/(RSR + 30)] - 80$, in kips/ ft2 (1.37a)

= $0.26\,(B + H)\,\{[8800/(RSR + 30)] - 80\}$, in kPa (1.37b)

The spacing pattern(s) for various diameter rock bolts can be determined as follows:[1.47]

1. For 1 in. diameter rock bolts having a working load of 24,000 lb: $s = (24/W_r)^{0.5}$ (1.38a)
2. For 5/8 in. diameter rock bolts: $s = (9.2/W_r)^{0.5}$ (1.38b)
3. For 3.4 in. diameter rock bolts: $s = (13.5/W_r)^{0.5}$ (1.38c)
4. For 1.25 in. diameter rock bolts: $s = (37.5/W_r)^{0.5}$ (1.38 d)

Table 1.64 Correlation of Rock Structure Rating (RSR) to Rock Load on Tunnel Arch (W_r) and Tunnel Diameter

Tunnel diameter (D, ft)	(W_r) Rock load on tunnel arch (K/sq ft^2)											
	0.5	1.0	1.5	2.0	3.0	4.0	5.0	6.0	7.0	8.0	9.0	10.0
	Corresponding values of rock structure ratings (RSR)											
10'	62.5	49.9	40.2	32.7	21.6	13.8						
12'	65.0	53.7	44.7	37.5	26.6	18.7						
14'	66.9	56.6	48.3	41.4	30.8	22.9	16.8					
16'	68.3	59.0	51.2	44.7	34.4	26.6	20.4	15.5				
18'	69.5	61.0	53.7	47.6	37.6	29.9	23.8	18.8				
20'	70.4	62.5	55.7	49.9	40.2	32.7	26.6	21.6	17.4			
22'	71.3	63.9	57.5	51.9	42.7	35.3	29.3	24.3	20.1	16.4		
24'	72.0	65.0	59.0	53.7	44.7	37.5	31.5	26.6	22.3	18.7		
26'	72.6	66.1	60.3	55.3	46.7	39.6	33.8	28.8	24.6	20.9	17.7	
28'	73.0	66.9	61.5	56.6	48.3	41.4	35.7	30.8	26.6	22.9	19.7	16.8
30'	73.4	67.7	62.4	57.8	49.8	43.1	37.4	32.6	28.4	24.7	21.5	18.6

If shotcrete lining is used as a means of tunnel support, its required minimum thickness (t) can be determined from: [1.47]

$$t = 1 + (W_r/1.25), \text{ in inches} \quad (1.39)$$

Support charts for 10, 24, and 30 ft-diameter tunnels relevant to the rock structure rating (RSR) is shown in Figures 1.17 to 1.19.

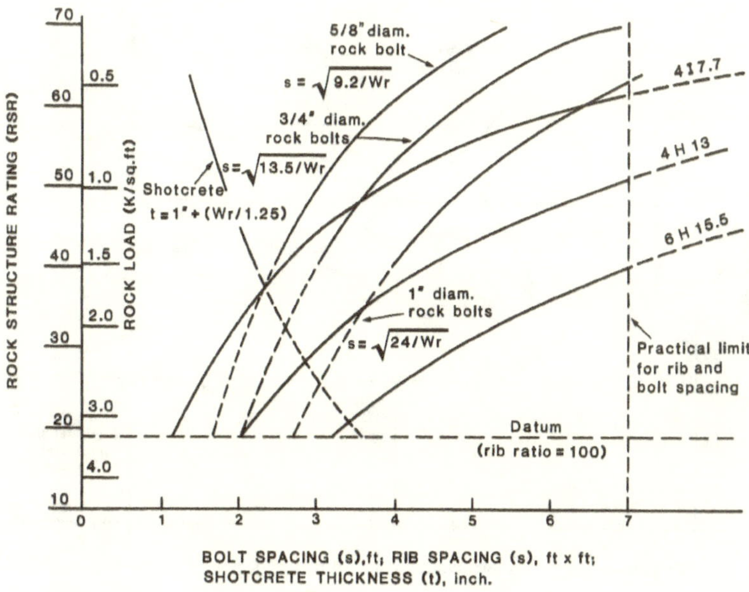

Figure 1.17. Support requirement chart for 10 ft diameter tunnel relevant to the RSR.[1.47]

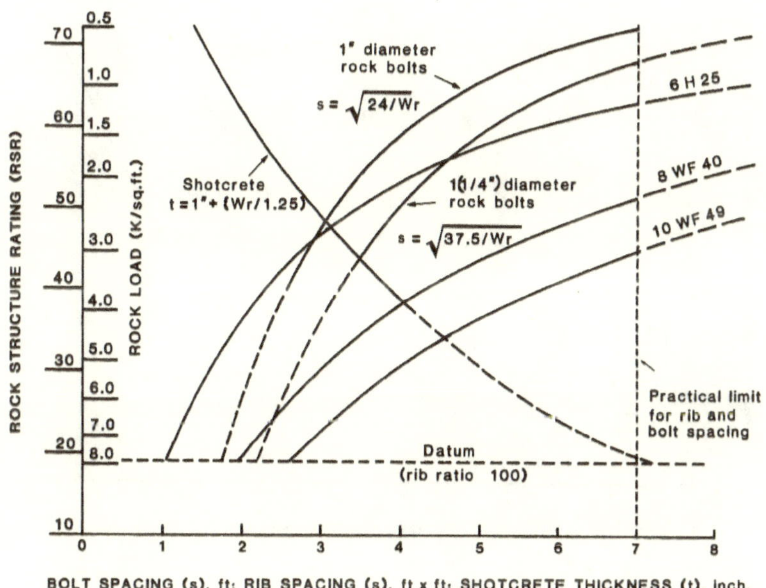

Figure 1.18. Support requirement chart for 24 ft diameter tunnel relevant to the RSR.[1.47]

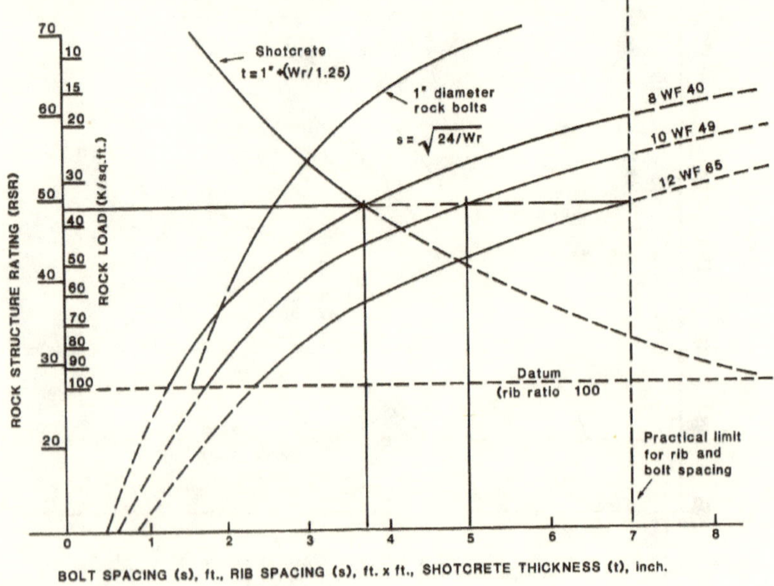

Figure 1.19. Support requirement chart for 30-ft diameter tunnel relevant to the RSR.[1.47]

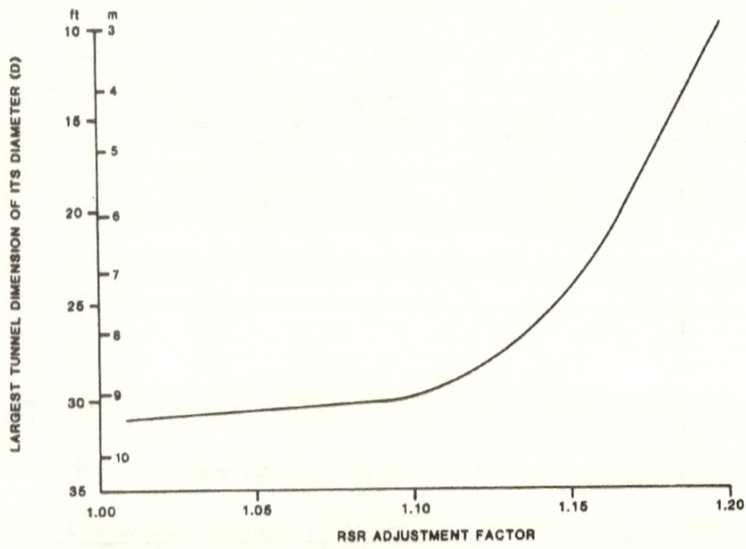

Figure 1.20. Wickhan et al. RSR Multiplication Adjustment for Machine-cut Operation.[1.47]

1.12.2 Effect of Tunnel Excavation Method on RSR Evaluation

It is commonly assumed that some rock structures penetrated by a boring machine would be more competent or suffer less damage than when excavated by a drill and blast methods. In instances where this is the case it is reasonable to assume a lesser amount of support would be required which in turn could be reflected by adjusting upward a predetermined RSR value. Suggested adjustment factors when considering the use of a boring machine is given in Figure 1.20.

1.13 MATHEW'S METHOD OF ROCK MASS CLASSIFICATION

This empirical design approach was developed by Mathews et al. in the *CANMET*.[1.48] It is applied to designs for stability of underground excavations and pillars. It is essentially plots of the rock mass stability number (N) vs. the excavation shape factor (S) as shown in Figure 1.21, in which:

N = rock mass stability number = Q' · A · B · C (1.40)

Q' = modified rock mass quality or NGI — rating (Q) = (Q) (stress reduction factor SRF)

A = rock stress factor, relating the ratio of uniaxial compressive strength of intact rock to the induced stress, (i.e., σ_c/σ_i), see Figure 1.22 (1.41)

B = factor relating the discontinuity or joint orientation relative to the excavation surfaces, see Figure 1.23

C = factor relating to the dip of excavation boundaries (θ), i.e., dip of roof, hangingwall, or footwall to the horizontal, in degrees, see Figure 1.24 = $8 - 7\cos\theta$ (1.42)

S = shape factor of the excavation = (surface area of the excavation at the roof, hangingwall, footwall)/(perimeter of the excavation roof, hangingwall, or footwall) (1.43)

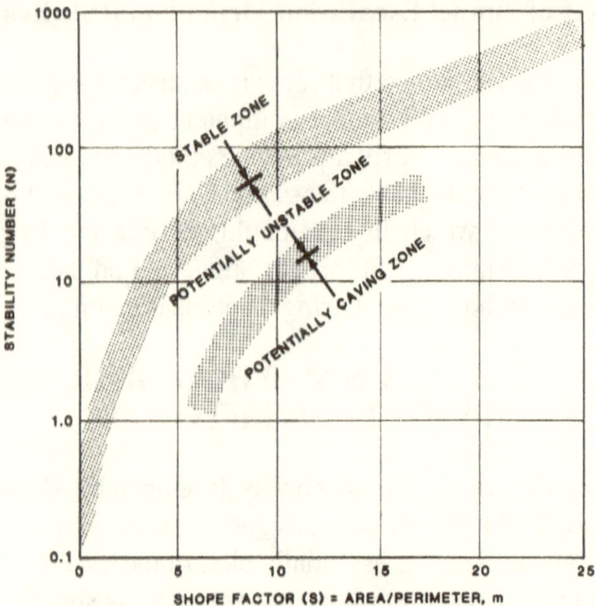

Figure 1.21 Mathews et al. Empirical design chart for rock mass classification.[1.48]

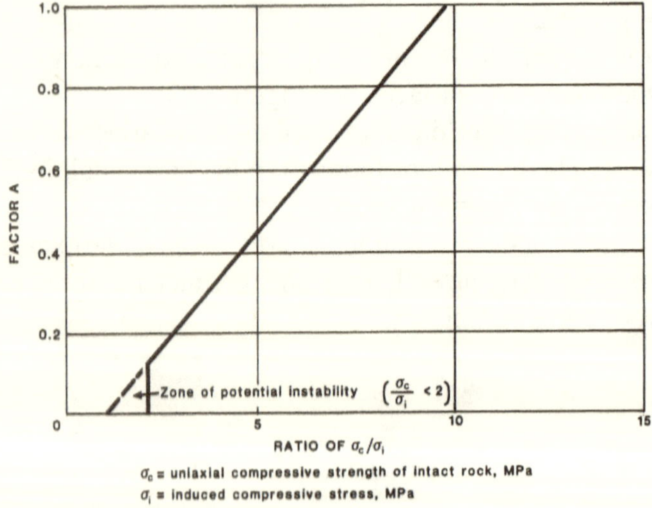

Figure 1.22. Mathews et al. rock stress factor A.[1.48]

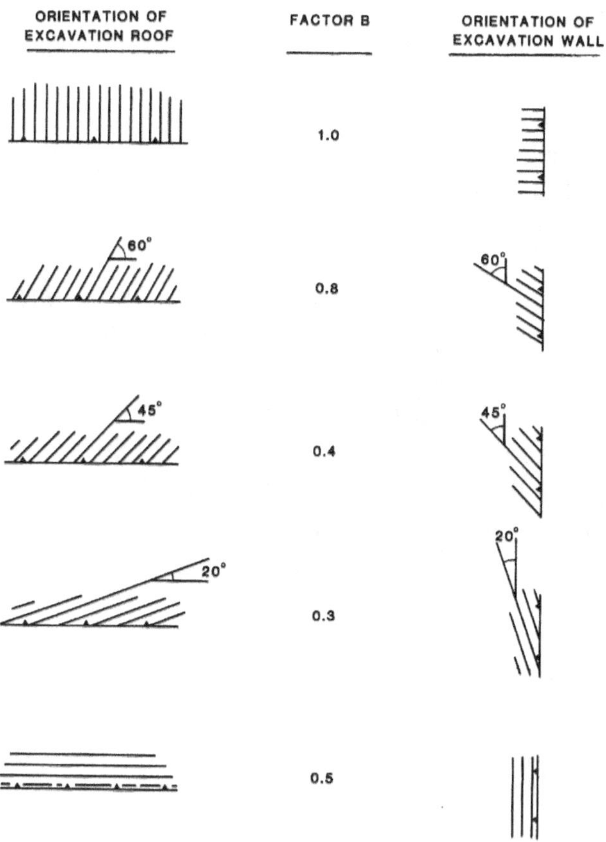

Figure 1.23. Mathews et al. rock discontinuity orientation factor B.[1.48]

1.13.1 Practical Example Utilizing Mathews' Method

Problem: A mining excavation in quartzite has the following rock mass specifications and ratings:

$J_n = 9$, from Table 1.7 ; $J_r = 1.5$, from Table 1.8 ; $J_w = 0.66$, from Table 1.10 ;
$J_a = 2$, from Table 1.9 ; $J_v = 1.10$; SRF = 0.8, from Table 1.13 ; $\sigma_c = 150.2$, MPa
$\sigma_i = 18.5$, MPa ; $\theta = 45°$; L = 100, m ; W = 20, m

Determine:

1. The rock mass quality (Q)
2. The rock mass rating (RMR)
3. The stability number of excavation roof (N)
4. The shape factor of excavation roof (S)
5. Check if the excavation needs any support.

Solution:

RQD, from Equation 1.3 = 115 – 3.3 J_v = 115 - 3.3 (10) = 82%, utilizing the NGI classification from Equation 1.4

$Q = (RQD / J_n)(J_r / J_a)(J_w / SRF)$

$= (82/9)(1.5/2)(0.66/0.8)$

= 5.6, fair rock mass

Q', modified Mathews rock mass quality = Q(SRF) = 5.6(0.8) = 4.5

RMR = 9 ln Q + 44 = 9 ln 5.6 + 44 = 9 x 1.72 + 44 = 59.5

Factor A (for σ_c / σ_i = 150 /18.5 = 8.1, from Figure 1.22) = 0.8

Factor B (for the discontinuities dipping at θ = 45° to horizontal in the excavation roof, from Figure 1.23) = 0.4

Factor C (for θ = 45° at the roof, from Figure 1.24) = 8 – 7 cos θ = 3.1

Therefore, N = stability number = Q' · A · B · C = 4.5 x 0.8 x 0.4 x 3.1 = 4.5

S = shape factor of the excavation roof = roof surface area/ roof perimeter

= L · W/2(L + W) = 100 x 20/2(100 + 20) = 8.3

Hence, from Figure 1.21, the excavation lies in the potentially unstable to caving zones, requiring ground reinforcement.

1.14 BROOK AND DHARMARATNE1.49 SIMPLIFIED ROCK MASS RATING (R)

In this method, four major parameters are utilized. These are:[1.49]

1. Intact rock strength (IRS) — which is the uniaxial compressive strength, though it is intended that the point load strength should be the parameter that is actually measured.[1.50, 1.51] The two alternative parameters that are given produce the same component rating percentage. Maximum rating for IRS is 30%.

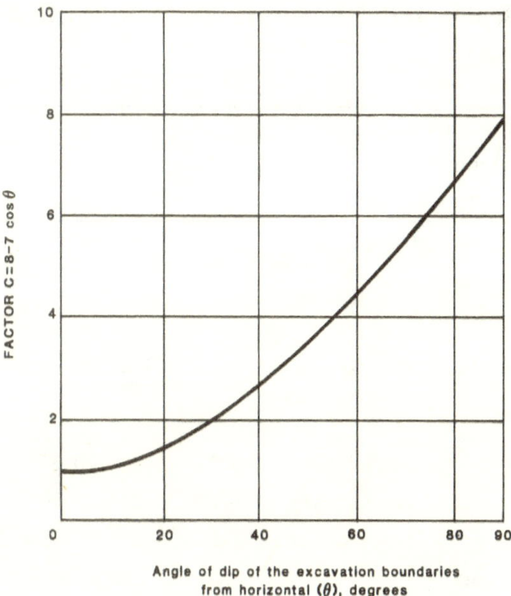

Figure 1.24 Mathews et al. design excavation surface orientation factor C.[1.48]

2. Joint spacing (J_s) — which is determined by calculating the average spacing relation to the typical size of the excavation. For example, a site has 2 distinct joint sets spaced at 0.25 and 0.76 m to an excavation of 2.8 m in width (plus some random joints). This gives an average relative spacing of (0.25 + 0.76)/(2 x 2.8) = 0.18 (this does not include random joints). According to Table 1.65, for 2 joint sets J_s = 20 to 25% (and for 3 joint sets J_s = 15 to 20%). In view of the presence of random joints, the selected rating is 20%. The maximum rating for J_s is 30%.
3. Joint type (J_t) — this rating is determined by multiplying maximum rating of 30% for J_t by adjustment factors given in Tables 1.65 and 1.66, until all aspects of joint surface, infilling, weathering, etc. are taken into account. The adjustment factors range from 0.5 to 1.0, and generally, two levels of influence are considered: a slight effect (0.75 – 1.0) and a more pronounced effect (0.5 – 0.75). For example, at the site noted for J_s rating, if the joints are continuous but undulating with less than 1 mm separation, the adjustment factor is 0.9.
4. Groundwater rating (GW) — which is allocated up to a maximum 10%, according to Table 1.65.

Table 1.65 Brook and Dharmaratne[1.49] Simplified Rock Mass Rating ®

Parameter	Maximum rating (%)	*In situ* values quantity	Rating				
1. Intact Rock Strength (IRS)	30	Compressive strength, MPa, usually from point load test	30% (IRS/200) {= 30% [I_S(50/9)]}				
2. Joint Spacing (J_s)	30	Spacing relative to excavation site	>0.3	0.3–0.1	0.1–0.03	0.03–0.01	<0.01
		One joint set	30%	25–30%	20–25%	15–20%	10–15%
		Two joint sets	25–30%	20–25%	15–20%	10–10%	0–5%
		Three joint sets	20–25%	15–20%	10–15%	5–10%	0–5%
3. Joint Type (J_t)	30		Exact value interpolated if necessary 30% x Adjustment factor				
			Adjustment			**Adjustment factor**	
		Expression and continuity	Discontinuous			1.0	
			Wavy			0.75–1.0	
			Straight			0.5–0.75	
		Surface if in contact	Rough			1.0	
			Slightly rough			0.75–1.0	
			Smooth to polished			0.5–0.75	
		Separation	< 1 mm			0.9–1.0	
			2–1 mm			0.8–0.9	
			5–2 mm			0.7–0.8	
			10–5 mm			0.6–0.7	
			>10 mm			0.5–0.6	
		Gouge properties	Hard pack			1.0	
			Sheared			0.75–1.0	
			Soft, clay			0.5–0.75	
4. Groundwater (GW)	10		Dry	Moist	Wet	Moderate pressure	High pressure
			10%	8%	5%	2%	0

Total rating for the rock mass (R) is obtained as follows:

$$R = IRS + J_s + J_t + GW \quad (1.44)$$

The *in situ* rating (R_a) and adjustment rating (R_b) for 11 sites is given in Table 1.67 for comparison.

The elimination of RQD utilized in other methods of rock mass rating does not affect the final result, and greatly simplifies the field procedure. Support requirements, using the simplified rock mass rating (R) is suggested to be according to Table 1.68.

Table 1.66 Adjustment Factors for *In Situ* Rating Components in Simplified Method[1.49]

	IRS	J_s	J_t
Weathering			
Slight	0.9	—	0.9
Moderate	0.75	—	0.75
High	0.5	—	0.5
Orientation of excavation			
Slightly unfavorable	—	0.9	0.9
Moderately unfavorable	—	0.75	0.75
Highly unfavorable	—	0.5	0.5
In situ or induced stress compared with IRS			
Causing slight failure	0.9	—	0.9
Moderate failure	0.75	—	0.75
General failure	0.5	—	0.5
Blasting			
Smooth	—	0.9	0.9
Moderately rough	—	0.75	0.75
Rough	—	0.5	0.5

Table 1.67 Comparison of *In Situ* Rating (R_a) and Adjusted Rating (R_b) using Brook and Dharmaratne[1.49] - Simplified Method

Site no	IRS	J_s	J_t	GW	R
1 (a)	30	20	30 x 1 x 1 x 0.9 x 1 = 27	10	87
(b)	30	20 x 0.75 x 0.75 = 11.25	27 x 0.5 x 0.75 = 10.12	10	61
2 (a)	17.25	27	30 x 0.95 x 0.95 x 0.85 x 0.75 = 17.26	10	72
(b)	17.25	27 x 0.9 x 0.75 = 18.2	17.25 x 0.9 x 0.75 = 11.64	10	57
3 (a)	21.6	18	30 x 1 x 0.95 x 0.85 x 0.5 = 12.1	10	62
(b)	21.6 x 0.9 = 19.44	18 x 0.9 x 0.75 = 12.15	12.1 x 0.5 x 0.9 x 0.75 = 4.08	10	46
4 (a)	21.5	20	30 x 0.6 x 0.6 x 0.65 x 0.6 = 4.2	5	51
(b)	21.5 x 0.9 x 0.75 = 14.51	20 x 0.75 x 0.9 = 13.5	4.2 x 0.75 x 0.75 x 0.9 x 0.9 = 1.91	5	35
5 (a)	21.5	20	30 x 1 x 0.6 x 0.75 x 0.6 = 8.1	8	58
(b)	21.5 x 0.5 x 0.9 = 9.68	20 x 0.5 x 0.9 = 9.0	8.1 x 0.5 x 0.5 x 0.9 x 0.9 = 1.64	8	28
6 (a)	25	16	30 x 1 x 1 x 0.9 x 1 = 27	8	76
(b)	25 x 0.9 = 22.5	16 x 0.75 x 0.9 = 10.8	27 x 0.9 x 0.9 x 0.75 = 16.4	8	58
7 (a)	21.9	25	30 x 1 x 1 x 0.9 x 1 = 27	10	84
(b)	21.9 x 0.9 = 19.71	25 x 0.5 = 22.5	27 x 0.9 x 0.9 = 21.87	10	74
8 (a)	23	15	30 x 0.8 x 0.75 x 0.9 x 0.9 = 14.58	8	61
(b)	23 x 0.9 = 20.7	15 x 0.5 x 0.9 = 6.75	14.6 x 0.9 x 0.5 x 0.9 = 5.91	8	41
9 (a)	22.8	21	30 x 0.75 x 0.75 x 0.85 x 0.75 = 10.76	8	63
(b)	22.8 x 0.75 x 0.9 = 15.39	21 x 0.9 = 18.9	10.8 x 0.5 x 0.9 x 0.9 = 4.37	8	47
10 (a)	22.6	15	30 x 0.65 x 0.75 x 0.75 x 0.75 = 8.2	8	54
(b)	22.6 x 0.5 x 0.9 = 10.17	15 x 0.5 x 0.9 = 6.75	8.2 x 0.5 x 0.5 x 0.9 x 0.75 = 1.38	8	26
11 (a)	10.5	30	30 x 1 x 1 x 0.95 x 1 = 28.5	10	79
(b)	10.5 x 0.9 = 9.45	30 x 0.9 = 27	28.5 x 0.9 = 25.65	10	72

Table 1.68 Support Requirements using Simplified Roof Mass Rating System (R)[1.49]

Adjusted Rating (R_b%)	Support requirement
>70	None
50–70	Local regions
40–50	Light, systematic
30–40	Medium, systematic with additional lagging or shotcreting
20–30	Increased level of systematic support and lagging, or steel mesh and shotcrete, or timber support with complete lagging
10–20	Extensive steel support and complete lagging; repair work is also expected
<10	Difficult to maintain an adequate tunnel profile, excavation size should be reduced

1.15 CORRELATION AND RELATIONSHIPS BETWEEN VARIOUS ROCK MASS CLASSIFICATION SYSTEMS

1.15.1 Correlation Between Rock Mass Rating (RMR) and Rock Mass Strength Parameters

Knowing the RMR- value, it will be possible to determine the unconfined compressive strength of discontinuous or jointed rock mass (σ_{cj}), cohesion of discontinuous or jointed rock mass (S_{oj}), internal friction angle of discontinuous or jointed rock mass (ϕ_j), and other relevant factors or constants as follows:[1.22, 1.37]

$$\sigma_{cj} = \sigma_c \cdot \exp[(RMR - 100)/18.75] = [(\sigma_1 - \sigma_3)^{1.25}/\sigma_3^{0.25}]/A_j^{1.25}$$
$$= 2\,S_{oj} \cdot \tan[45 + (\phi_j/2)] \quad (1.45)$$

$$S_{oj} = 0.5\ \tan[45 + (\phi_j/2)]\ \{\sigma_c\ .\ \exp[(RMR - 100)/18.75]\} \quad (1.46)$$

A_j = material constant for the rock mass

$$= A \cdot \exp[(RMR - 100)/75.5] \quad (1.47)$$

where: A = material constant for the intact rock (see Table 1.69)

= $\tan^2[45 + (\phi/2]$, using Mohr-Coulomb failure criteria

$= (1 + \sin \phi) /(1 - \sin \phi) = [(\sigma_1 - \sigma_3) / \sigma_3] (\sigma_3 / \sigma_c)^a$ (1.48)

α = slope of the failure curve on a log-log scale between parameters σ_c /σ_3 and $(\sigma_1 - \sigma_3) /\sigma_3$

E_m = estimated elastic modulus of the rock mass,[1.52] in GPa

$= 2(RMR) - 100$ (1.49)

Table 1.69 Empirical Criterion and Equation Constants for Triaxial Strength of Various Rocks[1.1]

Rock type	Criterion $[(\sigma 1 /\sigma_c) = A (\sigma_3 / \sigma_c)^{0.75} + 1]$		Criterion 2 $[(t_m /\sigma_c) = B (\sigma_m /\sigma_c)^{0.9} + 1]$
Norite	A = 5.0	Error: 3.6%	B = 0.8 Error: 1.8 %
Quartzite	A = 4.5	Error: 9.2%	B = 0.78 Error: 3.2 %
Sandstone	A = 4.0	Error: 5.8%	B = 0.75 Error: 2.3 %
Silstone	A = 3.0	Error: 5.6%	B = 0.7 Error: 4.2 %
Mudstone	A = 3.0	Error: 6.1%	B = 0.7 Error: 6.6 %
Average for all rock types	A = 3.5	Error: 10.4%	B = 0.75 Error: 8.3 %

For predominantly horizontal discontinuities:

$$E_m/E=\exp(0.0217RMR-2.17) \quad (1.50)$$

For inclined discontinuities predominantly at 45 to 60° horizontal:

$$E_m/E=\exp(0.0564RMR-5.64) \quad (1.51)$$

where, E = elastic modulus of the intact rock, GPa

The Hoek and Brown m and s values can be related to the RMR by the following expressions:[1.53]

1. For undisturbed or interlocked rock mass, where mechanical excavation, perimeter blasting techniques, or in some cases good blasting practice are used:

$$m /m_i = \exp [(RMR - 100) / 28] \quad (1.52)$$

where, m_i = value of m for the intact rock.

$$s = \exp [(RMR - 100) / 9] \quad (1.53)$$

2. For disturbed rock mass, where poor blasting is practiced, or for waste dumps:

$$m/m_i = \exp[(RMR - 100)/14] \quad (1.54)$$

$$s = \exp[(RMR - 100)/6] \quad (1.55)$$

where mi = value of m for the intact rock.

1.15.2 Correlation Between Bieniawski's Rock Mass Rating (RMR) and Barton's Rock Mass Quality (Q) Systems

Relationship between Bieniawski's RMR[1.40] and Barton's Q-values [1.15] can be modified in the following general form:

$$RMR = x^a \ln Q + y^b \quad (1.56)$$

where: x, y, a, and b = constants dependent upon the rock type and the state of its fracture, in such a way that:

x^a = 5 to 13.5 (see Reference 1.54)

y^b = 26 to 62 (see Figure 1.25)

Correlation between RMR and Q-systems is given in Table 1.70.

Table 1.70 Correlation Between Bieniawski's Rock Mass Rating (RMR) and Barton et al. Rock Mass Quality (Q) Systems [a]

Class	Rock mass description	RMR rating	Q rating
0	Extremely-exceptionally good	—	100–1000
1	Very good	81–100	40–100
2	Good	61–80	10–40
3	Fair	41–60	4–10
4	Poor	21–40	1–4
5	Very poor	0–20	0.1–1
6	Extremely–exceptionally poor	—	0.001–0.1

Note: a Modified from References 1.15 and 1.42.

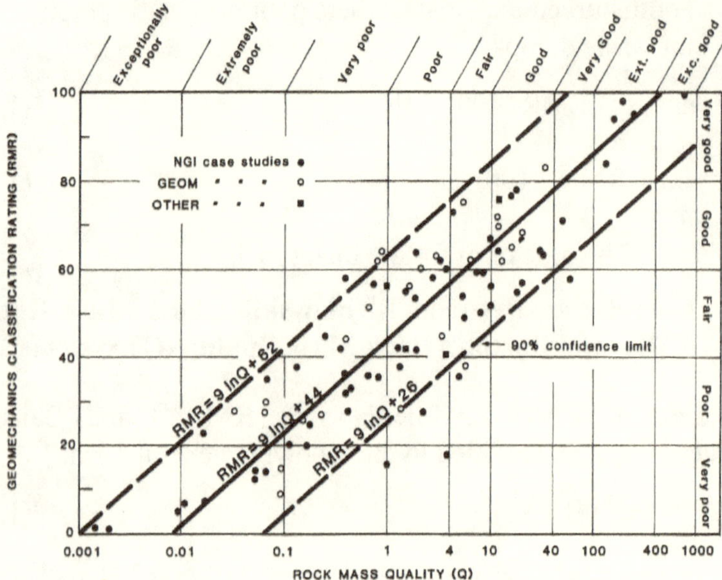

Figure 1.25. Correlation between rock mass ratio (RMR) and rock mass quality (Q) systems.[1.40]

1.15.3 Correlation Between Rock Quality Designation (RQD) and Barton et al. Rock Mass Quality (Q) System

A comprehensive comparison between RQD and Q, for various rock and discontinuities types, is given in Table 1.71. Moreover, relevant results from five bored-core logs totaling 160 m long, driven to the roof, floor, and face of a prospective tunnel in Norway, was given in Table 1.15.

Table 1.71 Correlation Between RQD, Joint Specifications, SRF, and Q[1.14]

Class	Rock type	J_n	J_r	J_a	J_w	SRF	RQD	Q
1	Hard and intact	≤2	4	1	1	1	100	≥200
2	Hard stratified or schistose	3	1	1	1	1	≥30	20–6
3	Massive, moderately jointed	6	≥1.5	1	1	1	100	50–25
4	Moderately blocky and seamy	9	1	≤3	0.66	1	80	6–2
5	Very blocky and seamy	12	1	≤3	0.66	1	50	1–0.4
6	Completely crushed but chemically intact	15	1	2	≤0.66	5	20	0.08–0.04
7	Squeezing rock, moderate depth	20	1	≥6	0.66	5–10	20	0.03–0.01
8	Squeezing rock, great depth	20	1	≥6	0.03	10–20	0	0.004–0.001
9	Swelling rock	20	1	12	0.66	10	0	0.003–0.001

1.15.4 Correlation Between Rocha's Rock Mass Classification (MR), Bieniawski's Rock Mass Rating (RMR), and Wickham et al. Rock Structure Rating (RSR)

The characteristics of rock masses investigated along four tunnels in Portugal and one in Macau,[1.46] resulted the following correlation

$$MR = 0.95(RMR) + 5.4 \tag{1.57}$$

Regression analysis of results obtained from several New Zealand case studies on tunnels,[1.52] modified by the author suggests the following expression:

$$RSR = a\,(RMR) + b \tag{1.58}$$

where: a = constant varies between 0.7 to 0.8
b = constant between 12 to 24

A summarized correlation between MR, RMR and RSR, obtained from 15 tunnel sections, is given in Table 1.72 and shown graphically in Figure 1.26.

Table 1.72 Correlations Between Rock Mass Rating (RMR), Rock Structure Rating (RSR), and Rocha's Classification (MR)[1.46]

Tunnel and rock type	Geotechnical Zone [a]	Ratings RMR	RSR	MR
Guia (Macau) —	I	21	48	21
Granite	II	37	50	53
	III	57	65	63
Beliche-GaFA —	I	17	34	18
Shale + greywacke	II	36	44	44
(flysch)	III	46	56	52
Odeleite —	I	28	34	24
shale + greywacke	II	45	44	48
(flysch)	III	62	66	59
Castelo do Bode —	I	17	36	25
Gneiss	II	36	50	45
	III	60	69	60
Alamos-Alqueva —	I	20	34	25
Phillite	II	45	50	43
	III	65	76	67

Note: a = see Table 1.58

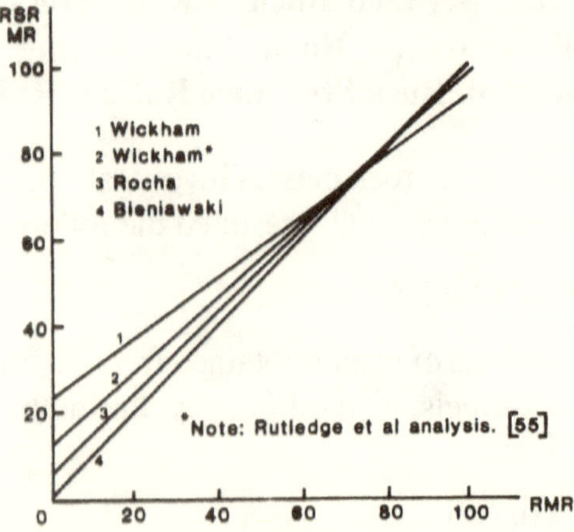

Figure 1.26. Correlations between Rock Mass Rating (RMR), Rock Structure Rating (RSR), and Rocha's Classification (MR).[1.46]

1.15.5 Correlation between Rock Structure Rating (RSR) and Rock Mass Quality (Q)

Result of rock mass analysis along tunnels in New Zealand is shown in Figure 1.27, which suggests the following relationship:[1.55]

$$RSR = 13.3 \log Q + 46.5 \tag{1.59}$$

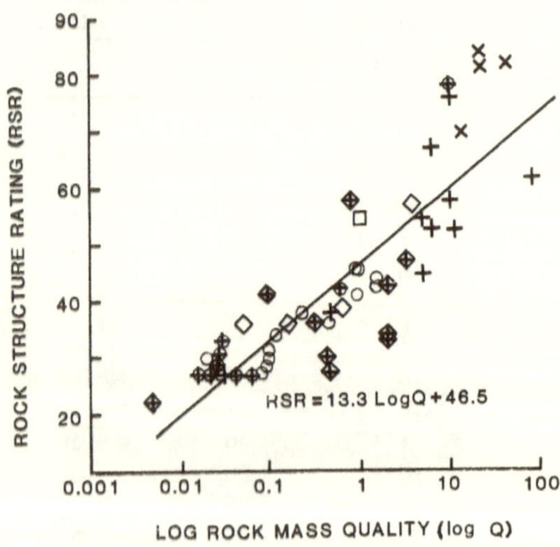

Figure 1.27 Correlation between rock structure rating (RSR) and rock mass quality (Q).[1.55]

Table 1.73 Geotechnical Details of Selected Sites along Tunnels[1.49]

Site no.	Location and rock type	RQD (%)	Intact rock strength (MPa)	Joint spacing (m)	Joint condition [a]	Infill thickness and type	Ground -water
			Kahatogaha-Kolongaha Mine				
1	Garnet biotite gneiss	90	206	0.25, 0.76 and random joints	C/R/HW/NG	0.1 mm, open	Dry
			Bogala Mine				
2	Quartzite	94	115	0.82. 1.5	C/SR/HW/WG	1–2 mm, with clay and sheared rock	Dry
3	Unweathered garnet biotite gneiss	99	144	0.25,0.35, 1.5	C/SR/HW/WG	1–2 mm, clay	Dry
4	Slightly weathered garnet biotite gneiss	92	143	0.30, 0.96	C/S/SW/WG	5–10 mm, mica schist	40 1/min
5	Highly weathered garnet biotite gneiss	25	143	0.25, 0.72 and random joints	C/S/SW/WG	2.5 mm, graphite plus calcite	Moist only
			Victoria Dam Project				
6	Quartzite	98	166	0.52. 0.60, 1.9	C/RH/HW/NG	<0.1 mm, stains only	Moist only
7	Unweathered garnet quartz gneiss	99	146	1.5 and random joints	C/R/HW/NG	<0.1 mm, stains only	Dry
8	Slightly weathered garnet quartz gneiss	98	153	0.66, 0.73 and random joints	C/S/SW/WG	1 mm, sheared rock	Moist only
9	Moderately weathered garnet quartz gneiss	60	152	0.39 and random joints	C/S/SW/WG	1–2 mm, sheared rock	Moist only
10	Highly weathered garnet quartz gneiss	52	152	0.45, 0.50 and random joints	C/S/SW/WG	2–5 mm, sheared rock	Moist only
11	Limestone	100	70	Random joints only	NC/R/HW/NG	Tightly healed	Dry

Note on the designation a: R = rough; SR = slightly rough; S = smooth; C = continuous; NC = not continuous; HW = hard wall; SW = soft wall; WG = with gouge; NG = no gouge.

1.15.6 Comparison of Simplified Rock Mass Rating Method (R) with Other Rock Mass Classification Systems

A summary of rock description and measurements at 11 selected tunnel sites in Sri Lanka[1.49] is given in Table 1.73. *In situ* and adjusted rock mass ratings obtained by various methods are obtained in Table 1.74. The support requirements according to the method of rock mass rating determination is given in Table 1.75 for comparison.

Table 1.74 Comparison of Rock Mass Ratings Obtained by Different Methods[1.49]

Site no.	Q	RMR (%)	Laubscher *In-situ* (IR) (%)	Laubscher Adjusted (AR) (%)	Simplified method (R) *In-situ* (R_a) (%)	Simplified method (R) *Adjusted* (Rb) (%)
1	30	80	85	60	87	61
2	20	70	73	54	72	57
3	9	65	67	50	62	46
4	4	51	59	34	51	35
5	1	35	55	28	58	28
6	25	72	71	58	76	58
7	297	84	86	75	84	74
8	21	62	72	46	61	41
9	15	53	66	45	63	47
10	2	33	54	23	54	26
11	711	92	84	71	79	72

1.16 ESTIMATION OF ROCK MASS DEFORMABILITY

Since rock mass deformation (E, in GPa) is sometimes as important as the rock mass failure, its estimation in terms of the Bieniawski's rock mass rating (RMR) can be given as follows:[1.56]

$$E = 10^{[(RMR - 10)/40]} \tag{1.60}$$

Equation (1.60) is particularly useful during the early stage of an excavation, when relatively little field data is available.

Table 1.75 Comparison of Tunnel Support Predictions Utilizing Various Rock Mass Classification Systems and the Actual Support Installed[1.49]

Site no.	CSIR (RM)	NGI (Q)	Laubscher's AR	Actual support	Remarks
1	General no support except occasional spot bolting	No support required	No support required	None	Stable rock mass
2	3.5 m long, 2.5 m spaced bolts locally in crown, occasional wire mesh and 50 mm shotcrete as required	No support required	No support required	None	Stable rock-mass
3	3 m long, 2.5 m spaced bolts in crown, occasional wire mesh and 50 mm shotcrete as required	No support required	Generally no support, but locally joint intersections may require bolting	None	Block failures observed; may be prevented by spot bolting
4	4 m long, 1.5–2 m spaced systematic bolting; crown: 50–100 mm shotcrete and wire mesh; sides: 30 mm shotcrete and wire mesh	No support required	Patterned grouted bolts at 1 m spacing and 50 mm shotcrete	Wall and roof: 600-mm thick concrete	Shotcrete may have been adequate, but not available
5	3 m long, 0.5–1 m spaced systematic bolting; crown and sides; 50 mm shotcrete and wire mesh	No support required	Patterned grouted at 0.75 m spacing and 100 mm shotcrete with wire mesh	Timber sets at 1 m spacing with lagging	
6	3 m long, 2.5 m spaced bolts in crown with occasional wire mesh and 50-mm shotcrete as required	No support required	Generally no support, but locally joint intersections may require bolting	Patterned grouted bolts at 1–1.5 m spacing	
7	No support except spot bolting	No support required	No support required	Patterned grouted bolts at 1–1.5 m spacing	
8	3 m long, 2.5 m spaced bolts locally in crown, occasional wire mesh	No support required	Patterned grouted bolts at 1 m spacing	Patterned grouted bolts at 1 m spacing	
9	4 m long, 1.5–2 m spaced bolts; shotcrete 50–100 mm on crown and 30 mm on sides	Bolts at 1 m spacing with 20–30 mm shotcrete	Patterned grouted at 1 m spacing	Patterned grouted bolts at 1 m spacing, occasional wire mesh	

10	4–5 m long, 1–1.5 m spaced bolts with wire mesh and shotcrete 100–150 mm on crown, 100 mm on sides	Bolts 1–1.5 m long; wire mesh with 150 mm shotcrete	Patterned grouted bolts at 0.7 m spacing; 700 mm reinforced concrete or yielding arches	Steel arches at 1 m spacing with lagging
11	Spot bolting	No support required	No support required	Patterned grouted bolts 1 m spacing

REFERENCES

1.1. Bieniawski, Z. T., Estimating the strength of rock material, *J. S. African Instit. Mining Met.,* March, 312–320, 1974.

1.2. Anderson, M. G. and Richards, K. S., *Slope Stability — Geotechnical Engineering and Geomorphology*, John Wiley & Sons, New York, 146–155, 1987.

1.3. Kendorski, F. S., Cummings, R. A., Bieniawski, Z. T., and Skinner, E.·H., Rock mass classification for block caving mine drift support, *Proc. 5th Int. Cong. Rock Mech.,* Melbourn, 1983, B51–B63.

1.4. Bieniawski, Z. T., Engineering rock mass classification — new aid for engineers and geologists, *Bull. Earth Miner. Sci.,* 49(1) 1–5, 1979.

1.5. Laubscher, D. H., Geomechanics classification of jointed rock masses — mining application, *Min. Eng. J. (U.K.),* 6(86) A1–8, 1977.

1.6. Bell, F. G., *Engineering Properties of Soils and Rocks,* 2nd ed., Butterworth, London, 1983, 149.

1.7. Deere, D. V., Geological considerations, *Rock Mechanics in Engineering Practice,* Stagg, M. G. and Zienkiewiez, O. C., Eds., Wiley, London, 1968, 1–19.

1.8. Barton, N., Recent experiences with the Q-system of tunnel support design, Vol. 1, Proc. Symp. Exploration Rock Engineering, Johannesburg, 1976, 107–117.

1.9. Goodman, R. E., *Introduction to Rock Mechanics,* J. Wiley & Sons, New York, 1980, 18–49.

1.10. Deere, D. V., Design of surface and near surface construction in rock and breakage of rock, *Proc. 8th Symp. Rock Mech.,* Fairhust, C., Ed., Am. Inst. Min. Eng., New York, 1967, 237–302.

1.11. Hoek, E. and Brown, E. T., Underground excavations in rock, *Inst. Mining Met.,* London, 1980, 14–35, 287–299.
1.11a. Hoek, E. and Brown, E. T., Hoek and Brown Failure Criterion — a 1988 update, in *Rock Engineering for Underground Excavations*, 15th Canadian Rock Mechanics Symposium, University of Toronto, 1988, 31–38.
1.12. Afrouz, A, Mechanical Behavior of Mine Floor, Ph.D. thesis, University of Wales, Caraliff, U.K., 1973.
1.13. Plamstrom, A., Karakterisering av oppsprekningsgrad og fjellmassers kvalild, Internal Report, Ing. A. B. Berdal A/S, Oslo, 1975, 1–26.
1.14. Barton, N. R., Lien, R., and Lunde, J., Engineering classification of rock masses for the design of tunnel supports, *Rock Mech.,* 6, 189–236, 1974.
1.15. Barton, N. R., Loset, F., Lien, R., and Lunde, J., Application of Q-system in design decisions concerning dimensions and appropriate support for underground installations, *Subsurf. Space,* 2, 553–561, 1980.
1.16. Barton, N., Lien, R., and Lunde, J., Analysis of Rock Mass Quality and Support Practice in Tunnelling, Norwegian Geotechnical Institute, Report No. 54206, June, 1974, 74.
1.17. Cecil, O. S., Correlations of rockbolt-shotcrete support and rock quality parameters in Scandinavian tunnels, Ph.D. thesis, University of Illinois, Urbana, IL, 1970, 414.
1.18. Benson, R. P., Conton, R. H., Merrit, A. H., Joli-Coeur, P., and Deere, D. U., *Rock Mechanics at Churchill Falls*, American Society of Civil Engineers, Proc. Symp. on Underground Rock Chambers, Phoenix, AZ, 407–486, 1971.
1.19. Glass, C. H., Seismic considerations in siting large underground openings in rock, Ph.D. thesis, University of California, Berkeley, 1973, 132.
1.20. Fourmaintraux, D., Characterization of rocks, Laboratory tests, in *La Mécanique Les Roches Appliquée aux ouvrages du génie civil*, Marc Panet, et al., Ed., École Nationale des Ponts et Chausées, Paris, 1976, chap. 4.

1.21. Williamson, D. A., Uniform rock classification for geotechnical engineering purposes, Transportation research record 783, S. Africa, 9–14, 1979.

1.22. Bieniawski, Z. T., Geomechanics classification of rock masses and its application in tunnelling, *Proc. 3rd Int. Congress Rock Mechanics,* Vol. 11A, Int. Soc. Rock Mech., Denver, 1974, 27–32.

1.23. Deere, D. U., Technical description of cores for engineering purposes, *Rock Mech. Eng. Geol.,* 1, 17–22, 1964.

1.24. Bieniawski, Z. T., Rock Classifications: State of the Art and Prospects for Standardization, S. A. Transport Research Record 783, S. A., 1979, 2–9.

1.25. Hoek, E. and Bray, J. W., Rock Slope Engineering, 3rd Ed., Inst. of Mining and Met., London, 358, 1981.

1.26. Brady, B. H. G. and Brown, E. T., *Rock Mechanics for Underground Mining,* Allen, G. & Unwin, U.K., 1985, 527.

1.27. Stillborg, B., *Professional Users Handbook for Rock Bolting,* Trans Tech., W. Germany, 1986, 36.

1.28. Bell, F. G., *Engineering Properties of Soil and Rocks,* 2nd ed., Butterworths, U.K., 1983, 76–122.

1.29. Underwood, L. B., Classification and identification of shales, Proc. American Society of Civil Engineers Soil Mech. Found. Engr. Div., 93, No. SM6, 1967, 97–116.

1.30. Franklin, J. A., Broch, E., and Walton, G., Logging the mechanical character of rock, Inst. of Mining and Met., January, London, 1971, A1–A9.

1.31. British Geological Society Working Party Report, The logging of rock cores for engineering purposes, *Q. J. Inst. Engr. Geol.,* 3, U.K., 1970, 1–24.

1.32. Barton, N., The shear strength of rock and rock joints, *Int. J. Rock Mech. Min. Sci.,* (Pergamon Press), 13, 255–279, 1976.

1.33. Patton, F. D., Multiple modes of shear failure in rock, *Proc. 1st Int. Congress of Rock Mechanics,* Vol. 1, Lisbon, 1966, 509–513.

1.34. Barton, N., Review of a new shear strength criterion for rock joints, *Engineering Geology,* Vol. 7, Elsevier, 1973, 287–332.

1.35. Ladanyi, B. and Archambault, G., Simulation of shear behaviour of a jointed rock mass, *Proc. 11th Symp. on Rock Mechanics,* Am. Inst. Min. Eng., New York, 1970, 105–125.
1.36. Fairhurst, C., On the validity of the Brazilian test for brittle materials, *Int. J. Rock Mech. Mining Sci.,* 1, 535–546, 1964.
1.37. Barton, N., A review of shear strength of filled discontinuities in rock, Norwegian Geotechnical Institute Publication, No. 105, 1974, 38.
1.38. Hobbs, N. B., Factors affecting the prediction of settlement of structures on rocks with particular reference to the chalk and triors, in settlement of structures, *British Geotechnical Society*, Pentech Press, London, 1975, 579–610.
1.39. Bieniawski, Z. T., Engineering classification of jointed rock masses, *Civil Eng. S. Africa*, 15, 335–344, 1973.
1.40. Bieniawski, Z. T., Rock mass classifications in rock engineering, *Proc. Symp. Exploration Rock Engineering*, Johannesburg, 1976, 97–106.
1.41. Bieniawski, Z. T., Tunnel design by rock mass classification, U.S. Army Waterways Experiment Station, Technical Report No. GL-79-19, 1979, 131.
1.42. Bieniawski, Z. T., *Rock Mechanics Design in Mining and Tunnelling,* A. A. Balkema, P. O. Box 1675, Rotterdam, Netherlands, 1984.
1.43. Bieniawski, Z. T., Case studies: prediction of rock mass behaviour by the geomechanics classification, *Proc. 2nd Australia-New Zealand Conference Geomechanics*, Brisbane, 1975, 36–41.
1.44. Goodman, R. E., *Methods of Geological Engineering in Discontinuous Rocks,* West, St. Paul, Minnesota, 1976.
1.45. Moreno Tallon, E., *Tunnelling*, Inst. Min. Eng., London, 1982, 6.
1.46. Costa-Pereira, A. S. and Rodrigues-Cavalho, J. A., Rock mass classification for tunnel purposes — correlations between the systems proposed by Wickham et. al., 1987.
1.46a. Bieniawski, Z. T. and Rocha, *Proc. of 6th ISRM Conference*, Vol. 2, A. A. Balkema, Montreal, 1987, 841–844. (A.

A. Balkema Publishers, Old Post Rd., Brookfield, VT 05036).

1.47. Wickham, G. E., Tiedemann, H. R., and Skinner, E. H., Ground support prediction model — RSR concept, *Proc. 1st Rapid Excavation and Tunnelling Conference*, Am. Inst. Min. Eng., New York, 1974, 691–707.

1.48. CANMET, Prediction of stable excavation spans for mining at depths below 1000 metres in hard rocks, DSS Serial No. OSQ80-00081, DSS File No. 17SQ.23440-0-9020, 1980.

1.49. Brook, N. and Dharmaratne, P. G. R., Simplified rock mass rating system for mine tunnel support, *Trans. Inst. Mining Met.,* London, A148–A154, 1985.

1.50. International Society for Rock Mechanics Commission on Testing Methods, Suggested Method for determining point load strength (to replace document published in 1972), *Int. J. Rock Mech. Mining Sci.,* (Pergamon Press), 22(2), 51–60, 1985.

1.51. Brook, N., The equivalent core diameter method of size and shape correction in point load testing, *Int. J. Rock Mech. Mining Sci.,* (Pergamon Press), 22(2) 61–70, 1985.

1.52. Bieniawski, Z. T., Determination of rock mass deformability — experience from case histories, *Int. J. Rock Mech. Mining Sci.*, (Pergamon Press), 15, 237–248, 1978.

1.53. Priest, S. D. and Brown, E. T., Probabilistic stability analysis of variable rock slopes, *Trans. Inst. Mining Met.,* 92, A1-A12, 1983.

1.54. Kaiser, P. K., MacKay, C., and Gale, A. D., Evaluation of rock classifications at B.C. rail tumbler ridge tunnels, *Rock Mech. Rock Eng.*, 19, 205–234, 1986.

1.55. Rutledge, T. C. and Preston, R. L., New Zealand experience with engineering classifications of rock for the prediction of tunnel support, *Proc. Int. Tunnel Symp.*, Tokyo, 1978, A3-1 to A3-7.

1.56. Serafim, J. L. and Pereira, J. P., Considerations of the Geomechanics Classification of Bieniawski, *Proc. Int. Symp. Eng. Geol. and Underground Construction*, Lisbon, Portugal, 1983, 1133–1144.

CHAPTER 2

MODES OF GROUND FAILURE

2.1 INTRODUCTION

A prerequisite to maintaining stability around an excavation by artificial means is to understand mechanics of movement and geometry of the failure or potential failure zones. The reason is that different failure mechanisms usually require different methods of stabilization.

The mechanism of ground failure is complex. It is influenced by complicated parameters, for example:

- geomechanical properties of the rock
- rock mass characterization
- mining method
- extent, spacing, and orientation of rock discontinuities
- slope and dimensions of excavation
- ratio of the discontinuity spacing to the excavation dimensions
- influence of the groundwater
- strata loading conditions
- rock fragmentation method
- seismic effects
- the rock-support interactions

However, the ground failure can be classified into one or a combination of the main types discussed in the foregoing sections.

2.2 PLANAR FAILURE

This type of failure involves sliding to occur on a single plane which strikes subparallel to and is undercut by one of the excavation boundaries. Conditions for development of planar failure are:[2.1]

1. The failure plane must strike parallel or near parallel, within approximately ± 20° to the nearest excavation boundary.
2. The angle of the failure plane to horizontal (α) must be smaller than that of the nearest excavation boundary (β), i.e., $\beta > \alpha$. In this case, the failure plane will "daylight" in the excavation boundary. A variation of this condition is shown in Figure 2.1 in which the failure daylights at the toe or at some height from the excavation boundary.
3. The angle α must be greater than or equal to the angle of internal friction of the rock (ϕ) in which the failure plane develops, i.e., $\alpha \geq \phi$. Therefore for the failure to take place: $\beta > \alpha \geq \phi$, as shown in Figure 2.1.

2.2.1 Potential Occurrence of Planar Failure in a surface excavation

Planar failure could occur under the following circumstances:

- □ Where the individual or overall slope of the excavated boundary is steeper than the bedding as shown in Figures 2.2a and 2.2b. In this case, the failure occurs along a parallel to the plane of the bedding which has the least cohesion and cementation strength to resist the driving forces due to the strata load.
- □ When a major joint, fault, unconformity, or similar through-going discontinuity is exposed by mining (Figure 2.2c), the failure progresses along and parallel to the plane of the discontinuity.

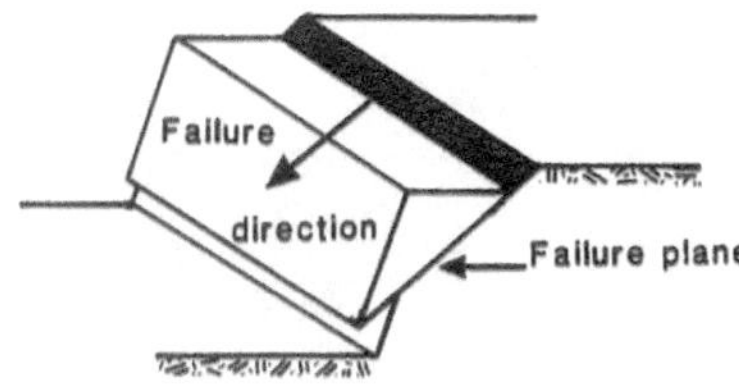

(a) A planar failure in three dimensions.

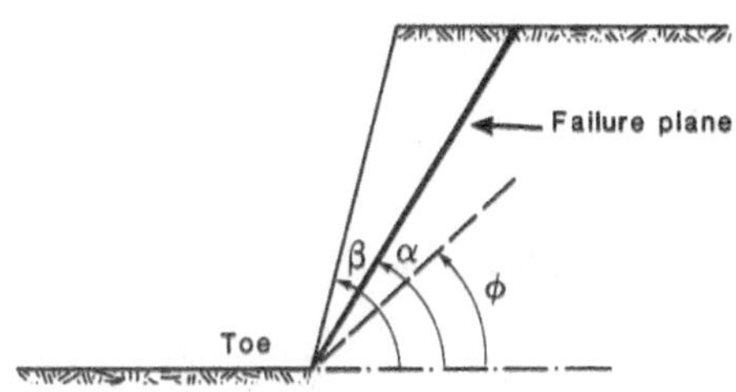

(b) Failure daylights at the toe.

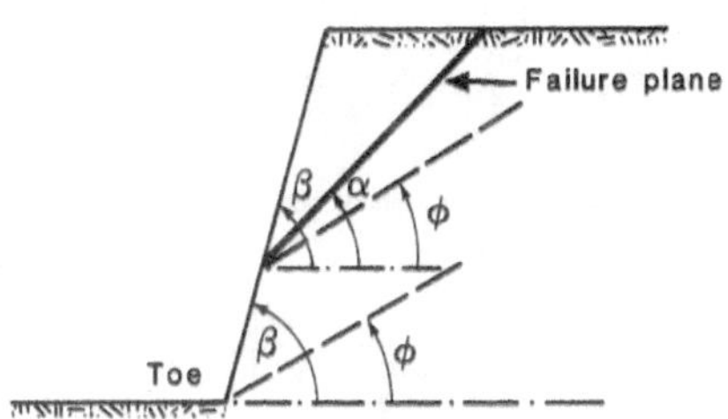

(c) Failure daylights above the toe.

Figure 2.1. General configurations of planar failure.

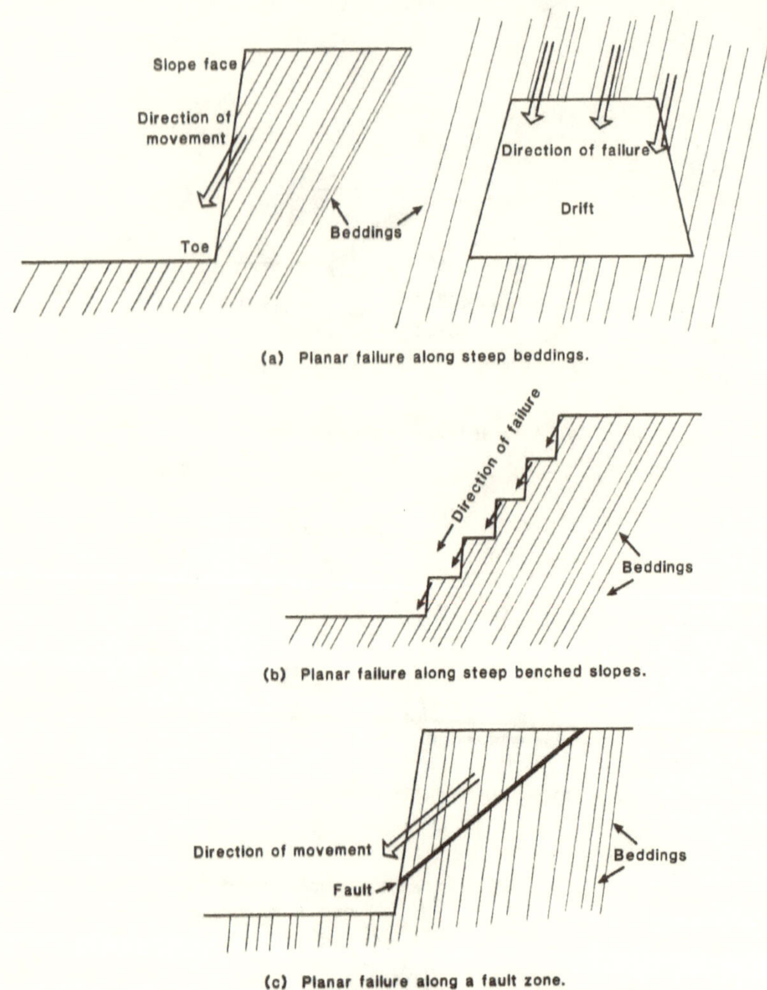

Figure 2.2. Potential occurrences of planar failure in mining conditions.

2.2.2 Development of Tension Crack During the Failure

Model studies on the failure of slopes in jointed rocks suggests that tension cracks generate as a result of small shear movements within the rock mass.[2.2] In practice, occurrence of tension cracks due to planar failure in underground excavations is not pronounced. However, this type of crack in the surface mine slopes can be seen in two locations.

1. Slope with tension crack in its face as shown in Figure 2.3a.
2. Slope with tension crack in its upper surface (Figure 2.3b).

The transition from one case to another occurs when:[2.1]

$$z/H = (1 - \cot\beta \cdot \tan\alpha) \tag{2.1}$$

where: z = Vertical depth of the tension crack from the slope crest, m
H = Height of the slope, m

In any case, presence of tension crack should be taken as an indication of potential instability.

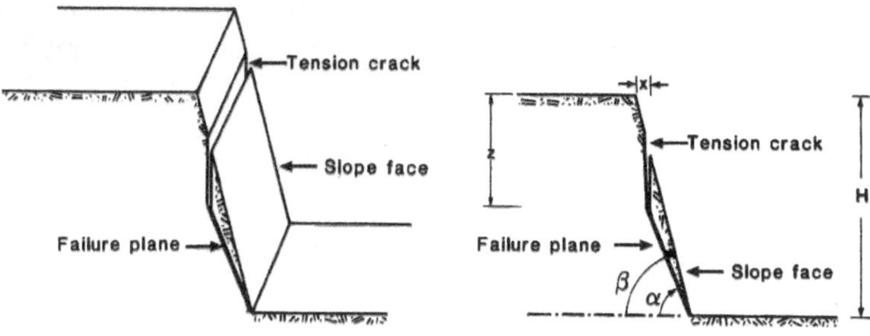

(a) Tension crack in a slope face.

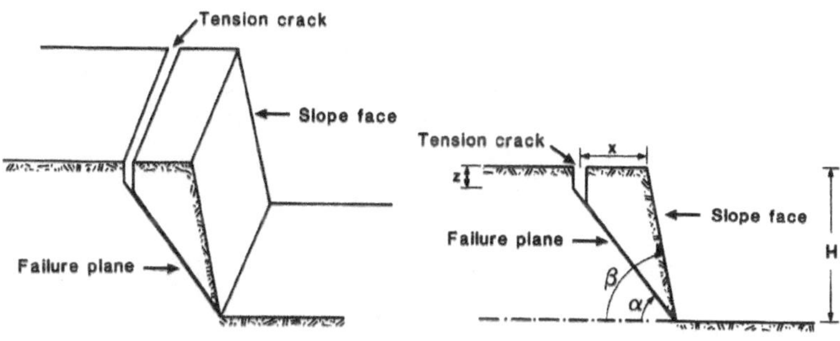

(b) Tension crack in upper slope surface.

Figure 2.3. Potential location of tension crack in the planar failure.

2.2.3 Critical Geometry of Tension Crack

Considering that the position of tension crack is known from its visible trace on or around the excavation walls. Under a dry or nearly dry condition, the critical depth (z_c) and critical horizontal location of the tension crack from the slope crest (x_c) can be expressed as follows:

$$z_c = H\,[1 - (\cot\beta \cdot \tan\alpha)^{0.5}] \quad (2.2)$$

$$x_c = H\,[\cot\beta \cdot \cot\alpha)^{0.5} - \cot\beta] \quad (2.3)$$

Field observation[2.1] suggests that tension cracks usually occur behind the crest of a slope in the manner shown as Figure 2.3b.

Depths and location of critical tension cracks in planar failure for a range of dry slopes (β) and the failure angle (α) are plotted in Figure 2.4. Influence of water in a tension crack formation is such that when it becomes water-filled, as a result of rain or surface water fillment, the Equations 2.2 and 2.3 will no longer hold. In these circumstances, the position and depth of tension crack should be determined empirically.

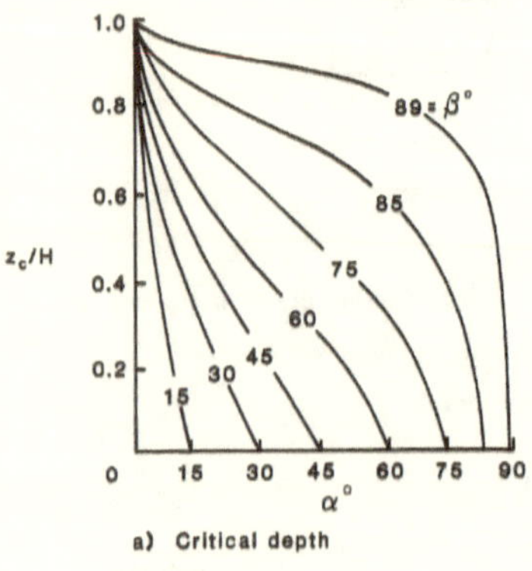

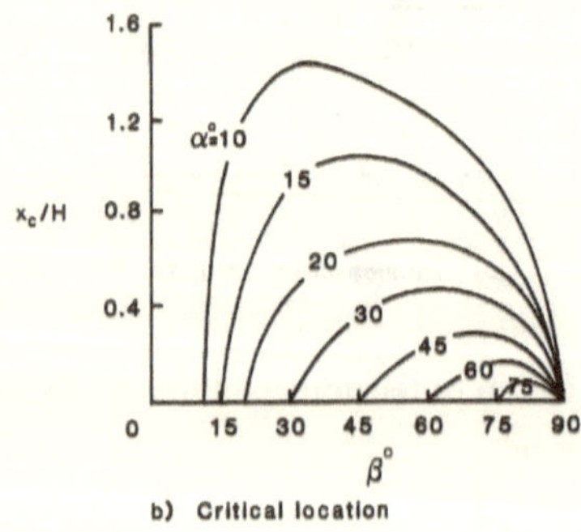

Figure 2.4. Critical geometry of tension crack in the planar failure and dry slope conditions.

2.2.4 Determination of the Safety Factor in Planar Failure Condition

Consider a surface excavation slope configuration where the groundwater table is above the toe of the slope or the excavation wall and a potential failure plane exists at an angle (α) to the horizontal less than the excavation slope (β). In this case, a minimum safety factor is considered to exist when the failure plane develops in the slope when no reinforcement is in place.[2.3] Figure 2.5 indicates a surcharge (q), blasting or seismic coefficient (K), phreatic surface, the weight of the failure wedge, and the water force acting through the centroid of the sliding mass without rotation.

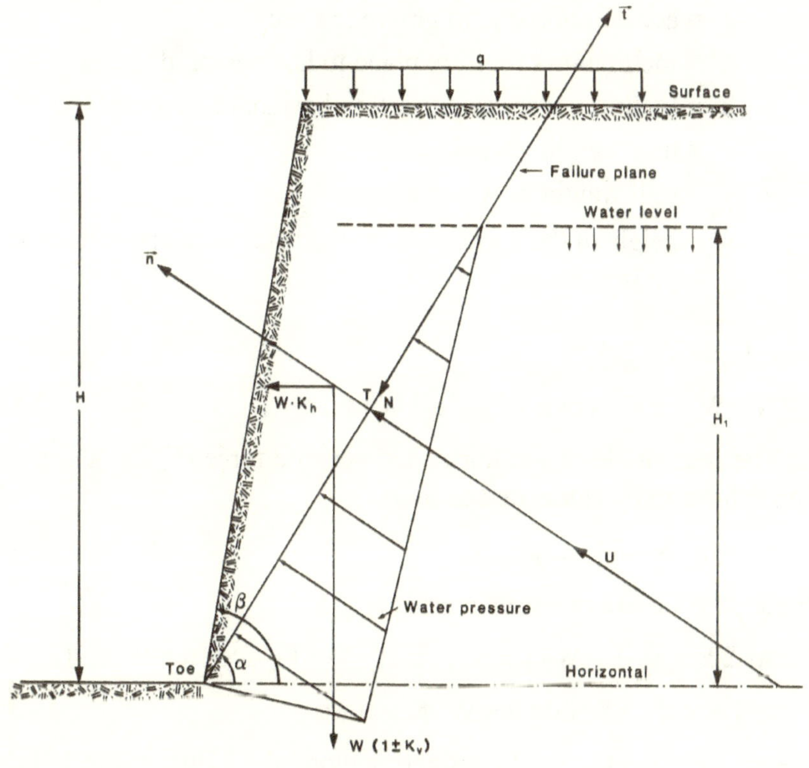

Figure 2.5. Detailed planar failure configuration.

Therefore, when $\Sigma F_n = 0$:

$$N + U - W(1 \pm K_v)\cos\alpha + W \cdot K_h \cdot \sin\alpha = 0 \tag{2.4}$$

where: F_n = sum of the forces acting normal to the failure plane, kN/m

N = reaction force acting normal to the failure plane, kN/m

U = total neutral force acting normal to the failure plane

$= H \cdot \gamma_w (\cot \alpha - \cot \beta) \text{ Sec } \alpha /2$, kN/m

W = weight of the sliding block, kN/m

K_v = vertical component of the seismic factor $\approx K_h /2$, in practice

K_h = horizontal component of the seismic factor = 0.10 to 0.33, for a = 0.1 to 0.33 g

a = seismic acceleration, m/sec^2

g = acceleration due to gravity, m/sec^2

α = inclination of failure plane to horizontal, degrees

β = inclination of the pit wall to horizontal, degrees

γ = unit weight of rock, kN/m^3

γ_w = unit weight of water, kN/m^3

H_1 = height of the horizontal groundwater table above the final pit bottom, m

From Equation 2.4:

$$N = W(1 \pm K_v) \cos \alpha - U - K_h \cdot W \cdot \sin \alpha \tag{2.5}$$

The magnitude of the horizontal seismic force (F_h) in kN/m of the anchor can be calculated as follows:

$$F_h = m \cdot a = W \cdot a /g = W \cdot K_h \tag{2.6}$$

where, m = failure mass, kN·sec^2/m^2

$$\text{At } \Sigma F_t = 0\text{: } T - W (1 \pm K_v) \sin \alpha - W \cdot K_h \cdot \cos \alpha = 0 \tag{2.7a}$$

$$\text{or } T = W (1 \pm K_v) \sin \alpha + W \cdot K_h \cdot \cos \alpha \tag{2.7b}$$

where: ΣF_t = sum of the forces acting along the failure plane, kN/m
T = Disturbing force along the failure plane, kN/m

The factor of safety (FS) can be calculated as follows: [2.3]

$$FS = \text{Resisting Force} / \text{Disturbing Force}$$
$$= (\text{Cohesional Force} + \text{Frictional Resistance}) / T$$
$$= \{(S_o \,.\, H / \sin \alpha) + \tan \phi\, [W(1 \pm K_v) \cos \alpha - U - W \cdot K_h \cdot \sin \alpha]\}/[W (1 \pm K_v) \sin \alpha + W \cdot K_h \cdot \cos \alpha] \quad (2.8)$$

Considering surcharge (q) and water level at H_1, then the total weight of rock above the failure plane (W) can be calculated as follows:

$$W = [q \cdot H(\cot \alpha - \cot \beta) + \gamma\, (H - H_1)]\ \{[H(\cot \alpha - \cot \beta) + H_1 (\cot \alpha - \cot \beta) / 2] + \gamma_{sat} .\, H_1^{\,2} [(\cot \alpha - \cot \beta) / 2]\}$$
$$= (\cot \alpha - \cot \beta)\ \{q.\, H + [\gamma\, (H - H_1^{\,2}) / 2] + (\gamma_{sat} .\, H_1^{\,2} / 2)\} \quad (2.9)$$

where: H = Total height of the slope, m

γ_{sat} = Unit weight of saturated rock, kN/m^3

$$(\cot \alpha - \cot \beta) = [\sin (\alpha - \beta) / \sin \alpha .\sin \beta]$$

Let ψ = Total weight factor, kN/m = $q.\, H\, [(H^2 - H_1^{\,2)}/2] + [(\gamma_{sat} + H_1^{\,2)}/2]$ (2.10)

Substituting Equation 2.10 into Equation 2.9 gives:

$$W = \psi\, [\sin (\beta - \alpha) / \sin \alpha .\sin \beta] \quad (2.11)$$

The water weight factor (ψ_1) can be calculated as follows:

$$\psi_1 = \gamma_w .H_1^{\,2}/2 \quad (2.12)$$

The natural force due to water uplift (U) is then:

$$U = [\psi_1 .\, \sec \alpha .\, \sin (\beta - \alpha)] / \sin \alpha .\sin \beta \quad (2.13)$$

Therefore, the safety factor (FS) in terms of ψ and ψ_1 from Equations 2.8 and 2.11 to 2.13 can be expressed as follows:

$$FS = \{S_o .\, H + [\sin (\beta - \alpha) / \sin \beta] \tan \phi\, [\psi\, (1 \pm K_v) \cos \alpha - \psi_1 .\, \sec \alpha - \psi .\, K_h \cdot \sin \alpha]\}/\{[\sin (\beta - \alpha) / \sin \beta].\ [\psi (1 \pm K_v) \sin \alpha + \psi .\, K_h \cdot \cos \alpha]\} \quad (2.14)$$

Let $(1 \pm K_v) = \Omega$ and $\psi_1 / \psi = \Omega_1$ (2.15)

Then Equation 2.14 in a dimensionless form can be expressed as follows:

$$FS = \{S_o . H . \sin\beta / [\psi. \sin(\beta - \alpha)] + \tan\phi\,(\Omega . \cos\alpha - \Omega_1 .\sec\alpha - K_h \cdot \sin\alpha)\} / (\Omega . \sin\alpha + K_h \cdot \cos\alpha) \qquad (2.16)$$

2.2.5 Inclination of the Critical Failure Plane (α)

The α - value depends largely on factors such as S_o, ϕ, γ, H, H_1, β, q, K_v, and K_h.

For any constant value of the above noted factors, the FS will be a minimum when $\partial FS/\partial\alpha = 0$. Thus differentiating equation 2.16 with respect to α and equating to zero gives:

$$\{S_o . H . \sin\beta / [\psi. \sin(\beta - \alpha)]\}[K_h \cdot \cos(\beta - 2\alpha) - \Omega \cdot \sin(\beta - 2\alpha)] - \tan\phi\,[\Omega^2 + \Omega \cdot \Omega_1 (\tan^2\alpha - 1) + 2\Omega_1 \cdot K_h \cdot \tan\alpha + K_h^2] = 0 \qquad (2.17)$$

For a particular case when there is no surcharge (i.e., q = 0), on a fully drained slope (i.e., $H_1 = 0$ and $K_h = K_v = 0$), then equations 2.9, 2.11, and 2.13 give:

$$\text{Sin}^2(\beta - \alpha) \cdot \tan\phi = (2 S_o . \sin\beta / \gamma . H) \qquad (2.18)$$

The presence of water in the tension crack will cause a 10% reduction in the α value, as a rough estimation.[2.1]

2.2.6 Analysis of a Rough Planar Failure

Most rock surfaces exhibit a nonlinear relationship between shear strength (τ) and the effective stress normal to the failure plane (σ_n). This relationship is defined differently by various investigators[2.2, 2.4–2.6] as follows:

$$\tau = [\sigma_n (1 - a_s)(v + \tan\phi) + a_s . \tau_r] / [1 - (1 - a_s) v. \tan\phi] \qquad (2.19)$$

$$\text{or } \tau = \sigma_n . \tan[\phi + JRC . \text{Log}_{10}(\sigma_j / \sigma_n)] \qquad (2.20)$$

where: a_s = The proportion of the discontinuity surface which is sheared through projections of intact rock material
$= 1 - [1 - (\sigma/\sigma j)]^L$

v =The dilation rate dv/dt at peak shear strength[2.5]
= $[1- (\sigma /\sigma j)]^{K\cdot}$ tan i

τ_r = Shear strength of the intact rock material[2.7]
= $\sigma_j\{[(1 + n) – 1 / n]^{0.5}\}[1 + (n. \sigma /\sigma j)]^{0.5}$

σ_j = uniaxial compression strength of the rock material adjacent to the discontinuity, which due to weathering or loosening of the surface, may be lower than the uniaxial compressive strength of the rock material within the body of an intact block

n = ratio of uniaxial compressive to uniaxial tensile strength of the rock material
= 10, for most hard rocks[2.8]

L and K = roughness constants = 1.5 to 4 for rough rocks

i = number of angles for second order projections

JRC = Joint Roughness Coefficient. It is evaluated as follows:
For rough joints and beddings = 20
Medium joints and beddings = 10
Smooth planar joints and beddings = 5

In order to utilize either Equation 2.19 or 2.20 it is necessary to determine the effective normal stress (σ_n) acting on the rough plane of failure. This quantity can be calculated from Equation 2.4 in the following form:

$$\sigma_n = W(1 \pm K_v)\cos\alpha - U - K_h \cdot W \cdot \sin\alpha \tag{2.21}$$

where, W and U can be calculated from Equations 2.9 and 2.13, respectively. Finally the factor of safety (FS) can be determined from either equations 2.8, 2.14, or 2.18.

2.2.7 Influence of Undercutting the Toe of a Slope with Potential Planar Failure

Undercutting the toe or base of excavation walls can occur by:

1. Action of water on the soft underlying rock in the toe
2. By weathering of underlying strata along the toe to cause caving

3. By excessive blasting or shovel cutting at the toe of a surface mine slope, as shown in Figure 2.6a.
4. By intentionally undercutting the toe or base along an underground roadway, to an appropriate depth (d) and thickness (t), in order to hold back the stress concentration at some distance from the excavation boundary (Figure 2.6b). The latter was found by the author to equalize the strata loading on both sides of an underground roadway to achieve a more uniform and symmetrical closure.[2.9] In any case, the undercutting forms a negative angle to the surface and underground excavation walls or slopes.

A previous failure is assumed to have left a face inclined at α_1 with an initial vertical tension crack depth z_1. An undercutting of depth, d, inclined at a negative angle, α_0, causes a new failure plane inclined at α_2 with a new tension crack of depth z_2, as shown in Figure 2.6.

Under the above noted condition, weight of the sliding block due to the undercutting (W_2) can be expressed as follows:

$$W_2 = (\gamma/2)\,[(H^2 -)\cot\alpha_2 - (H - z)\cot\alpha_1 + (H + H_1)\,d] \qquad (2.22)$$

where: d = depth of the undercut, m $= (H - H_1)\cot\alpha_o$ (2.23)

z = critical depth of the new tension crack for dry undercut, m

$= (S_o \,.\, \cot\phi) / [\gamma .\, \cos\alpha_2 .\, \sin(\alpha_2 - \phi)]$ (2.24)

α_2 = inclination of the new failure plane caused by the undercut, degree.

$= (½)\ \{\phi + \arctan[(H - z_2{}^2) / [(H^2{}_1 - z1\,^2)\cot\alpha_1] - (H + H_1)d\}$ (2.25)

Thus the factor of safety (FS) under the final condition can be determined by utilizing either Equations 2.8, 2.14, or 2.18.

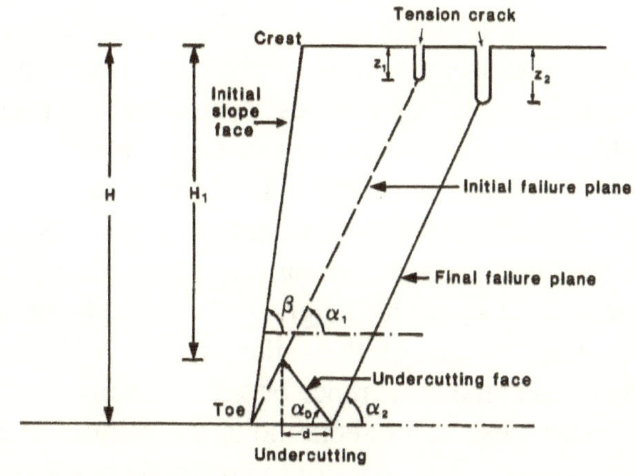

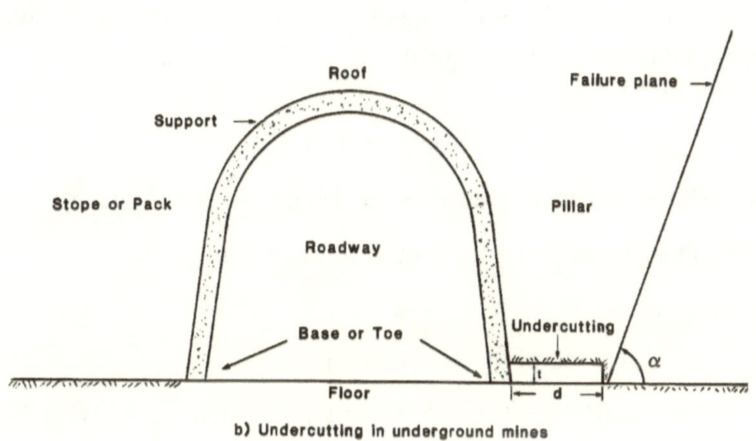

Figure 2.6. Configuration of undercutting in the surface and underground mines.[2.9]

2.3 BILINEAR SLAB FAILURE

2.3.1 Potential Occurrence

This type of failure occurs when rock slabs sliding along a primary bedding-plane discontinuity or split bedding in combination with sliding along a secondary shallow dipping discontinuity which undercuts the slope as shown in Figure 2.7a. If such a preexisting discontinuity does not occur, bi-linear slab failure could occur by crushing and shearing through the rock mass at the toe of the slope (Figure 2.7b). In some cases, occurrence of rolls or thinning of beddings or veins in the footwall may contribute to development of bi-linear slab failure as shown in Figure 2.7c.

Main factors in development of bi-linear slab failure are: the slab thickness (t); its physical and mechanical properties — slope inclination (β) and its height (H) or height of the underground excavation, weight of the failure mass (W), groundwater, strata pressure; and the occurrence of rolls or thinning of the slab.

Under these failure conditions, the failure mass can be divided into two parts. These are driving or active block and resisting block.

2.3.2 Driving or Active Block

This is the main mass, specifications of which are shown in Figure 2.8a. For limiting equilibrium of the driving block:

$\Sigma F_x = 0$ and $\Sigma F_y = 0$

W_d = weight of the driving block $= \gamma \cdot 1_d \cdot t$ (2.26)

τ_d = shear force at the bottom of the driving block $= N_d \cdot \tan \phi_d$ (2.27)

τ_b = shear force along the length of the driving block

$= N_b \cdot \tan\phi_b + S_b \cdot 1_d$ (2.28)

N_d = driving force of the active block

$= \{[W_d \sin (\beta - \cos \beta \,.\, \tan \phi_d) - S_d \cdot 1_d]\} / (1 - \tan \phi_b \,.\, \tan \phi_d)$ (2.29)

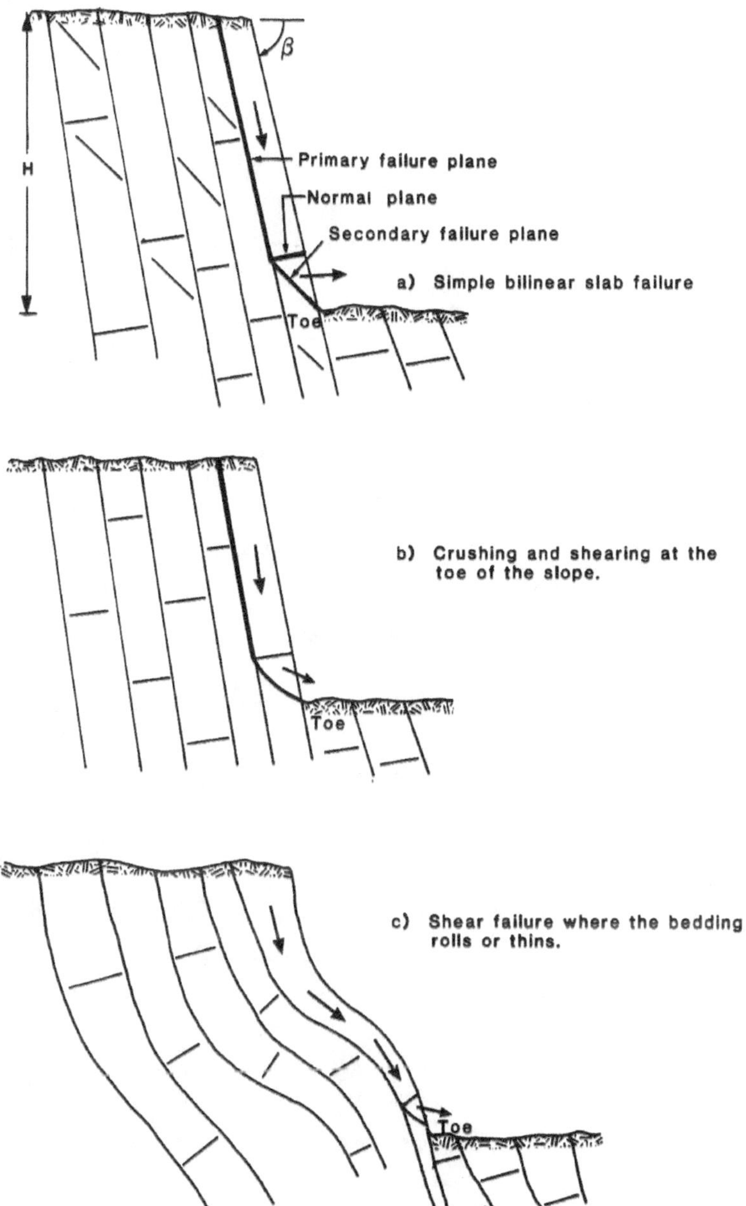

Figure 2.7. Occurrence of bi-linear slab failure.

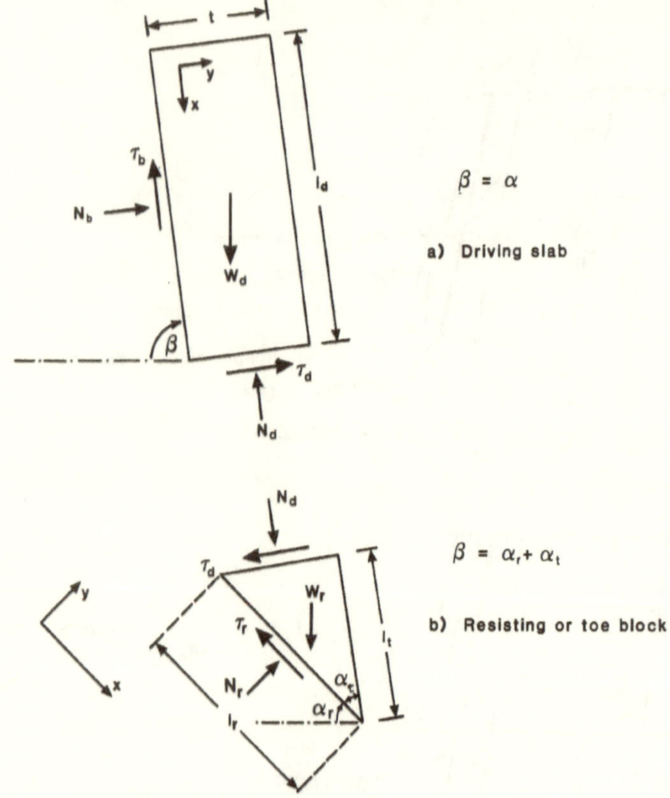

Figure 2.8. Free-body diagram of a bi-linear slab failure.

2.3.3 Resisting Block

The resisting block exists at the toe or base of the excavation with specifications given in Figure 2.8b. For limiting equilibrium of the resisting or toe block:

$\Sigma F_x = 0$ and $\Sigma F_y = 0$

W_r = weight of the resisting block $= \gamma \cdot 1_r \cdot t/2$ (2.30)

τ_r = shear force along the base of the resisting block

$= N_r \cdot \tan \phi_r + S_r \cdot 1_r$ (2.31)

N_r = resisting force of the toe block $= FS \cdot N_d$

$= FS \{[Wd(\sin \beta - \cos \beta . \tan \phi_b) - S_d \cdot 1_d] / (1 - \tan \phi_b . \tan \phi_d)\}$ (2.32)

2.4 PLOUGHING SLAB FAILURE

2.4.1 Potential Occurrence

Ploughing occurs when slab sliding along a basal plane (i.e., a bedding or a fault) combines with sliding along a secondary steeply dipping discontinuity (e.g., joint) which strikes at an angle opposite to the slopes as shown in Figure 2.9a. This occurrence is opposite to that discussed in the case of bi-linear slab failure where shear failure predominates. In the case of ploughing, a toe block will be separated from the rock mass and rotated out of the slope. This type of failure can also be initiated on bedding rolls (Figure 2.9b).

The failure mass in the ploughing slab failure can be divided into two parts much similar to that of the bi-linear slab failure. These are described in the next two sections.

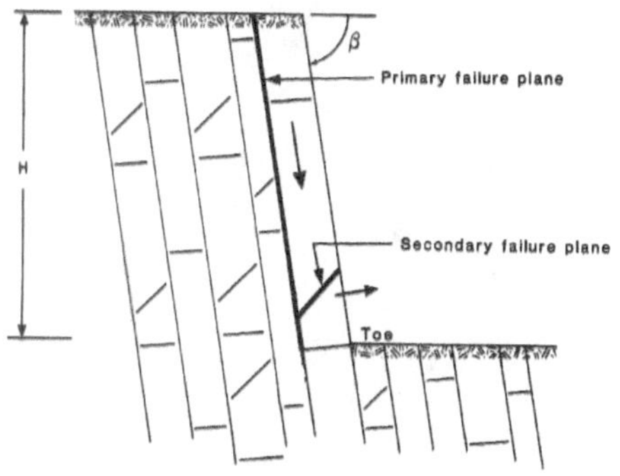

a) Simple ploughing slab failure

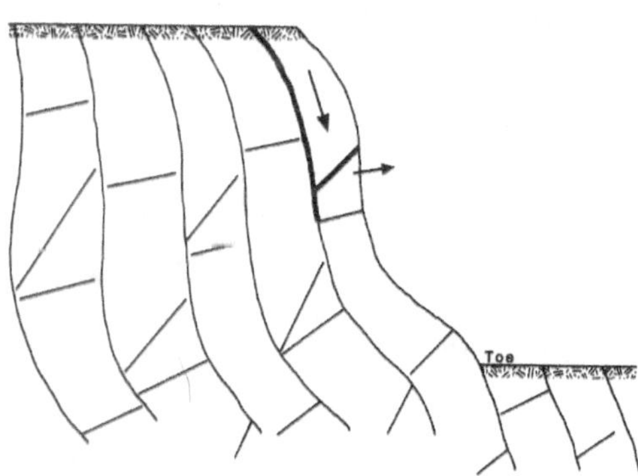

b) Ploughing initiated on bedding roll.

Figure 2.9. Configuration of ploughing slab failure.

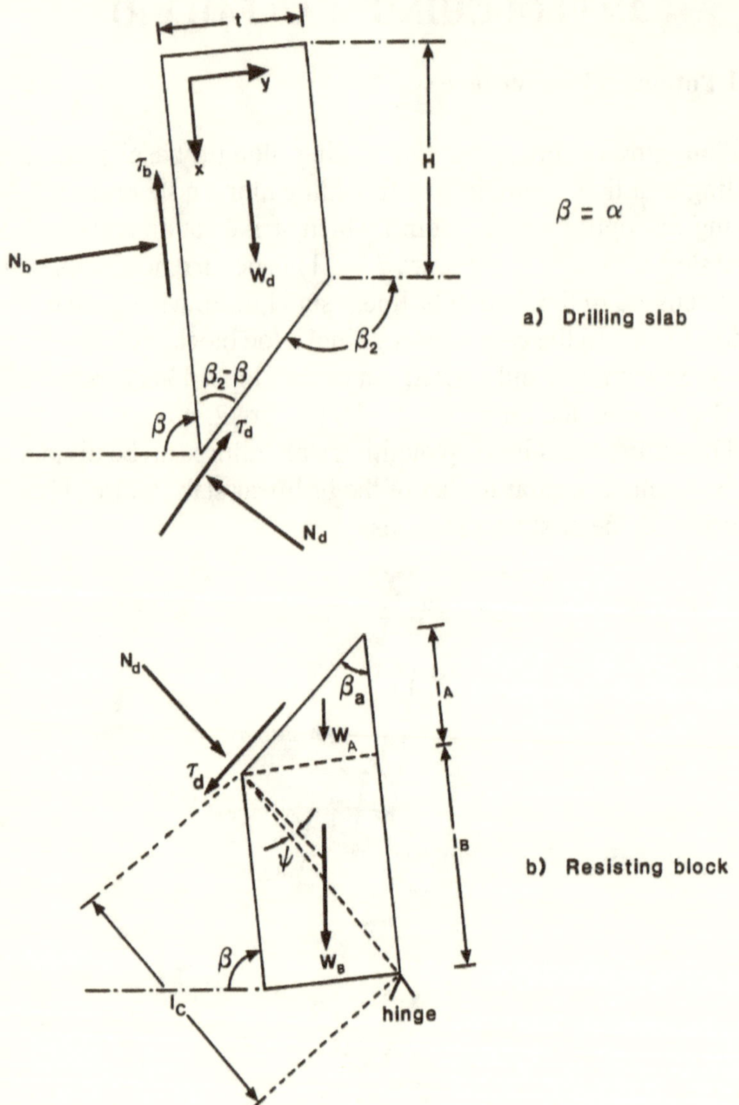

Figure 2.10. Free-body diagram of a ploughing slab failure.

2.4.2 Driving Block or Slab

Driving block or slab have specifications as shown in Figure 2.10a. For limiting equilibrium of the driving slab:

$\Sigma F_x = 0$ and $\Sigma F_y = 0$

H_{max} = maximum allowable slope height = $H_1 + H_2$

$$= \{[(W_d - Wa)\sin\beta]/\gamma . t\} + (1_A + 1_B)\sin\beta \qquad (2.33)$$

where: $W_A = \gamma \cdot t \cdot 1_A/2$

$W_B = \gamma \cdot t \cdot 1_B$

W_d = weight of the driving block

$$= \{\{N_d . \tan\phi_d [\tan\phi r. \sin(\beta_2 - \beta)] - \cos(\beta_2 - \beta)\}/ (\cos\beta. \tan\phi d - \sin\beta)\} - \{N_d [\tan\phi r .\cos(\beta_2 - \beta) + \sin(\beta_2 - \beta)] / (\cos\beta. \tan\phi d - \sin\beta)\} \qquad (2.34)$$

M_d = driving or overturning moment = $N_d \cdot 1_c \cdot \tan\psi$ (2.35)

2.4.3 Resisting or Toe Block

A resisting or toe block forms at the base or toe of the failure mass, specifications of which is shown in Figure 2.10b. For limiting equilibrium:

$M_d = M_r$

M_r = resisting moment

$$= \tau d . 1_c + W_A \{\cos\beta [(1_A/3) + 1_B] + \sin\beta (t/3)\} + W_B [\cos\beta(1_B/2) - \sin\beta (t/2)] \qquad (2.36)$$

N_d = driving force on the resisting or toe block

$$= W_B [\cos\beta [(1_A/3) + 1_B + \sin\beta (t/3)] + W_B [\cos\beta(1_B/2) + \sin\beta (t/2)] \qquad (2.37)$$

As is evident from Equations 2.33 to 2.37, the analysis of results are sensitive to the length of the toe block. In this regard, a number of potential toe block geometries should be assessed to determine the critical length of the resisting block. Also, maximum weight of the driving slab to maintain limiting moment equilibrium can be converted to maximum allowable or critical slope height, by geometry. In addition, parameters such as dip of the discontinuities, slope curvature or rolls, cementation between the beddings and joints, and groundwater conditions may also be of critical importance.

2.5 STEPPED FAILURE

2.5.1 Potential Occurrence

This type of failure occurs where the length of discontinuities is less than the length of the slope or the excavation wall as shown in Figures 2.11a and b. In this case, failure surface may be formed along a stepped path, by rock mass sliding along bedding and joints in combination with release along steeper secondary discontinuities. Secondary discontinuity of this type are generally observed in bedded deposits which have undergone folding.

Where significant sliding may occur on two sets of discontinuities involved in stepped failure, an assessment of the strength parameters of both discontinuity sets and the relative contribution of each set to the failure path is required. Rigorous analysis of this type of failure is not generally practical. Therefore, stability analysis are often conducted by assuming a single continuous failure surface inclined at the average dip of the stepped failure path (α), as shown in Figure 2.11c, and by estimating the average shear strength along the failure path. This approach is conservative, as sliding along two sets of discontinuities at different orientations may induce slippage and rotation between adjacent discrete blocks within the mass. Hence, the actual net vector direction of movement of the failure mass could be somewhat shallower than the assumed overall failure dip (α). Although numerous sizes and configurations of potential stepped failure surfaces are possible, if the spacing and lengths of the discontinuities are known, the most probable or critical failure path may be determined.

2.5.2 Analysis of Stepped Failure

Stability analysis of stepped failure may be carried out using a modified plane failure analysis technique in which the weight or size of the failure block is calculated from the average dip of the failure surface. Driving and resisting forces are determined by resolving the weight force into components normal and parallel to the dip of the primary discontinuity set (Figure 2.11c). The failure block is assumed to remain intact because sliding is assumed to occur primarily along one discontinuity set with little rotation or internal readjustment of the block. This approach results in a higher FS than a conventional plane failure analysis, because the size of the failure block is reduced as a result of stepped failure surface. Under the above noted conditions and Figure 2.11c:

Figure 2.11. Stepped failure configuration.

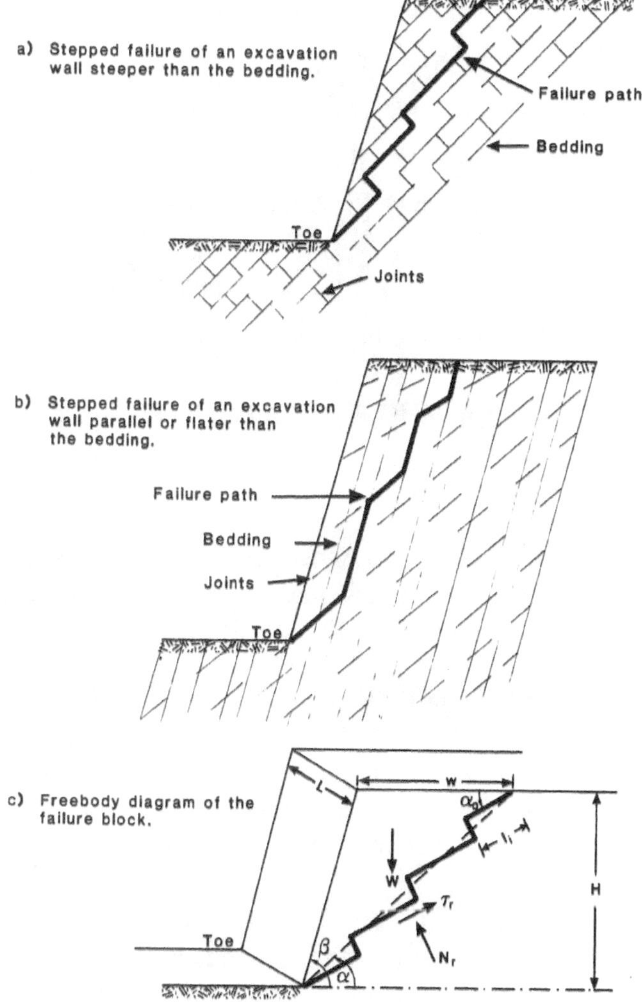

F_d=drivingforce=W·sinα_0 (2.38)

where: W = [γ .H.L.W/2] [(1 /tan α) - (1 /tan β)], mN

α_0 = dip of the primary discontinuity (or bedding) to the horizontal, degree

L = length of the failure block (Figure 2.11c)

w = width of the failure block (Figure 2.11c)

F_r = resisting force = τ_r

= W · cosα_0 · tanϕ_r + S_r Σbetween i = 1 to n of l_i (2.39)

FS = factor of safety = F_r/F_d

= cot α_0 · tan ϕ_r + [S_r Σbetween i = 1 to n of l_i / W. sinα_0] (2.40)

ϕ_r = internal friction angle of the bedding material, degree

S_r = cohesion of the bedding material, MPa

2.6 WEDGE FAILURE

2.6.1 Potential Occurrence

Wedge failure may occur where two or more discontinuities intersect, and that intersection is undercut by the slope or excavation boundary as shown in Figure 2.12.

2.6.2 Analysis of Wedge Failure

The geometry of the wedge for the purpose of analyzing its basic failure mechanism is defined in Figure 2.13. In this analysis the following assumptions were made:[2.1]

1. Failure condition is defined by: B > ψ_i > ϕ
   where: β = inclination of the slope face (Figure 2.13a)
   ψ_i = dip of the line of interception from the two planes of discontinuities to the horizontal
   ϕ = internal friction angle of the rock mass
2. The flatter of the two planes of wedge failure is called plane A and the steeper one is called plane B, as shown in Figure 2.13b.

3. The wedge failure occurs by sliding and this is resisted by friction only.
4. The angle of internal friction (ϕ) is the same for both failure planes.

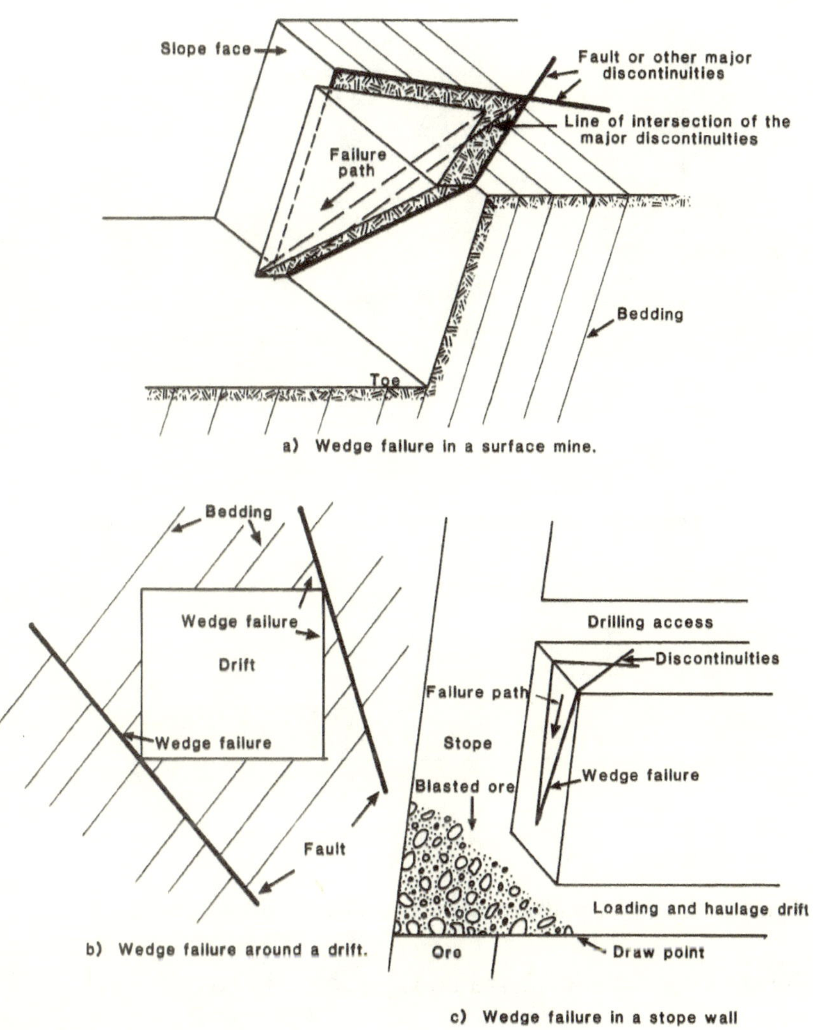

Figure 2.12. Configuration of wedge failure.

Under the above noted and dry conditions, FS can be determined as follows:[2.1]

$$FS = (R_A + R_B) \tan \phi / W. \sin \psi_i \tag{2.41}$$

Where: R_A = the normal reaction provided by plane A

R_B = the normal reaction provided by plane B

$$R_A + R_B = W \cdot \cos\psi_i \sin\alpha_A / \sin\Delta \quad (2.42)$$

Δ = half included angle of the wedge (Figure 2.13b)

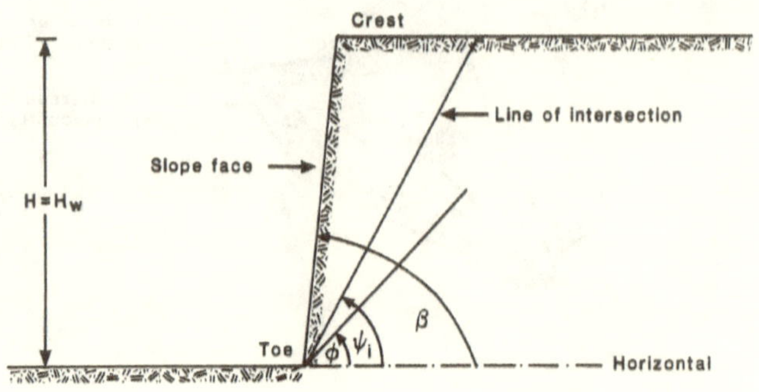

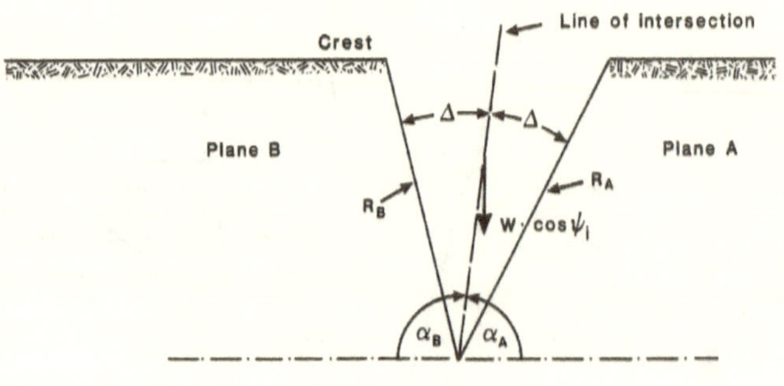

Figure 2.13. Geometry of a simple wedge failure in dry condition.

Substituting Equation 2.42 into Equation 2.41 gives:

$$FS = \sin\alpha_A \cdot \tan\phi \cdot \cot\psi_i / \sin\Delta = K \cdot \tan\phi \cdot \cot\psi_i \quad (2.43)$$

where: K = the wedge factor = $\sin\alpha_A / \sin\Delta$ (2.44)

Equations 2.43 and 2.44 suggest the steeper failure planes or decreases in the included angle, decrease the wedge factor which in turn decrease the safety factor.

The average pressure under dry wedge condition $= \gamma \cdot H_W / 6$ (2.45)
where: H_W = height of the wedge

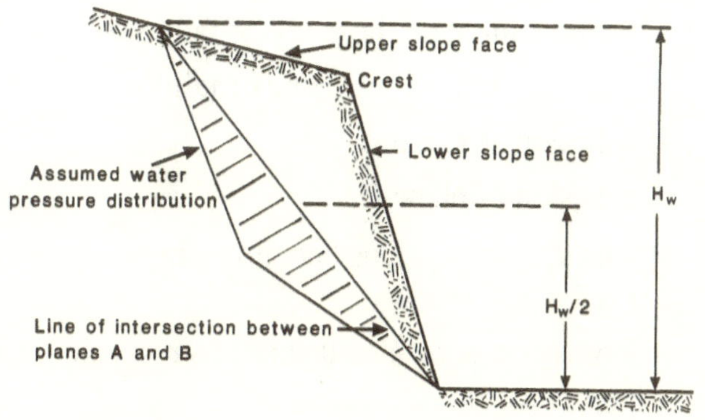

a) Wedge failure without tension crack

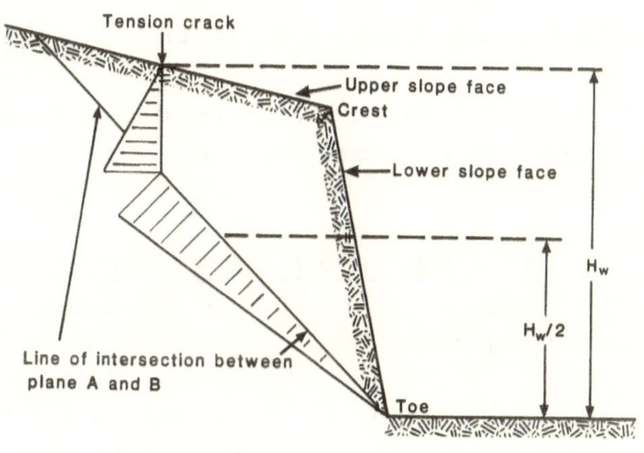

b) Wedge failure with tension crack.

Figure 2.14. Water pressure distribution in wedge failure.

Under wet conditions and when different cohesion and friction angles are taken into account, determination of the wedge safety factor is given as follows:[2.10]

$$FS = [3\, y\, (S_A \,.\, x + S_B) / \gamma.\, H_W] + [(2A.\, \gamma - \gamma w.\, x) \tan \phi_A / 2\,\gamma] + [2B.\, \gamma - \gamma w.\, y) \tan \phi_B / 2\gamma] \quad (2.46)$$

where: γ = unit weight of the rock mass

γw = unit weight of the water

H_w = total height of the wedge (Figures 2.14a and b)

ϕ_A = friction angle on the plane of failure A

ϕ_B = friction angle on the plane of failure B

S_A = cohesion of plane A

S_B = cohesion of plane B

A = $(\cos \alpha_a - \cos \alpha_b . \text{Cos } \theta na.nb) / (\sin \psi_5 . \sin^2 \theta na.nb)$ (see Figure 2.15)

B = $(\cos \alpha_b - \cos \alpha a. \text{Cos } \theta na.nb) / \sin \psi_5 . \sin^2 \theta na.nb)$ (see Figure 2.15)

x = $\sin \theta 24 / \sin \theta 45. \cos \theta 2na$ (see Figure 2.15)

y = $\sin \theta 13 / \sin \theta 35. \cos \theta 1nb$ (see Figure 2.15)

The average pressure in a wet wedge condition $= \gamma \cdot H_t/3$ (2.47)

where: H_t = depth of the bottom vertex of the tension crack below the upper slope face (see Figure 2.15)

2.7 CIRCULAR FAILURE

2.7.1 Potential Occurrence

This type of failure occurs when a segment of the excavation boundary fails by rotation on a curvilinear path of finite length offering least resistance to sliding. Variations of circular failure are shown in Figure 2.16. Circular failure is more predominant in soil, crushed or milled rock, and altered or weathered rock formations.[2.11] A tendency to circular failure may be attributed to several factors such as: the cohesion of failed material; shape and interlocking characteristics, the material size with respect to the excavation height; and groundwater and seismic activity. Circular failure conditions prevail mainly in the overburden and alluvial formations around a surface mine and in the waste or tailing dumps.

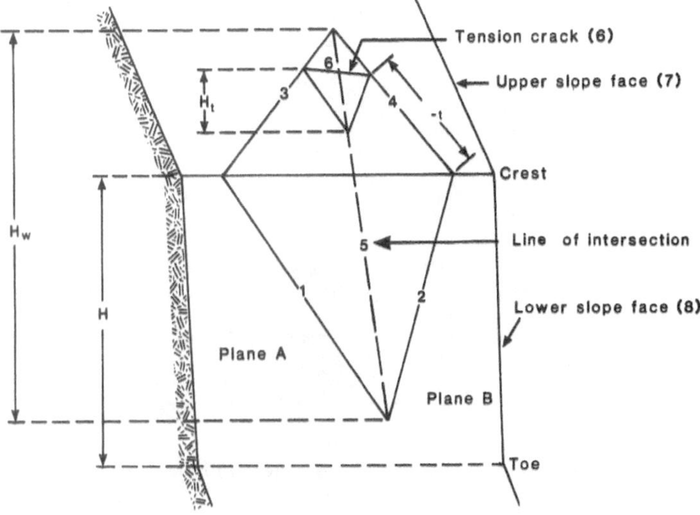

a) Pictorial view of a wedge failure with a tension crack showing number of intersection lines and planes used in equation 4-46

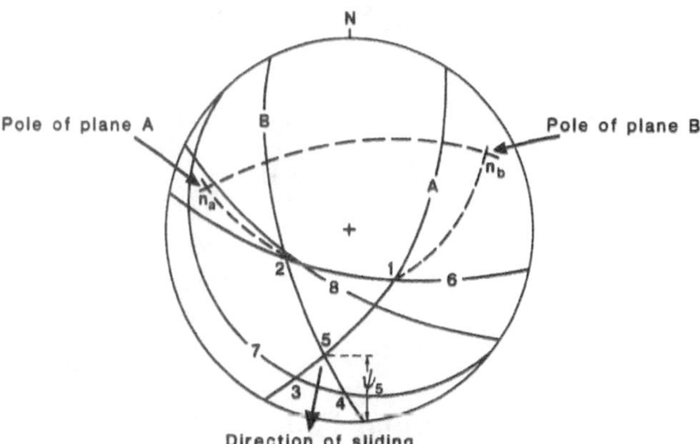

b) Sterioplot of a wedge failure showing numbering used in equation 4.46

Figure 2.15. Geometry of a wedge failure with tension crack.

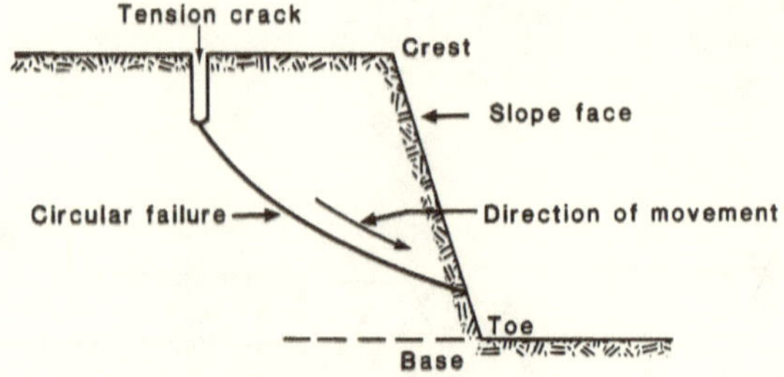

a) Failure ending above the toe in relatively hard base and variable strength in the top soil

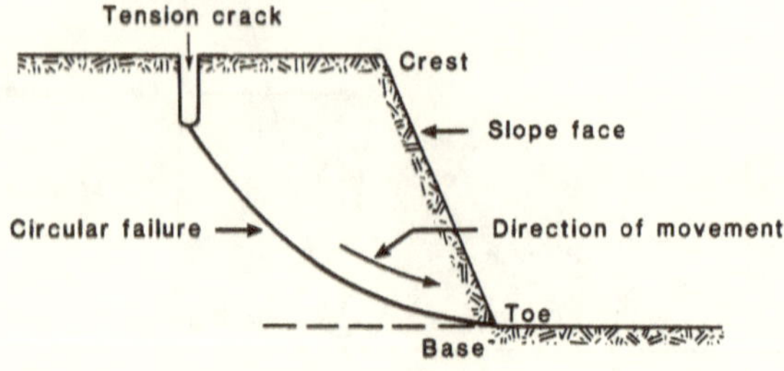

b) Failure ending at the toe in relatively hard base and uniform strength top soil.

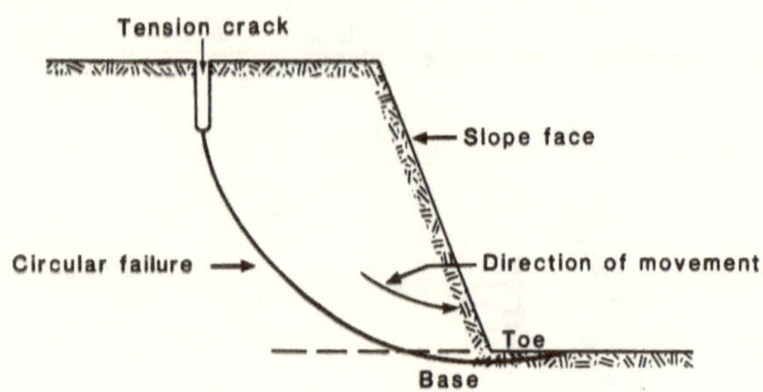

c) Failure ending below toe in relatively soft base.

Figure 2.16. Variations of circular failure.

2.7.2 Stability Analysis of Circular Failure

Available circular failure charts[2.1,2.12,2.15] are utilized mainly as a guide or rough check during the preliminary assessment of the slope stability. An accurate determination can be performed based on the critical sliding surface.[2.16] Assumptions made are as follows:

1. Homogeneous medium
2. Shear strength of the material (τ) is characterized by the Mohr-Coulomb failure criterion, expressed as follows:

$$\tau = S_0 = \sigma_n \tan \phi \qquad (2.48)$$

where: S_0 = average cohesion of the material in MPa
= shear strength of the material at zero normal stress, i.e. $\sigma_n = 0$

σ_n = σ_α = normal stress acting on the failure surface in MPa

ϕ = average internal friction angle of the material in degree

In equation 2.48, it may be necessary to consider:

1. Total stress analysis based on the undrained shear strength (τ), then ($S_0 = S_u$) and ($\phi = \phi_u$) usually utilized to determine short-term stability during or at the end of slope construction.
2. Effective stress analysis based on the drained shear strength (τ^1) commonly called effective shear strength parameters (S), the effective friction angle on the failure plane (ϕ^1), and effective normal stress (σ). The major difference between a total stress analysis and effective stress analysis is that the former does not require a knowledge of the pore pressure, in contrast to effective stress analysis.
3. Slope failure occurs on a circular surface passing through the slope toe. This was suggested by Taylor[2.13] as being especially true for $\phi > 3$ and the slope angle $\beta > 10$.

A number of methods are available to determine the slope FS.[2.12,2.17–2.22] All of these methods are based on the limit equilibrium concept, and in turn based on the slip surface theory. Some other methods consider log-spiral failure surface.[2.13,2.23] However, their results are similar to those obtained using a circular arc surface.[2.17,2.19,2.21,2.22] Common to all methods is a free body under gravitational force loaded by external forces (i.e., pore water and seismic), with a normal and shear stress distribution along the failure surface.

The equilibrium condition with regard to the slices of the total sliding mass can be expressed in a general form as follows:

$$(d\,Q_i/dx) + \gamma\,[F(x) - D(x)] + d\,S_i/dx - \sigma_a - \tau_a\,[d\,F(x)/dx] = 0$$

Therefore, $-S_i - E_i\,[dG(x)/dx] - (dE_i/dx)\,F(x) + (dE_i/dx)\cdot G(x) = 0$ (2.49)

Hence, $(d\,E_i/dx) - \tau_a + \sigma_a\,[d\,F(x)/dx)] = 0$ (2.50)

Definition for the above noted symbols is given in Figure 2.17.

Slope stability charts given by Janbu[2.20] relate the minimum stability number N to the slope angle, β, in terms of dimensionless factor, $\lambda_{So}\phi$. For a special case where there is no tension crack and ground water pressure, the slope FS can be expressed as follows:

$$FS = N \cdot S_o/\gamma \cdot H \quad (2.51)$$

$$\lambda_{So} \cdot \phi = \gamma \cdot H \cdot \tan\phi/S_o \quad (2.52)$$

where : γ = unit weight of the sliding material, kN/m^3
H = slope height, m

Slope stability charts to determine the location of the center of the critical circle and the minimum slope safety factor are given in Figures 2.18 and 2.19.

2.7.3 An Analytical Approach to Determine Circular Failure Parameters

The popularity of the method of slices suggested by Fellenius[2.18] relates to its ability to accommodate complex and irregular geometries, including variable soil and water pressure conditions. In the ordinary or Fellenius method the solution is simplified by assuming the forces on the two sides of a slice are parallel to the failure surface, at the bottom of the slice. Therefore, they do not affect the force normal to the failure surface. This produces a small inaccuracy in calculations, especially where the forces are not parallel to the failure surface. In order to express a theoretical solution to the above noted approach, a simple analytical solution by integration is suggested here under circular failure conditions as follows:

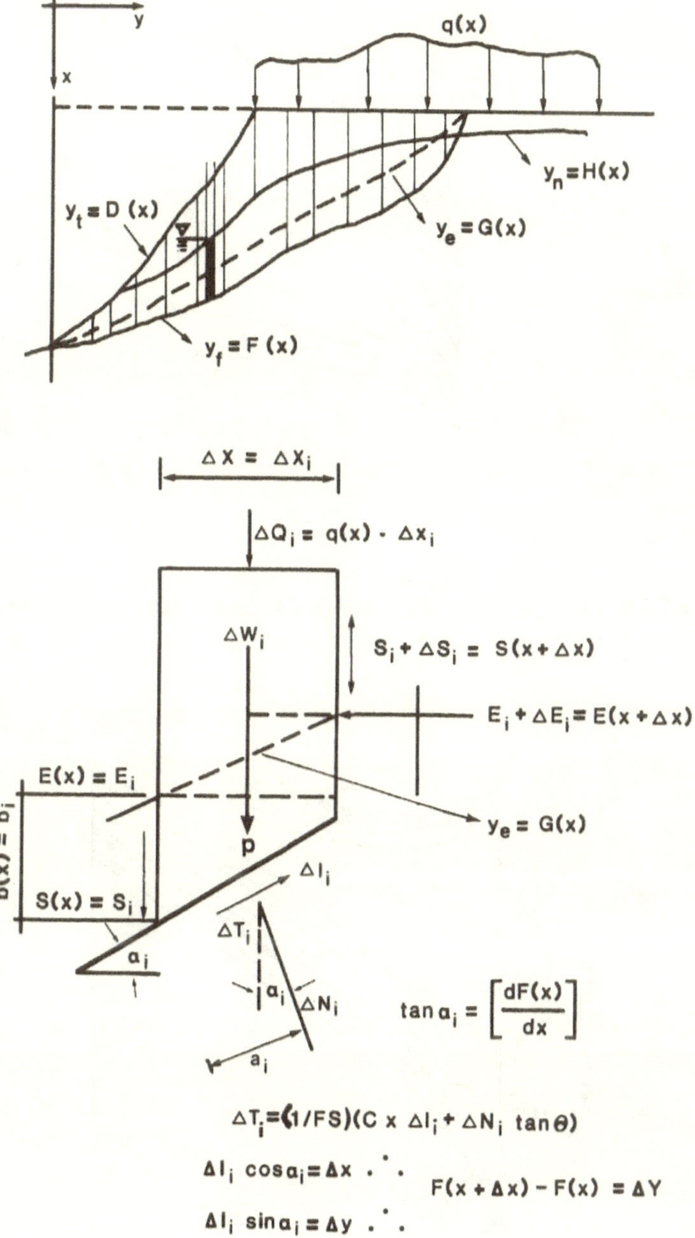

Figure 2.17. Graphical presentation of the forces acting as a potential failure slice.[2.16]

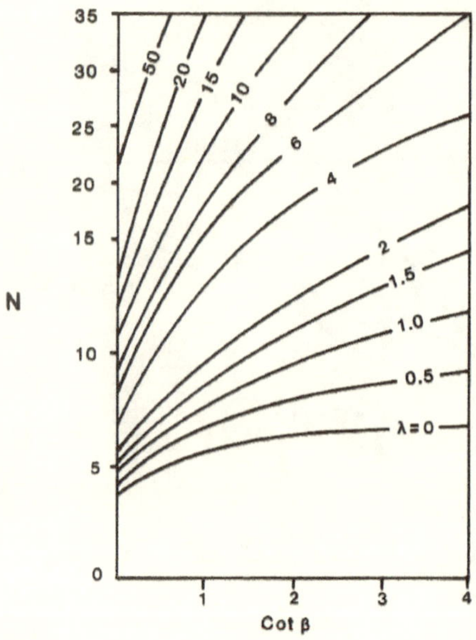

Figure 2.18. Variation of stability number (N) as a function of the slope angle (β).[2.20]

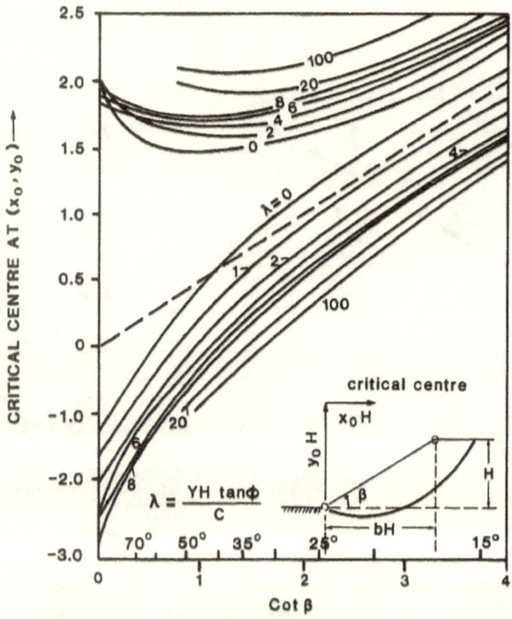

Figure 2.19. Location of the critical circle for failure through the toe.[2.20]

2.7.3.1 Determination of the Total Failed Weight of Soil /Rock (W) in mN

$$W = \int \text{Integral between } \theta_1 \text{ and } \theta_b \text{ of } (R.\ \sin\theta - a)\ R.\ \sin\theta.\ d\,\theta + \int \text{Integral between } \theta_1 \text{ and } \theta_b \text{ of } [R^2 (\sin\theta + \cos\theta.\ \tan\beta) - b]\sin\theta\ .\ d\,\theta$$

$$= (\gamma \cdot R/2)\{R(\theta_2 - \theta_1) - R[(\sin 2\theta_2 - \sin 2\theta_1)/2] + 2[a(\cos\theta_b - \cos\theta_1) + b(\cos\theta_2 - \cos\theta_b)] + R \cdot \tan\beta\,(\sin^2\theta_2 - \sin^2\theta_b)\} \quad (2.53)$$

where: γ = average weight of unit volume of the failing soil or rock mass, mN/m^3

θ_1 = angle of the radius vector (R) at the top of the potential failure surface on the bench with the horizontal, degrees

θ_2 = angle of the radius vector (R) at the toe of the slope with the horizontal, degrees

θ_b = angle between the horizontal and the radius vector (R) which intercepts the vertical projection of the slope face crest on the potential failure surface, degrees (Figure 2.20)

R = radius of the circular failure, m

θ = angle between the radius vector (R) and the horizontal, degrees

β = angle of slope face to the horizontal, degrees

a = ordinate of the potential failure curve at the top of the bench, m (Figure 2.20)

b = ordinate intercept of the line that defines the slope faces (see Figure 2.21)

2.7.3.2 Determination of Total Normal Force Acting on the Failure Surfacc (N) in mN/m

$$N = \gamma \int \text{Integral between } \theta_1 \text{ and } \theta_b \text{ of } (R.\ \sin\theta - a)\ R.\ \sin^2\theta\ d\theta + \gamma \int \text{Integral between } \theta_b \text{ and } \theta_2 \text{ of } [R^2 (\sin\theta + \cos\theta.\ \tan\beta) - b]\ \sin^2\theta\ d\theta$$

$$= (\gamma \cdot R/2)\big\{R[2R\tan\beta\,(\sin^3\theta_2 - \sin^3\theta_b)/3] - a[(\theta_b - \theta_1) - (\sin 2\theta_b - \sin 2\theta_1)/2] - b[(\theta_2 - \theta_b] - (\sin^2\theta_2 - \sin^2\theta_b)/2] - 2R[\cos\theta_2(\sin^2\theta_2 + 2) - \cos\theta_1(\sin^2\theta_1 + 2)]/3\big\} \quad (2.54)$$

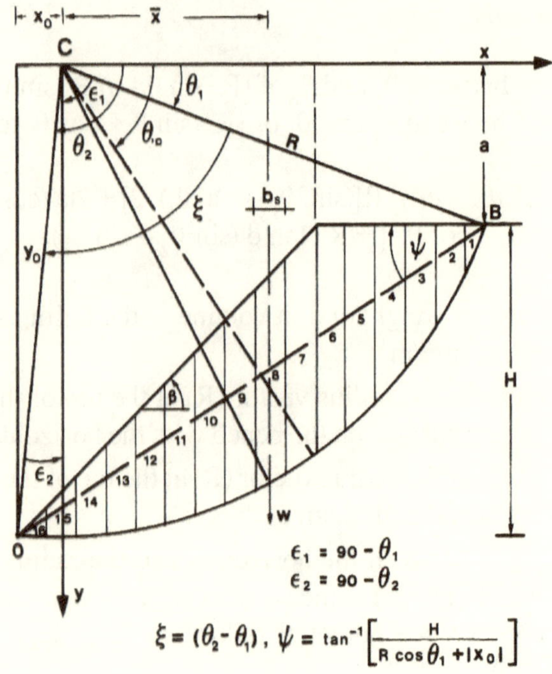

Figure 2.20. Method of dividing the potential failure mass into slices.

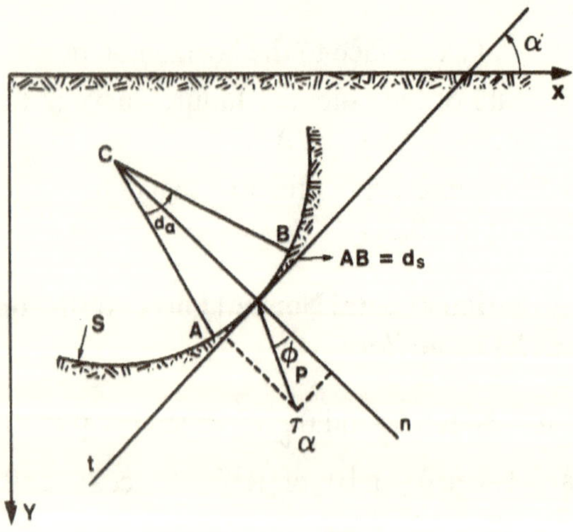

Figure 2.21. Schematic of the reaction pressure (P) acting on the potential failure surface.

2.7.3.3 Determination of Total Tangential Force Acting on the Failure Surface (T) in mN/m

$T = \gamma \int$ Integral between θ_1 and θ_b of (R. sinθ - a) R. sinθ. Cosθ dθ + $\gamma \int$ Integral between θ_b and θ_2 of [R^2 (sinθ + cosθ. tan β) - b] sinθ. cosθ dθ

$$= (\gamma \cdot R/2)\{[R(\sin^3\theta_2 - \sin^3\theta_1)/3] - a(\sin^2\theta_b - \sin^2\theta_1) - b(\sin^2\theta_2 - \sin^2\theta_b) - [2R \cdot \text{tab}\, \beta\, (\cos^3\theta_2 - \cos^3\theta_2 - \cos^3\theta_b)/3]\} \quad 2.55)$$

2.7.3.4 Determination of the Driving Moment (M_d) with Respect to the Critical Center (C)

$$M_d = (\gamma \cdot R^2/6)\{2R[\sin^3\theta^2 - \sin^3\theta_1) - \tan\beta\, (\cos^3\theta_2 - \cos^3\theta_b)] - 3[a(\sin^2\theta_b - \sin^2\theta_1) - b(\cos^2\theta_2 - \cos^2\theta_b)]\} \quad (2.56)$$

A more compact equation for the total weight (W) and driving moment (M_d) was suggested by Taylor[2.13] for the circular failure surfaces. This is expressed as follows:

$$W = \gamma\{R^2[(\xi/2) - \sin(\xi/2)] + H^2(\cot\psi - \cot\beta)/2\} \quad (2.57)$$

$$M_d = (\gamma \cdot H^3/12)[1 - 2\cot^2\beta + 3\cot\beta \cdot \cot\psi - 3\cot\beta \cdot \cot(\xi/2) + 3\cot(\xi/2) \cdot \cot\psi] \quad (2.58)$$

where: $\xi = \theta_2 - \theta_1 = \tan^{-1}(H/R \cdot \cos\theta_1 + X_0)$

H = height of the slope bench, m

X_0 = the abscissa of the critical center with reference to the coordinate of the slope toe, m

It is to be noted that the accuracy of both methods is very similar.

2.7.3.5 Determination of the Resisting Moment (M_r)

In order to determine the slope FS in terms of moments, it is necessary to obtain the resisting moment of the soil or fractured rock mass (M_r). This is done by using Kötter's differential equation[2.24,] for a homogeneous soil, as follows:

$$(dp/ds) - 2p \cdot \tan\phi\, (d\alpha/ds) - \gamma \cdot \sin(\alpha - \phi) = 0 \quad (2.59)$$

where: $d\,p$ = differential reaction pressure on the failure surface, kPa
$d\,s$ = differential length of arc of circular failure surface, m
p = reaction pressure on the failure surface, mPa
$d\,\alpha$ = differential angle at any segment on the potential failure surface, degree.

The differential equation of the shear strength along the circular failure surface τ_a is:

$$d\tau_\alpha/d\alpha = 2\tau_a \cdot \tan\phi - \gamma \cdot R \cdot \sin\phi \cdot \sin(\alpha - \phi) \quad (2.60)$$

Hence: $\tau_\alpha = [m \cdot f \cdot \sin(\alpha - \phi)/[1 + f^2)] - [m \cdot \cos(\alpha - \phi)/(1 + f^2)] + (\Omega_1/e^{f\alpha})$ (2.61)

where: $m = -\gamma \cdot R \cdot \sin\phi$
$f = -2\tan\phi$
Ω_1 = integration constant to be evaluated from the boundary conditions at
σ_α = 0 and from the Mohr-Coulomb equation $\tau_\alpha = S_o$ at the top of the bench.

To obtain a more realistic result, it is necessary to account for the tensile stress or if possible to observe the presence of the tension cracks on the top of the slope. Under the above noted conditions:

$$\Omega_1 = \{S_o + [m/(1 + f^{2)}]\cos(\epsilon_1 - \phi) - [m \cdot \sin(\epsilon_1 - \phi)/(1 + f^{2)}]\}e^{f\epsilon 1} \quad (2.62)$$

where: $\epsilon_1 = 90 - \theta_1$, (see Figure 2.22)

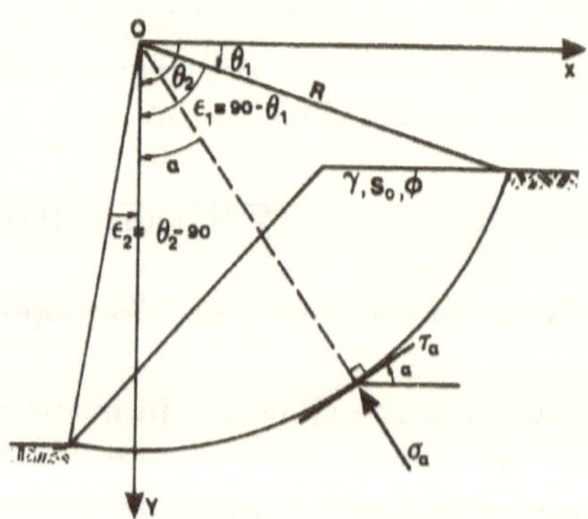

Figure 2.22 Schematic of tangential and normal stresses acting on the potential failure surface—Kotter's Differential Equation.[2.24]

If the shear strength of soil or rock mass is defined by the Mohr-Coulomb failure criterion, then normal stress (σ_a) can be determined as follows:

$$(d\tau_\alpha/d\alpha)=(d\sigma_\alpha/d\alpha)\tan\phi \tag{2.63}$$

Substituting equation 2.63 into equation 2.60 gives:

$$(d\sigma_\alpha/d\alpha)\tan\phi = 2\tan\phi\,(S_o + \sigma_\alpha \cdot \tan\phi) - \gamma \cdot R \cdot \sin\phi \cdot \sin(\alpha - \phi) \tag{2.64}$$

The differential equation of the normal stress (σ_a) on the circular failure surface is:

$$\sigma_a = (2S_o/f) + [B.\,f/(1 + f^{2)}]\sin(\alpha - \phi) - [B/(1 + f^{2)}]\cos(\alpha - \phi) + \Omega_2/e^{-f\cdot\alpha} \tag{2.65}$$

where: $B = -\gamma \cdot R \cdot \cos\phi$

e = base of naperian log = 2.17

Ω_2 = the integration constant at $\sigma_\alpha = 0$ and at $\alpha = \epsilon_1$ on the top of the bench

Then, Ω_2 value is determined as follows:

$$\Omega_2 = -\{(2S_0/f) + [B \cdot f/(1 + f^{2)}]\sin(\epsilon_1 - \phi) - B \cdot \cos(\epsilon_1 - \phi)/(1 + f^2)\}e^{-f\epsilon 1} \tag{2.66}$$

where: $\epsilon_1 = 90 - \theta_1$

Finally, from the geometry of Figure 2.22, the resisting moment by Kötter (M_r) can be determined as follows:

$$\begin{aligned} M_r &= \int \text{Integral between } \epsilon_1 \text{ and } \epsilon_2 \text{ of } \tau_\alpha\,.\,R^2\,.\,d\alpha \\ &= R^2 \int \text{Integral between } \epsilon_1 \text{ and } \epsilon_2 \text{ of } \{[m.\,f.\,\sin(\alpha - \phi)/(1 + f^{2)}] - [m \cdot \cos(\alpha - \phi)/(1 + f^{2)}] + (\Omega_1/e^{f\epsilon 1})\}d\alpha \\ &= [R^3 \cdot \gamma \cdot \sin\phi/(1 + 4\tan^2\phi)]\{\sin(\epsilon_1 - \phi) + 2\tan\phi\,[\cos(\epsilon_2 - \phi) - \cos(\epsilon_1 - \phi)] - \sin(\epsilon_2 - \phi)\} + (R^2\Omega_1/2\tan\phi)(e^{-f\epsilon 1} - e^{-f\epsilon 2)}) \end{aligned} \tag{2.67}$$

where: $\epsilon_2 = \theta_2 - 90$ (see Figure 2.22).

2.7.3.6 Determination of the Slope Safety Factor (FS)

Utilizing Fellenious and Bishop methods, the slope safety factor before reinforcement is given below in terms of moments:

$$FS = M_r / M_d = R(S_o L_f + N \cdot \tan\phi) / M_d \qquad (2.68)$$

where: L_f = length of the perimeter of failure surface, m
N = total normal force acting on the failure surface

2.7.3.7 Determination of Critical Circle and the Tension Crack Location

Hoek and Bray charts, shown in Figures 2.23 and 2.24 can be utilized to estimate the above noted locations.[2.1] Such determination serves as a first approximation, after which interactive methods should be used for more accurate determinations.

As an example of the application of these charts, consider the case of a slope face where: $\beta = 30^\circ$ to horizontal in a drained soil of $\gamma = 16\ kN/m^3$, with a friction angle of $\phi = 20^\circ$, cohesive strength of $S_o = 45$ kPa.

Using Figure 2.23b, at the interception of β and ϕ curves:

1 - Projected on the X axis gives the horizontal position of center of the failure circle at 0.2H.
2 - Projected on the Y axis gives the vertical position of center of the failure circle at 1.85H.

Using Figure 2.23c, at the interception of the β axis with the ϕ curve, projected on the vertical axis (b/H) gives:

b = horizontal position of the tension crack measured from the slope crest = 0.1H

Overall configuration of the above noted example is shown in Figure 2.25a.

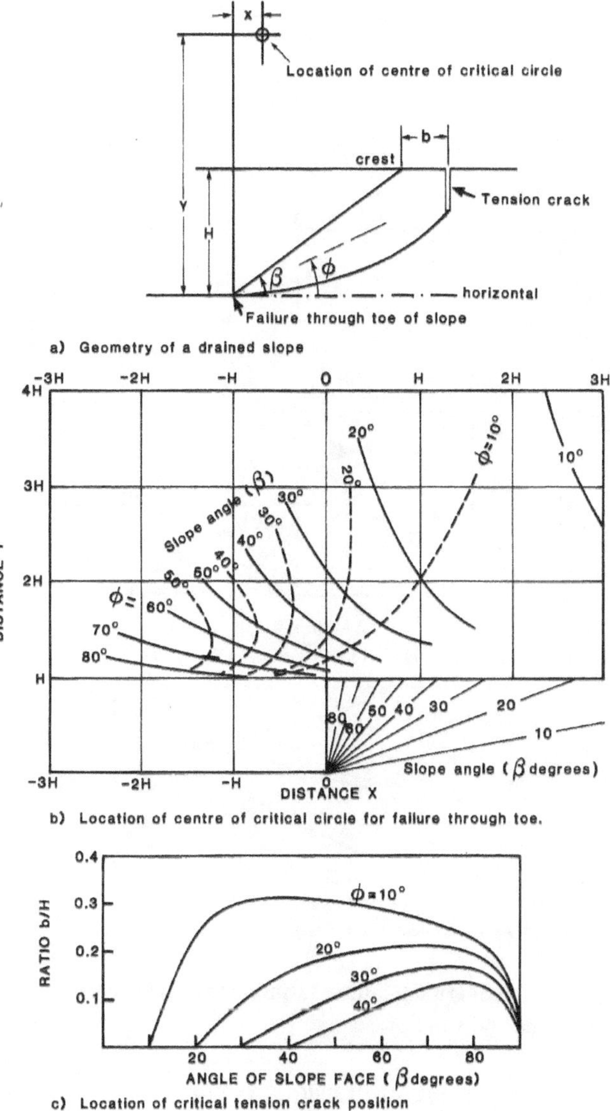

Figure 2.23. Location of critical failure surface and critical tension crack for drained slopes.[2.1]

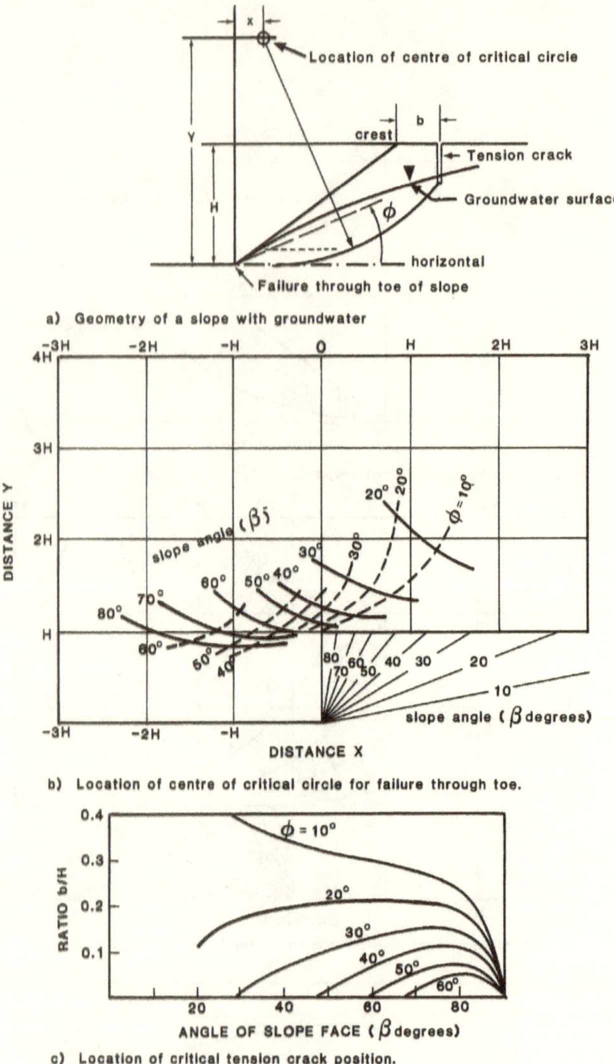

Figure 2.24. Location of critical failure surface and critical tension crack for slopes with groundwater present.[2.1]

In the above example, if the slope is partially saturated in such a way that the seepage of the groundwater is at H/4 above the slope toe, as shown in Figure 2.24a, and with all other factors remaining the same, then, using Figure 2.24b at the interception of β and ϕ curves:

(A) Projected on the X axis gives the horizontal position of center of the failure circle at 0.5H.

(B) Projected on the Y axis gives the vertical position of center of the failure circle at 1.5H.

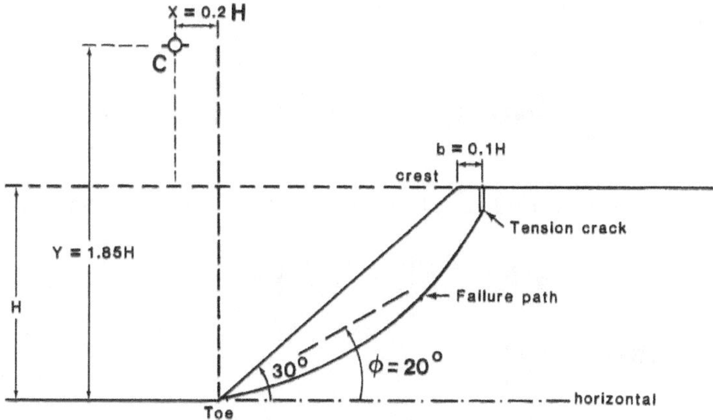

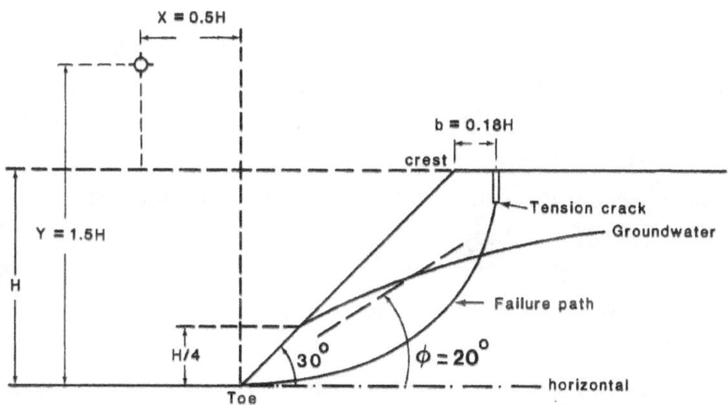

Figure 2.25. Comparative location of critical circular failure and critical tension crack in drained and water logged slopes.

In the above example, if the slope is partially saturated in such a way that the seepage of the groundwater is at H/4 above the slope toe, as shown in Figure 2.24a, and with all other factors remaining the same, then, using Figure 2.24b at the interception of β and ϕ curves:

1. Projected on the X axis gives the horizontal position of center of the failure circle at 0.5H.
2. Projected on the Y axis gives the vertical position of center of the failure circle at 1.5H.

Using Figure 2.24c, at the interception of β axis with ϕ curve, projected on the vertical axis (b /H) gives:

b = horizontal position of the tension crack measured from the slope crest = 0.18H

Overall configuration of the latter case is shown in Figure 2.25b.

2.8 DEEP-SEATED FAILURE

2.8.1 Potential Occurrence

This type of failure may occur in rotation, where steep footwall slopes or the footwall side from underground excavation, in fractured or weak rocks with no dominant geologic structural discontinuities exist as shown in Figure 2.26. In such case, analytical techniques which can model non-circular or logarithmic spiral failure surfaces, composed of curvilinear and planar segments, with different shear strength characteristics are required. Janbu,[2.19,2.20] Spencer,[2.25] and Chen[2.26] have provided the necessary techniques by incorporating calculus of variation, limit analysis, and upper-bound solution.

Regardless of the analysis technique chosen, an assessment of the strength characteristics of the rock mass is required. In most cases the confidence with which the *in situ* rock-mass strength can be predicted determines the accuracy of the stability analysis, not the analysis technique used. Assessment of rock mass strength may be based on semi-empirical techniques or back analysis of existing slopes. Any estimate of rock-mass strength must recognize the highly anisotropic nature of the rock due to the presence of discontinuities.

2.8.2 Analysis of Deep Seated Failure

Considering randomly fractured rock mass, unable to resist tension, no energy is dissipated in the formation of a simple tension crack, and the rate of external work done by the gravity is equal to the dissipation of internal energy. Then, according to Figure 2.27, the lower-bound and upper bound solutions for the critical slope height (H_c) are[2.27]:

$$(k.\ c/\gamma) \tan[(\pi/4) + (\varphi/2)] \succcurlyeq H_c \succcurlyeq (k.\ c/\gamma) \tan[(\pi/4) - (\varphi/2)] \quad (2.69)$$

where, k = Proportionality constant =2 to 4,[2.27, 2.28]

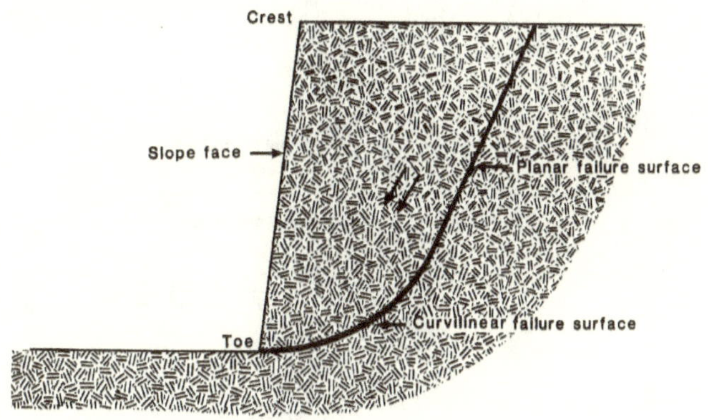

a) Surface slope condition.

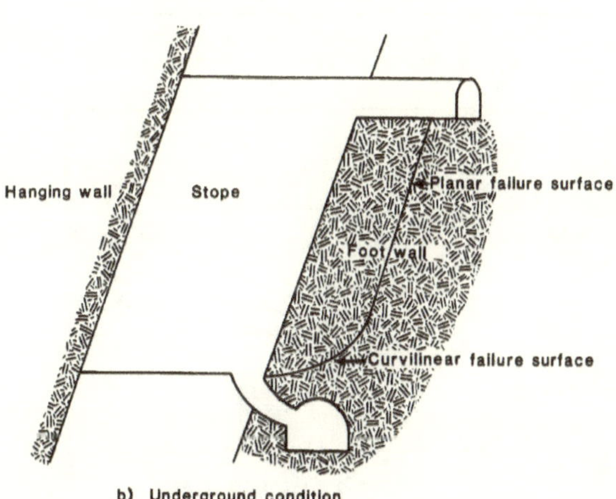

b) Underground condition

Figure 2.26. Configurations of simplified deep seated failure with logarithmic spiral failure path.

The stability factor (N_s) for the logarithmic spiral failure path passing through the slope toe (Figure 2.27a) or below the toe (Figure 2.27b) can be determined as follows:

$$N_s = H_c \cdot \gamma / c \tag{2.70}$$

Variation in the stability factor as a function of the slope angle (β) and the internal friction angle at the failure surface for each of the above noted conditions, where α = 0, is shown in Figure 2.27c.

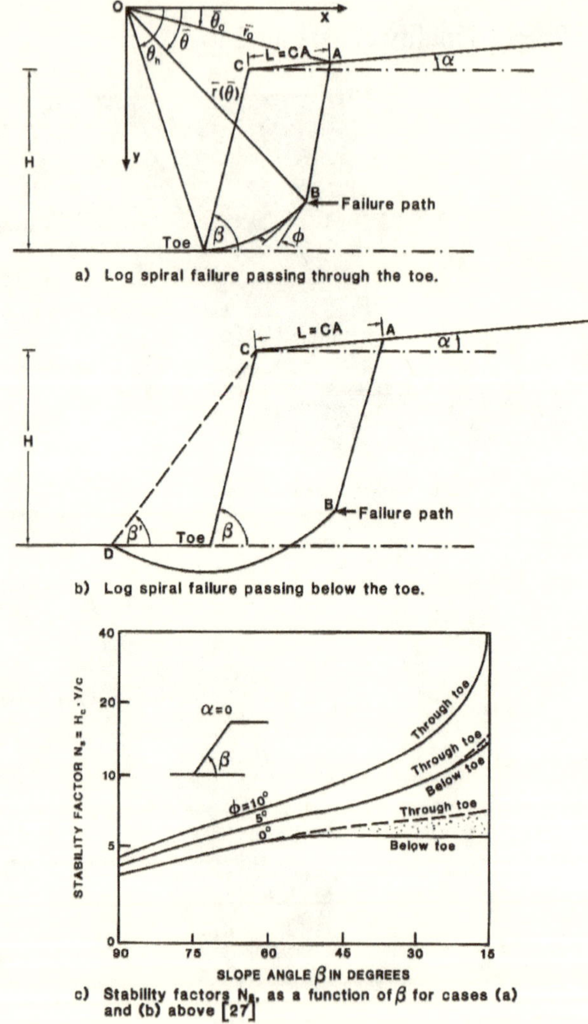

Figure 2.27. Mechanism of the deep seated failure.

σ(θ) = normal stress at any point B along the logarithmic spiral failure path, with regard to the notations given in Figure 2.27a:[2.27]

$= \gamma \cdot H\{A_1 + \{3 \tan \phi / [(H / r_0) (1 + 9 \tan^2 \phi . \exp ((\theta_0). \tan \phi)]\}$
$\{\cos(\theta). \exp ((\theta). \tan \phi) + [\sin (\theta). \exp ((\theta). \tan \phi) / 3\sin \phi\} +$
$[A_2 . \exp (-2(\theta). \tan \phi)]$ (2.71)

where: $A_1 = 1 / \{\{[[u_3(\theta)]$between θ_0 to $\theta_h]/[u_3(-\theta)]$between θ_0 to $\theta_h] +$
$[u_1(\theta)]$between θ_0 to $\theta_h]/[u_1(-\theta)]$between θ_0 to $\theta_h]\}\}$
$\{\{[[-3(1 + \tan^2 \phi) u_4(2\theta) - 3 \tan \phi\ u_1(2\theta) -$
$u_3(2\theta)]]$ between θ_0 to $\theta_h\}/$
$\{[4 (H / r_0) (1 + 9 \tan^2 \phi) u_4(\theta_0) u_1(-\theta)]$between θ_0 to $\theta_h\}\}$
$+ \{\{(\cos \phi + 17 \tan \phi) [u_4(2\theta)]$between θ_0 to $\theta_h\} +$
$[u_1(2\theta)]$ between θ_0 to θ_h - $3\tan \phi.u_3(2\theta)]$
between θ_0 to $\theta_h\}/$
$\{(4H / r_0) (1 + 9 \tan^2 \phi)[u_4(\theta_0) u_3(-\theta)]$between θ_0 to $\theta_h\}$ -
$\{\{[\tan \phi\ u_1(\theta + u_3(\theta)]$ between θ_0 to $\theta_h) / [f (1 + \tan^2$
$\phi) [u_1(-\theta)]$ between θ_0 to $\theta_h\}$ - $\{[u_1(\theta) + \tan \phi.u_3(\theta)]$
between θ_0 to $\theta_h\}/$
$\{f (1 + \tan^2 \phi) [u_3(-\theta)]$between θ_0 to $\theta_h\} - \{(L / r_0)$
$\sin(\theta_0). u_4(\theta_0) - \sin(\theta_h).\cos(\theta_0) + (L / r_0) \sin(\theta_h) -$
$\sin(\theta_o).\cos(\theta_h)] u_4(\theta_h)\}/$
$2 (H / r_0) [u_3(-\theta)]$ between θ_0 to θ_h

$A = \{[\tan \phi\ u_1(\theta) - u_3(\theta)]$ between θ_0 to $\theta_h\}/\{f (1 + \tan^2 \phi)$
$[u_1(-\theta)]$between θ_0 to $\theta_h\} - \{(A_1)[u_1(\theta)]$between θ_0 to θ_h /
$[u_1(-\theta)]$between θ_0 to $\theta_h\}$- $\{[3(1 + \tan^2 \phi) u_4(2\theta) +$
$[u_3(2\theta)]$between θ_0 to $\theta_h] / 4 (H / r_0) (1 + 9 \tan^2 \phi) u_4(\theta_0)$
$[u_1(-\theta)]$between θ_0 to $\theta_h\}$

In which the functions $u(\theta)$ are defined as:

$$u_1(\theta) = (\tan\phi. \cos\theta + \sin\theta) \exp(\theta. \tan\phi)$$

$$u_2(\theta) = (\tan\phi. \sin\theta + \cos\theta) \exp(\theta. \tan\phi)$$

$$u_3(\theta) = (\tan\phi. \sin\theta - \cos\theta)\exp(\theta. \tan\phi)$$

$$u_4(\theta) = \exp(\theta. \tan\phi)$$

2.9 TOPPLING FAILURE

2.9.1 Potential Occurrence

This type of failure occurs under gravity, where bedding dips steeply away from the excavation walls and the rock mass is jointed, as shown in Figure 2.28a. The failure results from propagation of the stress release discontinuities in the form of slabbing or spalling. Toppling can also occur where the bedding dips in the same direction as steep slope or the excavation wall (Figure 2.28b). In this case, failure results from high peizometric pressures in the rock mass. Complex toppling failure may also develop, involving some combination of toppling induced by deep seated block movements, or ploughing slab sliding where rolls or minor folds are exposed by mining (Figure 2.28c). The failure mechanisms discussed above, all involve bedding or joints or both as dominant discontinuities.

Conditions in which toppling failure can occur are illustrated in Figure 2.29a, where a block of rock having a height, h, and a base length, t, is resting on a plane with inclination of β to horizontal. It is assumed that the force resisting downward movement of the block is due to friction only, i.e. $S_0 = 0$. In this case, when the weight vector (W) of the block falls within the base, b, and the inclination of the plane, β is greater than the internal friction angle between the block and the plane (ϕ), the sliding of the block will occur. When the block is tall and slender (h > t), the weight vector (W) can fall outside the base length, t, which causes the block to topple, i.e., rotate about its lowest contact edge. The conditions for sliding or toppling are defined in Figure 2.29b. The four main regions in this diagram are as follows:[2.1]

- Region 1: β < ϕ, and t/h > tanβ — the block is stable. It will not slide or topple.
- Region 2: β > ϕ, and t/h > tanβ — the block will slide but it will not topple.
- Region 3: β < ϕ, and t/h < tanβ — the block will topple but it will not slide.
- Region 4: β > ϕ, and t/h < tanβ — the block can slide and topple simultaneously.

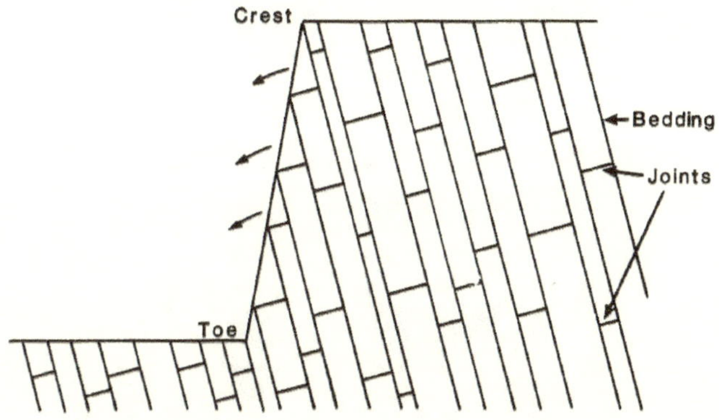

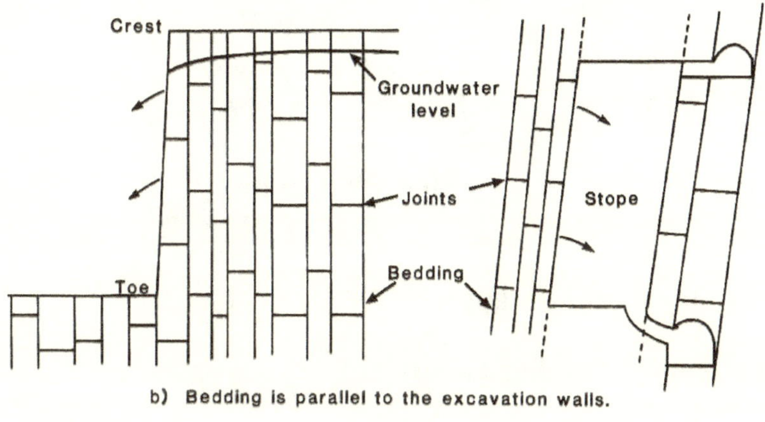

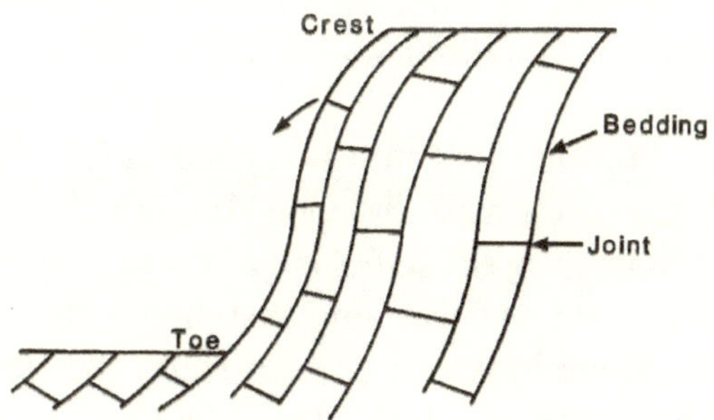

Figure 2.28. Modes of toppling failure.[2.1]

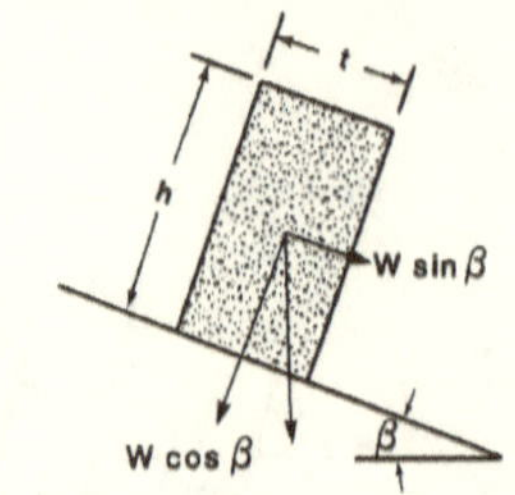

a) Geometry of block on inclined plane.

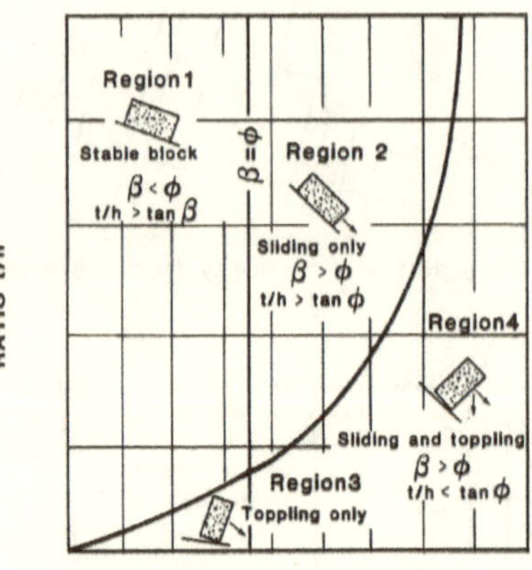

b) Conditions for sliding and toppling of a block on an inclined plane.

Figure 2.29. Conditions for toppling failure.[2.1]

2.9.2 Types of Toppling Failure

Following types of toppling failure have been encountered in the field and reported by Goodman and Bray:[2.29]

1. **Flexural toppling** — in which continuous columns of rock, which are separated by well-developed steeply dipping discontinuities, break in flexure as they bend forward (Figure 2.30a). Erosion, undermining or sliding of the toe of the slope, or underground excavation wall starts the toppling process with formation of crack along the discontinuities. These cracks widen towards the opening causing columns of rock to separate from the mass and topple. The separation movement of each column produces an inter-layer slip exposing a portion of the upper surface of each plane in a series of back facing

or ob-sequent scarps as shown in Figure 2.30b. The latter is visible in the surface excavations.

2. **Block toppling** — occurs when, in addition to the bedding, the individual columns of hard rock are divided by widely spaced orthogonal joints (Figure 2.31). The short columns forming the toe or base of the excavation wall are pushed forward by the load from the longer overturning columns behind the short columns. This sliding of the toe allows further topping to develop higher up the excavation wall. The base of this type of failure consists of a stairway rising from one cross-joint to the next. Block toppling is very similar to the bi-linear slab failure discussed in Section 2.3, except that in the block toppling all the failure occurs along the bedding and cross-joints and the top column of rock topples rather than slides down the slope.

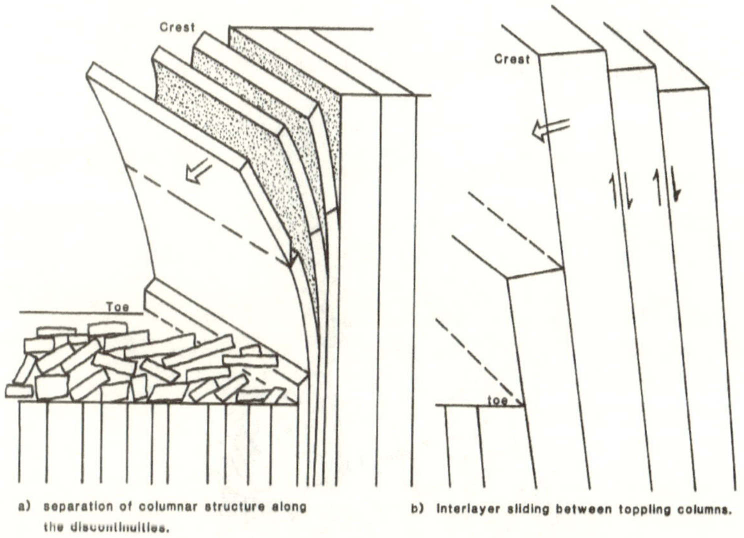

Figure 2.30. Flexural toppling in hard rock.

3. **Block-flexure toppling** — is characterized by pseudo-continuous flexure along long columns which are divided by numerous cross-joints. Instead of the flexural failure of continuous columns, resulting in flexural toppling, the toppling of columns in this case results from accumulated

displacements on the cross-joints (Figure 2.32). This is a combination of flexural and the block toppling. However, because of the large number of small movements in this type of topple, there are fewer cracks than in flexural toppling and fewer edge-to-face contacts and voids than in block toppling.

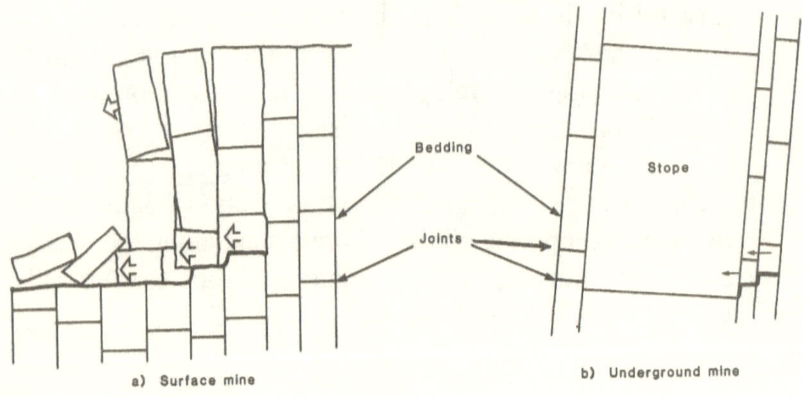

Figure 2.31. Block toppling in hard jointed rock.

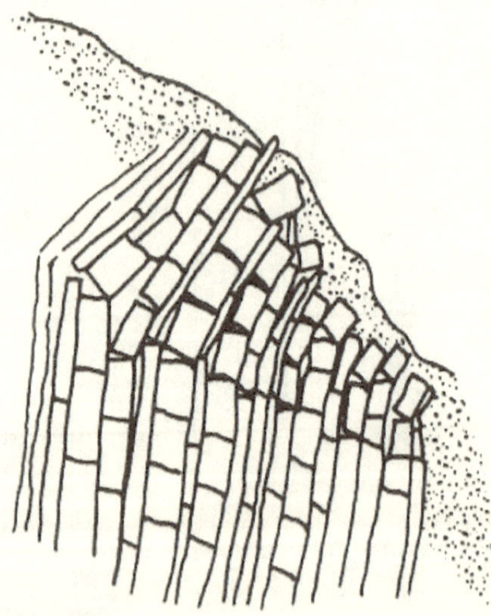

Figure 2.32. Block flexure toppling characterized by continuous accumulated motions.[2.1,2.29]

4. **Induced toppling by primary failure** — is initiated by natural erosion or weathering, or undercutting of the toe of the slope due to mining activity. In all cases, the primary failure mode involves sliding or physical breakdown of the rock and toppling is induced in some part of the slope as a result of this primary failure (Figure 2.33).

2.9.3 Analysis of Toppling Failure

Goodman and Bray[2.29] considered a regular system of blocks in a rock mass with bedding inclination at $90 - \alpha°$, in which a slope angle θ is excavated as shown in Figure 2.34. The failure base is stepped upwards with an overall inclination β. The constants a_1, a_2 and b shown in the figure are given by:

$$a_1 = \Delta x \cdot \tan(\theta - \alpha)$$

$$a_2 = \Delta x \cdot \tan(\alpha - \theta_u)$$

$$b = \Delta x \cdot \tan(\beta - \alpha)$$

where: Δx = width of each block ; and θ_u = Inclination of the upper ground surface

Height of the *nth* block in a position below the crest of the slope $(Y_n) = n(a_1 - b)$. Height of the *nth* block in a position above the crest of the slope $(Y_n) = Y_{n-1} - a_2 - b$.

When a system of blocks, having the form shown in Figure 2.34 commences to fail, it is generally possible to distinguish three separate groups according to their mode of behavior:

1. A set of sliding blocks in the toe region
2. A set of stable blocks at the top
3. An intermediate set of toppling blocks

With certain geometries, the sliding set may be absent in which case the toppling set extends down to the toe.

Figure 2.35a shows a typical block (n) with the forces developed on the base (R_n, τ_n) and on the interfaces with adjacent blocks $(P_n, Q_n, P_{n-1}, Q_{n-1})$.

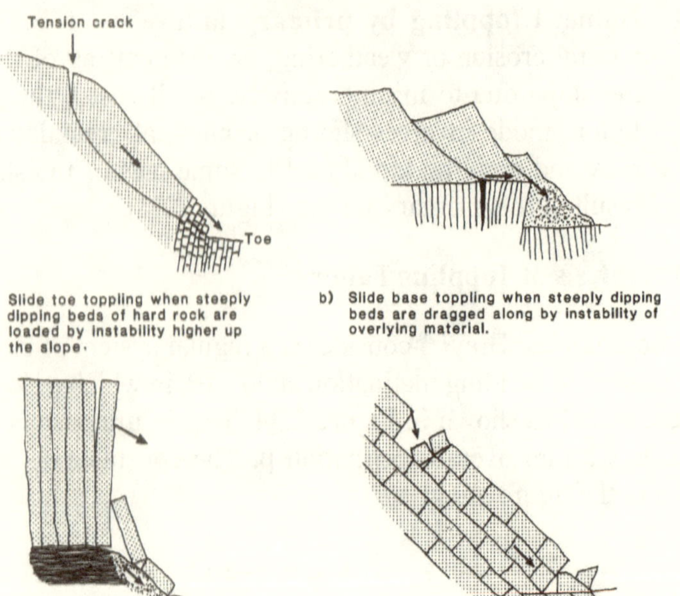

Figure 2.33. Various mechanisms of induced toppling by primary failure suggested by Goodman and Bray.[2.24]

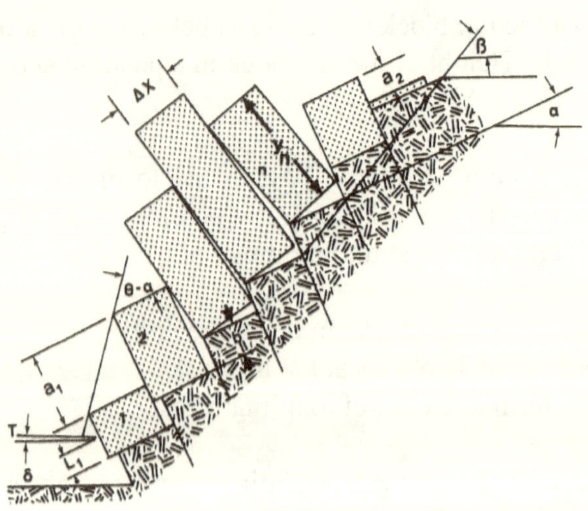

Figure 2.34. Model for limiting equilibrium analysis of toppling on a stepped base.[2.29]

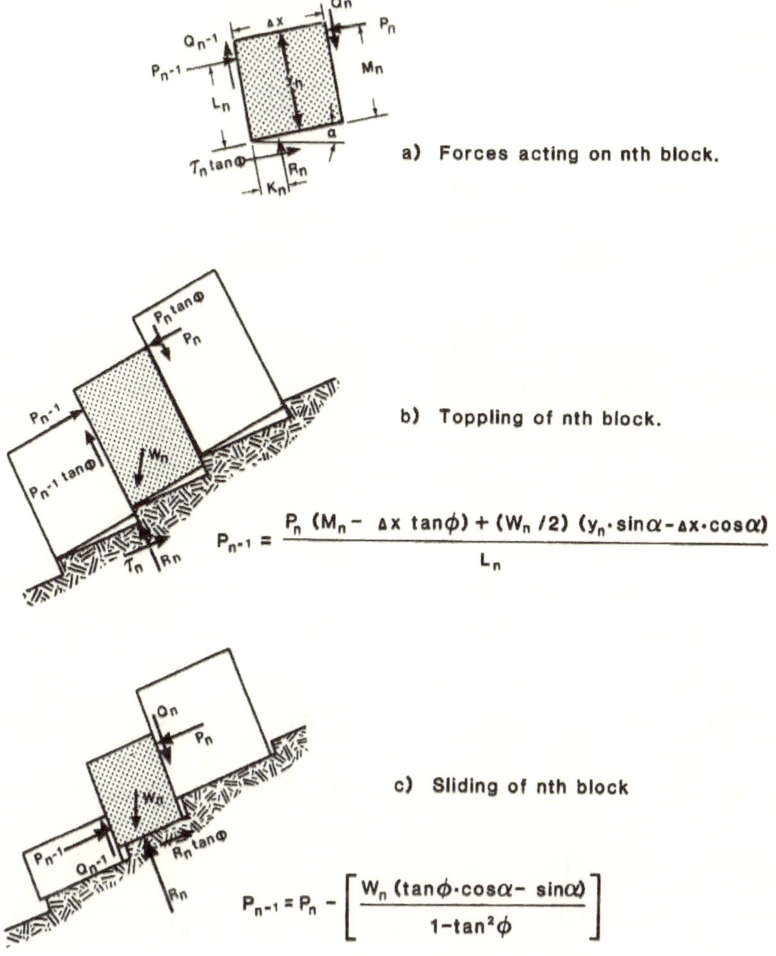

Figure 2.35. Limiting equilibrium conditions for toppling and for sliding of the n*th* block.[2.29]

When the block is one of the toppling set, the points of applications of all forces are known, as shown in Figure 2.35b. If the n*th* block is below the slope crest:

$$M_n = Y_n; \text{ and } L_n = Y_n - a_1$$

If the nth block is the crest block: $M_n = Y_n - a_2;$ and $L_n = Y_n - a_1$

If the n*th* block is above the slope crest: $M_n = Y_n - a_2;$ and $L_n = Y_n$

In all cases: $K_n = 0$

For an irregular array of blocks, Y_n, L_n, and M_n can be determined graphically. For limiting friction on the sides of the block: $Q_n = P_n \cdot \tan \phi$; and $Q_{n-1} = P_{n-1} \cdot \tan \phi$

By resolving perpendicular and parallel to the base:

$$R_n = W_n \cdot \cos \alpha + (P_n - P_{n-1})\tan \phi \tag{2.72}$$

$$\tau_n = W_n \cdot \sin \alpha + (P_n - P_{n-1}) \tag{2.73}$$

Considering rotational equilibrium, it is found that the force P_{n-1} which is just sufficient to prevent toppling has the value of:

$$P_{n-1}, t = [P_n (M_n - \Delta x. \tan \phi) + (W_n /2) (Y_n \cdot \sin \alpha - \Delta. \cos \alpha)] / L_n \tag{2.74}$$

When the block under consideration is one of the sliding set,

$$\tau_n = R_n \cdot \tan \phi \tag{2.75}$$

However, the magnitudes and points of application of all the forces applied to the sides and base of the block are unknown. The procedure suggested here[2.29] is to assume that, as in the toppling case, conditions of limiting equilibrium are established on the side faces so that equations 2.72 and 2.73 apply. Taken in conjunction with equation 2.75, these show that the force P_{n-1} which is just sufficient to prevent sliding has the value:

$$P_{n-1,s} = P_n - [W_n (\tan \phi . \cos \alpha - \sin \alpha)] / (1 - \tan^2 \phi) \tag{2.76}$$

The assumption introduced here is quite arbitrary, but a little consideration will show that it has no effect on calculations of the overall stability of the slope. Any other reasonable assumption would produce the same results.

Calculation Procedure[2.29]

Let: n_1 = uppermost block of the toppling set ; and
n_2 = uppermost block of the sliding set.

A. To determine the value of ϕ for limiting equilibrium:

1. Assume a reasonable value of ϕ, such that $\phi > \alpha$.
2. Establish n_1 by determining the uppermost block of the whole group which satisfies the condition: $Y_n/\Delta x > \cot \alpha$.
3. Starting with this block, determine the lateral forces $P_{n-1,t}$ required to prevent toppling, and $P_{n-1,s}$ to prevent sliding.

   a. If $P_{n-1,t} > P_{n-1,s}$, the block is on the point of toppling, and P_{n-1} is set equal to $P_{n-1,t}$

   b. If $P_{n-1,s} > P_{n-1t,}$ the block is on the point of sliding and P_{n-1} is set equal to $P_{n-1,s}$

   For this particular block, and all other tall blocks of the system, it will be found that the toppling mode is critical, and this check is purely a matter of routine. It is required at a later stage to determine n_2 which defines the upper limit of the sliding section. Further checks should be carried out to ensure that:

$$R_n > 0; \qquad \text{and} \qquad |\tau_n| < R_n \cdot \tan \phi$$

4. The next lower block $(n_1 - 1)$ and all the lower blocks are treated in succession, using the same procedure.
5. Eventually a block may be reached for which $P_{n-1,s} > P_{n-1,t}$. This establishes block n_2, and for this and all lower blocks, the critical state is one of sliding. If the condition $P_{n-1,s} > P_{n-1,t}$ is not met for any of the blocks, the sliding set is absent and toppling extends down to block 1.
6. Considering the toe block 1:

   a. If $P_0 > 0$, the slope is unstable for the assumed value of ϕ. It is necessary to repeat the calculation for an increased value of ϕ.

   b. If $P_0 < 0$, repeat the calculations with a reduced value of ϕ.

   c. When P_0 is sufficiently small, the corresponding value of ϕ can be taken as that for limiting equilibrium.

B. To determine the cable force required to stabilize a slope:

Suppose that a cable is installed through block 1 at a distance L_1, above its base. The cable is inclined at an angle δ degrees below horizontal and anchored a safe distance below the base. The tension in the cable required to prevent toppling of block 1 is:

$$T_t = [(W_1 /2)(y_1.\sin\alpha - \Delta x.\cos\alpha) + P_1 (y_1 - \Delta x.\tan\phi)] /[L_1.\cos(\alpha + \delta)] \quad (2.77)$$

While the tension in the cable to prevent sliding is:

$$T_s = [P_1 (1 - \tan^2 \phi) - W_n (\tan\phi.\cos\alpha - \sin\alpha)] /\tan\phi . \sin(\alpha + \delta) + \cos(\alpha + \delta)] \quad (2.78a)$$

The normal and shear force on the base of the block are, respectively:

$$R_1 = P_1 \cdot \tan\phi + T \cdot \sin(\alpha + \delta) + W_1 \cdot \cos\alpha \quad (2.78b)$$

$$\tau_1 = P_1 - T \cdot \cos(\alpha + \delta) + W_1 \cdot \sin\alpha \quad (2.78c)$$

The procedure in this case is identical to that described above apart from the calculations relating to block 1. The required tension is the greater of T_t and T_s defined by equations 2.77 and 2.78.

Example[2.29]

An ideal example is illustrated in Figure 2.36. A rock slope 92.5 m high is cut on a 56.6° slope in a layered rock mass dipping at 60° into the hill. A regular system of 16 blocks is shown on a base stepped at 1 m in every 5 (angle $\beta - \alpha = 5.8°$). The constants are $a_1 = 5.0$ m, $a_2 = 5.0$ m, b = 1.0 m, $\Delta x = 10.0$ m and = 25 kN/m³. Block 10 is at the crest which rises 4° above the horizontal. Since cot α = 1.78, blocks 16, 15, and 14 comprise a stable zone for all cases in which $\phi > 30°$ ($\tan\phi > 0.577$).

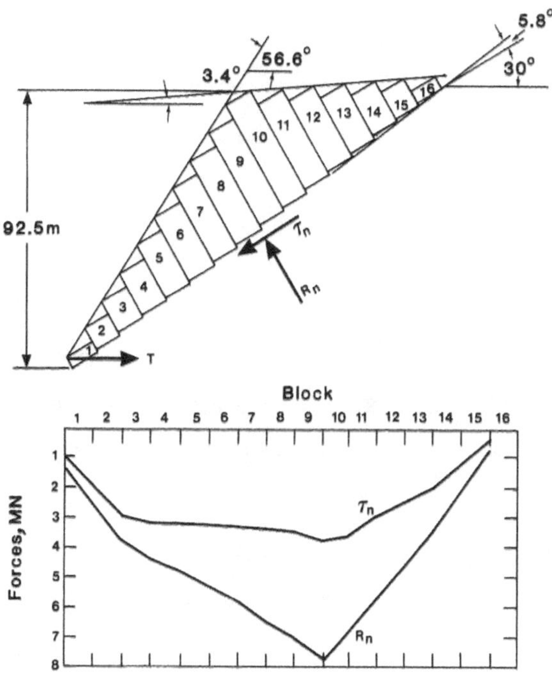

Figure 2.36. Limiting equilibrium of a toppling slope with tan ϕ = 0.7855.[2.29]

In this example, tan ϕ is set as 0.7855, P_{13} is then equal to O and P_{12} calculated as the greater of $P_{12,t}$ and $P_{12,s}$ given by Equations 2.74 and 2.75, respectively. As shown in Table 2.1, $P_{n-1,t}$ turns out to be the larger until a value of n = 3, where upon $P_{n-1,s}$ remains larger. Thus blocks 4 to 13 constitute the potential toppling zone and blocks 1 to 3 constitute a sliding zone.

The force required to prevent sliding in block 1 tends to zero which indicates that the slope is very close to limiting equilibrium. The installed tension required to stabilize block 1 is 0.5 kN/m of slope crest length, as compared with the maximum value of P (in block 5) equal to 4837 kN/m.

Table 2.1 Calculation of Forces for Example Shown in Figure 2.36[2.29]

No	Y_n	$Y_o/\Delta x$	M_n	L_n	P_{n-t}	P_{n-s}	P_n	R_n	τ_n	T_n/R_n	Mode
16	4.0	0.4			0	0	0	866	500	0.557	
15	10.0	1.0			0	0	0	2165	1250	0.577	STABLE
14	16.0	1.6			0	0	0	3463	2000	0.577	
13	22.0	2.2	17	22	0	0	0	4533.4	2457.5	0.542	
12	28.0	2.8	23	28	292.5	–2588.7	292.5	5643.3	2966.8	0.526	T
11	34.0	3.4	29	34	825.7	–3003.2	825.7	6787.6	3520.0	0.519	O
10	40.0	4.0	35	35	1566.0	–3175.0	1556.0	7662.1	3729.3	0.487	P
9	36.0	3.6	36	31	2826.7	–3150.8	2826.7	6933.8	3404.6	0.491	P
8	32.0	3.2	32	27	3922.1	–1409.4	3922.1	6399.8	3327.3	0.520	L
7	28.0	2.8	28	23	4594.8	156.8	4594.8	5872.0	3257.8	0.555	I
6	24.0	2.4	24	19	4837.0	1300.1	4837.0	5352.9	3199.5	0.598	N
5	20.0	2.0	20	15	4637.5	2013.0	4637.5	4848.1	3159.4	0.652	G
4	16.0	1.6	16	11	3978.1	2284.1	3978.1	4369.4	3152.5	0.722	
3	12.0	1.2	12	7	2825.6	2095.4	2825.6	3707.3	2912.1	0.7855	
2	8.0	0.8	8	3	1103.1	1413.5	1413.5	2471.4	1941.3	0.7855	SLIDING
1	4.0	0.4	4	---	–1485.1	472.2	472.2	1237.1	971.8	0.7855	

If tan ϕ is reduced to 0.650, it will be found that blocks 1 to 4 in the toe region will slide while blocks 5 to 13 will topple. The tension in a bolt or cable installed horizontally through block 1, required to restore equilibrium, is found to be 2013 kN/m of slope crest. This is not a large number, demonstrating that support of the "keystone" is remarkably effective in increasing the degree of stability. Conversely, removing or weakening the keystone of a slope near failure as a result of toppling can have serious consequences. The support force required to stabilize a slope from which the first n toe blocks have been removed can be calculated from equations 2.77 and 2.78 substituting P_{n+1} for P_1.

Now that the distribution of P forces has been defined in the toppling region, the forces R_n and S_n on the base of the columns can be calculated using equations 2.72 and 2.73; and assuming $Q_{n-1} = P_{n-1} \cdot \tan \phi$, R_n and τ_n can also be calculated for the sliding region. Figure 2.36 shows the distribution of these forces throughout the slope. The conditions defined by $R_n > 0$ and $|\tau_n| < R_n \cdot \tan \phi$ are satisfied everywhere.

2.9.4. Factor of Safety for Limiting Equilibrium Analysis of a Toppling Failure

The FS for toppling can be defined by dividing the tangent of the friction angle believed to apply to the rock layers (tan ϕ available) by the tangent of the friction angle required for equilibrium with a given support force T(tan ϕ required).

$$F = \tan\phi_{available} / \tan\phi_{required} \quad (2.79)$$

If, for example, the best estimate of $\tan\phi = 0.800$ for the rock surfaces sliding on one another, the FS in the example, with $\tan \phi_{required} = 0.7855$ and with a 0.5 kN support force in block 1, is equal to 0.800/0.7855 = 1.02. With tan ϕ required = 0.650 and a support force of 2013 kN, the FS is 0.800 /0.650 = 1.23.

Once a column overturns by a small amount, the friction required to prevent further rotation increases. Hence, a slope just at limiting equilibrium is metastable. However, rotation equal to $2(\beta - \alpha)$ will convert the edge to face contacts along the sides of the columns into continuous face contacts and the friction angle required to prevent further rotation will drop sharply, possibly even below that required for initial equilibrium. The choice of FS, therefore, depends on whether or not some deformation can be tolerated.

The restoration of continuous face-to-face contact of toppled columns of rock is probably a very important arrest mechanism in large scale toppling failures. In many cases in the field, large surface displacements and tension crack formation can be observed and yet the volumes of rock which detach themselves from the rock mass are relatively modest.

2.9.5 General Comments on Toppling Failure

The analysis presented on the previous pages can be applied to a few special cases of toppling failure and it is obviously not a rock slope design tool at this stage of development. However, the basic principles which have been included in this analysis are generally true and, with suitable additions, will probably provide a basis for further developments of toppling failure analysis.

The reader is strongly advised to work through the example given and to try examples of his own since this can be a very instructive exercise. The calculations are relatively simple to program on a desk top calculator or a computer and the availability of such a program will enable the user to explore a number of possibilities, thereby gaining a better understanding of the sensitivity of the toppling process to changes in geometry and material properties.

2.10 BUCKLING FAILURE

2.10.1 Potential Occurrence

Buckling failure can occur in moderate to steep dipping, thinly bedded rocks, resting on a plane having a low friction angle. This type of failure occurs over limited areas, particularly in the vicinity of rolls or undulations, and are usually followed by planar slab sliding of material up-slope of the buckled area. Two possible models for buckling failure are three-hinge buckling and Euler buckling. These are shown in Figure 2.37. In both cases, buckling occurs over the lower portion of the slope. The upper part of the slope acts as the driving force.

In the three-hinge buckling, it is assumed that the buckling segment is composed of two discrete blocks (Figure 2.37a). Failure is initiated when sufficient pore pressure exists along the base plane to induce simple rotation of the two blocks about three hinge points, or rotation combined with shearing at the hinge points, or shearing along cross-joints between blocks causing one or both blocks to be thrust upward. This type of buckling occurs mostly in hard rocks.

In the Euler buckling, failure occurs when the axial force exceeds a certain critical buckling load causing inelastic deformation or heave of the buckling segment (Figure 2.37b). This type of buckling occurs mostly in rocks with soft to medium strength ranges.

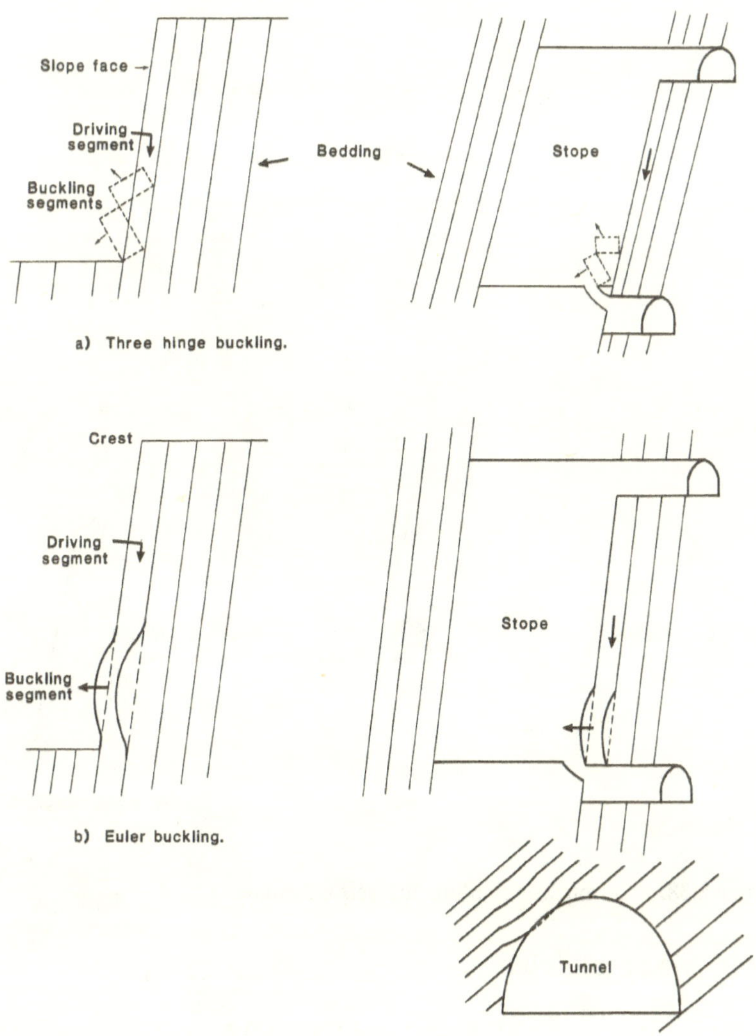

Figure 2.37. Types of buckling failure.

2.10.2 Analysis of Buckling Failure

The buckling length of the slab is assumed to be weightless subject to axial compression. Utilizing limit equilibrium techniques and the Euler buckling solution, the maximum load which can be carried by the column, can be converted to an equivalent maximum slope length or height of the driving slab as shown in Figure 2.38a.

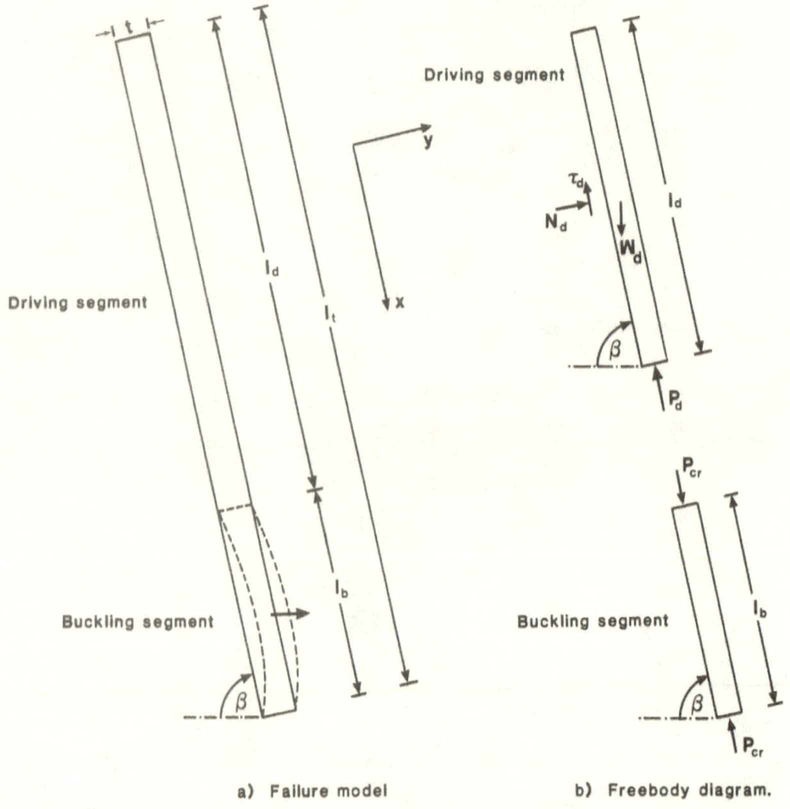

Figure 2.38. Euler buckling failure analysis technique.

For limiting equilibrium:

$$\Sigma F_x = 0 \text{ and } \Sigma F_y = 0$$

τd = shear resistance to the driving force (Figure 2.38b)

$= N_d \cdot \tan\phi_d + S_d \cdot 1_d$ (2.80)

P_d = driving force = $W_d (\sin\beta - \cos\beta \cdot \tan\phi_{d)} - S_d \cdot 1_d$ (2.81)

P_{cr} = critical buckling load = $K \cdot \pi^2 \cdot E \cdot I / 1_b^{\ 2}$ (2.82)

where: K = 1, for column with pinned ends

E = elastic modulus of the rock

I = moment of inertia of the rock slab = $b \cdot t^3/12$

b = width of the slab

t = thickness of the slab

FS= factor of safety = P_{cr}/P_d

$$= (K_r \pi^2 \cdot E \cdot I)/\{l_b^2 [W_d (\sin\beta - \cos\beta \cdot \tan\phi_{d)}) - S_d \cdot l_d]\} \quad (2.83)$$

2.11 SAGGING OR BED SEPARATION FAILURE

2.11.1 Potential Occurrence

This type of failure occurs due to excessive bending of the roof and floor beddings surrounding an excavation. Its occurrence is accompanied with shear or slip, leading to separation of the bedding (Figure 2.39). It is more predominant in the excavation roof, especially where the span of the excavation (s) is large in comparison with the thickness of the bedding (t). In addition to the above noted factors, parameters such as state of horizontal and vertical stresses underground, cohesion and internal friction angle of the rock, groundwater, and vibration affect development of failure.

2.11.2 Analysis of Sagging Failure for a Roof Load

Considering a beam of narrow rectangular cross section to have a length (or span) of s = 2L, thickness of t = 2c and unit width. The beam, as a roof bedding, is supported at the ends by pillars and is bent under a uniformly distributed load of intensity q, as shown in Figure 2.40.

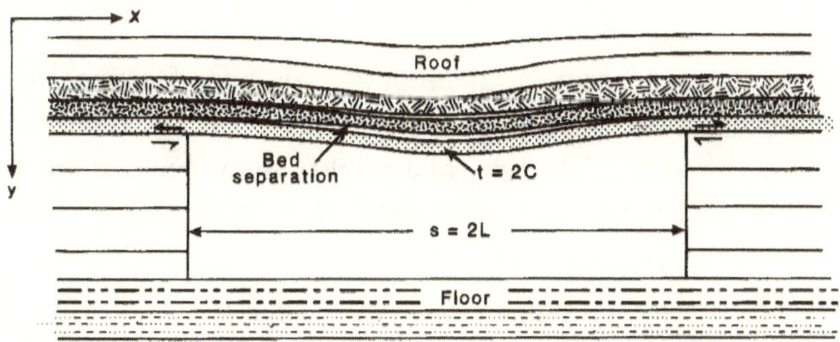

Figure 2.39. Sagging failure due to bending and bed separation in hard rock excavations.

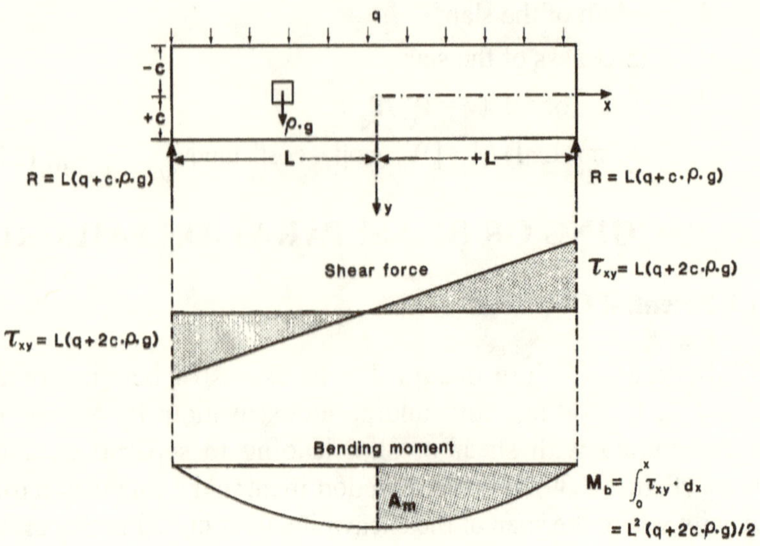

Figure 2.40. Schematic of a horizontal mine roof beam.[2.9]

Distribution of the horizontal, vertical, and shear stresses in the beam, and its corresponding horizontal and vertical displacement, shear strain, and the deflection of the roof beam, utilizing Airy Stress Function, is given as follows:[2.9]

σ_x = Horizontal stress

$$= [(q + 2c.\ \rho\ .g)/\ I]\ /\ \{[(L^2 - x^2)\ y/2] + \{[y^3 - (3\ c^3 - (y/5)]\ /3\}\} \quad (2.84)$$

where: I = moment of inertia of the beam = b $(2c)^3/12$

b = width of the beam, assumed unity in this plane stress condition = 1

x = horizontal distance of a point from the center of the beam

y = vertical distance of a point from the center of beam, in the direction of the loading

σ_y = vertical stress

$$= \{[(q + 2c.\rho.g)\ y^3] + [(3q + 2c.\rho.g)\ c^2.y - 2\ c^3.q\}/\ 6I \quad (2.85)$$

τxy = shear stress = $(q + 2c.\rho.g)x\ (c^2 - y^{2)}\ /2I$ (2.86)

u = horizontal displacement

$= (1/6E.I)\{(q+2c.\rho.g)\{3[L^2.x - x^3/3\}y + 2y[y^2 - 3c^2/5]\} + \upsilon.x[(q+2c.\rho.g)y^3 - (3q+2c.\rho.g).y + 2.q]\}$ (2.87)

v = vertical displacement

$= (1/6E.I)\{\{[(q+2c.\rho.g)y^4/4] - (3q + 2c.\rho.g)c^2.y^2/2] + 2c^3.q.y\} + \upsilon(q+2c.\rho.g)\{[3(L^2 - x^{2)}y^2/2] + y^2[(y^2/2) - (3c^2/5)]\}\} + (c^4/120 E.I)[(q+2c.\rho.g)(5-2\upsilon) - 10(3q+2c\cdot\rho\cdot g) + 40q]$ (2.88)

γxy = shear strain $= [2(1-\upsilon)(q+2c.\rho.g)(c^2 - y^{2)}x]/2E.I$ (2.89)

θ = slope of the roof beam at its ends

$= [L^3(q+2c.\rho.g)/3E.I]$ (2.90)

η_{max} = maximum sagging or deflection of the beam

$= 15L^4(q+2c.\rho.g)/48E.c^3$ (2.91)

η = general sagging or deflection of the beam

$= x(q+2c.\rho.g)(L^3 - 2L.X^2 + x^3)/16E.c^3$ (2.92)

2.11.3 Analysis of Sagging Failure in Roof Beams Under Combined Axial and Transverse Stresses

Horizontal *in situ* stresses are generally present in underground mines. These will load roof beams axially which increases the moment in the beam by so-called secondary bending effect, as shown in Figure 2.41. This results from the axial force multiplied by a moment arm equal to the deflection. Under this condition:[2.30]

Q = total vertical load on the roof beam $= q \cdot s = 2q \cdot L$

P = total horizontal load on the sides of the roof beam $= p \cdot t = 2p \cdot c$

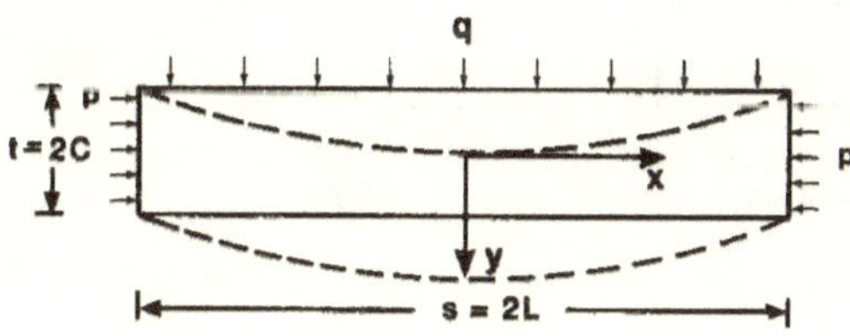

Figure 2.41. Schematic of combined vertical and horizontal loads on a roof beam.

M_{center} = moment at the center of the roof beam, with fixed ends

$$= (4q.\ L^2/24)\ [6(v - \sin v)/v^2.\sin v]$$

$$= q.\ L^2\ [(v - \sin v)/v^2.\sin v] \qquad (2.93)$$

M_{ends} = moment at the roof beam ends, with fixed ends

$$= -\ (4q.\ L^2/12)\ [3\ (\tan v - v)/v^2.\tan v]$$

$$= -\ q.\ L^2\ [(\tan v - v)/v^2.\tan v] \qquad (2.94)$$

where: $v = s\ (3P/E.\ t^3)^{0.5} = 1.73L\ (P/E.\ c^3)^{0.5}$

If the beam ends are not fully fixed, or one allowed to rotate, the secondary bending effect becomes larger. Therefore, Equations 2.93 and 2.94 are not conservative. The extreme condition is a uniformly loaded beam simply supported at the center. In the later case:

M_{max} = movement at the center of roof beam with free ends

$$= (4q.\ L^3/8)\ [2(1 - \cos v)/v^2.\cos v]$$

$$= q.L^3[(1 - \cos v)/v^2.\cos v] \qquad (2.95)$$

$\sigma_t = \sigma_{max}$ = total stresses on the beam $= p \pm \sigma_{bending}$

$$= p \pm (6M_{max}/t^{2)} = p \pm (1.5M_{max}/c^{2)} \qquad (2.96)$$

τ_{max} = maximum shear stress at the neutral axis and the beam ends $= 3\ V_x/2t = 1.5V_x/c$ (2.97)

where, V_x = shear at the beam ends $\gamma \cdot s \cdot t/2$

Substituting the allowable M_{max} and V_x values in the expressions 2.96 and 2.97 and solving for the beam span (s) gives:

S_{max} = maximum allowable, or safe, beam span

$$= t\ (2\sigma_{max}/q)^{0.5} = 1.33\ t \cdot \tau_{max}/q \qquad (2.98)$$

Note: The maximum total stresses in tension is obtained by dividing the beam modulus of rupture by the safety factor, and for compression and shear by dividing the beam compression and shear strength by the safety factor.

Fs_{sq} = safety factor of the roof beam as a square plate of sides s (see Figure 2.42a).[2.31]

$$= 2\sigma_c \cdot n \tan \phi/\gamma \cdot s = s.\ \tan \phi/6z \qquad (2.99)$$

where: $n = 3[1 - (z/t)] /2$

$z = \eta_{max}$ = maximum sagging or deflection of the roof beam

FS_{rect} = safety factor of the roof beam as a rectangular plate of sides s x b (see Figure 2.42b).[2.31]

$$= \sigma_c \cdot n.s. \tan \phi / [\gamma (b (s - y))] = [b(3s - 4y. \tan \phi] / 12 z (s - y) \quad (2.100)$$

where: $y = b[(k^2 + 3)^{0.5} - k] /2$ and $k = 12\sigma_y /7\sigma_c = 1.71\sigma_y /\sigma_c$

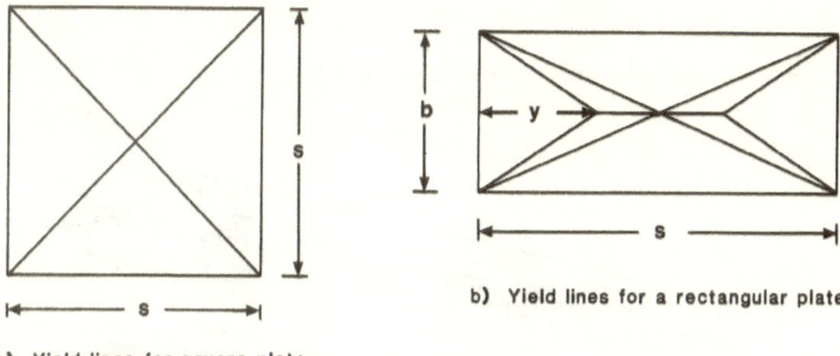

Figure 2.42. Plan of the roof beam in an underground excavation.

2.11.4 Effect of Multiple Beds

Where the immediate roof is composed of more than one bed and the upper, less rigid beds are partially supported by the bottom bed (Figure 2.43), the load on the lowest roof bed (q_1) can be calculated by the following formula:[2.30]

$$q_1 = E_1 .t_1^{.3} [(\gamma_1 . t_1 + \gamma_2 . t_2 + ... + \gamma_n . t_n)] / (E_1 .t_1^{.3} + E_2 .t_2^{.3} + ... + E_n .t_n^{.3}) \quad (2.101)$$

where, n = number of deflected roof beds.

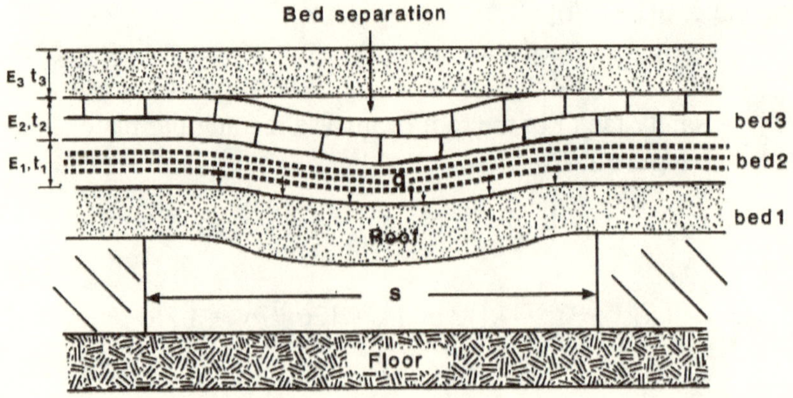

Figure 2.43. Horizontal roof with multiple beds.[2.30]

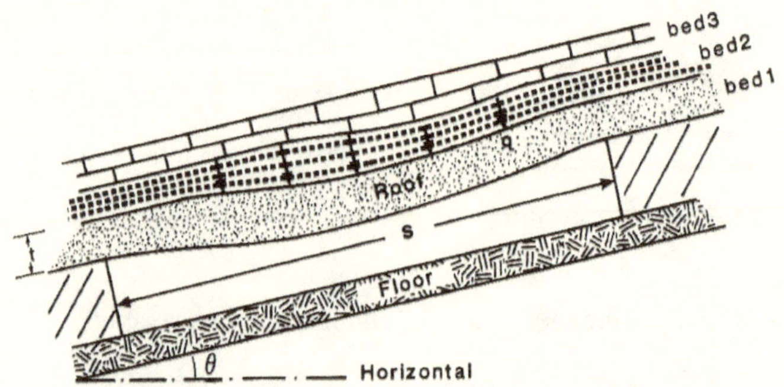

Figure 2.44. Excavation driven along strike of inclined beddings.[2.30]

2.11.5 Effect of the Bedding Inclination

For inclined bedding (Figure 2.44) the unit transverse load causing flexure of the roof beam (q_{inc}) is equal to the value of q as determined for horizontal beds (equation 2.98) times cosine of the angle of dip (θ), in the following form:[2.30]

$$q_{inc}=q_{hor}\cdot\cos\theta \qquad (2.102)$$

The influence of angle of dip on the value of q is generally ignored for beds dipping less than 10 to 15°. Correcting for = 20°, reduces q by 6%.

2.11.6 Effect of Cracks

The presence of the crack influences sagging of the roof beam and consequently its failure dependent upon the orientation of the cracks and the coefficient of friction of the crack surfaces.

In a long room, steeply dipping cracks or fracture with a strike parallel to the span of the room, do not affect a beam analysis. This is because such cracks merely cut the roof slab into a number of parallel beams, a condition which was assumed in the Section 2.11.2. If the strike of the cracks is inclined moderately to the direction of the room span (Figure 2.45), the roof can be approximated by beams with spans equal to the distance along the crack between abutments.[2.30] For greater angles between the directions of the room span and the cracks, the roof should be analyzed both by beam and arch theory and judgment used as to which analysis is most applicable.

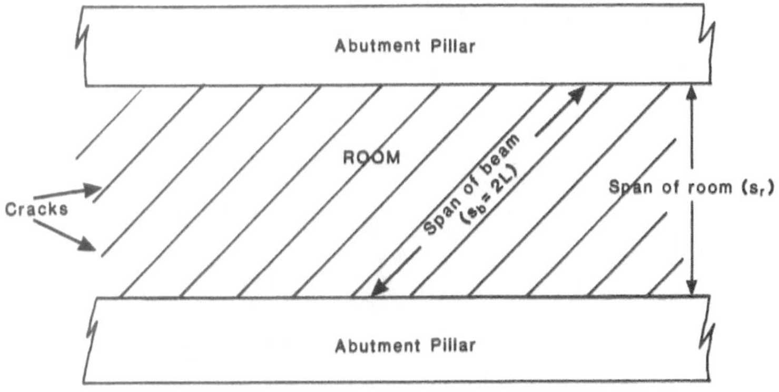

Figure 2.45. Plan view of parallel roof cracks occurring across a long room.

In a long room, if the steep cracks strike in the direction that the room is driven and the axial force (P) is not large enough to overcome all tensile stresses in the beam, the beam theory is not applicable and arch theory must be used to determine the stability of the roof.[2.30] If only a single open crack close to one pillar removes the support provided at one abutment, the roof will behave as a cantilever beam. The maximum (-ve.) movement will be 6 times as great and the maximum shear will be twice as great as for a beam fixed at both ends.

If steep cracks or other planes of weakness strike in the direction that the room is driven and the beam also is loaded by an axial force (P), which is large enough to overcome all tensile stresses in the beam, the roof can be treated as a beam under combined axial and transverse loads and the stresses can be calculated as in Section 2.11.2. However, the FS for such a beam may depend on resistance to sliding along the crack surfaces rather than on the compressive or shear strength of the rock.[2.30]

The most critical crack is one near and dipping toward an abutment, as shown in Figure 2.46. A simple method to analyze sliding along the crack surface is to resolve the resultant of the axial force (P) and the shear (τ) into components normal and parallel to the crack surface.

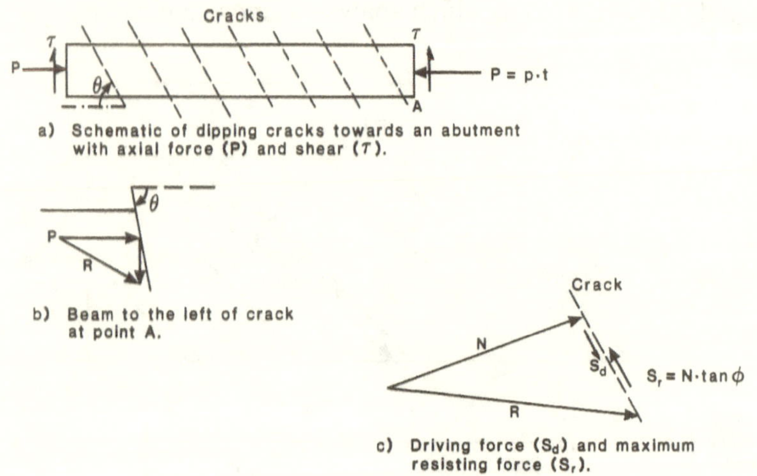

Figure 2.46. Method to determine driving and resisting forces along a crack in a beam.[2.30]

The component parallel to the crack surface is the driving force (S_d) that tends to cause sliding. It is determined as follows:

$$S_d = P \cdot \cos\theta + \tau \cdot \sin\theta \qquad (2.103)$$

This force is resisted by an opposite force (S_r) that can have a maximum value equal to the normal component (N) times the coefficient of internal friction of the crack surface (tan ϕ_{μ}), as follows:

$$S_r = N \cdot \tan(\phi_\mu + i) \tag{2.104}$$

where: $N = P \cdot \sin\theta - \tau \cdot \cos\theta$

i = inclination of the asperity in the crack surface, to a common tangent to the base of the asperities.

and finally: $FS = S_r / S_d$

$$= \{[\tan(\phi_\mu + i)\,(P \cdot \sin\theta - \tau \cdot \cos\theta)]\} / (P.\cos\theta + \tau \cdot \sin\theta) \tag{2.105}$$

2.12 KEYBLOCK FAILURE

2.12.1 Potential Occurrence

Key-blocks occur on the surface of an excavation and one or more of their faces are created by the excavation. A key-block is potentially critical to the stability of an excavation because it is finite, removable, and potentially unstable. Goodman and Shi[2.32] have classified five types of blocks, named in Table 2.2 and illustrated in Figure 2.47. Among these five types of rock blocks, type 1 is a true key-block (Figure 2.47a). It is removable and oriented in an unsafe manner, unless restraint and reinforcement is provided before the excavation has completely isolated the block.

Table 2.2 Types of Rock Blocks[2.32]

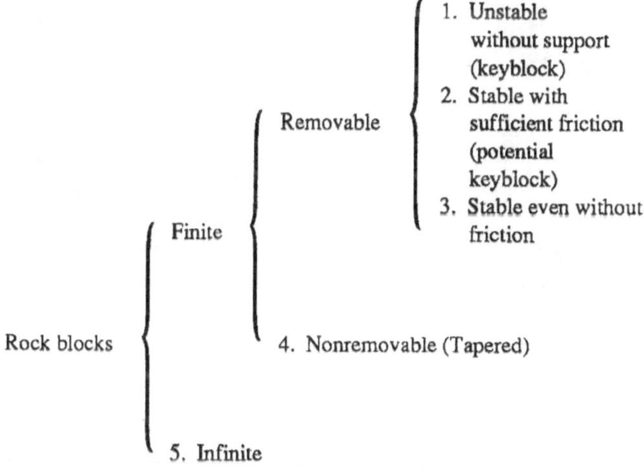

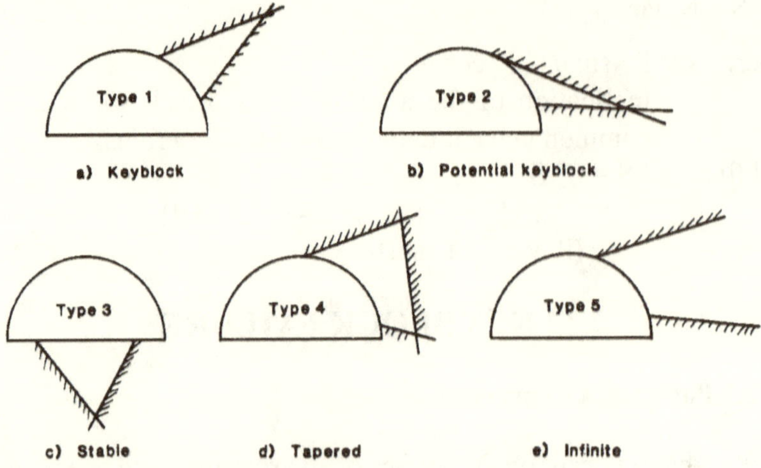

Figure 2.47. Schematic of rock block types.[2.32]

Two measures of relative importance of key-blocks are:

1. Block size, as measured by the block volume.
2. Net shear force of the block, as measured by the difference between sliding and resisting forces per unit volume.

Configuration of a key-block with adjoining blocks in an excavation and possible sequence of failure is illustrated in Figure 2.48.

2.12.2 Analysis of Sliding in Key-block Failure

Denoting a specific removable block by B, ignoring rotation, and considering every part of the B to undergo sliding motion described by the same vector $\hat{s}$. At limiting equilibrium the sliding is without acceleration. Under a given set of forces, B cannot be expected to be exactly in a condition of limiting equilibrium. To bring B to a limiting state, a fictitious force $(-F\cdot \hat{s})$ is added, as shown in Figure 2.49. When F is positive the block tends to slide unless artificial support is added.

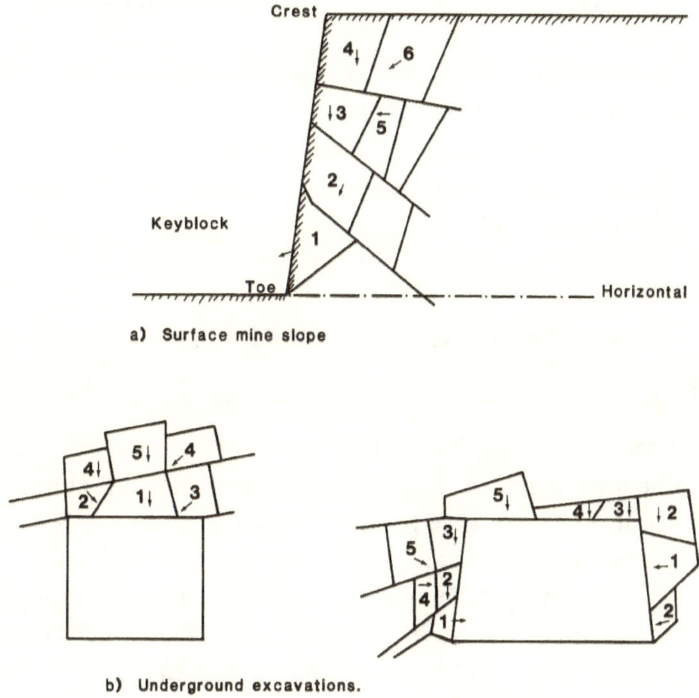

Figure 2.48. Configuration of key-blocks and possible sequence of failure in various excavations.

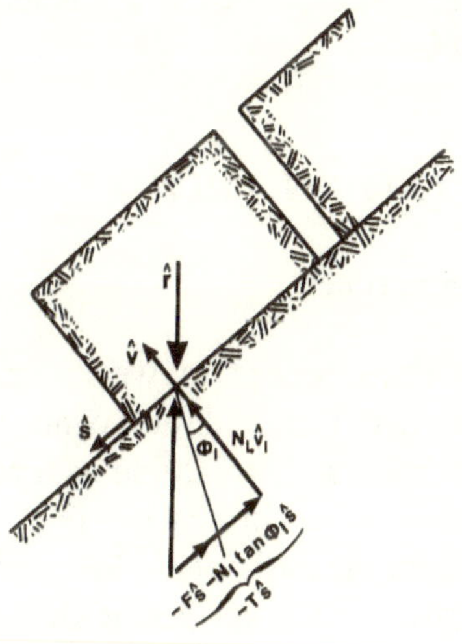

Figure 2.49. Schematic of a key-block sliding.[2.32]

There are three contributions to the forces acting on block B.[2.32]

1. The resultant (N) of the normal components of the reaction forces on the sliding planes. Let be a unit vector normal to joint plane L, directed into block B. Then the normal reactions, assuming there is no tensile strength across the joint (so: $N_L \geq 0$), are:

$$N = \Sigma_L N_L \cdot v_1 \qquad (2.106)$$

2. The resultant (T) of the tangential frictional forces are:

$$-T \cdot \hat{s} = -\Sigma_L T_L - F \cdot \hat{s} = -\Sigma_L N_L \cdot \tan \phi_L \cdot \hat{s} - F \cdot \hat{s}$$

   Therefore: $T = \Sigma_L N_L \cdot \tan \phi_L \cdot \tan \phi_L + F$ (2.107)
   For a potential or a real key-block, by design, sliding will occur if $\phi_L = 0$ since positive F implies sliding, $T \geq 0$.

3. The active resultant (r) of all other forces acting on block B, including weight, seepage forces or external water pressures, inertia forces, and support forces from cable or rock bolts. The conditions of equilibrium for a potential or real key-block B is: $r + \Sigma_L N_L \cdot v_1 - T \cdot \hat{s} = 0$ (2.108)

With $T > 0$ and $N_L \geq 0$, from equation 2.108, if $N_L = 0$, $r = T \cdot \hat{s}$ and for $T \geq 0$ then $\hat{s} =$. From the theorem of block removability,[2.32] the sliding direction of $\hat{s}$ removable block B belongs to the joint plane (JP) of block B, that is:

$$\hat{s} \subset JP \qquad (2.109)$$

2.13 RAVELING AND ROCKFALL FAILURE

2.13.1 Potential Occurrence

This type of failure occurs in weak, heavily fractured, or degradable rocks, or where sides and walls of an excavation are subject to the effects of mechanical degradation as a result of weathering, freeze-thaw action of water in cracks, general deterioration of materials which cement the individual blocks or adjustments due to unloading. On long footwall of surface or underground excavations or embankments which are inclined steeper than the angle of repose of the rock mass, rockfalls could bounce or roll down the slopes.

2.13.2 Analysis of Raveling Failure

Consider a surface slope or footwall of an underground excavation with a gradient β to horizontal and height H, initially at OT, as shown in Figure 2.50a. At some subsequent time, raveling occurs and the position of slope changes to PQ, projected down to N. The curve OP shows the form of the rock core, protected beneath the rockfall surface PM, inclined at constant angle α. The development of the rock core is traced using inclined axes as indicated with origin at O, to represent the point P (x,y). In retreating a horizontal distance dx, the rock core advances from P to P′. The cross section PQQ′P′ is removed from the slope cross section, and after swelling by a factor of k, added to the rockfall cross section. At the same time a thickness of m · dx (measured horizontally) is removed from the rockfall either by basal undercutting or by weathering of the rockfall material.

For the cross section PNM, the input from the slope allowing for raveling is k (H - y) dx· sinβ, and the output from weathering is m· dx· sinβ where m is the weathering constant. The resulting change is cross section:

$$PMM'P' = (PM \cdot \cos\alpha)\, PL \qquad (2.110)$$

where: PL = the vertical height change (Figure 2.50b)

$$= dy \cdot \sin\beta - (dx + dy \cdot \cos\beta)\tan\alpha \qquad (2.111)$$

The overall raveling equation for PNM is then:[2.28]

$$[k\,(H - y) - m \cdot y]dx \cdot \sin\beta = y \cdot \sin\beta \cdot \cot\alpha[dy \cdot \sin\beta - (dx + dy \cdot \cos\beta)\tan\alpha] \qquad (2.112)$$

Grouping terms in x and y in the equation 2.112 gives:

$$y.\, dy / [k.\, H - y(k + m - 1)] = dx / (\sin\beta.\, \cot\alpha.\, \cos\beta) \qquad (2.113)$$

For small dx and dy, differential equation 2.113 yields the following solution:

$$(k + m - 1)\, x / [k.\, H(\sin\beta.\, \cot\alpha.\, \mathrm{Cos}\beta)] = \ln[1 - y(k + m - 1) / k.\, H] - [y(k + m - 1) / k.\, H] \qquad (2.114)$$

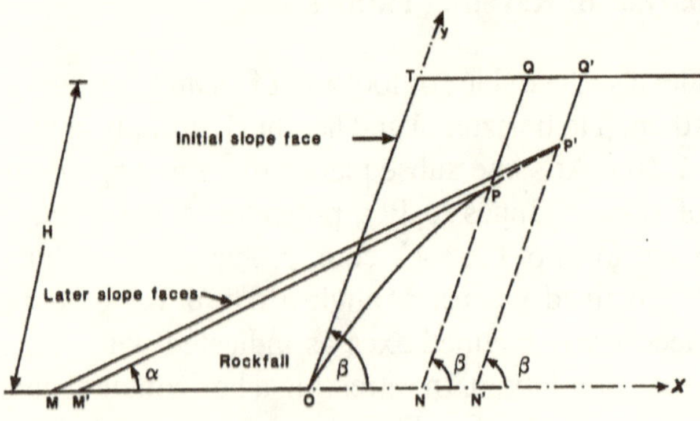

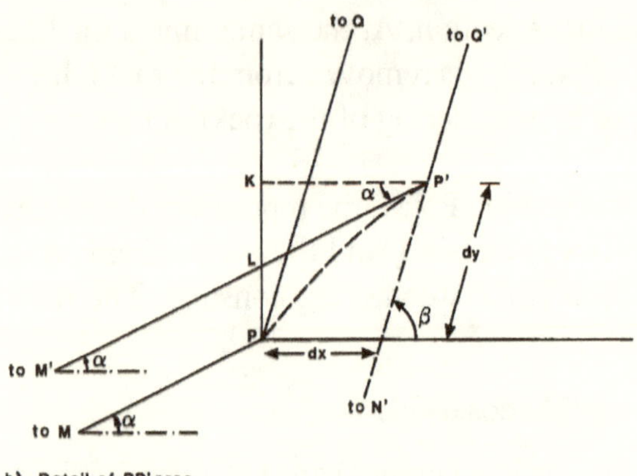

Figure 2.50. Schematic of a raveling or rockfall.[2.28]

There are two important special cases, derived from Equation 2.113. These are:

(1) For $k = 0$: $y = (m - 1)\,x / (\sin\beta.\ \cot\alpha - \cos\beta)$ (2.115)

(2) For $k + m = 1$: $y = [2\,k.\ H.\ x / (\sin\beta.\ \cot\alpha - \cos\beta)]^{0.5}$ (2.116)

2.14 STRESS-INDUCED FAILURE

2.14.1 Potential Occurrence

Rock is very weak in tension. In hard rock where the cementation is weak, even groundwater pressure may be sufficient to induce failure near the excavation free surfaces. Moreover, change in the virgin *in situ* stress distribution due to mining excavation is greatly affected by the stratigraphy and tectonics in the rock mass. Local stress concentration may reach as high as three times the virgin *in situ* stress and may exceed the strength of the rock mass. Spalling or rock-burst may then occur. This type of failure does not produce a distinct sliding segment but rather it is characterized by internal deformation. The boundary of failure surface is not likely to be either planar or circular, and in its preliminary stage, it cannot be predicted accurately. Some geological conditions that can result in the stress induced failure is illustrated in Figure 2.51.

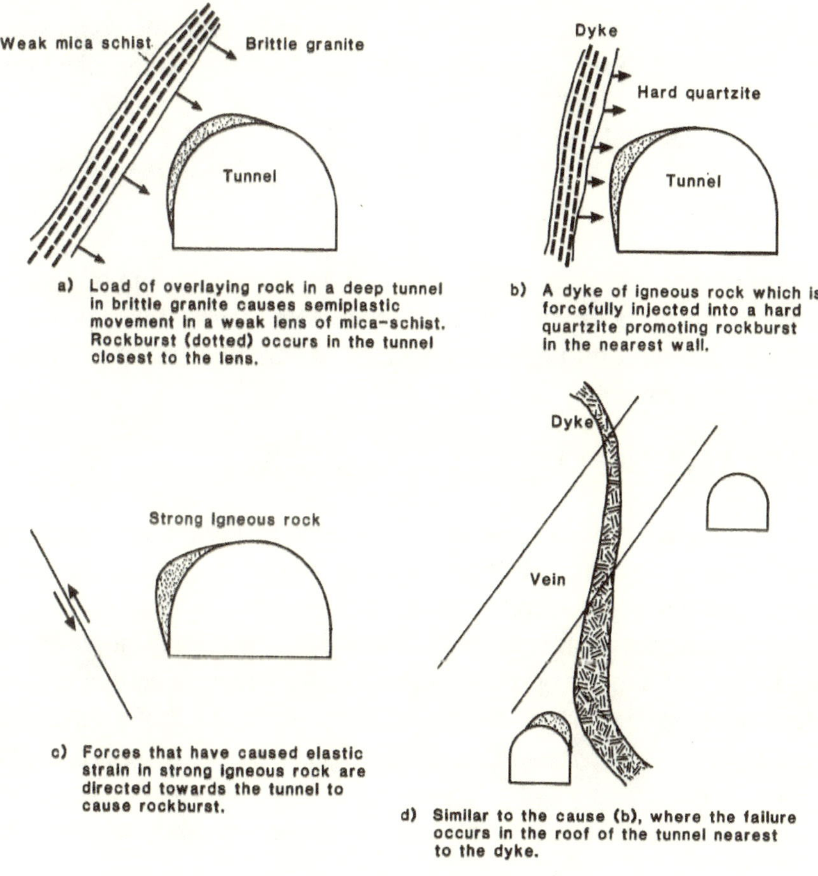

Figure 2.51. Stress induced failure and rock-burst under various geological conditions.[2.33]

2.14.2 Analysis of Stress-Induced Failure

Assuming the deflection of the crest of the slope after excavation is a valid measure of critical conditions, the following equation has been derived:[2.34]

$$\tan\beta_2 = \tan\beta_1 \,(K_1 .H_1 / K_2 .H_2)^{0.5} \tag{2.117}$$

where: β = slope angle of the excavation

k = ratio of horizontal to vertical stresses, i.e., σ_h/σ_v

H = height of the slope;

Subscript 1 = the existing or initial slope which is either stable or somewhat unstable due to stress induced failure.

Subscript 2 = The new slope in the same rock with the same wall direction for which the same stability can be tolerated.

2.15 CHIMNEY CAVING OR FAILURE

2.15.1 Potential Occurrence

In some cases, overall failure could develop by caving of a large trapezoidal mass of rock, formed by intersection of three or more sub-vertical discontinuities. Figure 2.52 illustrates some of the forms of chimney caving or failure.

2.15.2 Analysis of Chimney Caving or Failure

Considering the general case illustrated in Figure 2.53 the base and top of a vertical wall block coincide with the hanging wall of a stope and with the ground surface, respectively.

It is assumed that rigid-body block caves vertically downwards under the influence of gravity when the shear resistance that can be developed on the vertical block boundary is exceeded. It is assumed that the vertical *in situ* stress is given by $\sigma_z = \gamma \cdot h$, where h is depth below surface and γ is the unit weight of the overburden, and the horizontal normal stress acting equally around the failure zone, i.e., $\sigma_x = \sigma_y = k \cdot \gamma \cdot h$, where k is a constant. Then:[2.31]

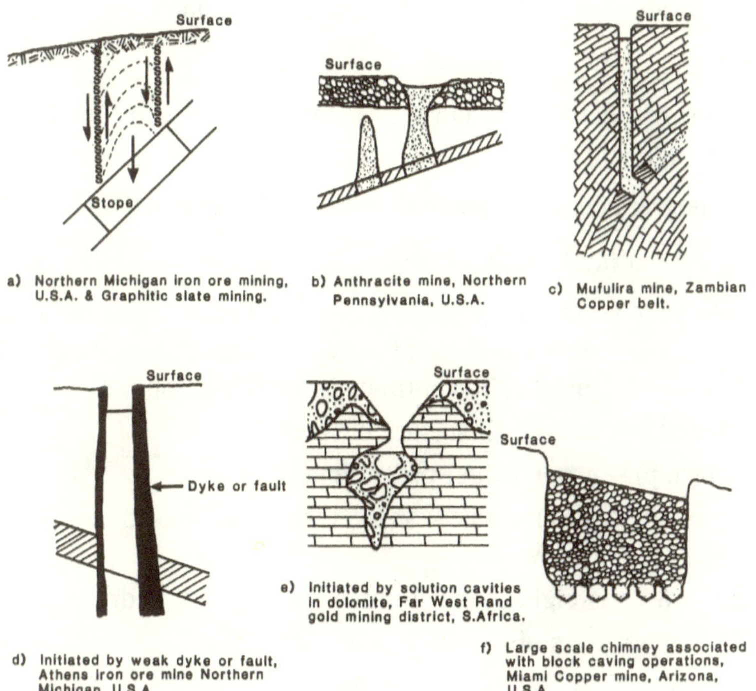

Figure 2.52. Various forms of chimney caving or failure.[2.31]

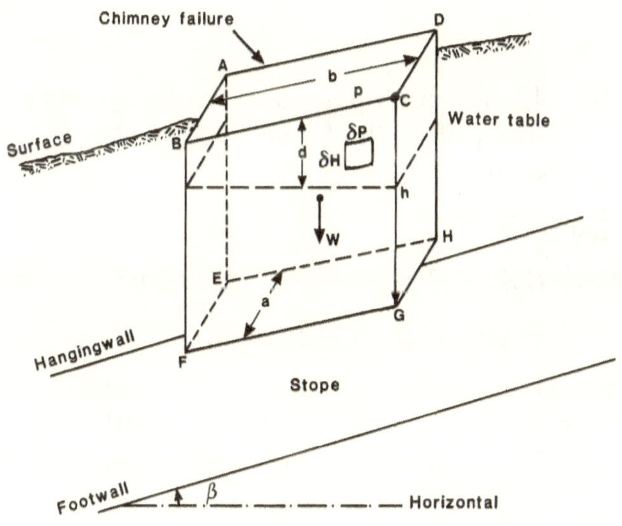

Figure 2.53. Schematic of a chimney failure for limiting equilibrium analysis.

Q = Available shear resistance for the entire surface

$= 2Q_{BCGF} + Q_{ABFE} + Q_{DCGH}$

= ∫Integral between 0 and p then∫Integral between 0 and h of τ. dh. dp

= ∫Integral between 0 and p then∫Integral between 0 and h of {c + [γ . h – u(h, p) tan ϕ] dh. dp (2.118)

where: τ = limiting vertical Coulomb shear stress that can be developed on a boundary element δp · δH to which an effective normal stress of σ is applied.

h and p = sum of all surface elements.

u(h, p) = groundwater pressure at an element.

FS = factor of safety on the vertical wall block

= Q /(W + w) (2.119)

where: W = weight of the chimney caving rock in dry condition

= γ · a · b · cosβ [h – (b. sinβ /2)] (2.120)

w = weight of water in the caving block

Substituting equations 2.118 and 2.120 into equation 2.119 leads to the following results for the three cases of groundwater level considered:

(a) For d ≥ h:

$(FS)_1$ = [2c'(a + b. cosβ) /γ · a · b · cosβ] + [k. tan ϕ' / (2h - b. sinβ)]. {[h^2+ (h - b. sinβ)^2 / b · cosβ} + {(2/a)[h (h - b. sinβ) + (b^2 . Sin^2 β) / 3]} (2.121)

(b) For h ≥ d ≥ – b · sinβ:

(FS) = $(FS)_1$ – [2 γ_w tan ϕ' (h – d)^2/2 γ. b (2h - b. sinβ)]. {secβ + [2 (h – d) / 3a. sinβ]} (2.122)

(c) For 0 ≤ d ≤ h – b · sinβ:

(FS) = $(FS)_1$ – 3γ_w tan ϕ'/ 3γ (2h - b. sinβ){[h^2 + (h - b. sinβ)^2 - 2d (2h - b. sinβ - d) / b ·cosβ] + 2 /3a [3h(h - b. sinβ) + b^2 · $\sin^2$β – 3d (2h - b. sinβ - d)]} (2.123)

2.16 SWELLING FAILURE

This is a common problem for rocks and joint fillings which swell up to three times their dry volume, in the presence of water. Bentonite and clay minerals such as illite, kaolinite, montmorillinite exhibit this property. Swelling failure generally occurs in rocks or joint fillings with low mass strength properties. The author has found that in laboratory and *in situ* conditions, the swelling failure is a result of time-dependent deformation characterized by creep, volume expansion, and plastic behavior, especially when the rock absorbs water under low stress level.[2.35]

Under these conditions, mechanically anchored cable bolts are inappropriate for the ground reinforcement. It is recommended to utilize doweling or rock bolts with wire or weld-mesh, or to use 5- to 10-cm thick shotcrete, cement grout, or resin injection.[2.36] The latter is more water resistant, its strength and curing time can be designed to match the ground condition to be reinforced. However, it is more expensive than the other types of reinforcement materials.

2.17 WEATHERING AND EROSION

Weathering and eventually erosion in rock is generally caused by wind, water, sun, and seasonal temperature variations. It changes the geotechnical properties of the rock mass, particularly its angle of internal friction and drainage characteristics. If weathering proceeds far enough, the direction of change is usually from properties of a high interlocking masses, such as quartzite, granite, limestone, and shale, towards leached material and eventually residual sand and clay. In this process, there are two main effects. These are:

1 - Normal association of the rock grain size with its mineralogy, from quartz and feldspar through mica to clay minerals in a descending sequence of grain size and internal friction angle[2.37] as shown in Figure 2.54.
2 - The increase in interlocking and dilation angles associated with mixtures of grain sizes.

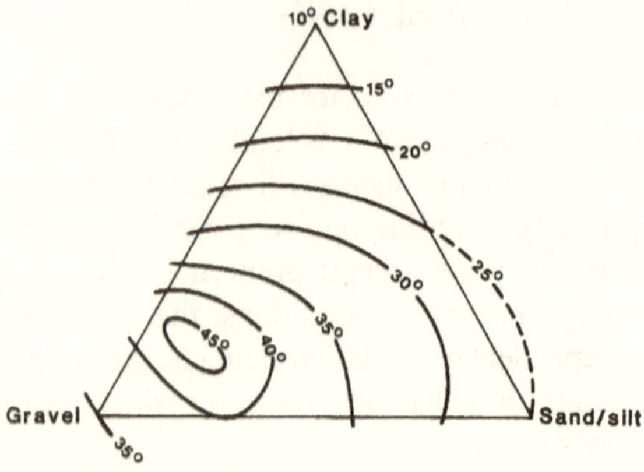

Figure 2.54. Generalized relationship between grain size and friction angle.[2.37]

Physical breakdown of rocks, partly saturated zone above the water table, will expose an ever increasing surface area to chemical weathering processes. Detailed studies highlighted the biogenic oxidation of ferrous sulfide minerals such as pyrite, marcosite, and pyrrhotite, as the most important chemical decomposition process, expressed as follows:[2.38]

Stage 1: $2\ FeS_2 + 7\ 0_2 + 2\ H_20 \rightarrow 2\ FeSO_4 + 2\ H_2SO_4$
Stage 2: (Bacteria as catalyst), $4\ FeSO_4 + O_2 + 2\ H_2S0_4 \rightarrow 2\ Fe_2(SO_4) + 2\ H_20$
Stage 3: $7\ Fe(SO_4)_3 + FeS_2 + 8\ H_20 \rightarrow 15\ FeS0_4 + 8\ H_2SO_4$

At some stage in the weathering and erosion process, enough fines will be produced to retard drainage of the rock mass, which will become critical during wet seasons. At this stage, the stable angle for the rock mass will drop to approximately half its dry value.[2.39]

A measure of overall weathering and its depth is called the total "rock deficit", defined as the total mass of material required to convert the existing rock back to its parent material. It is expressed in volumetric terms as an equivalent depth of bedrock. The rock deficit may be augmented directly by solution and depleted somewhat less directly through mechanical erosion of the surface. The mass balance for rock deficit may be expressed as follows:[2.28]

$$\delta w / \delta t = \delta / \delta x\ [V - (1 - P_s) / P_s \cdot S] \qquad (2.124)$$

where: w = the rock deficit

P_s = near surface proportion of bedrock remaining

V = the solute transport at distance x down-slope

S = the mechanical sediment transport at x

The coefficient of S in the final term is obtained by considering the equivalent depth of rock eroded by removing S (equal to S/P_s), and the loss of rock deficit per unit depth below the free face (equal to $1 - P_s$). For low rates of mechanical removal, this expression describes indefinite thickening of the disintegrated rock through weathering. For high rates of weathering, the erosion is associated with equilibrium between mechanical and chemical removal. For this equilibrium $P_s = S/(S + V)$.

2.18 DYNAMIC-INDUCED FAILURE

2.18.1 Potential Occurrence

This type of failure occurs due to vibration, rock-burst, and cyclic dynamic loading on rock mass. Mining activities such as blasting, rock fragmentation by air hammer, tunneling machines or impulsive water jets, excessive vibration of machinery and equipments can be a cause. In addition, natural phenomenon such as earthquake may produce dynamic failure. An apparent feature of this type of ground failure is comparatively more ragged failure surface.

2.18.2 Analysis of Dynamic Failure

Considering the displacement of rock particles (y) during a period of vibration (t) to be sinusoidal, then:[2.40]

$$y = A \cdot \sin 2\pi .\ f(t - x/V) \tag{2.125}$$

where: A = decaying amplitude of the vibration, m = $(A_0 \cdot e^{-k \cdot x})/x$

A_0 = initial amplitude of the vibration, m

k = proportionality constant, depending to the rock type and ground conditions

x = distance from the source of vibration to the potential failure surface, m

f = frequency of vibration, c/sec
V = velocity of the stress wane, m/sec

Particle velocity of the stress wave (v) can be obtained by differentiating equation 2.125 with respect to time (t), as follows:

$$v = dy/dt = 2\pi \cdot a \cdot f \cdot \cos 2\pi .\ f(t - x/V) \tag{2.126}$$

where: a = acceleration of the stress wave particles, m/sec^2

$$= dy^2/dt^2 = -A(2\pi \cdot f)^2 \sin 2\pi .\ f(t - x/V) \tag{2.127}$$

KE = kinetic energy of the stress wave having particle mass m

$$= m(dy/dt)^2/2 \tag{2.128}$$

PE = potential energy of the stress wave having particle mass m

$$= \int m\,(d^2y/d^2t)\,dy \tag{2.129}$$

$$TE = \text{total energy} = KE + PE = 2m\,(\pi .\ A .\ f)^2 = 2\gamma\,(\pi \cdot A \cdot f)^2/g \tag{2.130}$$

$(FS)_d$ = dynamic factor of safety = P_s / W_{gd}

$$= 2T_c \cdot \sin\theta\,(\sigma_d \cdot D)/0.3g \cdot W_{gs} \tag{2.131}$$

where:
- P_s = capacity of the support or reinforcement under dynamic condition per unit linear length, kN/m
- W_{gd} = the ground dynamic loading on the support or reinforcement per unit length, kN/m
- T_c = thrust coefficient at failure
- θ = abutment angle, generally between 0 and 30°
- σ_d = tensile or compressive strength of the support or reinforcement under dynamic loading, whichever is applicable *in situ*, kPa.
- D = diameter or thickness of the support or reinforcement system, m
- W_{gs} = the ground static loading on the support or reinforcement per unit length, kN/m

Dynamic force at the rupture surface (F) can be determined as follows:[2.41]

$$F = (\pi \cdot C \cdot D)^a \cdot \sigma_t \cdot W^b \tag{2.132}$$

where: C = constant related to the type of dynamic loading
= 4.07, for direct impact and less for indirect impacts
D = diameter of the impact pin or blasthole, m
σ_t = dynamic tensile strength of rock, kPa
W = shortest distances between the source of impact to the rupture surface, m

a and b are constants related to the effect of rate of impact on rock fragmentation as well as rock type in such a manner that a + b = 2. The values of a and b were found to be 0.96 and 1.04 for limestone blocks, and 0.72 and 1.28 for concrete blocks, respectively.

Distance (W) can be calculated from the following expression:[2.42]

$$\log W \cdot v = 3.95 + 0.57 M_L \quad (2.133)$$

where: M_L = magnitude of the event = 0 to 5
v = velocity of the stress wave, m/sec

The force (F), in terms of strain energy and wave propagation velocity can be expressed as follows:[2.43]

$$F = 8.1\, U / V_L \,.\, T_r \,.\, \epsilon = 8.1\, U / K.\,.\, T_r \,.\, \epsilon^{1/3} \quad (2.134)$$

where: U = critical fracture strain energy, kN · m
V_L = p-wave propagation velocity for the longitudinal dynamic loading, m/sec
T_r = rise time to fracture, sec
ϵ = critical fracture strain necessary to cause tensile fracture in rock, μ cm /cm
$= K \cdot \epsilon'^{1/3}$
K = time dependent constant varying between 206 and 756 (average for quartz monzonite rock is 360)
ϵ' = straining rate, μ cm/cm/sec

The peak radial stress in rock (σ_r) is calculated as follows:[2.44]

$$\sigma_r = 9.8\, \gamma \cdot \epsilon \cdot V^2_L / g = 9.8\, \gamma \cdot v \cdot V_L / g \quad (2.135)$$

where: γ = weight per unit volume of the rock, kN/m^3
g = acceleration due to gravity, m/sec^2
v = velocity of the stress wave, m/sec
E_d = dynamic modulus of elasticity of rock, kPa
V_L = longitudinal P-wave propagation, m/sec = $(g \cdot E_d/9.8\,\gamma)^{0.5}$
$= \{[g. E_d (1 - \upsilon] / \gamma (1 - \upsilon - 2\,\upsilon^2\}^{0.5}$ (2.136)

The initial wave front reaching the potential failure surface consists of longitudinal vibrations in which the rock particles oscillate to the line of wave front. This disturbance travels with a velocity of (V_L) expressed mathematically as equation 2.136. It follows a secondary wave consisting of transverse or shear vibrations, in which the rock particles oscillate in the plane of the wave front. This disturbance travels with a lesser velocity of (V_T) expressed as follows:

$$V_T=[g.E_d/2\gamma(1+\upsilon) \quad (2.137)$$

Mass of rock slabs which can be supported under dynamic conditions (m_r) is estimated as follows:[2.42]

$$m_r = F_L \cdot c / v^2 \quad (2.138)$$

where: F_L = potential failure load, kN
c = maximum elongation of rock before failure, mm

2.19 REMEDY FOR THE FAILURE CONDITIONS

The following actions can be considered to eliminate or at least reduce the possibility and extent of the failure in mining or other excavations:

1. Consider smaller excavation slopes down to the slope of the bedding. This may cost a high stripping ratio.
2. Develop slope without benches where possible, or to adopt an excavation design which avoids undercutting parallel to the beddings.
3. Keep the slope and the excavation as dry as possible.
4. Support the potential failure areas around the excavation.
5. Vegetation on the overburden slopes, or alluvial material, soil, and waste damps can add to the slope stability.

6. Utilize ground reinforcement techniques such as cable bolting, rock bolting, doweling, wire meshing, resin or cement grout injection.

REFERENCES

2.1. Hoek, E. and Bray, J. W., *Rock Slope Engineering*, 3rd Ed., The Institution of Mining and Metallurgy, London, 1981, 150-269.

2.2. Barton, N. R., A model study of the behaviour of excavated slopes, Ph.D. Thesis, University of London, Imperial College of Science and Technology, 1971, 520 pages.

2.3. Ucar, R., Hassani, F. P., and Afrouz, A., Design of Anchoring in Rock with Planar Failure, *Int. J. Surf. Min.,* 1(2), 199–208, 1987. (A. A. Balkema, Old Post Road, Brookfield, VT 05036).

2.4. Ladanyi, B. and Archambault, G., Simulation of shear behaviour of a jointed rock mass, *Proc. 11th Symposium on Rock Mechanics,* AIME, New York, 1970, 105–125.

2.5. Ladanyi, B. and Archambault, G., Evaluation de la résistance au cisaillement d'un massif rocheux fragmenté, *Proc. 24th International Geological Congress*, Montreal, Section 13D, 1972, 249–260.

2.6. Barton, N. R., A relationship between joint roughness and joint shear strength, *Proc. International Symposium on Rock Fracture*, Nancy, France, 1971, 1–8.

2.7. Fairhurst, C., On the validity of Brazilian test for brittle materials, *Int. J. Rock Mech. Min. Sci.,* 1, 535–546, 1964.

2.8. Hoek, E., Brittle failure of rock, *Rock Mechanics in Engineering Practice*, Stragg, K. G. and Zienkiewicz, O. C., Eds., J. Wiley, London, 1968, 99–124.

2.9. Afrouz, A., Mechanical Behaviour of Mine Floor, Ph.D. Thesis, University of Wales, Cardiff, U.K., 1973.

2.10. Hoek, E., Bray, J. W., and Boyd, J. M., The stability of rock slope containing a wedge resting on two intersection discontinuities, *Q. J. Eng. Geol.,* 6(1), 1973.

2.11. Coutes, D. F., McRorie, K. L., and Stubbins, J. B., Analyses of pit slides on some incompetent rocks, Trans. AIME, 226, 94–101, 1963.

2.12. Cousin, B. F., Stability charts for simple earth slopes, Proceedings of American Society of Civil Engineers, *J. Geotech. Eng. Div.*, 104(GT2), 267–279, 1978.

2.13. Taylor, D. W., *Fundamentals of soil mechanics*, J. Wiley & Sons, London, 1948, 455–461.

2.14. Terzaghi, K., *Theoretical soil mechanics*, J. Wiley, New York, 1943.

2.15. Lamb, W. T. and Whitman, R. V., Soil mechanics, J. Wiley, New York, 1969.

2.16. Ucar, R., Hassani, F. P., and Afrouz, A., A design analysis for anchoring soil and fractured rock masses prone to circular failure, *Int. J. Surf. Min.*, 1(2), 267–276, 1987. (A. A. Balkema, Old Post Road, Brookfield, VT 05036).

2.17. Bishop, A. W., The use of slip circle in the stability analysis of slopes, *Geotechnique*, 5, 7–17, 1955.

2.18. Fellenius, W., Calculation of the stability of earth dams, *Proc. 2nd Congress on Large Dams*, 4, 445–463, 1936.

2.19. Janbu, N., Stability analysis of slopes with dimensionless parameters, Ph.D. dissertation, Harvard University, Cambridge, Massachusetts, 81, 1954.

2.20. Janbu, N., Slope stability computation, Embankment — Dam engineering casagrande volume, Hirschfeld, R. C. and Poulos, S. J., Eds., J. Wiley & Sons Ltd., New York, 1973, 47–86.

2.21. Morgstern, N. R. and Price, V. E., The analysis of stability of general slip surfaces, *Geotechnique*, 15, 70–83, 1965.

2.22. Spencer, E., A method of analysis of the stability of the embankments assuming parallel interslice forces, *Geotechnique*, 17, 11–26, 1967.

2.23. Baker, R. and Garber, M., Variational approach to slope stability, *Proc. 9th Int. Conf. Soil Mechanics and Foundation Engineering*, 1977, 9–12.

2.24. Kötter, F., Bestimmung des Drucks an gekrummten Geleitfleichen, ueine Aufgabe ans der Lebre von Erddrucks, *Sitzungsberichte der Königlichen Preussischen Akademic der Wissenscherften,* Berlin, 1903, 229–233.

2.25. Spencer, E., Circular and logarithmic spiral slip surfaces, *J. Soil Mech. Found. Div. ASCE,* 95(SM1), 227–234, 1969.

2.26. Chen, W. F., Discussion on circular and logarithmic spiral slip surfaces, *J. Soil Mech. Found. Div.,* ASCE, 96(SM1) 324–326, 1970.

2.27. Chen, W. F., Limit analysis and soil plasticity, *Developments in Geotechnical Engineering,* Elsevier Scientific Publishing, 1975, 399–420.

2.28. Anderson, M. G. and Richards, K. S., Slope Stability, Geotechnical Engineering and Geomorphology, John Wiley & Sons, 1987.

2.29. Goodman, R. E. and Bray, J. W., Toppling of rock slopes, *Proc. Speciality Conference on Rock Engineering for Foundation and Slopes*, Vol. 2, ACSE, Boulder, Colorado, 1976, 201–234.

2.30. Wright, F. D., Roof control through beam action and arching, *Mining Engineers Handbook, AIME,* Vol. 1, Mud Series, Given, C., Ed., 1973, 13.85–13.87.

2.31. Brady, B. H. and Brown, E. T., *Rock Mechanics for Underground Mining,* George Allen & Unwin Publishers, Sydney, 1985, 214–222.

2.32. Goodman, R. E. and G. Shi, *Block Theory and Its Application to Rock Engineering*, Prentice-Hall, New Jersey, 1985, 98–120.

2.33. Stillborg, B., *Professional Users Handbook for Rock Bolting*, Trans. Tech. Publications, W. Germany, 1986, 145.

2.34. Coates, D. F. and Gyenge, M., Incremental Design in Rock Mechanics, Mining Research Centre, Mines Branch Monograph 880, Department of Energy, Mines and Resources, Ottawa, 1973, 2.9.

2.35. Afrouz, A. and Harvey, J. M., Rheology of rocks within the soft to medium strength range, *Int. J. Mech. Min. Sci. Geomech.,* 11, 281–290, 1974.

2.36. Afrouz, A., Floor behaviour along Longwall Roadways, *Int. J. Rock Mech. Min. Sci. & Geomech.*, 12, 229–240, 1975.

2.37. Kirkby, M. J., Landslide and weathering rates, Geologia Applicata e Indrogeologia, *Bari*, 8(1), 171–183, 1973.

2.38. Russell, D. J. and Parker, A., Geotechnical mineralogical and chemical interrelationships in weathering profiles in an overconsolidated clay, *Q. J. Eng. Geol.,* 12, 107–116, 1976.

2.39. Kenney, T. C., The influence of mineral composition on the residual strength of natural soils, *Proc. Geotechnique Conf.*, Oslo, 1967, 123–129.

2.40. Cording, E. J., Henderson, A. J., and Deere, D. E., Rock Engineering for Underground Caverns, *Proc. of Symp. on Underground Rock Chambers, ASCE*, New York, 1981, 567–600.

2.41. Afrouz, A., Rupturing mechanism of rock under dynamic loading, M.Sc. thesis, University of Wisconsin, Madison, U.S., 1969.

2.42. McGarr, A., Green, R. W. E., and Spottiswoode, S. M., Strong ground motion of mine tremors: some implications for near source ground motion parameters, *Bull. Sysmp. Soc. U.S.*, 71(1), 295–319, 1981.

2.43. Birkimer, D. L., A possible fracture criterion for the dynamic tensile strength of rock, *Proc. 12th Symp. on Rock Mechanics,* Univ. of Missouri, AIME, 1971, 573–590.

2.44 Rinehardt, J. S., Dynamic fracture strength of rocks, *Proc. VII Symp. on Rock Mech.,* Penn. State Univ., AIME, 1965.

CHAPTER 3

PRACTICAL EXAMPLES

3.1 APPLICATION OF ROCK MASS CLASSIFICATION SYSTEMS TO WEAK AND WEATHERED ROCK

By: Andy. A. Afrouz and Jamal Rostami

ABSTRACT

The paper encompasses utilization of the rock mass classification systems and failure modes in the design of excavation in weak ground and its support under static and dynamic conditions. The paper provides a brief review of the rock types, its geomechanical properties, preexisting discontinuities, development of fractured zone around the excavation, modes of rock mass failure and methods for designing ground control measures pertaining to the purpose, type, and useful life of the excavation. This is to ensure that large, uncontrolled or unacceptable displacements and ultimately ground failure does not occur. The process involves appropriate use of rock mass classification, deformation and failure criterion.

Special attention to the application of rock mass classification systems to weak and weathered rock masses was made. Some case studies, comparison of the results and suggested modifications is also presented.

INTRODUCTION

Geotechnical investigation pertinent to rock mass characterization is prerequisite to construction or mining excavations and removal of rock. The purpose of the investigation is to examine the ground behavior during and after the excavation, need of support or reinforcement and to identify the potential problems. This is to develop appropriate measures to ensure that the structure will serve its purpose. However, ground behavior can be considered from two different aspects, one is the excavation operation and the second is the ground control issues. In hard and massive rock formations excavation is the primary challenge and poses a major limit to the operation, in soft and weak rock formations the ground control issues dominate the design and operational criterion. Therefore, in soft and weak rocks, specific approaches have to be developed to address the ground control issues and provide a sound, practical, and comprehensive basis to characterize the rock formation.

The early rock mass classifications (Terzagi, 1943) were fairly simple and utilized the basic rock features such as the joint system to classify the rock. Obviously, the application of these methods were limited to specific rocks in a region and could not be used in variety of rock types encountered in nature with different genesis, formation, physical and mechanical properties. Advancement was made as more rocks, discontinuity filling and rock mass properties were identified to influence the ground behavior and support requirements (Deere 1968, Barton et. al. 1974, Bieniawski 1979). These classification methods included more features such as rock strength, joint systems, joint characteristics, joint filling, stress system in the ground and ground water information. With establishment of sufficient database of case histories, the rock mass classification evolved to a higher level of development through statistical analysis of data and subsequent interpretations to serve as knowledge base and expert systems. The present day rock mass classification systems, in addition to providing a universal means to characterize and report the rock mass properties, offer means for estimation of ground support requirements. They provide the engineers with the tools to quantify the rock mass behavior and use it towards estimation of the affected zones, load on the support system, rock mass failure criteria, rock mass elastic and plastic properties. All these features are important in rock engineering and geotechnical investigations, especially in weak rock formations.

This paper presents and discusses some of the issues involved in rock mass classifications with special emphasis on the weak and weathered rock masses. The content of this paper is kept very general. Readers are referred to the appropriate reference material for more detailed information and analysis, tabulated ratings on specific classification systems, the ground failure, support requirement and area of application for each method with examples (see Afrouz, 1992).

ROCK CHARACTERISTICS AND PROPERTIES

The rock characteristics can be divided into two categories:

a) intrinsic which reflects the inherent properties of the intact rock,
b) the overall rock mass characteristics.

There are several basic rock properties known to affect its engineering characteristics and behavior. Some of the well known and most frequently measured rock properties include unconfined and confined compressive strengths, tensile strength, shear strength, internal angle of friction, joint sets and frequency, joint filling, cementation, porosity, grain size, moisture, texture, fabric, plasticity and rock toughness. Among these characteristics, compressive, shear tensile strengths and plasticity are most common. These properties may also be represented by specially developed indices such as point load, toughness, and hardness and plasticity indices.

These features are addressed by various rock mass classification systems to provide a basis for uniform ranking and characterization of different rock masses in the following sections:

(A) Parameters Defining The Properties Of The Intact Rock

Parameters defining the properties of the intact rock are those, which can be typically measured in a laboratory on the samples of rock, represented by different size cores. In most geotechnical investigations, the exploration of the subsurface is via core samples obtained by drilling or from drifts and adits, which can be used to identify the different formations and geological measures. Whereas the surface rock samples can be obtained from outcrop of the formation, trenches, other type of exposures and augured or cored to provide samples for performing the mineralogical, physical, and mechanical property tests. Testing of these samples provide. quantitative measurements of the rock properties. The geomechanic tests on rock samples must be performed to obtain measurement of rock physical and mechanical properties before the residue or discard of sample is made into samples for other purposes such as chemical analysis.

The tests have to be performed in accordance with the uniform standards such as the American Standards for Testing Materials (ASTM) to warrant utilization of the results in the subsequent analysis and design. This is especially important when using the information for design of a project in an environment, which has different set of standards. Therefore the compatibility of the test procedures and the results has to be assured. The frequency and number of tests depend on the variation in the geology of the region. This can be optimized using geostatistical methods to attain a desirable level of confidence in the results and be able to extend the measurements to the entire area of interest.

The most frequently measured intact rock properties are listed in Table 1. However, few issues should be remembered when planning, performing and applying the physical and mechanical property tests to rock mass classifications. The first issue is that when a rock mass contains clay filling or shale parting, the characteristics of the intact rock does not represent the mass behavior. In this case, the rock mass behavior is controlled mainly by the degree of moisture and time. In addition, the tests should be performed under the conditions encountered during the excavation and operating life of the structure. Further, the rocks, which deteriorate when exposed to the air, gasses, radiation or ground water, should be tested for its durability and time dependent behavior.

The shear strength of the intact rock sample has much lesser effect on the rock mass behavior than the shear properties of the discontinuities. Directional properties of the intact rock should also be considered in the testing. This means that some rock samples have to be prepared and tested for the physical and mechanical properties in different directions if the rock is isotropic. This case often rises in weak and weathered ground where grain orientation, special cleavage of composing minerals, sedimentation and cleat shows more weakness in particular directions.

Table 1. The rock physical and mechanical properties measured in testing programs.

• Density • Porosity • Strength (static Loading): Unconfined Compressive Strength (UCS) Confined Compressive Strength (CCS) Tensile Strength Point Load Index Shear strength Bending Hardness • Rock Strength Under Partial and Full Saturation	• Elastic Properties Static Elastic Modulus & Poissons Ratio Dynamic Elastic Modulus & Poissons Ratio • Durability Slake Test • Moisture Content • Plastic Limit • Time Dependent Parameters (Creep) • Directional Properties and Anisotropy in each of the above noted tests

(B) Parameters Controlling Behavior of the Rock Mass

The parameters dominating the physical and mechanical behavior the rock mass are mainly its discontinuities, their cementation and in filling. This refers to the frequency of the planes of weakness, their surface characteristics, filling material, orientation with respect to the excavation, and finally presence of ground water. This is the main reason that early rock mass classifications started by focusing on the joints and their properties. A quick look at the classification systems available to date show that there are a few features of the discontinuities that are commonly used in all classification systems although with different ratings. For example, presence of ground water has a major impact on the behavior of the discontinuities. This is due to a variety of reasons such as reducing the effective normal stresses (and hence shear resistance) between the joints, erosion of the joint surface, reacting with the clay minerals and effectively reducing or eliminating the cohesion, friction between the joint surfaces and finally removing the filling which results in creation of open voids. Table 2. is the summary of the required information for rock mass classification purposes.

Table 2. The features of discontinuities used in rock mass classifications.

• Number of Joint Sets	• Bedding/Fault/Cleat
• Frequency/Spacing between the Joint Sets	• Surface Characteristics: Continuity Smoothness/Undulations Weathering
• Joint Orientation	
• Aperture (Open/Closed Joints)	• Degree of Alteration/Decomposition
• Filling Material Mineral Composition Moisture Content washout	• Groundwater: Source and Variation Pressure and Volume Chemical Characteristics

Some one of these parameters can be expressed differently. For example the number and frequency of joint sets can be expressed as Rock Quality Designation (RQD), joint density, or joint spacing. Likewise, the ground water can be reported in different terms depending on the geotechnical tests performed. This includes water pressure, flow rate, and permeability and so on. Nonetheless, these different expressions are interrelated and there are some formulae defining the relationship between them (Afrouz, 1992). The rating assigned to these parameters depend on the scale and arithmetic used in the specific classifications. Details of the rating of joint properties for each classification system are given in the literature and are not discussed here.

BASIS AND EXAMPLES OF ROCK MASS CLASSIFICATIONS

Rock mass classifications are algebraic sum or multiplication of some ratings assigned to the rock properties and the joint characteristics. Added to these factors is the state of in situ stresses in the ground surrounding the structure. This parameter is represented by the term such as Stress Reduction Factor (SRF). Table 3. was developed by Barton et. al. (1980) in the Norwegian Geotechnical Institute (NGI) to account for the in situ stresses and their impact on behavior of the rock mass and the excavation.

Table 3. Values of SRF for different ground conditions (after Barton et. al. 1980)

Category	Ground Condition	SRF
	I. Weakness zones intersecting excavation, which may cause loosening of rock mass when tunnel is excavated	
A	Multiple occurrences of weakness zones containing clay or chemically disintegrated rock, very loose surrounding rock (any depth)	10
B	Single weakness zones containing clay or chemically disintegrated rock (depth of excavation <50 m)	5
C	Single weakness zones containing clay or chemically disintegrated rock (depth of excavation >50 m)	2.5
D	Multiple shear zones in competent rock (clay-free), loose surrounding rock (any depth)	7.5
E	Single shear zones in competent rock (clay-free) (depth of excavation <50 m)	5.0
F	Single shear zones in competent rock (clay-free) (depth of excavation >50 m)	2.5
G	Loose open joints, heavily jointed or "sugar cube" etc. (any depth)	5.0

Note: (1) Reduce these values of SRF by 25-50% if the relevant shear zones only influence but do not intersect the excavation.

II. Competent rock, rock stress problems

	$\sigma c/\sigma 1$	$\sigma t/\sigma 1$	*(SRF)*
H. Low stress, near surface	>200	>13	2.5
J. Medium stress	200-10	13-0.66	1.0
K. High stress, tight structure	10-5	0.66-0.33	0.5-2
L. Mild rock burst (massive rock)	5-2.5	0.33-0.16	5-10
M. Heavy rock burst (massive rock)	<2.5	<0.16	10-20

Note: (2) For strongly anisotropic virgin stress field (if measured): when $4 \geq \sigma_1/\sigma_{2 \geq 10,}$ reduce σ_c and σ_1 to 0.8 σ_c and 0.8 σ_t. When $\sigma_1/\sigma_3 > 10$, reduce σ_c and σ_t to 0.6 σ_c and 0.6 σ_t, where: σ_c = unconfined compression strength, and σ_t = tensile strength (point load) and σ_1 and σ_3 are the major and minor principal stresses.

Note: (3) Few case records available where depth of crown below surface is less than span width. Suggest SRF increase from 2.5 to 5 for such cases (see H).

III. Squeezing rock plastic flow of incompetent rock under the influence of high rock pressure

N. Mild squeezing rock pressure	5-10
O. Heavy squeezing rock pressure	10-20

IV. Swelling rock chemical swelling activity depending on presence of water

P. Mild swelling rock pressure	5-10
R. Heavy swelling rock pressure	10-15

In addition, the type, dimensions, application, and service life of the structure have an impact on the classification systems. This parameter is represented by Excavation to Support Ratio (ESR), (Barton et. al., 1974). Table 4 shows the ESR values for some typical applications.

The main rock mass classifications published to date are listed in Table 5. The limitation for application of each classification system depends on the conditions they were developed. There are classification systems specially developed for mining applications and there are some others specially fitted for civil constructions. Most the classifications were based on the data and the experiences in underground excavations and were later adapted to surface excavations.

Table 4. Values of ESR for different applications (Barton et. al. 1974).

Category	Type of excavation	ESR	No. of cases
A	Temporary mine openings, etc.	3–5	2
B	Vertical shafts:		
	(I) Circular section	2.5	1
	(ii) Rectangular/square section	2.0	1
C	Permanent mine openings, water tunnels for hydropower (exclude high pressure penstocks), pilot tunnels, drifts and headings for large excavations, etc.	1.6	83
D	Storage rooms, water treatment plants, minor road and railway tunnels, surge chambers, access tunnels, etc. (hemispherical caverns?)	1.3	25
E	Power stations, major road and railway tunnels, civil defense chambers, portals, intersections, etc.	1.0	79
F	Underground nuclear power stations, railway stations, sports and public facilities factories, etc.	0.8	2

Table 5. The Main Rock Mass Classification Systems (after Afrouz, 1992).

1. Terzagi 2. Protodiaganov 3. Classification of Degree of Weathering	11. Laubscher's Adjusted RMR rating (AR) 12. Rocha's Quality of Rock Mass (MR) 13. Wickham et. al. Rock Structure Rating (RSR)
4. Classifi cation of Discontinuities 5. Deere's Rock Quality Designation (RQD) 6. Barton et. al. Rock Mass Quality (NGI or Q/System) 7. Rock Quality Index (QI) 8. Strength Properties of Rock Mass 9. Rock Mass Factor (j) 10. Bieniawski's Rock Mass Rating (RMR)	14. Mathew's Rock Mass Classification (Adjusted Q for block caving evaluation-CANMET) 15. Brook et. al. Simplified Rock Mass Rating (R) 16. Palmstrom's Rock Mass Index (RMi) 17. Rock Mass Deformability

In general, the rock masses are divided into several categories by the classification systems based on the estimated rating in each classification. This classes range from extremely weak to moderate, good and extremely good or excellent ground conditions. Table 6 shows a comparison of rock mass classification based on the RQD, RMR and NGI (Q) systems of rating. The classifications offered by different systems are often similar in the category of the ground or rock mass, but their numerical value of rating is different. This is because the controlling parameters are almost identical but was given different weight due to various grounds encountered. Therefore, the ratings in one classification system can be translated to or estimated from another system. Formulae exist to define the relationship between different rock mass classification ratings and can be found in the literature (Afrouz, 1992). Figure 1 shows the plot of RMR versus Q values from NGI system for some cases and the conversion formulae.

Table 6. Comparison of the RQD, RMR and NGI (Q system) of Rating (Modified From Afrouz, 1992).

Rock Mass Classifications		Rating		
RQD%	RMR	RQD System	Q System	Condition of the Rock Mass or Ground
-	-	-	0.001-0.01	Exceptionally Poor
0-1	-	<0.1	0.01-0.1	Extremely Poor
1-25	0-20	3	0.1-1	Very Poor
25-50	21-40	8	1-4	Poor - Weak
51-75	41-60	13	4-10	Fair - Moderate
76-90	61-80	17	10-40	Good
91-100	81-100	20	40-100	Very Good
-	-	-	100-400	Extremely Good
-	-	-	400-1000	Exceptionally Good

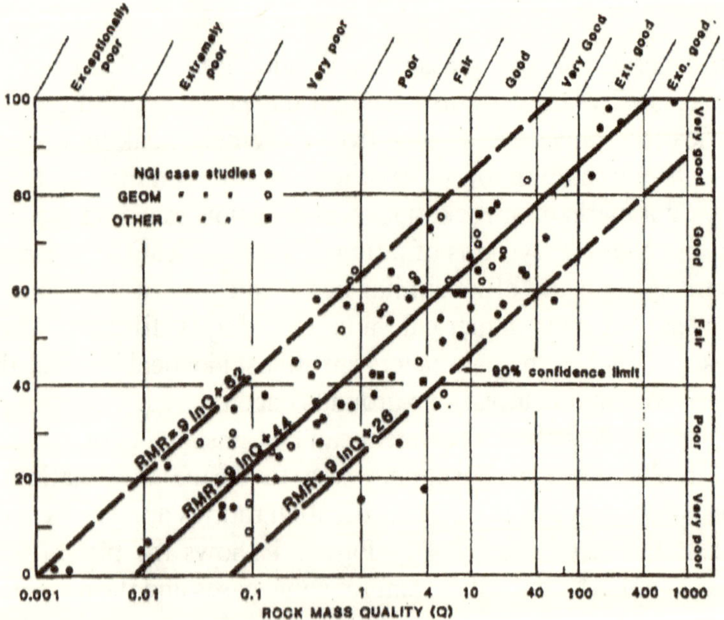

Figure 1. The relationship between RMR and Q ratings (Bieniawski, 1979).

ESTIMATION OF STAND UP TIME FROM ROCK MASS CLASSIFICATION

The classification systems typically provide an estimation of the stand up time of the excavation depending on the size of the opening and the ground conditions. Stand up time represents the period of time that the excavation is supported by the rock without any need for additional ground support or reinforcement. Figure 2 shows the estimated span and stand up time for different rock classes offered by NGI system. If the stand up time is longer than the service life of the excavation, it means that no ground support is required and the excavation is self-supporting.

Several graphs have been developed by researchers to show the stand up time for different classes of rock mass as a function of the effective excavation dimensions. An example for this is shown in Figure 3, which covers very weak to very strong rocks.

Rock mass classifications often offer some measures for estimation of the load on the support system. This is accomplished by providing formulae to estimate the size of the affected zone and the load of the fractured zone exerted on the support system. The size and extent of the broken ground depends on the dimensions of the excavation, type of the rock mass, method of excavation and its utilization with time. The method of excavation has an important effect on this parameter. Structures excavated by mechanized methods, as compared with drill and blast operation, tend to have smaller fractured zone and broken ground. Therefore, the magnitude of the load and the support requirements for such an excavation is less. This phenomenon is more pronounced as the size of the excavation gets larger. Figure 4 shows the adjustment factor (to be multiplied) for machine excavated openings as a function of the excavation dimension in RSR system.

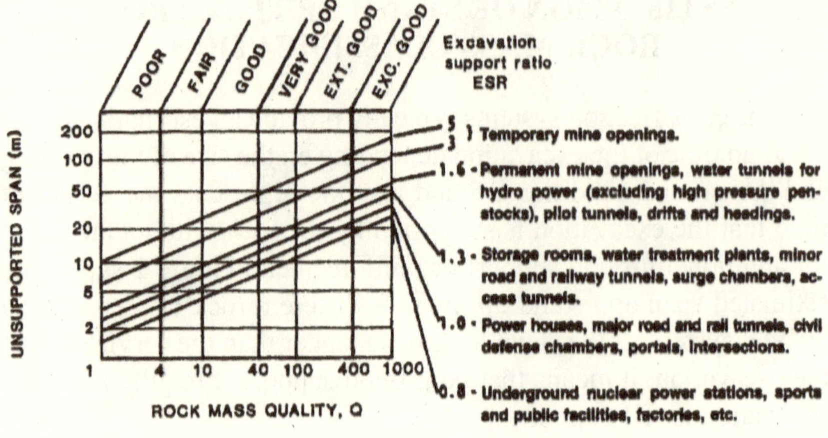

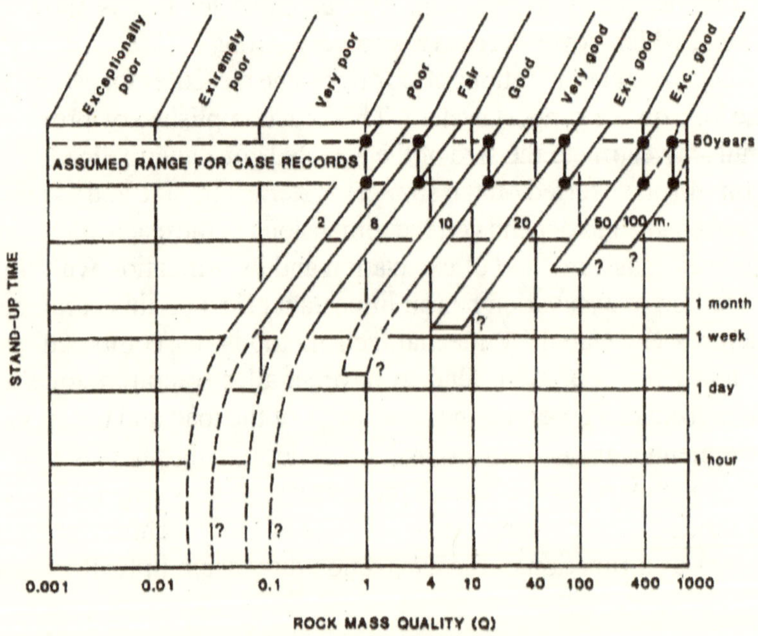

Figures 2. (a) Unsupported span and (b) Stand up time for different rock mass classes according to NGI system (Barton et. al 1980).

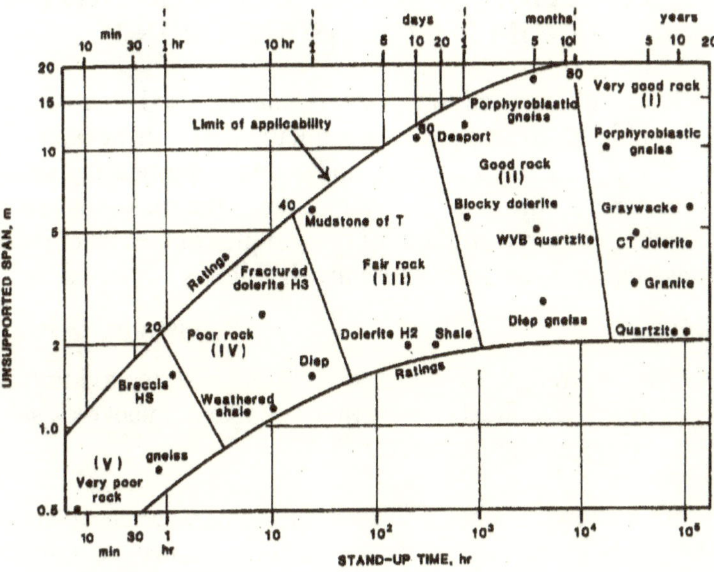

Figure 3. Geomechanics classification of rock masses or rock mass rating (RMR) applied to predicting Tunneling conditions (Bieniawski, 1979).

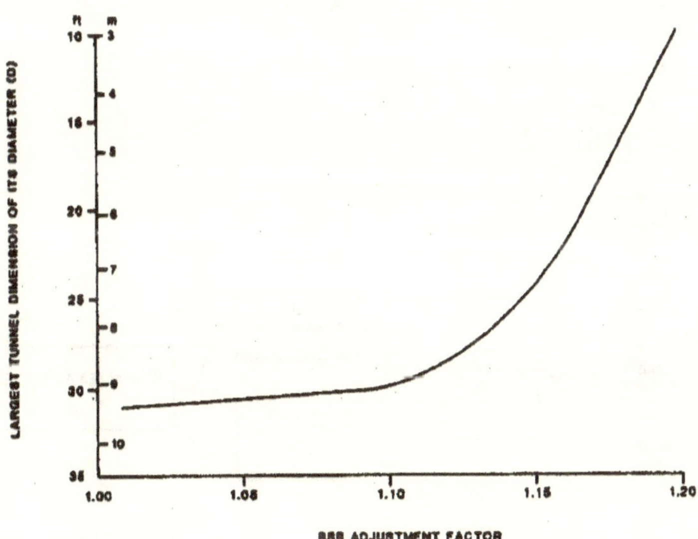

Figure 4. The adjustment factor (multiplier) for machine excavated openings in Rock Structure Rating (RSR) system (Wickham et. al. 1974).

ESTIMATION OF ROCK MASS STRENGTH FROM ITS CLASSIFICATION

Rock mass classification systems can be used to estimate its strength. This is an important parameter that can be applied for design of the structures in rock and related engineering analysis. There are several formulae for estimation of the rock mass strength based on the measured strength of the samples of the intact rock and rock mass features. Some of these are as follows:

$$RMi = \sigma_c \cdot JP \qquad \text{(Palmstrom, 1996)}$$

Where: Rmi = strength of the rock mass reduced due to jointing
σ_c = unconfined compressive strength of intact rock samples
JP = jointing parameter

Figure 5 shows a schematic view of the parameters influencing the RMi. This graph is very typical of the rock mass classification systems. Brooks et. al. (1985) offered the following equation as a simplified rock mass rating:

$$R = IRS + Js + Jt + GW$$

Where: R = Simplified or total rock mass rating
IRS = strength of intact rock = UCS
Js = jointing spacing
Jt = Joint type
GW = Ground water rating.

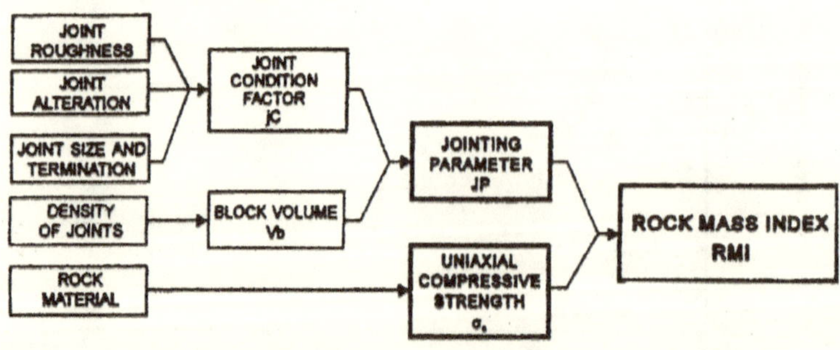

Figure 5. Factors influencing rock mass strength. (Palmstrom, 1996)

Adjustments for the in situ rating for R-value are shown in Table 7 and comparison of various classifications for strength of intact rock is given in Table 8.

Table 7. Adjustment Factors for In situ Rating Components for R-Value (Brook et. al. 1985).

In Situ Conditions	IRS	Js	Jt
Weathering:	-	-	
Slight	0.90	-	0.90
Moderate	0.75	-	0.75
High	0.5	-	0.50
Orientation of Excavation:			
Slightly Unfavorable	-	0.90	0.90
Moderately Unfavorable	-	0.75	0.75
Highly Unfavorable	-	0.50	0.50
In Situ or Induced Stress for IRS:			
Causing slight Failure	0.90	-	0.90
Moderate Failure	0.75	-	0.75
General Failure	0.50	-	0.50
Blasting:			
Smooth	-	0.90	0.90
Moderate	-	0.75	0.75
Rough	-	0.50	0.50

Note: For detailed practical information refer to Afrouz, 1992, pp 78-80.

Hoek and Brown (1988) simplified equations for failure criteria of the rock mass has widely been accepted and used to estimate rock mass behavior. These formulae allow for estimation of the maximum principal stress as a function of the confining stresses and some empirically developed factors ("s" and "m"). These factors are determined from the rock mass rating offered by the classification systems. It can also be used to estimate the angle of internal friction, "Φ", of the rock mass. This can be utilized to estimate the shear strength of the rock mass under given set of principle in situ stresses. Other rock mass properties such as modulus of deformation and elastic properties of the rock mass can also be estimated from the rock mass classification ratings using empirically developed equations. One such equations is given as follows:

Rock Mass Deformability, $E = 10^{[(RMR - 10)/40]}$ (Sarafim et. al. 1983)

Table 8. Comparison of Various Classifications for Strength (σ_c) of intact rock (Bieniawski, 1979).

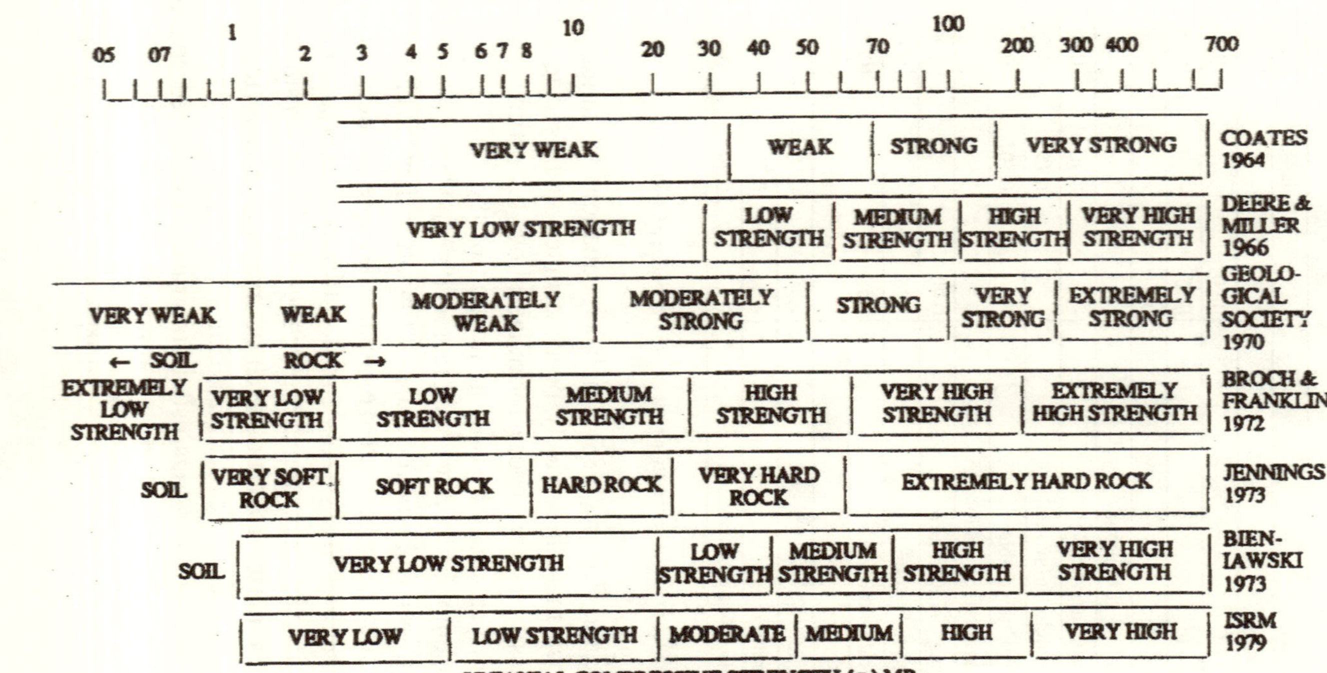

Scale (MPa): 05, 07, 1, 2, 3, 4, 5, 6, 7, 8, 10, 20, 30, 40, 50, 70, 100, 200, 300, 400, 700

Classification	Classes (increasing strength, left to right)
COATES 1964	VERY WEAK · WEAK · STRONG · VERY STRONG
DEERE & MILLER 1966	VERY LOW STRENGTH · LOW STRENGTH · MEDIUM STRENGTH · HIGH STRENGTH · VERY HIGH STRENGTH
GEOLOGICAL SOCIETY 1970	VERY WEAK · WEAK · MODERATELY WEAK · MODERATELY STRONG · STRONG · VERY STRONG · EXTREMELY STRONG (← SOIL / ROCK →)
BROCH & FRANKLIN 1972	EXTREMELY LOW STRENGTH · VERY LOW STRENGTH · LOW STRENGTH · MEDIUM STRENGTH · HIGH STRENGTH · VERY HIGH STRENGTH · EXTREMELY HIGH STRENGTH
JENNINGS 1973	SOIL · VERY SOFT ROCK · SOFT ROCK · HARD ROCK · VERY HARD ROCK · EXTREMELY HARD ROCK
BIENIAWSKI 1973	SOIL · VERY LOW STRENGTH · LOW STRENGTH · MEDIUM STRENGTH · HIGH STRENGTH · VERY HIGH STRENGTH
ISRM 1979	VERY LOW · LOW STRENGTH · MODERATE · MEDIUM · HIGH · VERY HIGH

UNIAXIAL COMPRESSIVE STRENGTH (σ_c) MPa

Note: 1 MPa= 145 lbf/in.^{2}

These parameters are essential in any type of modeling and numerical analysis of the structure. The shear strength of rock mass also can be estimated from the classification rating. For example the shear strength of the rock mass featuring planar unfilled discontinuity can be estimated from the Mohr-Culomb or Barton formulae. Hoek and Brady have proposed equations for estimation of the shear strength of rock mass with smooth inclined unfilled joints. Also, equations have been given for estimation of the shear strength of rock mass with rough joint surfaces by Ladany and Archambolt, Hoek and Brown (using s and m), Barton, and Mohr-Culomb. Shear strength of the discontinuities which are filled are a function of the filling material and its properties. The estimated strength of the rock mass can then be utilized to examine the stability of the excavation with a given shape under the in situ stress conditions (see Afrouz 1992 for more details).

ESTIMATION OF SUPPORT REQUIREMENTS FROM ROCK MASS CLASSIFICATION

Most of rock mass classification systems offer suggestions for ground support requirement. These suggestions are based on the past experience in a number of case studies and are in the form of system of support for different ground conditions, excavation and dimensions for various applications. Figure 6 shows a typical ground support chart suggested by the NGI system.

The full description of the support systems can be found in the literature. Some additional information is also available for support design through developing the ground and support reaction curves. This curve represents the ground loading as a function of the deformation or closure of the opening. These curves can be used to predict the ground behavior over the time and choose the appropriate type of support to attain the required factor of safety. The type, density, and time of installation of the ground support measures can be optimized for each individual site. Figure 7 shows a typical ground reaction curve. Some of the rock mass properties and strength values estimated by the above noted formulae are used to develop the ground reaction curves.

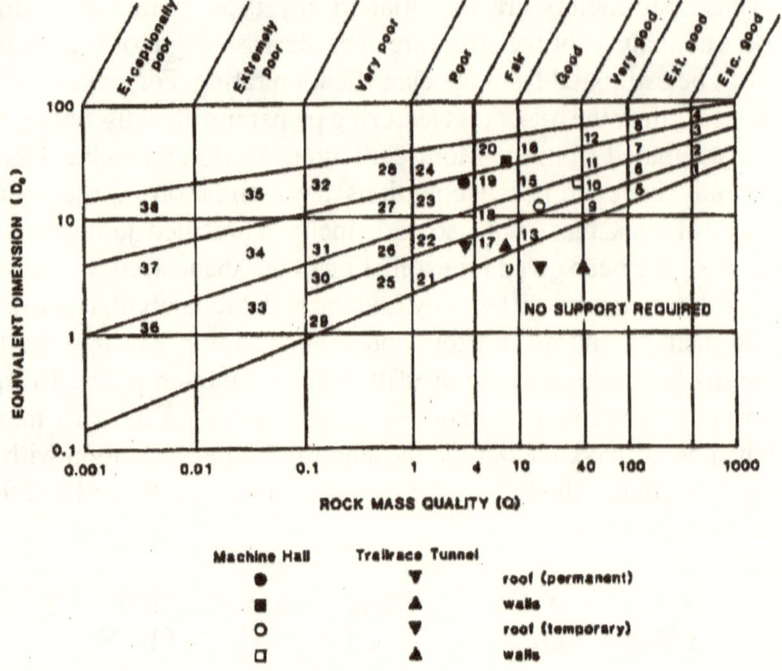

Figure 6. Support systems suggested by NGI system for various Q ratings and excavation dimensions (See Chapter 1 of this book or Barton, 1980 for more details).

SENSITIVITY OF ROCK CLASSIFICATION

If an error is made in the estimation of the rock mass quality, hopefully the combination of the favorable and the unfavorable judgments will cancel each other and minimize the impact on the final results. One of the most serious errors is failure to anticipate a clay-filled weakness zone. This might have a significant effect on the results specially if the clay is of the swelling variety. The difference in the rock quality index might reach up to 100 times. In such cases the ground support measures have to be redesigned and upgraded to meet the requirements of the lower rock quality. This is likely to happen in weak rocks and strong formations with clay parting, especially when time dependent behavior of the formations containing clay is not quantified. This would impact almost all the factors used in determination of the index.

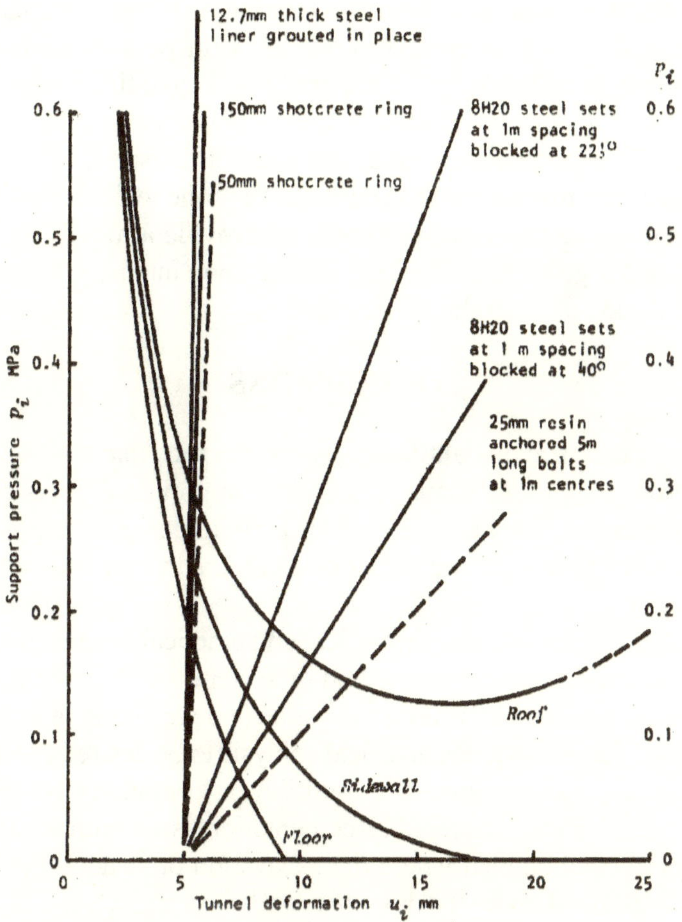

Figure 7. An example of a ground and support reaction curve (Hoek & Brown, 1988)

SPECIAL RECOMMENDATION FOR APPLICATION OF THE ROCK MASS CLASSIFICATION TO WEAK AND WEATHERED ROCK MASSES

Weak and weathered rock masses typically involve finely bedded sedimentary rocks, washout, highly weathered igneous or metamorphic rocks, clay bearing rocks, or clay filling in the discontinuities of the rock mass. Previously noted suggestions pertaining to clay bearing rock masses apply to these conditions. In addition, the provisions for performing physical and mechanical property test and measurements under in situ moisture and saturated conditions is recommended for weak rocks. An important part of dealing with weak rocks is study of the long term behavior of the mass and quantification of the time dependent characteristics of the components and the overall rock mass. Under dynamic conditions, such as blasting, vibration of equipment and machinery or in earthquake prone regions, rock mass rating should be reduced by 30-60% to provide a higher factor of safety for design of structures, depending upon intensity, frequency, and amplitude of the induced vibration.

CONCLUSIONS

Qualification and quantification of in situ rock mass in weak and weathered ground are some of the most important aspects of site characterization. They are utilized for input to design of excavation or structure in the rock and its support pertinent to the utilization of such activity.

With all the impressive theoretical and practical advancements in this field, there is no single method or system which could provide 100% accuracy in rock mass classification and prediction of its timely behavior. More knowledge and field analysis to be desired especially where weak and weathered or multilayer formations containing soft, poor or water susceptible cementation is encountered. This problem is compounded where the excavation or structure in weak ground is utilized under the dynamic conditions and when budgetary restrictions for detailed investigation apply.

To overcome this complexity, cross investigation with two or more of the pertinent classification systems together with keen field observations may go a long way to obtaining fairly accurate and representative results.

REFERENCES

Afrouz, A.A., 1992, "Practical Handbook of Rock Mass Classifications and Modes of Ground Failure". CRC Press, Florida, p-195.

Barton N.R., Lien, R., & Lunde, J., 1974, "Engineering Classification of Rock Masses for Design of Tunnel Supports, Rock mechanics, Vol. 6. pp 189-236.

Barton N.R., Loset F., Lien, R., & Lunde, J., 1980, "Application of Q-System in Design Decisions Concerning Dimensions and Appropriate Support for Underground Installations", SubSurface Space, 2, pp 553-561.

Bieniawski, Z.T., 1979, "Rock Classifications: State of the Art and Prospects for Standardization", S. A. Transport Research Record, 783, South Africa, pp 2-9.

Bieniawski, Z.T.,, 1979, "Tunnel Design by Rock Mass Classification", U.S. Army Waterways Experiment Station, Technical Report No. GL-78-19, p131.

Brook, N., and Dharmaratne, P.G.R., 1985, "Simplified Rock Mass rating System For Mine Tunnel Support", Trans. Inst. Mine & Met., London, U.K., pp A-148-A152.

Deere, D., 1968, "Geological Considerations, Rock Mechanical in Engineering Practice", J. Wiley, London, pp. 1-19.

Hoek E., Brown, E.T., 1988, "Updated Failure Criterion, Rock Engineering for Underground Excavations". 15th Canadian Rock Mechanics Symposium, Univ. of Toronto, pp.31-38.

Palmstrom, A., 1996, "Characterizing Rock Masses by RMi for Use in Practical Rock Engineering" Part 1, Tunneling and Underground Space Technology, Vol. 11, No. 2, pp 175-188. (also Part 2 in Vol. 9, No. 3, pp 287-303).

Sarafim, J.L. and Pereria, J.P., 1983, "Considerations of Geomechanics Classification of Bieniawski", Proc. Int. Symp. Eng. Geol. & Underground Construction, Lisbon, Portugal, pp. 1133-1144.

Terzagi, K. 1943. "Theoretical Soil Mechanics", J. Wiley. U.S.

Wickham, G.E., Tidemann, H.R., & Skinner, E.H., 1974, "Ground Support Prediction Model- RSR Concept", Proc. Rapid Excavation & Tunneling Conference. 1974, SME-AIME, New York pp 691-707.

3.2 BLAST DESIGN

3.2.1 AN INVESTIGATION INTO ROCK BREAKING BY DIRECT IMPACT

By: Andy A. Afrouz and F.P. Hassani

ABSTRACT

An empirical approach towards analysis of rock breaking under direct impact and laboratory conditions is suggested utilizing limestone and concrete blocks. Characteristics and mechanism of rupture in the test blocks are discussed. It is found that the effect of reflected shock waves is less significant than that of the stress relief in rupturing the rocks at the potential fracture zone. The test variables are defined and their relationships are expressed mathematically.

1. INTRODUCTION

In the nineteenth century, the fracture mechanism of rock with explosives was thought to be due to expansion of the confined products of detonation. About 1909, A.W. Daw and Z.W. Daw postulated that the mechanism of failure is by shear [1]. The Daws' experiment was performed in quarries using one shot hole which led to the empirical relationship:

$$Q = C \cdot V \tag{1}$$

where: Q = explosive weight;
C = charge or blasting coefficient (g/cm^3); and
V = fractured volume of rock (cm^3).

Daws then supplemented the above expression by loading blocks of ice under laboratory conditions and suggested:

$$F = K \cdot S \cdot W \tag{2}$$

where: F = impact force at rupture (kN)

K = proportionality constant (kN/mm^2)
S = periphery of the impact hole = π. D *(mm)*
$\pi = 3.14$
D = diameter of the impact hole (mm); and
W = thickness of the block or burden (mm).

Investigations [2] into breaking coal by wedge action suggested the following empirical relationship:

$$F \propto d^a \cdot S^b \tag{3}$$

where: F = mean grooving force
d = depth of cut
S = size of scale of the wedge
a = constant varying between 1 and 1.35; and
b = constant varying between 0.1 and 0.8.

Saluja, in his research conducted in collaboration with Clark [3], has shown that the equations (2) and (3) are compatible for rocks in the form:

$$F = K \cdot S^a \cdot W^b \tag{4}$$

where K, a and b are constants dependent on the type of rupture and physical properties of rock.

Pariseau and Fairhurst [4], in a theoretical plasticity analysis to characterize force-penetration of wedges into rock, reached the following conclusion:

$$F = [(bh\sigma_0 . \tan\beta)/(\tan\phi \,.\, \tan\mu)]\{[1 + \sin\phi\,(\cot\beta \,.\, \tan\mu - 1)]. \\ \exp(2\beta \,.\, \tan\phi - \tan^2\mu)\} \tag{5}$$

Where: F = Force on the bit;
b = width of the bit cutting edge;
h = depth of penetration;
σ_0 = unconfined compressive strength of rock;
β = half-wedge angle of the bit ($0 \leq \beta \leq \pi/2$);
ϕ = angle of internal friction of the rock; and
$\mu = (\pi/4 - \phi/2)$.

However, the above expression does not make a satisfactorily close fit to the laboratory and in situ measurements, necessitating further research. There are numerous other published investigations [5–15] relating the crack propagation, compressive strength of rocks, impact velocity, penetration and size of broken rocks under various dynamic loadings. Some have substantiated the validity of equation (4) for various rock types and dynamic loadings [5-10]. However, experiments to measure and to relate accurately the amount of impact force at rock breaking point to the mechanical properties of the rock, as well as to find the meaning, magnitude and definition of the constants involved in the relevant expressions, are much to be desired.

The present paper is mainly concerned with:

(a) The characteristic and mechanism of rupture on blocks of rocks made of two different constituents with two different dimensions;
(b) The effect of reflected shock waves initiated by the impact; and
(c) The relationship of the test variables to the volume and exposed area of rupture.

The basic application of this investigation is in drilling, blasting, coal cutting and mechanized impact tunneling machines.

2. SAMPLE PREPARATION

Two types of rectangular samples were selected, as follows:

(a) Concrete blocks as artificial samples, the physico-mechanical properties and geometry of which can be controlled at room temperature (20 ± 1°C) and atmospheric pressure;
(b) Limestone blocks as natural massive samples with no visible joints, pronounced bedding structure within the samples, and with controlled geometry, tested under the same environmental conditions as noted above.

Concrete blocks

The concrete blocks were made of a 3.6 : 2 : 1 mixture, by weight, of arrowhead-screened foundry sand, cement and water, respectively. The mix was poured slowly into wooden frames of internal dimensions 175 × 175 × 100 mm. The concrete thus obtained had a dynamic tensile strength of 4137 kPa after a total of 4 months setting and curing time. Holes were drilled on the face center of the blocks by a carbide-tipped masonry drilling bit with an Asarco powermatic drilling machine, and using an air-powered drill-press feed to ensure a constant rate of pressure on the bit. To ensure uniform loading during the test, the bottom of each hole was made smooth by means of a drill ground flat at the tip. The holes were made such as to leave burden W between the hole bottom and free face of the block. The burden W was 12.5–25.0 mm, covering all sizes within the range. The diameter of the hole was varied within the range 15.6–25.0 mm, and almost all possible sizes were included within these limits.

Limestone blocks

The rock samples were cut by a wire saw to form blocks of dimensions 450 × 450 × 150 mm. The average dynamic tensile strength of the blocks was 6205.5 kPa. The hole drilling was similar to that described for concrete blocks, with diameters of 15.6 and 25.0 mm. The larger ranges of 37.5 and 50.0 mm diameter holes were drilled utilizing clipper diamond core bits. In the latter case, a small water pump circulated water to the face of the drilling bit as a coolant and lubricant. The holes were made to leave smooth and flat burdens W of 12.5–87.5 mm including almost all sizes within the range.

3. INSTRUMENTATION

To simulate the actual dynamic impact loading conditions, a testing machine was designed on the drop-weight principle. The weight was raised by means of 0.250 HP electric motor through a rope drive, which then released automatically on hitting the adjustable stop which could be set at any desired drop-height position. In this way, the required impact force could be obtained by means of changing the height of drop to suit the condition for rupturing the rock.

Water-hardened tool-steel cylindrical rods, grade W4 steel, were utilized, having the following specification:

C = 1.00; Mn = 0.30; Si = 0.50; P = 0.02;
S = 0.02; Cr = 0.25; and Ni = 0.05.

The impact pins were 175.0 mm long, with diameters from 15.6 to 50.0 mm to close-fit the holes drilled in the testing blocks.

In order to attach two strain gauges longitudinally on the mid-length of each pin for strain measurements, two opposite sides of the pins were carved to provide surfaces and the necessary room for the strain gauges to be mounted, forming a Wheatstone Bridge with two external dummy gauges, as shown in Fig. 1. The modulus of elasticity of the pins thus made was 203.75×10^6 kPa.

Electrical resistance SR-4 wire gauges, with gauge factor (G.R.) of $2.00 \pm 2\%$, and an effective strain-gauge area of the 20.0×5.0 mm, were mounted on the pins and on the rock samples using Eastman-910 cement.

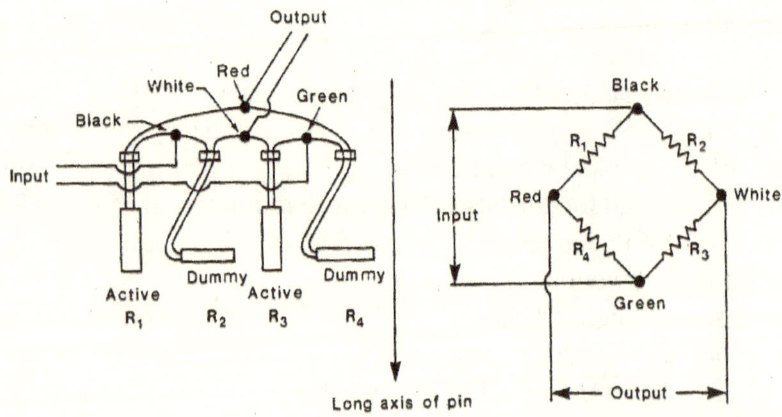

Fig. 1. Gauge circuit forming a Wheatstone Bridge around the impact pin.

For dynamic strain gauge applications, a dual-beam oscilloscope was used in conjunction with two Q-type plug-in units. The *X*-axis on the oscilloscope screen represented the time base, while the voltage change from the Wheatstone Bridge was displayed on the *Y*-axis, which was calibrated to represent the strain changes in the strain gauges thus utilized. Hence, the strain gauge record, showing strain as a function of time, was obtained by photographing the oscilloscope display.

4. TEST PROCEDURE

Three strain gauges were mounted at the base of the test blocks (see Fig. 2) after sanding and cleaning the selected regions with acetone. This was done at least 24 hours prior to the test.

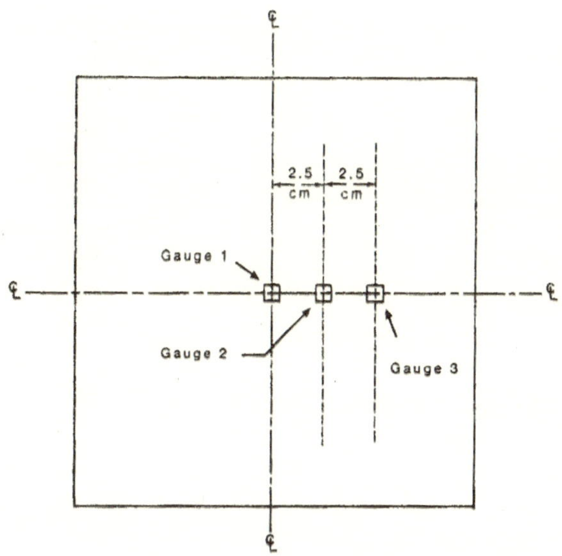

Fig. 2. Strain gauge arrangements on bottom of the test blocks.

The test blocks were supported along all four sides of their bases, with a 37.5 mm margin, using four steel I-sections as a frame on which to place the blocks, as illustrated in Fig. 3(a). This arrangement offered a free face at the bottom of the blocks, allowing 37.5 mm clearance above the base of the impact testing machine. The impact pin was placed in the hole and centralized with respect to the cylindrical impact weight attached to the cross-head of the testing machine, as shown in Fig. 3(b).

The dynamic recording circuit was so arranged as to trigger the circuit when the impact weight was released from the predetermined height. In this process, signals from the strain gauges mounted on the impact pins and on the bottom of the blocks were transmitted via two Q-type plug-in units to the CA-type plug-in unit incorporated within the oscilloscope. The circuit arrangement is shown in Fig. 4. The assembly was calibrated every time before testing, and a calibration factor was obtained [8].

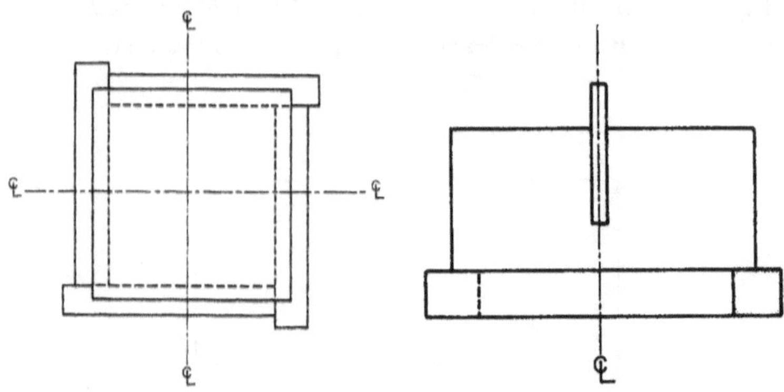

(a) Plan View (b) Elevation

Fig. 3. Method of supporting the bottoms of the blocks.

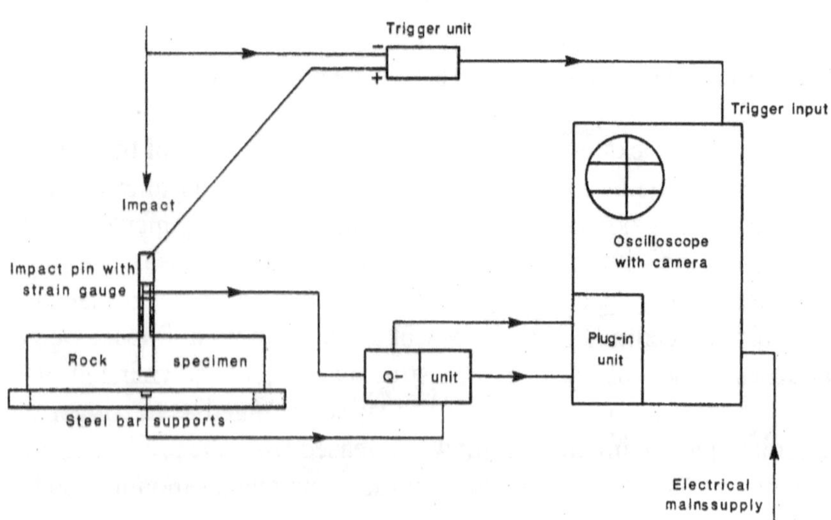

Fig. 4. Block diagram of the strain recording circuit utilized.

A Polaroid camera, fitted with high-speed 300/type-47 film and set for 2 seconds exposure, was mounted on the oscilloscope to record the signal corresponding to strains caused by the impact at the point of application of the strain gauge. A typical strain–time record thus obtained is illustrated in Fig. 5. The strain-time graph were then interpreted using the preset scanning scale of the oscilloscope. It was then corrected by its calibration factor. The outcome indicated the actual strain change at the point of interest.

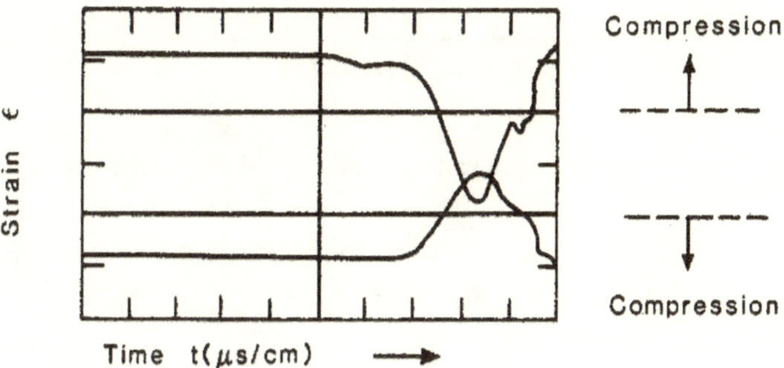

Fig. 5. A typical dynamic strain/time record obtained.

Knowing the actual strains ϵ under any particular set of conditions, the corresponding stress σ and force F to cause rupture in the block is determined by use of equations:

$$\sigma = E \cdot \epsilon \text{ and } F = \sigma \cdot A \qquad (6)$$

Where A = cross-sectional area of the pin.

5. ANALYSIS AND DISCUSSION OF RESULTS

5.1 Characteristics of rupture

Rupture of the blocks under direct impact appeared to be always conical (Fig. 6).

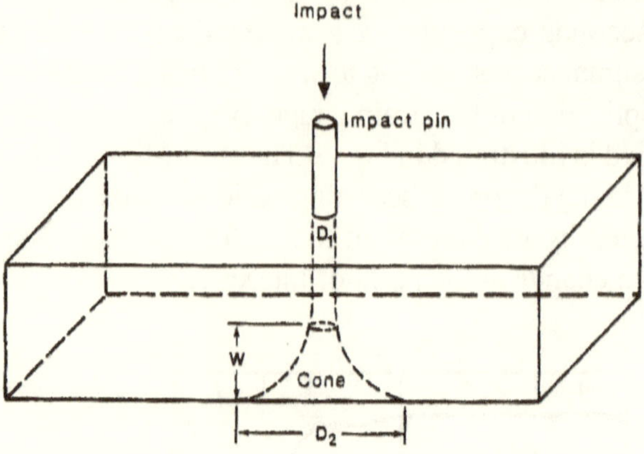

Fig. 6. Schematic diagram of the conical rupture produced in the laboratory.

The apex of the cone was directly beneath the impact pin. The conical rupture is initiated from the periphery of the pin base, and progresses downwards and radially outwards towards the bottom of the blocks under test. The slope of the fractured cone increases non-linearly with an increase in burden *W*, impact-hole diameter *D* and/ or impact-hole periphery *S*. The volume of rupture *V* is non-linearly, but directly, proportional to the burden *W* and hole diameter *D*, as illustrated in Fig. 7.

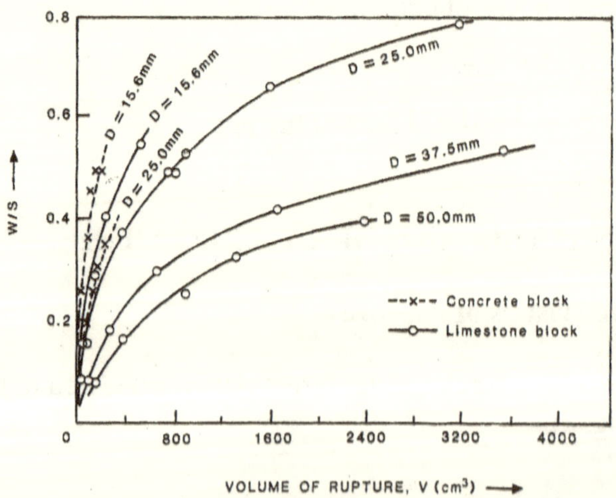

Fig. 7. Relationship between the ratio *W/S* and volume *V* of rupture.

5.2 Mechanism of rupture

It is believed that the effective energy imposed by the direct impact on the pin is mainly a result of two kinds of stress waves:

(1) Shear waves—acting outwards with resulting radial fracture;
(2) Longitudinal waves—acting downwards and creating tensile fracture.

The strain development caused by the impact, as measured on the rock surface, consisted of three stages. In the order of occurrence, these were:

(1) Rapidly increasing strain (Fig. 5);
(2) Peak strain coinciding with initiation of rupture;
(3) Gradual strain relief, longer in duration than that of the stage (1).

Correlation of the readings obtained from sets of three strain gauges placed one inch (2.5 cm) apart at the bottom of the limestone blocks (Fig. 2) indicates that the rock undergoes bending during the impact [8]. This suggests that the rock failure may be due to high-velocity compression leading to tensile stress, originating from stress concentration at micro-cracks which may be present in the rock before the impact.

Localizing of strain energy at the point of impact may quickly initiate fractures that would propagate through the impacted material along with the traveling stress waves. As the cracks form, the initially absorbed strain energy would be released and parts of it would travel ahead of the crack as stress waves derived from the impact, and would produce high localized strain conditions at the potential fracture zone. The potential fracture zone would propagate and the specimen fail.

5.3 Effect of the reflected shock waves

For a given rock type, the distance covered by the imposed shock wave in traveling over the free surface of the rock and then being reflected back at some angle, is directly proportional to the time for the reflected wave to reach the potential conical fracture zone, and is inversely proportional to the strain energy of the reflected wave. This was measured in the test blocks, which indicated that the reflected waves alone had little influence in causing fragmentation along the potential fracture zone (Fig. 8).

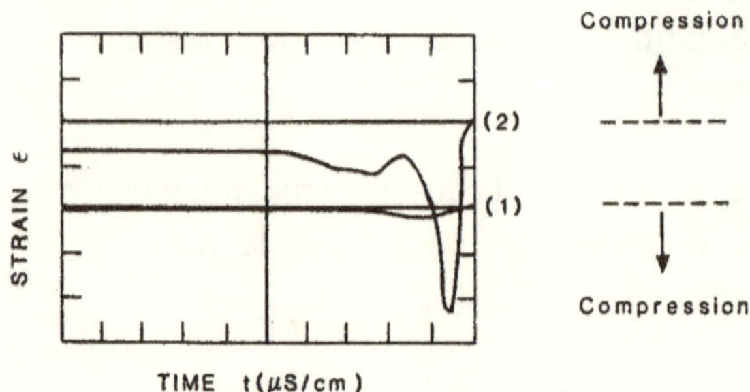

Fig. 8. Comparison between strain recorded at the side of a limestone block (1) with that at the center bottom gauge (2).

Theoretical scale-model studies on reflected shock waves in the test blocks, drawn to each extreme case of maximum and minimum diameter D of the impact pin used against thickest and thinnest burden W utilized, also indicated the above noted effect [8]. This suggests that the phenomena of spalling in rocks may be due mainly to the effect of stress relief rather than reflection of impact stress.

5.4 Relationship amongst the test variables:

5.4.1. The strain development on the impact pin at rupture ϵ_p was directly proportional to the burden W and indirectly proportional to the periphery of the impact hole, $S = \pi \cdot D$, as shown in Fig. 9. This was further supported by the linear relationship between the dimensionless ratios of W/S versus ϵ_p/ϵ_g expressed in the following form for the Valders limestone blocks:

$$W.\epsilon_g = 2.23D.\epsilon_p \tag{7}$$

Where ϵ_g is the strain reading at the time of rupture from the gauge mounted on the rock burden below the impact pin.

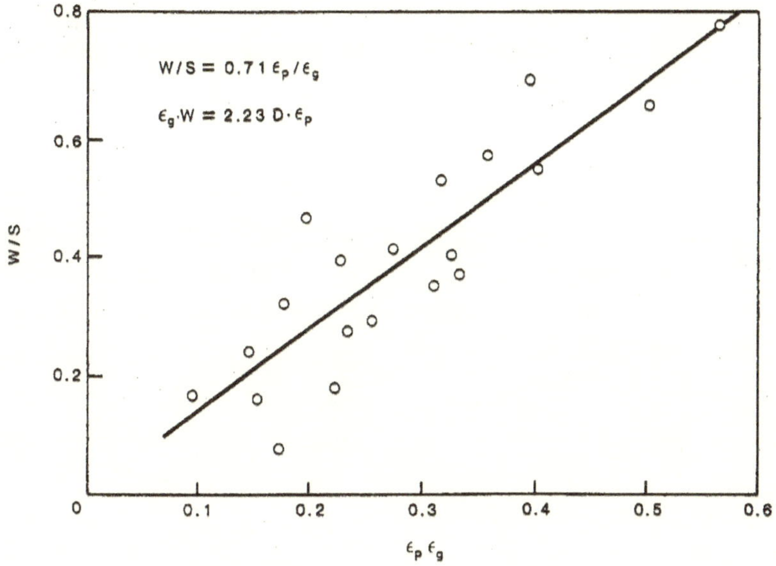

Fig. 9. Linear relationship between dimensionless ratios of ϵ_p/ϵ_g and *W/S* for the limestone.

5.4.2 The impact force at rupture *F* is the total force applied and is not an absolute quantity. It is measured from the peak strain ($F = E \cdot \epsilon_p \cdot A$) and is therefore a peak force. As the force is linearly proportional to the strain at rupture for impact hole diameter *D*, shown in Fig. 10, it can logically be employed in the development of the blasting equations and utilized for comparisons.

Dimensionless graphs illustrated in Fig. 11 suggest a general form mathematically expressed as follows:

$$W/S = C\,(F/S_t \cdot W^2)^n \qquad (8)$$

Where: *C* – A constant related to the type of loading; for the direct impact applied it was found to be 4.07;

St = the dynamic tensile strength of rocks (kPa);

n = a constant related to the effect of rate of loading on the breaking properties of the rock; it was found to be 1.04 and 1.39 for the limestone and concrete blocks, respectively.

Equation (5), modified in the following form, is valid for the rock types tested in all ranges of loading applied under the laboratory conditions:

$$F = (\pi \cdot C \cdot D)^a \cdot S_t \cdot W^b \qquad (9)$$

Where a and b are constants related to the rate of loadings employed. They vary for rock types as well as rate of loading in such a manner that $a + b = 2$ and $a = 1/n$. The values of a and b were found to be 0.96 and 1.04 for the limestone blocks, and 0.72 and 1.28 for the concrete blocks.

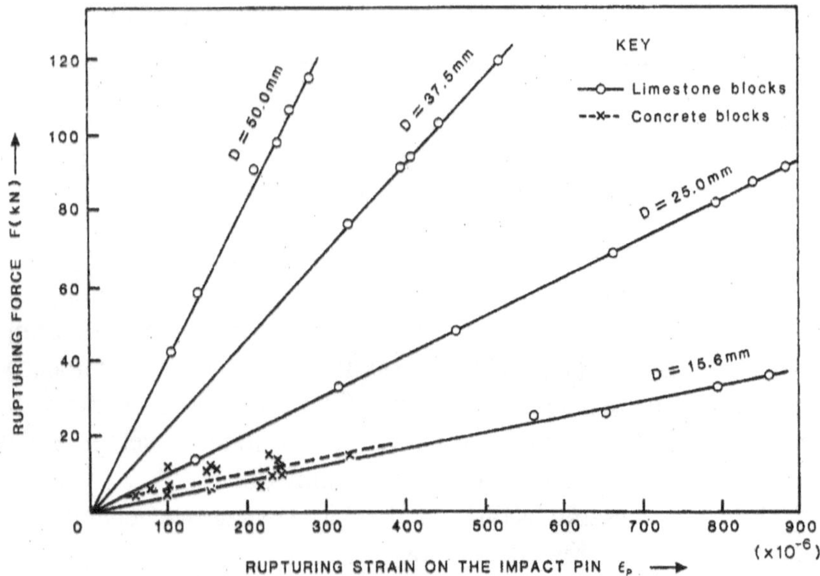

Fig. 10. Linear relationship between the strain and force at rupture.

6. CONCLUSIONS

(a) Rupture of rocks under direct impact was observed to be conical, the slope of the cone being dependent on the burden W and on the diameter D of the impact hole.

(b) The volume V of ruptured rock was directly but non-linearly proportional to the burden and to the impact hole diameter.

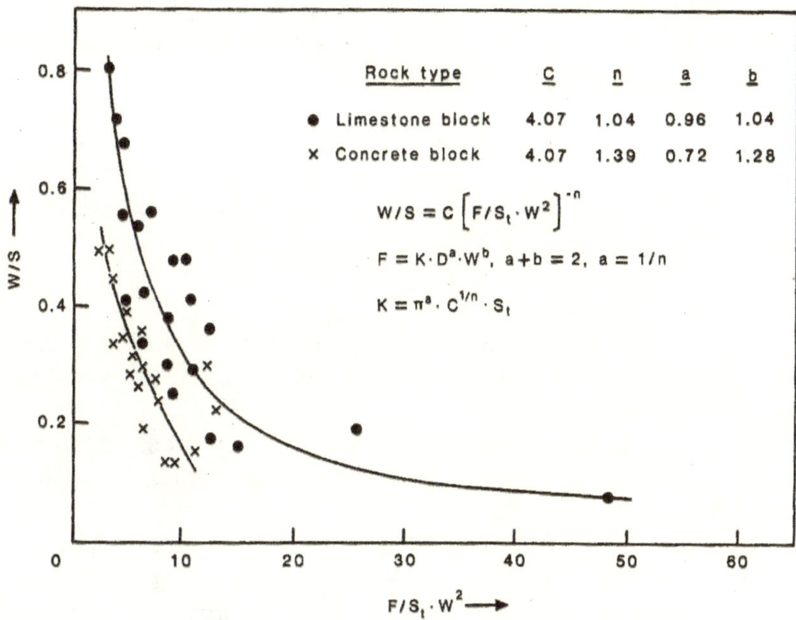

Fig. 11. Dimensionless relationship between the test variables utilized.

(c) The mechanism of rupture appeared to be due to combined shear and longitudinal waves, resulting in radial and tensile fractures, respectively, at the potential fracture zone.

(d) The reflected strain wave did not appear to have a significant effect in rupturing rock specimens at the potential fracture zone.

(e) Strain development in the impact pin at rupture was directly proportional to the burden and inversely proportional to the impact hole diameter, expressed in the form:

$$W \,.\, \epsilon_g = 2.23D \,.\, \epsilon_p.$$

(f) The force of impact at rupture F was linearly proportional to the strain developed at rupture. The general equation relating the rupture force to the physical properties of rock, the geometry of the blocks and the rate of loading, was found to be:

$$F = (\pi \cdot C \cdot D)^a \cdot S_t \cdot W^b.$$

(g) a and b are constants related to the rate of impact, and are employed in such a way that: $a = 1/n$ and $a + b = 2$,

Where *n* is a constant related to the effect of rate of loading on the breaking properties of the rock. Its value for Valders limestone was 1.04 and for the concrete blocks 1.39.

REFERENCES

1. A.W. Daw and Z.W. Daw, Blasting Rock, 2nd Ed. McGraw-Hill, London/New York, 1909.
2. C.D. Pomeroy, The breakage of coal by wedge action. Colliery Guardian (Nov. 21) (1963) 642–677.
3. S.S. Saluja, Study of the mechanism of rock failure under the action of explosives. Ph.D. Thesis, Univ. of Wisconsin, Madison, USA, 1964.
4. W.G. Pariseau and C. Fairhurst, The force–penetration characteristic for wedge penetration into rock. Int. J. Rock Mech., 4 (1967) 165–180.
5. K.K. Wu, An investigation of physical properties of rock under impact using a relationship of rupturing force to hole diameter and burden. M.Sc. Thesis, Univ. of Wisconsin, Madison, USA, 1965.
6. S.K. Khaderi, A study of rock rupturing under hydrodynamic impact loading. M.Sc. Thesis, Univ. of Wisconsin, Madison, USA, 1967.
7. R.J. Jones, Laboratory scale measurements of rock rupturing under impact and explosive loading. Ph.D. Thesis, Univ. of Wisconsin, Madison, USA, 1967.
8. A. Afrouz, Rupturing mechanism of rock under dynamic loading. M.Sc. Thesis, Univ. of Wisconsin, Madison, USA, 1969.
9. L.D. Clark, R.J. Jones and R.C. Howell, Blasting mechanics. A.I.M.E. and S.M.E. Trans., 252 (1972): 371–375.
10. J.S. Rinehard, Impact on rock, Q. Colo. Sch. Mines, 55(4) (1960): 201–208.
11. J. Schimazek and H. Knatz, The influence of rock structures on the cutting speed and pick wear of heading machines (in German). Glückauf, 106 (March) (1970) 275–278.
12. H.M. Hughes, Some aspects of rock machining. Int. J. of Rock Mech. Min. Sci., 9 (1972) 205–211.
13. I.W. Farmer, Engineering Behavior of Rock, 2nd Ed. Chapman and Hall, London, 1983, 224 pp.
14. H.L. Ewalds and R.J. Wanhill, Fracture Mechanics. Edward Arnold, London, 1984, 304 pp.
15. H.M. Hughes, The relative cuttability of Coal-Measures stone. Min. Sci. and Technol., 3(2) (1986) 95–109.

3.2.2 AN INVESTIGATION INTO BLASTING DESIGN FOR MINING EXCAVATIONS

By: Andy A. Afrouz, F.P. Hassani and R. Ucar

ABSTRACT

Blastability of rock is defined utilizing Bieniawski's rock mass classification with Hoek and Brown m *and* s *values. A new method is introduced to determine optimum blasthole diameter, burden, spacing, stemming length, ratio of maximum burden to the bench height and relationship between the blastability, rock strength factor and specific consumption of explosive. Moreover, sets of empirical equations are suggested to determine the amount of explosive per blasthole or chamber under various surface and underground mining conditions. This was done by the use of evaluated blasting coefficients for ranges of ground conditions.*

1. INTRODUCTION

Fragmentation of rocks by blasting in mines and quarries depends upon several factors, most important of which are: rock structure and its classification; mechanical properties and blastability of rock; blasthole diameter; its length and inclination to the free face; burden and spacing between blastholes; blasthole dryness and smoothness of its wall; ground pressure in the vicinity of the blastholes; type, density, blasting power and detonation velocity of the explosive; charge length and diameter; design of the charge; quality of placing and packing the explosive in the blasthole; type and length of stemming in the blasthole; staggering design; characteristic impedance of explosive and rock; point of initiation and decoupling of the blast; maximum available energy of explosive and its interaction with the surrounding rock; and degree of fragmentation. Hence, proper blasting design under various conditions is complex, difficult, challenging and yet highly desired.

This investigation is concerned with an approach to: (a) utilize Bieniawski's rock mass classification and Hoek and Brown *m* and *s* empirical failure criterion to evaluate the blastability of rock in relation to its specific explosive consumption and the rock strength parameter; and (b) to determine the diameter of blasthole, burden, spacing between blastholes, weight of explosive for various ground conditions and blast design.

There have been numerous investigations to achieve efficient blasting in mines, by formulating some of the above noted factors.

Anderson [1] has developed the following empirical equation:

$$B = K(D_h \cdot H)^{0.5} \tag{1}$$

Where: B = burden (m)
K = a proportionality constant
Dh = blasthole diameter (mm)
H = bench height (m).

In the above equation, for a good fragmentation: $H/B \leq 4$. Fraenkel [2] suggested the following more sophisticated equation:

$$B = K \cdot D_h^{\,0.8}(H \cdot h)^{0.3}/50 \tag{2}$$

Where: K = experimental constant (between 1 to 6 for most rock types)
h = length of the charge in the blasthole (m).

Lambooy and Jones [3] expressed the following formula for determination of burden:

$$B = h.W_e/S.H.q \tag{3}$$

Where: S = spacing between the blastholes (m)
We = weight of the explosive in kg/m run in a blasthole
q = weight of explosive to break unit volume of rock (kg/m^3)

Langefors and Kihlstrom [4] suggested a basic equation to calculate explosive quantity required for a single shot cratering of one meter height, as follows:

$$Q = K_2 \cdot B^2 + K_3 \cdot B^3 + K_4 \cdot B^4 = K_3 \cdot B^3 \tag{4}$$

Where: Q = weight of the explosive in a blasthole (kg)
B = burden = 1 to 15 m.
$K2$, $K3$ and $K4$ are factors defined in Table 1 [4,5].

TABLE 1. Definition and range of Langefors' variables [1,2]

K_2 = factor relating to the strength properties of the rock = 0.07 to 0.10 kg/m^2

K_3 = factor relating to the lifting of one m^3 of rock mass by explosive = 0.4 kg/m^3

K_4 = factor relating to the throw of one m^3 of rock mass 1 m by explosive = K_3/100
= 0.004 kg/m^4

B = burden = $(D_h / 33)[f_p \, . \, S_w / c \, . \, f(S / B)]^{0.5}$

f_p = factor relating quality of placing and packing the explosive in blasthole (g/cm^3 of V):

For tamping pole, nonmatched size	= 0.8 –1.0
For tamping pole, matched size	= 1.0 –1.4
For pneumatic loader	= 1.0 –1.6
For granular ANFO, without loader	= 0.9

v = volume of the blasthole (cm^3)

S_w = relative weight strength of the explosive:
For dynamite with 35% nitroglycerine = 1.0
For blasting gelatin = 1.27
For ANFO = 0.87

c = rock strength factor (kg/m^3) = $(0.07/B) + K_3$
For $B > 4$ m; = K_3 = 0.4 kg/m^3
c for beginning design value, single hole = 0.45
c for beginning design value, one row of holes = 0.36
c for horizontally bedded sedimentary rock = 0.2 0.45
c for empirical range = 0.2–1.2

f = fixity of hole or degree of relief.
for free bottom = 0.75
for fixed slope 2 : 1 (h : v) = 0.85
for fixed slope 3 : 1 (h : v) = 0.9
for fixed vertical = 1.0
for tunneling-reliever holes = 1.25 –1.5

S/B = spacing/burden ratio:
for bench shots = 1.25
for smooth blasting = 0.7

Hansen [6] extended the Langefors' equation in the following form:

$$Q = 0.028B(H+1.5B) + 0.4B^2C(H+1.5B) \quad (5)$$

Where: C is the rock strength factor, kg/m^3, given in Table 1.

Pearse [7] derived an expression that accounts for the characteristic of the explosives and the strength of the rock mass, given as follows:

$$B = 10^{-3} K . D_e (P_e/\sigma_t)^{0.5} \quad (6)$$

Where: K = proportionality constant = 0.8 (for most rocks)
D_e = diameter of explosive (mm)
P_e = explosion pressure (MPa) = $P_d/2$
P_d = detonation pressure (MPa) = $(p_e \cdot v_d \cdot v_g) \times 10^{-3}$
ρ_e = explosive density (gr/cm^3)
υ_d = detonation velocity (m/sec)
υ_g = particle velocity of gases (m/sec) = $v_d/4$
σ_t = tensile strength of the rock (MPa).

In most cases it is difficult to evaluate the ratio P_e/σ_t. Ash [8, 9] has simplified the equation (6) in the following form:

$$K_B = K(P_e/\sigma_t)^{0.5} \quad (7)$$

Where: K_B = burden constant = between 20 to 40
$= 30\ (2.56\,\rho_e/1.3\,\rho_r)^{1/3} = 37.6(\rho_e/\rho_r)^{1/3}$

ρ_r = rock density (gr/cm^3)

Substituting eqn. (7) into eqn. (6) gives: $B = 10^{-3} K_B . D_e$ (8)

Russians suggested [10] a variety of equations to relate burden and blasthole diameter. Amongst the most predominantly used are the ones as follows:

$$D_h = d/D_e = B/[7.85B . \rho_e (2x) / r . q]^{0.5}$$
$$= 0.25\,[B . r . q/\rho_e . x]^{0.5} \quad (9)$$

and $B = (Dh/d)\ 0.28\ [B. \rho e. fp/q. S]^{0.5}$ (10)

Where: $2x$ = length of the charge in the blasthole (m)

r = radius of the fractured zone in rock (m)

f_p = charge packing factor (see Table 1)

d = decoupling = D_h/D_e.

Afrouz [11] presented an empirical formula to determine the burden in terms of a single impact force to cause rupture (F) and the dynamic tensile strength of rock (σ_{td}) as follows:

$$B=\{F^{[n/(2n-1)]}\}/g\ \{\sigma td^{[2/(2n-1)]}.[\pi.c.d.D_h]^{[1/(2n-1)]}\} \quad (11)$$

Where: n = a constant related to the effect of rate of explosion on the braking properties of the rock = 1.04 for limestone, and 1.39 for concrete.

c = constant related to the type of loading, for direct impact it was evaluated to be 4.07.

Hino [12] based on the propagation of the shock waves and its reflection at a free face suggested the following equation:

$$B=10^{-3}D_h(P_d/\sigma_t)^{(1/n)}/4 \quad (12)$$

Where: n = a constant = 1.5, on average.

Rinehart [13] has developed an equation to relate the spall thickness (δ) to the explosion wave length (λ) as follows:

$$\delta=\lambda.\sigma_n/2\sigma_c=\lambda/2N=v_e.t/2 \quad (13)$$

Where: σ_n = normal stress at the time of fracture (MPa)

σ_c = maximum compressive stress developed by a saw tooth explosion wave (MPa)

v_e = velocity of explosion propagation (m/sec)

t = time taken for the explosion wave to travel to the free face and back (sec)

N = number of spalls that an explosive wave may produce = σ_c/σ_n.

Cook [14] has found that the pressure developed in the borehole to blast the rock (P_h) is related to the explosion pressure (P_e) in the following form:

$$P_h=P_e.\Delta^a \quad (14)$$

Where: P_e = explosion pressure (MPa) = $10^{-3}\,\rho_e \cdot v^2/8$
v = detonation velocity (m/sec)

a = exponential constant = 2.5
Δ = the loading density or fraction of the blasthole occupied by the explosive
= volume of explosive/volume of blasthole.

Ucar [15] has found that for range of a = 2.5, the cylindrical loading density (Δ) can be determined as follows:

$$\Delta = V_e / V_b = (D_e / D_h)^2 = (1/d)^2 \tag{15}$$

Where: V_e = volume of the explosive in blasthole (cm^3)
V_b = volume of the blasthole (cm^3)
d = decoupling = D_h / D_e

Substituting eqn. (15) into eqn. (14) gives:

$$P_h = P_e (1/d)^{2a} = P_e \cdot d^{-5} \tag{16}$$

2. VALUATION OF ROCK BLASTABILITY USING *m* AND *s* VALUES AND BIENIEWSKI'S ROCK MASS RATING

Tensile and compressive strength properties of rocks are the main parameters in determining their blastability. All other geological, physical and mechanical factors in rocks lead to a variation in the tensile and compressional strengths. An important criterion in the efficiency of fragmentation process is the ratio of compressive strength (σ_c) to tensile strength (σ_t) of the rock [16]. This ratio is called blastability coefficient (ϵ). It ranges from 10 to 500, except for very poor quality rock mass or soil for which it approaches infinity [17], and for the intact rock which is seldom encountered around mining excavations.

According to Mohr–Coulomb and Hoek et al. [17] failure criterion the relationship between major principal stress (σ_1) and minor principal stress (σ_3) in a rock at failure is follows, respectively:

$$\sigma_1 = \sigma_3 \times \tan^2 [45 + (\phi/2)] + 2C \cdot \tan [45 + (\phi/2)]$$
$$= \sigma_3 + \sigma_c [(m \times \sigma_3/\sigma_c) + s]^{0.5} \tag{17}$$

Where: C = cohesion of the rock (MPa) = shear strength (when normal stress in rock is zero)
ϕ = internal friction angle in the rock (degree)
m and s = constants dependent upon the rock mass properties and its state of fracture (noted in Table 2).

TABLE 2. m, s, $\xi = -\sigma_t/\sigma_c$, A and B values [17]

Empirical failure criterion, $\sigma_1 = \sigma_3 + [m.\sigma_c.\sigma_3 + s.\sigma_c^2]^{0.5}$ $\tau_\alpha = A\sigma_c(\sigma_\alpha/\sigma_c - \xi)^B$, and $\xi = -\sigma_t/\sigma_c$	Q^*	Dolomite, limestone, and marble	Mudstone, siltstone, shale & slate (normal to cleavage)	Sandstone and quartzite	Andesite, dolerite, diabase, and rhyolyte	Amphibolite, gabbro, gneiss, granite, norite, quartz-diorite
(1)	(2)	(3)	(4)	(5)	(6)	(7)
Intact rock samples Laboratory size specimens free from joints	500	$m = 7.0$ $s = 1.0$ $A = 0.816$ $B = 0.658$ $\xi = -0.140$	$m = 10.0$ $s = 1.0$ $A = 0.918$ $B = 0.677$ $\xi = -0.099$	$m = 15.0$ $s = 1.0$ $A = 1.044$ $B = 0.692$ $\xi = -0.067$	$m = 17.0$ $s = 1.0$ $A = 1.086$ $B = 0.696$ $\xi = -0.059$	$m = 25.0$ $s = 1.0$ $A = 1.220$ $B = 0.705$ $\xi = -0.040$
Very good quality rock mass Tightly interlocking undisturbed rock with unweathered joints at ±3m.	100	$m = 3.5$ $s = 0.1$ $A = 0.651$ $B = 0.679$ $\xi = -0.028$	$m = 5.0$ $s = 0.1$ $A = 0.739$ $B = 0.692$ $\xi = -0.020$	$m = 7.5$ $s = 0.1$ $A = 0.848$ $B = 0.702$ $\xi = -0.01$	$m = 8.5$ $s = 0.1$ $A = 0.883$ $B = 0.705$ $\xi = -0.012$	$m = 12.5$ $s = 0.1$ $A = 0.998$ $B = 0.712$ $\xi = -0.008$
Good quality rock mass Slightly weathered and disturbed with joints at 1–3 m	10	$m = 0.7$ $s = 0.004$ $A = 0.369$ $B = 0.669$ $\xi = -0.006$	$m = 1.0$ $s = 0.004$ $A = 0.427$ $B = 0.683$ $\xi = -0.004$	$m = 1.5$ $s = 0.004$ $A = 0.501$ $B = 0.695$ $\xi = -0.003$	$m = 1.7$ $s = 0.004$ $A = 0.525$ $B = 0.698$ $\xi = -0.002$	$m = 2.5$ $s = 0.004$ $A = 0.603$ $B = 0.707$ $\xi = -0.002$
Fair quality rock mass Several sets of moderately weathered joints spaced at 0.3–1m	1	$m = 0.14$ $s = 0.0001$ $A = 0.198$ $B = 0.662$ $\xi = -0.007$	$m = 0.20$ $s = 0.0001$ $A = 0.234$ $B = 0.675$ $\xi = -0.0005$	$m = 0.30$ $s = 0.0001$ $A = 0.280$ $B = 0.688$ $\xi = -0.0003$	$m = 0.24$ $s = 0.0001$ $A = 0.295$ $B = 0.691$ $\xi = -0.0003$	$m = 0.50$ $s = 0.0001$ $A = 0.346$ $B = 0.700$ $\xi = -0.0002$
Poor quality rock mass Numerous weathered joints at 30–500 mm	10^{-1}	$m = 0.04$ $s = 0.00001$ $A = 0.115$ $B = 0.646$ $\xi = -0.0002$	$m = 0.05$ $s = 0.00001$ $A = 0.129$ $B = 0.655$ $\xi = -0.0002$	$m = 0.08$ $s = 0.00001$ $A = 0.162$ $B = 0.672$ $\xi = -0.0001$	$m = 0.09$ $s = 0.00004$ $A - 0.172$ $B = 0.676$ $\xi = -0.0001$	$m = 0.13$ $s = 0.00001$ $A - 0.203$ $B = 0.686$ $\xi = - -0.0001$
Very poor quality rock mass Numerous heavily weathered joints spaced < 50 mm	10^{-2}	$m = 0.007$ $s = 0$ $A = 0.042$ $B = 0.534$ $\xi = 0$	$m = 0.010$ $s = 0$ $A = 0.050$ $B = 0.539$ $\xi = 0$	$m = 0.015$ $s = 0$ $A = 0.061$ $B = 0.546$ Б = 0	$m = .017$ $s = 0$ $A = 0.065$ $B = 0.548$ $\xi = 0$	$m = 0.025$ $s = 0$ $A = 0.078$ $B = 0.556$ $\xi = 0$

Note: * = Quality index obtained from the NGI (Norwegian-Geotechnical Institute) classification.
ϕ = internal friction angle in the rock (degree)
m and s = constants dependent upon the rock mass properties and its state of fracture (noted in Table 2).

In case where there is no lateral confinement, i.e. $\sigma_3 = 0$, the uniaxial compressive strength of rock is from eqn. (17) as follows:

$$\sigma_c = \sigma_1 = 2C\,tan\,[45 + (\phi/2)] \text{ or } \sigma c = \sigma_1/(s)^{0.5} \quad (18)$$

when $\sigma_1 = 0$ and $\sigma_3 = \sigma_t$, then eqn. (17) becomes:

$$\sigma_t = -\,2C\,.\tan\,[45 + (\phi/2)]\,/\tan^2\,[45 + (\phi/2)]$$
$$= -\,2C\,\tan\,[45 + (\phi/2)] = -\,2C\,\tan\,[45 - (\phi/2)]$$
$$= \sigma c[m - (m^2 + 4\,s)^{0.5}]\,/\,2 \quad (19)$$

Therefore, from equations (18) and (19), the ratio of σ_c/σ_t is as follows:

$$\epsilon = \sigma_c/\sigma_t = (1 + \sin\phi)/(1 - \sin\phi) = f(\phi)$$
$$= 2/[m - (m^2 + 4s)^{0.5}] \quad (20)$$

Finally, eqn. (17) can also be expressed according to Fig. 1 as follows:

$$(\sigma t/\sigma c) + (\sigma 3/\sigma 1) = 1 \text{ or } \epsilon = \sigma c/\sigma t = \sigma 1/(\sigma t - \sigma 3) \quad (21)$$

Priest and Brown [18] recommend following expressions relating the *m* and *s* values to the Bieniawski's rock mass rating [19]:

Category (a): In good blasting practice, perimeter blasting technique, or where the rock was previously excavated by machine rather than blasting, where the rock has no discontinuity and low fracture intensity:

$$R = m/m_i = \exp[(\text{RMR} - 100)/28] \quad (22)$$

$$s = \exp[(\text{RMR} - 100)/9] \quad (23)$$

where: R = a parameter dependent on the fracture intensity of rock mass
= 0.002 for heavily broken rock up to 1 for intact rock.

M_i = value of *m* for intact rock = between 7.5 to 25, dependent upon the rock type and its mineralogy (noted in Table 2)

RMR = Bieniawski's rock mass rating [19] = 20 for weak, heavily fractured rock mass, up to 100 for high strength intact rock = 9 ln Q + 44.

Q = Barton et al. [20] rock mass quality (noted in Table 2)

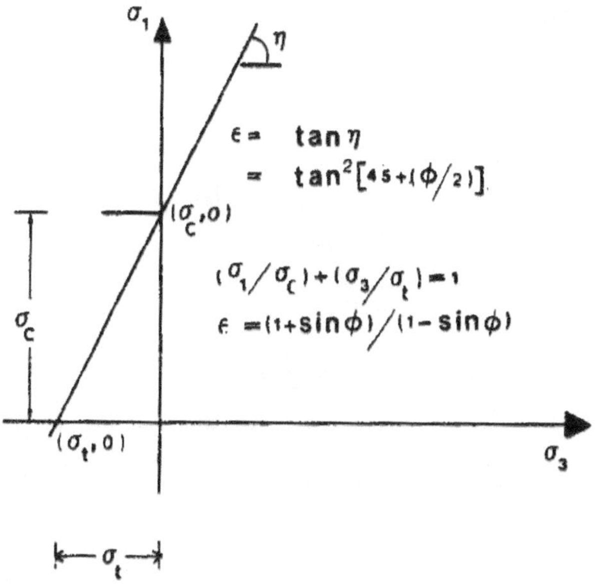

Fig. 1. Relationship between the major, minor, tensile and compressive stresses in rock at failure.

Category (b): In disturbed rock mass mainly occurring around the surface mine slopes and underground excavations, which have been loosened or damaged by poor blasting practice:

$$R = m/m_i = \exp[(\mathrm{RMR} - 100)/14] \tag{24}$$

$$s = \exp[(\mathrm{RMR} - 100)/6.3] \tag{25}$$

Hence, for condition of Category (a):

Substituting equations (22) and (23) into eqn. (20) in terms of *m* and *s* yields:

$$\varepsilon a = 2/\{\{m_i \exp[(\mathrm{RMR} - 100)/28]\} - \{\{m_i \exp[(\mathrm{RMR} - 100)/28]\}^2 + 4\exp[(\mathrm{RMR} - 100)/9]\}^{0.5}\} \tag{26}$$

and for condition of Category (b):

Substituting equations (24) and (25) into eqn. (20) in terms of *m* and *s* gives:

$$\varepsilon b = 2/\{\{m_i \exp[(\mathrm{RMR} - 100)/14]\} - \{\{m_i \exp[(\mathrm{RMR} - 100)/14]\}^2 + 4\exp[(\mathrm{RMR} - 100)/6.3]\}^{0.5}\} \tag{27}$$

3. DETERMINATION OF THE BLASTHOLE DIAMETER (D_h)

Volume of explosive in each blasthole (V_e) with respect to Figs. 2(a) and (b) can be determined as follows:

$$V_e = Q/\rho_e \tag{28}$$

$$V_e = k.\pi.D_h^2.H/4 = 0.79k.D_h^2 \tag{29}$$

where: k = packing factor = 0.6 to 1.00 (see Table 3).

TABLE 3. Explosive packing factor (k)

Type of blasthole	Type of explosives	k
Deep blastholes	Slurry and Gelatineous explosives	1.00
Deep blastholes	Powdered ANFO loaded pneumatically	0.98
Deep blastholes	Powdered ANFO loaded free flow down-hole	0.90
Deep blastholes	Pellet ANFO loaded free flow down-hole	0.80
Deep blastholes	Large diameter dynamites	0.90
Deep blastholes	Small diameter dynamites	0.80
Deep blastholes	Granular black powder	0.85
Potholes and chambers	Large diameter dynamites	0.80
Potholes and chambers	Small diameter dynamites	0.70
Potholes and chambers	Granular black powder	0.70
Potholes and chambers	Explosives in rectangular boxes	0.70
Potholes and chambers	Explosives in cylindrical boxes	0.60
Potholes and chambers	ANFO in bags	0.80

From equations (28) and (29):

$$D_h = (V_e/0.79k.h)^{0.5} \tag{30}$$

However, total weight of explosive per hole (Q) is evaluated as follows:

$$Q = q.V_r \tag{31}$$

where: q = specific consumption of explosive, g/m^3 of rock
V_r = volume of rock to be blasted.

Two important cases to note as follows:

Case (1) In the case where the inclination of bench to horizontal (α) is the same as that of the blasthole, i.e they are parallel to each other (Fig. 2a), then:

$$Q = q \, . \, B \, . \, S \, (l - S_d) \tag{32}$$

where: l = length of the blasthole (m)

S_d = subdrilling, m = 20 to 35% of B, or:

For blastholes with $D_h > 60$ mm, $5D_h < S_d < 15D_h$

For blastholes with $D_h < 60$ mm, $S_d > 10D_h$

For chambers, $S_d < 2D_h$

Substituting equations (28) and (32) into eqn. (30):

$$D_h = 1.13[q \times B \times S(1 - S_d)/\rho_e \times k \times h]^{0.5} \tag{33}$$

Case (2) In the case where the bench inclination is (α) and the blasthole is vertical (Fig. 2(b)) then:

$$Q = q \, [B \, . \, S \, . \, H + (S \, . \, H \, . \, l_t/2)]$$

$$= q \, . S \, . H[B + (H \cdot \cot\alpha/2)] \tag{34}$$

where: $l_t = H \, . \, \cot \alpha$ (from Fig. 2(b)).

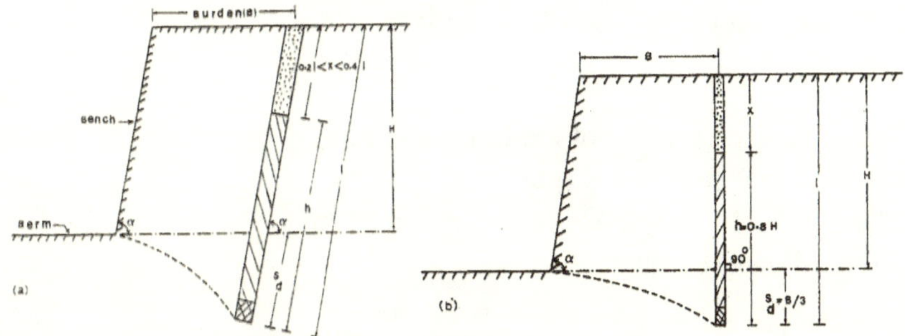

Fig. 2. Variety of blasthole inclination relative to that of the bench: (a). Inclination of bench to horizontal with parallel blasthole; (b). Bench inclination to horizontal is $\alpha°$, where the blasthole is vertical.

Substituting equations (28) and (34) into eqn. (30):

$$D_h = 0.8[q \, . \, H \, . \, S \, (2B + H \times \cot \alpha) \, / p_e \, . \, k \, . \, h]^{0.5} \tag{35}$$

4. DETERMINATION OF THE BURDEN (*B*)

A quadratic equation in which $B = f\,(H, h, W_e, S, q)$ can be considered. According to Fig. 2(b), length of charge (h) is determined as follows:

$$h = H - x + S_d \tag{36}$$

According to Langefors and Kihlstrom [4], the weight of explosive used per metre of blasthole (W_e) can be derived as follows:

$$W_e = f_p (D_h/36)^2 \tag{37}$$

where:

f_p = factor relating quality of placing and packing the explosive in blasthole (in gr/cm^3 of volume of blasthole) = 0.8 to 1.6 (gr/cm^3)

D_h = diameter of the blasthole (mm).

Therefore, the total explosive weight per hole (Q) can be determined as follows:

$$Q = W_e \times h = W_e (H - x + S_d) \tag{38}$$

Also, $Q = q \,.\, H \,.\, S \,.\, B = q \,.\, H \,.\, n \,.\, B^2$ (39)

where: $n = S/B$ = between 1 to 2.

Equating equations (38) with (39) gives:

$$W_e (H - x + S_d) = q \,.\, H \,.\, n \,.\, B^2 \tag{40}$$

For surface mining of medium strength to competent rocks such as limestone, sandstone, quartzite and granite with $H \leq 30$ m, $q = 0.4\ Kg/m^3$, $x = B$, $S_d = B/3$ and $n = 1.25$, eqn. (40) can be simplified in the following form:

$$1.5H \,.\, B^2 + 2W_e \,.\, B - 3W_e \,.\, H = 0 \tag{41}$$

Equation (41) is also applicable for high and low density explosives, and for small bench heights (H) and burden (B) calculations, occurring in underground conditions, and where $1 \leq B \leq 2$ meters and $5 \leq H/B \leq 6$. It gives a close fit to the empirical expressions suggested by Langefors [4] as shown in Table 4. For example, considering: $H = 1.2$ m; $D_h = 33$ mm; $f_p = 1.27\ gr/cm^3$; and $n = S/B = 1.25$, then:

W_e = 1..27(33/36)^2 = 1.07 *kg/m; and B* = 0.98 m = 1.0 m

The accuracy of this calculations can also be seen from the Table 4, which indicated a maximum burden of 1.1, and also a practical burden of 1.0 m.

Equation (40) can be simplified for the case $h = 5H/6$ to give:

$$5W_e/6 = q \cdot n \cdot B^2 \text{ or } B = 0.91\,(W_e/q \cdot n)^{0.5} = \psi \cdot W_e^{\,0.5} \quad (42)$$

where: $\psi = 0.91\,/\,(q\,.\,n)^{0.5}$ for $q = 0.4$ kg/m^3 and $n = 1.25$:

$$\psi = 0.91/(0.4 \times 1.25)^{0.5} = 1.3.$$

Therefore, eqn. (42), for $\psi = 1.3$ becomes: $B = 1.3$ (43)

TABLE 4. Multiple-row bench blasting: Diameter of the drill bits 34–29 mm (1.35–1.16 in.), $f_p = 1.27$; Length of the rod 0.8–4.8 m (2.6–16 ft); Slope of the holes 3:1–2:1 and $S = 1.25\,B$ [4]

Bench height (H, m)	Depth of the hole (l, m)	Diameter at the bottom (D_h, mm)	Max burden (B_{max}, m)	Practical burden (B, m)
0.3	0.67	34	0.60	0.55
0.45	0.80	34	0.70	0.65
0.6	1.00	33	0.77	0.7
0.9	1.35	33	0.92	0.85
1.0	1.45	33	1.00	0.95
1.2	**1.65**	**33**	**1.10**	**1.00**
1.5	1.0	32	1.20	1.10
1.8	2.4	32	1.30	1.20
2.1	2.7	31	1.35	1.25
2.4	3.1	31	1.40	1.25
2.7	3.4	30	1.40	1.25
3.0	3.8	30	1.40	1.25
3.3	4.2	30	1.40	1.20
3.6	4.5	29	1.35	1.15
4.0	4.9	29	1.35	1.15

Comparison of average burden from eqn. (43) with maximum burden from Langefors [4] is given in Table 5 which shows close similarities. It is to be noted that at $\psi = 1.5$ in eqn. (43) the results will be the same in both the approaches.

Substituting eqn. (37) into eqn. (43) gives:

$$B = 0.04\,D_h \quad (44)$$

where: B = burden (m)

D_h = blasthole diameter (mm).

TABLE 5. Relationship between burden (B) and weight of charge per metre of the blasthole (W_e) for bench blasting and stoping (f_p = 1.27 (tamping pole); $S = 1.25\ B$; $x = B$; $S_d = 0.3\ B$; Slope 2:1–3:1 and weight strength of explosives = 1.0)

Blasthole diameter (D_h, mm)	W_e (kg/m)	Langefors [4] ($B_{max.}$ mm)	Burden (B, m) (from eqn. 37)
14	0.18	0.60	0.55
16	0.25	0.70	0.65
19	0.36	0.85	0.78
22	0.48	1.00	0.90
25	0.63	1.15	1.03
29	0.84	1.30	1.20
32	1.00	1.50	1.30
38	1.40	1.80	1.54
44	1.90	2.00	1.80
50	2.50	2.30	2.10
63	4.00	3.00	2.60
75	5.60	3.50	3.10
88	7.70	4.10	3.60
100	10.00	4.70	4.11
125	15.50	5.90	5.12
150	22.50	7.00	6.20
175	30.00	8.20	7.12
200	40.00	9.40	8.22
225	50.00	10.60	9.20
250	62.00	11.80	10.23
300	90.00	14.00	12.33

Another method to determine burden (B) is to consider the ratio $H/B = \lambda$. Thus eqn. (41) becomes:

$$1.5\lambda \,.\, B^3 + 2W_e \,.\, B - (3\lambda \,.\, W_e \,.\, B) = 0 \qquad (45)$$

Therefore, $B = [(3\lambda - 2)/1.5\lambda]^{0.5}$ (46)

Comparing eqn. (42) with eqn. (46) results:

$$\psi = [(3\lambda - 2)/1.5\lambda]^{0.5} \qquad (47)$$

5. RELATIONSHIP BETWEEN MAXIMUM BURDEN (B_{max}) WITH MAXIMUM BENCH HEIGHT (H_{max})

In bottom initiated, long charge columns a pear shape explosion wave will be developed. The actual shape and symmetry of the wave propagation depends upon the charge length, the hole inclination, occurrence of bedding planes and their inclination, and finally occurrence of other discontinuities and in-fillings, their orientation, extent and quality. However, the complex wave propagation can reasonably be simplified in vertical blastholes, for calculation purposes, as being of conical shape as shown in Fig. 3.

For the explosive with a characteristic detonation velocity (v_d) exceeding that of the longitudinal or *P*-wave velocity of the rock (v_p) with the stress cone angle of (θ), following relationship exists [8]:

$$\sin\theta = vp/v_d \tag{48}$$

The travel time for explosion compressive waves to reach the free face and the time for the resulting tension waves to travel back (t_1) can be expressed as follows:

$$t_1 = 2B/v_p \tag{49}$$

Equation (49) is important in order to choose a proper delay interval in the blasting design. As a rule of thumb, the optimum delay is about 5 to 10 milliseconds per metre of burden.

The travel time for detonation velocity (v_d) to pass the column charge (h) can be denoted as (t_2) and expressed as follows:

$$t_2 = h/v_d \tag{50}$$

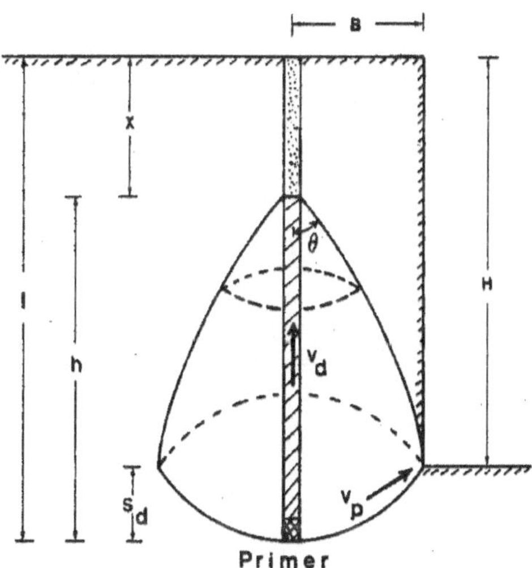

Fig. 3. Schematic of explosion wave propagation in the bottom initiated, vertical, long charge columns.

In the critical condition, for maximum burden (B_{max}), the following relationship exists:

$$t_1=t_2 \text{ and } 2B/v_p=h/v_d \tag{51}$$

In cases where $h = 0.8H$, eqn. (51) can be expressed as: $2B/v_p = 0.8H/v_d$. Therefore,

$$v_d/v_p=0.4H/B=0.4\lambda \tag{52}$$

The P-wave velocity in various rocks (v_p) is noted in Table 6 [4,21] and the detonation velocity for various explosives (v_d) is given in Table 7 [21]. For practical purpose,

$$v_p < v_d < 3v_p.$$

In case where $D_h > 150$ mm; $H \leq 30$ m; and $v_d = 2v_p$ eqn. (46) gives:

$$\lambda=H/B=2/0.4=5 \tag{53}$$

TABLE 6. Average values for the longitudinal P-wave velocities (v_p) in various rocks

Rock type	Specific gravity (gr/cm³)	v_p (m/sec)
Unconsolidated soil	1.5–2.1	< 900
Clay stone	1.8–2.3	1000–2000
Conglomerate	2.0–2.5	1200–2100
Soft shale	2.0–2.6	1200–2400
Hard shale	2.2–2.7	1800–3000
Chalk	2.1–2.6	2800–3000
Sandstone	2.2–2.7	2800–3300
Marlstone	2.2–2.9	2800–3800
Limestone	2.1–3.0	2500–5000
Basalt	2.2–2.8	2500–4000
Gneiss	2.2–3.1	2800–5500
Granite	2.5–3.2	4000–6000

TABLE 7. Average detonation velocity (v_d) for various explosives

Explosive type	Constituent	Specific gravity (gr/cm^3)	v_d (m/sec)
Ammonium nitrite	93% NO_3 NH_4 + 7% C	0.80	3600
Ammonium nitrite	93% NO_3 NH_4 + 7% C	1.00	4000
10% Nitroglycerine	10% Nitroglycerine + 80% NO_3 NH_4 + 10% Cellulose	0.98	4300
50% Nitroglycerine	50% Nitroglycerine + 41.5% NO_3 NH_4 + 5.5% Cellulose + 7% Combustible material + 2.3% Cotton wool	1.50	6800
Nitroglycerine	Pure $(NO_3)C_3H_5$	1.60	8500

Substituting eqn. (53) into eqn. (47) gives:

$$\psi = [(3 \times 5 - 2)/1.5 \times 5]^{0.5} = 1.3$$

The latter determination checks with ψ-value obtained from equations (42) and (43).

6. DETERMINATION OF SPACING BETWEEN BLASTHOLES (S)

The following suggestions [4,22] work well under field conditions, considering a staggering pattern:

$$\text{If } 2B < H < 4B, \text{ then: } S = (B \,.\, H)^{0.5} \quad (54)$$

$$\text{If } H > 4B, \text{ then: } S = 2B \quad (55)$$

For square pattern, smaller spacing, S / B ratio is recommended.

7. DETERMINATION OF THE STEMMING LENGTH (x)

Length of the stemming is a function of the burden (B), length of the blasthole (l), type of the stemming material, method of placement, and compaction of the stemming in blasthole and type of explosive utilized.

For most blasting conditions, the stemming should have enough length and compaction to permit the explosion gases to perform the necessary fracturing of solid rock before there is rock movement. Generally, the compressional wave travels much faster in solid rock than in the stemming. The critical condition occurs when travel time of the compressive explosion wave to reach the free face is equal to that of the wave to pass through the column of stemming, to reach the top of the hole. In this case:

$x(v_p)_{stemming} = B/(v_p)_{rock}$; Therefore,

$$x = B(v_p)_{stemming}/(v_p)_{rock} \tag{56}$$

In practice, for the rock strength factor (C) of 0.3 to 0.45, kg/m^3, and delay time of 3 to 5 millisecond/m of burden, a length of stemming (x) between 0.6 B to B is satisfactory.

8. RELATIONSHIP BETWEEN THE ROCK STRENGTH FACTOR BLASTABILITY (*E*) AND SPECIFIC CONSUMPTION OF EXPLOSIVE (*q*)

According to Langefors and Kihlstrom [4], the amount of explosive per blasthole (Q) can be derived from the following equation:

$$Q = q\,.\,B\,.\,S\,.\,H = [1.4\,C.\,B^3 + 0.4C.\,B^2(H-2B)] \tag{57}$$

where: Q at the bottom of blasthole = 1.4 C. B^3

Q in the blasthole column = 0.4 C. $B^2(H-2B)$

$C = 0.50 + 2.60(\sigma_t/\sigma_c)^{0.5} + 13\,\sigma_t/\sigma_c$, kg/m^3

From eqn. (57):

$$q = Q/B.\,S.\,H = (1.4\,C\,.\,B^3 + 0.4\,C.\,B^2\,(H-2B))\,/\,n\,.\,H\,.\,B^2$$
$$= 0.4\,C\,(1 + (1.5B/H))\,/n$$
$$= 0.4[0.50 + 2.60\,(\sigma_t/\sigma_c)^{0.5} + 13\sigma_t/\sigma_c]\,.\,[1 + (1.5B/H)]\,/n$$
$$= [0.20 + 1.04(1/\epsilon)^{0.5} + (5.2/\epsilon)]\,.\,[1 + (1.5B/H)]\,/n \tag{58}$$

TABLE 8 - Empirical determination of explosive weight per blasthole or chamber (Q) for surface and underground mining

Condition	Q (kg/hole)	Parameters (m)					Types of blastholes and chamber								Remarks
		H	l	B	S	D_h	Chamber		Blasthole						
							T-shape	L-shape	Horizontal	Vertical	$15° < \alpha < 75°$	$45° < \alpha < 85°$	Deep	Shallow	
1	$q \times B \times S \times H$	$2B$–$4B$	$\geq 2B$	$1/2 < B < 3$	$(B \times H)^{0.5}$	< 0.06						×		×	
2	$q \times B \times S \times H$	$2B$		$2 < B <$	$(B \times H)^{0.5}$	< 0.06			×					×	
3	$q \times C_t \times B \times S \times H$	$> 3B$	$> 2B$	> 3	$2B$	> 0.06						×	×		
4	$q \times G_e \times B \times S \times H/2$	$< B$	< 2	< 2	$(B \times H)^{0.5}$	< 0.06						×	×		Ground resistance is more at the back of the face
5	$q \times G_e B \times S \times H$	$> 2B$		> 3	$2B$	< 0.06						×	×		Ground resistance is more at the back of the face
6	$q \times G_r \times C_t \times B \times S \times H$		< 2	< 3	$(B \times H)^{0.5}$	< 0.06					×			×	Ground resistance is more at the back of the face
7	$q \times G_r \times B \times S \times H$		< 2	> 3	$(B \times H)^{0.5}$	< 0.06				×				×	Quarrying with black powder
8	$q \times G_r \times C_t \times B \times S \times H$	$> 2B$		> 4	$2B$	> 0.06				×			×		Quarrying with black powder
9	$q \times G_r \times C_t \times B \times S \times H/2k$	$\leq 3B$	$< 4B$	> 4	$2B$	> 0.06						×	×		
10	$q \times C_t \times B \times S \times H/2k$	$> 3B$	$< 4B$	> 4	$2B$	> 0.06				×			×		Quarrying
11	$q \times C_t \times B \times S \times H$	$> 2B$		$3 < B < 4$	$1.5B$	< 0.1			×				×		
12	$q \times B \times S \times H/2$	$> 2B$		< 3	$1.5B$	< 0.1			×				×		
13	$q \times C_t \times B \times S \times H/2$	$2B$		< 3	$2B$	< 0.06			×					×	
14	$1.3q \times C_t \times B \times S \times H/k$	$2B$		$H/2$	$(B \times H)^{0.5}$	< 0.1		×					×		Quarrying
15	$1.6q \times C_t \times B \times S \times H/k$	$< 3B$		$< H/2$	$2B$	> 0.1		×					×		Quarrying
16	$q \times C_t \times B \times S \times H/k$	$1.5 < H < B$		$0.3H < B < 0.7H$	$(B \times H)^{0.5}$	< 0.1	×						×		Quarrying
17	$q \times_e \times C_t \times B \times S \times H/2$	$1.5B < H < B$		$< S_c$	$(B \times H)^{0.5}$	< 0.1	×						×		Quarrying

Equation (58) is applicable within the range of $1.25 \leq n \leq 1.5$ and $H \leq 5B$.

Definitions of the factors and coefficients used in the expressions of Table 8 are follows:

Q = weight of explosive per blasthole or per chamber (kg/hole)

q = specific consumption of explosive, gram/m^3 or rock = $P_1 \times S_p \times B_i \times F_f \times S_q$

P_1 = plastic coefficient of rock = E_e/E_p (see Table 9)

E_e = energy consumption to fracture the rock within its elastic range

E_p = energy consumption to break the rock into the plastic range

S_p = coefficient for shattering power of explosive (Table 10)

B_i = blasting index due to rock physical properties (Table 11)

F_f = coefficient relating to number of free faces (n) (Table 12)

S_q = coefficient related to the quality of stemming the blasthole or chamber (Table 13)

B = burden (m)

S = spacing between blastholes (m)

H = height of the bench (m)

l = length of the blasthole (m)

D_h = diameter of the blasthole (m)

α = angle of blasthole to horizontal (degrees)

S_c = spacing centre to centre between adjacent chambers (m)

C_f = corrective factor related to the burden (Table 14)

G_r = coefficient related to the resistance of the ground to blasting (Table 15)

k = explosive packing or charging factor (Table 3)

9. EMPIRICAL DETERMINATION OF THE EXPLOSIVE AMOUNT PER BLASTHOLE OR CHAMBER (*Q*)

Experimental results in various surface and underground blasting conditions for soft to hard rocks such as coal, limestone, sandstone, quartzite, magnetite, hematite, copper ore, shale, siltstone, underclay, mudstone, magnesite, Alaskite, galena, marble, chronite and phosphate deposits are summarized in categories noted in Table 8. The percentage error was seen to be within ± 15% of that indicated in Table 8. This was largely dependent upon the ground conditions, rock type and its blastability.

TABLE 9. Plastic coefficient of various rocks (P_1)

Group no.	Rock hardness	Rock type	$P_1 = E_c / E_p$
1	Very hard and intact	Diabase, basalt, gneiss, gabro, diorite, norite, amphibolites, chert	> 160
2	Hard and intact	Hard black shale, quartzite, granite, magnetite, dolomite, hematite, marble, hard limestone,	80–200
3	Medium hard and intact	Ordinary limestone, ordinary sandstone, shale, slate, siltstone, hard underclay or fireclay	50–100
4	Medium hard weathered and intact	Ordinary limestone, ordinary sandstone, shale, slate, siltstone, hard underclay or fireclay	30– 70
5	Soft and intact	Potash, rock salt, coal gypsum, chalk, underclay	20– 40
6	Soft weathered and fractured mass	Potash, rock salt, coal gypsum, chalk, underclay	10– 30
7	Soil	Aluvial deposites and wastes, sand and clay	3– 15

TABLE 10. Coefficient for shattering power of explosive (S_p)

Group no.	Explosive type	Detonation velocity (m/sec)	S_p
1	Blasting gelignite, startex A, gelatine donarite no. 1, Jel-anon, acid pycric, tetryl	6000–8000	1.1
2	Dynamite, gelex no. 3, delobel, dubel, monobel, T.N.T., melinite, ammonium dynamite, ammonium gelatine	5000–6000	1.2
3	Liquid oxygen, donarite, reolite, detonite, ammonite no. 1, dynamex A & B, pantry, lead azide, silver azide	4500–5000	1.3
4	ANFO, gurite, pabdite, donarite no. 2, reomex, nitrites, mercury fulminate	3500–4500	1.4
5	Decamon, chlorites, dynalite, granular dynamite	3000–3500	1.5–1.6
6	Nobel dynamite no. 1235, prilite A & B	1500–2500	1.6–1.7
7	Black powder, gun powder	200– 800	1.7–2.0

TABLE 11. Blasting index due to the ground conditions (B_i)

Group no.	Ground conditions	Drilling pattern	Type of charging	B_i
1	Sloped bedding with weak cementation and joints; footwall blasting	(a) Vertical or horizontal holes	spread	0.6
		(b) Vertical or horizontal holes	concentrated	0.7
		(c) Angular deep holes and chambers	concentrated	0.9–1.1
2	Near horizontal bedding with weak cementation and joints	(a) Vertical or horizontal holes	spread	0.7
		(b) Vertical or horizontal holes	concentrated	0.9
		(c) Angular deep holes and chambers	concentrated	1.1
3	Slopes bedding with strong cementation and joints; hanging wall blasting	Angular deep holes and chambers	spread or concentrated	1.2–1.5
4	Less than one metre thin bedding with weak cementation, joints and clay partings; footwall blasting	(a) Angular deep holes	spread	1.8–2.0
		(b) Vertical or horizontal holes	spread	0.7

5	Less than one metre thin beddings with strong cementation and joints; footwall blasting	(a) Angular deep holes (b) Vertical or horizontal holes or chambers	spread concentrated	0.9 1.2
6	Same as group 4; hangingwall blasting	(a) Angular holes (b) Vertical or horizontal holes or development works	spread concentrated	1.2 1.5
7	More than one metre thick beds without joints; footwall blasting	(a) Vertical or horizontal holes (b) Deep holes of chambers	spread concentrated	0.8 1.0
8	More than one metre thick beds without joints; hangingwall blasting	Angular deep holes or chambers	spread or concentrated	1.2
9	Same as group 4	Vertical deep holes	spread or concentrated	1.4
10	More than one metre thick massive beddings with weak cementation and high slope; footwall blasting	(a) Horizontal holes (b) Vertical holes	spread concentrated	0.7 0.9
11	As that of group 10 with near horizontal beddings	(a) Vertical holes (b) Angular deep holes	spread concentrated	0.8 1.0–1.2
12	As that of group 10; hangingwall blasting	Angular deep holes	spread	1.3
13	More than one metre thick near vertical beddings; perpendicular to the direction of advancement	Vertical or angular deep holes	spread	1.6
14	Massive thick limestones and sandstones	Angular deep holes	concentrated	1.5–1.8
15	Massive horizontal beddings and fine grained, without joints	(a) Vertical or angular deep holes (b) Vertical or angular deep holes	spread concentrated	0.7–1.0 0.8
16	Same as group 15 with joints	Vertical or angular deep holes	spread	1.4

17	Hard, coarse grained rocks	(a) Vertical holes (b) Angular deep holes or chambers	concentrated concentrated	1.2 1.5
18	Hard, fine grained rocks	(a) Vertical holes (b) Angular deep holes or chambers	spread concentrated	1.0 1.2
19	Dykes and sills; footwall blasting	Angular deep holes or chambers	concentrated	1.2–1.3
20	Dykes and sills; hangingwall blasting	Angular deep holes or chambers	concentrated	1.3–1.5
21	Basalts	(a) Vertical holes (b) Angular deep holes	spread concentrated	1.5 1.8
22	Consolidated soil, clay and aluminal deposits together with rock beddings	(a) Vertical holes (b) Angular holes	spread concentrated	1.5 2.0
23	Thick beds of conglomerates, shale, siltstone, tuff and dolomite; footwall blasting	(a) Vertical or horizontal holes (b) Angular holes or chambers	spread concentrated	1.5 2.0
24	As that of group 23, but thin beds and hangingwall blasting	Angular holes or chambers	concentrated	2.5–3.0
25	Moderately weathered and jointed shale	(a) Vertical or horizontal holes (b) Angular deep holes or chambers	spread concentrated	2.2–2.8 2.7–3.5
26	Hard and dry underclay or fireclay	(a) Vertical or horizontal holes (b) Angular holes or chambers	spread concentrated	1.5 1.8
27	Soft underclay and clay rocks with	Angular holes or	spread	2.0–3.0
28	Unconsolidated soil, sand and clay	Any type of hole	spread	3.0–4.0

TABLE 12. Coefficient relating to the number of free faces (n) in each blast (F_f)

Group no.	Type of blasthole	Types of charging	F_f		
			n = 1	n = 2	n = 3
1	Shallow; angular to vertical holes	(a) Spread	1.2	1.0	0.8
		(b) Concentrated	1.0	0.8	0.6
2	Deep; angular to vertical holes	Concentrated	1.6	1.4	1.2
3	Horizontal holes; one metre up the floor	(a) Spread	1.4	1.2	1.0
		(b) Concentrated	1.2	1.0	0.8
4	Angular holes 30 to 40° to horizontal drilled into an advancing face and development of an underground work, or bench of an open pit	(a) Spread	1.3	1.0	0.9
		(b) Concentrated	1.1	0.9	0.7
5	Chambers driven about one metre up the floor	Concentrated	1.6	1.4	1.1
6	Chambers driven at the floor level or in stressed ground	Concentrated	1.8	1.6	1.4
7	Chambers driven below the floor level (this type of blasting is accompanied by high ground vibration)	Concentrated	2.2	2.0	1.8
8	Tunneling and development work	(a)Pyramidal center	1.2		
		(b) Conical center	1.1		
		(c) Roof holes	0.7		
		(d) Floor holes	0.9		
		(e) Wall holes	0.8		
		(f) Middle holes	1.1	0.9–1.0	

TABLE 13. Coefficient related to the quality of stemming the blasthole or chamber (S_q)

Group no.	Types of blasthole, charging and stemming	S_q
1	Deep angular to vertical holes or chambers	1.1–1.2
2	Shallow 0.3 to 0.7 m horizontal holes or 0.7 to 1.1 m vertical holes	1.2–1.4
3	Shallow horizontal holes of 1 ≤ 0.3 m	1.4–1.5
4	Blasholes with good stemming	0.9–1.0

TABLE 14. Correction factor related to the burden (C_f)

Burden (B, m)	C_f	Burden (B, m)	C_f	Burden (B, m)	C_f
2.8	0.36	5.5	0.23	8.5	0.19
3.0	0.33	6.0	0.22	9.0	0.18
3.5	0.30	6.5	0.22	9.5	0.18
4.0	0.28	7.0	0.21	10.0	0.18
4.5	0.26	7.5	0.21	10.5	0.17
5.0	0.24	8.0	0.20	11.0	0.17

TABLE 15. Coefficient related to the resistance of the ground to blasting (G_r)

Group no.	Type of ground and blasthole	G_r
1	Ground resistance to the back of the faceline is little concentrated charging	1.0
2	Ground resistance to the back of the faceline is high and blastholes are vertical or chambers of T-type	1.1–1.3
3	Same as group 2 where it is necessary to throw the blasted rock well off the face area	1.2–1.5
4	Chambers of L-type	1.2–1.3
5	Ground heavily jointed or fractured	1.1–1.3
6	Ground with 2 free faces	0.7–0.9
7	Ground with 3 free faces	0.6–0.7
8	Leveling the floor or berm	0.2–0.3
9	Deep angular holes of 45° to 55° to horizontal	0.5–0.6
10	Deep angular holes of 60° to 65° to horizontal	0.6–0.7
11	Deep angular holes of 70° to 80° to horizontal	0.7–0.8
12	Spread charging and high ground resistance to the back of the faceline	0.6–0.8
13	Quarry blasting with large blocks of rocks	0.3–0.5
14	Same as group 13 with high ground resistance at the back of the faceline	0.3–0.6
15	Same as group 14 with large blocks of rocks just to be detached from the face	0.2–0.5

10. CONCLUSIONS

The Bieniawski's rock mass rating can successfully be utilized in determination of rock blastability. Therefore, specific consumption of explosive can be determined as a function of the rock blastability using Bieniawski's RMR values. Other factors of prime importance in rock fragmentation such as blasthole and explosive diameters, decoupling, bench height, subdrilling, burden, spacing, specific consumption of explosive, its type and length of stemming can be determined accurately using the relevant expressions given in this investigation.

REFERENCES

1 O. Anderson, Blasthole burden design: Introducing a formula. Proc. Australian Inst. Min. Metal., (1952): 166–167.

2 K.H. Fraenkel, Factors influencing blasting results. Manual on Rock Blasting, Atlas Diesel and S.J. Sweden, 1(2) (1952): 15 pp.

3 P. Lambooy and R.C. Esplay–Jones, Practical considerations of blasting in open cast mines. Proc. Open Pit Mining Symp., Johannesburg, S. Africa, (Aug. 29–Sept. 4, 1970) Balkema, 227–234.

4 U. Langefors and B. Kihlstrom, The Modern Technique of Rock Blasting. John Wiley and Sons, New York, (1976) 405 pp.

5 C.H. Dowding, Blast Vibration Monitoring and Control. Prentice–Hall Inc., (1985), 246–247.

6 D.W. Hansen, Drilling and blasting techniques for Morrow Point Power Plant. 9th Symp. on Rock Mechanics, Colorado School of Mines, Golden, CO, (1967), 347–360.

7 G.E. Pearse, Rock blasting—Some aspects on the theory and practice. Mine and Quarry Eng., 21(1) (1955): 25–30.

8 R.L. Ash, The mechanics of rock breakage. Pit and Quarry, 56(2–5) (1963): 98–143.

9 R.L. Ash, Class Notes on Explosives. University of Missouri–Rolla, (1974).

10 J.A. Otano, Fragmentation de Rocas Con Explosives. Editorial Pueblo Y Educacion, (1980): 174–175.

11 A. Afrouz, Rupturing mechanism of rock under dynamic loading. M.Sc. Thesis, University of Wisconsin, Madison, (1969), 150 pp.
12 K. Hino, Theory and Practice of Blasting. Nippon Kayacu Co., Asa, Japan, (1959) 189 pp.
13 J.S. Rinehart, On fractures caused by explosions and impact. Q. Colorado School of Mines, 55(4) (1960): 150 pp.
14 M.A. Cook, The Science of High Explosives. Reinhold Publ. Corp., NY, (1958), 440 pp.
15 R. Ucar, Decoupled explosive charge effects on blasting performance. M.Sc. Thesis, University of Missouri–Rolla, 1975, 59 pp.
16 T.C. Atchison, Surface Mining–Fragmentation Principles. Amer. Inst. Min. Metal. Pet. Eng., (1st ed.), (1968): 355–372.
17 E. Hoek and T. Brown, Empirical strength criterion for rock masses. J. Geotech. Eng. Div., ASCE, 106 (GT9) (Sept. 1980): 1013–1035.
18 S.D. Priest and T. Brown, Probabilistic stability analysis of variable rock slopes. Trans. Inst. Min. Metal. Section A: Mining Industry, 92 (1983): A1–A12.
19 Z.T. Bieniawski, Rock mass classification in rock engineering. Proc. Symp. on Exploration for Rock Engineering, Vol. 1, A.A. Balkema, Rotterdam, (1976): 97–106.
20 N. Barton, R. Lien and J. Lunde, Estimation of support requirements for underground excavations. Proc. 16th Symp. on Design Methods in Rock Mechanics, University of Minnesota and Amer. Soc. Civil Engineers, (1977): 163–177.
21 A. Afrouz, Blasting in Mines. Mighat Publishing Co., (1st ed.), 1985, 482 pp.
22 J. Pugliese, Designing Blast Patterns Using Empirical Formulae. Information Circular 8550, U.S. Dept. Interior, Bureau of Mines, 1972.

3.2.3 SEMI-THEORETICAL DESIGN OF BLASTHOLE SPACING FOR VARIABLE GROUND CONDITIONS

By: Andy A. Afrouz and Jamal Rostami

SYNOPSIS

This paper presents findings to determine the average spacing between parallel blastholes. It is based on the previously established empirical expressions to relate spacing and burden in the design of blasts, using dynamic strength of rock. This is then utilized to calculate the spacing by accounting for variation in the rock mass rating and blastability of the ground. Use of this technique is in the design of blasts for parallel holes in open pits as well as in underground cut-and-fill and longhole stopes.

INTRODUCTION

Spacing between blastholes is a function of several parameters, the most important of them being:

- Rock type, its degree of saturation, fissuration and ground conditions.
- Blastability of rock.
- Burden.
- Diameter and length of holes.
- Orientation of blastholes relative to each other and the free face.
- Size of rock fragments required.
- Type of explosive to be utilized.
- Method of explosive charging, stemming and blasting.
- Cost of drilling, explosives, charging, stemming and blasting.

Permafrost and freeze-thaw conditions also affect the mechanical properties and blastability of the rock mass and, therefore, the blast design. Each of the above mentioned parameters comprises numerous site-dependent factors, some of which are interdependent. Hence, optimum spacing between blastholes is site dependent and its accurate determination for variable ground conditions is complicated. There are two ways to determine blasthole spacing in rocks. These are:

- By trial and error, in which one of the factors such as explosive type, amount or blasthole diameter is changed a little at a time and the rock is blasted until it produces the desired fragmentation for a particular site. This method is the oldest way of blast design, requiring frequent fine tuning to achieve satisfactory results. In the case of frozen ground, hot water or steam is injected into the frozen rock mass via the blasthole to thaw the ground. The blastholes are then quickly dewatered, charged and blasted in water saturated conditions, in accordance with the conventional blast design formulae given later in the paper. In this case, blasting should be carried out before the ground freezes again (12). This is a costly method of blasting.
- By inputting the actual condition of the rock mass and variation in its blastability into the design calculation, which is covered in this paper. With the latter technique, less fine tuning in the field will be required, resulting in cost reduction. This investigation is a continuing effort to achieve more accurate blast designs incorporating variations in ground conditions. It does not claim to have solved all the interacting problems but offers a start to quantify and formulate a design chart.

Therefore, more co-operation between the explosive manufacturers, blast designers and mining operators is essential in order to obtain adequate field data required for advances in the theoretical and practical research in this subject.

PRACTICAL EXAMPLE AND COMPARISON WITH OTHER TECHNIQUES

A mass of metal bearing quartzite is to be mined by open pit drilling and blasting. The blastholes are to be parallel to each other and to the bench. The objective is to determine the average spacing between the blastholes (S). The following information is available:

- Quality of rock mass in good and stable conditions, $Q_a = 20$.
- Quality of rock mass in fractured conditions, $Q_b = 4$.
- Dynamic tensile strength of quartzite, $\sigma_{td} = 20$ MPa.
- Diameter of blasthole, $D_h = 130$ mm.
- Angle of blasthole to horizontal, $\alpha = 70°$.
- Length of blasthole, L = 13.5 m.
- Charge length, h = 9 m.
- Length of stemming, x = 3 m.
- Subdrilling, $s_d = 1.5$ m.
- Detonation velocity of metalized Ammonium Nitrate and Fuel Oil (ANFO), $v_d = 3637.4$ m/sec.
- Volumetric specific consumption of explosive, $q_v = 0.4$ kg/m^3.
- Explosive density, $p_e = 1.0$ g/cm^3.
- Decoupling, $d = D_h / D_e = 1.1$.
- Charge placing and packing factor, $f_p = 1.2$.
- Constant related to the type of loading, C = 4.0.
- Constant related to the effect of the explosive impact velocity on the breaking property of the rock, N = 1.2.
- Constant for the intact quartzite and its state of fracture, $m_i = 15$.

Solution:

Utilizing the equation RMR =9.ln(Q) + 44 ; (ref.5)

Where:

RMR = rock mass rating ; (ref.6)

Q = rock mass quality or Norwegian Geomechanics Institute (NGI) rock quality.

RMR= in good condition = 9 ln 20 + 44 = 71.

RMR= in poor condition = 9 ln 4 + 44 = 56.

P_e = maximum explosion pressure in the blasthole = $10^{-3}\ p_e.v_d^{\ 2}/8$; (ref. 11)

= 10^{-3} x 1.0 x 3 637.4^2/8 = 1653.8, MPa

P_h = pressure developed in the blasthole to break the rock,

= $P_e.d^{-5}$ (modified from Cook 1958 ; (ref. 7)

= 1653.8 x 1.1^{-5} = 1 026.9,MPa.

F= impact force to cause rupture in the blasthole, MN = $\pi . D_h . p_e . d^{-5} . [(D_h/4) + h]$

$= 3.14 \times 0.13 \times 1\,653.8 \times 1.1^{-5} \times [(0.13/4) + 9] = 3\,786.17$, MN.

B_e= effective burden under dynamic conditions (m).

$= \{F^{[n/(2n-1)]}\}/g\ \{\sigma t d^{[2/(2n-1)]} . [\pi . c . d. D_h]^{[1/(2n-1)]}\}$ (ref. 4)

$= 6.03 \sim 6.0$, m.

ε_a= Blastability of the quartzite in good ground conditions

$= 2 /\{\{m_i \exp [(RMR - 100) / 28]\} - \{\{m_i \exp [(RMR - 100) / 28]\}^2 + 4 \exp [(RMR - 100) / 9]\}^{0.5}\}$ (ref. 2)

$= 2 / - 0.015 = - 134$

ε_b= Blastability of the quartzite in poor ground conditions

$= 2 /\{\{m_i \exp [(RMR - 100) / 14]\} - \{\{m_i \exp [(RMR - 100) / 14]\}^2 + 4 \exp [(RMR - 100) / 6.3]\}^{0.5}\}$ (ref. 2)

$= 2 / - 0.00286 = - 700$

S= Blasthole spacing in variable ground conditions,

$= \{B\ [0.7 + (3.6 /- \varepsilon)^{0.5} + (18.2 / \varepsilon] + (H - 2 B_e)\ [0.2 + (1.04 / - \varepsilon)^{0.5} + (5.2 / \varepsilon)\}/[q\ (x + h - s)]$ (modified from ref. 3)

S_a= Blasthole spacing in good conditions,

$= 6\ \{6\ [0.7 + (3.6 / 134)^{0.5} + (18.2 / -134) + (15 -12)\ [0.2 + (1.04 /134)^{0.5} + (5.2 / - 134)]\} / [0.4\ (3 + 9 - 1.5)] = 7.3$, m

S_b= Blasthole spacing in poor conditions,

$= 6\ \{6\ [0.7 + (3.6 / 700)^{0.5} + (18.2 / -700) + (15 -12)\ [0.2 + (1.04 /700)^{0.5} + (5.2 / - 700)]\} / [0.4\ (3 + 9 - 1.5)] = 7.4$, m

Other rock mass classifications such as rock mass strength parameters, rock quality designation (RQD), and Laubscher's adjusted RMR rating (AR), Rocha's rock mass ratio (MR), rock structure rating (RSR) and Brook Mathew's modified rock mass quality (Q') can also be utilized in the blastability equations instead of RMR or Q. This can be facilitated by using the relationships between the above noted classification systems (1). Comparison of the above finding with methods of determination published prior to this investigation are as follows:

M_e = Mass of explosive in kg/m run in the blasthole = 21.6, kg/m.

From Lambooy and Jones (ref.8): $S = h.\ M_e / B\ .\ H\ .\ qv$
$= (9 \times 21.6) / (6 \times 12 \times 0.4)$

$= 194.4 / 28.8 = 6.8$, m (a conservative value)

From Longefors and Kihlstrom (ref. 9): $S = 0.96\ h^2/B = 0.96 \times 9^2/6$
$= 12.96$, m (this is too large).

From Otano (ref.10): $S = 7.9 \times 10^{-3}\ [(\rho e.\ fp\ .\ Dh^2) / (qv\ .\ B^2\ .\ d^{2)}$

$= 7.9 \times 10^{-3}\ [(1.0 \times 1.2 \times 130^{2)} /(0.4 \times {}_6{}^2 \times {}_{1.1}{}^{2})$

$= 9.2$, *m* (this is large).

Modified from Afrouz (ref. 3):

Case (A) For packing factor (k) = 0.9, tight formations and angled holes to be blasted parallel to the free face:

$$S = 1.28\ [(k.\ h.\ \rho e.\ Dh^{2)} / qv\ .\ B\ (L - s)]$$
$$= 1.28\ [(0.9 \times 9 \times 1000 \times 0.130^2) / 0.4 \times 6\ (13.5 - 1.5)] = 6.1, m$$

(This value is very near to a square pattern; too small and utilized for tight formations).

Case (B) For packing factor (k) = 0.9, tight formations and vertical holes to be blasted to the free face at an inclination angle of "α" to the horizontal plane:

$$S = 3.5\ \{(k.\ h.\ \rho e.\ Dh^2) / qv\ .\ H\ [2B + H.\ \text{cotan}\ (\alpha)]\}$$

$$= 3.5\ \{(0.9 \times 9 \times 1000 \times 0.13^2) / 0.4 \times 12\ (2 \times 6 + 12\ \text{cotan}\ 70)\} = 6.2, m$$

(This value is for tight formations)

Figure 1 shows a more pronounced decrease in blastability of the poor and heavily fractured or loose ground in comparison with that of a good, tight and more stable ground. Comparison of various methods indicates that the new method provides an average or medium range design for blasthole spacing under variable rock mass conditions.

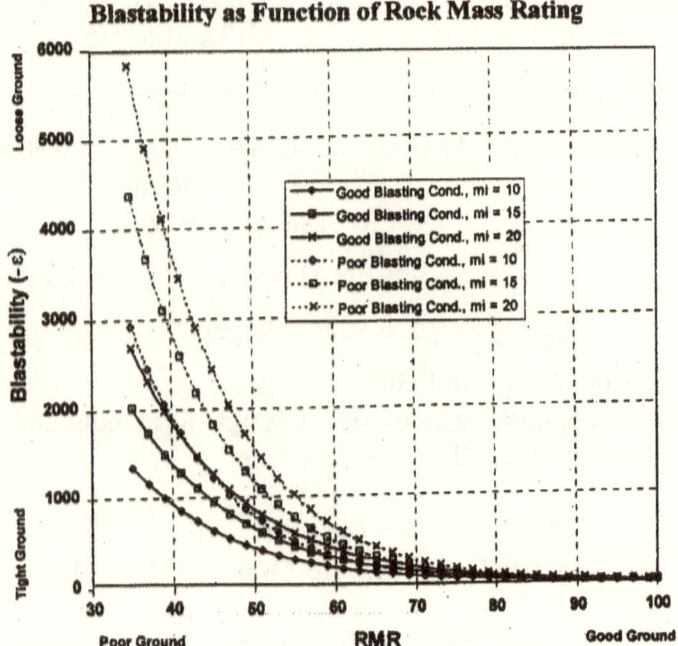

Figure 1. Variation in the Ground Blastability (-ε) as a Function of the Rock Mass Rating (RMR) for Rocks with Different Intact Strength (m_i)

CONCLUSIONS

The advantage of the suggested method over the others is its incorporation of:

- dynamic tensile strength of rock, and
- variation in the rock mass and ground condition in the blasting formulae, which leads to a more representative blastability and spacing optimization.

REFERENCES

1. AFROUZ, A (1992). 'Practical Handbook of Rock Mass Classification Systems and Modes of Ground Failure'. CRC Press, USA, April, pp 80-89.
2. AFROUZ, A, HASSANI, F P and UCAR, R (1988). 'An Investigation into Blast Design for Mining Excavation'. *Mining Science*

& Technology, **Volume 7**, Elsvier Science Publishers, B V, Amsterdam, February, pp 45-62.

3. AFROUZ, A (1985). 'Blasting in Mines'. Mighat Publishing Co, Tehran, Iran, **Volume 1,** August 1985, p 482.
4. AFROUZ, A (1969). 'Fracture Mechanizm of Rocks Under Dynamic Loading'.

MSc Thesis, University of Wisconsin, Madison, USA.

5. BARTON, N, LIEN, R and LUNDE, J (1977). 'Estimation of Support Requirements for Underground Excavations'. Proceedings of the 16th Symposium on Design Methods in Rock Mechanics, Univ of Minnesota, American Society of Civil Engineers, **Volume 1**, May, pp 163-177.
6. BIENIAWSKI, Z T (1989). 'Engineering Rock Mass Classification'. J Wiley & Sons Publishing Co, **Volume 1**, September, pp 52-55.
7. COOK, M A (1958). 'The Science of High Explosives'. Reinhold Publishing Corp, NY, **Volume 1**, September, USA, p 40.
8. LAMBOOY, P and JONES, R C E (1970). 'Practical Considerations of Blasting in Open Cast Mines'. Proceedings of the Open Pit Mining Symposium, Johannesburg, S Africa, 29 August - 4 September, Balkema, **Volume 1**, pp 227-234.
9. LONGEFORS, U and KIHLSTROM, B (1978). 'The Modern Technique of Rock Blasting'. J Wiley & Sons Publishing Co, New York, **Volume 1**, March, p 405.
10. OTANO, J A (1980). 'Fragmentacion de Rocas Con Explosices'. Editorial Poeblo Y Educacion, Spain, **Volume 1**, March, pp 174-175.
11. PEARSE, G E (1955). 'Rock Blasting - Some Aspects on the Theory and Practice'. *Mine and Quarry Engineering*, **Volume 21** (1), June, pp 25-30.
12. TAYBASHEV, V N and YEGUPOV, A A (1983). 'On Blastability of Frozen Placer Deposits in N Eastern Siberia, USSR'. **Volume 1**, August, p 15.

3.2.4 DETERMINATION OF BLASTHOLE SPACING IN ROCKS UNDER ARCTIC CONDITIONS

By: Andy A. Afrouz

ABSTRACT: This investigation reports on the theoretical and empirical findings to determine optimum spacing between parallel blastholes for open pit and surface excavations in frozen conditions. Determination of the burden is facilitated using dynamic tensile strength of the rock. This is then utilized to calculate the spacing by taking account of variation in the rock mass rating and blastability of frozen ground.

Comparisons between the suggested method and those published previously is made in the form of a practical example.

KEYWORDS: Blasthole spacing, Blasting, Rock Fragmentation.

1- INTRODUCTION

Spacing between blastholes is a function of several parameters, the most important of which are:

- Rock type, its degree of saturation, fisuration and ground conditions;
- Blastability of rock.
- Burden.
- Diameter and length of the blasthole.
- Orientation of the blastholes relative to each other and to the main free face.
- Degree or size of the rock fragments required.
- Type, physical and mechanical properties of the explosive to be utilized.
- Method of explosive charging, stemming and blasting.
- Cost of drilling, explosive, charging, and blasting.

Under arctic conditions, an additional parameter should also be considered which relates change in the physical and mechanical properties of the rock due to continuous permafrost or discontinuous freeze-thaw in the rock mass. This change affects blastability of the rock and, therefore, the blast design.

Each of the above noted parameters comprise numerous site dependent factors, some of which are interdependent. Hence, optimum spacing between the blastholes is site dependent and its accurate determination in the arctic conditions is complicated. There are three ways to determine blasthole spacing in rocks. These are:

(A) By trial and error - in which step by step one of the factors such as: the explosive type, amount of explosive and blasthole diameter is changed and the rock blasted until it produces the desired fragmentation for a particular site. This method is the oldest way of blast design, requiring numerous fine tunings to satisfaction.

(B) By thawing the frozen ground - in which hot water or steam is injected into the frozen ground via the blastholes to thaw the rock mass. The blastholes are then quickly dewatered, charged and blasted in water saturated condition, according to the conventional blast design formulae given in forgoing literature review. Blasting should be carried out before the ground freezes again [Taybashev & Yegupov, 1983]. This is a more elaborate but costly method of blast design.

(C) By taking the actual frozen condition of rock mass, as is, into the design calculation and input the variations in blastability of rock mass due to the permafrost, which is the scope of this paper. With the latter technique less fine tuning in the field will be required, resulting cost reduction.

This investigation is a continuing effort to formulate the effect of freeze - thaw on the ground conditions for blasting purposes. It does not claim to have solved all the interacting problems other than a start to quantify and formulate a curve, by considering the main affecting parameters under arctic conditions. Therefore, co-operation of the explosive manufacturers and mining establishments is highly encouraged to advance this work on both theoretical and the field fronts.

2- LITERATURE REVIEW

Popular formulae for determination of spacing between blastholes are as follows:

[Lambooy & Jones, 1970]: $S = h.W_e / B.H.q_v$ (1)

[Pugliese, 1972 and Langefors & Kihlstrom, 1978]:

$$S = (B.H)^{0.5,} \text{ if } 2B < H < 4B \quad (2)$$
$$S = 2B, \text{ if } H > 4B \quad (3)$$
$$S = 1.25\, h^2/1.3B = 0.86\, h^2/ B \quad (4)$$

In general, $B \leq s \leq 1.5B$, for square or rectangular blasthole patterns (5)

[Otano, 1980]:

$$S = 7.9 \times 10^{-3} [(\rho e.\, fp\, .\, Dh^2)/(qv\, .\, B^2\, .\, d^2) \quad (6)$$
$$= 7.9 \text{ x } 10^{-3} [(\rho e.\, fp\, .\, De^2)/(qv\, .\, B^2)$$

Where:

d = decoupling = $D_h/D_e = \Delta^{-5} = 1$ to 1.2 [Ucar, 1975] (7)

f_p = charge placing and packing factor [Anderson, 1952 and Frankel, 1952].

= 0.8 to 1.4 using tamping pole.

= 1.0 to 1.6 using pneumatic loader.

= 0.9 for granular ANFO without loader.

$S = 80 \text{ x } 10^3\, D_h^{\ 2}\, .\, \rho_e / B\, .\, q_v$, for 3<L<5 (8)

$S = W_e\, (L - 22D_h)\, /B\, .\, q_v\, .\, H$, for $3 < L < 5$, m. (9)

[Afrouz, 1985]:

$$S = Q_e / q_v\, .\, Be\, .\, H(n\text{-}1) = q_v\, .\, Bs\, .\, \sin\alpha\, .\, H(n\text{-}1)$$
$$= Q_e\, .\, \rho r / qm\, .\, Be\, .\, H(n\text{-}1)$$
$$= Qe\, .\, \rho r / qm\, .\, Bs\, .\, \sin\alpha\, .\, H(n\text{-}1) \quad (10)$$

[Afrouz, et al, 1988]:

$$S = 1.28\, k.\, h.\, \rho_e.\, D_h^{\ 2}/q_v\, .B\, (L - s_d) \quad (11)$$

where: k = packing factor

= 0.6 to 0.8 for cartridge explosives

= 0.8 to 0.9 for powdered explosives

= 0.9 to 1.0 for slurry explosives

For vertical holes, to blast free face with angle α to horizontal:

$$S=3.5k.h.\rho_e.D_h^2/q_v.H(2B+H.\cot\alpha) \quad (12)$$

Although useful in general ground conditions, none of the above noted equations account for parameters such as: variation in the state of rock mass under freeze - thaw, its strength and blastability under dynamic conditions.

3- THEORETICAL TREATMENT

Considering part of an open pit configuration shown in Figure 1.

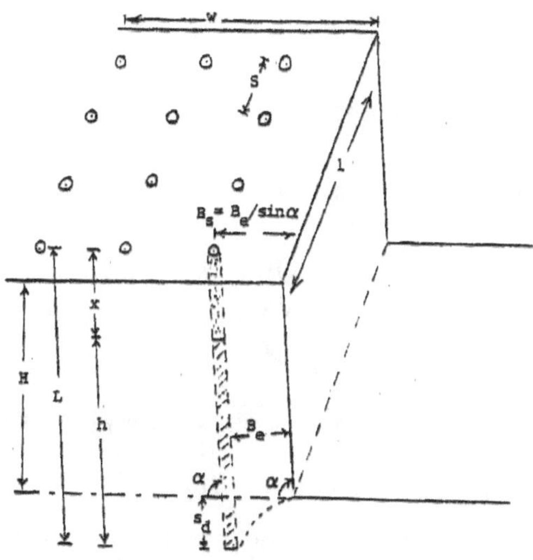

Figure 1 - Schematic of parallel blastholes in a surface mine.

The optimum burden (B_e) under dynamic conditions can be determined as follows [Afrouz, 1969]:

$$B_e=F^{[N/(2N-1)]}/g\{\sigma_{td}^{[N/2N-1)]}\ (\tau.C.d.D_h)^{[1/(2N-1)]}\} \quad (13)$$

Where: $F\ = P_h\,(A + A_e) = \pi.\,D_h.\,P_h\,[(D_h/4) + h]$, MN (14)

$P_h = P_e\,.\,d^{-5}$, MPa [modified from Cook, 1958] (15)

$P_e\ = P_d/2 \sim 10^{-3}\,\rho_e\,.\,v_d\,.\,V_g/2$, MPa [Pearse, 1955] (16)

In general, $v_g \sim v_d/4$, m/second (17)

Equation (13) can be expressed in various forms by use of equations (14) to (17).

Substituting equations (7) and (16) into equation (13) gives:

$$B_e = (1/g)\ \{3.14D_h\ .\ P_e\ [(D_h/4) + h]/d^5\ .\ \sigma_{td}\}^{[N/(2N-1)]}\ [0.32/C.d\ .D_h]^{[1/2N-1)]},\ m. \qquad (18)$$

On the other hand a relationship between the spacing (S), volumetric specific consumption of explosive (q_v), rock blastability (ϵ) and rock mass rating can be expressed as follows [Afrouz, et.al, 1988]:

$$S = Q_e/q_v\ .\ B_e\ .\ H = (Q_b + Q_c)/q_v\ .\ B_e\,(x + h - s_d)$$
$$= B_e\ \{B_e\ \{B_e[0.7 - (3.6/\epsilon^{0.5}) + (18.2/\epsilon)] + (H - 2B_e)\ [0.2 - (1.04/\epsilon^{0.5}) + (5.2/\epsilon)]\}\ /q_v\,(x + h - s_d),\ m. \qquad (19)$$

Equation (19) is applicable within the range of: $1.0 \leq S/B_e \leq 1.5$ and $H < 5B_e$

Blastability of rock mass (ϵ) can be given in two categories as follows [Afrouz, et.al., 1988]:

Category a - in a good blasting practice, stable rock mass and continuous permafrost condition, with low discontinuity and fracture intensity:

$$\epsilon_a = 2\ /\{\{m_i\ .\ \exp\ [(RMR - 100)\ /28]\}\ -\ \{\{m_i\ .\ \exp\ [(RMR - 100\ /28]\}^2 + 4\exp\ [(RMR - 100\ /9]\}^{0.5}\} \qquad (20)$$

Category b - in poor blasting practice, disturbed rock under freeze - thaw condition and fractured rock mass:

$$\epsilon_b = 2\ /\{\{m_i\ .\ \exp\ [(RMR - 100)/14]\}\ -\ \{\{m_i\ .\ \exp\ [(RMR - 100)\ /14]\}^2 + 4\exp\ [(RMR - 100)\ /6.3]\}^{0.5}\} \qquad (21)$$

Where: m_i = Constant for intact rock dependent upon the rock type, its properties and the state of fracture = 7 to 25, [Hoek and Brown, 1980]

RMR = Rock mass rating = 9 ln Q + 44 [Bieniewski, 1989] (22)

= 0 to 20, for unconsolidated to consolidated soil.

= 20 to 100 for weak, heavily fractured rock mass to high strength intact rock.

Q = Rock mass quality, [Barton, et.al., 1977]
= 0.01 to 500, for poor quality rock mass to intact strong rock.

4- PRACTICAL EXAMPLE & COMPARISONS WITH OTHER TECHNIQUES

(A) ***Problem:***

A mass of metal bearing quartzite is to be mined by open pit drilling and blasting. The blastholes are to be parallel to each other and to the bench. Determine optimum spacing between the blastholes (S). Following information is available:

- Quality of rock mass in continuous permafrost condition (Q_a) = 20
- Quality of rock mass in freeze - thaw condition (Q_b) = 4
- Dynamic tensile strength of quartzite (σ_{td}) = 20, MPa
- Diameter of blasthole (D_h) = 130, mm.
- Angle of blastholes to horizontal (α) = 70°
- Length of blasthole (L) = 13.5, m.
- Charge length (h) = 9, m.
- Length of stemming (x) = 3, m.
- Sub drilling (s_d) = 1.5, m.
- Detonation velocity of metalized ANFO (v_d) = 3637.4, $m.\ s^{-1}$.
- Volumetric specific consumption of explosive (q_v) = 0.4, $kg.m^{-3}$.
- Explosive density (ρ_e) = 1.0, $g.cm^{-3}$.
- Decoupling (d) = D_h/D_e = 1.1
- Charge placing and packing factor (f_p) = 1.2
- Constant related to the type of loading (C) = 4.0
- Constant related to the effect of the explosive impact velocity on the breaking property of the rock (N) = 1.2
- Constant for the intact quartzite and its state of fracture (m_i) = 15

(B) ***Solution:***

From equation (22), RMR = 9 ln Q + 44

RMR in continuous permafrost condition = 9 ln 20 + 44 = 71

RMR in freeze - thaw condition = 9 ln 4 + 44 = 56

P_e = maximum explosion pressure in the blasthole = $10^{-3} \cdot \rho_e \cdot v_d^2/8g$

$= 10^{-3} \times 1.0 \times 3637.4^2/8 = 1653.8$, MPa

P_h = Pressure developed at the blasthole wall to break the rock

$= P_e \cdot d^{-5} = 1653.8 \times 1.1^{-5} = 1026.9$, MPa

From equations (13) to (17):

F = Impact force to cause rupture in the blasthole, MN.

$= \pi \cdot D_h \cdot P_e \cdot d^{-5}[(D_h/4) + h]$

$= 3.14 \times 0.13 \times 1653.8 \times 1.1^{-5}[(0.13/4) + 9] = 3786.17$, MN

From equation (18):

The effective burden ($B_{e)}$ = (1/9.81) {3786.17/20} $^{[1.2/(2.4-1)]}$

$[0.32/4 \times 1.1 \times 0.13]^{[1/(2.4-1)]}$

$= 0.102\ (189.31)^{0.857}\ (0.559)^{0.714}$

$= 0.067 \times 89.441 = 6.03 \rightarrow 6.0$, m.

From equation (20):

Blastability of continuously frozen quartzite mass (ϵ_a)

$= 2/\{\{15 \exp[(71-100)/28]\} - \{\{15 \exp[(71-100)/28]\}^2 +$

$4 \exp[(71-100)/9]\}^{0.5}\} = 2/-0.025 = -80$

From equation (21):

Blastability of quartzite mass in freeze - thaw condition (ϵ_b)

$= 2/\{\{15 \exp[(56-100)/14]\} - \{\{15 \exp[(56-100)/14]\}^2 +$

$4 \exp[(56-100)/6.3]\}^{0.5}\} = 2/-0.007 = -286$

Substituting ϵ_a and ϵ_b into equation (19) gives, respectively:

Blasthole spacing in the continuous permafrost condition (S_a)

$= 6\{6\ [0.7 - (3.6/-80^{0.5}) + (18.2/-80)] + [12 - (2x6)]\ [0.2 -$

$(1.04/-80^{0.5}) + (5.2/-80)]\}/0.4\ (3 + 9 - 1.5) = 7.5$, m.

Blasthole spacing under freeze - thaw condition (S_b)

$= 6\{6\,[0.7 - (3.6 / {-286}^{0.5}) + (18.2 / {-286})] + [12 - (2x6)]\,[0.2 - (1.04/{-286}^{0.5}) + (5.2/{-286})]\}/0.4\,(3 + 4 - 1.5) = 7.2$, m.

Comparisons of the above findings with previously published methods of the spacing determination is as follows:

From equation (1):

We = Weight of explosive charged in the blasthole = 14.4, kg/m.

$S = h.W_e / B \,.\, H \,.\, q_v = 9 \times 14.4 / 6 \times 12 \times 0.4 = 4.5$,m.

It is too small, and more useful for underground condition, also it does not distinguish between a frozen and unfrozen ground.

From equation (2): $S = (B \,.\, H)^{0.5} = (6 \times 12)^{0.5} = 8.5$, m.
It is OK, but rather large for permafrost conditions.

From equation (4): $S = 0.86\, h^2 / B = 0.86 \times 9^2/6 = 11.61$, m. ; It is too large.

From equation (6): $S = 7.9 \times 10^6 D_h^{\,2} . \rho_e\, f_p / d^2 \,.\, q_v \,.\, B^2$

$= 7.9 \times 10^{-3} \times 0.13^2 \times 1.0 \times 1.2 / 1.1^2 \times 0.4 \times 6^2$

$= 9.2$, m.

It is O.K., but large for permafrost conditions.

From equation (9): $S = W_e\,(L - 22\, D_h) / B \,.\, q_v \,.\, H$

$= 14.4\,[13.5 - (22 \times 0.13)] / 6 \times 0.4 \times 12 = 5.4$, m.

It is small and useful for underground conditions, also it does not distinguish between frozen and unfrozen ground.

From equation (11):

k = Packing factor = 0.9

$S = 1.28\, k \,.\, h.\, \rho_e \,.\, D_h^{\,2} / q_v \,.\, B_e\,(L - s_d)$

$= 1.28 \times 0.8 \times 9 \times 1000 \times 0.13^2 / 0.4 \times 6\,(13.5 - 1.5) = 6.1$, m.

From equation (12):

$S = 3.5\, k.\, h.\, \rho_e . D_h^{\,2} / q_v .\, H(2B + H.\, \mathrm{Cot}\, \alpha)$

$= 3.5 \times 0.9 \times 9 \times 1000 \times 0.13^2 / 0.4 \times 12\,(2 \times 6 + 12\, \mathrm{Cot}\, 70°)$

$= 6.1$, m.

It gives very near a square pattern and can be maximized.

It can be seen that the new method gives smaller values than those of equations (2), (4), (6), and (9) in the arctic conditions. It gives larger values than equations (1), (11) and (12). Hence, it provides a medium range optimized design under continuous permafrost and freeze - thaw conditions.

Therefore, in very poor, and freeze - thaw ground, the spacing should be reduced from the normal average values. Similarly, in very tight and compacted continuous permafrost conditions the spacing should be reduced from the normal conditions.

Variation in the ground conditions, blastability and the optimum blasthole spacing for the above noted problem is given in Table 1 and illustrated in Figure 2.

TABLE 1 - SPACING BETWEEN THE BLASTHOLES (S) WITH VARIATIONS IN THE GROUND FREEZING CONDITIONS (RMR) AND ITS BLASTABILITY (ϵ)

Rock Quality Q	Rock Mass Rating RMR	Blastability ϵ	Spacing S, m	Ground Condition
0	25	-1000	6.8	Freeze-Thaw
1	44	-531	7.0	Freeze-Thaw
4	56	-286	7.2	Freeze-Thaw
10	65	-156	7.4	Permafrost
20	71	-80	7.5	(Maximum)
40	77	-50	7.2	Permafrost
100	85	-36	6.8	Permafrost
200	92	-25	5.9	Continuous freezing and tight ground
500	100	-15	3.6	Very tight ground

NOTE: The above values are applicable for burden (B) = 6 m and specific consumption of explosive (q_V) = 0.4, kg/m^3

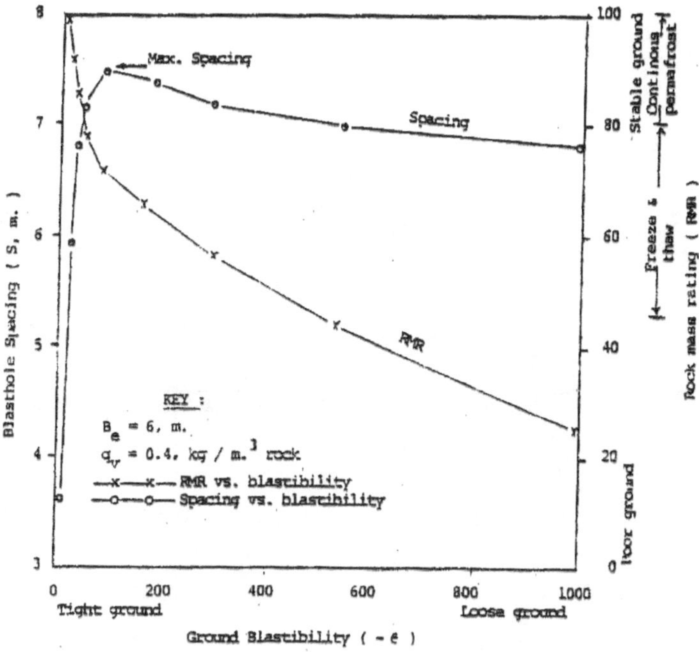

Figure 2 - Determination of blasthole spacing (S) with variation in the rock mass rating (RMR) and ground blastability (ϵ) due to freeze - thaw conditions.

5- CONCLUSION

For open cut operations in permafrost, the proposed technique is comparable to those suggested by Pugliese, Langefors and Kihlstrom, the equation (6) of Otano, and two equations suggested by the author previously. The advantage of this technique over the others is its considerations for the dynamic tensile strength of rock and variable ground freezing conditions which leads to a more representative blastability and blasthole spacing optimization.

REFERENCES

Afrouz, A., (1985), Blasting in Mines, Mighat Publishing Company, 482p.

Afrouz, A., Hassani, F.P. and Ucar, R., (1988), An Investigation into Blasting Design for Mining Excavations, Mining Science and

Technology, 7, Elsevier Science Publishers, B.V., Amsterdam, p 45-62.

Anderson, O., (1952) Blasthole burden design: Introducing a Formula, Proc. Australian Inst. Min. Met., p.166-167.

Barton, N., Lien, R. and Lunde, J., (1977), Estimation of Support Requirements for Underground Excavations, Proc. 16th Symposium on Design Methods in Rock Mechanics, University of Minnesota and American Society of Civil Engineers, p163-177.

Bieniawski, Z.T., (1989), Engineering Rock Mass Classification, J. Wiley and Sons, p52-55.

Cook, M.A., (1958), The Science of High Explosives, Reinhold Publishing Corp., N.Y., 40p.

Frankel, K.H., (1952) Factors Influencing Blasting Results, Manual on Rock Blasting, Atlas Diesel and S.J. Sweden, 1(2), 15 p.

Hoek, E., and Brown, E.T., (1980), Empirical Strength Criterion for Rock Masses, J. Geotech. Eng. Div., ASCE, 106(GT9) p1013-1035.

Lambooy, P., and Jones, R.C.E., (1970), Practical Considerations of Blasting in Open Cast Mines, Proc. Open Pit Mining Symposium, Johannesburg, S. Africa, Aug. 29-Sept. 4, Balkema, p 227-234.

Langefors, U. and Kihlstrom, B., (1978), The Modern Technique of Rock Blasting, J. Wiley and Sons, New York, 405p.

Otano, J.A., (1980), Fragmentacion de Rocas Con Explosives, Editorial Poeblo YEducacion, p 174-175.

Pearse, G. E., (1955), Rock Blasting - Some Aspects on the Theory and Practice, Mine and Quarry Engineering, 21(1), p 25-30.

Pugliese, J., (1972), Designing Blast Patterns Using Empirical Formulae, Information Circular 8550, U.S. Department of Interior, Bureau of Mines.

Taybashev, V.N. and Yegupov, A. A. (1983), On Blastability of Frozen Placer Deposits in N. Eastern Siberia, U.S.S.R.

Ucar, R., (1975), Decoupled Explosive Charge Effects on Blasting Performance, M.Sc. Thesis, University of Missouri at Rolla, Missouri, U.S.A.

NOMENCLATURE:

A = Cross sectional area of blasthole = $\pi.D_h^2/4$, m^2.

A_e = Effective area of the blasthole wall = $\pi.D_h.h$, m^2.

$B = B_S$ = Apparent burden, from the free face to blasthole, m.

B_e = Effective or real optimum burden, from the center of the charge to the nearest free face determined under dynamic conditions, m.

c = Constant related to the type of loading = 4.07 for direct impact.

d = Decoupling = D_h / D_e = 1 to 1.2.

D_e = Diameter of explosive placed in the blasthole, m.

D_h = Diameter of blasthole, m.

F = Impact force to cause rupture in the blasthole, MN.

fp = Charge placing and packing factor = 0.8 to 1.6.

g = Acceleration due to gravity = 9.81, m.s^{-2}.

H = Bench height, m.

h = Total length of the explosive charged in the blasthole, m.

k = Packing factor = 0.6 to 1.0.

L = Length of the blasthole, m.

M_e = Mass of explosive in kg/m. charged in the blasthole = W_e/g.

m_i = Constant for intact rock properties = 7 to 25 [Hoek and Brown, 1980].

n = Number of blastholes in one blast round.

N = Constant related to the effect of impact velocity on the breaking properties of rock = 1 to 1.4.

P_e = Maximum explosion pressure, MPa.

P_d = Detonation pressure, MPa.

P_h = Explosion pressure to break the rock, MPa.

q*m* = Mass of explosive to break one tonne of the ground with acceptable fragmentation size range, kg/tonne of rock.

qυ = Mass of explosive to break unit volume of the ground with acceptable fragmentation size range, kg/m^3 of rock.

Q = Rock mass quality = 0.01 to 500, [Barton et al, 1977].
Q_b = Total mass of explosive used at the bottom of blastholes for one blast round, kg.
Q_c = Total mass of explosive used at the column of blastholes for one blast round, kg.
Q_e = Total mass of explosive for one blast round, kg. = $Q_b + Q_c$.
RMR = Rock mass rating = 0 to 100, [Bieniawski, 1989].
S = Blasthole spacing in general, m.
S_a = Blasthole spacing in good ground and continuous permafrost conditions, m.
S_b = Blasthole spacing in poor ground and freeze - thaw conditions, m.
s_d = Sub-drilling, m.
v_d = Detonation velocity, $m.s^{-1}$.
vg = Particle velocity of explosion gases, $m.s^{-1}$.
W_e = Weight of explosive, kg/m. charged in the blasthole = g. M_e
x = Length of the stemming in blasthole, m
α = Angle of free face (bench) to horizontal, degrees.
= Angle of blasthole to horizontal, degrees.
ρe = Mass per unit volume of explosive, kg. m^{-3}.
ρr = Mass per unit volume of rock mass, tonne. $m.^{-3}$.
Δ = Explosive loading density, g/cm^3 of the blasthole volume.
ϵ = Blastability of rock mass.
ϵ_a = Ground blastability in good and continuous permafrost conditions.
ϵ_b = Ground blastability in poor and freeze - thaw conditions.
σtd = Dynamic tensile strength of rock, MPa.

3.3 MECHANICAL BEHAVIOR OF ROCK

3.3.1 FRACTURE CRITERIA OF SOFT ROCKS

By: Andy A. Afrouz, Ph.D., M.S., Mining
& Rock Mechanics Specialist

ABSTRACT

Scanning electron micrography was used as a technique to observe surface fracture of the lower four feet coal, ironstone nodule and its host underclay of the UK South Wales coalfield, fractured under the uniaxial compression. It was found that the fractured surfaces of these soft rocks exhibit a combination of brittle and ductile fracture with an overall behavior classified as elasto-plastic.

The surface structure of rocks, whether broken naturally or deliberately, exhibit markings and topography which are characteristic of the deformation and fracture mechanisms during the initiation and propagation of deformation and/or fracture.

The physical measurements of soft rocks by conventional methods such as strength, strain and compaction does not always provide a decisive answer on whether a particular fracture in the rock was developed as a result of brittle or ductile behavior. In porous rocks and rocks with primary cracks and other discontinuities, the overall behavior, specially in the initial stages of loading, appears to be exclusively plastic, by measurement. The stress-strain graphs generally obtained in the physical properties measurements of rocks show this behavior to commence as a curve start which accounts generally for the closure of the pores and cracks in the rock samples. Although the overall mechanical behavior of rock at this stage appear plastic, the true induced micro-fractures may or may not be ductile. A simple and accurate method of determining the development of deformation and fracture in rock is to observe its strained and/or failed surface structure by instruments which possess great magnification.

INTRODUCTION

The low depth of the light microscope makes photomicrography of non-planar fractured surfaces almost impossible. The Transmission Electron Microscope (TEM) with a considerably larger depth of field, a high resolving power and a large range of magnification is a better but not ideal tool to study deformed and fractured surfaces. Since, the TEM requires specimens thin enough for the penetration of electrons, rough fracture surfaces cannot be viewed directly. Hence, replicas should be made using either cellulose acetate or a direct evaporation of carbon. However, the replication of extremely rough fractured surfaces requires considerable expertise as well as being a time consuming method.

The Stereo Scanning Electron Microscope (SEM) facilities direct examination of the surface structure of thick and even the roughest samples without preparation of the replicas.

INSTRUMENTATION

An electron beam is generated by a heated tungsten filament (F) and is accelerated down to a vacuum column by a voltage usually in the range of 1 to 30 KV. The electrons are formed into a narrow probe by three electricity activated magnetic lenses (L_1, L_2 and L_3), and the final electron spot is focused onto the surface of the specimen (P), as shown in the accompanied figure.

The spot is made to scan across the specimen so as to give a square raster (in a manner similar to that in a television set). This is achieved by placing two pairs of deflecting coils between lenses L_2 and L_3, one pair (A_1 and A_2) being at right angle to the other pair (A_3 and A_4). Currents from a sawtooth generator (G) pass through the coils, one pair giving the line scan and the other frame scan. The same sawtooth currents also pass through the corresponding coils of cathode ray tube (T) so as to produce an identical but much larger raster on the viewing screen.

Schematic diagram of the Scanning Electron Microscope (SEM) with an X-ray micro-probe attachment

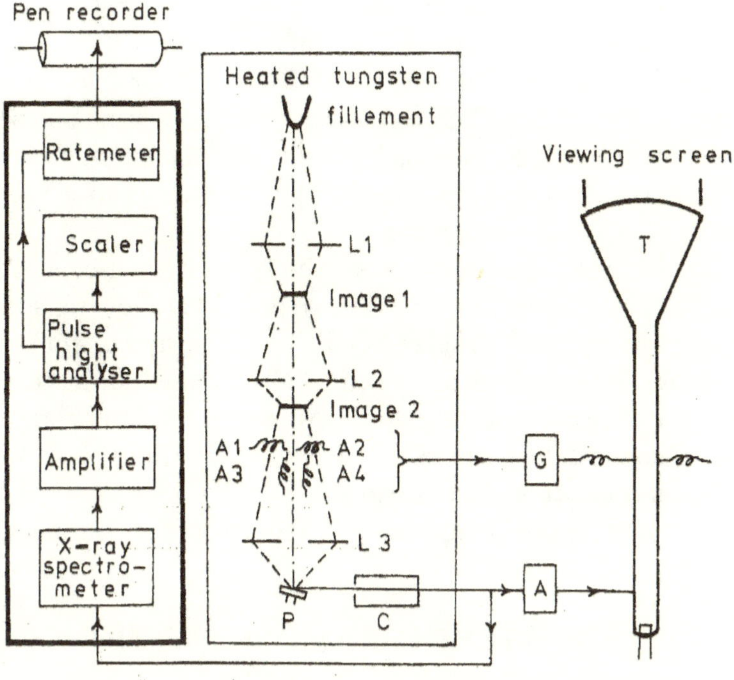

The electron spot incident on the specimen surface causes the electrons to leave the surface. These later electrons are attracted to a positively charged collector (C) adjacent to the specimen. The resulting electron current signal is amplified by an electron/ scintillator photomultiplier/amplifier chain (A), and used to control the brightness of the cathode ray tube. This is an exact point-to-point correspondence between the raster on the specimen surface and the raster on the screen of the cathode ray tube, and the brightness of each point on the latter is a direct measure of the number of electrons leaving the corresponding point of the specimen surface. Hence, a map of the surface of the rock specimen is obtained. As the illuminating electron beam is very narrow, the image displays a large depth of focus about 500 times greater than the light optical microscope, and the displayed topographical image clearly shows the three-dimensional nature of the fracture surface of the rock specimen at a magnification of up to 100,000 times depending upon operating conditions and the nature of the rock specimen. The lower four feet coal, underclay and ironstone nodules of the South Wales Coalfield, was investigated. Uni-axial compressive loads were applied to the irregularly and cylindrical shaped rock samples up to their failure under the laboratory condition. The failure surfaces thus obtained were cut down to approximately 1×1×0.5 cm., and air dried in room temperature of 20 deg. C. for one week before mounting on special aluminum stubs. They were left for one day for the mounting to harden, then fitted directly into the specimen stage in a vacuum chamber within the micro-probe.

The underclay specimens were shown to be non-conducting. Therefore, a thin conducting layer of gold palladium alloy was sprayed on these specimens for higher conduction and, hence, emissivity. Aluminum and/or carbon sprays could also be used to improve the emissivity of non-conducting rocks. Although, the latter sprays are more economical, they can readily be oxidized in time which would have adverse effect on the emission.

The test procedure was according to Thornton[1] and was conducted in room temperature of 20 deg. C.

MACROSTRUCTURAL OBSERVATION

Failure of soft rocks under compressive and indirect tensile pressures appeared to be accompanied with single and compound fractures leading to crush compaction. Balla[2] has shown that stresses in cylindrical samples subjected to triaxial compression are dependent on the ratio of the cylinder length to diameter and not on absolute dimensions. He has shown that if the boundary conditions were such that they prevented lateral movement of the end faces of the samples during compression, then the stress difference and the strain energy would have their maximum values at the perimeters of the plane ends. Thus, failure is most likely to be initiated at a point on the sample perimeter, this theory was found to be consistent with the observation made on the crack development of the rock samples.

The fracture was mainly docile and was along one or more irregular planes inclined to the direction of applied stress. Natural bedding and artificially induced layering bonded with Durafix resin in the rock samples have shown to influence the direction of fracture progression and consequently the fracture plane. The angle of fracture

(θ) was considered to be the angle between the fracture plane and the direction of the axial applied stress (vertical) in the rock. this angle was measured directly from the specimens after completion of the tests by a protractor to within ± 1 degree accuracy.

The average measured angle of fracture for the coal and underclay was 32 and 27 degrees producing an internal friction angle (ϕ == 90-2 θ) of 26 and 36 degrees, respectively. Since, the coal exhibited a lower average uniaxial strength, perpendicular to the bedding plane, of 5.8 ± 0.2 MN/sq m. in comparison to 18.1 ± 0.9 MN/sq. m. of the underclay, it can be suggested that the angle of fracture is indirectly proportional to the strength of the rocks. Moreover, the angle of internal friction is directly proportional to the strength of the rocks.

MICROSTRUCTURAL OBSERVATION

Commonly, under stresses, whether tensile, compressional, or shear the molecules, grains and/or crystals in a rock may deform and eventually separate by pulling apart or sliding past one another. In soft rocks, the deformation and failure processes stem from a combination of these two theoretical mechanisms.

Usually, rocks especially dry and/or hard ones are inherently brittle. But brittleness is a normal property of most solids at low temperature ranges (between –273 deg. C. and zero deg. C.). Most of the knowledge about brittleness has come from studies on glass, ice and structural steel, which exhibit plastic behavior, especially at medium and high temperature ranges.

Observed fracture and/or rupture in the rocks, under the room temperature of 20 degrees Celsius, can be categorized into either: (A) Brittle, (B) Ductile, or (C) combination of Brittle and Ductile. In any of the above noted categories, a fracture may be:

(1) Inter-crystalline — where failure occurs by the separation of crystal aggregates from one another along the grain boundaries utilizing certain amount of energy for the crack formation and propagation.
(2) Transcrysalline — where the fracture passes through or across the crystals or grains utilizing more energy than the case (1) for the crack development.

A — Brittle Fracture: This is sometimes known as cleavage fracture. Each crystal and/or grain of rock tends to break on a cleavage plane. Since, crystals and grains are at random orientations, the failure process results in a fracture surface that is flat, but rough and jagged, and lustrous when rotated in the hand. In general, if the resistance of the rock to sliding (or shearing) is greater than the resistance to separation (or tension) the rock may be expected to behave as brittle.

Failure of rock caused locally at microscopic defects and/or crack formations in the direction normal to its largest separation distance. The development of a crack is considered to involve slip and tension. The micro-cracks join together and create cracks which, in an idealized case of brittle fracture, is normal to the greatest tension. Parallel brittle fractures were observed in the ironstone nodule with the lower four feet underclay exposing its layered structure, in a dry condition. These fractures may join together as a large cleavage fracture (C) as observed in the ironstone nodule. The steps between the different planes of fracture give rise to characteristic "River-line" markings on the fracture surface (C). These "River-lines" tend to run together, so giving the local direction of mean crack propagation. They can also be seen to run along and across the main cleats or any other per-existing plane of weakness. A brittle fracture may also be as straight fibrous and have a ripple like surface.

B — Ductile Fracture: This type of fracture is caused by shear stress. The fracture surface appears glossy with plastic deformation and/or curved fibers. In general, if the resistance of the rock to separation (tension) is greater than the resistance to sliding (shearing), the rock behaves in a ductile manner. This fracture was usually accompanied with void like surfaces. This type of fracture, in underclay, has a close resemblance to dimple fracture of metals under moderate to high temperature ranges.

C — Combination of Brittle and Ductile Fractures: In general, fracture of the soft mine rocks encountered rarely appears to be exclusively brittle or ductile. Variation in structure of layers comprising these sedimentary rocks causes variation in deformation and consequently fracture surface of the rocks. This was observed especially in coal where a fibrous fusain layer (B) was sandwiched between two layers of clarain (A and C). Magnification reveals the fracture characteristics of the fusain (B) to be ductile and that of the clarain (C) to be brittle creating an overall behavior which is elasto-plastic.

CONCLUSION

The Scanning Electron Micrography have shown that:

1. Due to predominant discontinuities such as cleats, bedding and other planes of weakness in coal, the overall behavior appears as plastic. However, the majority of fracture developed in coal samples observed to be brittle in nature comprising fusain bands exhibiting plastic deformation. The average fracture angle of Coal is 31 ± 2 degrees to the direction of applied load.
2. Behavior of the underclay is observed to be elasto-plastic leading to predominately ductile fractures with small pockets of brittle cracks when the rock is in a dry state. The wet rock displays exclusively plastic behavior leading to ductile fractures oriented at an average angle of 26 ± 2 degrees to the direction of applied load.
3. The ironstone nodules, comprising of higher metal and quartz content than its host underclay, have a higher elastic property and fracture in a brittle manner.
4. The overall behaviour of the multimedia soft mine rocks may be classified as elasto-plastic when dry and plastic when wet. The majority of fracture characteristics observed was a combination of brittle and ductile.
5. The velocity of crack propagation and the speed of rock fracture appeared to be very high. Experimentally, it was not possible to stop a fracture and study early stages of its growth. But it is believed that the yield point is the point where several micro-cracks form or open up. A small amount of plastic deformation may also take place before initiation of the cracks. These micro-cracks then join to form a larger crack leading to the rock failure application of a higher load.

REFERENCES

1. THORNTON, J., Scanning Electron Microscopy, Chapman and Hall, (1968).
2. BALLA, A., A new solution to the stress conditions in triaxial compression, Acta. Tech. Hung., Vol. 28, (1960), pp. 349-388.

3.3.2 RHEOLOGY OF ROCKS WITHIN THE SOFT TO MEDIUM STRENGTH RANGE

By: Andy A. Afrouz and John Harvey

This paper contains a description of laboratory tests conducted at room temperature and atmospheric pressure, on dry and saturated rocks exhibiting uniaxial compressive strengths within the soft to medium strength range. A computer program was developed to obtain and output mathematical expressions to the creep of rocks in order of best fit. The time-dependent equations thus chosen were compared with the expressions obtained by graphical methods. A comparison has also been drawn between the laboratory results and in-situ *time-dependent measurements.*

1. INTRODUCTION

One of the main difficulties in interpreting and relating the results of laboratory tests to *in-situ* behaviour is the time involved in each type of analysis. While laboratory experiments take only a few minutes to perform, under *in-situ* conditions, time is an important factor and deformations may take many days or months to complete.

In multilayer strata, time-dependent rock deformation manifests itself under natural conditions of geological processes, such as compression and shear produced by tectonic forces within the earth's crust, leading to plastic states covering large areas. Engineering activities, such as mining operations also cause local plastic zones from the yielding of pillar bases and the permanent deformation of mine openings. The state of stress in the plastic zone and the shape of the zone in mine openings depends, to a large extent, on the type of deformation, degree of roughness of the contact surface between rock and mine supports, moisture and boundary conditions, thus indicating the complexity of the problem.

Time-dependent deformation can generally be described either by a stress–relaxation law, which expresses the dependence of stress on the strain history, or by a creep law, which expresses the dependence of strain on the stress history. The experimental investigation of these constitutive relations generally involves stress relaxation at constant strain and creep tests at constant stress (or at constant load). In stress relaxation tests, a rock specimen is brought suddenly to a state of strain, which is held constant, and the time-dependent stress required to support this constant deformation is measured. In a creep test, the specimen is brought suddenly to a state of stress, which is held constant, and the resulting time-dependent strain is measured. Although each of the above noted tests offer advantages and simulate different conditions, the latter method is readily adaptable because of the simpler instrumentation, test procedure and control of variables involved, and therefore, was considered a more accurate and feasible method to use under laboratory conditions.

2. EXPERIMENTAL TECHNIQUE

Laboratory investigations were limited to creep tests upon air-dried and water saturated specimens of cylindrical shape observed at room temperature (20 ± 4°C) and atmospheric pressure (1·02 ± 0·05 kgf/cm^2). Consideration was given to the measurement of time-dependent, load-deflection behaviour of rectangular strips of rocks as beams or as cantilevers. Apart from difficulty in accurately obtaining symmetrical beams of soft rocks, the compressional load on the beams or cantilevers would induce a non-uniform stress distribution throughout the sample under test. Therefore, it was decided to use cylindrical specimens under direct uniaxial compression. Moreover, the loading technique suggested itself as being a simpler and inexpensive way in which a rock may be loaded with a reasonably homogeneous stress over the load-application surface of the specimen.

The cylindrical specimens were air dried over a period of 6 days, after which time strain gauges were mounted at the mid-length of the samples, one parallel and the other perpendicular to the specimen axes. A light paraffin coating was applied to prevent ingress of moisture. In cases when the rocks were tested in a saturated condition, the samples were positioned in the middle of a water tight steel cylindrical jacket filled with distilled water. The jacket was then placed on the spherical seat of the creep rig.

For application of constant stress to the rock samples, a small creep rig capable of applying compressive loads ranging from 45 up to 2270 kg was found to be adequate. The arrangement and test procedure using the creep rig is described elsewhere [1]. However, the rig was used in conjunction with an accurate dial gauge and strain meter rather than the telescopic arrangement used by the previous researchers. The theoretical mechanical advantage of the rig was 125 to 1, but the mechanical advantage obtained by calibration was found to be 100 to 1. Calibration was performed using a load cell constructed of strain gauges mounted on a cylindrical brass specimen with known physical properties and of the same geometry as that of the test specimens. The strain gauges were connected to a strain meter and the circuit balanced before applying load to the creep rig. Upon application of the instantaneous load on the specimen, its corresponding strain ($\varepsilon_1 \times 10^{-3}$), ($\varepsilon_3 \times 10^{-3}$) and overall shortening ($\Delta l \times 10^{-3}$), with respect to time (t/hr), was noted and illustrated graphically.

3. THEORETICAL CONSIDERATION

A cylindrical rock sample of volume (v) compressed axially by a constant load (p) under laboratory conditions (Fig. 1) was considered.

Upon application of the load, the initial instantaneous stress (σ_0) will develop as:

$$\sigma_0 = \text{Load } (p) / \text{Area } (A) = 4p / \pi D_0^2 \quad (1)$$

where D_0 is the original diameter of the sample. If the loading is continued over a time $0 < t < \infty$, the specimen geometry tends to change, creating a new time-dependent diameter (D_t).

So, the stress at any time $(0 < t < t_f) = \sigma_t = 4p / \pi \,.\, Dt^2$ (2)

where: t_f = the time taken for the rock to fail under the creep load (p), and $Dt^2 = Do^2 + \Delta D^2$, as $A_t = A_0 + \Delta A$.

From equations 1 and 2: $\sigma_t = \sigma_0 . Do^2 / Dt^2$ (3)

According to the creep definition [2]:

$$v = \text{constant} = L_0 \times A_0 = L_t \times A_t = \pi.\ L_0 \,.\ Do^2/4 = \pi\ .\ L_t \,.\ Dt^2/4,$$
$$\text{for } 0 < t < t_f.$$

Hence, $Dt^2 = L_0 . Do^2 / L_t.$ (4)

$$\sigma_t = \sigma_0 D_0^2 / D_t^2.$$

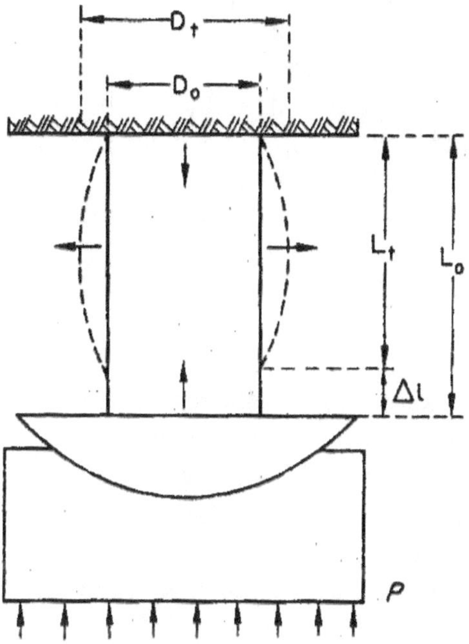

Fig. 1. Simplified deformation of rock sample under a uniaxial creep rig.

According to the strain definition: $\varepsilon_t = (L_0 - L_t)/L_0 = \Delta l / L_0,$ for $0 < t < t_f$

Therefore, $L_t = L_0 - \Delta l = L_0 (1 - \varepsilon_t)$ (5)

From equations 4 and 5: $Dt^2 = L_0 . Do^2 / L_0 (1 - \varepsilon_t) = Do^2 (1 - \varepsilon_t)$ (6)

Substituting equations (6) into equation (3) yields:

$$\sigma_t = \sigma_0 . Do^2 / D0^2 (1 - \varepsilon_t) = \sigma_0 (1 - \varepsilon_t) = E . \varepsilon_0 (1 - \varepsilon_t) \quad (7)$$

At any given time period, ($t_1 < t_f$, $t_2 < t_f$), the average strain (ε_t) may be obtained by integrating the rate of change of the strain with respect to time ($d\varepsilon/dt$) within the time limits (t_1, t_2). Hence, equation (7) can be expressed in the following form:

σ_t = σ0 |1- ∫Integral between t1to t2 of (dε /dt) dt|

= E . ε0|1- ∫Integral between t1to t2 of (dε /dt) dt| (8)

Equation (8) suggests that, as strain in the specimen varies with time, the effective stress on the specimen (σ_t) varies monotonically and independently of the applied constant load (p). This equation also suggests that it is possible to determine or predict the stress level in a rock at any time prior to its failure, if the initial stress imposed upon the rock and its time-dependent deformation (or the strain rate) is known.

4. ANALYSIS OF DATA

There have been a number of empirical equations introduced [2–5] in an effort to establish and define a relationship between deformation and time that would fit experimental results. However, these equations do not adequately describe the full range of experimental results and deformation mechanism of rock. As a result, many investigators [6–8] have recommended that the observed behaviour be expressed in terms of mechanical models and it is about such models, together with polynomial equations, that a computer program has been constructed. The equations chosen for the analysis are:

(1) $\varepsilon = \sigma t / \mu$
(2) $\varepsilon = \sigma(1/E + t/\mu)$
(3) $\varepsilon = \sigma / E + (\sigma - \sigma_y) t / \mu$
(4) $\varepsilon = A + B.\ \ln t$
(5) $\varepsilon = A + B.\ t^n$
(6) $\varepsilon = (\sigma / E)[1 - \exp(- Et / \mu)]$
(7) $\varepsilon = \sigma\{1/E + (1/E_2)[1 - \exp(- t\ .\ E_2 / \mu)]\}$
(8) $\varepsilon = \sigma / E\ \{1 - [E_1 / (E_1 + E_2)\ \exp_1(-E_1\ .\ E_2\ .\ t) / \mu\ (E_1 + E_2)]\}$
(9) $\varepsilon = \sigma / (E_1 + E_2) + \sigma\ \{[(1 / E_2) - [1 / (E_1 + E_2)]\}$
(10) $\varepsilon = A + B \ln t + ct^n$. This equation has been applied to the time–strain results for rocks [2, 3]

(11) $\varepsilon = \sigma\{1/E + t/\mu + (1/E_2)[1 - \exp(-E_2 t/\mu_2)]\}$
(12) $\varepsilon = \sigma\{1/E + (1/E_2)[1 - \exp(-E_2 t/\mu_1)]\} + (\sigma - \sigma_y) t/\mu_2$
(13) $\varepsilon = \sigma\{1/E + (1/E_2)[1 - \exp(-E_1 t/\mu_2\}] + t/\mu_1(1 - P/\sigma)\}$
(14) $\varepsilon = A + B.\ t^m + c.\ t^n$. This equation has been found to represent closely the creep curve for metallic alloys [9].
(15) $\varepsilon = A + B[1 - \exp(mt)] + ct^n$. This equation is of the Berger Model type and has been applied to the strain–time results of rocks [2, 10, 11].

For simplicity of analysis and to obtain a uniformity between the models and polynomials, the coefficients related to the physical properties of the rocks, such as rock viscosity (μ) and modulus of elasticity (E), were regarded as constant. It is thus possible to categorize each of the above equations as a basic model. From this, nine different basic models were generated:

(1) $\varepsilon = p_1 t$, representing equation (1)
(2) $\varepsilon = p_1 + p_2 t$, representing equations (2) and (3)
(3) $\varepsilon = p_1 + p_2 . \ln t$, representing equation (4)
(4) $\varepsilon = p_1 + p_2$, representing equation (5)
(5) $\varepsilon = p_1 + p_2 \exp(p_3 t)$, representing equations (6), (7), (8) and (9)
(6) $\varepsilon = p_1 + p_2 \ln(t) + p_3$, representing equation (10)
(7) $\varepsilon = p_1 + p_2 t + p_3 \exp(p_4 t)$, representing equations (11–13)
(8) $\varepsilon = p_1 + p_2 + p4$, representing equation (14)
(9) $\varepsilon = p_1 + p_2 + p_4 \exp(p_5 t)$, representing equation (15)

It can be seen from the above that these basic models represent a progression of complexity with the use of the standard functions upon time, with respect to deformation. It was apparent, even before actual data was used in the analysis, that the more complex basic models would yield the better approximations to the experimental work, although it was not apparent as to which combination of functions of time would yield the optimum expression.

With respect to various samples of experimental data, each of the basic models was subjected to an unweighted least squares fit by means of stepwise Gauss–Newton iterations [12] on the parameters, p, minimizing the error sum square with each iteration;

Ess = Σ between i=1 to n of $\{[\varepsilon_1 - f(t_1, p_1, \ldots\ldots\ldots, p_{np})]^2\}$

where: ε_i is the ith value of the deformation of the sample,

t_i is the corresponding value of time,

f is the function of time, with respect to deformation, corresponding to the basic model under analysis,

n is the sample size, and

np is the number of parameters corresponding to the basic model under analysis.

Exit from the least squares analysis for each basic model is accomplished when comparison of the error sum squares from two consecutive iterations,

$$ERR = |(E_0 - E_1)/E_1|$$

where E_0 and E_1 represent the error sum squares of the ultimate and penultimate iterations, respectively, diminishes past a criterion of convergence (for this analysis, set at $1 \times 10^{-6)}$.

After a least squares analysis has been completed on each of the basic models, they are sorted, with respect to their error sum squares, in order of best fit. Analysis of a number of rock type samples enabled a comparison of the best fitting models of each sample to be performed. The results for the eight rock samples thus analyzed are given in Table 1, and Fig. 2 displays a graphical representation of the best fitting model of each sample.

5. DISCUSSION OF RESULTS

The rheology exhibited by rocks tested under constant uniaxial compression is qualitatively similar to those obtained by some of the previous investigators [1–6, 11]. In general, complete time-dependent behaviour comprised of the four stages: the instantaneous elastic, primary creep, secondary creep and tertiary creep, leading to the eventual failure of the rock sample, as shown in Figs. 2–4.

5.1. *Computer analysis*

Analysis of the resulting 'best fits' produces the term of the power of time, t^p, where p is a parameter, as the most dominant term (Table 1). This can be observed by the relative positions of models 4, 7 and 9: model 4, apart from in the case of saturated underclay with clay band, always precedes model 7, and model 9 always precedes model 4, while the only difference between model 7 and model 9 is the power of time in the second term of each. Furthermore, other terms, apart from the intercept, accompanying a power of time within a model are, in general, greatly suppressed, either from within the term or by the multiplicative parameter of the term. For example, for the air-dried coal, the best fit is model 9, expressed as: $\varepsilon = p_1 + p_2 \times t + p_4 \exp(p_5 \times t)$. The parameter within the exponential term, p_5, is almost unity ($9{\cdot}926875{\times}10^{-1}$) while the multiplicative parameter to that term, p_4, is almost zero ($1{\cdot}631667{\times}10^{-22}$).

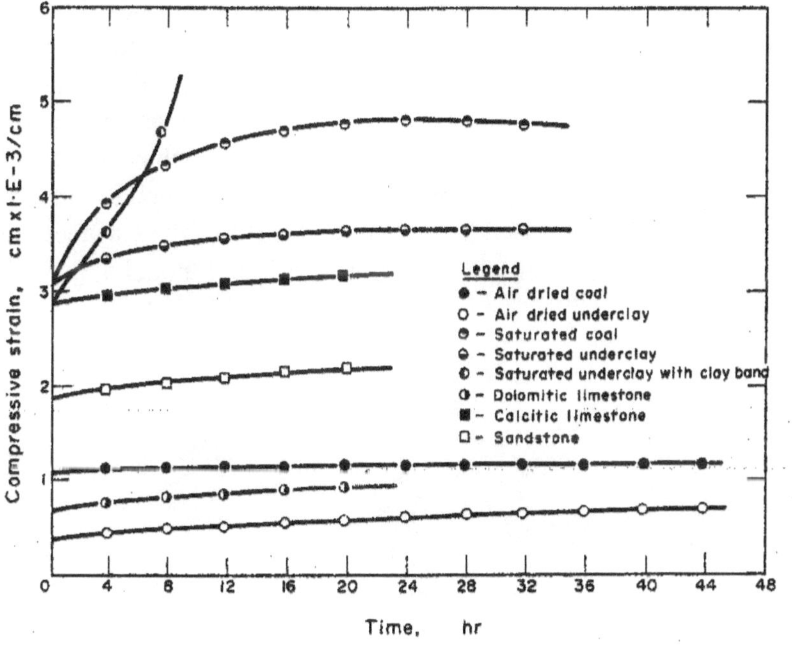

Fig. 2. Comparison of time dependent deformation of some soft and medium rocks.

Alternatively, for the sandstone, it is the parameter within the term that suppresses the exponential (i.e. $p_5 = 19{\cdot}999995 \times 10^2$). The behaviour of a rock sample, indicates that, as the time becomes large, the deformation and, consequently, development of strain in the sample becomes almost linear in its progression, prior to its tertiary creep range, in much the same way as the Maxwell-Kelvin (i.e. Bunger) model predicts. It is this that the parameters of these three terms of time suppress the great increase in the time value so as to restrict the projection of the strain/time graph to near linear in behaviour. Model 2, ($\varepsilon = p_1 + p_2 \cdot t$), does have an advantage over the other models comprising only two terms in as much as it is linear throughout, but it does have the disadvantage in that it cannot compensate for the large, nonlinear deformation that occurs during the primary creep of the rock samples. Model 8 comprises an intercept and two terms composed of powers of time. This model is the best fit for the saturated coal and saturated underclay samples, expressing elasto–plastic behaviour, while it was a constant position of third for the other rock types. However, powers of time appear to be the dominant terms in this analysis so, as model 8 is composed mainly of such terms, its positioning of third in the majority of samples analyzed appears slightly low. It is possible that the two terms of powers of time are in conflict with each other—each could take the role that the other plays—and combine to form a model similar to model 4, slightly modified to account for a little of the variances due to the rock type and sample data from the line of model 4.

TABLE 1. RESULTS OF THE COMPUTER PROGRAM, CREEP, UPON THE ROCK SAMPLES PARAMETER VALUES OF BEST FIT

Rock type	Models, in order of best fit									Parameter values of best fit
	1	2	3	4	5	6	7	8	9	
Air-dried coal	9	6	8	4	7	2	3	5	1	$p_1 = 1{\cdot}027867$ $p_2 = 5{\cdot}860112 \times 10^{-2}$ $p_3 = 2{\cdot}521902 \times 10^{-1}$ $p_4 = 1{\cdot}631667 \times 10^{-22}$ $p_5 = 9{\cdot}926875 \times 10^{-1}$
Air-dried underclay	6	9	8	4	7	2	3	5	1	$p_1 = 3{\cdot}625824 \times 10^{-1}$ $p_2 = 3{\cdot}852710 \times 10^{-3}$ $p_3 = 3{\cdot}216117 \times 10^{-2}$ $p_4 = 6{\cdot}284337 \times 10^{-1}$
Saturated coal	8	6	9	4	7	3	2	5	1	$p_1 = 2{\cdot}768859$ $p_2 = -1{\cdot}393853$ $p_3 = 8{\cdot}154296 \times 10^{-1}$ $p_4 = 1{\cdot}974527$ $p_5 = 7{\cdot}385941 \times 10^{-1}$
Saturated underclay	8	6	9	4	7	2	3	5	1	$p_1 = 2{\cdot}972680$ $p_2 = -8{\cdot}247645 \times 10^{-1}$ $p_3 = 7{\cdot}383078 \times 10^{-1}$ $p_4 = 1{\cdot}024724$ $p_5 = 6{\cdot}936851 \times 10^{-1}$
Saturated underclay with clay band	6	9	7	5	8	4	2	3	1	$p_1 = 3{\cdot}094857$ $p_2 = 2{\cdot}008622 \times 10^{-1}$ $p_3 = 1{\cdot}315394 \times 10^{-2}$ $p_4 = 2{\cdot}225700$
Calcitic limestone	6	9	8	4	7	2	5	3	1	$p_1 = 2{\cdot}822289$ $p_2 = 1{\cdot}613669 \times 10^{-3}$ $p_3 = 4{\cdot}790967 \times 10^{-2}$ $p_4 = 6{\cdot}511509 \times 10^{-1}$
Dolomitic limestone	9	6	8	4	7	2	5	3	1	$p_1 = 6{\cdot}484839 \times 10^{-1}$ $p_2 = 5{\cdot}632693 \times 10^{-2}$ $p_3 = 4{\cdot}891489 \times 10^{-1}$ $p_4 = 7{\cdot}083883 \times 10^{-5}$ $p_5 = 4{\cdot}901808 \times 10^{-1}$
Sandstone	9	6	8	4	7	2	5	3	1	$p_1 = 1{\cdot}857982$ $p_2 = 4{\cdot}096582 \times 10^{-2}$ $p_3 = 6{\cdot}869447 \times 10^{-1}$ $p_4 = 5{\cdot}804307 \times 10^{-2}$ $p_5 = 9{\cdot}999995 \times 10^{-2}$

The best fit obtained by the computer analysis for the rocks tested is as follows:

(a) for the air-dried coal:

$$\varepsilon_t = [1{\cdot}028 + 4{\cdot}368 \times 10^{-22} \times \exp(0{\cdot}971t) + 0{\cdot}059t^{0{\cdot}252}] \times 10^{-3}$$
$$= [1028 + 4{\cdot}368 \times 10^{-19} \times \exp(0{\cdot}971t) + 59t^{0.252}] \times 10^{-6} \quad (9)$$

(b) for the air-dried underclay:

$$\varepsilon_t = (0{\cdot}363 + 0{\cdot}004 \ln t + 0{\cdot}032t^{0{\cdot}628}) \times 10^{-3}$$
$$= (363 + 9 \log t + 32t^{0{\cdot}628}) \times 10^{-6} \quad (10)$$

(c) for the saturated coal:

$$\varepsilon_t = (2769 - 1393t^{0{\cdot}815} + 1975t^{0{\cdot}739}) \times 10^{-6} \quad (11)$$

(d) for the saturated underclay:

$$\varepsilon_t = (2973 - 825t^{0{\cdot}738} + 1025t^{0{\cdot}694}) \times 10^{-6} \quad (12)$$

(e) for the saturated underclay with clay band:

$$\varepsilon_t = (3095 + 463 \log t + 13t^{2{\cdot}226}) \times 10^{-6} \quad (13)$$

(f) for the air-dried calcitic limestone:

$$\varepsilon_t = (2822 + 5 \log t + 48t^{0{\cdot}651}) \times 10^{-6} \quad (14)$$

(g) for the air-dried dolomitic limestone:

$$\varepsilon_t = [648 + 56t^{0{\cdot}489} + 0{\cdot}7 \exp(0{\cdot}490t)] \times 10^{-6} \quad (15)$$

(h) for the air-dried sandstone:

$$\varepsilon_t = [1858 + 410t^{0{\cdot}687} - 58 \exp(0{\cdot}010t)] \times 10^{-6}. \quad (16)$$

5.2 *Graphical analysis*

The mathematical expression for the time-dependent behaviour of air-dried and water-saturated coal and underclay samples, under an initial stress of 25·7 kg/cm^2, using graphical method. (Fig. 5), is as follows:

(a) for the air-dried coal:

(i) loading:

$$\varepsilon_t = \varepsilon_0 + \varepsilon_1 + \varepsilon_2 + \varepsilon_3$$
$$\varepsilon_t = (1018 + 8{\cdot}25 \log t + 2{\cdot}48t) \times 10^{-6} + \varepsilon_3 \quad (17)$$

(ii) unloading (or recovery):

$$\varepsilon_r = \varepsilon_{r0} + \varepsilon_{r1} + \varepsilon_{r2}$$
$$\varepsilon_r = (1108 - 0{\cdot}63 \log t) \times 10^{-6} + \varepsilon_{r2} \quad (18)$$

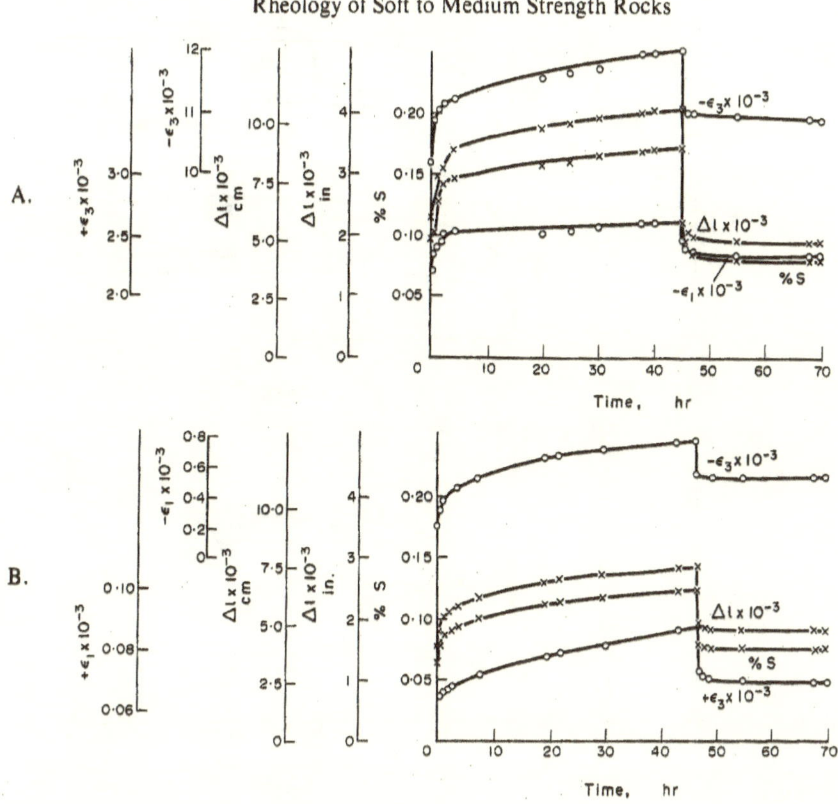

Fig. 3. Time–dependent deformation of coal and underclay due to 25·7 Kg/Cm² loading and unloading. (A) Lower four feet coal: (B) lower four feet underclay.

(b) for the air-dried underclay:

(i) loading:

$$\varepsilon_t=(274+21{\cdot}6\log t-4{\cdot}22t)\times 10^{-6}+\varepsilon_{r3} \quad (19)$$

(ii) unloading (or recovery):

$$\varepsilon_r=(545-0{\cdot}42\log t)\times 10^{-6}+\varepsilon_{r2} \quad (20)$$

(c) for the saturated coal:

$$\varepsilon_t=(3250+25{\cdot}7\log t+9{\cdot}2t)\times 10^{-6}+\varepsilon_3 \quad (21)$$

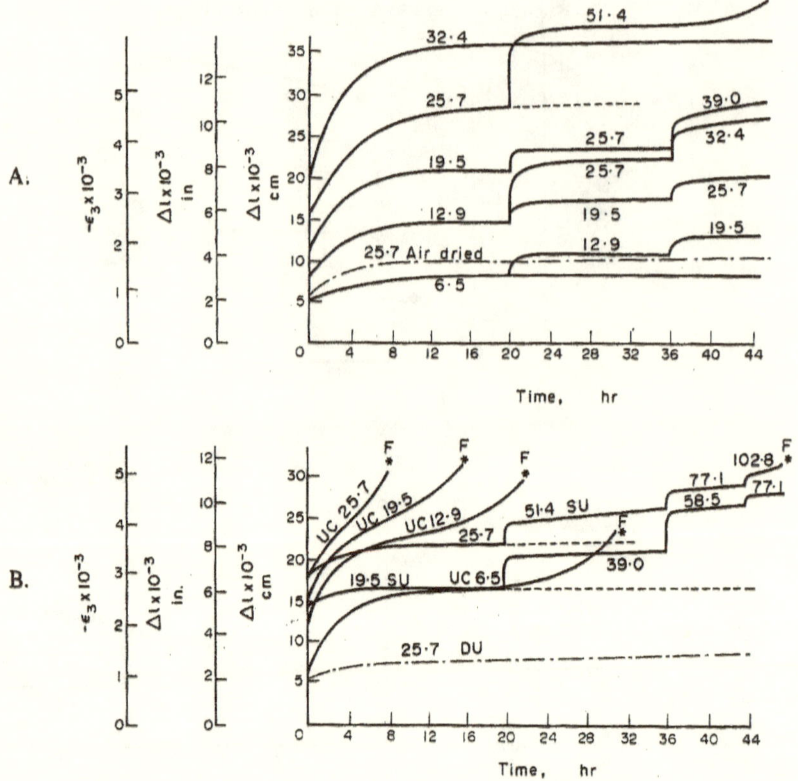

Fig. 4. Creep of water saturated Coal and Underclay under various stress levels. (A) Water saturated coal; (B) water saturated underclay. *Legend*: DU = Air-dried underclay; SU = saturated underclay; UC = underclay with clay band. Numbers on each curve indicate the compressive stress on sample in Kg/Cm².

(d) for the saturated underclay:

$$\varepsilon_t = (2890 + 172{\cdot}8\log t + 29{\cdot}5t) \times 10^{-6} + e_3 \quad (22)$$

(e) for the saturated underclay with clay band:

$$\varepsilon_t = (2883 + 864{\cdot}5\log t + 189{\cdot}7t + 385{\cdot}2t^{1{\cdot}5}) \times 10^{-6}. \quad (23)$$

The precise expression for the tertiary creep period (ε_3) and the long term recovery (ε_{r2}) was not determined, since all the tests were conducted under short periods of time to study the rock behaviour up to the tertiary creep range.

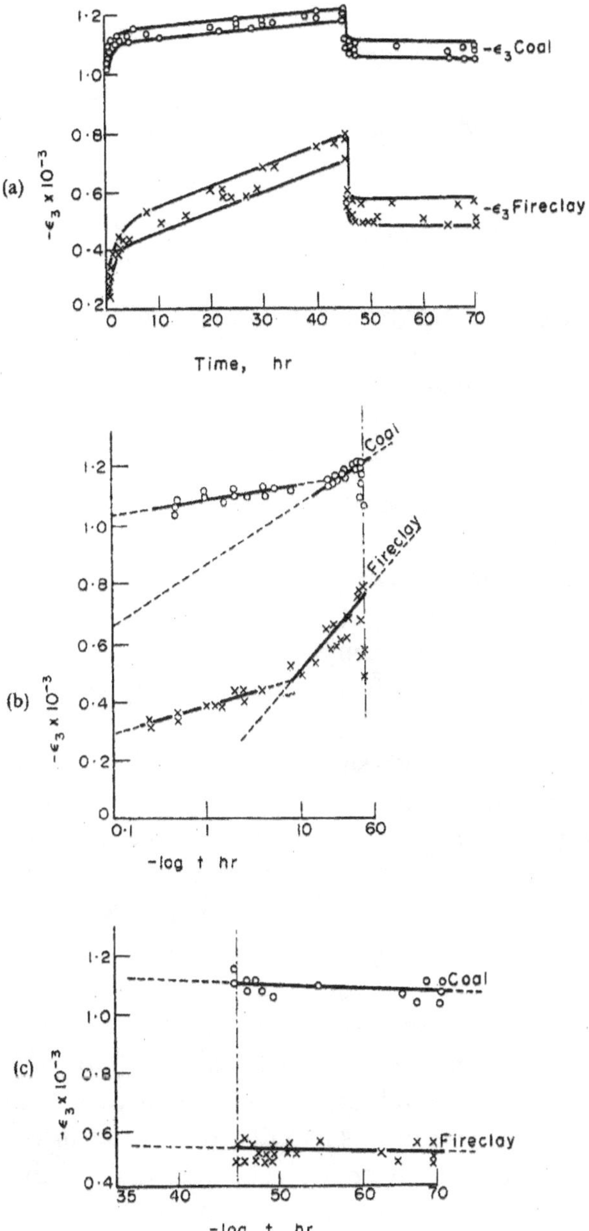

Fig. 5. Graphical analysis of loading and unloading creep curves for the coal and underclay. (a) Comparison of axial strains in the coal and fireclay; (b) loading creep curves for the coal and fireclay; (c) unloading creep curves for the coal and fireclay.

Close similarity exists between the experimental data and that calculated utilizing equations 9–16. The percentage error between the two above noted methods for the rocks tested was analyzed to average between 0·1 and 4·1. However, comparatively larger instantaneous strain was indicated by the computer analysis than that obtained experimentally.

5.3 *Rheological classification of the rocks*

According to the test results, the rocks can be placed into three categories relative to their strength and state of wetness. The best three equations to fit each of the categories investigated is averaged to be as follows, respectively:

(a) air-dried, soft rocks: coal and underclay; equations (15), (10) and (5)
(b) water-saturated: coal, underclay and underclay with clay band; equations (14), (10) and (15)
(c) air-dried medium strength rocks: calcitic limestone, dolomitic limestone and sandstone; equations (15), (10) and (5).

The above classification shows that the graphical method of curve fitting provided the second best fit (equation 10) to express the time-dependent behaviour of the rock samples. Moreover, the succession of the best fit to creep of the soft and medium strength air-dried rocks (categories A and B) is the same, on average, whereas that of the water-saturated rocks (category B) differs, indicating more plasticity in the rocks due to the effect of water.

Comparison of the expressions obtained reveal that the introduction of water to an air-dried coal merely trebles its overall creep. Similarly, an overall creep of approximately 8 times higher was observed when water was introduced to the air-dried underclay. Under water-saturation, overall creep is increased 5 times when a clay band was present in the underclay specimen (i.e. 13 times that of the air-dried underclay).

5.4 *Development of instantaneous and secondary strain*

A number of strain-time relationships were obtained for single specimens subjected to various levels of stress. The strain–time data was illustrated in Figs. 4(A) and (B). Assuming the linear portion of the curves thus obtained represent the secondary creep range in each stress level, the average rate of secondary creep ($d\varepsilon_2/dt$) and the average instantaneous strain (ε_0) for the various stress levels (σ_0) in the rocks tested in relation to their uniaxial compressive strength (σ_c), is shown in Table 2.

It appears that deformation of the rocks at low levels of stress has affected the subsequent strain–time data obtained at a higher stress level. Plotting a graph of the average instantaneous strain (ε_0) against the induced stress (σ_0) shows a direct proportionality between the two parameters (Fig. 6). This is expressed in the general form as:

$$\varepsilon_0 = m.\sigma \tag{24}$$

where, $0 < m < 1$ and $0 < n < 1$ are constants changing with the rock type, time void ratio, porosity and the state of moisture in the rock as well as temperature and the stress rate.

A similar behaviour prevails with the effect of the induced stress (σ_0) on the average secondary creep rate (Fig. 7) expressed in general as:

$$d\varepsilon_2/dt = (m \times \sigma) \tag{25}$$

Equation (25) is a differential equation, the solution of which yields:

$$\varepsilon_2 = (m\,._{\sigma 0}{}^{n}.t) + C \tag{26}$$

where, C is the integration constant.

Equations (24–26) suggest that an increase in the stress level would give a corresponding increase in the rate of secondary creep, although the latter to a lesser degree. The decrease in the rate of proportionality may be explained in terms of the progressive closing of initially swelled rock, voids and pore spaces within the samples. Thus, initially the effective stress in the portion of rock between the void spaces is greater than the nominal applied stress (σ_0). This higher effective stress gives rise to relatively higher instantaneous strain and rate of creep, which in turn gradually brings about the closure (or partial closure) of the pore spaces. As the pores close, a local redistribution and general lowering of the effective stress (σ_t) occurs, which is responsible for the reduction in creep rate with time. From this explanation, it is readily clear that porosity, discontinuity and planes of weakness play important parts in strain development and the *in-situ* creep rate. They are not only responsible in the development of regions of high internal stresses, but also as pockets where water or any wetting agent may enter. Therefore, dislocation movements result according to the efficiency of the wetting agent and its interaction with the crystal boundaries of the rock. Since the coal tested was observed to have a higher void ratio and porosity than the underclay, it was expected to exhibit a higher incremental instantaneous strain and creep rate over incremental induced stresses, as seen in Figs. 4(A)

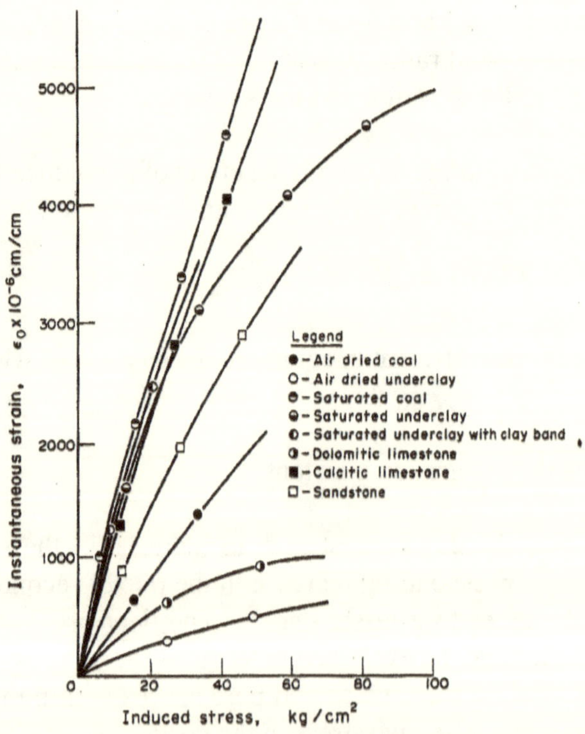

Fig. 6. The effect of stress on the instantaneous strain.

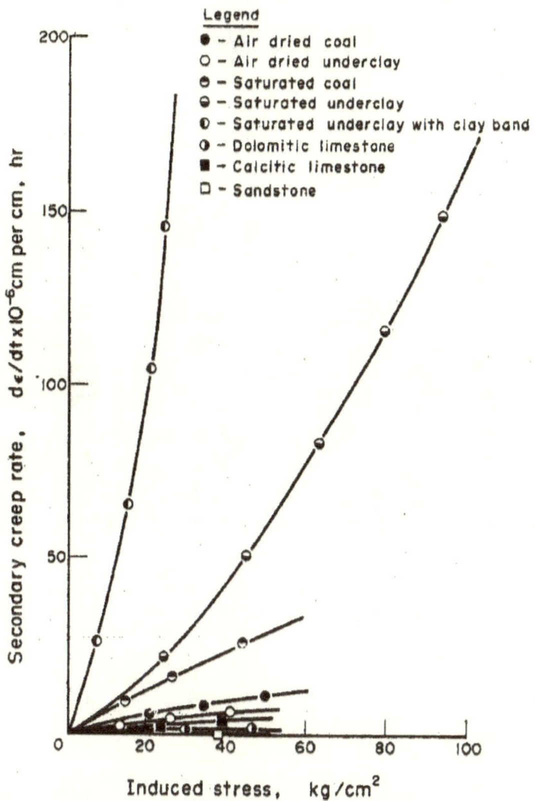

Fig. 7. The effect of stress on the secondary creep rate.

and 4(B). The time-dependent behaviour of the saturated underclay, especially those samples that contain clay bands (Fig. 4B), validates the above suggestion. Hence, in the latter case, the constants *m* and *n* in equations (24–26) may change between unity and infinity, revealing a highly plastic behaviour accelerated with time.

It can be observed from the deformation–time graphs (Figs. 2–5, 8, 9) that the majority of deformation under laboratory conditions occur in the initial 12 hr and, practically, all deformations approach a linear rate within the secondary creep range by the first 24 hr under stress.

TABLE 2. AVERAGE DEVELOPMENT OF THE INSTANTANEOUS STRAIN AND THE SECONDARY CREEP RATE IN ROCKS AT VARIOUS STRESS LEVELS

Rock type	σ_c (kN/cm²)	δ_0 (kN/cm²)	$\varepsilon_0 \times 10^{-6}$	$D\epsilon_2/dt$ ($\times 10^{-6}$/hr)
Air-dried coal	0·56	25·7	1120	6·6
		51·4	1850	11·9
Air-dried undeclary	1·90	25·7	274	5·4
		51·4	450	10·0
Water-saturated coal	0·33	6.6	900	5·5
		12·9	1650	8·5
		19·5	2290	11·6
		25·7	3300	13·5
		32·4	3600	19·8
		39·0	4330	22·9
		51·4	5600	30·1
Water-saturated underclay	1·14	19·5	2100	15·4
		25·7	2690	20·0
		39·0	3220	41·1
		51·4	3880	65·5
		58·5	4110	78·2
		77·1	4600	115·0
		102·8	4880	166·6
Water-saturated underclay with clay band	0·95	6·6	1050	28·0
		12·9	1700	50·0
		13·4	1800	54·5
		19·5	2400	102·2
		25·7	2880	195·0
Calcitic limestone	4·55	25·7	2800	2·0
		51.4	4850	2·5
Dolomitic limestone	6·88	25·7	660	1·9
		51·4	940	2·3
Sandstone	9·42	25·7	1800	0·7
		51·4	3300	1·0

In cases where the pressure on rock is constant, as in the laboratory condition, there is no difficulty in obtaining and/or predicting the rate of deformation of the samples with time by utilizing equations (8) and/or (26). However, stress constantly changes under mining conditions, producing time-dependent deformation under varying stress levels similar to the steps illustrated in Figs. 4(A) and 4(B). The latter case may be simplified by fitting a curve to pass through all the steps and by considering points on the curve thus obtained as the average overall time-dependent behaviour of the rock under *in-situ* conditions.

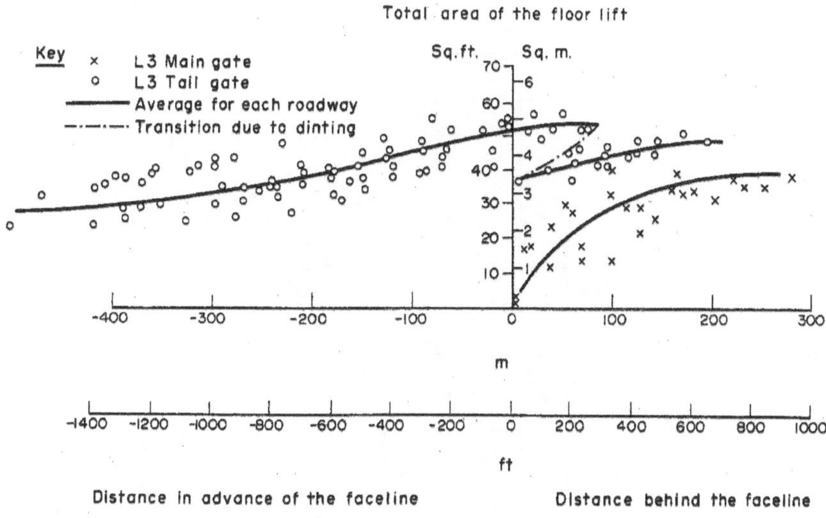

Fig. 8. Behavior of the lower 4 ft underclay floor along the L3 roadways, Britannia Colliery, S. Wales, U.K.

5.5 In-situ *investigations*

Measurements of time-dependent deformation of the underclay along the floor of mine roadways (Figs. 8 and 9) have shown the following general expression:

$$\varepsilon = [A + Bt^{C} + D\exp(-Et)] \times 10^{-6} \qquad (27)$$

where, ε = strain developed on the floor surface = $\Delta l/l$

Δl = the floor lift (in cm)

l = the thickness of the upper most floor layer (in cm)

t = time (per hr)

A, *B*, *C*, *D* and *E* are constants depending on the rock type state of moisture in the rock, dimensions of the opening floor, rate of face advance and the mining methods. The average values *A* and *D* found graphically to vary between 220–500 cm/hr; *B* is between 200–6000; *C* is between 0·21 to 0·61; and *E* is between 1·1–27.

A comparison of the expressions obtained from the laboratory with *in-situ* results show close similarity in behaviour of the underclay under the two conditions.

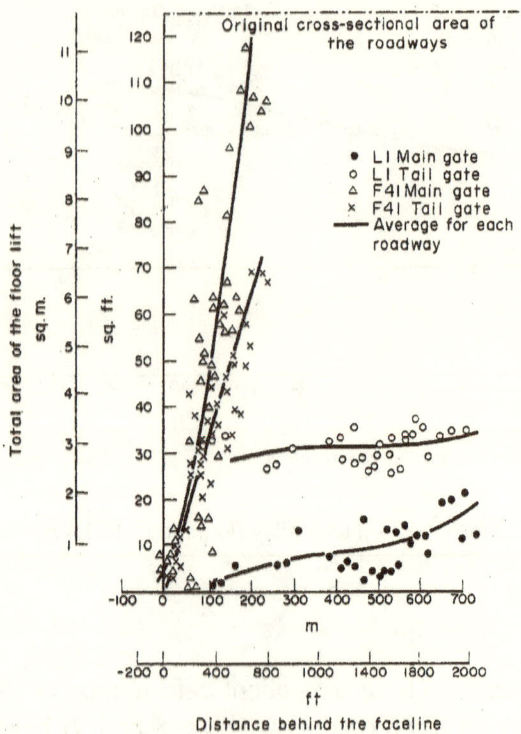

Fig. 9. Behavior of the lower 4 ft (L1) and standard 6 ft (F41) underclay floor along roadways, S. Wales area, U.K..

Empirical equation (27), derived graphically from the *in-situ* mine floor lift measurements, is of the form described as equation (15) of the model 9, which is the best fit for the laboratory results. Since the testing conditions between the laboratory and underground measurements varied from one another, close correlation between the magnitude of the constants involved in the above noted equations was not achieved. However, close similarity of the general behaviour is thought to facilitate correlation and prediction of the time-dependent deformation of rocks investigated under laboratory and *in-situ* conditions.

6. CONCLUSIONS

Close similarity was achieved between laboratory and *in-situ* time-dependent behaviour, enhancing the possibility of predicting, within reason, the *in-situ* creep of these rocks. The results indicate that air-dried soft to medium strength rocks behave in an elasto-plastic manner and very close to the Maxwell-Kelvin model, expressed in the general form:

$$\varepsilon = A + (B \,.\, t^{c)} + D[1 - \exp(E \,.\, t)].$$

The saturated soft rocks generally exhibit plastic behaviour expressed as:

$$\varepsilon = A + (B \,.\, t^{c)} + (D \,.\, t^{E}).$$

In general, introduction of water to air-dried coal and underclay increases their overall creep rates three- and eight-fold, respectively. The presence of clay bands in the saturated underclay further increase the creep five-fold.

The non-linear primary creep range of these rocks is within 12 hr of their exposure or application of pressure. After the primary creep range, the linear rate of secondary creep occurs during a further 12 hr, if the pressure stays unchanged. Any increase in pressure would also increase the rate of secondary creep in a non-linear form expressed by:

$d\varepsilon_2/dt = (m \,.\, \sigma \,.\, t_2)$ or $\sigma_t = \sigma 0$ [1- ∫ between t1 and t2 of ($d\varepsilon/dt$) dt]

Upon removal of the pressure, time dependent recovery of the rocks are of the form: $\varepsilon r = \varepsilon r0 + \varepsilon r1 + \varepsilon r2 = A + (B.\log t) + (C.t)$

REFERENCES

1. Price N. J., A study of the Time-Strain behavior of Coal measure Rocks, NCB. MRE Report, No. 2232 (1963).
2. Griggs D.J., Creep of rocks, Journal of Geol. XLVII, No. 3 (1939).
3. Griggs D.J., Deformation of rock under high confining pressures, Journal of Geol. 44 (1963).
4. Robertson E. C., Viscoelasticity of rocks: State of the stress in Earth's crust, Elsevier, New York (1964).

5. Handin J., Strength and Plasticity, Handbook of Physical Constants, 2nd edition, Geological Society of USA, New York (1957)
6. Eirich F.R., Rheology, Volumes. I, II and III, Academic Press, New York, (1957).
7. Bland D.R., The theory of Linear-Elasticity, p. 125, Pergamon Press, New York (1969).
8. Zener C., Elasticity and Anelasticity of Metals, p. 140-163, Chicago University of Chicago Press (1948).
9. Johnson A. E. and Henderson J., Complex Stress Creep, Relaxation and Fracture of Metallic Alloys, DSIR, HMSO (1962).
10. Terry N. B., The Elastic Properties of Coal - VI: Some measurements of Internal Damping and some consideration of Viscoelastic Behavior, NCB, MRE Report No. 2080 (1960).
11. Hardy H.R. Jr., Determination of the Inelastic Parameters of Geologic Materials from incremental creep experiments, American Inst. of Min. Met. and Pet. Engr., Paper No. PSE 1707 (1967).
12. Hartley H.O., The modified Gauss-Newton method for the fitting of non-linear regression functions by least squares, Technometrics, 3, p. 269-280 (1961).

3.3.3. A NUMERICAL & LABORATORY INVESTIGATION INTO THE BEHAVIOUR OF A DRY UNDERCLAY BEAM

By: ANDY A. AFROUZ AND F. P. HASSANI

Department of Mining and Metallurgical Engineering, McGill University, Montreal, Canada, H3A 2A7

SUMMARY

A simple and inexpensive laboratory device has been developed and used to simulate the behaviour of a dry underclay beam as the floor of an underground roadway subjected to a uniformly distributed uplift pressure. The Airy stress function method proved to be a useful means of expressing and predicting of the uplift of such beams. This method has been related to both classical beam theory and laboratory investigations. It was found that the Airy stress function method, utilizing polynomial equations, provides a closer fit to the mechanism of beam deformation and stress developments under dry laboratory conditions.

INTRODUCTION

Deformation and, ultimately, the failure of any mine floor in dry conditions is largely affected by loads imposed by the geological media. These loads can be broadly divided into the following categories:

1. The cover load, acting downwards, is transmitted to the floor through the mine supports. The magnitude of this load

depends largely on the depth from the surface, the dimensions of the opening, the mechanical properties of the overlying strata, and the geometry of the support base.

2. Side pressures act on the sides of a mine opening and are caused by the cover load. The resultant direction of these pressures is towards the centre of the opening, causing its gradual closure.
3. Bottom pressure is related to the cover load. For a weak underclay material at depth, it is considered to be a uniformly distributed load beneath the floor layer causing floor lift.

Terzaghi[1] quoted empirical evidence indicating that bottom pressures in tunnels and deep foundations are approximately one-half of the roof load, while lateral thrusts are one-third of the roof (or vertical) load intensity.

A number of attempts have been made to predict, on a theoretical or semi-theoretical basis, the magnitude of strata movements likely to develop at a particular site.[2–5] Theories developed for mine subsidence have stressed the importance of the thickness and rigidity of the layers comprising the strata. Grant[6] and Berbower[7] noted the importance of the above, and suggested that the strata behave as a thick plate which bends to maintain continuity with all its constituent layers as a result of compression. Yam[8] and Chapman[9] modeled simply supported composite beams of steel and concrete to explore the effects of beam stiffness, connector distribution and type of loading. Wright[10] investigated the behaviour of cracked roof beams and concluded that those that have developed steeply inclined fractures behave as flat arches unless the side pressure in the formation is much larger than any developed horizontal tensile stress in the beam. Hardy *et al.*[11] studied failure and crack propagation in Plexiglas and granite beams under three-point loading, and suggested that the behaviour of a beam with no notch (or crack) is equal to the sum of the behaviors of an infinite number of beams with notches (or cracks) at different depths.

Despite such work on beams, there seems to have been no previous work on the behaviour of dry underclay as a beam. The significance of this investigation is that a relatively dry and strong underclay in the mine floor causes beamlike elastic deformation leading to fracture, rather than plastic flow. The particular underclay studied in this work is a grey, fine grained, unlaminated argillaceous rock, with apparent and true densities of 2·51 and 2·72 g/cm^3, respectively. The uniaxial compressive strength of the underclay as dry beams of various thicknesses and lengths was between 40 and 200 kN/m^2. This is, however, within the range of soft rocks. The major mineralogical constituents of the relatively hard underclay are quartz, iron pyrite, calcite, chlorate, kaolinite, montmorilinite and illite, with small quantities of haematite and carbonaceous materials.

In this study, a cross-section of a dry mine roadway was simulated under a uniaxial stress field in laboratory conditions. The floor of the roadway was considered as a beam resting on a uniformly distributed load acting upwards. Point loading based on one to four points was considered as an alternative loading condition, but was abandoned. The thinner underclay beam sample failed under a very low upward load (less than 0·22 kN at the bottom of the beam) which would have necessitated the use of a more sensitive hydraulic loading device than that used. Moreover, it was considered that a uniform loading was more realistic in the case of relatively soft floor strata in mine roadways.

INSTRUMENTATION

The laboratory testing assembly is shown in Figure 1, and consisted of the following:

1. A 12 tons capacity hydraulic press (1) with appropriate pressure transducers (2) having an accuracy of ± 0·5 per cent of the measured pressure.
2. A cylindrical vulcanized rubber bag (3), The bag was made of rubber tubing 41·9 cm long and 6·1 cm in diameter sealed at both ends with 1·3 cm thick hard steel disks. The maximum length of the beams used was 24·8 cm,

most beams having a constant width of 4·3 cm. A pressure transducer (4) with a range of 0 to 4·17 MPa was mounted on one of the sealed ends of the cylindrical rubber bag to record the average hydraulic pressure inside the bag, which was filled with distilled water during the test. This pressure recording was used to determine the effective applied pressure at the rubber bag/rock beam interface. The bag assembly was placed inside an exact-fit steel channel (5) whose dimensions were 41·9 × 6·1 × 5·08 cm. This was done in order to eliminate distortion in the rubber bag under sustained pressure, and to direct the pressure towards the least resistant part of the bag, i.e. its top surface, where the underclay beam rested.

3. A hard-steel rectangular frame (6) simulated the legs of an arch sitting on the beam. The frame was 5·08 cm wide and 1·3 cm thick, and was adjustable in order to accommodate prepared beams of 14·8 to 24·8 cm length. In order to make line contact, with the surface of the beam, the bases of the frame's legs were rounded. Two to three LVDTs (7) were placed at the half, quarter and one-eighth points of the beam to measure its vertical movements when bent upwards due to the bottom loading pressure. The LVDTs were mounted on the steel frame and could be adjusted by sliding them along a steel rod (8) for positioning on beams of various lengths. A lightweight spirit level accurate to 2 minutes of arc (9) was used to ensure that the legs of the frame, when positioned on the horizontal beam, were perpendicular to the surface of the beam.
4. A strain meter (13) and multichannel connector (12) were used to measure the strain due to deformation of the beam to which the strain gauges (11) were cemented. The maximum sensitivity of the instrument was 2·5 microstrains, and the overall accuracy for the static measurement was ± (0·5% + 2) microstrains.

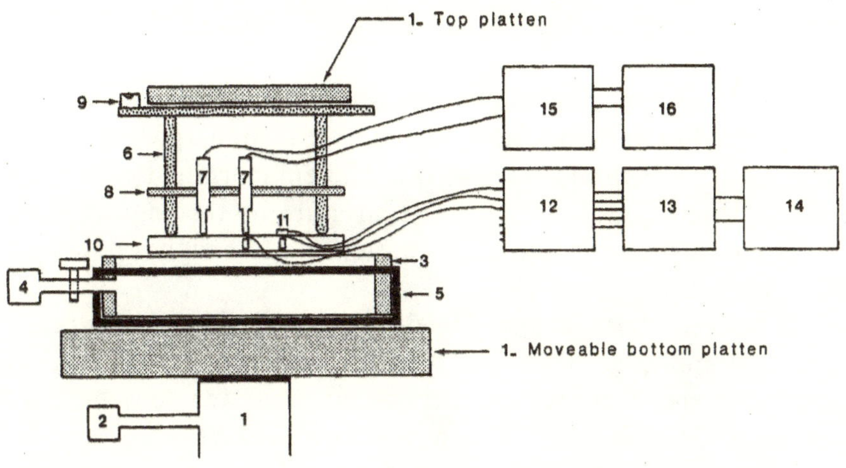

KEY

1- Hydraulic Press
2- Pressure transducer for the press
3- Rubber bag
4- Pressure transducer on the rubber bag
5- Steel channel
6- Adjustable rectangular frame
7- L.V.D.Ts.
8- Silver steel rod
9- Spirit level
10- Underclay beam
11- Strain gauges
12- Multichannel connector
13- Strain meter
14- Data acquisition system
15 Signal conditioner and amplifiers
16- Data acquisition system for (7)

Figure 1. Schematic diagram of the experimental arrangement

EXPERIMENTAL PROCEDURE

It was decided to simulate the roadway floor by using samples of underclay cut into beams. The samples ranged in length-to-width ratio and length-to-thickness ratio from 3·4 : 1 to 6·5 : 1 and from 7·4 : 1 to 24·8 : 1, respectively (Table I).

The rubber bag was completely filled with distilled water at atmospheric pressure, with care taken to remove all air bubbles. It was then positioned firmly in the steel channel and placed on the platen of the hydraulic press.

The underclay beam samples were cut and prepared and left for one week to air-dry. Some of the beams were then fitted with strain gauges and left for a further day for the strain gauge adhesive to cure. The level of the assembly was checked as accurately as possible with the spirit level, as any departure from the horizontal at this stage was found to affect the test result and consequently introduce considerable error.

The length of the rectangular steel frame was adjusted to correspond to the beam length to be tested, so that, when placed on the beam, it would leave only a 0·25 cm margin at either end of the beam. The LVDTs were secured to the rectangular steel frame in positions at the half, quarter and one-eighth points of the beam.

The steel frame was then placed on the beam within the hydraulic press assembly, and the upper platen of the press was lowered to the top of the steel frame and tightened. This was done by bolts incorporated in the press. The spirit level was used to check that the frame was vertical, and adjustments made by loosening or tightening the bolts securing the top platen on the frame.

Where strain gauges were used, equal lengths of lead wires from the gauges were connected to the strain meter through the multichannel connector. A half-bridge circuit with one dummy gauge was used.

Incremental loading was applied up to beam failure, and the corresponding uplift and strain resulting from the load change were noted. The specifications of the test samples, together with the positions of the strain gauges on the samples, are given in Table 1.

FLOOR BEAM THEORY

In order to compare observed laboratory results with theory, two mathematical models were used. One was based on the Airy stress function (ϕ), and the other was simple beam theory. The rock beams all had a fixed width of either 4·3 or 3·8 cm, and rested horizontally upon a uniformly distributed load of intensity q acting upwards on them. They had a mass density of p and thickness of $2C$. The rigid arch leg plates restricted upward movement of the beam at its ends (Figure 2).

Table 1. Specification of the underclay beam samples

Sample No.	Dimensions (cm)	Failure Angle to Vertical (θ^o)	Failure Occurred at	Strain Gauge Arrangement
1	24.8x4.3x2.0	18	2/5 of the length endwise	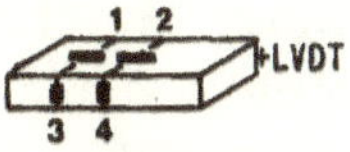 1 2 +LVDT 3 4
2	24.8x4.3x2.0	22	1/3 of the length	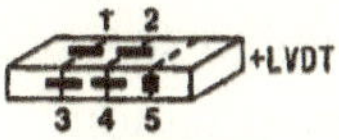 1 2 +LVDT 3 4 5
3	24.8x4.3x1.0	25	1/4 of the length	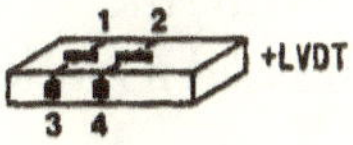 1 2 +LVDT 3 4
4	24.8x3.8x2.0	16	1/2 of the length	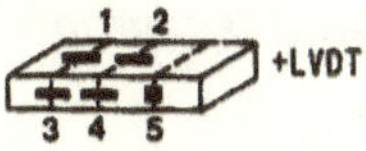 1 2 +LVDT 3 4 5
5	24.8x3.8x1.0	20	1/3 of the length	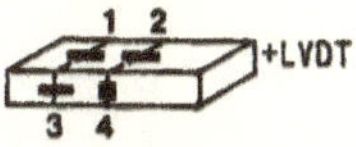 1 2 +LVDT 3 4
6	24.8x3.8x1.0	22	1/2 of the length	LVDT
7	19.8x4.3x2.0	28	1/4 of the length	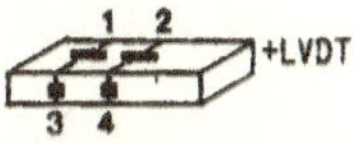 1 2 +LVDT 3 4
8	19.8x4.3x2.0	19	1/3 of the length	LVDT
9	19.8x4.3x1.0	26	1/2 of the length	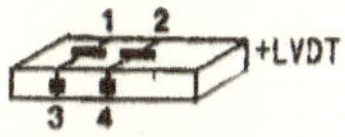 1 2 +LVDT 3 4
10	19.8x3.8x2.0	24	1/2 of the length	LVDT
11	19.8x3.8x2.0	17	1/3 of the length	LVDT
12	19.8x3.8x1.0	27	1/2 of the length	LVDT
13	14.8x4.3x2.0	20	1/3 of the length	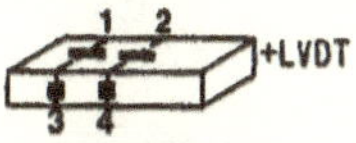 1 2 +LVDT 3 4
14	14.8x4.3x1.0	28	1/4 of the length	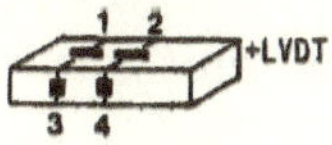 1 2 +LVDT 3 4
15	14.8x3.8x2.0	23	1/2 of the length	LVDT
16	14.8x3.8x1.0	25	1/3 of the length	LVDT

The neutral axis of the beam was considered to be at mid-thickness. It was recognized that the neutral axis deviates from the cent-reline by up to one-sixth of the beam thickness, the exact magnitude depending on the beam thickness and loading conditions. However, Grant[6] and Berbower[7] suggested that the neutral axis was effectively at mid-thickness. This is the case for a normal symmetrical beam, and there has been some evidence, in the form of measured depths to sheared oil well casings, to indicate that it is a reasonable approximation.[12]

With the above noted assumptions, the simulation can be expressed as follows.

Boundary condition

With reference to Figure 2:

at $y = +C$: $\sigma_y = 0$ and $\tau_{xy} = 0$

at $y = -C$: $\sigma_y = -q$ and $\tau_{xy} = 0$

at $x = +l$: net horizontal force = ∫Integral between -c to +c of $\sigma x\, dy = 0$

net moment (M_b) = ∫Integral between -c to +c of y. $\sigma x\, dy = 0$

total shear force = ∫Integral between -c to +c of $\tau_{xy\, dy}$

= -l (q -2cρg)

at $x = -l$: net horizontal force = ∫Integral between -c to +c of $\sigma x\, dy = 0$

net moment (M_b) = ∫Integral between -c to +c of y. $\sigma x\, dy = 0$

total shear force = ∫Integral between -c to +c of $\tau_{xy\, dy}$

= +l (q - 2cρg)

Stress and strain equations:

From classical stress equations of equilibrium, the strain–displacement and stress–strain equations, a compatibility equation in terms of strain in the x–y plane and plane stress, are respectively as follows:

$$\partial^2 \varepsilon_{xy} / \partial x\, \partial y = (\partial^2 \varepsilon_x / \partial_y{}^2) + (\partial\, \varepsilon_y / \partial x^2) \tag{1}$$

$$[(\partial^2/\partial x^2) + (\partial^2/\partial y^2)]\,(\sigma_x + \sigma_y) = \nabla^2(\sigma_x + \sigma_y) = 0 \tag{2}$$

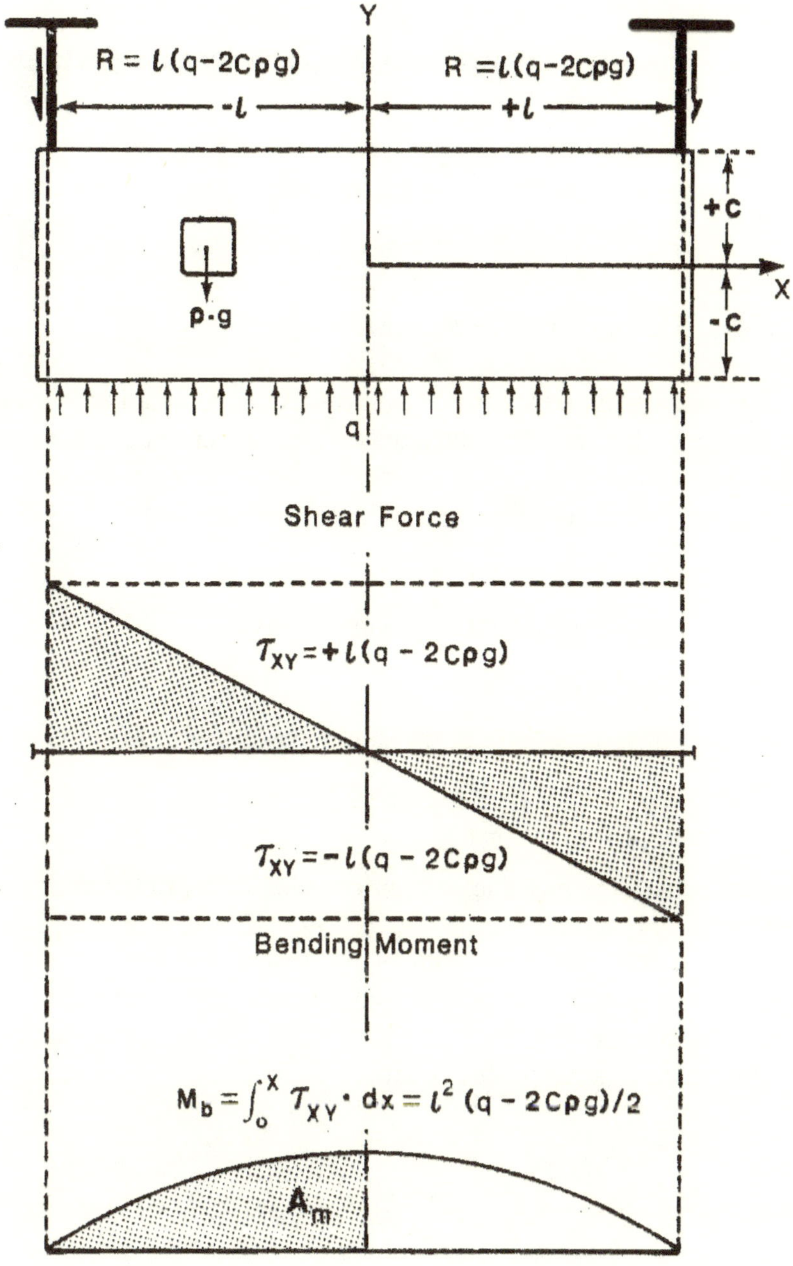

Figure 2. Model conditions of uniform beam bottom loading with top end constraint

Airy stress function:

The stresses in terms of the Airy stress function (ϕ) are as follows (ref.13):

$$\sigma_x = \partial^2\phi / \partial y^2 \quad (3)$$

$$\sigma_y = \partial^2\phi / \partial x^2 \quad (4)$$

$$\tau_{xy} = -\partial^2\phi / \partial x\, \partial y + K \quad (5)$$

where, K is a constant to be evaluated.

Differentiating equation (4) with respect to y, equation (5) with respect to x and substituting the products into the classical equilibrium equation yields: $\partial K/\partial x = +pg$

Therefore, for the equilibrium condition to be satisfied:

$$K = \text{Integral between 0 to x of } \rho.g\; dx = +\rho.g.x \quad (6)$$

Substituting equation (6) into equation (5) gives:

$$\tau_{xy} = -\partial^2\phi / \partial x\, \partial y + \rho g x \quad (7)$$

The Airy stress function for the solutions of the problem can be written as:

$$\phi = \phi_1 + \phi_2 + \phi_3 + \phi_4 + \phi_5 \quad (8)$$

Note that it is possible to continue the stress function to any degree of polynomial (i.e. to ϕ_6, ϕ_7, ϕ_8, . . .) to achieve higher accuracy, within the sensitivity of the instrumentation. It is also legitimate to add stress functions and delete terms within individual functions to achieve a desired accuracy. The following polynomials have been considered for the equation (8):

$$\phi_1 = a_1 x + b_1 y$$
$$\phi_2 = a_2 x^2/2 + b_2 xy + c_2 y^2/2$$
$$\phi_3 = a_3 x^3/6\; b_3 x^2 y/2 + c_3 xy^2/2 + d_3 y^3/6$$
$$\phi_4 = a_4 x^4/12 + b_4 x^3 y/6 + c_4 x^2 y^2/2 + d_4 xy^3/6 + e_4 y^4/12$$
$$\phi_5 = a_5 x^5/20 + b_5 x^4 y/12 + c_5 x^3 y^2/6 + d_5 x^2 y^3/6 + e_5 xy^4/12 + f_5 y^5/20$$

From equations (3), (4) and (5): $a_1 = b_1 = 0$, because there is a double differentiation in each of the equations, and a first-degree polynomial will be zero in the second differentiation. Evaluation of the other constants in terms of q, p, C, l and g gave the following results:

$$a_2 = -q/2,\ b_2 = 0,\ c_2 = 0,$$
$$a_3 = 0,\ b_3 = (3q - 2C\rho g)/4c,\ c_3 = 0$$
$$d_3 = 3(2C\rho g - q)\ (l/5 - l^2/2C^2)\ /2C$$
$$a_4 = 0,\ b_4 = 0,\ c_4 = 0,\ d_4 = 0,\ e_4 = 0$$
$$a_5 = 0,\ b_5 = 0,\ c_5 = 0,\ d_5 = 3(2C\rho g - q)\ /4C^3$$
$$e_5 = 0, f_5 = -\ (2C\rho g - q)\ /2C^3$$

Substituting all the evaluated coefficients and the polynomials into equation (8) yields:

$$\phi = (-qx^2/4) - [(2C\rho g - 3q)x^2 y/8C] + [(2C\rho g - q)\ (l/5 + l^2/2C^2)\ y^3/4C] + [(2C\rho g - q)x^2y^3/8C^3] - [(2C\rho g - q)\ y^5/40C^3] \quad (9)$$

Substituting equation (9) into equations (3), (4) and (7) gives:

$$\sigma_x = 3(q - 2C\rho g)\ (l^2 - x^2)\ y/4C^3 + (q - 2C\rho g)[y^3 - (3C^2y/5)]/2C^3 \quad (10)$$

$$\sigma_y = [-(q - 2C\rho g)\ y^3 + (3q - 2C\rho g)\ (C^2y - 2C^3q)]\ /4C^3 \quad (11)$$

$$\tau_{xy} = -3(q - 2C\rho g)(C^2 - y^2)x/4C^3 \quad (12)$$

In general, the moment of inertia (I) of the beam is b $(2C)^3/12$. Where b is the width of the beam (assumed to be unity for plane stress) and 2C is total thickness of the beam. Therefore,

$$I = 8C^3/12 = 2C^3/3 \text{ or } C^3 = 3I/2 \quad (13)$$

Substitution of equation (13) into equations (10) t0 (12) yields:

$$\sigma_x = [(q - 2C\rho g)\ (l^2 - x^2)\ y\ /\ 2I] + \{(q - 2C\rho g)\ [-(3C^2y/5)]/\ 3I\} \quad (14)$$

$$\sigma_y = [-\ (q - 2C\rho g)\ y^3 + (3q - 2C\rho g)\ (C^2y - 2C^3q)]\ /\ 6I \quad (15)$$

$$\tau_{xy} = -\ 3(q - 2C\rho g)\ (C^2 - y^2)\ x\ /\ 2I \quad (16)$$

Equation (14) can be written as:

$$\sigma_x = (M_b \cdot y\ /I) + (q - 2C\rho g)[y^3 - (3C^2y/5)]\ /\ 3I \quad (17)$$

Where M_b = Bending moment of the beam = $(q - 2C\rho g)\ (l^2 - x^2)\ /2$ (18)

The positive sign in equation (17) indicates that the bending moment is in the positive direction of the y-axis, and the parameter x is the horizontal distance from any point of interest to the mid-length of the beam (i.e. To the origin of the co-ordinate axis). The magnitude of x can vary between zero and *l* such that:

(a) At the beam ends, for x = L:

$$\min(M_b) = (q - 2C\rho g)(l^2 - x^{2)}/2 = 0 \quad (19)$$

(b) At middle of the beam, for x = 0:

$$\max(M_b) = (q - 2C\rho g)(l^2 - 0)/2$$
$$= l^2 (q - 2C\rho g)/2 \quad (20)$$

Displacement in the beam:

(a) Horizontal displacement (u) – Substituting equations (14) and (15) into the classical stress-strain relationship gives:

$$\varepsilon x = \partial u/\partial x = \{(q - 2C\rho g)[15(l^2 - x^{2)}y + 2(5y^3 - 3C^2 y)] + 5v[(q - 2C\rho g)y^3 - (3q - 2C\rho g)C^2.y + 2C^3.q]\}/30EI \quad (21)$$

Integrating equation (21) with respect to x provides the horizontal displacement u:

$$u = \int \varepsilon x \,.\, dx = \{(q - 2C\rho g)[5x.y(3l^2 - x^{2)} + 2x.y(5y^2 + 3C^2)] + 5v.x[(q - 2C\rho g)y^3 - (3q - 2C\rho g)C^2.y + 2C^3.q]\}/30EI + K_1 \quad (22)$$

Where K1 is the integration constant, to be determined. Theoretically, the horizontal displacement u at the central co-ordinates of the beam (i.e. At x = 0,

y = 0) is zero. Therefore, from equation (22) and this bondary condition, K1 = 0.

(b) Vertival displacement (v) – Substituting equations (14) and (15) into the classical stress-strain relationship yields:

$$\varepsilon y = \partial v/\partial y = \{5[(q - 2C\rho g)y^3 - (3q - 2C\rho g)C^2.y + 2C^3.q] + v[y(q - 2C\rho g)[15(l^2 - x^{2)}y +$$

$$2(5y^2 - 3\,C^2)]\}/\,30\,\mathrm{EI} \qquad (23)$$

Integrating equation (23) with respect to y provides the vertical displacement (v):

$$v = \int \varepsilon y\,.\,d_y = \{150[(q-2C\rho g)\,y^4 - C^2y^2(3q-2C\rho g) + 8C^3yq] +$$

$$2y^2(q-2C\rho g)[45(l^2-x^2)+4(5y^2-9C^2)]\}/3600EI+K_2 \qquad (24)$$

Where K_2 is the integration constant, to be determined. Theoretically, the vertical lift (v) under the reactionary supports at the beam ends (i.e. at $x = l$, $y = C$) is zero.

Therefore, from equation (24) and this boundary condition:

$$K_2 = C\;[(q - 2C\rho g)\;(150 - 60v) - 300(3q - 2C\rho g) + 1200q]\;/2400E \qquad (25)$$

Shear strain (γ_{xy}) can be obtained either by differentiating equation (22) with respect to *y or by d*ifferentiating equation (24) with respect to x and adding the results, or by substituting equation (16) into the classical stress–strain relationship, which will give:

$$\gamma_{xy} = \pm\,(1-v)\,(q-2C\rho g)\,(C^2-y^2)x\,/EI \qquad (26)$$

Slope and deflection

From Figure 2, the area A_m under the shaded parabolic bending moment diagram is determined as follows:

$$A_m = 2M_b\,.l/3 = l^3/(q-2C\rho g)\,/3 \qquad (27)$$

$$= \text{slope of the beam at its ends} = |\,A_m\,/EI\,|$$

$$= |\,l^2(q-2C\rho g)\,/3EI\,| \qquad (28)$$

x = distance of the centroid of the shaded area from the end of the beam = $5l/8$

Therefore, the maximum deflection is η_{max}, the deflection of the ends relative to the centre of the beam:

$$\eta_{max} = A_m/EI = 5l^4(q-2C\rho g)\,/24EI \qquad (29)$$

The general deflection η is given by:

$$\eta = x(q-2C\rho g)(l^3 - 2l.\,x^2 + x^2)/24EI \qquad (30)$$

The radius of curvature R is given by:

$$R = - \left[1 + (\partial^2 \eta / \partial x^2)\right]^{2/3} / \left[\partial^2 \eta / \partial x^2\right]$$
$$= - 24E.I \left\{1 + [(q - 2C\rho g)(l^3 - 6l.\, x^2 + 4\, x^3) / 24E.I]^2\right\}^{2/3} \qquad (31)$$

A software was written to compare the analysis from the above equations with that of the classical equations for uniformly loaded beams. The laboratory test data were also compared graphically with each of the above theoretical methods for their predictions of the behaviour of underclay beams.

DISCUSSION OF RESULTS

The stress–strain behaviour of the underclay beams due to bottom pressure revealed the fact that thinner beams have less overall strength before failure (Figure 3). Increased beam length also resulted in reduced overall strength of the beam.

In all cases except sample 13 (Figure 3(b)), the horizontal strain gauge placed at the quarter-span of the beam surface (ε_1) indicated negative values, corresponding to the development of compressional strain in that region. This indicated that the beam ends had undergone a slight compression, leading to shear stresses and ultimately to shear failure at the ends.

All the surface strain readings at the mid-span (ε_2) were positive, corresponding to the development of indirect tension in this region of the beam. This tension is caused by upward lift and bending due to the bottom pressure. Theoretically, failure should occur at the point where the highest stress is developed. For beams under the action of uniformly distributed loads, this point of failure is expected to be at the mid-span. However, in practice, the failure developed at various points, mostly between one-third and one-half of the beam length as measured from the simulated arch legs (Table 1).

The strain gauges placed along the centre of the beam depth at the quarter-span (ε_3) and mid-span (ε_4) indicated the average strain development under their area of contact with the beam. This area lies partly above and partly below the beam's neutral axis. Therefore the overall strain reading depended on the algebraic sum of the tensile and compressional strain concentration in the contact areas above and below the neutral axis. The position of the neutral axis was found to change with variation in load along the beam's mid-depth.

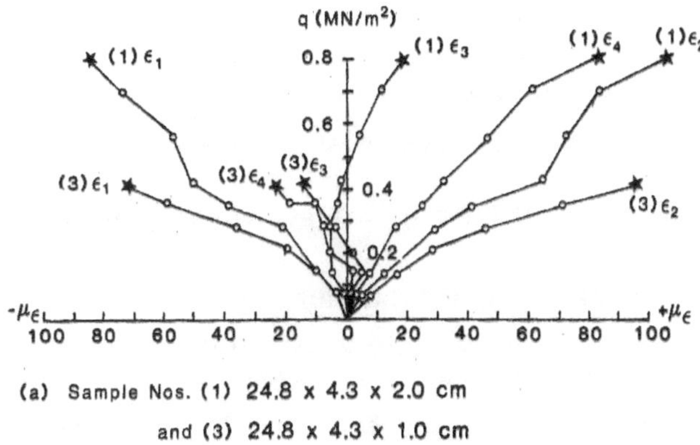

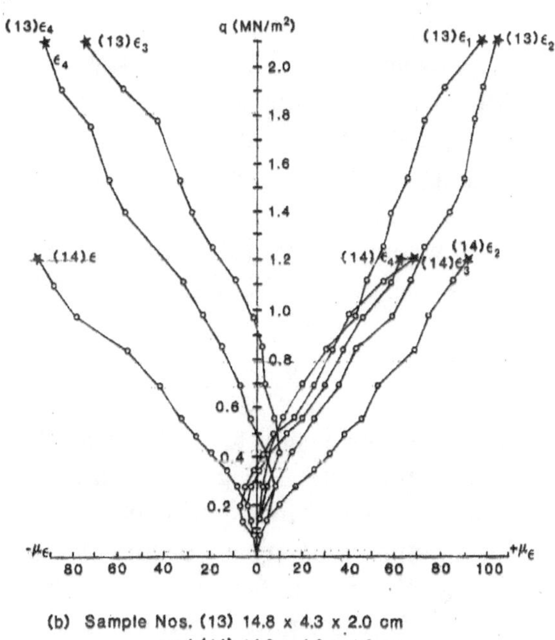

Figure 3. Stress-strain response of underclay beams to uniformly applied bottom loading for (a) samples 1 and 3, and (b) samples 13 and 14 (see Table I)

It appears from Figure 3 that the neutral axis tends to lie towards the tensile zone of the beam under strain. Therefore, the position of the neutral axis changes from the lower half of the beam depth to the upper half as the applied load and beam geometry change. However, the magnitude of this change was very small and, in terms of the strain change, it was no more than 10 microstrains ($\mu\varepsilon$), which led to a maximum differential stress (σ) of less than:

$$E\,.\varepsilon = 6269 \times 10/10^6 = 0{\cdot}063 \text{ MN/m}^2 \text{ } (=9{\cdot}10 \text{ psi})$$

The vertical lift at the quarter-span just before failure was generally 7–8, 20–30 and 17–18 per cent lower than that of the mid-span for the beams of length 24·8, 19·8 and 14·8 cm, respectively. As is apparent, anisotropy of the rock may have masked any definite systematic variation in the vertical behaviour of the beam. However, differences in the vertical readings became larger as the beam length was decreased.

Comparison of the results obtained with the Airy stress function with those from the classical beam theories and laboratory data for selected beam dimensions, through all ranges of loading up to failure, is illustrated in Figure 4. It is evident from these graphs, that the Airy stress function provides a closer fit to the laboratory data. Some of the inconsistency in the curves obtained from the experimental results was due to heterogeneity of the beam material, as would be expected. Figure 4 also reveals that a decrease in the beam length results in an increase in its strength before failure, while a decrease in the beam thickness (or depth) decreases the strength of the beam.

A comparison of the average vertical lift (v) at mid-span of various types and geometry of rock beams tested by one of the authors for a different purpose[14] is illustrated in Figure 5, as a check on the results obtained from the underclay beam experiments. It shows that the unit pressure used to fracture a given rock beam is larger for one-point loading than for any other multipoint loading, and that it decreases as the number of points increases towards a uniformly distributed load. However, the total load required to fracture a given beam is much smaller in one-point loading than in subsequent multipoint loading. Figure 5 also illustrates the weakness of the underclay in comparison to more competent rocks.

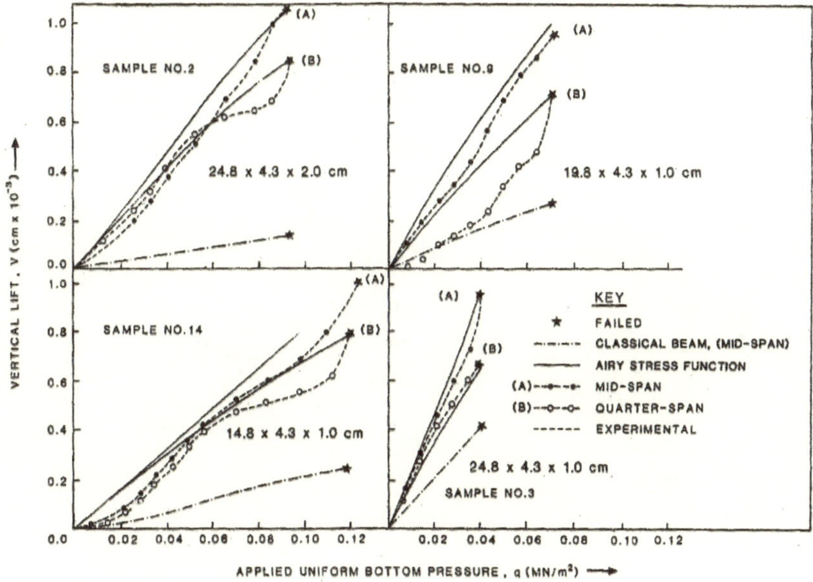

Figure 4. Comparison of Airy stress function and classical beam theory curves with observed response of underclay beams.

There was little or no measurable horizontal movement *(u)* at the mid-span of the beam. At each location along the beam surface, the direction of horizontal movements as indicated by the strain gauge readings was towards the region of maximum vertical lift, i.e. the mid-span.

As shown in Figure 6, the magnitude of the beams' horizontal movement under a uniformly distributed load was relatively high at the beam ends. It diminished towards the central region of the beam, as indicated by LVDTs and strain gauges positioned at various points along the beam length. The same phenomena have been observed along the soft underclay floor of the F41 roadways in Windsor Colliery, South Wales Coalfield, where the floor—especially in the west side of the roadways—flowed towards the middle of the roadway, distorting the arch legs.[5] Figure 6 illustrates that, under a uniformly distributed pressure, increase in the beam length increases the horizontal movement, whereas increase in the thickness of the beam decreases the horizontal movement.

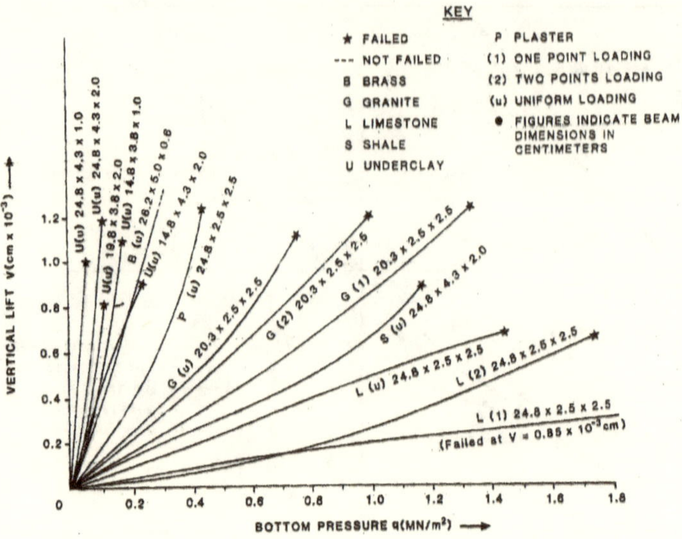

Figure 5. Comparative failure data for model floor beams of various material under point and uniform bottom loads.[14]

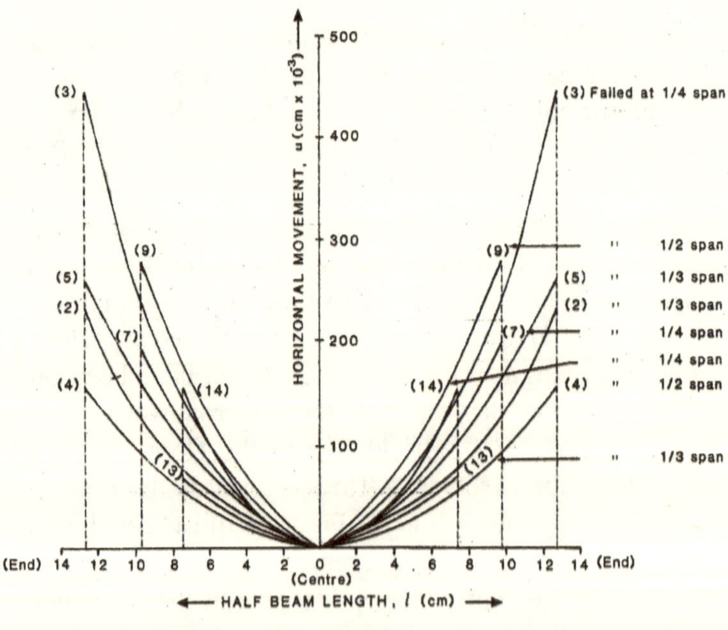

Figure 6. Beam horizontal movement due to uniformly distributed pressure, from equation (22) for various sample numbers (see Table 1)

The magnitude of the associated horizontal stress (σ_{x}) developed appears to be maximum at the mid-span of the beams, and to diminish to zero at the beam ends (Figure 7). The horizontal stress is shown to be directly proportional to the beam length and inversely proportional to its thickness. The theoretical expression derived as equation (14) shows the horizontal stress development (σ_{x}) to be, on average, 7 per cent larger than that derived using the classical beam method.

The vertical stress distribution (σ_{y}) obtained by the Airy stress function and classical beam methods are both close to the experimental results. The vertical stresses were found by the Airy stress function method to be directly proportional to the beam thickness, as expressed in equation (15). However, the test results revealed that, in addition, (σ_{y}) is inversely proportional to the beam length (Figure 8). At the beam surface (σ_{y}) reaches zero, and it is equal to the applied pressure at the base of the beam.

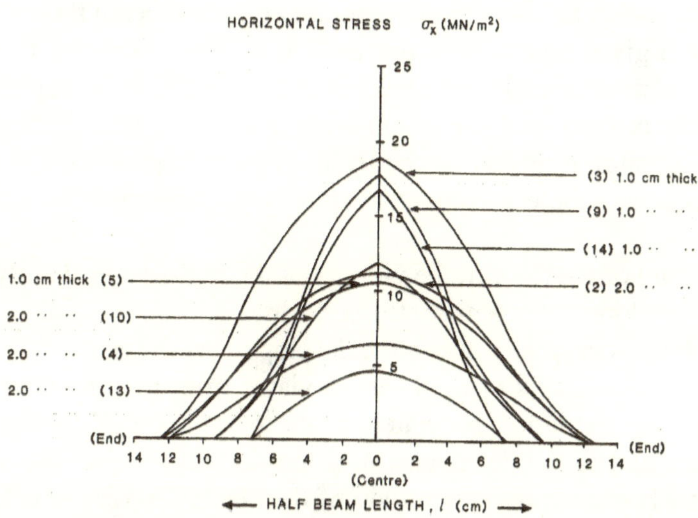

Figure 7. Calculated horizontal stress at beam surface due to horizontal strain, from equation (21) for various sample numbers (see Table 1)

The shear stress τ_{xy} development, expressed by equation (16), was on average 50 per cent lower at the mid-span than beneath the simulated arch legs. This indicates that the failure of the underclay beams at the mid-span is due mainly to tension rather than to shear, which predominates in the floor failure beneath supports.

All other parameters derived from the Airy stress function gave similar results to classical beam theory, except for the radius of curvature of the beam *R* which was, generally, two to three times greater with the Airy stress function than with the classical beam equation. This reflects the fact that rock beams are not as elastic as metals, for which the classical elastic theory was developed. It would be desirable to substantiate these mathematical derivations by *in situ* measurements on borehole plugs with extensometers at various floor depths and distances from each of the arch legs along the underground roadways. A correlation with the variation in pressure measurements obtained from load cells beneath supports at various time intervals dictated by the mining operations could then be evaluated.

CONCLUSIONS

The Airy stress function provides a better model of the behaviour of a dry underclay floor beam than does classical beam theory.

For a given rock type, the overall strength of the mine floor as a dry beam is directly proportional to its thickness (or depth) and inversely proportional to its length (Figures 4 to 8). This relationship is important in establishing the width of the roadways for a desired performance where a competent roof is accompanied by a softer floor.

Failure of the floor in the vicinity of the mid-span is mainly due to the development of tensile stresses, while within the vicinity of the supports it is generally due to the development of shear.

The neutral axis of the beam deviates from the mid-thickness plane. This deviation is related to the magnitude of the applied pressure and to the development of tensile and compressive stress regions. However, the magnitude of the change is very small, and for mine floors of variable thickness it is considered to be effectively negligible *in situ.*

The vertical lift at the mid-span of floor beams is 7 to 18 per cent greater than at its quarter-span. The longer the beam length (or the width of the underground roadway), the smaller the difference between vertical lift at the mid-span and quarter-span. The harder the rock, the lower the magnitude of the vertical lift before failure.

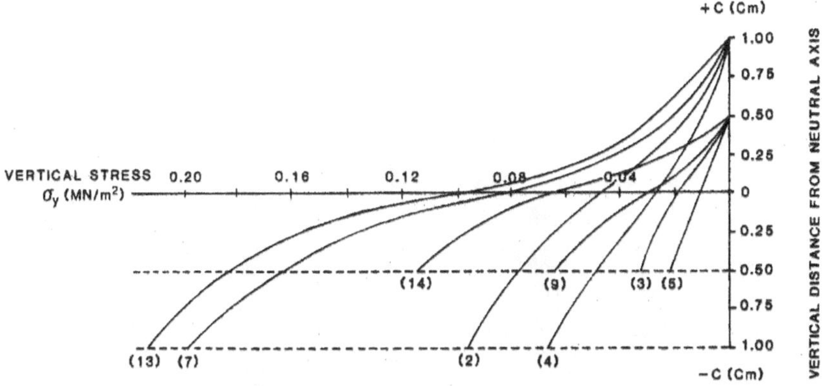

KEY: 1 - All beams had 4.3 cm width, except beams numbered 4 and 5 which had 3.8 cm width.
2 - Beam lengths were as follows:

Sample No:	2	3	4	5	7	9	13	14
Length, cm:	24.8	24.8	24.8	24.8	19.8	19.8	14.8	14.8

Figure 8. Theoretical vertical stress at failure versus underclay beam thickness and length at failure under uniform bottom loading for various sample numbers (see Table 1)

The horizontal movement of the floor beam is zero at the mid-span, increasing towards the support base. The direction of horizontal movement is towards the region of maximum vertical lift (i.e. the mid-span).

The horizontal stress developed along the beam is directly proportional to the beam's length and inversely proportional to its thickness. The maximum horizontal stress occurs at the mid-span, the stress diminishing to zero at the beam ends.

The shear stress and strain are, on average, 50 per cent lower at the mid-span than beneath the simulated arch legs.

References

1. K. Terzaghi, *Theoretical Soil Mechanics*, Wiley, New York, 1943, pp. 118–143.

2. H. Hoffmann, 'The effects of direction of working and rate of advance on the scale-deformation on a self-loaded stratified model of a large body of ground', in Proc. 4th Int. Conf. on Strata Control and Rock Mechanics, Columbia University, New York, 1964, pp. 397–411.
3. O. Rellensmann, 'Rock mechanics in regard to static loading caused by mining excavation', *Q. Colorado School Mines*, **52**(3), 387–395 (1957).
4. S. Litwiniszyn, 'On certain linear and non-linear strata theoretical models, in Proc. 4th Int. Conf. on Strata Control and Rock Mechanics', Columbia University, New York, 1964, pp. 384–396.
5. A. Afrouz, 'Floor behaviour along Longwall roadways, *Int. J. Rock Mech. Mining Sci.*, **12**, 229–240 (1975).
6. U. S. Grant, 'A further analysis of horizontal earth movements in the Long Beach Harbour area, The elliptical plate analogy', Report, University of California, Berkeley, 1948.
7. R. F. Berbower, 'Subsidence problems in Long Beach Harbour district, *Waterways Harbours Div., ASCE*, **85**, 43–80 (1959).
8. L. C. P. Yam, 'Ultimate load behaviour of composite T-beams having inelastic shear connection', Ph.D. Thesis, University of London, 1966.
9. J. C. Chapman, 'The behaviour of composite beams in steel and concrete', *Struct. Eng.*, **4**, 115–125 (1964).
10. F. D. Wright, 'Arching erection in cracked roof beams', in Proc. 5th Int. Strata Control Conf., 1972, Section 29.
11. M. P. Hardy, J. A. Hudson and C. Fairhurst, 'The failure of rock beams', Parts I and II, *Int. J. Rock Mech. Mining Sci.*, **10**, 52–82 (1973).
12. U. S. Grant, 'Subsidence of Wilmington Oil Field, California', Division of Mines Department of Natural Resources, San Francisco, CA, Bulletin 170, Chapter 10, 1954.
13. L. Obert and W. Duval, *Rock Mechanics and Design of Structures in Rock*, Wiley, New York, 1967.
14. A. Afrouz, 'Mechanical behaviour of mine floor', Internal Report No. 112, Mining Department, McGill University, 1986.

3.3.4 FRICTIONAL PROPERTIES OF ROCK APPLIED TO MINING EXCAVATIONS

By: J. Hadjigeorgiou, F.P. Hassani and Andy A. Afrouz

Department of Mining and Metallurgical Eng.,
McGill University, Montreal, Canada

ABSTRACT

Results of direct shear tests on rock discontinuities and triaxial compression tests on intact rocks are presented and discussed. The main parameters considered are: stress functions, angle of internal friction and cohesion variations. The Primary objective of this work is to determine relationships that adequately describe the rock behaviour for design of mining excavations, while still maintaining a certain degree of simplicity.

INTRODUCTION

The essence of stability analysis, applied to mining excavations, lies with the accurate estimation of the shear strength, internal friction and cohesion of the rock mass.

In surface and shallow underground excavations the stability of the mining structure tend to be mostly influenced by the presence of structural discontinuities, weight of the overburden, groundwater regime and the degree of weathering. Excavations at greater depths, however, tend to primarily be controlled by the rock mass response to the in situ stress variation in the vicinity of the openings, the tectonic stresses and state of fractured-zones surrounding the openings.

Under high stresses and in the absence of major weakness planes, the intact strength of rock in the immediate vicinity of an opening is much higher than the broken or heavily jointed rock masses. However, the fissured, fractured or jointed rock is capable of supporting considerable loads, especially in the presence of a lateral restrain. Therefore, an understanding of the complete stress-strain characteristics of a confined rock mass have application in the mining field, especially where deep underground structures are developed in rocks which fail due to the induced ground stresses.

The purpose of this investigation is to present the laboratory results performed on discontinuities and intact sedimentary rocks. Use of a stiff-testing machine allowed the determination of complete stress-strain curves. It is believed that the above program allows for the determination of the frictional properties of the investigated rocks, for mining excavation design considerations.

SHEAR STRENGTH OF ROCK DISCONTINUITIES

Variety of models have been suggested to adequately predict the frictional behaviour of discontinuities in rock (1-15). The mere proposal of so many different models is a testimony to the underlying complexity in the determination of rock mass frictional behaviour. Patton's (2,3) pioneering work laid the foundations for most of the recent studies. The proposed bi-linear criterian (2) recognized the effects of surface roughness and dimensions on the frictional properties of rock discontinuities. While mechanically correct, his model was limited due to the assumption of a single asperity inclination. Ladanyi and Archambault (4) have suggested a model that provides a physically appropriate simulation of the shearing process. Its main restriction, however, lies in the accurate collection of the required input parameters, which is greatly hindered by currently available laboratory and in situ methodology. Heuze and Barbour (10) have proposed a model that has the capacity for strain softening and dilatancy effects. Swan (15) has analyzed the contact mechanics of a number of rock joints by obtaining information on the detailed topography of joint surfaces. Barton (6,7) and his co-workers have (8,11,12) provided valuable contributions that improved our understanding of the behaviour of rock joints. The Barton-Bandis model (11) is a most comprehensive design tool, backed by an abundance of experimental data. Its major premise is the use of simple index tests in order to determine the input parameters. As such the overall reliability of the model can not be any greater than the quality of the derived input parameters.

In recent years the use of a Power Curve has been applied as a means of representing failure envelopes of discontinuities. Hassani (13) and Udd et al. (14) have used direct shear test results and obtained Power Law relationships that adequately described the frictional behaviour of the rock joints.

ANALYSIS OF DIRECT SHEAR RESULTS ON ROCK DISCONTINUITIES

A large direct shear testing machine was used to determine the shear strength of rock joints. The testing procedure complied with the suggested ISRM guidelines (16). Table 1 provides a summary of the peak failure envelopes for the tested samples. A minimum of five points were used to determine the failure envelope. Table 2 indicates the ultimate failure envelope of the tested samples. The term "ultimate" was chosen over "residual" in that the authors tend to agree with Krahn and Morgenstern (17) view that rock joints do not have a fixed residual value as the case for soils. It is quite probable that the ultimate frictional resistance of hard unweathered rock depends on the initial surface roughness and the type of alteration occurring during shearing.

The power law equation is: $\tau=A+\sigma n^{B}$ (1)

where: τ = shear stress, MPa

σ_n = normal stress, MPa

A, B = unit dependent coefficients, MPa, given in tables 1 and 2.

The relationships presented in Tables 1 and 2 were obtained by a statistical analysis of previously conducted tests (18,19). For comparative purposes the linear Mohr-Coulomb relationships were also derived.

The relationships presented in Tables 1 and 2 were obtained by a statistical analysis of previously conducted tests (18,19). For comparative purposes the linear Mohr-Coulomb relationships were also derived.

Table 1. Peak Failure Envelopes for Rock Joints.

Specimen	Peak Failure Envelopes					
	Power Law			Mohr-Coulomb		
	A	B	r	c	φ	r
Sandstone No. 1	0.586	0.928	0.9991	0.015	29.41	0.9972
No. 2	0.647	0.929	0.9998	0.020	31.75	0.9999
No. 3	0.763	0.905	0.9999	0.043	34.99	0.9995
No. 4	0.759	0.876	0.9997	0.063	33.64	0.9876
Mustone No. 1	0.912	0.796	0.9897	0.127	35.61	0.9976
No. 2	0.682	0.786	0.9992	0.226	24.40	0.9985
No. 3	0.864	0.840	0.9850	0.132	34.51	0.9949
No. 4	0.844	0.819	0.9904	0.367	26.69	0.9985
Siltstone No. 1	0.668	0.852	0.9987	0.102	28.18	0.9999
No. 2	0.847	0.872	0.9977	0.144	34.47	0.9884
No. 3	1.000	0.827	0.9977	0.167	38.14	0.9998

Table 2. Ultimate Failure Envelopes for Rock Joints.

Specimen	Ultimate residual Failure Envelopes					
	Power Law			Mohr-Coulomb		
	A	B	r	c	Φ	r
Sandstone No. 1	0.491	0.938	0.9956	0.000	27.33	0.9919
No. 2	0.530	0.943	0.9971	0.003	27.76	0.9929
No. 3	0.549	0.921	0.9988	0.007	28.50	0.9985
No. 4	0.608	0.897	0.9997	0.042	28.61	0.9981
Sandstone saw-cut	0.723	0.967	0.9986	0.054	32.80	0.9974
Mudstone saw-cut	0.527	0.989	0.9894	0.000	27.87	0.9952
Siltstone saw-cut	0.575	0.994	0.9945	0.009	29.53	0.9998

While equation (1) adequately describes the failure envelope, it is sometimes necessary to determine the angle of internal friction (ϕ) and cohesion (*c*) so that they can be used as input parameters for mine design excavations. An additional requirement being that these parameters should be easily determined. The traditional design approach is to take unique *c* and ϕ values and substitute for them in the design equations. This however is in direct contradiction with the curvilinear nature of the failure envelope i.e. their dependence on the applied normal stress. A way to overcome this difficulty has been proposed by Hadjigeorgiou (18).

Equation (1) can also be represented by: $\tau = \sigma_n . \tan(\phi_T)$ (2)

where ϕ_T is the total angle of internal friction. Barton (7) has suggested that ϕ_T comprises of the basic friction angle (ϕ_b), the peak dilation angle (d_n) and the shear component (S_n), i.e.

$$\phi_T = \phi_b + d_n + S_n \quad (3)$$

Hassani and Scoble (20) have examined the relationship of the ϕ_T (which they referred to as ϕ_P) with the residual angle of friction (ϕ_r) and the inclination of the surface asperities (i). In their work they found that: $\phi_P > \phi_r + i$

The reason for this discrepancy was attributed by the authors to a deformation of the asperities in the direction of shear. This reduces the effective inclination of the asperities, consequently resulting in lower frictional resistance.

To overcome these difficulties and in order to determine a simple criterion suitable for design considerations the following modified, curvilinear form of the Mohr Coulomb criterion could be more appropriate.

$$\tau_i = c_i + \sigma_n \cdot \tan \phi_i \quad (4)$$

where: τ_i = instantaneous shear strength, MPa
ϕ_i = instantaneous angle of friction, degrees
c_i = instantaneous apparent cohesion, MPa

The difference between the total friction angle (ϕ_T) and the instantaneous angle of friction being that ϕ_T is secant to the failure envelope while ϕ_i is a tangential value

(Fig. 1). Assuming the validity of the Power Law relationship then by differentiating eqn. (1):

$$\phi_i = \tan^{-1} AB\sigma n^{B-1)} \quad (5)$$

The instantaneous angle of internal friction can then be equated to:

$$\phi_i = (\phi_b + d_n) \quad (6)$$

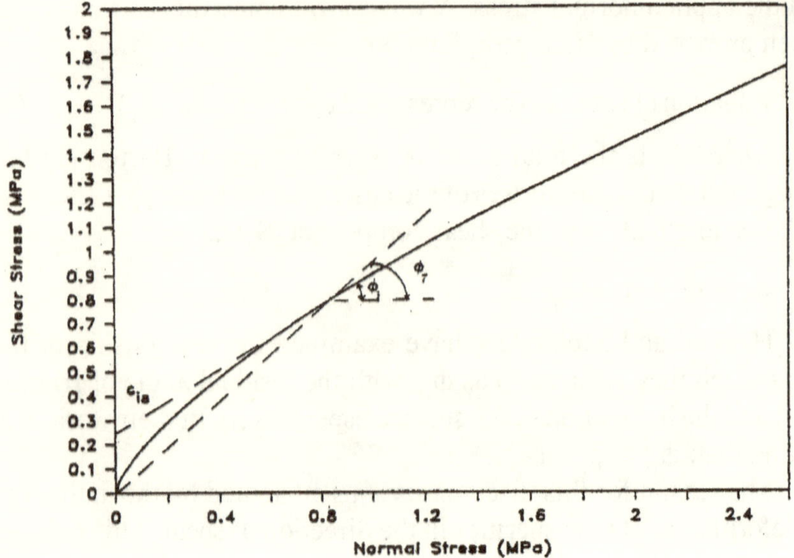

Figure 1. Modified Failure Criterion for Rock Joints

The instantaneous apparent cohesion which has been introduced in equation (4) is equivalent to the shear component S_n. This value is highly dependent on the surface roughness of the discontinuities in that surfaces with more asperities produce more gouge (Hassani and Scoble (20)) and on the rigidity of the asperities. For design purposes it can be shown that a good estimate of the instantaneous apparent cohesion is obtained by:

$$c_i = A\,(1 - B) \tag{7}$$

Figure 2 demonstrates the relationship of ϕ_T, ϕ_i, and c_i with respect to the applied stress for mudstone specimen No. 2. It can be seen that ϕ_i decreases as the normal stress increases; similarly the c_i is increasing with increased normal stress. This is attributed to the collected grout created by the shearing of the asperities.

It is hence possible to determine the appropriate c and ϕ values at a range of normal stresses based on the A and B coefficients of the power curve. The basis of this theorem is that the strength envelope can be predicted by means of a Power Law relationship.

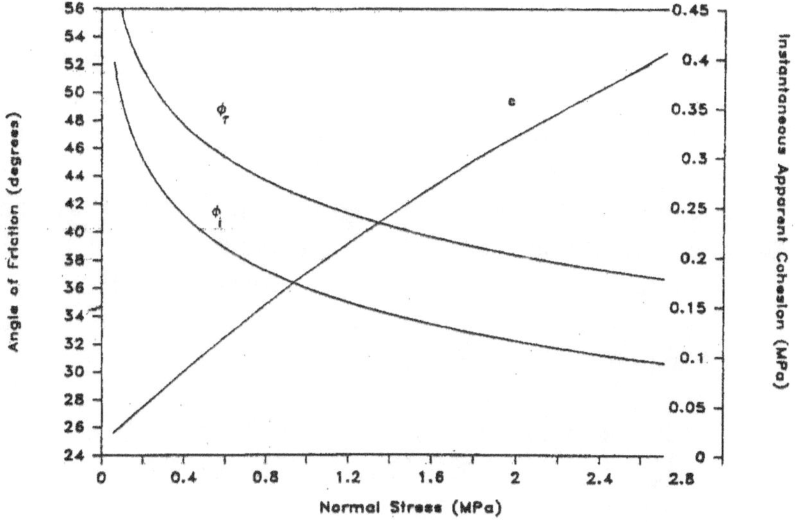

Figure 2. Variation of the Frictional Characteristics of Rock Joints with respect to Applied Normal Stresses (Mudstone Sample).

PEAK AND POST-PEAK TRIAXIAL COMPRESSIVE STRENGTH OF ROCK

Since the early work of Coulomb (21), several criteria have been developed in order to predict the behaviour of rock in compression. Theoretical criteria are generally an inadequate fit to data obtained by experimentation on rock over a wide range of applied compressive stress conditions. In modifying these theoretical criteria to better fit experimental data, some loss of simplicity is incurred. Alternatively an empirical criterion can be adopted, having little or no theoretical foundation, selected only to fulfill a specific practical requirement for accurate strength prediction and simplicity of use. This approach has been followed by Franklin (22), Hoek (23), Bieniawski (24), and Hoek and Brown (25) among others.

Hudson et al. (26) have described the development of mechanical testing of materials and testing machines. The two major advances they describe are the development of stiff frames and the use of feedback control systems. Rummel and Fairhurst (27), Wawersik and Brace (28) have used servo-controlled testing machines to determine the post-failure behaviour of brittle rock.

In this work, the experiments were performed in a "conventional" triaxial cell. This is characterized in that $\sigma_1 > \sigma_2 = \sigma_3$. The principal stress, $_1$, was applied by means of an electro-hydraulic servo-controlled testing machine and the confining stress was maintained constant by means of a hydraulic pump unit. The rock specimens of 50 mm diameter and 100 mm height were accommodated in a Hoek and Franklin cell (29). The confining fluid pressure was kept constant and specimen dilation was recorded. In order to minimize the influence of the testing procedure on the material stress-strain characteristics, the ISRM standardized procedure was followed (16).

GEOTECHNICAL DESCRIPTIONS OF INVESTIGATED ROCK SPECIMENS

The major classification of sedimentary rocks in their division by origin into respective chemical and/or physical derivatives, i.e., sandstone, siltstone, mudstone, seat-earth and coal. Major subdivisions of such rock types is on the basis of the average specimen grain size (Blythe and Feritas (30). In this study, the investigated rock specimens were:

(a) Fine-grained sandstone: This sandstone has a quartz content of greater than 60% by volume and a median grain size of between 0.2 and 0.06 mm. Individual grains were just distinguishable with a x10 hand lens. The unit weight, γ, was equal to 25.1 KN/m^3, the void ratio e = 6.5% and the porosity n = 6.10%.

(b) Coarse-grained sandstone: Defined as having a quartz content of greater than 60% and a median grain size of between 0.6 and 2.0 mm. Individual grains are distinguished by the naked eye. Unit weight, γ = 219 KN/m^3, e = 15.07% and n = 13.10%.

(c) The quartz content for siltstones is between 30 and 60%. Siltstones are commonly grey and the reflectance of the surface is not increased by the self-polishing test and remains matt or granular. Rock has a gritty feel when scraped against a knife. γ = 26.7 KN/m^3, e = 7.87% and n = 7.30%.

(d) Mudstone: The quartz content of mudstone is less than 30%. Mudstone has a graying-black color and when the self-polish test is applied, sticky or soapy patches are found and the rock surface increases in reflectance. When scratched against a knife, mudstone exhibits a greasy, smooth feel. $\gamma = 24.5$ KN/m^3, e = 16.14% and n = 13.90%.

(e) Seatearth: This is a genetic term which refers to the fossil soil that underlies coal seams. Seatearths are identified by their "earthy" texture and highly organic content, frequently present as fossil rootlets. In the strict sense the term includes siltstones and sandstones with high organic content. $\gamma = 23.5$ KN/m^3.

ANALYSIS OF TRIAXIAL TEST RESULTS

Tables 3 and 4 provide a summary of the statistical analysis performed on the results of the compressive triaxial test program. It was evident that the investigated four criteria all gave acceptable descriptions of the σ_1 vs. σ_3 relationships. The four criteria for peaks tress conditions were:

$$\sigma_1 = A\sigma_3^B + C \quad (8)$$

$$\sigma_1 = A\sigma_3 + C \quad (9)$$

$$\sigma_1/\sigma_c = A(\sigma_3/\sigma_c)^B + C \quad (10)$$

$$\sigma_1/\sigma_c = \sigma_3/\sigma_c + [m(\sigma_3/\sigma_c) + s]^{1/2} \quad (11)$$

Criterion (8) is a curvilinear criterion while criterion (9) is the traditional linear criterion commonly used for design purposes. While they both gave satisfactory results over the range of stresses applied, they were subsequently rejected. The reasons being in, that both equations rely on unit dependent coefficients A, B and C, and to the more serious underlying consideration the fact that these equations were not normalized. The advantages in normalizing the experimental results by dividing each measured value by the uniaxial compressive strength of that particular material is that it permits the direct comparison of a number of tests on the same plot in a dimensionless form. An additional case for normalizing is that the influence of testing techniques are minimized since they are common to both numerator and denominator of the dimensionless ratios.

The normalized criteria (10) and (11) which were selected are the Bieniawski (15) and Hoek and Brown (16) stress functions. Criterion (10) which gave the best results, appears to be the more satisfactory of the four, for the investigated rocks. It differs, however, from the original Bieniawski criterion in that coefficient B is allowed to vary for rocks of the same type. This is similar to the approach followed by Brook (31). Bieniawski had opted to keep B as constant for each rock type, i.e. 0.75 for sandstone, not allowing for different size grains of the same material. The coefficients m and s for the Hoek and Brown criterion were derived by menas of their proposed procedure (32). It should be noted that Hoek and Brown assume a value of s = 1 for intact rock material and a value of s = 0 for completely broken down rock. It was interesting to observe that the Bieniawski criterion consistently provided a better fit to the experimental data for intact rock than Hoek and Brown's more widely used criterion (Table 3).

The following criteria were used to describe the residual triaxial stress functions:

$$\sigma_1 = A.\sigma_3^{B} \quad (12)$$

$$\sigma_1 = A\sigma_3 + C \quad (13)$$

$$\sigma_1/\sigma_C = A(\sigma_3/\sigma_c)^{B} \quad (14)$$

$$\sigma_1/\sigma_C = \sigma_3/\sigma_c + [m(\sigma_3/\sigma_c) + s]^{1/2} \quad (15)$$

All criteria provide an adequate description of the residual stress functions for the tested rock specimens (Table 4). For the same reasons described above, the non-normalized criteria (12) and (13) were subsequently rejected in favor of criteria (14) and (15). It should be noted that when equation (15) results in a negative value of coefficient s, s is assumed to be equal to zero and coefficient m is recalculated.

Table 3. Regression Analysis of Peak Triaxial Stress Functions.

Equation		F.G. Sandstone	C.G. Sandstone	Mudstone	Siltstone	Seatearth
1. $\sigma_1 = C + A.\sigma_3^B$	A.	7.1575	16.5981	12.3383	13.0050	10.1708
	B.	0.8893	0.6869	0.6953	0.6008	0.5600
	C.	10.809	46.62	20.69	20.69	20.17
	r.	0.9843	0.9965	0.9440	0.9440	0.986
2. $\sigma_1 = A\,\sigma_3 + C$	A.	3.8295	5.0.307	3.1071	3.1274	2.4386
	C.	118.0947	62.7890	38.4757	60.3203	28.8195
	r.	0.9766	0.9874	0.9549	0.9840	0.9867
3. $\sigma_1/\sigma_c = C + A\,(\sigma_3/\sigma_{c)}B$	A.	4.1475	4.9381	4.8983	2.7565	2.6967
	C.	0.8766	0.6819	0.6952	0.5987	0.5591
	r.	0.9839	0.9974	0.9425	0.9894	0.9890
4. $\sigma_1/\sigma_c = \sigma_3/\sigma_c + [m(\sigma_3/\sigma_c) + s]^{1/2}$	m.	7.650	17.253	7.6215	6.6998	3.3449
	S.	1.000	1.000	1.000	1.000	1.000
	r.	0.9559	0.9950	0.9343	0.9617	0.9598

Table 4. Regression Analysis of Residual Triaxial Stress Functions.

Equation		F.G. Sandstone	C.G. Sandstone	Mudstone	Siltstone	Seatearth
1. $\sigma_1 = C + A.\sigma_3^B$	A.	21.1039	18.9684	11.7299	14.0414	11.9293
	B.	0.4931	0.4876	0.6144	0.5762	0.5474
	r.	0.9907	0.9941	0.9884	0.9865	0.9986
2. $\sigma_1 = A.\,\sigma_3 + C$	A.	3.9768	2.9224	3.3333	3.8314	2.5199
	B.	17.7174	18.6942	9.4407	9.0112	11.2129
	r.	0.9919	9.9777	0.9987	0.9987	0.9861
3. $\sigma_1/\sigma_c = A(\sigma_3/\sigma_{c)}B$	A.	1.9255	2.6266	3.6488	2.7043	3.0614
	C.	0.4842	0.4825	0.6150	0.5738	0.5464
	r.	0.9897	0.9945	0.9877	0.9864	0.9985
4. $\sigma_1/\sigma_C = \sigma_3/\sigma_c + (m(\sigma_3/\sigma_c)+s)^{1/2}$	m.	2.727	3.050	4.460	3.940	2.1650
	S.	0.	0.0668	0.	0	0.3884

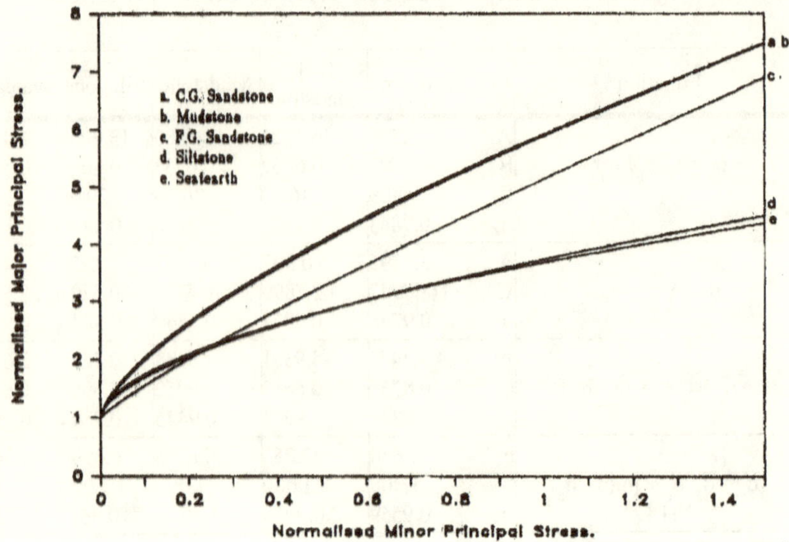

Figure 3. Bieniawski Peak Failure Criteria for all Rock Types

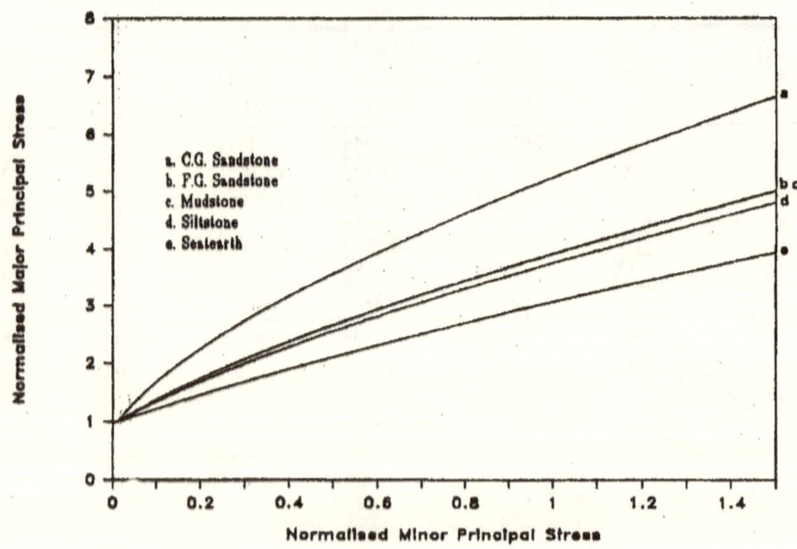

Figure 4. Hoek and Brown Peak Failure Criteria for all Rock Types

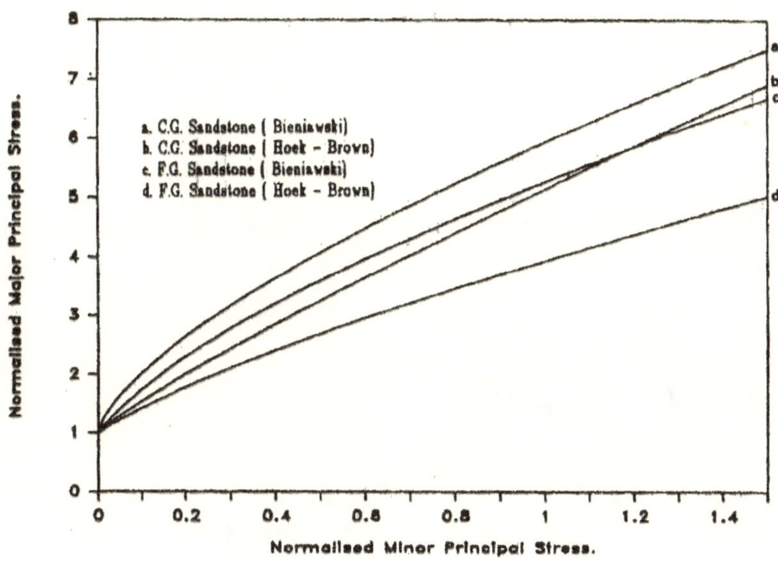

Figure 5. Comparison of Predicted Peak Values for the Sandstone Specimens

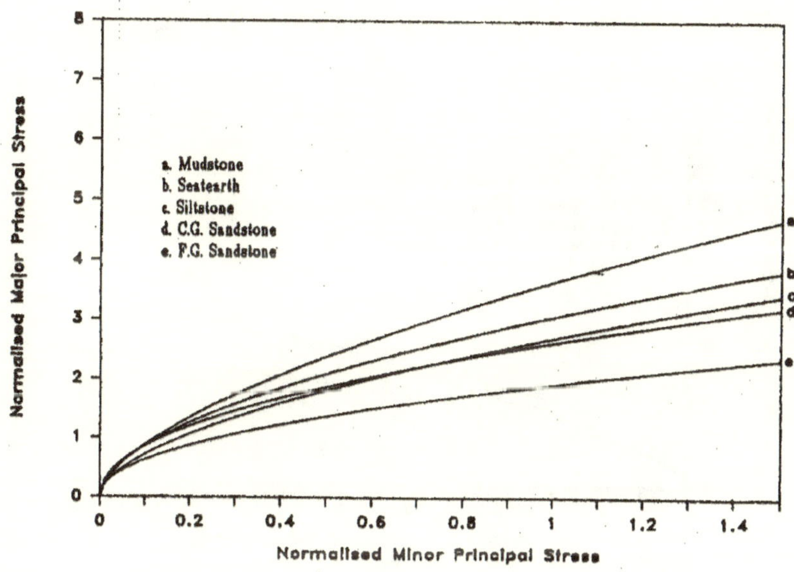

Figure 6. Power-Law Residual Failure Criteria for all Rock Types

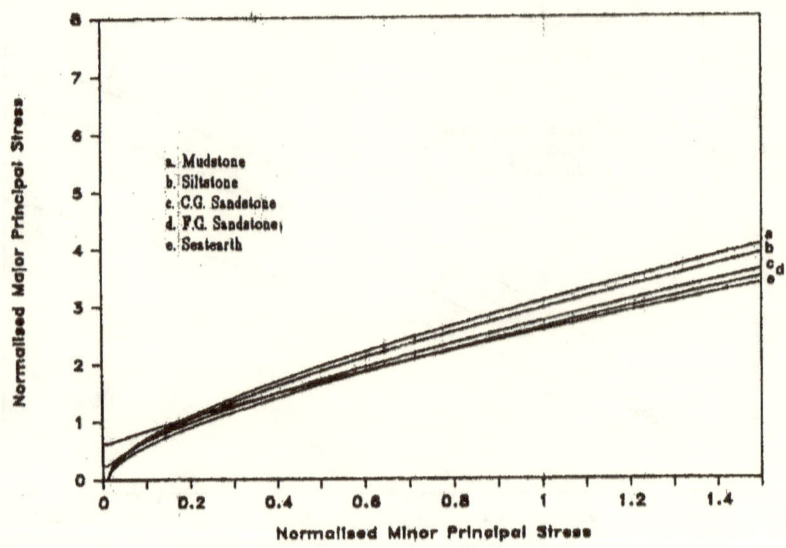

Figure 7. Hoek and Brown Residual Failure Criteria for all Rock Types

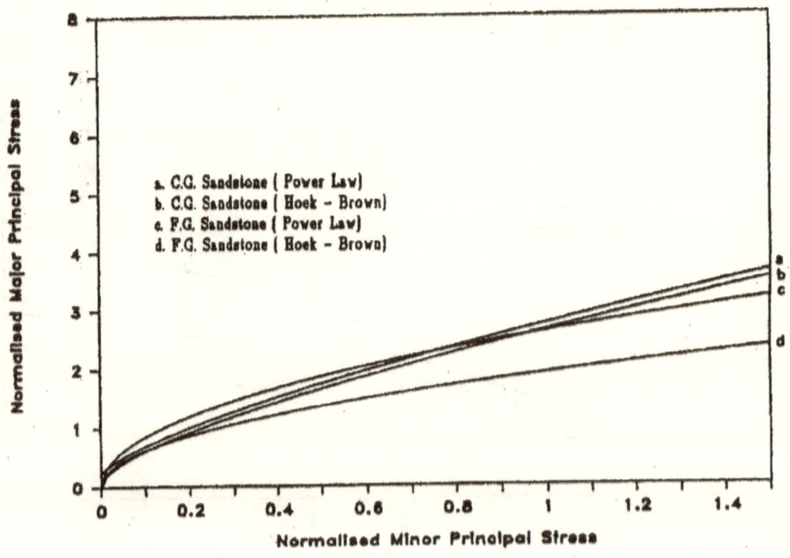

Figure 8. Comparison of Predicted Residual Values for the Sandstone Specimens

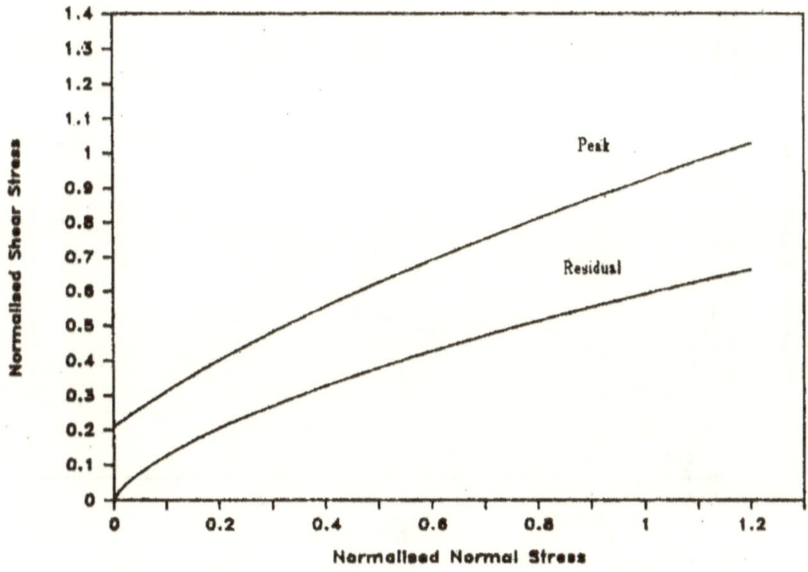

Figure 9: Peak and Residual Failure Envelopes for Fine Grained Sandstone

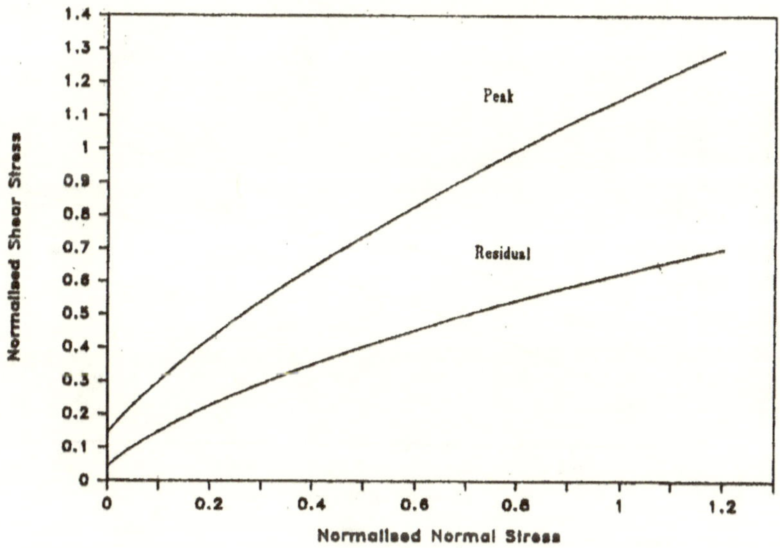

Figure 10. Peak and Residual Failure Envelopes for Coarse Grained Sandstone

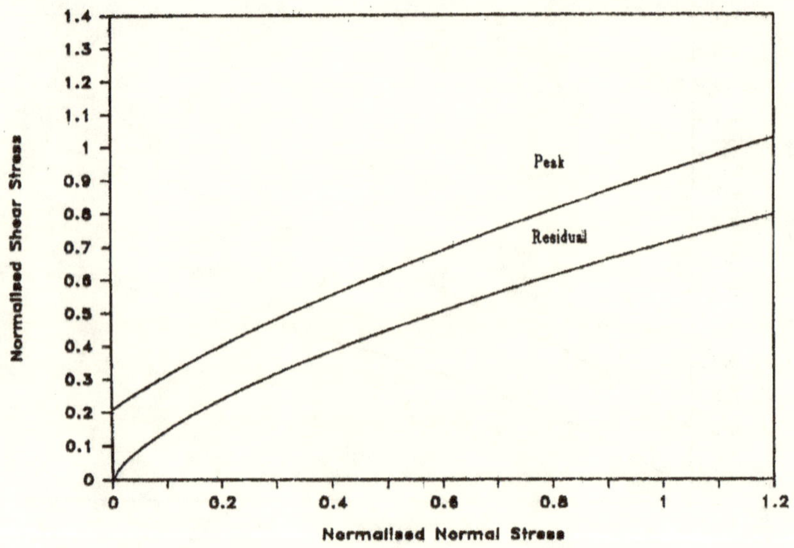

Figure 11. Peak and Residual Failure Envelopes for Mudstone

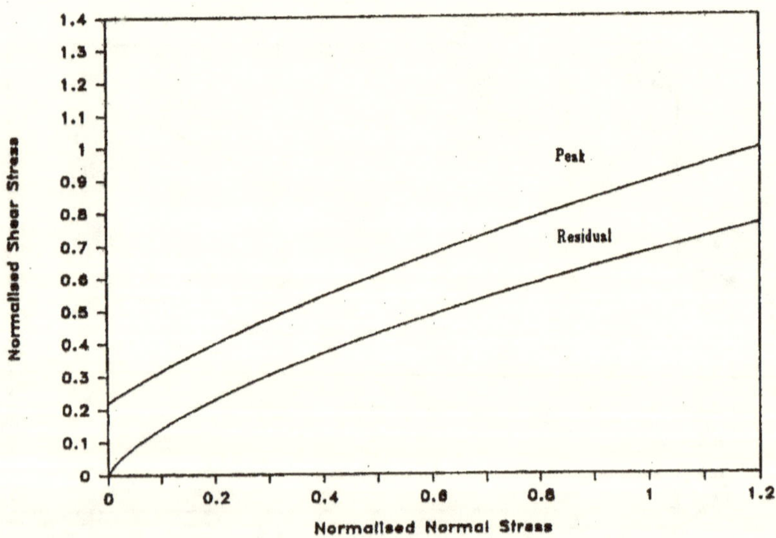

Figure 12. Peak and Residual Failure Envelopes for Siltstone

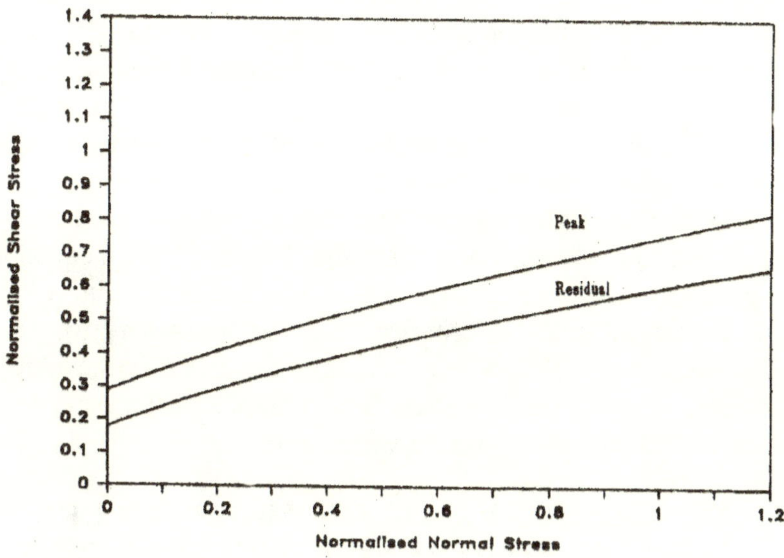

Figure 13. Peak and Residual Failure Envelopes for Seatearth

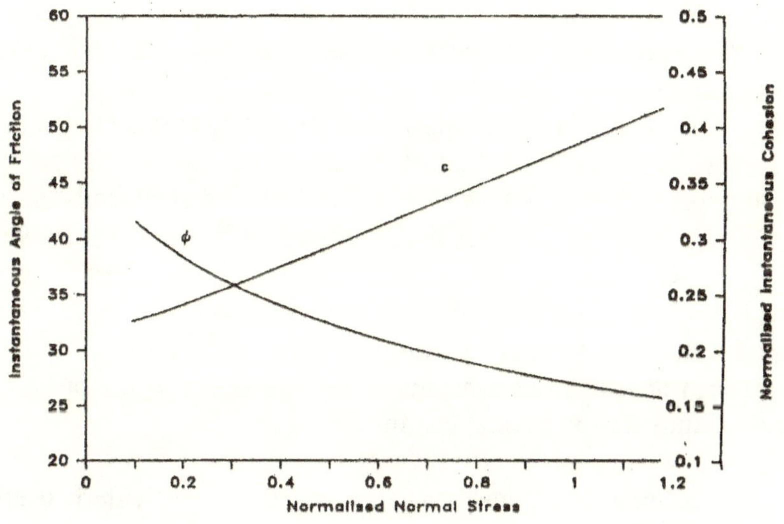

Figure 14. Variation of ϕ_i and c_i with respect to Normalized Normal Stress (Mudstone)

Figure 3 illustrates the predicted peak behaviour of all rock types as represented by the Bieniawski criterion. It is interesting to observe the deviation in peak stress for the two sandstone samples. The Hoek and Brown representation of the peak stresses for the same specimen is shown in Figure 4, while the range of predicted peak stresses for the sandstone specimens, as predicted by the two criteria, is shown in Figure 5. It is evident that application of the Bieniawski criterion leads to higher predicted values.

Figure 6 and Figure 7 show the predicted residual stress functions for all rock types by the Power Law and Hoek and Brown equations, respectively. The closer cluster of the Hoek and Brown curves may tend to suggest that this criterion is not as stress-sensitive as the Bieniawski criterion. This is again exemplified in Figure 8, where the predicted range for residual stress behaviour of sandstone specimens is contrasted with the Power Law functions.

DETERMINATION OF THE MOHR FAILURE ENVELOPE FOR THE INVESTIGATED ROCKS

The principal stress functions that have been investigated are more appropriate for underground excavation design. However, for surface design purposes, it is traditional to express the failure criteria in τ. σ coordinates. If the direct shear test is used on the jointed rock specimens, then the experimental data are readily represented in τ, σ space. The use of triaxial testing apparatus leading to σ_1 and σ_3 values calls for further manipulation of the experimental data. By far the easiest approach is to plot the σ_1, σ_3 data points and graphically derive the failure envelope. This method, however, suffers in the somewhat arbitrary selection of the failure envelope, which is inherent in most graphical techniques. Furthermore, this process has to be repeated for all tested rock specimens. Alternatively, several different criteria have been developed as a means of representation of the principal stress results for the triaxial test in τ, σ space.

(a) Bieniawski's Criterion: This is based on a particularly useful criterion proposed by Hoek (18) which stated that:

$$(\tau m - \tau 0) / \sigma_C = A (\sigma_m / \sigma_C)^B \qquad (16)$$

where: $m = 1/2 (\sigma_1 - \sigma_3)$ and B, C and τ_0 are constants. For practical purposes $\tau_0 = \sigma_c$ (uniaxial compressive strength).

Bieniawski (24) found that results could be obtained within an accuracy of ±1% if the uniaxial tensile strength was assumed to be one tenth of the uniaxial compressive strength. Therefore equation (16) can be written as:

$$(\tau m / \tau c) = A(\sigma_m / \sigma_C)^B + 0.1 \tag{17}$$

(b) Hoek and Brown Criteria: Hoek and Brown (32) have shown that their empirical principal stress criterion could be expressed in terms of shear strength, τ, and normal stress, σ_n, as:

$$(\tau / \tau c) = [A(\sigma n / \sigma_C) - (\sigma_t / \sigma_C)]^B \tag{18}$$

$$\text{where: } (\sigma_t / \sigma_C) = 0.5\,m - (m^2 + 4\,s)^{0.5} \tag{19}$$

Furthermore, Hoek and Brown have suggested that A and B are functions of the m and s coefficients.

(c) Power Law: This has also been applied to triaxial test data in order to derive the failure envelopes. Brook (17) and Hassani (12) have both reached the conclusion that Power Law between normalized τ_m and σ_m is always, if only marginally, a more accurate relationship. They recommend th following expression:

$$(\tau m / \tau c) = A(\sigma_m / \sigma_C)^B \tag{20}$$

(d) Analytical Solutions:

Ucar (33) has taken the empirical m and s coefficients derived by the Hoek and Brown method and has proceeded to derive the failure envelope by the analytical rather than the empirical solution proposed by Hoek and Brown. The final form of his proposed equation is:

$$\tau = (m.\,\sigma_C / 8)[(1 - \sin\phi)(1 - \sin^2\phi) / \sin\phi \tag{21}$$

$$\text{where: } \phi = \sin^{-1}\{(k/3)\,2\cos(1/3)\cos^{-1}[1 - (27/4k^3)] + 4\,\pi/3) + 1\} \tag{22}$$

$$k = (8 /)(m.\,\sigma / \sigma_C) + s + 1.5 \tag{23}$$

In an earlier work, Bray (34) has suggested that the Mohr failure envelope for the Hoek and Brown non-linear failure criterion can be described by:

$$\tau = (\cot \phi_i - \cos \phi_i) \cdot m \cdot \sigma_C / 8 \tag{24}$$

where: $\phi i = \tan^{-1}\{4h \cos^2 [30 + (1/3) \sin^{-1} h^{-2/3}] - 1\}^{0.5}$ (25)

$$h = 1 + [16(m + s.\sigma_C)/(3m^2 . \sigma_C)] \tag{26}$$

Under closer scrutiny, it is observed that both methods lead to the same analytical solution. It can, in fact, be shown that: h = 2k / 3 (27)

REPRESENTATION OF THE MOHR FAILURE ENVELOPE FOR THE INVESTIGATED ROCK TYPES

In table 5 the derived peak failure functions for the investigated rocks are presented while table 6 shows the residual failure functions. The difference between the definition of the τ_m/σ_c peak failure envelope and the actual normalized failure envelope, as obtained by using analytical techniques has been shown by Bieniawski (24). The τ_m and σ_m quantities are given physical representations as the maximum shear stress acting in the specimen at failure and the normal stress acting on the plane of the maximum shear stress. Bieniawski (24) and the Power Law failure criteria define the locus of points at the top of Mohr failure envelopes and are referred to as the maximum shear stress locus. An argument in favor of the shear stress locus representation is that the Mohr envelope can be constructed by a simple and reliable graphical representation. On the other hand, Hoek and Brown's solution as well as the analytical techniques provide a more accurate representation of the "true" Mohr failure envelope.

Table 5. Peak Failure Envelopes for Rock.

Equation	F.G. Sandsone	C.G. Sandstone	Mudsone	Siltstone	Seatearth
1. $(\tau m/\sigma_C) = A(\sigma_m/\sigma_C)^B + 0.1$	A.0.7245 B.0.8349 r.0.9967	0.7914 0.9268 0.9979	0.7895 0.8656 0.9905	0.7233 0.7949 0.9931	0.7236 0.7304 0.9919
2. $(\tau m/\sigma_C) = A(\sigma_m/\sigma_C)^B$	A.0.8233 B.0.7048 r.0.9976	0.9016 0.8231 0.9992	0.9027 0.7743 0.9931	0.8287 0.6899 0.9945	0.8306 0.6475 0.9937
3. $(\tau/\sigma c) = [A(\sigma n/\sigma_C) - (\sigma_t/\sigma_C)]^B$	A.0.832 B.0.661	1.096 0.696	0.802 0.654	0.839 0.663	0.641 0.593

Table 6. Residual Failure Envelopes for Rock.

Equation	F.G. Sandsone	C.G. Sandstone	Mudsone	Siltstone	Seatearth
1. $(\tau m/\sigma_C) = A(\sigma_m/\sigma_C)^B$	A.0.6007 B.0.8295 r.0.9988	0.6366 0.7827 0.9957	0.6774 0.8303 0.9976	0.6257 0.8384 0.9991	0.6326 0.7597 0.9983
2. $(\tau/\sigma c) = [A(\sigma n/\sigma_C) - (\sigma_t/\sigma_C)]^B$	A.0.649 B.0.714	0.630 0.681	0.631 0.703	0.710 0.709	0.546 0.619

Table 7 compares the derived normalized peak failure envelopes for the intact fine-grained sandstone samples. For the range of normalized stresses investigated, it can be seen that the difference in prediction of the normalized shear stress between the semi-empirical Hoek and Brown relation (equation 18) and the analytical solutions of Bray and Ucar (equations 21 and 24) is never more than 2%. In contrast, the linearized Mohr-Coulomb relation could differ from the analytical solution by as much as 12%, this percentage number increasing for higher normal stress. Similar results were obtained fro the other rock types.

Table 7. Comparison of Derived Normalized Failure Envelopes.

F.G. Sandstone σ/σ_c	Normalized Shear Stress		
	Analytical	Hoek & Brown	Mohr-Coulomb
0.1	0.312	0.316	0.33
0.2	3.401	0.401	0.4
0.3	0.482	0.479	0.48
0.4	0.556	0.550	0.55
0.5	0.625	0.617	0.62
0.6	0.691	0.680	0.7
0.7	0.753	0.741	0.77
0.8	0.813	0.799	0.85
0.9	0.870	0.855	0.92
1.0	0.925	0.909	1
1.1	0.978	0.962	1.07
1.2	1.029	1.013	1.14

Figures 9-13 illustrate the peak and residual failure envelopes for the investigated rock types. These were derived based on the analytical solutions and are believed to be representative of the behaviour of the rocks.

Table 8. Instantaneous Angle of Friction and Cohesion Values for Fine-Grained Sandstone.

σ/σ_c	Analytical		Hoek & Brown		
	Φ_i	σi /σc	Φ_i	σi /σc	Mohr-Coulomb
0.1	43.59	0.22	42.21		
0.2	40.19	0.23	38.73		
0.3	37.66	0.25	36.24		
0.4	35.65	0.27	34.32		
0.5	33.98	0.29	32.77		Φ = 36.59
0.6	32.56	0.31	31.48		σ /σc = 0.25
0.7	31.34	0.33	30.38		
0.8	30.26	0.35	29.42		
0.9	29.29	0.37	28.60		
1.0	28.43	0.38	27.83		
1.1	27.64	0.40	27.15		
1.2	26.92	0.42	26.54		

FRICTIONAL AND COHESIVE PROPERTIES OF ROCK

Based on the shear failure envelopes, it is possible to quantify the angle of internal friction and cohesion of the investigated rocks. The angle of internal friction, ϕ, describes the rate of increase of peak strength with normal stress and cohesion, c, is the shear strength intercept. Since the shear failure envelope, however, is not linear, then it is evident that the parameters ϕ and c are not constant but vary with respect to the applied stress. Therefore, it is inappropriate to quote singular and c values for rock unless the specific stress for which they were derived is also defined.

Determination of the instantaneous angle of friction, ϕ_i, and instantaneous cohesion, c_i, is possible and is dependent on the equation used to describe the failure envelope. Hoek and Brown (32) suggest the following relationship:

$$\phi_i = \tan^{-1} AB(\sigma/\sigma_c - \sigma_t/\sigma_c)^{B-1} \quad (28)$$

$$c_i = \tau - \sigma \tan \phi_i \quad (29)$$

where all parameters are based on their defined shear envelope.

Ucar (33) recommends the use of equation (22) while Bray (34) has suggested equation (25) for the calculation of the instantaneous angle of friction, while the instantaneous cohesion is defined by equation (29).

Table 8 provides a summary of the c_i and ϕ_i values for the fine grained sandstone under varying stresses. Similar results were obtained for all investigated rock type. It is thus demonstrated that use of unique values of angle of friction and cohesion, based on the linear Mohr-Coulomb failure criterion, can lead to misleading results. Another observation is the close agreement between analytically derived c_i and ϕ_i values and values derived by Hoek and Brown's approximation.

Figure 14 illustrate the behaviour of the ϕ_i and c_i under a range of normal stress values for the mudstone specimen. It can be seen that the angle of internal friction decreases as the normal stress increases, while the instantaneous cohesion increases as the normal stress increases.

CONCLUSIONS

The design of structures in rock requires a solid knowledge of the mechanical properties of the rock mass. In this work a modified criterion is introduced which is hoped to aid in the prediction of the frictional characteristics of rock discontinuities. This is based on the validity of the Power Law as a suitable description of the failure envelope. Its inherent simplicity makes it an attractive and more appropriate alternative to the linear Mohr-Coulomb criterion.

Based on the analysis of triaxial compressive tests on intact rock specimens the validity of the Hoek and Brown, Bieniawski and Power Law relationships was investigated. The results would indicate that the above criteria are powerful design tools that adequately describe the behaviour of the rock in compression. Furthermore, due to the established non-linearity of the failure envelopes, instantaneous angle of friction and cohesion should be used in subsequent analysis.

REFERENCES

1. Horn, H.M. and Deere, D.C. (1962), Frictional Characteristics of Minerals, Geo-technique, 12, pp. 319–335.
2. Patton, F.D., (1966), Multiple Modes of Shear Failure in Rock, Proc. 1st Congress, Int. Soc. Rock Mech., Lisbon, Vol. 1, pp. 509–513.
3. Patton, F.D., (1966), Multiple Modes of Shear Failure in Rock and Related Materials, Ph.D. Thesis, Univ. of Illinois, p. 282.
4. Ladanyi, B. and Archambault, G., (1970), Simulation of Shear Behavior of a Jointed Rock Mass, Proc. of the 11th Symp. on Rock Mech., (AIME), pp. 105-125.
5. Jaeger, J.C., (1971), Friction of Rocks and Stability of Rock Slopes, 11th Rankine Lectures, Geo-technique, Vol. 21, No. 2, pp. 97–134.
6. Barton, N., (1971), A Relationship between Joint Roughness and Joint Shear Strength, Proc., Symp. Int. Soc. Rock Mech., Nancy, Paper I-8.
7. Barton, N., (1973), Review of New Shear-Strength Criterion for Rock Joints. Engineering Geology, Elsevier, Vol. 7, pp. 287–332.
8. Barton, N. and Choubey, V., (1977), The Shear Strength of Joints in Theory and Practice, Rock Mech., Vol. 10, pp. 1–54.
9. Heuze, F.E., (1979), Dilatant Effects on Rock Joints, Proc. 4th Congress Intl. Soc. Rock Mech., Montreaux, Switzerland, Vol. 1, pp. 169–175.
10. Heuze, F.E. and Barbour, T.G., (1981), Models for Jointed Rock Structures, Proc. 1st Int. Conf. on Computing in Civil Engr., N.Y., ASCE, pp. 811–24.

11. Bandis, S., Lumsden, A.C. and Barton, N., (1983), Fundamentals of Rock Joint Deformation, Int. J. Rock Mech. Sci. and Geomech. Abst., Vol. 20, pp. 249–268.
12. Barton, N.R., Bandis, S.C. and Bakhtar, K., (1985), Strength Deformation and Conductivity of Rock Joints, Int. Jour. Rock Mech., Min. Sci. & Geomech. Abst., Vol. 229, pp. 121–140.
13. Hassani, F.P., (1980), Study of Physical and Mechanical Properties of Rocks and their Discontinuities Associated with Opencast Coal Mining Operations, Ph.D. Thesis, University of Nottingham, p. 750.
14. Udd, J.E., Leinberger, P. and Kozarycz, R., (1981), The Stability of Cut Slopes in Chazy Limestone in a Montreal area Quarry, CIM Bulletin, Vol. 74, No. 829, pp. 65–72.
15. Swan, G., (1983), Determination of Stiffness and other Joint Properties from Roughness Measurements, Rock Mech. and Rock Engr., Vol.1 6, pp. 19–38.
16. Brown, E.T., (1981), Rock Characterization Testing and Monitoring, ISRM Suggested Methods Pergamon Press, Oxford.
17. Krahn, J. and N.R. Morgenstern, (1979), The Ultimate Frictional Resistance of Rock Discontinuities, Int. J. Rock Mech., Min. Sci. and Geomech. Abstr., Vol.1 6, pp. 127–133.
18. Hadjigeorgiou, J., (1987), A Study of Frictional Properties of Rock Masses, M.Eng. Thesis, McGill University, pp. 176.
19. Denby, B., (1983), Shear Assessment in Mine Slope Design, Ph.D. Thesis, University of Nottingham, pp. 349.
20. Hassani, F.P., and M.J. Scoble, (1985), Frictional Mechanism and Properties of Rock Discontinuities, Proc. of the Intl. Symp. on Fundamentals of Rock Joints, Bjorkliden, Sweden, pp. 185–204.
21. Coulomb, C.A., (1776), Essai sur une application des regles de maximis et minimis a quelques problemes de statique, relatifs a l'architecture, Memoires de Mathematique et de Physique, l'Academie Royale des Sciences, 7, pp. 343–382.
22. Franklin, J., (1970), Classification of Rock According to its Mechanical Properties, Ph.D. Thesis, University of London, pp.1 56.

23. Hoek, E., (1968), Brittle Failure of Rock, Rock Mechanics in Engineering Practice, Stagg, K.G. and O.C. Zienkiewicz (Eds.), John Wiley and Sons.
24. Bieniawski, Z.T., (1974), Estimating Strength of Rock Materials, J.S. Afr. Inst. Min. Metall., 74, pp. 312–320.
25. Hoek, E. and E.T. Brown, (1980), Empirical Strength Criterion for Rock Masses, Journ. of the Geotech. Engr. Divsn., ASCE, Vol. 106, No. GT6, pp. 1013–1035.
26. Hudson, J.A., S.L. Crouch and C. Fairhurst, (1972), Soft, Stiff and Servo-Controlled Testing Machines: A Review with Reference to Rock Failure, Engineering Geology, 6, pp. 155–189.
27. Rummel, F. and C. Fairhurst, (1970), Determination of the Post-Failure Behavior of Brittle Rock using a Servo-Controlled Testing Machine, Rock Mechanics, 2, pp. 189–204.
28. Wawersik, W.R. and W.F. Brace, (1971), Post Failure Behaviour of a Granite and Diabase, Rock Mechanics, 3, pp. 61–85.
29. Franklin, J.A. and E. Hoek, (1970), Developments in Triaxial Testing Technique, Rock Mechanics 2, pp. 223–228.
30. Blyth, F.G.H. and M.H. Feritas, (1974), Wentworth and Atterberg Fragments and Grain Size Classifications, Geology for Engineers, 6th Edn., Edward Arnold, London, pp. 177.
31. Brook, N., (1979), Estimating the Triaxial Strength of Rocks. Rock Mech. Min. Sci. and Geomech. Abstracts, Vol. 16, pp. 261–264.
32. Hoek, E. and E.T. Brown, (1980), Underground Excavations in Rock, London Inst. Min. Met., pp. 527.
33. Ucar, R., (1986), Determination of Shear Failure Envelope in Rock Masses, Journal of the Geotechnical Engineering Division, ASCE, Vol. 112, No. GT3, pp. 303–315.
34. Hoek, E., (1984), Strength of Jointed Rock Masses, Rankine Lecture, Geotechnique 33, No. 33, pp. 137–223.

3.3.5 GEOTECHNICAL ASSESSMENT OF THE BEARING CAPACITY OF COAL MINE FLOORS

By: Andy A. Afrouz, F.P. Hassani and M.J. Scoble

Summary

This investigation concerns 3 longwall faces having variable strata and mining conditions, but supported by 202 tonnes hydraulic 6-leg chocks. Bearing capacity tests along the faces were conducted to evaluate the factors influencing floor deformation and failure. The effect of size, shape and perimeter of the base plates, thickness of the floor layer, time and moisture on the ultimate bearing capacity of the floor was measured and discussed. Application of this work is the prediction of stability and support performance of face ends, as well as the design of support systems and ground control on production faces.

Keywords: Longwall coal mining; hydraulic supports; coal mine floors; bearing capacity

Introduction

Floor reaction to any kind of support installed along or behind longwall faces significantly influences strata stability. If support design is to be based on an acceptable rate of closure or deformation along a longwall face and its ends, then in order to ensure support balance and stability the strata pressure within the face region should be controlled. This requires:

(a) uniform pressure and deformation distribution along the face;
(b) a floor bearing capacity in excess of the effective strata pressure exerted upon it through the supports.

The discontinuous and anisotropic nature of Coal Measure rock masses indicates the suitability of *in situ* testing for the determination of strength or bearing capacity underground (see Afrouz, 1976; Coats, 1965; Denby *et al.*, 1983; Hassani and Scoble, 1982; Jenkins, 1960). In plate loading tests it is preferable to utilize full scale base plates as large as those of the mechanized longwall face supports. In this way the measured bearing capacity should be representative of actual values beneath the supports. In these studies however, floor space along the face was limited and only a 21 kN capacity prop dynamometer was available which limited the base plate size.

Prior bearing capacity research has tended to be limited to civil engineering (see for instance Meyerhof and Purkayastha, 1985), with some reported work on mine roof and pillar structures (Hsuing and Peng, 1985) and weathered and jointed rock masses (Rowe, 1982). In addition, variation in modulus of elasticity with mine floor thickness under load is reported (Afrouz and Hassani, 1986).

This paper reports on investigations directed towards characterizing the effect of floor thickness as well as base plate shape and size on deformation and bearing capacity of mine floors along three longwall coal faces with variable strata and mining conditions. A summary of the site specification is given in Table 1 as well as in Figs 1 and 2.

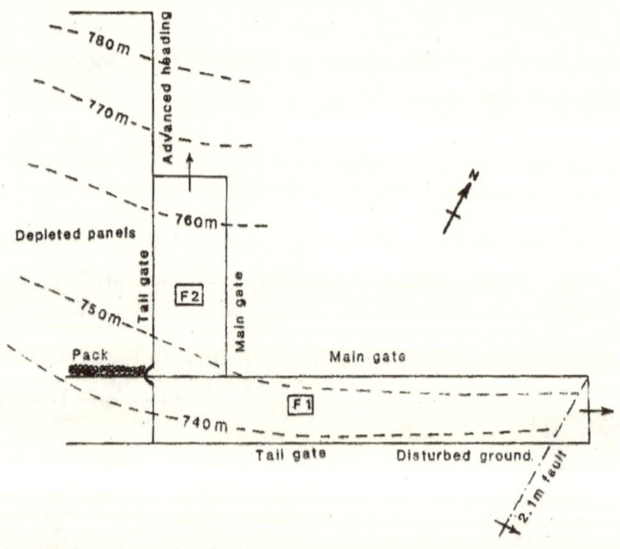

Fig. 1. Plan view of F[1] and F[2] panels relative to each other and the general contour of the coal seam.

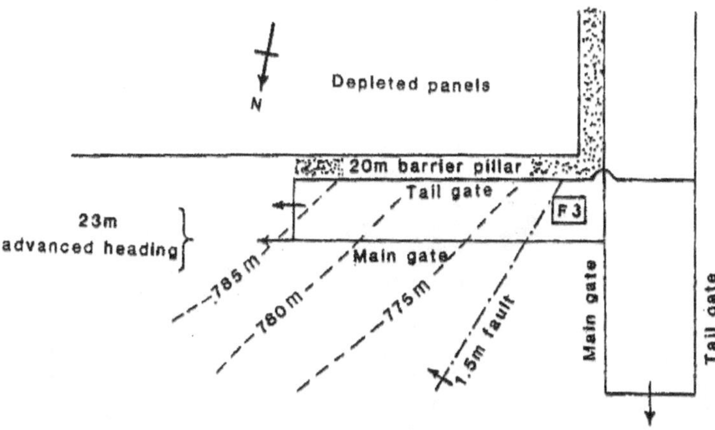

Fig. 2. Plan view of F^3 panel relative to the general contour of the coal seam.

Instrumentation

The equipment used consisted of the following:

(a) Hydraulic prop dynamometer with 21 kN load capacity, similar to design to a standard hydraulic prop and fitted with a 30 kN capacity pressure gauge on the hydraulic main of the prop.
(b) Extension tubes – in lengths from 10 up to 76 cm and diameter equal to that of the hydraulic prop. These were used to extend the effective height of the prop dynamometer to suit all conditions along the face.
(c) A dial gauge was used to record floor penetration and its rate with incremental increase or decrease in applied pressure.
(d) Base plates were manufactured from 1.3 cm thick hardened steel, cut to circular, square and rectangular shapes. The base plates consisted of per-selected sizes within each shape group as shown in Table 2.
(e) A top plate was made of the same material as that of the base plates. Its dimensions were 45.7 × 45.7 × 1.3 cm and so chosen for two purposes:
    (i) to make the ratio of top plate and base plate area as large as possible, in order to reduce and possibly eliminate roof penetration by the prop, as far as practicable.
    (ii) to fit between the mechanized roof supports along the faces.

Table 1- Physical character of the gate roads and faces investigated.

Longwall faces	F1		F2		F3	
Roadways investigated	Main	Tail	Main	Tail	Main	Tail
State of moisture along roadways	99% dry	60% wet	98% dry	98% dry	85% dry	85% dry
Size of arches (in metres):						
(i) Main Gate	4.5 × 3.5		4.5 × 3.5		Originally 4.5 × 3.5 + 0.7 m stilts, the dinted 2.5 m and double skinned with 4.5 × 3.0 m arches	
(ii) Tail Gate	4.5 × 3.5		Originally 4.5 × 3.5, doubled skinned at 15 m in advance of the face with 2.8 × 2.5 arches		Originally 4.5 × 3.5 + 0.7 m stilts, then dinted 2.5 m and doubled skinned with 4.5 × 3.0 m arches	
Spacing of arches	1 m		1 m		1 m	
Other support material used:						
(i) Wood blocks under arches	15 × 15 × 10 cm		15 × 15 × 10 cm		30 × 30 × 15 cm	
(ii) Concrete blocks along face side	50 × 25 × 25 cm		50 × 25 × 25 cm		50 × 25 × 25 cm	
(iii) Road side packs (cm)	130 H × 50 W		130 H × 50 W		130 H × 50 W	
(v) Barrier pillar (m)	25–30		25–30		25–30	
Dinting along main gate (m)	None		First dint 0.81 coal floor at 13 behind the face, then, dint 0.86 underclay at 50 behind face		First dint 1.0 underclay at 10 behind the face then dint 1.5 at 100 behind the face	
Dinting along tail gate (m)	None		0.86 m of underclay at the faceline		1.5 m of underclay at 180 m behind the face	
Position of main gate relative to faceline:	Conventional		Conventional		23 m advanced	

Table 1 continued.

Position of tail gate relative to faceline:	Conventional	Fully advanced	Convention
Depth below surface (m)	740	770	785
Average thickness of the seam (m)	2.8	2.3	2.4
Height of cut (m)	1.6	1.6	1.6
Total length of face (m)	118	192	192
Dip of seam	1:10	1:12	1:161
Average inclination of roadways	1:54	1:10 to 1:15	1:40
Direction of roadways relative to the dip of seam (degree)	87	172	90
System of face support		← 202 tonnes, 6 leg Dowty chocks →	
Number of chocks along face	104	175	116
System of waste control	Total caving, except at the face ends where 3 m long stone and concrete packs are used.		
Method of coal breaking along face		← Uni-directional shearing →	
System of stable control		← 3.5 m long H bars on Dowty duke hydraulic props. →	
Spacing of stable supports		← 0.95 m centre to centre →	
Method of coal breaking in stables		← Conventional shotfiring →	
Averate rate of advance (metres/week)	10	5	20
Type of coal		← Steam bituminous →	
Average uniaxial strength of coal (MN/m^2)	5	5	6
Average thickness of underclay (m)	0.9	0.9	0.6+2 m of shale and underclay
Average uniaxial strength of underclay (MN/m^2)	18	18	10
Type of roof rocks	S. St. & M. St.	S. St. & M. St.	S. St., Clift & Clod.
Average uniaxial strength of roof (MN/m^2)	66	59	83
State of bonding between coal and roof	Good	Moderate	Poor
State of bonding between coal and underclay	Good	Good	Moderate
Type of sidewall rock	Coal, S. St. & M. St.	Coal, S. St. & M. St.	Coa, Clift & S. St.
Working above	Five 9 m pillars 24 m above	Three 9 m pillars 25 m above	None
Working below	None	None	None

Table 2. Specifications of the base plates utilized

Base Area (A)	Circular Plates (C)		Square Plates (S) (a = b or a /b = 1:1)		Rectangular Plates (R) (a = 2b or a/b = 2:1)	
(cm²)	No.	Diameter (cm)	No.	Diameter (cm)	No.	Diameter (cm)
81.01	C1	10.16	–	–	--	--
126.61	C2	12.70	S1	11.25 x 11.25	R1	15.85 x 7.95
182.28	C3	15.24	S2	13.51 x 13.51	R2	19.10 x 9.55
248.13	C4	17.78	S3	15.75 x 15.75	R3	22.26 x 10.15
324.05	C5	20.32	S4	18.31 x 18.31	R4	25.45 x 12.72
411.66	C6	22.86	S5	20.24 x 20.24	R5	28.75 x 14.38
506.33	C7	25.40	S6	22.50 x 22 50	R6	31.85 x 15.93
612.69	C8	27.94	S7	24.77 x 24.77	R7	35.00 x 17.50
729.11	C9	30.48	S8	27.05 x 27.05	R7	38.20 x 19.10

Test procedure

The prop assembly was calibrated under laboratory conditions by a hydraulic compression testing machine, up to its yield point, prior to *in situ* tests. These were conducted at 5 locations selected at 45 m intervals along the three longwall faces, including the main stable, mid-face and tail stable in each case.

Prior to any testing, the roof and floor surfaces were cleared of loose debris and planed flat to achieve uniform loading over the entire contact surface between floor and the base plate. Several tests were carried out on each location, where test points were spaced at about one metre, in order to eliminate the influence of tests upon each other.

Loads were applied incrementally at an average rate of 1 kN/min. During the tests, each incremental load increase or decrease was left for sufficient time to permit the corresponding deformation to become reasonably steady, before the representative deformation was taken from the dial gauge. The next incremental load was then applied. Loading was terminated when either 5 cm floor penetration, (the maximum capacity of the dial gauge), or the yield point of the prop was reached. If the floor failed, the load at initial failure was noted to be the floor bearing capacity. At this point, the penetration rate started to increase rapidly while being accompanied by a drop in the pressure reading. During all unloading, the load was incrementally decreased and the corresponding floor recovery was noted on the dial gauge.

Analysis of results

The load–deformation relationship of the floor observed along the face, beneath various base plates (Fig. 3) suggests:

$$\delta^1 = K_1 P^n \qquad (1)$$

where: δ^1 = deformation or settlement of floor, cm.
P = total vertical load applied on base plate, MN.

K_1 and n are proportionality constants varying between zero and one, depending upon the size and perimeter of the base plates, strength and elastic moduli of the floor.

In order to calculate the strain values from the floor deformation measurements, it is necessary to know the total depth to which the floor is strained. Particularly, the most affected part of the floor, which is the upper floor layer. According to Figure 4, the average percentage vertical strain in the floor beneath the base plate (ϵ_3) can be expressed as: $\epsilon 3 = 100\, \delta^1 / l = K2.\ \sigma 3^n$ (2)

where: l = thickness of the floor layer, cm.

K_2 and n are proportionality constants which vary between zero and one, depending upon the size and perimeter of the base plates, strength of the floor and degree of contact between the base plate and the floor.

$\sigma 3$ = vertical applied pressure at the base plate-floor contact surface, Pa = P/A

A = area of the base plate, cm^2

Equation (2) can also be expressed in the following forms:

$100\, \delta^1 / l = K2\ (P/A)^n$; or $\delta^1 = K2\ .\ l.\ P^n / 100\, A^n$ (3)

According to Hook's Law, $\epsilon = {}_{\delta}{}^{1} / 1 = \sigma / E = P/AE$ (4)

From equations (1) and (4): K1. P^n = 1. P/ AE ; or 1 = K1. A.E. P^{n-1} (5)

In equation 5, since A is constant throughout the experiment and all other parameters change with load and deformation, the value of 1 changes proportionally. This 1-value is the actual floor depth by which the measured deformation should be divided to obtain the true strain developed within the floor.

Equations 1 and 3 provide the relationship between the two proportionality constants K_1 and K_2 as follows:

$$K_1 / K_2 = 1/100A^n \text{ or } K_1 = K_2 \,.\, 1/100A^n \qquad (6)$$

Substituting Equation 5 into Equation 6 yields:

$$K_2 = 100\,\mathrm{P}^{1-n} / \mathrm{E}.A^{1-n} \qquad (7)$$

Equations 5 to 7 suggest that K_1 and K_2 are directly proportional to each other and their value varies non-linearly with the applied load (P). The magnitude of the constants and parameters involved in the above noted equations is averaged in Table 3.

Statistical analysis of the ultimate bearing capacity results is noted in Table 4. The error involved in the *in situ* measurements was between 4.4 and 26.0%. The variation in bearing capacity encountered is considered to have been due to the following factors.

Table 3. Average Values for the Constants and Parameters involved in the equations 1 to7

Floor Rock	K_1	$K_2 \times 10^{-3}$	n
Medium strength Underclay	0.10 – 0.15	0.07 – 0.11	0.8 – 1.0
Coal layer on the Medium Underclay	0.20	1.45	0.95
Coal layer on the Soft Underclay	0.30	2.75	0.50

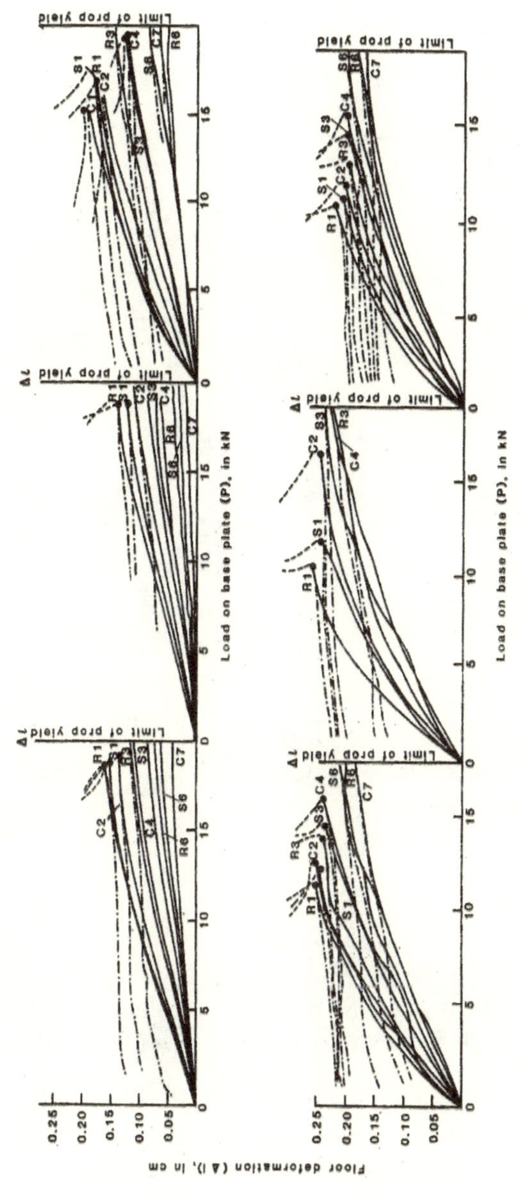

Key: —— loading, —·— unloading, fracture, ----- loading after fracture.

Fig. 3. Load/deformation characteristics of coal and underclay floor (a) F^1 Main stable underclay. Floor thickness = 8.00 cm, (b) F^1 Mid face underclay. Floor underclay thickness = 80.0 cm, (c) F^2 Tall stable underclay. Floor underclay thickness = 86.4 cm, (d) F^3 Main stable coal. Floor coal thickness = 76.2 cm, (3) F^3 Mid face coal. Floor coal thickness = 60.0 cm, (f) F^3 Tall stable coal. Floor coal thickness = 45.7 cm.

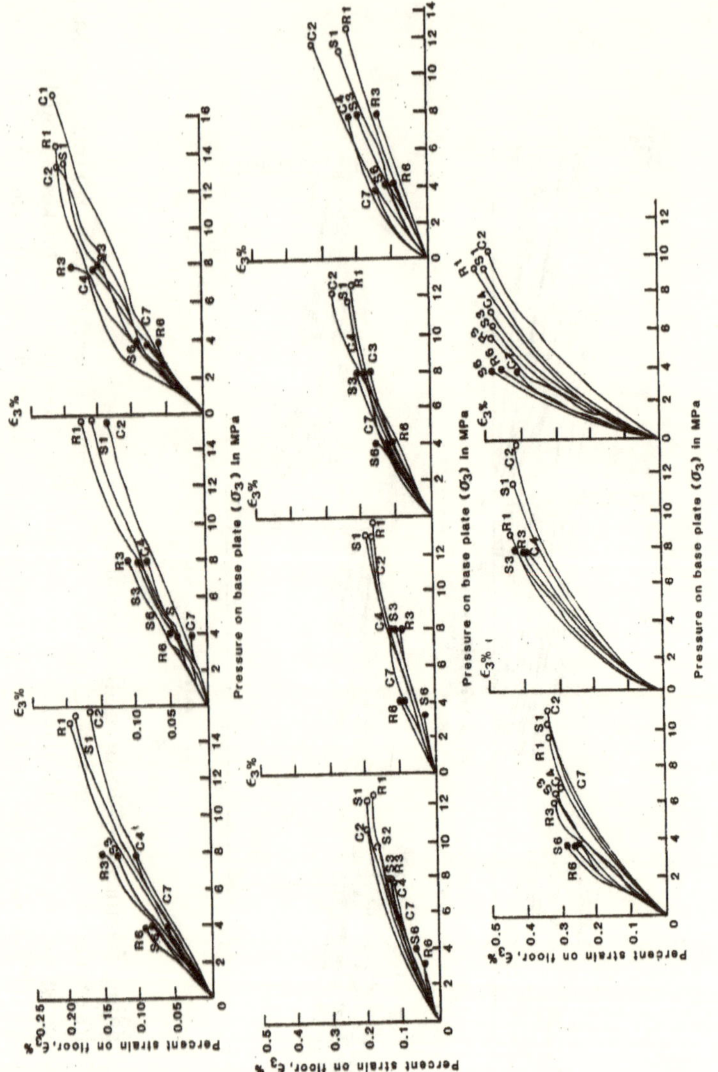

Key: C = circular base plate; S = square base plate; R = rectangular base plate; numbers 126.6 cm^2 and 248.1 cm^2 refer to the area of base plates.

Fig. 4. Stress/strain characteristics of coal and underclay floor. (a) F^1 main stable underclay. Underclay floor thickness 80 cm σ_{mean} = 15.25 MPa, E_{mean} = 6318 MPa, (b) F^1 mid-face underclay, underclay floor thickness = 80 cm, E_{mean} = 9024 MPa, (c) F^2 Tail stable underclay, underclay floor thickness = 86.4 cm, σ_{mean} = 12.91 MPa E_{mean} = 6323 MPa, (d) F^2 main stable coal, coal floor thickness = 81.3 cm, σ_{mean} = 10.43 MPa E_{mean} = 4683 MPa, (e) F^2 meters from main stable coal. Coal floor thickness = 73.7 cm, σ_{mean} 13.33 MPa, E_{mean} = 8219 MPa, (f) F^2 mid face coal. Coal floor thickness = 55.9 cm, σ_{mean} = 12.14 MPa, E_{mean} = 4450 MPa, (g) F^2 45 meters from main stable coal. Coal floor thickness = 45.7 cm, σ_{mean} = 10.76 MPa, E_{mean} = 4450 MPa, (h) F^3 main stable coal. Coal floor thickness = 76.2 cm σ_{mean} = 8.42 MPa, E_{mean} = 3636 MPa, (i) F^3 mid face coal. Coal floor thickness = 61 cm, E_{mean} = 2667 MPa, σ_{mean} = 11.36 MPa, (j) F^3 Tall stable coal. Coal floor thickness = 45.7 cm σ_{mean} = 7.87 MPa, E_{mean} = 1600 MPa. Key: o = fracture, • = maximum pressure applied.

The thickness of floor coal: Observation revealed that there exists a critical thickness for soft rocks at which the bearing capacity is a maximum. It is shown in Fig. 5 that this critical thickness is at 73.7 cm for the F^2 coal floor and at 61.0 cm for the F^3 coal floor.

Geotechnical assessment of the bearing capacity of coal mine floors

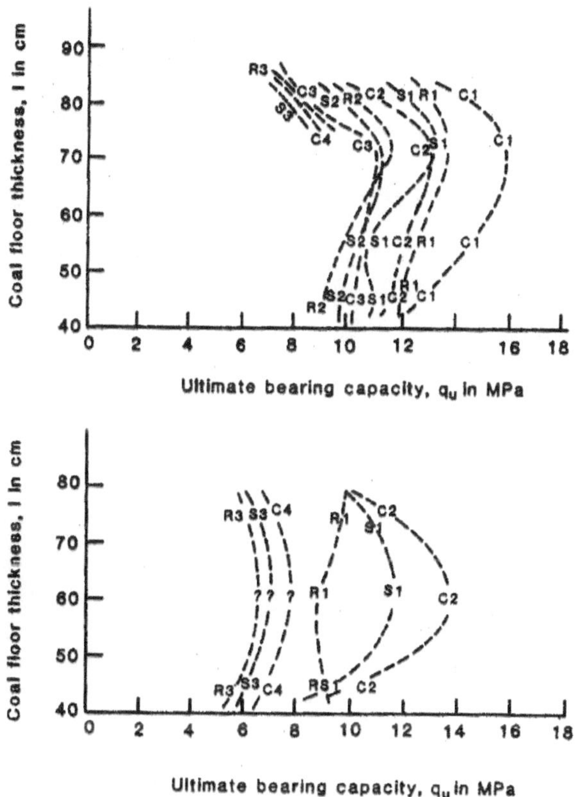

Fig. 5. Thickness effect of coal floor on the floor leaving capacity (a) along F^2 coalface, (b) along F^3 coalface.

Table 4- Statistical analysis of the bearing capacity test results.

	F^1			F^2						F^3						
	Underclay			Underclay			Coal			Coal						
	C	S	R	C	S	R	C	S	R	C 126.6 cm²	248.1 cm²	S 126.6 cm²	248.1 cm²	R 126.6 cm²	248.1 cm²	All together
Number of observations (n)	1	1	1	3	2	1	12	8	7	3	2	3	2	3	2	15
Degree of freedom^{n-1}	0	0	0	2	1	0	11	7	6	2	1	2	1	2	1	14
Sum of ultimate bearing capacity (UBC) (Σq_u)	15.46	15.28	14.97	52.71	24.17	14.41	142.63	84.52	77.55	34.64	13.81	31.38	12.64	27.58	11.80	131.83
Mean of UBC ($\Sigma q_u/n$)$=m$, in MPa	15.46	15.28	14.97	14.23	12.09	14.41	11.89	10.57	11.06	11.55	6.90	10.46	6.32	9.19	5.9	8.79
Square of UBC (m^2)	237.2	234.1	225.0	202.5	146.2	207.7	141.3	111.8	122.4	133.4	47.6	109.5	39.9	84.5	34.8	77.3
Sum of squares of UBC (Σq_u^2)	237.2	234.1	225.0	648.5	296.3	207.7	1741.10	914.0	887.4	407.3	95.3	331.0	79.9	254.0	69.7	1226.0
Mean square of UBC ($\Sigma q_u^2/n$)$-m^2$	237.2	234.1	225.0	216.2	148.1	207.7	145.1	114.3	126.8	135.8	47.7	110.3	40.0	84.7	34.9	85.1
Variance of results (V)$=(\Sigma q_u^2/n)-m^2$	—	—	—	13.7	1.9	—	3.8	2.5	4.4	2.4	0.1	0.8	0.1	0.2	0.1	7.8
Standard deviation $(s)=\sqrt{v}$	—	—	—	3.7	1.4	—	1.9	1.6	2.1	1.5	0.3	0.9	0.3	0.4	0.3	2.8
Per cent error$=100s/m$	—	—	—	26.0	11.5	—	15.9	15.1	19.0	12.9	4.4	8.6	4.7	4.9	5.1	31.6

The effect of base size: Tests showed that the larger the base area then the lower would be the effective vertical pressure on the floor for a given load and hence the lower would be the floor settlement (Fig. 3). The magnitude of this effect was such that when the base area was doubled, keeping all other factors constant, the K_1 value was decreased by 20 to 35% in the case of underclay, and 40 to 57% in the case of coal floor. At the same time the *n*-value was increased by 4 to 7% and 6 to 10% for the underclay and coal floor, respectively. However, the ultimate bearing capacity of floor was decreased by 5 to 20% and 16 to 35% in the underclay and coal floor, respectively.

The effect of base shape: Length of the base perimeter was observed to be an influencing factor. Fig. 6 demonstrates substantial increase in the ultimate bearing capacity (q_u) for coal and underclay floors with increase in the ratio of the base periphery (*P*) to its area (*A*). This can be expressed empirically as follows:

$$q_u = -K + m(p/A)^n \tag{8}$$

where: K = compressive stress in the floor directly beneath the base = – 2.6 to – 2.8,

MPa. (the negative sign indicates compression).

m = proportionality constant, related to the perimeter of the base plates = –0.2 to +0.2, MPa.

n = constant related to the shape and size of the base plates.

The average values of the constants and parameters introduced above are noted in Table 5. In nearly all cases, the floor failure was initiated from the periphery of the base plates and propagated mainly outwards with abrupt fracture surfaces. The mode of fracture beneath the base plates varied with the shape of the plates, and can be categorized as follows:

(i) For the circular bases, fracturing was radial with lengths of the main fractures from 4 to 10 cm.

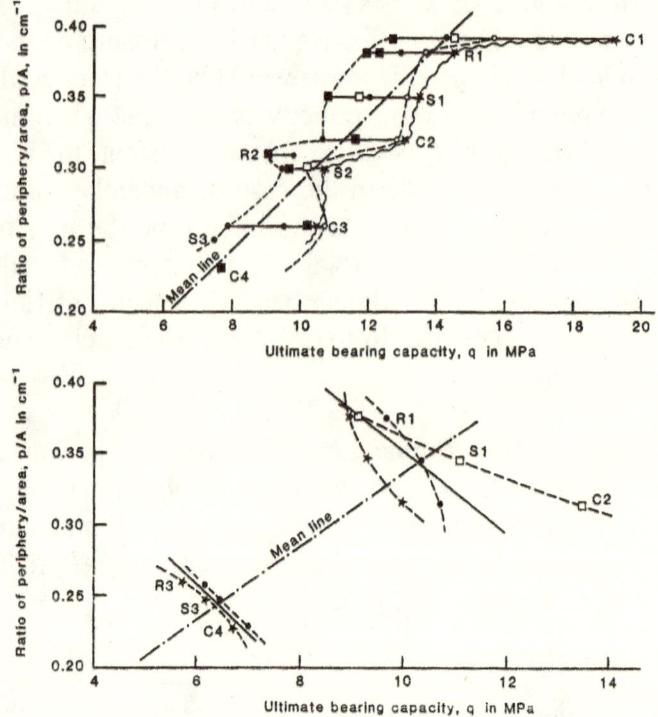

Key: o = 81.3 cm (32 in) thickness coal, • = 73.7 cm (29 in) thickness coal., = 55.9 cm (22 in) thickness coal, ▪ = 45.7 cm (18 in) thickness coal, * = 86.4 cm (34 in) thickness coal. (B) F^3 coal floor. General expression, $q_u = K + m(p/A)$, mean $q_u = 2.6 - 0.2p/A$. Key: • = 76.2 cm (30 in) thickness coal,. = 61.0 cm (24 in) thickness coal, * = 45.7 cm (18 in) thickness coal.

Fig. 6. Effect of base plate size and shape on beaming capacity of floor (a) F^2 coal and underclay. General expressions, $q_u = K + m(p/A)$, mean $q_u = 2.8 + 0.2p/A$.

(ii) For the square and rectangular bases, it tended to propagate outwards from the corners with 5 to 15 cm for the main fracture lengths.

Base edges and sharp corners were thus apparent to exert an adverse effect on the stress concentration in the floor which leads to the tertiary fracture type *in situ*.

Table 5. Average values of factors involved in Equation 8

Floor Rock	K, in MPa	m, in MPa. Cm.	n
Medium strength Underclay	– 2.85	0.25	1
Coal	– 2.62	0.23	1

The effect of rate of loading: During the *in situ* bearing capacity test, the loads were increased far more rapidly than is generally the case under actual mining conditions. It would be more appropriate if the roof and floor convergence by its continual closure is employed to provide the loading on the base plate. While this is justifiable, it gives rise to an additional non-linear variable. It is a fact that the rate of convergence of roof and floor (both in the vertical and horizontal directions) is not constant (see Afrouz and Harvey, 1974). It varies according to the physical state of the rock in the region under study and to the type of mining operation being carried out at the time. Another disadvantage of natural loading is the long time duration for each measurement. For an actual simulation of the floor loading, a prop once set would normally be expected to remain in position for between 5 to 24 h during weekdays and between 48 to 60 h during weekends. Thus each recording dynamometer prop can not be expected to yield more than one or two results per day during weekdays and more than one result during the weekends. The solution would be to install a larger number of instruments on the face at one time, in order to obtain sufficient data.

A time-dependent study on penetration of chock bases along the underclay floor of F^2 face and the coal floor of F^3 face was conducted utilizing a dial gauge mounted on the side of the chock near the base. Due to sensitivity of the instrument, mining operations and vibrations affected the gauge settings. However, during quiet periods the method proved effective. At such times, the chock base settlement into the underclay and coal floor was noted to be time-dependent at an average rate of 2×10^{-3} and 4×10^{-3} cm/hour, respectively. In order to appreciate the magnitude of these settlements, it was found for example that during a six week stoppage of the F^3 face, the average non-linear vertical closure of roof to floor along the face was 2.5 cm/week (2.5×10^{-3} cm h).

The effect of mining method: Shotfiring and the use of hydraulic props in the face ends were observed to develop secondary and tertiary fractures in the floor. The magnitude and concentration of these fractures and the state of stress within the floor of the face ends differ from one another and that of the mid face. This was evident to be mainly due to the influence of stress and time on the gate roads leading to their respective faces. Hence, it is felt that analysis relying solely on floor strength measurement is somewhat simplistic, especially where variable thickness of floor layers is encountered. However, the floor bearing capacity in areas where the face major fractured zones exist should be lower than in the other regions along any face. In general, three fracture zones have been distinguished along the floors of producing longwall faces. These are as follows:

(i) Maximum fractured zones – these tended to develop at face ends, mainly due to effects of hydraulic props and gate roads. About 3 to 4% of the face length may fall within this zone. The average floor bearing capacity (q_u) within this zone along the faces investigated was found to be: q_u for medium strength underclay = 13.2 to 14.6 MPa; q_u for coal floor = 8.3 to 10.3 MPa.

(ii) Minimum fractured zone – this is generally observed to lie along the mid-face where support is hydraulic chocks and the coal is cut by machine. The extent of this zone depended mainly on the face length, rate of advance and number of gate roads leading to the face. 80 to 83% of the face length along the mid-face floor conformed to this minimum fractured state. The average bearing capacity of floor (q_u) in this region of the faces investigated was: q_u for medium strength underclay = 15.8 MPa; q_u for coal floor = 10.8 MPa. It appeared that the average q_u of floor along the minimum fractured zone was 12 to 14% higher than that along the maximum fractured zones.

(iii) Transition zones – these exist between the maximum and minimum fractured zones. Their extent was observed to be between 4 to 8% of the face length. The average bearing capacity of the floor (q_u) within the transition zones was as follows: q_u for medium strength underclay = 14.9 MPa; q_u for coal floor = 9.6 MPa. The average q_u of floor along transition zones was 6 to 11% higher than that along the maximum fractured zones.

The effect of water. Laboratory creep tests on coal and underclay samples (see Afrouz and Harvey, 1974) revealed that the average instantaneous deformation (δ^1) of air dried coal and underclay under pressure was increased by 62 to 71% when water saturated. The effect was even greater when clay bands was present in the underclay. *In situ* observations indicated that introduction of water to the floor along the faces decreased bearing capacity by an average of 18 and 30% for coal and underclay, respectively.

The ultimate floor bearing capacity beneath chocks: Utilizing empirical Equation 8 and Table 5, to calculate the average bearing capacity of coal and underclay floors beneath the full size hydraulic chocks, yields:

A = Base area of the hydraulic chocks = 8130 cm^2

p = Perimeter of the hydraulic chocks = 855 cm

K = – 2.61 and –2.85 MPa for the coal and underclay floor, respectively.

m = 0.23 and 0.25, MPa.cm for the coal and underclay respectively

n = 1

Therefore:

q_u for the coal beneath the chocks = 2.61 + 0.23 (855/8130) = 2.63 MPa

q_u for the underclay beneath the chocks = 2.85 + 0.25 (855/8130) = 2.88 MPa

The designed capacity of the chocks encountered was 202 tonnes. However, their actual loading capacity was measured to be 180 tons, and their yield point was at 2.2 MPa. The actual floor loading beneath the chocks measured along the faces averaged 1.4 MPa. Hence the bearing capacity of coal and underclay floor was adequate to withstand the pressure beneath the chocks, under uniform loading conditions.

Comparison of the results obtained by the graphical method of Equation 8, noted above, and the bearing capacity tests from the base plates (Table 4) indicates that the graphical method provides a lower and more conservative figure. Hence, the graphical method is more applicable to face-ends and vicinity of gate road junctions, and locations where acute and concentrated loading is experienced, e.g. beneath arch legs.

Conclusions

(1) Floor deformation and consequently bearing capacity beneath full-size hydraulic chocks was observed to be time-dependent. The chock base settlement into the dry coal and underclay floors was measured *in situ* to be at an average rate of 4×10^{-3} and 2×10^{-3} cm hour, respectively.

(2) Thickness of the floor along the longwall faces (l) affected by strata loading (P) could be expressed as: $1 = K_1 . A . E . P^{(n-1)}$

where: $0.10 < K_1 < 0.30 = K_2 . 1/100A^n$
$0.07 < K_2 < 2.75 = 100\ P^{(1-n)}/E . A^{(n-1)}$
$0.50 < n < 1.0$

(3) Deformation of the floor (δ^1) and strata loading (P) were found to be related as follows: $\delta^1 = K_1 . P^n = K_2 . 1 . P^n/100A^n$

(4) Stress/strain characteristics of the floor beneath full-size chocks utilizing a base plate test comprised the following: $\varepsilon_3 = K_2 . \sigma = \varepsilon_a . C_b . Q_b/n^3 . C_a . q_a$

where: $q_b = C_a . A_b . q_a/n . C_b . A_a$

(5) There was observed to exist a critical thickness at which the stability of mine floors comprising soft rock is a maximum. This critical thickness was found to be 73.7 cm and 61.0 cm for the F^2 and F^3 coal floor, respectively.

(6) Floor bearing capacity was observed to decrease from mid-face towards face-ends.

(7) Water reduced bearing capacity of the coal and underclay floors by 18 and 30% on average, respectively.

(8) The size effect of base plates on measured floor bearing capacity was such that doubling the base areas (A) causes 16 to 35% decrease in the coal floor bearing and 5 to 20% decrease in the underclay floor bearing capacity.

(9) The length of the base perimeter (p) and its area (A) were found to be additional factors influencing the ultimate floor bearing capacity (q_u) beneath full-size chocks, expressed empirically as follows: $q_u = -K + m(p/A)^n$

where the negative sign for K indicates compression. The magnitude of q_u predicted from the above expression, is less than that obtained from the base plate tests, Table 4. This is in accordance with the conclusion concerning base size effects. It provides more conservative values for design purposes.

References

Afrouz, A. (1976), The lower Four Feet floor rocks of south Wales, *Colliery Guardian*, **224,** No. 12.

Afrouz, A. and Hassani F.P., (1986), Theoretical analysis of variation in elastic moduli of mine floor under load with variable thickness, *Internal Report A*-86-3, *Mining and Metallurgical Engineering Department*, McGill University, p. 8.

Afrouz, A. and Harvey, J., (1974), Rheology of rocks within soft to medium strength range, *International Journal of Rock Mechanics & Mining Science*, **11,** Pergamon Press, U.K., pp. 281–290.

Coates, D.F., (1965), Rock mechanics principles, Canadian Department of Mines and Technical Surveys, Ottawa, Mines Branch Monograph 874, pp. 2–15.

Denby, B., Hassani, F.P. and Scoble, M.J., (1983), The shear strength of weak zones in coal measure rocks, *2nd International Symposium on Surface Mining & Quarrying*, Bristol, U.K. October, pp. 171–181.

Hassani, F.P. and Scoble, M.J., (1982), An investigation into parameters affecting the frictional properties of rock discontinuities, *Ground Engineering Journal,* **15,** No. 7.

Husiung, S.M. and Peng, Syd S., (1985), Chain Pillar Design for U.S. Longwall Panels, *Mining Science and Technology*, **2,** pp. 279–305.

Jenkins, J.D., (1960), Laboratory and underground study of the bearing capacity of mine floors, *Proceedings of 3rd International Conference on Strata Control*, Paris, May 16–20.

Meyerhof, G.G. and Purkayastha, R.D., (1985), Ultimate pile capacity in layered soil under eccentric and inclined loads, Technical note, *Canadian Geotechnical Journal*, **22,** N. 3, Aug., pp. 399–402.

Rowe, R.K. (1982), The determination of rock mass modulus variation with depth for weathered or jointed rock, *Canadian Geotechnical Journal,* **19,** No. 1, February, pp. 29–43.

3.3.6 DETERMINATION OF ROCK MASS MODULUS NONLINEAR VARIATION WITH LOADING AND DEPTH

By: Andy A. Afrouz

Introduction

The strata load (*P*) exerted upon the floor at the base of the support produces stress and strains in the floor varying with:

(1) Physical and mechanical properties and type of the support, e.g. rigid support, hydraulic support, yielding support, pillar or pack.
(2) Depth (*z*) below the support/floor contact.
(3) Physical and mechanical properties of the floor rock.
(4) Number of bedding planes, joints and other discontinuities in the floor beneath the support.

The pressure thus developed in the floor, beneath the support base, may be shown schematically in the form of contours of equal stress distribution, or pressure bulbs. The following conditions may occur:

(a) If the floor layer under a given size of support base (*A*) and load (*P*) is relatively hard, and thicker than the diameter of the support base, all the pressure bulbs created will be formed within the layer (Fig. 1a).
(b) When the floor layer beneath the support is relatively hard but much thinner than the diameter of the support base, the lower part of the pressure bulb may extend into successive floor layers (Fig. 1b). In this case, the stress distribution will be distorted due to the discontinuity encountered.

(c) If the floor rock is relatively soft and the support, pillar or pack is harder, punching of the support into the floor may occur. This type of response will be accompanied by floor heave or roof collapse.

(d) Where two different sizes of support base transmit equal loads (*P*) to floor layers of equal thickness (*z*) (Fig. 2), the pressure bulb below the wider base covers a larger area than that of the narrower base.

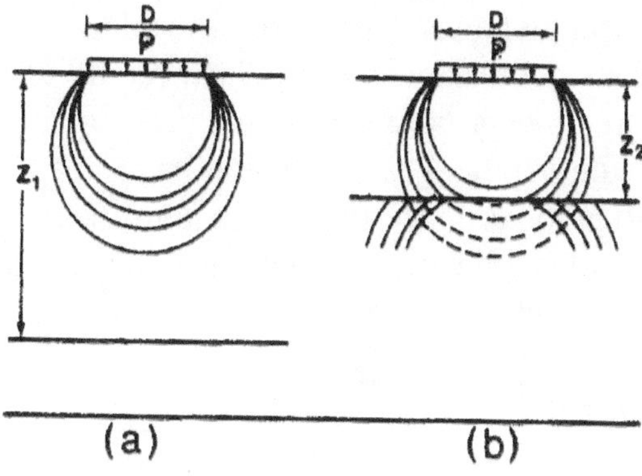

Fig. 1. Variation in the stress distribution in mine floors under a support with variation in the thickness of the floor layer. a. Thickness of the floor layer greater than the diameter of the support base. b. Support base diameter greater than the layer thickness.

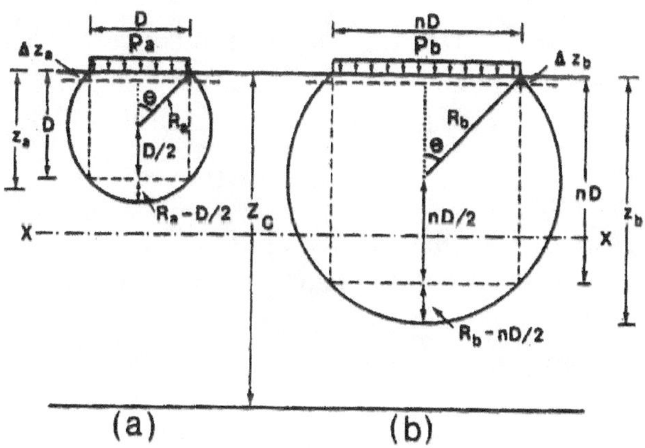

Fig. 2. Effect of the size of the support base on the stress development in the mine floor. Load transmitted to floor in (a) and (b) is equal.

Hence, there can be a depth (X–X in Fig. 2) at which the pressure due to the narrower base is insignificant compared to that of the wider base. Therefore, depending on the condition of the ground and on the supports, similar rock masses may exhibit different stress/strain characteristics. This results in the variable rock mass modulus (E_v).

The concept of the variable modulus of rocks in mines was first reported from studies of the floor of South Wales coal mines, U.K. [1]. Theoretical analysis and mathematical treatment of the above-noted phenomenon at depths below the floor surface has been given in previous publications [2–6]. However, all of the published mathematical treatments assumed a linear variation in the rock mass modulus (E_v) with depth (z) below the footing.

This investigation is concerned with a non-linear variation in the rock mass modulus with variation in vertical load (P) on the support (or pillar or pack) and depth (z).

Theoretical considerations

If a vertical strata load (P) is applied by a support (or pillar or pack) on the floor directly beneath, it develops stress (σ_z) deformation (Δz) and strain (σ_z) in the floor at a depth (z) (Fig. 2). Analysis of the results obtained from the plate load tests on the underclay and coal floors [1–3] provides the following expressions:

Δz = settlement or deformation = ϵ_z Integral between 0 to z of z dz

$= K_1 P^n$, on average (1)

ϵ_z = vertical strain in the floor = $P/A.E_v = \Delta_z$ / Integral between 0 to z of z dz

$= K_2 \sigma$, on average (2)

σ_z = vertical stress development = $K_3 . P/A_e$, MPa (3)

R = radius of the pressure bulb beneath the support, m = $D/2 \sin \theta$

$\approx D/2 \sin 27° = D/0.9$ (4)

z_c = critical depth below which the rock mass modulus remains constant, m

$\approx D + R - (D/2) = D + (D/0.9) - (D/2) \approx 1.6D$ (5)

q_u = ultimate bearing capacity of the floor beneath the support, MPa

$= -K_4 + m\,(p/A)^n$ (6)

(Afrouz empirical [3]. The negative sign indicates compression).

Or:

$q_u = \gamma w\ [1 - 0.4(w/l)][1.5 \tan\phi \cdot e^{\pi \tan}\phi \cdot \tan^2(0.25\pi - 0.50\phi)] +$

$S_0 \cdot \cot\phi\ [e^{\pi \tan}\phi \cdot \tan^2(0.25\pi - 0.50\phi)]\ [1 + \sin\phi\ (w/l)] -$

$S_0 \cdot \cot\phi$ (7)

(Hansen theoretical [7])

FS = factor of safety of the floor = q_u/σ_z (8)

Where: A_e = effective area of the pressure bulb beneath the support (m^2) (Fig. 2)

= πR^2 – [(πR^2 – D^2)/4] = 0.19π + 0.25D^2 in plane geometry

= $4\pi R^2 - \pi[2R^2 - (D^2/2) + D^2]/2$

= 0.19π(Integral between 0 to z_c of z dz) – 0.25πD^2 in spherical geometry

D = diameter or width of the support, m

n = constant relating to the ratio of p/A = 0.5–1 (Table 1)

p = perimeter of the support base, m

A = contact area of the support with the floor,m^2

K_1, K_2 and K_3= empirical proportionality constants varying between 0 and 1 depending on the rock type; state of fractures or other discontinuities in the rock mass; the critical depth (z_c); or thickness of the rock below which the rock mass modulus remains constant (Table 1)

K_4 = compressive strength of the floor directly beneath the support (or pillar or pack), MPa = –2.50 to –2.85, for soft underclay (Table 1)

m = proportionality constant related to the perimeter of the support base and Poison's Ratio of the rock mass beneath the support (v) = 0.20–0.25 (Table 1)

w = width of the support base

θ = angle from vertical of the line connecting the center of the pressure bulb and perimeter of the support (degrees) (Fig. 2)

γ = weight per unit volume of the floor rock, MN/m^3 = 0.015–0.025

S_0 = cohesion of the rock beneath the support, MPa

E_v = variable rock mass modulus which increases nonlinearly with depth beneath the support, up to a critical depth (z_c) after which it remains relatively constant, MPa (Fig. 3)

E_0 = rock mass modulus immediately beneath the support (or pillar or pack) (MPa)

= $q_u(1 - v^2)/\epsilon_0 = \sigma_z(1 - v^2/\epsilon_0$; for $FS = 1$ and $z = 0$

E = constant or ultimate rock mass modulus, MPa

= $\sigma_z(1 - v^2)/\epsilon_z$ for $z < z_c = E_0 + E_v$

v = Poisson's ratio of the rock in situ = 0.15–0.35

ϕ = internal friction angle of the rock (degrees)

σ_c = unconfined compressive strength of the intact rock samples under laboratory conditions.

TABLE 1. Average empirical values for the parameters used in equations (1)–(9) (modified from [1, 2 and 3])

Type of floor bedding	σc (MPa)	K_1	$K_2 \times 10^{-3}$	K_3	$-K_4$ (MPa)	m	n
Medium strength underclay	18	0.10–0.15	0.07–0.11	1.98	2.85	0.25	0.8–1.0
Soft underclay	10	0.20–0.28	1.30–2.20	1.10	2.75	0.24	0.70
Bituminous coal	5–6	0.25–0.32	1.30	0.95	2.62	0.23	0.60
45.7–81.3 cm Bituminous coal–on the medium strength underclay		0.20	1.45	0.98	2.70	0.24	0.75
45.7–81.3 cm Bituminous coal–on the soft underclay		0.30	2.75	0.90	2.65	0.24	0.50.

Note: Underclay or seat earth is a grayish clay occurring immediately beneath the coal seams. It contains silica, aluminum, magnesium and other impurities such as pyrites and fossilized plant roots.

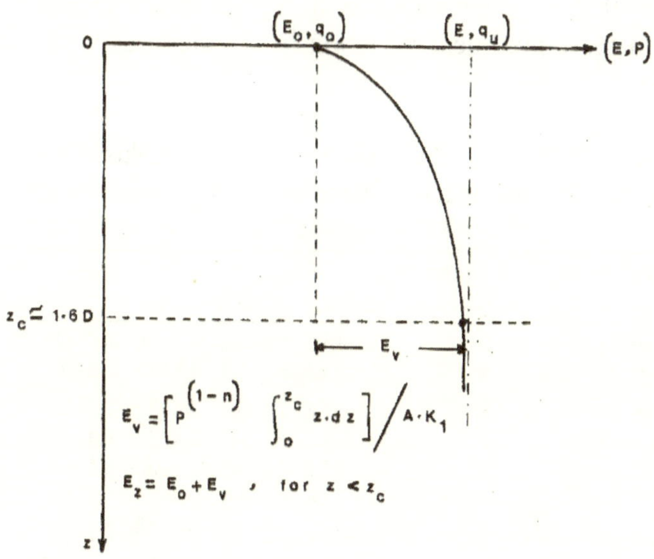

Fig. 3. Nonlinear variation of the rock mass modulus (E_v) with depth below the footing (z) and the strata loading (P).

To determine the variable rock mass modulus (E_V), substitute equations (1) and (5) into eqn. (2). Calculating E_V where $0 < z < z_c$, gives:

$$E_V = [P^{(1-n)} \int \text{Integral between 0 to z c of z dz}] / A.\ K_1$$

$$\approx [P^{(1-n)} \int \text{Integral between o to 1.6D of z dz} / A.\ K_1 \qquad (9)$$

Substituting equations (2) and (5), in terms of P, into eqn. (3) gives:

$$\sigma_z = K_3 A.E_{V.}\ \Delta z / 0.19\ [\int \text{Integral between o and 1.6D of } z^2\ d^2\ z\ -$$
$$1.32\ D \int \text{Integral between 0 and 1.6D of z dz}] \qquad (10)$$

Equations (9) and (10) prove the nonlinear increase of the rock mass modulus. This increases continuously to a critical depth (z) of 1.6 times the support base diameter (D), or the width of the support base (w), as shown in Fig. 3. In situ E_0 and E_V values for the composite underclay and coal floors, using the plate load test results (with ± 15.1% error), are given in Table 2 and their modulus variation is illustrated in Fig. 4. Comparison of the results of this investigation with that obtained from the softer weathered and jointed rocks [6] suggests that stronger rocks, with higher initial modulus values, exhibit a more nonlinear variation in the modulus with increasing load.

TABLE 2. Values of the in situ rock mass modulus at the base plate/ floor contact (E_0) and the variable modulus (E_v) (modified) from [3])

Type of floor layer	Thickness (cm)	q_u (MPa)	E_0 (GPa)	E_v (GPa)	E_v/E_0	E (GPa)
Medium strength	80.0	7.5–15.5	6.0–9.2	4.1–5.0	0.6–0.9	10.1–14.2
underclay	86.4	7.3–13.8	6.0–7.3	4.1–4.8	0.6–0.7	10.1–12.1
Bituminous coal on	81.3	6.5–11.4	4.5–5.3	2.4–3.1	0.5–0.6	6.9–8.4
medium underclay	73.7	6.7–13.5	5.0–6.5	2.4–2.9	0.4–0.5	6.4– 9.4
	55.9	6.2–12.5	4.1–4.7	2.4–2.8	0.5–0.6	6.5– 7.5
	45.7	6.0–11.6	4.1–4.9	2.4–2.6	0.5–0.6	6.5– 7.5
Bituminous coal on soft	76.2	5.5–10.1	2.2–2.6	0.5–0.7	0.1–0.3	2.7– 3.3
underclay	61.0	4.9–11.4	2.0–2.7	0.5–0.7	0.2–0.3	2.5– 3.4
	45.7	4.0–9.8	1.2–1.8	0.5–0.6	0.3–0.4	1.7– 2.4

DETERMINATION OF ROCK MASS MODULUS NONLINEAR VARIATION

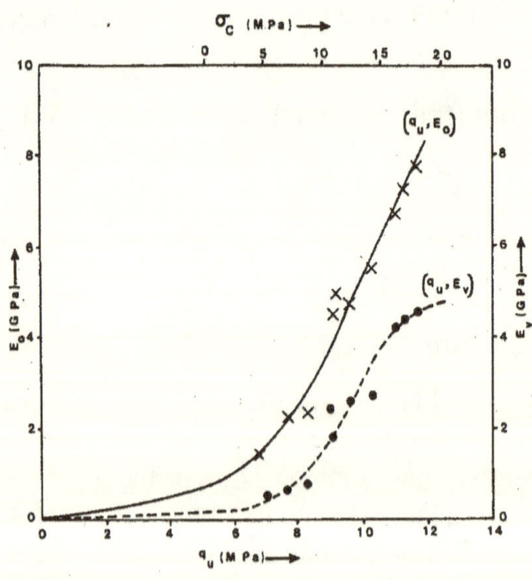

Fig. 4. Comparison of the nonlinear variation of the rock mass modulus at the support base/floor contact (E_0) with that at depths below the support base (E_v).

Conclusions

Under a given load (P), the rock mass modulus increases nonlinearly with depth below the support footing. This increase is continuous until a critical depth (z_c) is reached, after which the rock behaves as an intact mass and its modulus remains relatively constant.

The magnitude of the critical depth (z_c) varies mainly with: the ratio of the support base perimeter (p) to its area (A), strata load exerted on the support base, rock type, and the degree of discontinuities or fractures in the rock mass beneath the base. It can be approximated to $z_c \approx 1.6D$, where D is the diameter, or the smallest dimension, of the support/rock contact.

Rock masses with higher bearing capacity (q) or initial modulus (E_0) exhibit a more nonlinear variation in modulus (E_v). This can be expressed in the following general form:

$$E_v = [P^{(1-n)}{}_{\int}\text{Integral between 0 and z of z dz}] / A.\ K$$

$$E = E_0 + E_v$$

References

1. Afrouz, A., Mechanical behaviour of mine floor. Ph.D. Thesis, Univ. Wales (1973).
2. Afrouz, A. and Harvey, J., Rheology of rocks within soft to medium strength range. Int. J. Rock Mech. Min. Sci., 11 (1974): 281–290.
3. Afrouz, A., Yield and bearing capacity of coal mine floor. Int. J. Rock Mech. Min. Sci., 12 (1975): 241–253.
4. Afrouz, A., Floor loading and movement along coal face. Rock Mech. Rock Eng., 13 (4) (1981): 199–213.
5. Brown, E.T. and Gibson, R.E., Surface settlements on a finite layer whose modulus increases linearly with depth. Int. J. Numerical. Anal. Meth. Geomech., 3 (1979): 37–47.
6. Rowe, R.K., The determination of rock mass modulus variation with depth for weathered or jointed rock. Can. Geotech. J., 19 (1982): 29–43.
7. Hansen, B.J., Bearing Capacity. Dan. Geotech. Inst. Bull., 28 (1970): 5–11.

3.3.7 FLOOR LOADING AND MOVEMENT ALONG THE COALFACE

By: Andy A. Afrouz

Summary — Zusammenfassung — Résumé

Floor Loading and Movement Along the Coalface. Total roof to floor convergence, floor loading, horizontal and vertical movements of floor along three selected longwall faces were investigated. Empirical equations expressing the above noted linearly time-dependent factors in strata mechanics are developed. It was observed that the contribution of floor to the total roof to floor closure is between 31 to 63%, dependent upon the strength of roof and floor layers encountered.

Sohlhebungskraft und Bewegung entlang einer Abbaufront in Kohle. Die gesamte Konvergenz zwischen Firste und Sohle, die Sohlhebungskraft, die horizontalen und vertikalen Bewegungen der Sohle wurden an drei ausgewählten Strebabbauen untersucht. Funktionen, die die Zeitabhängigkeit der oben genannten Größen erfassen, wurden empirisch ermittelt. Es wurde beobachtet, daß der Anteil der Sohlhebung an der gesamten Konvergenz zwischen Firste und Sohle in Abhängigkeit von den Festigkeiten des Hangenden und Liegenden zwischen 31 und 63% beträgt.

Mouvements du fond d'une galerie en excavation dans une mine charbonnière. La convergence entre la calotte et le fond de la galerie, l'effet de la force montant le fond, ainsi que les mouvements horizontaux et verticaux du fond ont été étudiés sur trois tronçons d'excavation. Des équations empiriques donnant la relation plus ou moins linéaire ont été développé en fonction du temps pour des milieux stratifiés. On a observé-que la contribution du mouvement du fond à la convergence total entre la calotte et le fond de la galerie est entre 31 et 63% en fonction de la résistance de la masse rocheuse au-dessus de la calotte resp. en-dessus du fond.

Introduction

Part of the mechanical energy utilized to develop a longwall face is absorbed in the rock within the immediate vicinity of the excavation. In this way, the affected region is energized, thus loosing its initial equilibrium and comparative stability. Ground movements result wherever the total energy overcomes the resistance of the rock to deformation. The result is eventual closure of the opening, particularly in certain cases by horizontal and vertical movements in the floor.

In this context, numerous investigations have been conducted to understand and quantify the strata behaviour surrounding longwall faces. Shepherd [1] found that measurement of loads obtained along the face supports is mainly dependent on the elasticity of the supports and, therefore, provides information merely on the support behaviour. Hence, the results thus obtained can be expected to bear little relation to the actual upward pressure within the floor, especially if the yield strength of the floor is exceeded. Schwartz and Dubois [2, 3] have shown from convergence measurements on about 100 faces in France, that the roof to floor convergence (C) is linearly proportional to the rate of face advance (A), and that up to 72% of the convergence along the face was influenced by the rate of advance and 28% by all other factors when the floor is hard. They recorded an average roof to floor convergence of 4.2 and 5.5 cm/m advance, for hard and soft floor, respectively. The effect of rate of advance on convergence of roof to floor along longwall faces in Britain was averaged [4] suggesting that an increase of 39% in the rate of face advance produced a 16% decrease in the roof to floor convergence. In general, the total roof to floor convergence along the face (C_c in mm) was expressed [5] as:

$$C_c = C_i + r_s \cdot t \tag{1}$$

where: C_i = The mean roof to floor convergence during the cutting period (mm).

R_s = The mean roof to floor convergence per unit time while the face is standing (mm/hour), and

t = Average time between one cut and the next (hour).

It is clear that Eq. (1) provides an overall, qualitative assessment of the total convergence between roof and floor. This assessment appears to be without regard to specific influence of support advancement on the convergence and non-linear time-dependency of the convergence during the standing period of the face.

In general, little investigation has been conducted to quantify Coal and Underclay floor movements along the face. This investigation is directed towards establishing a method of measuring floor movements and loading along longwall faces irrespective of the roof and coalface deformations, and to obtain percentage contribution of the floor in the total roof to floor convergence along three faces in South Wales, UK.

Instrumentation and Test Procedure

Three measuring stations were chosen on each face, behind the armored conveyor. The physical character of the three faces is outlined in Table 1. Since it was found that convergence is more or less uniform across the face/waste line [6], the position of the stations along this line was not expected to alter the results obtained in any way. The measuring stations were at convenient locations, 5 m from the main stables, the mid-faces and 5 m from the tail stables. It would have been desirable to set up stations in the main and tail stables as well as along the faces, for comparison, but due to congestion this was impractical.

The stations were instrumented by a rigid prop which was modified from a friction prop. A suitable top and bottom plate was incorporated to reduce prop penetration into roof and floor as far as possible.

A load cell was placed between the base of prop and the bottom base plate to record any pressure development, as shown in Fig. 1 a.

Table 1. Physical Character of the Faces

Longwall faces investigated	L1	L3	F41
Location of the faces	Britannia Colliery	Britannia Colliery	Windsor Colliery
Name of the seam	Lower four feet	Lower four feet	Four feet
Depth below surface (m)	734	652	505
Average thickness of the seam (m)	2.8	2.3	2.4
Height of cut (m)	1.6	1.6	1.6
Total length of face (m)	118	192	192
Dip of seam	1:10	1:12	1:161
Direction of face advance relative to the dip of seam (degree)	87	179	90
System of face support	202 Tonnes, 6-legs Dowry chocks		
Spacing of supports along face	1 metre centre to centre		
Number of chocks along face	104	175	116
System of Waste control	Total caving, except at the face ends where 3 m long packs are used.		
Method of Coal breaking along face	Unidirectional Shearing		
Method of Coal breaking in stables	Conventional shotfiring		
Number of production shifts per day	1 cutting and 1 backing		2 cuttings
Average rate of advance (m/week)	10	5	20
Type of Coal	Steam Bituminous		
Average thickness of floor coal (m)	zero	0.7	0.8
Average uniaxial strength of coal (MN/m^2)	5	5	6 Underclay
Average uniaxial strength of Underclay (MN/m^2)	18	18	10
Type of roof rock	Coal, S. st., M. st.	S. st., M. st.	Clift, Clod, S. st.
Average uniaxial strength of roof (MN/m^2)	66	59	83
State of bonding between coal and roof	Good	Moderate	Poor
State of bonding between coal and Underclay	Good	Good	Moderate
Working below	None	None	None

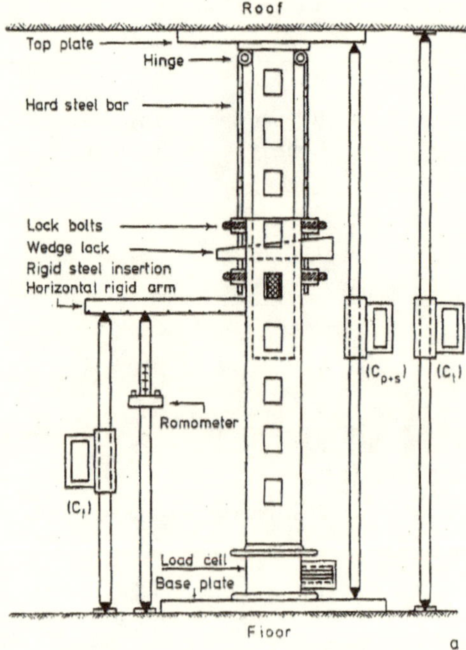

Key: (C_t) = Total roof to floor convergence recorder; (C_f) = Floor lift recorder;
(C_{p+s}) = Rigid prop penetration and sliding recorder.

Fig. 1 a. Schematic diagram of the rigid prop assembly utilized to record floor behaviour along the longwall faces

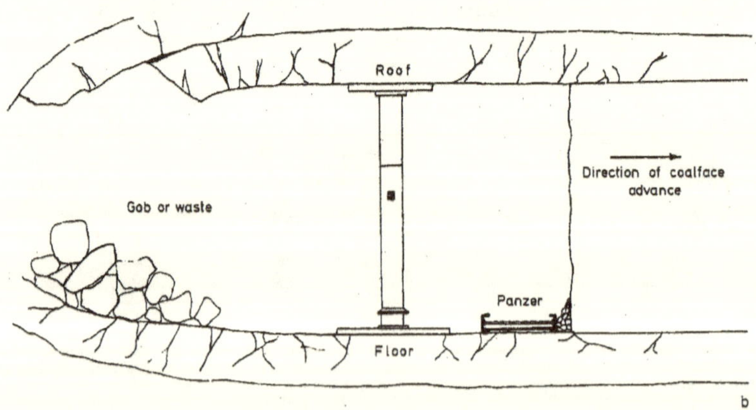

Fig. 1 b. Simplified diagram of the prop assembly along the direction of face advance.

Convergence recorders were taken in three positions:

(i) Between the top and bottom plates to record any vertical movement on the prop slide (s_p in mm).
(ii) Between the rigid horizontal arm (projecting 60 cm out from the prop) and floor, to record floor lift (C_f in mm).
(iii) Between roof and floor to record the overall convergence (C) in mm).

A romometer was placed between the prop horizontal arm and floor to record direction of the floor movements (Figs. 1a and 1b). In all cases, the recorders were placed well away from the area of influence of the base plate and other supports on the face. In this way the readings obtained would be representative of the roof and floor behavior.

It was possible to continuously determine plate penetration and prop sliding, and to correct for the actual floor lift and its direction irrespective of the roof behavior. The system provided movements of the roof and floor separately was reasonably accurate and provided a continuous record for comparison with survey methods.

Analysis and Discussion of Results

A. Floor Loading

Measurements taken from the load cell along the three faces (Table 1) indicated a general pattern of load increase at a decreasing rate. The graph of floor loading versus time is reasonably smooth and linear during the face standing period as shown in Figs. 2, 3, and 4. The advance of chocks and the cutting operation along the face induced additional load onto the floor. The time taken to advance a support, and the cutter to pass by a measuring station is very small, being less than a minute. These parameters have been treated as constants to be algebrically added for an empirical expression as follows:

$$\sigma = S + A_1 - A_2 + C + n \cdot t \qquad (2)$$

where: σ = Vertical pressure on floor beneath a support (in MN/m^2).

S = Constant pressure imposed on floor by setting and securing the instrument (in MN/m^2).

A_1 = Additional pressure due to release of pressure on the surrounding chocks (in MN/m^2).

A_2 = Relief of pressure due to advance and re-setting of load on the surrounding chocks (in MN/m^2).

C = Additional pressure due to the increase in the opening dimensions by cutting operation (in MN/m^2).

n = Constant (less than unity) depending on the physical properties of the floor and rate of loading during the standing period (in MN/m^2/hour).

t = Standing time of the face (in hour).

No measurable influence of the panzer advance on floor loading was detected. The average values of A_1, A_2 and C for the L1 Underclay floor (Table 1) were 10, zero and 10 MN/m^2, respectively. These values for the L3 Coal floor observed to be 12, 10 and 20 MN/m^2 and for the F41 Coal floor (above a soft Underclay) were 7, zero and 25 MN/m^2, respectively.

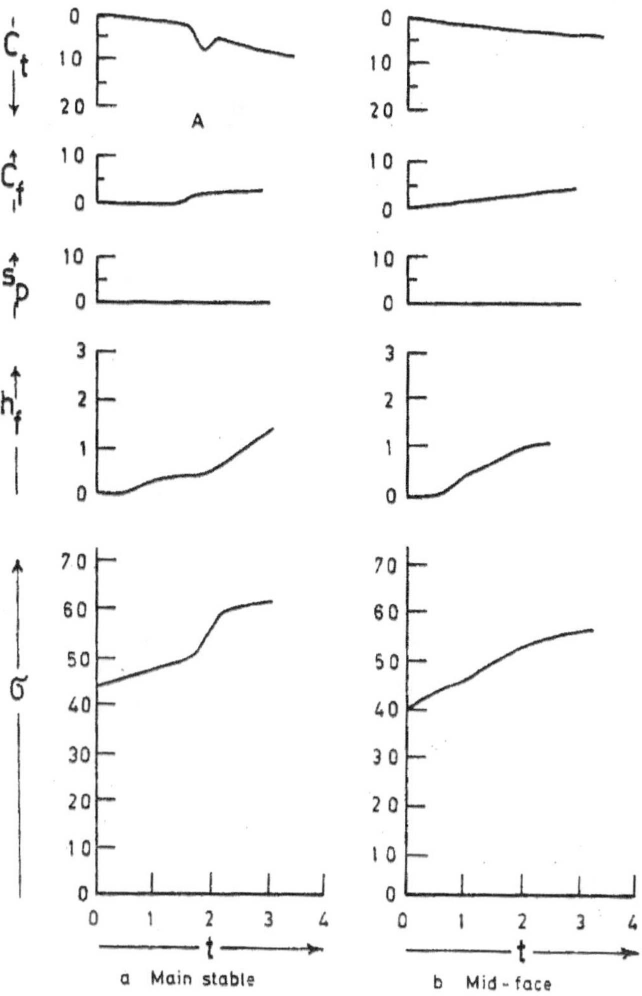

Key: C_t = Total roof to floor convergence (mm); C_f = Floor lift (mm); s_p = Prop slip (mm); h_f = Horizontal movement of floor towards waste (mm); σ = Floor loading (MN/m$^{2)}$; t = Time (hour); A = Chock advanced.

Fig. 2. Floor movements and loading along the L1 longwall face, Britannia Colliery, South Wales, United Kingdom

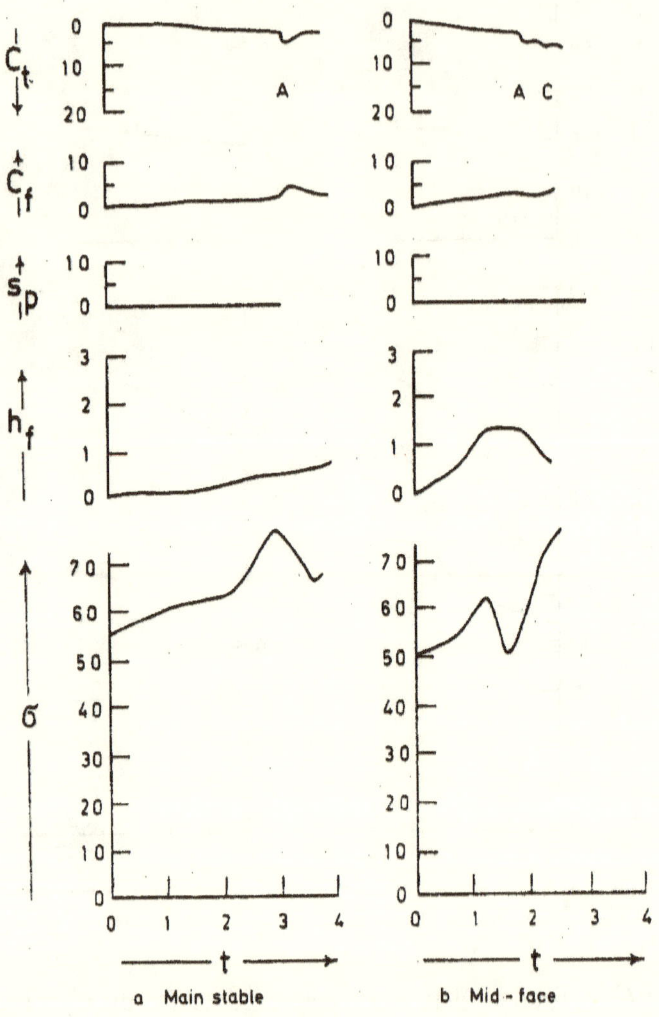

Key: *A* = Chocks advanced; *C* = Cutting

Fig. 3. Floor movements and loading along the L3 longwall face Britannia Colliery, South Wales, United Kingdom

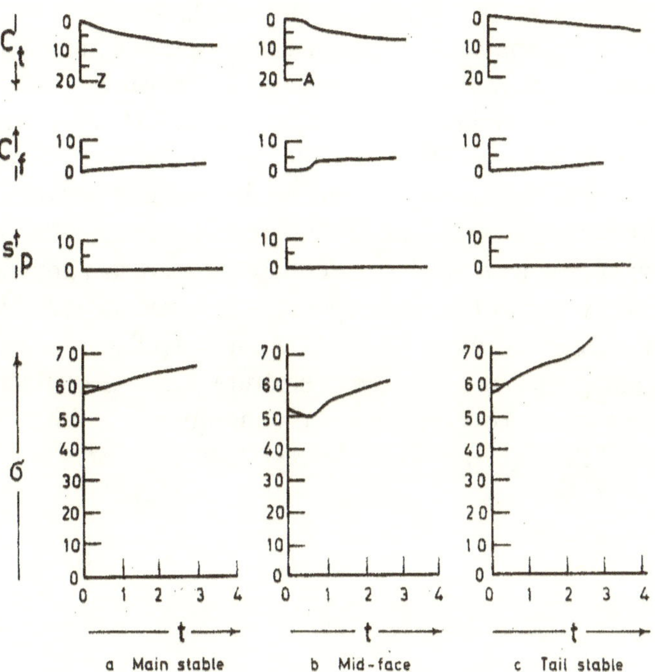

Key: A = Chock advanced; Z = Panzer advanced.

Fig. 4. Floor movements and loading along the F41 longwall face, Windsor Colliery, South Wales, United Kingdom

The n-values along the L1, L3 and F41 faces averaged at 0.40, 0.20 and 0.18 MN/m^2/hour, respectively. It appears that the softer the floor the smaller is its n-value. Comparison of the A-values suggest that no correlation exists between strength of the floor and advancement of the support base, as the A-values largely depend upon the strata pressure and the hydraulic feed pressure of the chocks.

An attempt was made to compare the results of floor loading beneath the rigid prop with that beneath the hydraulic chocks along the L3 face. It was assumed that the hydraulic pressure readings of the support circuits represented the pressure imposed on the floor beneath the chock. The hydraulic pressure readings were taken on 4 to 6 adjacent chocks along the L3 face over continuous periods of 20 to 30 hours at intervals corresponding to the face operations. Graphs of average chock pressure on each set of legs versus time is illustrated in Fig. 5. It shows that ploughing and advancement of the panzer conveyor has no appreciable influence on the pressure readings of the chock legs. The main factors influencing the pressure — time graphs appear to be the operations of chock advancement and cutting. The average *n*-value for all the chock legs extrapolated to the apparent floor loading along the L3 face during the standing period was 0.13 $MN/m^2/hour$. This value is about 35 percent lower than that obtained by the rigid prop on a much smaller base/floor contact area than that of the chock. Some of the discrepancy is believed attributable to the fact that there was about 7.5 to 22.9 cm of spillage beneath most chocks along the face creating premature yielding of floor which resulted in lower load readings and consequently smaller *n*-values than that of the rigid prop on a cleaned floor.

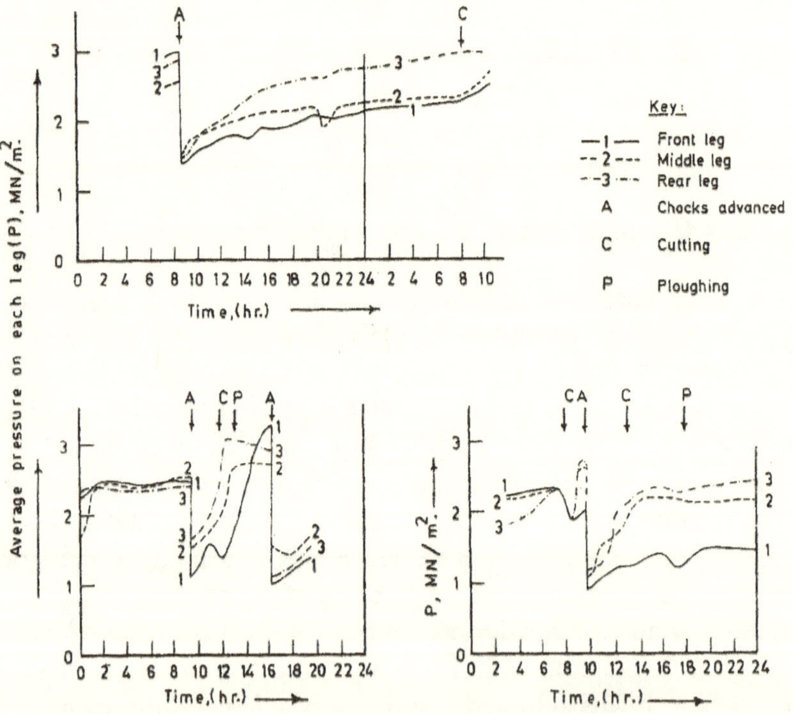

Fig. 5. Effect of mining operations at various times on the development of pressure in the hydraulic chock legs of the L3 mid-face, Britannia Colliery, United Kingdom.

B. Total Roof to Floor Convergence

An effect of strata pressure along the face was convergence of the roof and floor. The record of roof to floor convergence versus time along the L1, L3 and F41 faces is shown in Figs. 2, 3, and 4, respectively. The general behaviour of roof to floor convergence (C_t) can be expressed as follows:

$$C_t = A_1 - A_2 + C + n.t \tag{3}$$

where: C_t = Total roof to floor convergence (mm).

A_1 = Instantaneous convergence due to release of advancing supports (mm).

A_2 = Instantaneous expansion due to re-setting of the advanced supports (mm).

C = Convergence due to cutting operation (mm).

n = Constant less than unity the value of which depends on the rate of roof to floor convergence during the standing period, quality of the roof and floor (mm/hour).

Eq. (3) is similar to Eq. (1) suggested by other investigators [5]. However, Eq. (1) only takes into account the roof to floor convergence due to cutting and standing period, but not that due to advancing operations of the face support. For these reasons, it is believed that Eq. (3) would give a closer fit to the roof to floor time-dependent convergence.

The A_1-values for the L1, L3 faces averaged to be 6.0, 4.3 and 2.0 mm, respectively. The A_2-values for the L1, L3 and F41 faces averaged to 2.5, 2.0 and zero mm, respectively. The resultant roof to floor convergence during the support advancement (i. e. $A = A_1 - A_2$) is tabulated in Table 2 together with other parameters making up the total convergence along the faces investigated. The results indicate that the convergence due to support advancement (A) to be higher along faces with Coal roof (3.5 mm for the L1 face), than that along faces with Coal floor (2.3 and 2.0 mm for the L3 and F41 faces, respectively). The differences in A-values between the L3 and F41 faces appear very small in comparison with that of the L1 face and the limited results available do not allow definite conclusions taking into account all factors contributing to the differences. One possible explanation may be the more competent roof along the F41 face resulted in a smaller convergence due to support advance. The state of floor along the F41 was more critical than that along the L3 face and some of the differences may have been masked due to variations in strata and rate of face advance.

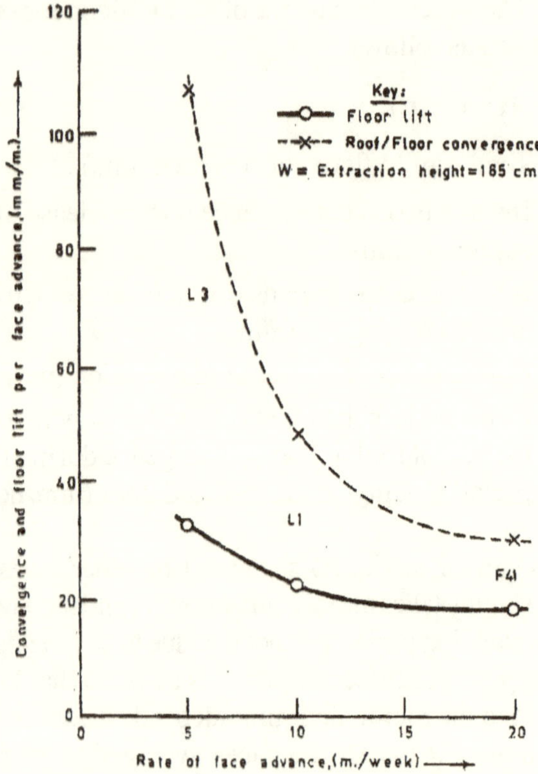

Fig. 6. Effect of the rate of the face advance on the total convergence and floor lift along longwall faces

The C-value for the L3 face was 2.0 mm. Since no cutting was performed along the L1 and F41 faces during the test period, their C-values were not obtainable. It is important that the recording be carried out on a 24-hours 7-days a week basis for at least two weeks on three to five stations along each face. In this way, a complete behaviour of roof to floor and a check on the repeatability of the test results could be achieved. However, practical difficulties together with the slow and erratic cycle of operation during the investigation period have lead to the recording being made within short time intervals.

Table 2. Vertical Movement Along the Longwall Faces Investigated

Location / Parameters:	Roof/floor convergence during cutting (mm)	Floor lift during cutting (mm)	Percentage contribution of floor to the total convergence during cutting	Resultant roof/floor convergence during chock advancement (mm)	Resultant floor lift during chock advancement (mm)	Percent contribution of floor to the total convergence during chock advance	Average rate of roof/floor convergence during the stand-still period (mm/hr)	Average rate of floor lift during the standing period in mm/hr	Percent contribution of floor to the total rate of convergence during standing period	Total roof/floor convergence per unit time (mm/hr)	Total floor lift per unit time (mm/hr)	Percent contribution of floor to the total convergence	Total roof/floor convergence per face advancement (mm/m)	Total floor lift per face advancement (mm/m)	Percent contribution of floor to the total convergence per face advancement
L1 Main stable	—	—	—	3.5	1.6	46	1.7	0.70	41	2.8	1.3	45	47	22	46
L1 Mid-face	—	—	—	—	—	—	1.7	0.86	80	—	—	—	—	—	—
Mean values for the L1 Face	—	—	—	3.5	1.6	46	1.7	0.78	46	2.8	1.3	45	47	22	46
L3 Main stable	—	—	—	2.0	1.0	50	0.7	0.33	47	3.4	0.8	34	115	27	24
L3 Mid-face	2.0	1.0	50	2.5	0.5	20	1.6	0.80	50	3.0	1.1	37	100	37	37
Mean values for the L3 Face	2.0	1.0	50	2.3	0.8	35	1.2	0.57	49	3.2	1.0	36	108	32	31
F41 Main stabel	—	—	—	—	—	—	2.3	0.71	31	—	—	—	—	—	—
F41 Mid-face	—	—	—	2.0	1.0	50	1.3	0.80	62	3.6	2.3	63	30	19	63
F41 Tail stable	—	—	—	—	—	—	1.8	1.20	66	—	—	—	—	—	—
Mean values for the F41 Face	—	—	—	2.0	1.0	50	1.8	0.90	53	3.6	2.3	63	30	19	63

Advancement of the armored conveyor and ploughing along the faces did not show measurable effect on the roof to floor convergence.

The *n*-values along the faces were variable averaging 0.77, 0.94 and 0.72 mm/hour for the L1, L3 and F41 faces, respectively. It is believed that since the extraction height of all three faces were equal (Table 1), the rate of face advance and strength of floor were two of the major influencing factors in producing different convergence values. The effect of rate of advance on the three faces, illustrated in Fig. 6, demonstrates that the larger the rate of advance (R_a) the lower would be the total convergence per unit of face advance (C_t/A), for a given height of extraction (W).

C. Floor Lift

The general characteristic of the vertical floor movement due to mining operations and time appeared much similar to that observed for the roof to floor convergence, but smaller is magnitude (Figs. 2, 3, and 4). Floor lift (C_f in mm) is expressed as follows:

$$C_f = A_1 - A_2 + \mathrm{C} + n.\,t \qquad (4)$$

where: A_1 = Instantaneous floor lift due to release of advancing supports (mm).

A_2 = Instantaneous compaction of floor due to re-setting of advanced supports (mm).

C = Floor lift due to cutting operation (mm).

n = Constant less than unity, dependent on the rate of floor lift during the standing period and the quality of floor rock (mm/hour).

t = Total time taken to complete one cycle of face operation (hour).

In equ. (4), the A_1-values for the L1, L3 and F41 faces averaged to 1.6, 1.8 and 1.0 mm, respectively. The A_2-values for the above noted faces averaged to zero, 1.0 and zero mm, respectively. Hence, the resultant floor lift due to support advancement (i. e. $A = A_1 - \mathrm{A}_2$) for the three faces averaged to 1.6, 0.8 and 1.0 mm, respectively. It appears that the Underclay floor along the L1 face has given a higher floor lift due to support advancement than the Coal floor along the other two faces. The percentage contribution of the floor to the total roof to floor convergence due to advancement of supports along the above noted faces was 46, 35 and 50, respectively.

The C-value on the L3 face was 1.0 mm which contributed up to 50 percent to the total roof to floor convergence. No record for the C-values along the L1 and F41 faces was available. There was no measurable effect due to advancement of the face armored conveyor and ploughing along the faces.

The *n*-values for the L1, L3 and F41 faces averaged to 0.72, 0.73 and 0.50 mm/hour, respectively. The average rate of floor lift during the standing period along the three faces were 0.78, 0.57 and 0.90 mm/hour, respectively. Hence, percentage contribution of the floor to the total roof to floor becomes 46, 49 and 53 percent. It is readily apparent that softer the floor the higher would be its percent contribution to the total closure along the face.

The effect of the rate of advance on the floor lift is illustrated in Fig. 6 as a comparison with that of the total roof to floor convergence. In general, the floor lift per unit face advance (C_f/A) decreases with increase in the rate of advance (R_a) for a given height of extraction (W), and the percent contribution of the floor to that of the roof to floor convergence increases with either softness of the floor and/or hardness of the roof as shown in detail in Table 2.

D. Horizontal Movement of Floor

Limited readings obtained from the romometer installed between the rigid arm of the prop assembly and floor along the L1 and L3 faces have indicated a variable degree of floor movement towards the waste (Figs. 2 and 3). Magnitude of the movement was affected by advancing the supports and the cutting operation. No measurable affect from advancing the armored conveyor and ploughing on the floor movement was recorded. Apart from unpredictable movement of floor accelerating due to the support advance and cutting operation on the floor movement, the rate of floor movement towards the waste appeared to be linear. This rate of floor movement during the standing period along the L1 main stable and mid-face was 2.0 and 4.5 mm/hour, respectively averaging to 3.2 mm/hour. The rate along the L3 main stable and mid-face was 1.7 and 6.0 mm/hour, respectively averaging to 3.8 mm/hour. Hence, the horizontal movement of floor (along the both faces), was observed to be higher along the mid-face than along the main stable. The accelerating effect of support advance and cutting operation on horizontal movement was not instantaneous as observed in the case of roof to floor convergence and floor lift. This effect was time-dependent and in a linear form averaged to 1.0 and 1.1 mm/hour along the L1 and L3 faces, respectively. Comparison of the results between the two faces suggests that the effect of advancement and cutting along the L1 face is a 31% increase in the rate of the Underclay floor movement towards the waste and that of the similar movement of the L3 face is 39% increase. Hence, the effect of face operations was more on the softer Coal floor than the harder Underclay floor.

Following empirical equation predicts the rate of horizontal movement:

$$Hf = mt + n.\,t_{(A+C)} \tag{5}$$

where:
- H_f = Floor movement towards waste (mm).
- m = Rate of floor movement during the standing period of the face (mm/hour).
- t = Total time for one cycle of face operation (hour).
- n = Rate of floor movement during support advance and cutting operation (mm/hour).
- $t_{(A+C)}$ = Time taken for the support advance and/or cutting operation (hour).

According to equ. (5), the horizontal movement of floor towards waste for the two faces recorded as:

1 - For the L1 Underclay floor: $H_f = 3.3\,t + 1.0\,t_{(A+C)}$ (6)

2 - For the L3 Coal floor: $H_f = 2.8\,t + 1.1\,t_{(A+C)}$ (7)

The time taken for support advance, cutting and total cycle of face operations depends greatly on the face length, height of extraction, method of support and type of cutter used. In this respect the only variable between the two faces was the face length (l), as shown in Table 1, which affects the total time per cycle of operation, i. e. for the specific faces investigated: $H_f - f(t) = f'(\text{l})$. It is anticipated that, apart from the fact that a softer floor exhibits larger horizontal movement than a harder floor since the length of the L3 face is larger than that of the L1 face, its (*t*)-value would be larger.

Conclusions

(a) Roof and floor movements and loading along the three longwall faces investigated was linearly time-dependent.

(b) Magnitude of the above noted movements and loading was inversely dependent upon the rate of advance, strength of roof and floor layers.

(c) Contribution of floor lift to the total roof to floor closure along the face having strong Sandstone roof, but Coal floor and soft Underclay (i. e. the F41 face), was 63% whereas for a face having a relatively moderate to strong roof but Coal floor and medium strength Underclay (i. e. the L3 face), was 31%. In the latter case, if a one metre thick Coal roof is left with Underclay as the immediate floor (i. e. the L1 face), the floor would contribute up to 46% to the total roof to floor closure. This means that a relatively better roof condition was achieved.

(d) A 72% decrease in strength of floor resulted in an overall 26% increase in the horizontal movement of floor towards the waste.

References

[1] Shepherd, R.: Measurement of Rock Pressure in Mines, Sheffield Univer. Mining Magazine, P. 72, 17th Nov. (1959).

[2] Schwartz, B., Dubois, R.: The Influence of Supports on the Movement of Roof and Floor in the Face, Proc. 2nd Int. Strata Control Congr. Essen, P 14 (1956).

[3] Dubois, R.: The Various Factors Governing Face Convergence, Proc. 3rd Int. Strata Control Congr. Paris, P. 443 (1960).

[4] Moore, J. F. A.: An Investigation of the Effect of Rate of Face Advance on Strata Behaviour at Bolsover Colliery, NCB, MRE Report No. 2349, Jan. (1969).

[5] Kenny, P., Wilson, A. H.: Strata Measurements on Faces Equipped with Conventional and Powered Supports, The Mining Engineer, P. 528, April (1963).

[6] Wilson, A. H.: Conclusions from Recent Strata Control Measurements Made by the Mining Research Establishment, The Mining Engineer, P. 369—373, April (1964).

3.3.8 YIELD AND BEARING CAPACITY OF COAL MINE FLOOR

By: Andy A. Afrouz*

Theoretical studies are combined with in situ *tests in order to investigate yielding and bearing capacity of mine floor along three longwall faces. Some of the parameters and constants related to deformation and failure of the floor are evaluated. A method is suggested to predict the bearing capacity of floor beneath full size chocks from base plate tests. The effect of size and shape of the base plates, thickness of the floor layer, water and time on the ultimate bearing capacity of floor is measured and discussed.*

INTRODUCTION

Equilibrium and reaction of the floor to mine supports may affect the stability of surrounding rocks. If the design of supports is to be based on an acceptable rate of closure or deformation within underground working spaces, to ensure balance and stability of supports, the strata pressure within that region should be such that the deformation of every part of the floor is uniform and the bearing capacity of the floor is larger than the effective strata pressure exerted upon it. Therefore, the relative contact deformation of floor under the influence of underground supports and the floor ultimate bearing capacity beneath the full size supports are important relationships.

In cases where the micro-structure of floor rock reveal the existence of plastic failure, fissures, slickensides, beddings and joints or other discontinuities, neither uniaxial nor triaxial tests or any other small scale laboratory investigation would accurately reveal the true *in situ* strength of the rock. In such a case it may be desirable to resort to *in situ* load tests for the determination of rock strength or bearing capacity. It is preferable to utilize base plates as large as possible so that the measured bearing capacity would represent an overall and a fair percentage of the floor area beneath the conventionally mechanized longwall face supports. However, capacity of the prop dynamometer to be used exerts limitation on the size of the base plates.

This investigation is directed towards obtaining the effect of thickness of the immediate floor, shapes and sizes of the base plates on deformation and bearing capacity of mine floor along three longwall coal faces in the South Wales coalfield.

INSTRUMENTATION

The equipment used consisted of the following:

(a) Hydraulic prop dynamometer—which was similar in design to the standard hydraulic prop. Its overall dimensions were 65 cm long by 10 cm diameter with a 15 cm hydraulic ram and 21 kN of load capacity. A 30 kN capacity pressure gauge calibrated to register directly the total load on the prop base, with 1-kN divisions, was mounted on the hydraulic main of the prop.

(b) Extension pieces—having various lengths of 10 up to 76 cm and the same diameter of that of the hydraulic prop. These were used to extend the effective height of the prop dynamometer to suit all conditions along the face.

(c) A dial gauge was used to record floor penetration and its rate with incremental load increase and/or decrease. The gauge had a range of 5 cm graduated to $2{\cdot}5 \times 10^{-3}$ cm with an overall accuracy of $0{\cdot}5 \times 10^{-3}$ cm. The gauge was firmly mounted by screw arrangement on the side of the prop dynamometer assembly via a 30 cm long × 0·6 cm dia silver stainless steel rod. This was to ensure that the measuring pivot of the dial gauge rested vertically on a fixed and stable floor surface as far from the base plate influence as possible. A schematic diagram of the assembly is shown in Fig. 1.

(d) Base plates were manufactured from 1·3 cm thick, hardened flat steel, cut to circular, square and rectangular shapes of per-selected sizes within each shape group as shown in Table 1. From an engineering standpoint, the most desirable test is one that requires a minimum of extrapolation to the full-scale application. Consequently, the largest possible sizes within the bounds of practicality were utilized in order to keep extrapolation to a minimum.

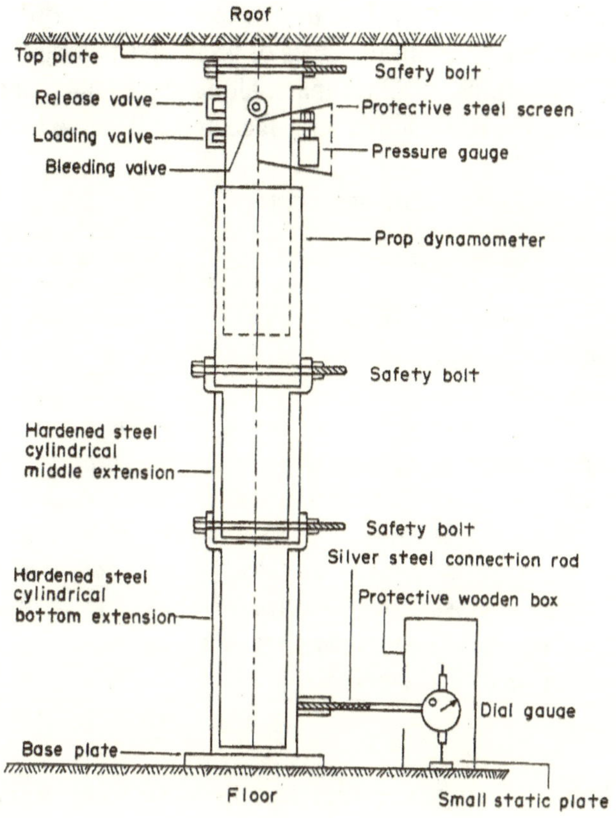

Fig. 1. Schematic side elevation of the prop dynamometer.

(e) A top plate was made of the same material as that of the base plates. Its 45·7 × 45·7 × 1·3 cm dimensions were chosen to make the ratio of 'top plate area/base plate area' as large as possible. This was in order to reduce or possibly eliminate roof penetration by the prop, as far as practicable. At the same time, consideration was given to the overall dimensions of the top plate to fit between the roof supports along the face.

TEST PROCEDURE

The prop assembly was calibrated under laboratory conditions by a 50 kN capacity hydraulic compression machine, up to its yield point, prior to *in situ* tests. The *in situ* tests were conducted at five locations selected at equal distances of 45 m apart along the L3 face and at three locations of 82 m apart, along the L1 face and the F41 face, including the main stable, mid-face and tail stable in each case.

TABLE 1. SPECIFICATION OF BASE PLATES

Base area (A)		Circular plates (C)			Square plates (S) ($a = b$, or, $a/b = 1{:}1$)			Rectangular plates (R) ($a = 2b$, or, $a/b = 2{:}1$)		
			Diameter			Dimension			Dimension	
(cm²)	(in²)	No.	(cm)	(in.)	No.	(cm)	(in.)	No.	(cm)	(in)
81·01	12·56	C1	10·16	4	—	—	—	—	—	----
126·61	19·63	C2	12·70	5	S1	11·25 × 11·25	4·43 × 4·43	R1	15·85 × 7·95	6.26 x 3.13
182·28	28·26	C3	15·24	6	S2	13·51 × 13·51	5·32 × 5·32	R2	19·10 × 9·55	7.52 x 3.76
248·13	38·47	C4	17·78	7	S3	15·75 × 15·75	6·20 × 6·20	R3	22·26 × 10·15	8.78 x 4.39
324·05	50·24	C5	20·32	8	S4	18·31 × 18·31	7·09 × 7·09	R4	25·45 × 12.72	10.02 x 5.01
411·66	63·59	C6	22·86	9	S5	20·24 × 20·24	7·97 × 7·97	R5	28·75 × 14·38	11.32 x 5.66
506·33	78·50	C7	25·40	10	S6	22·50 × 22·50	8·86 × 8·86	R6	31·85 × 15·93	12.54 x 6.27
612·69	94·99	C8	27·94	11	S7	24·77 × 24·77	9·75 × 9·75	R7	35·00 × 17·50	13.78 x 6.89
729·11	113·04	C9	30·48	12	S8	27·05 × 27·05	10·65 × 10·65	R8	38·20 × 19·10	15.04 x 7.52

Just before testing, the roof and floor surfaces were cleared of loose material and were made flat to achieve uniform loading throughout the entire contact surface between floor and base plate. The spacing between test points was kept at about one metre in order to eliminate the area of influence of the test points upon each other.

Loads were applied incrementally at an average rate of 1 kN min. During the test, each incremental load increase (or decrease) was left for sufficient time to permit the corresponding deformation to become reasonably steady before the representative deformation was taken from the dial gauge. The next incremental load was then applied. Loading of the prop was terminated when either a 5 cm floor penetration (maximum capacity of the dial gauge) or the yield point of the prop was reached. If the floor failed, the load at initial failure was noted to be the floor bearing capacity. At this point the penetration rate started to increase rapidly while being accompanied by a drop in the pressure reading.

During all unloading, the load was incrementally decreased and the corresponding floor recovery was noted on the dial gauge. Generally, when a load *P* is applied to a base plate/floor interface of a known area *A*, it develops a vertical stress equal to *P/A* causing an apparent floor settlement Δl_q. Since at zero vertical stress there could be no floor penetration under the base plates, the graphs of load vs apparent deformation for each base plate was corrected to pass through the origin of the coordinate axis, as shown in Fig. 2. The result thus observed was then used in the analysis.

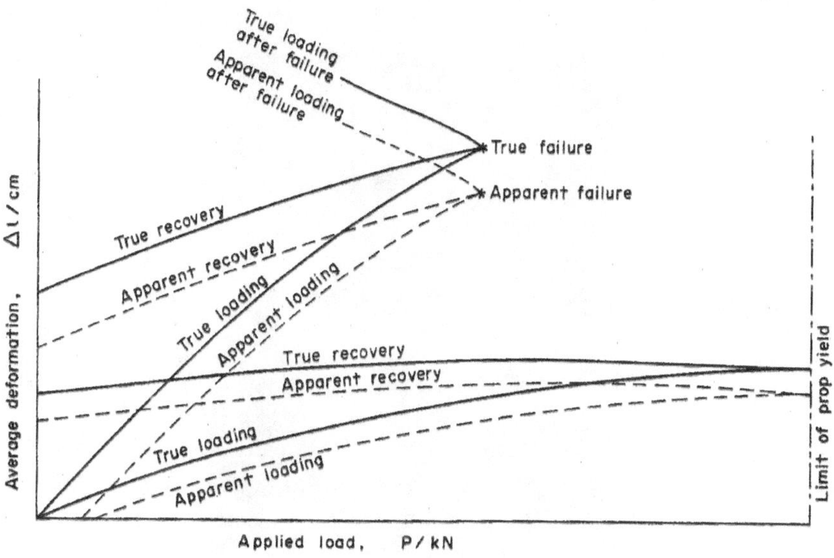

Fig. 2. Graphical technique to correct the *in situ* load/deformation test data.

THEORETICAL CONSIDERATIONS

Stress evaluation

Considering two support bases of equal size *D* and similar loading conditions. The pressure *q* exerted on the support base-floor contact surface produces stresses and strains in the floor varying with depth *z* below the support base-floor contact. This may be shown in the form of pressure bulbs or contours of equal stress distribution in the floor. Figures 3(a) and 3(b) were determined utilizing the scaled model photo-elastic technique.

If the floor layer under a given size of support base and load q is thicker than the dimensions of the created pressure bulbs, the whole of the bulbs will be formed within the layer (Fig. 4(a)). Whereas in the case of a much thinner layer or where a relatively thin layer of soft clay lies within a thick underclay floor (under the same above noted conditions) the bottom part of the pressure bulbs may enter the successive floor layers beneath, as shown in Fig. 4(b). In this process, the pressure distribution will be distorted due to the discontinuity encountered. This may result in different settlements under apparently similar support and loading conditions.

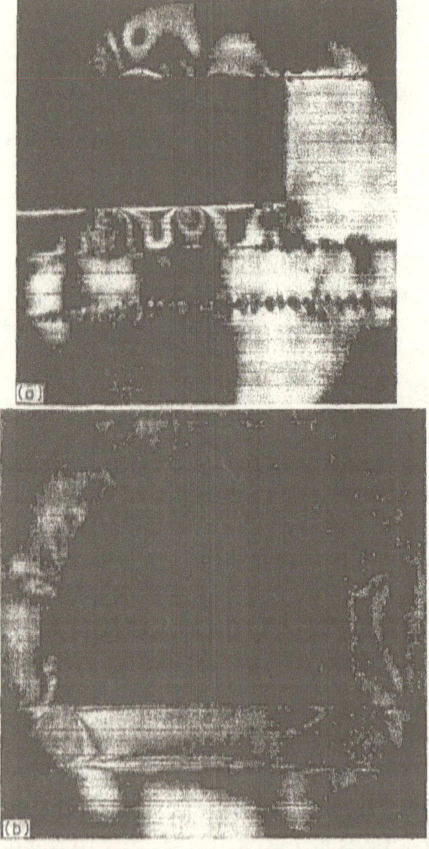

Fig. 3. Laboratory simulation of stress development around the immediate vicinity of underground supports. (Geometric scale factor model/actual = 1/50.) (a) Isochromatic development around longwall face supports

under strata pressure. (b) Isochromatic development around arched roadway under strata pressure.

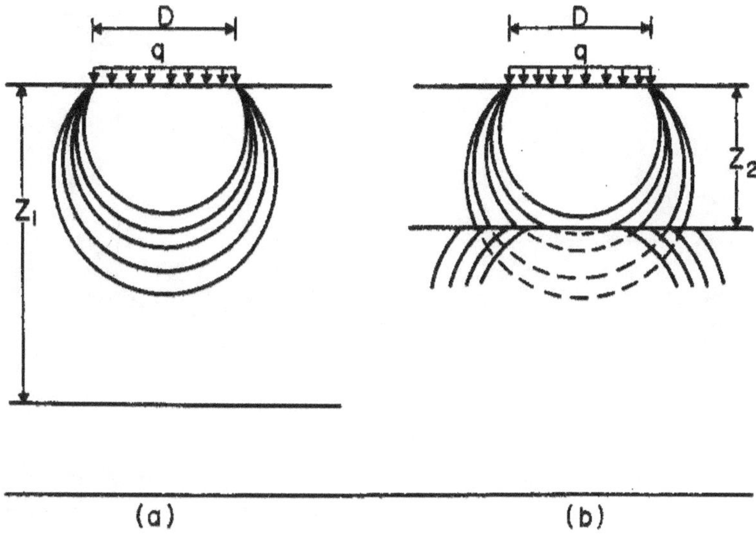

Fig. 4. Simplified comparison of pressure distribution in mine floors of different thickness.

A similar effect may occur when two different sizes of support base transmit equal loads q on to floors of equal thickness z as shown in Fig. 5. In this case, the pressure bulb below a larger base covers a larger area than that of the smaller base. Hence, there could be a depth (X–X in Fig. 5) at which the pressure due to the smaller plate is insignificant compared with that of the larger plate. Since the *in situ* tests were limited to base areas smaller than the hydraulic chocks used along the faces, an understanding of the size effect is essential if the test results are to be extrapolated to predict the bearing capacity of mine floor under the full-size supports.

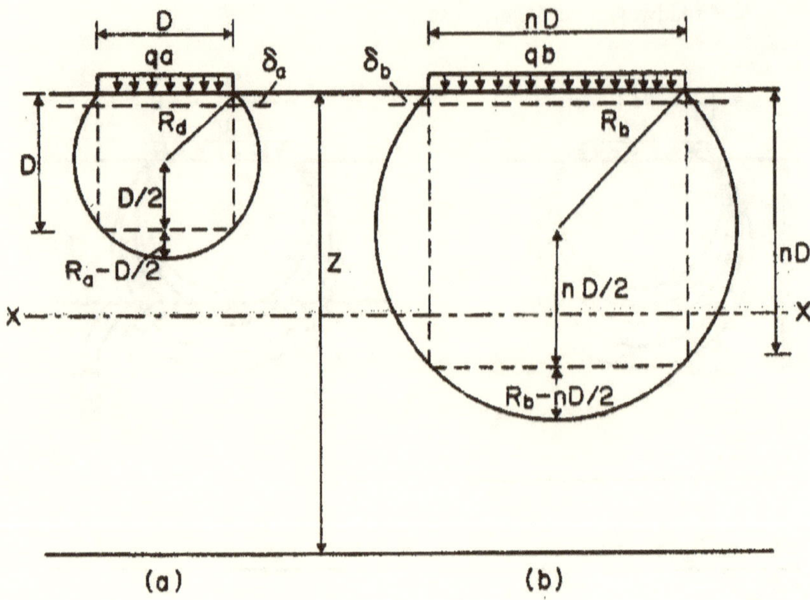

Fig. 5. Idealized size effects of support base on mine floor.

The problem may be simplified by assuming the pressure bulbs to be circular in vertical cross-section having radius *R*, in such a way that they encircle a square of equal dimensions to the support base, as shown in Fig. 5. Let one of the pressure bulbs represent an average pressure in the floor beneath a support, and the modulus of elasticity of the floor *E* remain constant with depth *z*.

From Hooke's Law ($\sigma = E \,.\, \epsilon$) one could write:

$$\epsilon_a = \sigma_a / E_a \text{ and } \epsilon_b = \sigma_b / E_b \qquad (1)$$

where: ϵ_a = train developed under base

a = $\Delta l_a / z$

ϵ_b = strain developed under base

b = $\Delta l_b / z$

Δl = average floor deformation under the base

z = the depth below floor surface;

σ_a = average stress under base *a* within all the pressure bulbs developed

= $C_a \cdot q_a K. D / A_a$ (2)

σ_b = average stress under base *b* within all the pressure bulbs developed: = $C_b \cdot q_b \cdot n. K. D / A_b$. (3)

C = constant pending on the rock type, base perimeter and quality of support base-floor contact;

K = proportionality constant between dimension *D* and area *A*

A_a = the effective area of the average pressure bulb beneath base *a*

A_a in plane geometry of Fig. 5(a) = $\pi Ra^2 - (\pi Ra^2 - D^2)/4$

= $(3\pi Ra^2 + D^2)/4$ (4)

A_a in spherical geometry of Fig. 5(a) = $4\pi Ra^2 - \pi(2 Ra^2 - D^2/2 + D^2)/2$

= $\pi(3 Ra^2 - D^2/4)$ (5)

A_b = the effective area of the average pressure bulb beneath base *b*.

A_b in plane geometry of Fig. 5(b) = $\pi Rb^2 - (\pi Rb^2 - n^2D^2)/4 = (3\pi Rb^2 + n^2D^2)/4$ (6)

A_b in spherical geometry of Fig. 5(b) = $4\pi\ {}_{Rb}{}^2 - (\pi 2 Rb^2 - n^2D^2/2 + n^2D^2)/2 = \pi(3 Rb^2 - n^2D^2/4)$ (7)

R_a = radius of the average pressure bulb beneath base *a*

= $D/2 \sin 45° = D/\sqrt{2}$. (8)

R_b = radius of the average pressure bulb beneath base *b*

= $n. D/2 \sin 45° = n. D/\sqrt{2} = n. R_a$. (9)

n = relative dimensional constant between different base plates which affect floor deformation = ratio of average base dimensions to ratio of square root of the base area of chock over area of base plate;

A = ratio of area of base plates.

At failure of floor, $\sigma_a = \sigma_b$, then equating equations (2) and (3) will give:

$$q_b = C_a \cdot q_a \cdot A_b / n \cdot A_a \cdot C_b. \tag{10}$$

Substituting equations (4), (6) and (9) into equation (10) for the plane circular consideration: $q_b = n \cdot C_a \cdot Q_a / C_b$. (11)

Substituting equations (5), (7) and (9) into equation (10) for the spherical consideration produces the same result. If Boussinesque's equation is considered in conjunction with the spherical pressure bulb [1]:

$\sigma_a = 3\,z^3\,(q/2\,\pi)$[∫ Integral between 0 and R ∫ Integral between 0 and 2 of

$$r.d.d_r/(r^2+z^2)^{5/2}] \tag{12}$$

Therefore, over the area of pressure bulb beneath the base *a*:

$$\sigma_a = q_a\,[1-(1+Ra^2/z^{2)-3/2}]\,/A_a \tag{13}$$

From the geometry of Fig. 5(a) and equation (8):

$$z = D + R_a - D/2 = 1.2D \tag{14}$$

Substituting equations (5) and (14) into equation (13) yields:

$$\sigma_a = 0{\cdot}09\,q_a / \mathrm{D}^2 \tag{15}$$

Also, considering equations (12), (9) and spherical geometry of Fig. 5(b):

$$z = n.\,D + R_b - n.D/2 = 1{\cdot}2\,n.\,D \tag{16}$$

$$\text{and } \sigma_b = 0{\cdot}09\,q_b / n^2 D^2 \tag{17}$$

Hence, at $\sigma_a = \sigma_{b'}$ the modified Boussinesque's equations (15) and (17) yield:

$$q_b = n^2 \times q_a \tag{18}$$

It is readily noticeable from the ratio methods of obtaining equations (11) and (18) that all lead to a similar general conclusion. However, equation (11) is more comprehensive than equation (18) as it covers the effect of support base perimeter as well as the size effect. Equation (11) suggests that the average bearing capacity beneath a full-size support base is equal to the bearing capacity tested by smaller size base plate multiplied by the ratio of dimensions between the support base and the test base plate. It is a theoretical approach slightly resembling the empirical equation suggested by Jenkins [2].

Effect of two- and three-dimensional analysis in strain evaluation

Substituting equations (2), (4) and (8) into equation (1), from the plane geometry point of view:

$$\epsilon_a = \Delta l_a / z = 4C_a \times q_a / (1.\ 5\,\pi + 1) E_a D^2$$
$$= 0.70\ \mathrm{Ca} \times \mathrm{qa} / \mathrm{D}^2 \times \mathrm{Ea} \qquad (19)$$

Substituting equations (2), (5) and (8) into equation (1), from the spherical geometry point of view:

$$\epsilon_a = \Delta l_a / z = 4C_a \times q_a / \pi\,(12\,Ra^2 - D^{2)} E_a$$
$$= 0{\cdot}25\ C_a \times q_a / D^2 \times E_a \qquad (20)$$

Similarly, the plane geometry consideration of equations (3), (6) and (9) substituted into equation (1) gives:

$$\epsilon_b = \Delta l_b / z = 4C_b \times q_b / n^2 \times D^2 (1{\cdot}5\pi + 1) E_b$$
$$= 0{\cdot}70\ C_b \times q_b / n^2 \times D^2 \times E_b \qquad (21)$$

and substitution of equations (3), (7) and (9) into equation (1) for the spherical geometry consideration gives:

$$c_b = \Delta l_b / z = 4C_b \times q_b / \pi (12\ {}_{Rb}{}^2 - n^2 \times D^2) E_b$$
$$= 0{\cdot}25\ C_b \times q_b / n^2 \times D^2 \times E_b. \qquad (22)$$

Comparison of equations (19) with (20) and (21) with (22) suggests that the shape factor between circular and spherical is important by a magnitude of 2·8(= 0·70/0·25). However, considering ratios of equations (19) and (21) (plane geometry) or equations (20) and (22) (spherical geometry) both yield the following expression:

$$\epsilon_b = C_b \times q_b \times E_a \times \epsilon_a / n^2 \times C_a \times q_a \times E_b \qquad (23)$$

Equation (23) demonstrates once more the advantage of the ratio method over individual absolute values in relating the test result to that of the actual support base.

Effect of increase in the modulus of elasticity of floor, by depth, on the strain development

Let the modulus of elasticity of floor increase with depth (z) and floor settlement (Δl) expressed as follows:

$$E' = E(z + \Delta l) \tag{24}$$

where: E' is a variable modulus of elasticity (i.e. E_a and/or E_b),
E is the constant modulus of elasticity.

The average modulus of elasticity of floor within the pressure bulb a is, therefore, from the geometry of Fig. 5(a), equations (8) and (24), as follows:

$$E_a = E\,[D/2 + (R_a - D/2)\,/2 + \Delta l_a] = DE(1{\cdot}1 + \Delta l_a/D) \tag{25}$$

Similarly from the geometry of Fig. 5(b), equations (9) and (24) yield:

$$E_b = E\,[nD/2 + (R_a - n\,D/2)\,/2 + \Delta l_b] = nD.\;E(1{\cdot}1 + \Delta l_b/nD) \tag{26}$$

Hence the vertical strain of the two cases, from the equation pairs (19), (25) and (21), (26) become:

$$\epsilon_a = 0{\cdot}70\,C_a \times q_a / D^3 E(1{\cdot}1 + \Delta l_a/D) \tag{27}$$

$$\epsilon_b = 0{\cdot}70\;C_b \times q_b /\, n^3\, D^3\, E\,(1{\cdot}1 + \Delta l_b/D) \tag{28}$$

Therefore:

$$\epsilon_b / \epsilon_a = C_b \cdot q_b\;(1.1 + \Delta l_a/\mathrm{D}) \,/\, C_a\, qa\, n^3\,(1.1 + \Delta\, l_b\,.\,\mathrm{n.\,D})$$

$$= C_b \cdot q_b\,(1.1\,\mathrm{D} + \Delta\, l_{a)}\,/n^2\, C_a \times q_a\,(1.1\,\mathrm{n.\,D} + \Delta\, l_b) \tag{29a}$$

or $\epsilon b = \mathrm{Cb}\,.\,\mathrm{qb}\,.\,\epsilon a\,(1.1\,\mathrm{D} + \Delta\,\mathrm{la}) \,/\, n^2\,\mathrm{Ca} \times \mathrm{qa}\,(1.1\,\mathrm{n.\,D} + \Delta\,\mathrm{lb})$ (29b)

Comparison of equations (23) and (29) shows the effect of variation in modulus of elasticity of the floor. An advantage of equation (29) over equation (23) is that it allows calculation of the ratio (ϵ_b/ϵ_a)

without having to determine the modulus of elasticity of the floor beneath the *in situ* support. Hence, knowing the dimensions of the support base and the test plates, *in situ* measurements of the average floor deformation beneath both of the bases, and the corresponding applied load on each base, it is possible to obtain an average ratio for the floor strain within the mean pressure bulb.

ANALYSIS AND DISCUSSION OF RESULTS

Statistical analysis of the ultimate bearing tests (Table 2) show the variance *V* in the results for individual size and shape of the base plates utilized to be between 0·1–13·7 MN/m^2, leading to an error of 4·4–26·0 per cent. In order to compare errors due to individual shape and size of the base plates with an overall result, the test results at failure of all the plates in the F41 floor coal were combined. The outcome was a high error of 31·6 per cent as shown in Table 2. This comparison clearly shows the additional variation due to effect of size and shape of the base plates. The considerable scatter in the test results necessitate a larger number of observations up to the failure of the floor, to be taken for each base size and shape. In practice, the larger the base plate, the more realistic would be the test result in expressing the behaviour of floor beneath large support bases. The suggested theoretical method does provide extrapolation of the ultimate bearing capacity of floor from the small base plates to larger support bases.

TABLE 2. STATISTICAL ANALYSIS OF THE BEARING CAPACITY RESULTS

	L1 Underclay			Underclay			L3		Coal	F41 Coal						
										C		S		R		
	C	S	R	C	S	R	C	S	R	126·6 cm²	248·1 cm²	126·6 cm²	248·1 cm²	126·6 cm²	248·1 cm²	All together
Number of observations (n)	1	1	1	3	2	1	2	8	7	3	2	3	2	3	2	15
Degree of freedom ($n-1$)	0	0	0	2	1	0	1	7	6	2	1	2	1	2	1	14
Sum of ultimate bearing capacity (UBC) (Σq_{μ})	15·46	15·28	14·97	52·71	24·17	14·41	142·63	84·52	77·55	34·64	13·81	31·38	12·64	27·58	11·80	131·83
Mean of UBC ($(\Sigma q_{\mu}/n) = m$, in MN/m²)	15·46	15·28	14·97	14·23	12·09	14·41	11·89	10·57	11·06	11·55	6·90	10·46	6·32	9·19	5·9	8·79
Square of UBC (m^2)	237·2	234·1	225·0	202·5	146·2	207·7	141·3	111·8	122·4	133·4	47·6	109·5	39·9	84·5	34·8	77·3
Sum of squares of UBC (Σq_{μ}^{3})	237·2	234·1	225·0	648·5	296·3	207·7	1741·0	914·0	887·4	407·3	95·3	331·0	79·9	254·0	69·7	1226·0
Mean square of UBC ($\Sigma q_{\mu}^{2}/n$)	237·2	234·1	225·0	216·2	148·1	207·7	145·1	114·3	126·8	135·8	47·7	110·3	40·0	84·7	34·9	85·1
Variance of results (V) = ($\Sigma q_{\mu}^{2}/n) - m^2$	–	–	–	13·7	1·9	–	3·8	2·5	4·4	2·4	0·1	0·8	0·1	0·2	0·1	7·8
Standard deviation (s) = $\sqrt{1}$	–	–	–	3·7	1·4	–	1·9	1·6	2·1	1·5	0·3	0·9	0·3	0·4	0·3	2·8
Per cent error = 100 s/m	–	–	–	26·0	11·5	–	15·9	15·1	19·0	12·9	4·4	8·6	4·7	4·9	5·1	31·6

Key: C = circular base plate; S = square base plate; R = rectangular base plate; numbers 126·6 cm² and 248·1 cm² refer to the area of base plates.

The use of high capacity prop dynamo meters capable of providing loads in excess of 75 tonnes has been suggested. Such instruments have the obvious disadvantage of being bulky and heavy to handle and transport along the face. To overcome this problem, it is possible to use a spare leg on a large-capacity hydraulic chock. A high capacity pressure gauge can be mounted on the hydraulic mains of the leg similar to that of the prop dynamometer, to indicate the applied load at any load level, regulated manually by a flow control valve. The assembly is similar to that used successfully by Barry and Nair [3] with an important modification in that the use of an independent portable hydraulic pump and the compressed air cylinders used in support can be eliminated. Instead it is possible to supply the hydraulic feed from the push–pull ram on the chocks via a flow control valve through an adequately long pressure hose, similar to those supplying fluid pressure to the conventional chock legs. The assembly would eliminate manual jacking-up of the prop, which is not efficient at high loads. It can be connected to any chock throughout the face.

The load-deformation of the floor along the face, beneath various base plates, is illustrated in Fig. 6, which suggests:

$$\Delta l = K_1 \times P^n \tag{30}$$

Where: Δl = Deformation on settlement of floor (cm),

P = Total vertical load applied on to base plate (kN),

K_1 and n are proportionality constants varying between zero and one, depending upon the size and perimeter of the base plates, strength and elastic moduli of floor.

The initial curved appearance of the graphs may be due to closure of cracks and pores existing in the floor rock prior to the test and incomplete contact between the base plate and floor. As higher loads were applied, the minute projections which caused this initial state gradually crushed and flattened, tending to a maximum contact which has produced a flatter curve. A similar characteristic was observed on the stress-strain curves. To calculate the strain values from the floor deformation measurements, it was necessary to know the total depth to which the floor was strained. However, an arbitrary vertical depth equal to the thickness of the upper layer of the floor was assumed. This thickness was shown by photoelastic analysis (Fig. 3) to be the most affected part of the floor and hence a reasonable approximation. According to Fig. 7, the average percentage vertical strain in floor beneath the base plate (ϵ_3) can be expressed as:

$$\epsilon_3 = \Delta l / 100 l = K_2 \times \sigma_3^{\ n} \tag{31}$$

where: l = thickness of the floor bed (cm), K_2 and n are proportionality constants which vary between zero and one depending on the size and perimeter of the base plates, strength of the floor and degree of contact between base plate and floor; σ_3 = vertical applied pressure at the base plate-floor contact surface in MN/m^2 = P/A; A = area of base plate (cm^2). Therefore, equation (31) can also be expressed as follows:

$$\Delta l / 100 l = K_2 (P/A)^n \text{ or } \Delta l = 100 K_2 \times l \times P^{\ n} / A^n \tag{32}$$

According to Hooke's Law:

$$\epsilon = \Delta l / l = \sigma / E = P / AE. \tag{33}$$

From equations (30) and (33):

$$K_1 \,.\, P^n = l \,.\, P/A \,.\, E \text{ or } l = K_1 \,.\, A \,.\, E \,.\, P^{(n-1)} \tag{34}$$

In equation (34), since A is constant throughout the experiment and all other parameters change with load and location, the value of l changes proportionally. This l-value is the actual floor depth by which the measured deformation should be divided to obtain the true strain developed within the floor. Equations (30) and (32) provides the relationship between the two proportionality constants K_1 and K_2 shown as follows:

$$K_1 \,.\, P^n = 100\, K_2 \,.\, l \,.\, P^{\ n} / A^n \text{ or } K_1 = 100\, K_2 \,.\, L / A^{\ n} \tag{35}$$

Substituting equation (34) into equation (35) yields:

$$K_2 = P^{(1-n)} / 100\, A \,.\, E \tag{36}$$

Equations (35) and (36) suggest that K_1 and K_2 are directly proportional to each other and their value varies non-linearly with the applied load (P).

The magnitude of K_1 from Fig. 6 was seen to be about 1·5–3 times lower in underclay than the coal floor. The average K_1-values were: 0·11 for the L1 underclay, 0·15 for the L3 underclay, 0·20 for the L3 coal and 0·30 for the F41 coal.

The K_2-values from Fig. 7 were generally 20–25 times lower in underclay than the coal floor. The average K_2-values were: $0{\cdot}07 \times 10^{-3}$ for the L1 underclay, $0{\cdot}11 \times 10^{-3}$ for the L3 underclay, $1{\cdot}45 \times 10^{-3}$ for the L3 coal and $2{\cdot}75 \times 10^{-3}$ for the F41 coal.

The n-values varied between 0·8 and 1·0 for underclay whereas it was generally between 0·50 and 0·95 for coal. The mean values were 0·90 for underclay and 0·75 for coal. It appears that the softer the rock the lower would be its n-value.

The load-deformation graphs (Fig. 6) appear to form a failure envelope. Although not enough points at floor failure under the large base plates were available, it is expected that the probable fracture envelope may be of the following form:

$$\Delta l = K_3 . P^{-m} \tag{37}$$

where K_3 and m are constants in such a way that $m < 1 < K_3$.

Combination of the equations (30), (35) and (37) yields:

$$K_1 . P^n = K_3 . P^{-m}$$

Therefore: $K_3 = K_1 \times P^{(n+m)} = 100\, K_2 . l . P^{(n+m)} / A^n$

$$= 100 l . P^{(1-n)} . P^{(n+m)} / 100\, E . A^{(n+1)}$$

$$= l . P^{(1+m)} / E . A^{(1+n)} \tag{38}$$

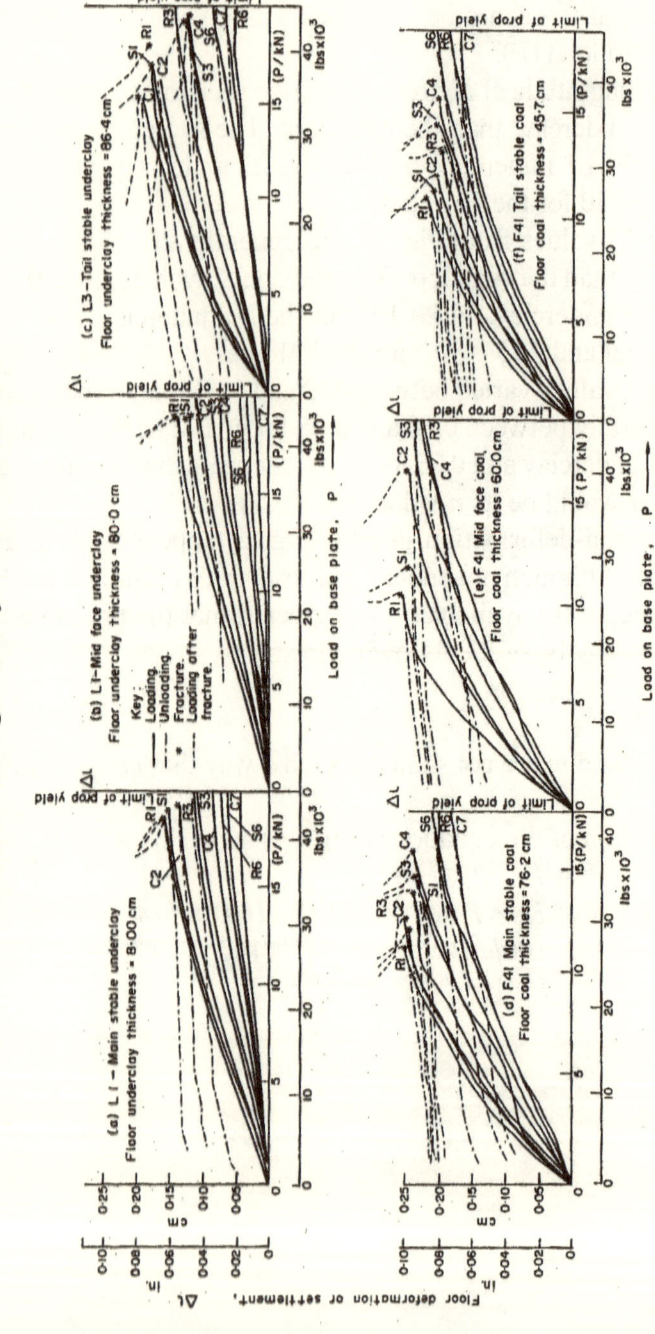

Fig. 6. Load-deformation characteristics of the coal and underclay floor.

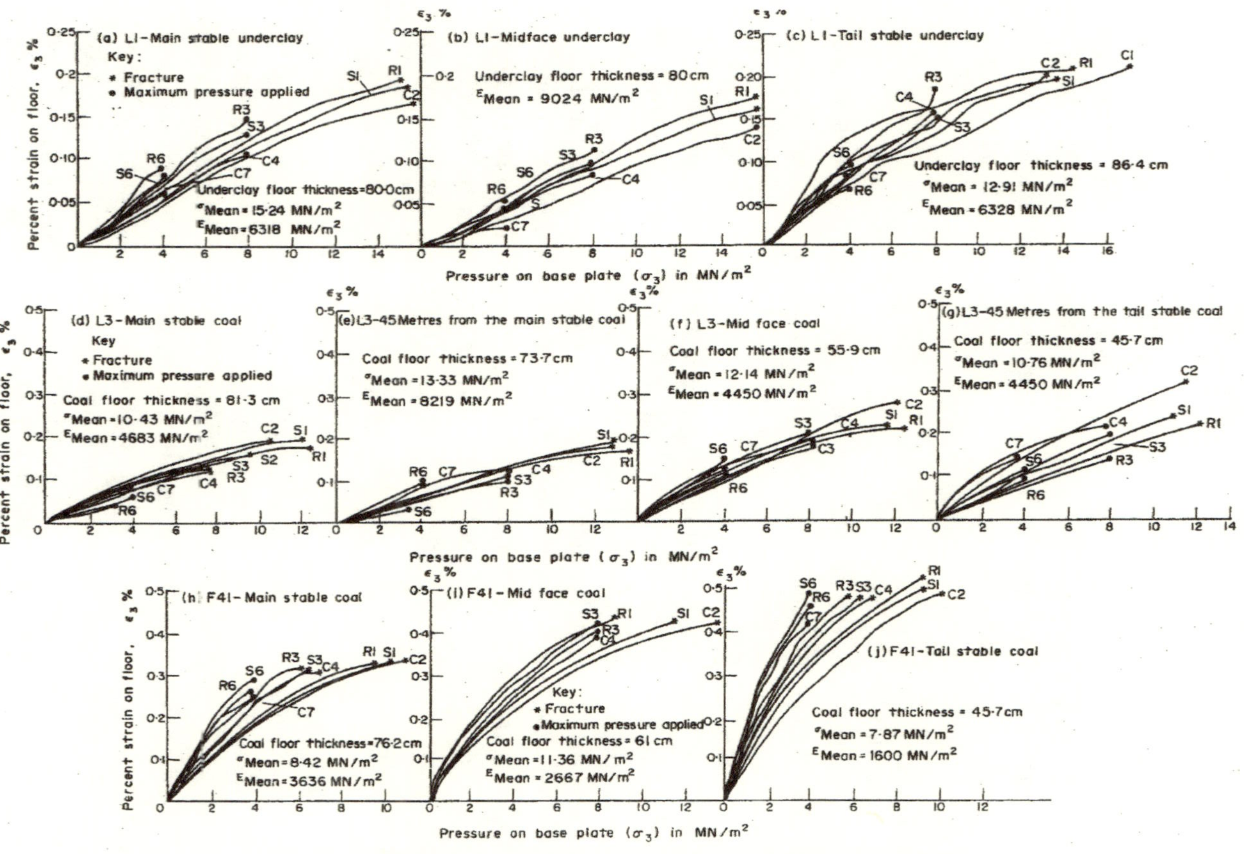

Fig. 7. Stress-strain characteristics of the coal and underclay floor.

If in equation (38) the values of m and n approach unity, then: $(P^{1+m}/A^{1+n}) \rightarrow \sigma^2$, and equation (38) becomes:

$$K_3 = l\sigma^2/E = l \cdot \epsilon \cdot \sigma \qquad (39)$$

Generally, the average *in situ* strain developed in underclay up to the point of failure was 0·5 per cent lower than that which occurred in coal. If all other parameters are equal, from equation (39) the K_3-value in underclay would be 0·4 per cent lower than that in coal.

The variations encountered in the proportionality constants may have been due to the following factors:

1- *Thickness of the floor coal*

Observations revealed that there exists a critical thickness for soft rocks at which the bearing capacity is a maximum. It is shown in Fig. 8 that this critical thickness is at 73·7 cm for the L3 coal floor and at 61·0 cm for the F41 coal floor.

Since the L3 floor coal was slightly harder and was lying over a harder underclay than that of the F41 coal and the rate of advance of the L3 face (5 m/week) was slower than the F41 face (20 m/week), a direct correlation and comparison is not possible. However, it appears that the stronger the coal the higher would be its critical thickness. In addition, the lower the rate of advance the longer would be the floor exposure, leading to higher fracture development in the floor, and this may have had the effect of masking the true behaviour of the two floors investigated.

2- *The effect of base size*

Tests have shown that the larger the base area, the lower would be the effective vertical pressure on the floor for a given load and hence the lower would be the floor settlement (Fig. 6). The magnitude of this effect was such that when the base area was doubled, keeping everything else unchanged, the value of K_1 was decreased by 20–35 per cent in the case of underclay and 40–57 per cent in the case of coal. At the same time the value of n was increased by 4–7 per cent for the underclay and 6–10 per cent for the coal. However, the ultimate bearing capacity of floor was decreased by 5–20 per cent underclay and 16–35 per cent in coal.

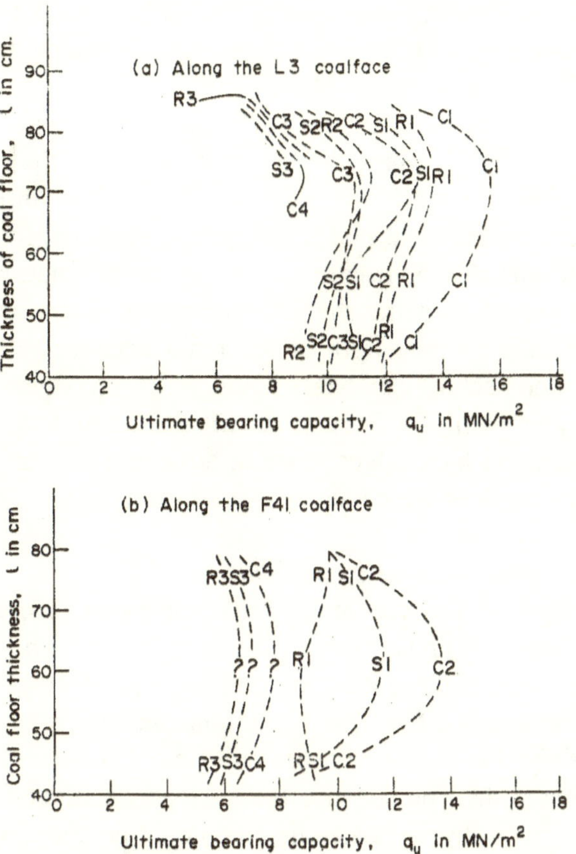

Fig. 8. Effect of thickness of the coal floor on the ultimate bearing capacity of floor along the face.

3- The effect of base shape

Due to variations encountered in the bearing capacity of floor, no definite and direct comparison between the effect of circular, rectangular and square base plates could be made. However, length of the base perimeter was observed to be a more influencing factor. Graph of the ratio of the base periphery to its area (*P/A*) against the ultimate bearing capacity (q_u) of floor coal along the L3 face (Fig. 9(a)) and the F41 face (Fig. 9(b)) indicate substantial increase in the bearing capacity with increase in the (*P/A*) ratio. It followed the general form:

$$q_u = K + m(p/A)^n \qquad (40)$$

In this case, the mean values along the L3 face are: $K = 0{\cdot}02$ MN/m^2; $m = 42{\cdot}2$ MN/m^2/cm; $n = 1$ and the mean value along the F41 face is: $K = 0{\cdot}08$ MN/m^2; $m = 33{\cdot}3$ MN/m^2/cm; $n = 1$.

It is possible that the thickness variations in floor coal throughout the faces have slightly masked the variations due to the base shape and/or periphery. In nearly all cases, the floor failure initiated from the periphery of the base plates and propagated mainly outwards with abrupt surfaces of fractures. The mode of fracture beneath base plates varied with the shape of the plates. For circular bases it appeared radial with length of the main fractures at 4–10 cm. For square and rectangular bases it tended to propagate outwards from the corners with 5–15 cm of the main fracture lengths. Hence sharp corners and base edges have an adverse effect on the stress concentration in floor which leads to *in situ* fractures.

4- Mining or other excavation methods

Shotfiring and the use of hydraulic props in face ends have developed secondary and tertiary fractures in the floor. The magnitude and concentration of these fractures and the state of stress within the floor of the face ends differ from one another and that of the mid-face. This is mainly due to the pressure and time influence on the gate roads leading to their prospective faces. Hence, a direct comparison between the floor strengths encountered could be open to criticism, especially where variable thickness of the floor layers is encountered. But observations illustrated in Fig. 7 have revealed that, generally, the mean bearing capacity and modulus of elasticity of both coal and underclay decrease as the approach is made from the mid-face towards the face ends. In general, three fracture zones can be distinguished along the floor of a producing longwall face. These are as follows:

(a) *Minimum fractured zone.* This occurs along the mid-face where the face is supported by hydraulic chocks and the coal is cut by machine. The extent of this zone depends mainly on the face length, rate of advance and number of gate roads leading to the face. About 80–83 per cent of the face length along the mid-face floor was in the minimum fractured state. The average bearing capacity of floor in this region of the faces investigated was: L1 underclay = 15·8 MN/m^2; L3 coal = 11·5 MN/m^2; F41 coal = 10·1 MN/m^2. These values vary greatly with shape and size of the base plates utilized as discussed in previous sections and illustrated in Figs. 6 and 7.

(b) *Maximum fractured zones.* These developed at the face ends mainly due to effects of hydraulic props and gate roads. About 2–4 per cent of the face length may fall within this zone. The average bearing capacity of floor within this zone was found to be: L1 underclay = 14·6 MN/m^2; L3 underclay = 13·2 MN/m^2; L3 coal = 10·3 MN/m^2; F14 coal = 8·3 MN/m^2. It appeared that the average bearing capacity of floor along the maximum fractured zone was 8–17 per cent lower than that within the minimum fractured zone.

(c) *Transition zones.* These existed between the minimum and maximum fractured zones, where the extent and concentration of fractures in the floor varied greatly from time to time. The extent of this zone was observed to be between 4 and 8 per cent of the face length. The bearing capacity of floor within the transition zones was 5–14 per cent lower than that of the maximum fracture zone and averaged to be as follows: L1 underclay = 14·9 MN/m^2; –L3 coal = 10·6 MN/m^2; F14 coal = 8·7 MN/m^2.

5- The effect of water

Stability and bearing capacity of the floor is affected by water. Although no effect was observed in a few isolated short time cases, water reduced the bearing capacity of coal floor by 10–35 per cent with an average around 18 per cent. The effect on underclay floor along the faces was between 20 and 50 per cent averaging to 30 per cent. Jenkins [2] has shown that introduction of water to underclay floor decreases its dry bearing capacity by 10–37 per cent with an average of about 26 per cent. Lee [4] indicated that the effect of water 35–69 per cent decrease in the *in situ* strength of underclay. Platt [5] suggested that introduction of water to underclay reduces its bearing capacity by 60–80 per cent. With such wide variations recorded, it appears that percentage saturation of the floor at the time of test, total time the floor was affected, quartz content of the floor rock and proximity of pre-developed cracks in the vicinity of the test plates are additional influencing parameters.

Laboratory creep tests on coal and underclay samples by the author [6] revealed that the average instantaneous deformation (Δl)

of air-dried coal under a pressure was increased by about 62 per cent when it was water saturated. Similarly, underclay samples have shown an increase of 71 per cent under the same load when saturated with water. The effect was even greater when clay bands were present in the underclay, where an overall creep rate five times higher than the saturated underclay was recorded. Since it is the deformation which leads to eventual failure, it is important to observe deformation pattern and its rate in the floor as well as the final rupture.

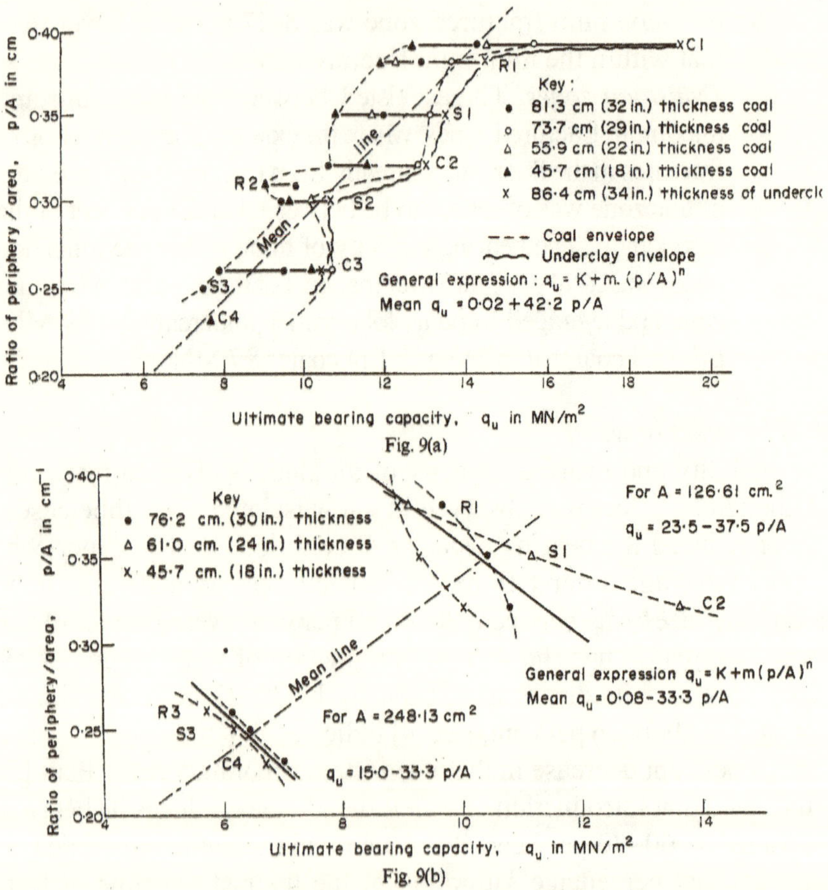

Fig. 9. (a) Effect of base plate size and shape on the bearing capacity of the L3 floor. (b) Effect of base plate size an shape on the bearing capacity of the F41 coal floor.

6- The effect of rate of loading

During the bearing capacity test, the loads were increased far more rapidly than is generally the case in practice. It would be ideal if the roof and floor convergence by its continual closure is employed to provide the loading on the base plate. While this is justifiable theoretically, it gives rise to an additional non-linear variable. It is a fact that the rate of convergence of roof and floor (both in the vertical and horizontal directions) is not constant, but varies according to the physical state of the rock in the region under study and to the type of mining operation being carried out at the time [6]. Another disadvantage of natural loading is the long time duration for each measurement. For an actual simulation of the floor loading, a prop once set would normally be expected to remain in position for between 5 and 24 hr during week days and between 48 and 60 hr during week-ends. Thus each recording dynamometer prop cannot be expected to yield more than one or two results per day during week days and more than one result during the weekends. The solution would be to install a larger number of dynamometer props on the face at one time, in order to obtain sufficient number of data.

A time-dependent study on penetration of chock bases along the coal floor of L3 face was conducted utilizing a dial gauge mounted on the side of the chock near the base. The assembly was similar to that shown in Fig. 1. Due to sensitive arrangement of instrumentation, the method, although useful, was not successful during mining operations because of vibration affecting the gauge settings. However, during quiet periods the method proved workable. At such times, the chock base settlement into the coal floor was noted to be time-dependent at an average rate of $3{\cdot}8 \times 10^{-3}$ cm/hr. The same experiment was conducted on the L1 underclay floor, the average recorded rate of penetration of the chocks was $1{\cdot}8 \times 10^{-3}$ cm/hr. To appreciate the magnitude of these settlements, it was found that during a six-week stoppage the average vertical closure of roof to floor along the L3 face was about 2·5 cm/week. This gives an average convergence of about $51{\cdot}2 \times 10^{-3}$ cm/hr along the L3 face. No direct relationship exists between the two parameters when yielding supports rather than rigid supports are used. There is a slight chock penetration into the floor, the magnitude of which is about twice in the lower four feet coal than in the underclay beneath the seam. Since there was generally between 2 and 18 cm thick spillage beneath the chocks throughout the face, it is thought that the majority of penetration may be caused by compaction of the spillage.

Ultimate bearing capacity of floor beneath the chocks

Extrapolation of the test results on the base plate to the full-size chock base is considered using Housel's perimeter shear theory [7] expressed as follows:

$$P = A \,.\, q_{\mu} = A \,.\, \sigma_c + p \,.\, m \quad \text{or} \quad q\mu = P/A = \sigma_c + p \,.\, m/A \tag{41}$$

where: P = total load applied (lb); A = base area = 1260 in^2 (8139 cm^2);

q_{μ} = ultimate bearing capacity (psi); m = perimeter shear (lb/linear in.);

σ_c = compressive stress in floor column directly beneath support base (psi);

p = periphery of the base plate (in.) = 4 (70 + 12) = 32·0 in. = 853 cm.

The result is shown in Table 3 which indicates a decrease of 12·5–53 per cent on the L3 coal from the measured values by the base plates. The percentage decrease for the F41 coal was between 8 and 45 and for the L3 underclay was between 60 and 63 per cent.

Utilizing empirical equation (40) revealed that on average the ultimate bearing capacity of floor beneath the chocks (q_{μ}) would be as follows:

(a) For the L3 coal floor: q_{μ} = 0·02 + 42·2 (0·11) = 4·66 MN/m^2. Percentage decrease in floor bearing = 58·5.
(b) For the F41 coal floor: q_{μ} = 0·08 + 33·3 (0·11) = 3·74 MN/m^2. Percentage decrease in floor bearing = 57·6.

The values obtained were under conditions of uniform loading beneath the chock base. Since the capacity of the chocks encountered were 180 tons, the floor loading at the chock yield point would only reach a maximum of 2·2 MN/m^2 by manufacturer's specifications.

In general, the approximate floor loading along the face (although much higher towards the face ends), measured by pressure gauge on various chock legs averaged to 1·4 MN/m^2. Hence, the lower yielding quality of the chocks is adequate to prevent the floor failure beneath the chocks under the condition of uniform loading.

TABLE 3. ULTIMATE BEARING CAPACITY OF MINE FLOOR ALONG LONGWALL FACES BENEATH THE CHOCKS CALCULATED FROM THE MEAN BEARING TEST RESULTS UTILIZING HOUSEL'S PERIMETER SHEAR THEORY

Location (district)	Base plate shape	Area A (in^2)	Perimeter p (in.)	Load P (lb/ft)	Pressure at failure (q_u)	m (lb/ft/in.)	σ_c (psi)	Ratio $x = p/A$ for the chock	Ultimate bearing beneath the chock (chock q_u) (psi)	(MN/m^2)
L1 underclay	Circular	C2 19·63	15·70	44,000	2241·5	—	—	0·26	—	—
	Square	S1 19·63	17·72	43,500	2216·0	—	—	0·26	—	—
	Rectangular	R1 19·63	18·78	42,600	2170·1	—	—	0·26	—	—
L3 underclay	Circular	C1 12·57	12·57	35,000	2786·6	5505	272·0	0·26	1648·2	11·53
		C2 19·63	15·70	33,000	1681·1					
	Square	S1 19·63	17·72	38,500	1961·3	2816	580·6	0·26	1284·6	8·99
		S2 28·26	21·28	43,600	1542·8					
	Rectangular	R1 19·63	18·78	41,000	2088·6	—	—	0·26	—	—
		R2 28·26	22·56	—	—					
L3 coal	Circular	C2 19·63	15·70	33,400	1702·8	529	1277	0·26	1409·3	9·85
		C3 28·26	18·84	41,000	1389·6					
	Square	S1 19·63	17·72	33,825	1723·1	200·4	1542	0·26	2042·1	14·28
		S2 28·26	21·28	40,233	1423·7					
	Rectangular	R1 19·63	18·78	36,175	1842·9	3850	1841	0·26	2803·5	19·61
		R2 28·26	22·56	38,700	1369·4					
F41 coal	Circular	C2 19·63	15·70	32,866·7	1630·6	2729	508·6	0·26	1190·8	8·33
		C4 38·47	21·98	38,500·0	1000·8					
	Square	S1 19·63	17·72	29,766·7	1516·4	2337	644·3	0·26	1228·5	8·60
		S3 38·47	24·80	32,250·0	916·3					
	Rectangular	R1 19·63	18·78	26,166·7	1666·3	1749	211·4	0·26	648·9	4·54
		R3 38·47	26·34	32,900·0	855·2					

Comparison of the results obtained by the perimeter shear method with the graphical method of equation (40) indicate that the latter method is simpler, quicker and closer to the manufacturer's specifications, for the maximum floor loading before chock yield. As far as the stability of floor is concerned, it is also safer, because it gives between 5 and 13 per cent lower bearing capacity than that deduced by Housel's method and, therefore, it is justifiable to use it where acute and concentrated loading, such as beneath arch legs and roadway junctions, is encountered.

CONCLUSIONS

(a) General deformation of floor (Δl) under load (P) can be expressed empirically as:

$$\Delta l = K_1 \,.\, P^{\mathrm{n}} = 100\, K_2 \,.\, L / A^{\mathrm{n}}$$

where: $|0 < K_1 < 1| = 100\, K_2 \,.\, L / A^n$,

$|0 < K_2 < 1| = P^{(1-n)} / 100\, A \,.\, E$,

$K_1 > K_2$, and $0 < n < 1$.

(b) Theoretically, thickness of the floor (l) affected by the applied load (P) is:

$l = K_1 \,.\, A \,.\, E \,.\, P^{(n-1)}$

(c) Apparent stress–strain characteristics of the floor beneath base plate (ϵ_3) can be expressed as: $\epsilon_3 = K_2 \,.\, \sigma_3{}^n$

(d) Theoretically, the vertical strain developed beneath a full-size chock (ϵ_b) can be obtained from that developed beneath a test plate (ϵ_a) as follows:

$$\epsilon_b = C_b \cdot q_b \cdot \epsilon_a (1{\cdot}1D + \Delta l_a) / n^2 \,.\, C_a \cdot q_a (1{\cdot}1n.\, D + \Delta l_b)$$

where: $q_b = n \,.\, C_a \cdot q_a / C_b$

(e) It is expected that the probable failure envelope of floor beneath base plates to be of the following general form:

$$\Delta l = K_3 \,.\, P^{-m}$$

where: $m < 1 < K_3$, and

$K_3 = l \,.\, P^{(1+m)} / E \,.\, A^{(1+n)} \rightarrow l \,.\, \epsilon \,.\, \sigma$

(f) There exists a critical thickness at which the stability of mine floor comprising soft rocks is a maximum. This critical thickness was found to be 73·7 cm for the L3 lower four feet coal floor and 61·0 cm for the F41 four feet coal floor.

(g) The size effect of the base plates on the bearing capacity of floor is such that doubling the base would decrease the coal floor bearing by 16–35 per cent and that of the underclay by 5–20 per cent.

(h) The length of the base perimeter p was observed to be an influencing factor on the ultimate bearing capacity of floor (q_μ) shown empirically as follows:

$$q_\mu = K + m(p/A)^n$$

(i) Under *in situ* conditions, the derived theoretical expression for the floor bearing capacity beneath a full-size chock provides a better approximation than the Boussinesque's equation. This is so because it covers the effect of support base size and perimeter.

(j) *In situ* results indicate that the strength of floor decreases from the mid-face towards the face ends. This is mainly due to three states of fracture along the face. These are classified as minimum, transition and maximum fractured zones as approach is made from the mid-face towards the face ends, respectively. The extent of each zone is mainly dependent on the face length, state and type of gate loads, mining methods and type of floor rock.

(k) Water reduced the bearing capacity of coal and underclay floors by 18 and 30 per cent, on average, respectively.

(l) Floor deformation and consequently the bearing capacity beneath hydraulic chocks is time-dependent. The chock base settlement into dry coal and underclay floors is observed to be at an average rate of $3{\cdot}8 \times 10^{-3}$ cm/hr and $1{\cdot}8 \times 10^{-3}$ cm/hr, respectively.

(m) *In situ* results indicate that the strength of floor decreases from the mid-face towards the face ends. This is mainly due to three states of fracture along the face. These are classified as minimum, transition and maximum fractured zones as approach is made from the mid-face towards the face ends, respectively. The extent of each zone is mainly dependent on the face length, state and type of gate loads, mining methods and type of floor rock.

(n) Water reduced the bearing capacity of coal and underclay floors by 18 and 30 per cent, on average, respectively.

(o) Floor deformation and consequently the bearing capacity beneath hydraulic chocks is time-dependent. The chock base settlement into dry coal and underclay floors is observed to be at an average rate of $3{\cdot}8 \times 10^{-3}$ cm/hr and $1{\cdot}8 \times 10^{-3}$ cm/hr, respectively.

REFERENCES

1. Coates D. F. Rock mechanics principles. Canadian Department of Mines and Technical Surveys, Ottawa. Mines Branch Monograph 874, pp. 2–15 (1965).
2. Jenkins J. D. Some investigations into the bearing capacities of floor in the Northumberland and Durham Coalfields. *Trans. Instn Min. Engrs* **117,** 730 (1958).
3. Barry A. J. and Nair O. B. *In situ* tests on bearing capacity of roof and floor in selected bituminous coal mines. U.S.B.M. Report No. 4706, Colorado (1970).
4. Lee R. D. Testing mine floors. *Colliery Engng* **38** (448), pp. 255–261 (1961).
5. Platt J. Underclays of South Wales coalfield and their influence on floor penetration. Ph.D. Thesis, University of Wales, Cardiff (1957).
6. Afrouz A. Mechanical behaviour of mine floor. Ph.D. Thesis, Mineral Exploitation Department, University of Wales, Cardiff (1973).
7. Housel W. S. Discussion, *Proc. Am. Soc. civ. Engrs* **91** (SM 1), 196–219.

3.3.9 FLOOR BEHAVIOUR ALONG LONGWALL ROADWAYS

By: Andy A. Afrouz

A simple and inexpensive method is adopted to measure penetration of arch legs into stilt and/or soft to medium strength floor, sides closure and floor lift along roadways of three longwall faces possessing variable strata and mining conditions. These conditions are compared in the light of the measurements taken in order to obtain the best positioning of the roadway drivage, relative to the strata conditions. Effect of producing face on its roadways is illustrated. Empirical time-dependent equations are developed to express and predict floor behaviour along the roadways. Contribution of floor to the total roadway closure is investigated and means of reducing floor lift along roadways discussed.

1. INTRODUCTION

Behaviour of mine floor along longwall roadways is a function of numerous geological, constructional and operational variables some of which are interdependent. The need for higher rate of production, adoption of advanced headings, using a roadway twice and retreat mining methods demand a longer life of roadways with no or minimum possible repairs. Hence, a greater understanding of the floor behaviour along the roadways is essential in order to ensure long term stability of strata in this region, so that more accurate and economical planning is facilitated.

There have been a number of investigations into strata behaviour along longwall roadways. Hobbs [1] suggested, from studies on scaled model mine roadways, that roof lowering, floor lift and sides closure to be directly proportional to the applied compressive load. The contribution to roadway closure of each of which at a given stress level is averaged to be 35, 42 and 23 per cent, respectively. However, in simulating the effect of induced stress on scaled models of circular roadways, the percentage contribution of the roof, floor and sides to the total closure of circular roadways was found to be equal [2]. Also, 47 per cent increase in diameter of a circular roadway resulted in an increase in closure of each of the three above noted regions by 100 per cent. Investigation on scaled models revealed [3] that reduction in the spacing of arched supports by 25 per cent reduced roof lowering by 34 per cent, but increased floor lift at 15 per cent. However, increasing width of road side packs has decreased the floor lift [4]. Moreover, packed ribsides resulted to less roadway closure than roadways with solid ribside [5]. Lawrence [6] investigated effects of horizontal and vertical stresses on models of mine roadways having rectangular cross-section. He indicated that horizontal stresses are a major cause of roadway failure underground, but a high vertical stress can reduce the effect of horizontal stress. Johnson [7] in studying gate road closure along the supply road of an advancing longwall face, planted bolts of 0·76 m long into roof and floor vertically below it and took measurements by surveying method. He concluded that approximately 50 per cent of the original roadway height was lost after a face advance of 550 m and noted that 40 per cent of this movement was due to roof lowering during the first 90 m of the face advancement. The remaining 60% of closure was contributed by floor at a more gradual and prolonged rate. He has also introduced the following empirical equation expressing floor heave (V) in terms of the distance from the faceline (d):

$$V = m . d + b(1 - e^{-c.r.d}) \tag{1}$$

where: m = Creep constant = 0·25

b = Total amount of floor heave due to causes other than creep = 6·1, m

c = Constant of asymptotic decay = 8·2

r = Conversion factor according to units used

= 1, if V is in meters and d is in kilometers, or

= $30{\cdot}5 \times 10^{-5}$ for V and d in feet.

It appears from equation (1) that the term for creep of floor was considered to be linear expressed as $m.\ d = 0{\cdot}25d$. The laboratory and *in situ* results on creep of floor coal and underclay, conducted under room temperature and atmospheric pressure by the author, revealed a nonlinear overall behaviour, especially during the primary and tertiary creep periods. Thomas [8] adopting a similar method of measurement to that of Johnson, noted above, concluded that, on average, roadway movements, especially floor lift and sides closure, in the advanced headings was greater than that along the conventional roadways at an equivalent distance behind the face.

In general, little *in situ* investigation has been conducted to quantify floor movements along the full cross-section of roadways, especially within the South Wales coalfield of U.K. This study is directed towards establishing a simpler and more representative method of measuring time-dependent penetration of arch legs, sides closure, floor lift and total roadway closure of three longwall faces details of which is given in Table 1.

TABLE 1. PHYSICAL CHARACTER OF THE ROADWAYS AND FACES

Longwall faces Roadway investigated	L1		L3		F41	
	Main	Tail	Main	Tail	Main	Tail
State of moisture along roadways	99% dry	60% wet	98% dry	98% dry	85% dry	85% dry
Size of arches (in metres): (i) Main Gate	4·5 × 3×5		4·5 × 3×5		Originally 4·5 × 3·5 + 0·7 m stilts, the dinted 2·5 m and double skinned with 4·5 × 3·0 m arches	
(ii) Tail Gate	4·5 × 3×5		Originally 4·5 × 3·5, double skinned at 15 m in advance of the face with 2·8 × 2·5 arches		Originally 4·5 × 3·5 + 0·7 m stilts, then dinted 2·5 m and double skinned with 4·5 × 3·0 m arches	
Spacing of arches	1m		1m		1m	
Other support material used: (i) Wood blocks under arches (ii) Concrete blocks along face side (iii) Road side packs (cm)	15 × 15 × 10 cm 50 × 25 × 25 cm 130 H × 50 W		15 × 15 × 10 cm 50 × 25 × 25 cm 130 H × 50 W		30 × 30 × 15 cm 50 × 25 × 25 cm 130 H × 50 W	

Dinting along main gate (m)	None	First dint 0·81 coal floor at 13 behind the face, then, dint 0·86 underclay at 50 behind face		First dint 1·0 underclay at 10 behind the face then dint 1·5 at 100 behind the face
Dinting along tail gate (m)	None	0·86 m of underclay at the faceline		1·5 m of underclay at 180 m behind the face
Name of the seam	Lower four	Lower four feet		Four feet
Depth below surface	feet	652		505
Average thickness of	734	2·3		2·4
the seam (m).	2·8	1·6		1·6
Height of cut (m)	1·6	192		192
Total length of face	118	1:12		1:161
(m)	1:10	1:10 to 1:15		1:40
Dip of seam	1:54			
Average inclination of roadways				
Direction of roadways relative to:	87	172		90
The dip of seam (degree)				
System of face support	←	202 tonnes, 6-legs Dowty chocks		
Number of chocks along face	104	175	116	116
System of waste control	Total caving, except at the face ends where 3 metres long sone and concrete packs are used			
Method of coal breaking along face	←	Uni-directional shearing		→
System of stable control	←	3·5 m long Hbars on dowty duke hydraulic props		→
Spacing of stable supports	←	0·95 m centre to centre		→
Method of coal breaking in stables	←	Conventional shotfiring		→
Average rate of advance (metres/week)	10	5	20	20
Type of coal	←	Steam bituminous		→
Average uniaxial strength of coal (MN/m^2)	5	5	6	6

Average thickness of underclay (m)	0·9	0·9	0·6 + 2 m of shale and underclay	0.6 + 2 m of Shale & Underclay
Average uniaxial strength of underclay (MN/m²)	18	18	10	10
Type of roof rocks	S. St. & M. St.	S. St. & M. St.	S. St., Clift & Clod.	S. stone; Clift & Clod
Average uniaxial strength of roof (MN/m²)	66	59	83	83
State of bonding between coal and roof	Good	Moderate	Poor	Poor
State of bonding between coal and underclay	Good	Good	Moderate	Moderate
Type of sidewall rock	Coal, S. St. & M. St.	Coal, S. St. & M. St.	Coal, Clift & S. St.	Coal, Clift & S,stone
orking above	Fine 9 m pillars 24 m above	Three9 m pillars 25 m above	None	None
Working below	None	None	None	None

2. INSTRUMENTATION AND TEST PROCEDURE

Permanent stations are chosen at 30 m interval along the L1 main and tail gates and the L3 tail gate, and at 10 meters interval along the L3 main gate and the F41 main and tail gates. The arch legs of these stations were nicked at convenient heights from the floor as permanent reference marks. The marks were placed at 1·5 and 0·8 m above the floor along the main and tail gates, respectively. However, due to rapid floor lift and arch penetration observed along the F41 roadways, the reference marks were placed at 1·5 m above the floor.

The method of floor lift measurement was a modification of the mid-ordinate system [9] requiring a 5 m long string marked at 0·3 m intervals, two bulldog clips, a light weight spirit level and a measuring tape. In addition to the mid-ordinate method of floor lift measurements, the arch legs was measured each time and their penetration into floor or stilt, when used, and deformation within the affected regions was noted and consequently deduced from the apparent floor lift readings. Similarly, where dinting operations were performed, dimensions were measured and corrections made to the measured values of the affected stations. The method was adopted because it provides the following information:

(i) Penetration and/or floor yield characteristics beneath the arch legs—this is obtained from differential measurements of distance between the nick on the arch leg and floor. Wherever distortion of the arch leg occurred the curved length of the arch legs was measured and consequently deduced from the original reading. The same technique was adopted to obtain penetrations due to use of stilts along the F41 roadways. In the latter case, to obtain the amount of floor penetration, the arch leg penetration into the stilt was deduced from the total shortening of the distance from the nick of the arch leg to floor. The penetration thus obtained is in effect a direct measurement of the resistance of floor to penetration due to strata pressure on supports at the floor level.

(ii) Sides closure, measured horizontally on the string clamped to the selected point nicked on the arch legs. Shortening of the horizontal distance between the arch legs, thus measured was considered to be the sides closure, which is a measure of the strata pressure upon sides of the roadways.

(iii) Floor lift across the roadways irrespective of the roof movements. The total cross-sectional area bounded between the fixed datum, floor and sides of the roadway computed originally as A_0. Successive cross-sectional areas, thus obtained at time intervals $t_1, t_2, t_3 ..., t_n$ were then deduced from the original area (A_0) to give the actual sectional area of the floor lift $A_1, A_2, A_3 ..., A_n$, respectively. The measurements were conducted on two monthly intervals along the L1 and L3 roadways and at monthly intervals along the F41 roadways.

3. ANALYSIS AND DISCUSSION OF RESULTS

In situ observations have led to the following distinct categories of floor behaviour within the immediate vicinity of longwall roadways.

3.1. *Penetration of roadway supports into the floor*

Penetration of arch legs into the L1 main and tail gates underclay floor appeared at an average rate of 0·32 and 0·50 cm/week, respectively. The relatively dry condition along the L1 main gate driven on strike almost parallel to the main cleat of coal seam resulted in a stable condition with low penetration of arch leg into floor. The effect of 60 per cent water along the otherwise similar condition of the L1 tail gate resulted in 36 per cent increase in floor penetration relative to the L1 main gate. The effect of setting arches originally on coal floor, then dinting the coal about 10 meters behind the face along the L3 main gate which is driven almost perpendicular to the main cleat of the coal seam in a dry condition, was 48 per cent increase in the rate of arch penetration compared with that of the L1 main gate.

Along the L3 tail gate driven on full dip the condition differs much from the above noted roadways, since this roadway was driven originally as the main gate of the previous district (L2), and majority of floor deformation and/or arch penetration occurred during that time. In general, the average rate of arch leg penetration along this roadway was 10 per cent lower than that of the L3 main gate. The effect of dinting at the faceline along the L3 tail gate, in addition to the above noted features, have caused a total of 43 per cent higher penetration to the floor than that of the L1 main gate.

The support penetration into the softer underclay floor along the F41 main and tail gates averaged to 5·85 and 5·08 cm/week, respectively. In addition, the arch penetration into stilt was at a rate of 2·85 and 1·40 cm/week, respectively. This yields a total penetration and/or shortening of arch legs along the F41 main and tail gates to be 8·70 and 6·48 cm/week, respectively. Effect of the dinting, was to non-linearly increase the rate of arch penetration due to reduction in the frictional resistance between the arch leg and floor material. Average of the measurements available along the F41 roadways, indicated 33 per cent increase due to dinting which was to be added to the above noted support penetration along the F41 main gate (i.e. on overall penetration rate of 11·5 cm/week).

Due to variable strata conditions and method of support along each roadway, detailed and accurate comparisons are not justifiable. However, taking the L1 main gate as the standard, it appears that the roadways driven along the main cleat of the coal seam would have the lowest arch penetration into the medium to hard underclay floor. Increase of 60 per cent wetness along such roadways increases the rate of arch penetration by 36 per cent. Roadways driven in a 98 per cent dry condition but perpendicular to the main cleat of the coal seam, with adoption of dinting, can be expected to have an increased arch penetration rate of 48 per cent of the standard L1 main gate. If such a roadway is to be used again to serve a semi-retreating face, the overall rate of arch penetration can be expected to be 43 per cent of the standard during the life of the second face (i.e. 48–43 = 5 per cent decrease in the original rate of arch penetration as the floor tends to stabilize). The effect of soft underclay floor in 85 per cent dry condition is to increase the penetration rate by a magnitude of 22·4 times that of the standard L1 main gate. Dinting operation in such condition together with the effect of advanced heading technique would give rise to the rate of arch penetration by a magnitude of 36·4 times that of the standard.

Penetration of arch leg into floor and stilt is graphed with respect to the distance from the F41 faceline shown in Fig. 1. The result indicates a higher support penetration into floor along the F41 main gate (Fig. 1a) than that of the F41 tail gate (Fig. 1b). The behaviour can be expressed empirically as follows:

(a) Along the F41 main gate:

$$p = d^{1\cdot05} + 7\cdot5(1 - e^{-0\cdot22d})$$
$$= 23\cdot24\,(t_2 - t_1)^{1\cdot05} + 7\cdot5\,[1 - e^{-4.40\,(t2 - t1)}] \qquad (2)$$

(b) Along the F41 tail gate:

$$p = d^{0\cdot74} + e^{+0.02d} = 9\cdot18(t_2 - t_1)^{0\cdot74} + e^{+0.40\,(t2 - t1)} \qquad (3)$$

where: p = Penetration of arch leg into floor (cm)
d = Distance of the arch from the faceline (m)
e = Constant of natural logarithm = 2·72
t_1 = Initial time of measurement (week)
t_2 = Final or subsequent time of measurement (week), and
r = Rate of face advance = $d/(t_2 - t_1)$ = 20 m/week.

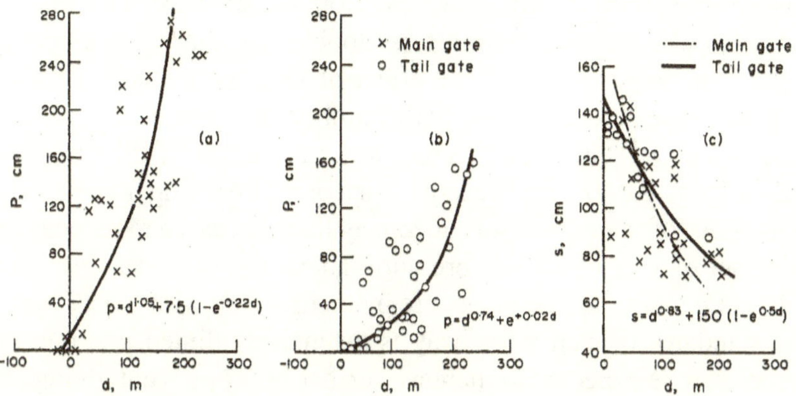

Fig. 1. Behaviour of arch legs with respect to the floor and stilt along the F41 roadways, Windsor colliery, S. Wales, U.K.

Penetration of arch leg into stilt(s) was also shown to be slightly greater along the F41 main gate than that of the tail gate (Fig. 1c). However, the difference is small by comparison with the wider scatter of results obtained. An overall sliding of arch into stilt along both of the roadways may be expressed as follows:

$$S = d^{0{\cdot}83} + 150\,(1 - e^{-0{\cdot}5d})$$
$$= 12{\cdot}01\,(t_2 - t_1)^{0{\cdot}83} + 150\,[1 - e^{-10\,(t2-t1)}] \quad (4)$$

None of the stilts used along the F41 roadways traveled its full distance by the time the face had passed, as is indicative in Fig. 1c.

3.2. *Sides closure*

The average rate of sides closure along the L1 main and tail gates was 0·15 and 0·20 cm/week, respectively. The average rate of sides closure along the L3 and F41 main and tail gates was 0·28, 0·41, 0·64 and 0·53 cm/week, respectively.

Considering the L1 main gate as a standard, the effect of 60 per cent increase in the wetness on the rate of sides closure is therefore to increase the closure rate by 33 per cent. Roadways driven in 98 per cent dry condition, but along the full dip and perpendicular to main cleat of coal seam with adoption of dinting would have an increase of 86 per cent sides closure. The average rate of sides closure along such a roadway to be used for a second time serving a semi-retreat face is thus expected to be 2·7 times increase from the standard. This means that sides closure would almost be double that during production of the first face. Therefore, the effect of face on the sides closure does not appear to diminish with number of times a face passes through a given station along a roadway, i.e. the rate of sides closure is directly proportional to the local strata pressure acting horizontally on the sides of roadways. Soft underclay floor appeared to reduce the resistance of arch legs to side pressure. Thereby, it caused an increase of 40 per cent in the rate of closure by comparison with the L1 main gate standard. If such a roadway is extensively dinted, the rate of closure can be expected to increase further by 27 per cent giving an overall rate of side closure equal to 67 per cent (=40 + 27) that of the standard. This shows the adverse effect of dinting on sides closure, since it creates larger side walls which would then undergo larger loading and hence exhibit larger closure rates.

The effect of face on the total sides closure along the roadways was such that the rate of closure tended to diminish as the distance of faceline from the measuring point along the roadway increased. This behaviour was observed in all the roadways investigated as illustrated in Fig. 2 which indicates the greatest rate of sides closure to be in the vicinity of the face. As the face distance is increased the sides closure steadily levels until a region is reached behind the face where the rate of closure is very small and the sides closure stabilized. This region, although not sharp and well defined, appeared graphically to be at 150 and 200 m behind the L1 face along the main and tail gates, respectively (Fig. 2a). The sides closure tended to remain steady at about 220 metres behind the L3 face along its main gate (Fig. 2b). The closure along the L3 tail gate was gradually accelerated at about 350 m in advance of the faceline. After the face passed, closure gradually decreased to a stable rate at about 75 m behind the face. Since, some arches along the L3 gate were already distorted due to the effect of previously worked L2 face and the time effect, much of their initial resistance had been overcome by the time the L3 face approached. Therefore, the resistance of these distorted arches to the pressure induced by approaching L3 face would be much less than that exhibited by the new arches along other roadways in the same seam (Fig. 2). Hence, the resultant front area of influence of the L3 face appeared to be much larger than its rear.

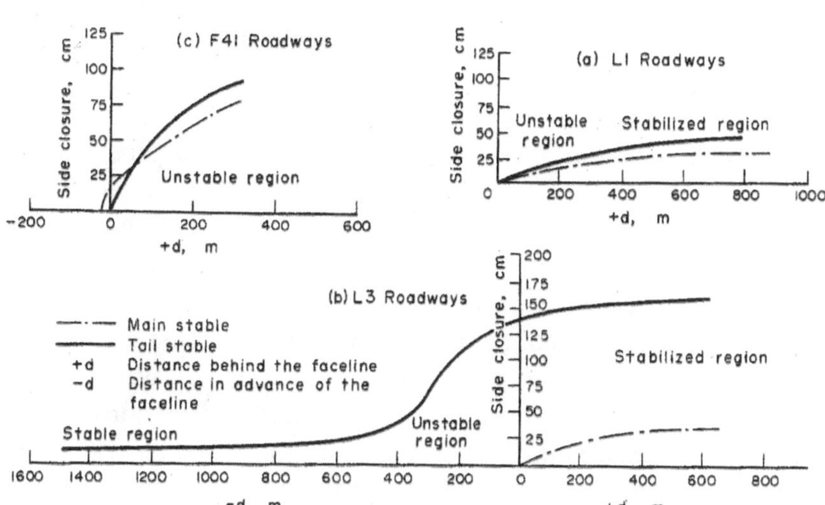

Fig. 2. Side closure along roadways of three faces in the S. Wales coalfield, U.K.

The limited data available were not adequate to locate the extent of influence of the F41 face on its roadways. It appears, from extrapolation of Fig. 2c, that the region lies at about 400 m behind the face.

To counteract sides closure in roadways use of splayed leg arches may be introduced, but their use has an adverse effect on the function of stilts. Where stilts are adopted, in order to provide yield for prevention of arch distortion, then it is better to use straight legs and to match the yield properties of the stilt to that of the arch, floor and the strata pressure.

3.3. *Floor lift*

A reasonably uniform rate of floor lift along the L1 main and tail gates is observed as illustrated in Fig. 3. The slight variation apparent is believed to be due to the presence of a 10 m coal pillar left in the overworked area 21 m above, a 1·5 m down-throw fault, about 33 m above, the plane of which is at 45 degrees to the axis of the L1 roadways, and a 28 m coal pillar left by the fault in the overworked area. The floor lift vertically below the 10 m coal pillar was, on average, 30 and 9 per cent lower than the overall lift along the L1 main gate and tail gate, respectively. However, at distances between 50–75 m either sides of the pillar centre (i.e. *ca* 66° to the vertical from the pillar edge projected towards the floor under study), the floor lift was 30–40 per cent higher than the average along the L1 roadways. In general, the average rate of floor lift along the L1 main and tail gates was 113·0 and 203·2 cm²/week, respectively. It appears that the rate of floor lift along the 60% wet L1 tail gate is 44 per cent higher than that along the 99% dry L1 main gate. The depth of water on floor along some sections of the L1 tail gate was measured to be 15 cm. It is possible that the actual floor lift occurring along these saturated sections is larger than that measured. Since, it was observed that the water tends to disintegrate the underclay floor forming a 5 cm crust of mud on floor this soft and unconsolidated crust is then easily washed away by constant flow of water from the abandoned overworking as more floor is lifted and disintegrated.

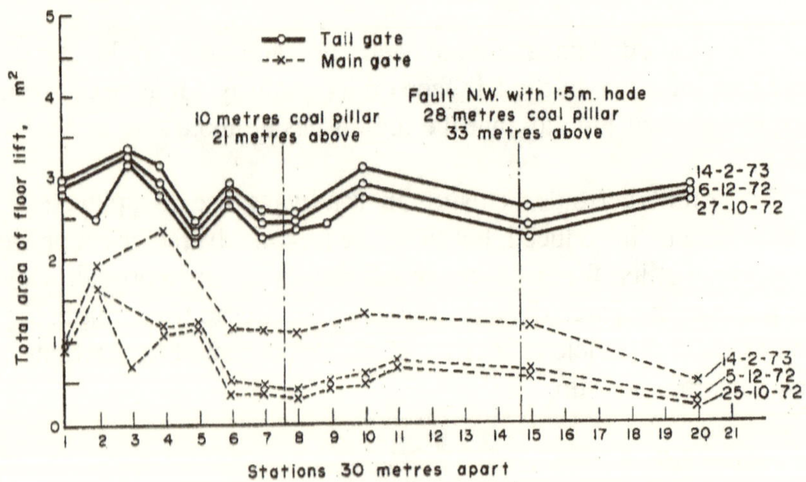

Fig. 3. Time-dependent floor lift along the L1 roadways.

The rate of floor lift along the L3 main gate was observed to decrease with time and increasing distance from the face (Fig. 4) with an average rate of 384 cm²/week. This roadway is driven on full dip and perpendicular to the main cleat of the seam with coal and dinting operation carried out which shows an increase of 54 per cent over the standard L1 main gate.

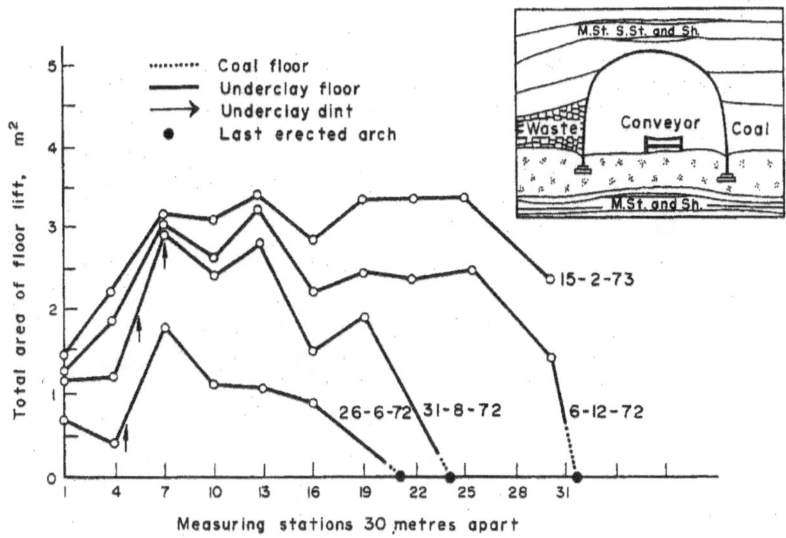

Fig. 4. Time-dependent floor lift along the L3 main gate.

The rate of floor lift along the L3 tail gate (Fig. 5) can be categorized as follows:

(a) About 180 m behind the face, the average rate of floor lift is 232·3 cm²/week and decreasing.
(b) Between 180 m behind the face and 160 m in advance of the faceline, the average rate of floor lift is constant at 245·1 cm²/week, but increasing as the face approaches.
(c) About 24 to 26 m vertically below the overworked 9 m wide pillars, the average rate of floor lift was observed to be less than that along the rest of the roadway at 139·3 cm²/week. The effect of the pillars appear to decrease the vertical downward pressure acting immediately below with a consequent decrease in stress, transfer to the floor.
(d) At distances 30 to 35 m either sides of the pillar centre of the overworked area (i.e. at 50 degrees to the vertical from the above pillar edge projected towards the floor understudy), the average rate of floor lift is 406·4 cm²/week. This is higher

than the case (c) because the peripheral stress of the pillar appears to be transferred downwards affecting the working below. This stress is theoretically larger than the central stress transformation which applies to the floor, thereby more energy is applied to move the floor. This behaviour is clearly apparent along the L1 roadways (Fig. 3) and the L3 tail gate (Fig. 5).

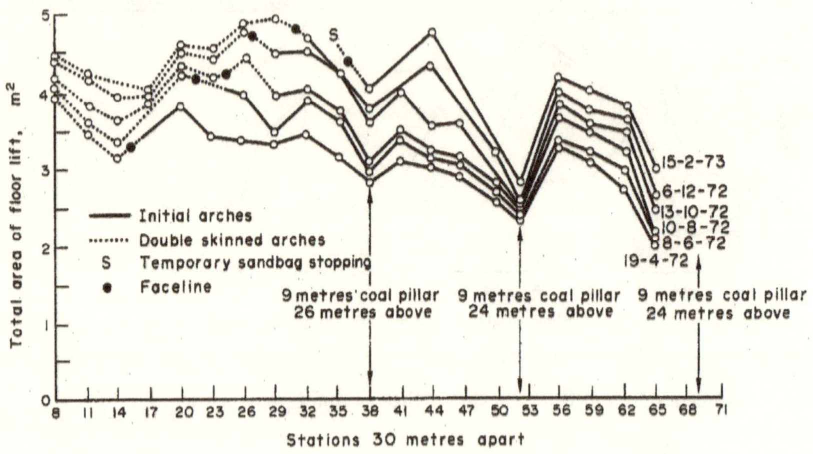

Fig. 5. Time-dependent floor lift along the L3 tail gate.

Hence, roadways driven in a 98 per cent dry condition along the full dip and perpendicular to the main cleat of the coal seam to be used second time (to serve a semi-retreat face), with dinting operation, the average floor lift is expected to be a 31 per cent increase over the standard.

The average rate of floor lift along the F41 main and tail gates are 5225 and 3483 cm ^{2}/week, respectively (Fig. 6).

This rate of floor lift is comparatively high due to softness of floor material and extensive dinting operation. The difference in the rate of floor lift between the F41 main and tail gates may be roughly considered as the effect of dinting on floor lift. With this assumption, the effect of dinting on average floor lift along the F41 roadways accounts for approx 33 per cent of the total rate of floor lift. Therefore, the effect of soft underclay floor in 85 per cent dry condition is to increase the floor lift by about 12 times that of the standard L1 main gate. Dinting operations in such conditions together with the effect of advanced heading would give rise to the floor lift by about 19 times that of the standard. Adverse effect of dinting measured indicates that although dinting would provide the required space in the closing roadway, it does not furnish a long term solution in floor stability. This is especially true for roadways having very soft floors. In fact, apart from the short term solution by dinting, the rate of floor lift was increased due to the less resistance to floor movement provided by eliminating the weight and internal friction of the removed layer.

Floor Behaviour Along Longwall Roadways

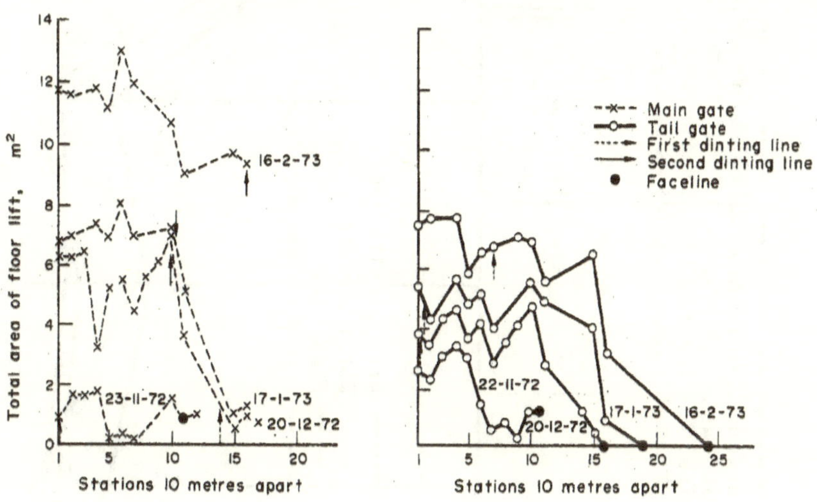

Fig. 6. Time-dependent floor lift along the F41 roadways.

Figure 7 is a summary of the above discussions and analysis as a comparison and contribution of arch penetration, side closure, floor lift and total of the arch penetration and floor lift along the roadways investigated. These are as average values relative to the standard L1 main gate.

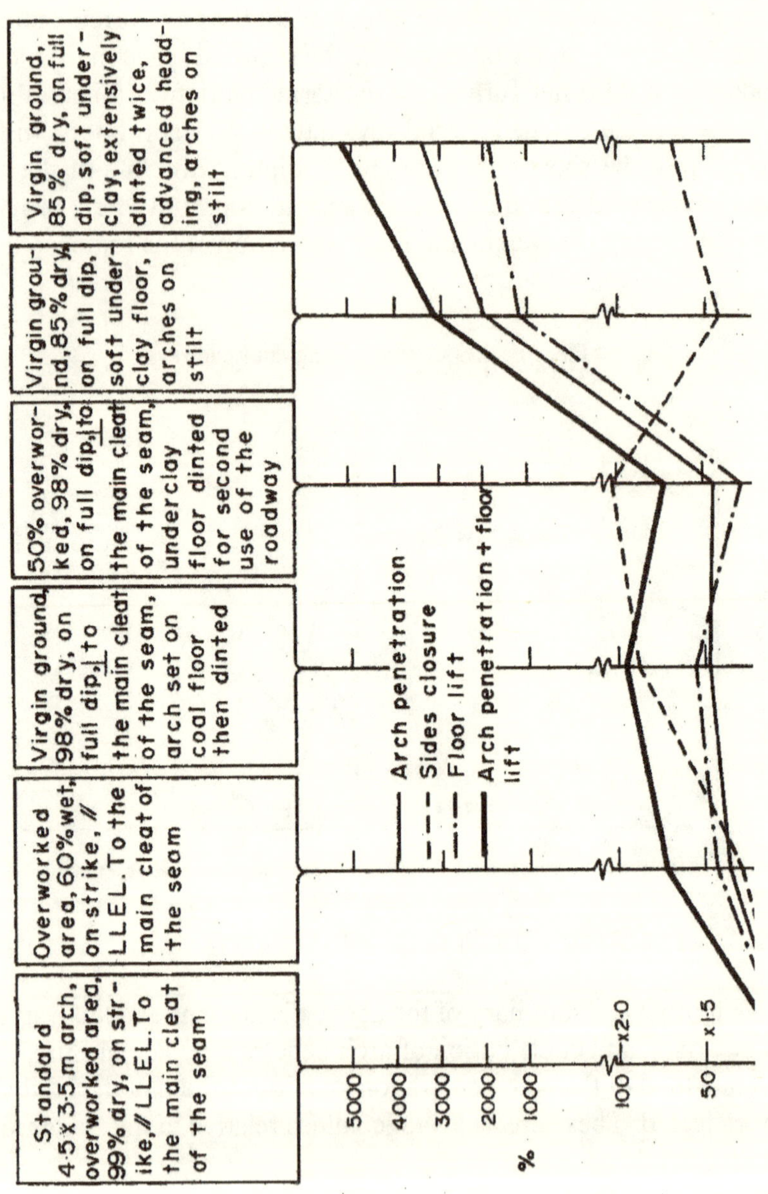

Fig. 7. Per cent contribution of each element of floor under roadway conditions relative to the standard L1 main gate.

3.4. *Prediction of floor lift along roadways*

The total area of floor lift along the roadways plotted against distance from the faceline in Figs. 8 and 9 which illustrate three distinct phases of floor behaviour. These are:

(a) A general slow floor lift due to the strata pressure and creep which predominates along the full length of the roadways. This behaviour is measurable along sections of roadways which lie outside the abutment pressure zones of their respective faces.
(b) A decreasing rate of floor lift at some distance behind the faceline (i.e. outside the back abutment pressure zone), until the roadway is either completely closed, repaired and kept open for further use, or an equilibrium is reached around the opening.
(c) A gradually increasing rate of floor lift, along the advanced heading and roadway serving a retreating face, as the face approaches any of the measuring stations.

Superimposing the above noted three phases of floor lift, the following equations is developed to express the area of floor lift (A) in terms of distance from the face (d) and time (t):

Along the L1 main gate: $A = d^{+0{\cdot}29} + 0{\cdot}25\,(1 + e^{-0{\cdot}26d})$

$$= 0{\cdot}51\,(t_2 - t_1)^{+0{\cdot}29} + 0{\cdot}06\,[1 + e^{-2.6\,(t2 - t1)}] \quad (5)$$

where: t_1 = Initial time of measurement, weeks
t_2 = Final time of measurement, weeks
r = Rate of the L1 face advance = 10, m/week
= $d/(t_2 - t_1)$, or $d = 10(t_2 - t_1)$.

Along the L1 tail gate:

$$A = d^{+0{\cdot}50} + 2{\cdot}25\,(1 + e^{-0{\cdot}32d})$$
$$= 0{\cdot}32(t_2 - t_1)^{+0{\cdot}50} + 0{\cdot}50\,[1 + e^{-3.2\,(t2 - t1)}] \quad (6)$$

Along the L3 main gate:

$$A = d^{+0{\cdot}38} + 0{\cdot}20\,(1 + e^{-5{\cdot}36d})$$
$$= 0{\cdot}54\,(t_2 - t_1)^{+0{\cdot}38} + 0{\cdot}04\,[1 + e^{-26.8\,(t2 - t1)}] \quad (7)$$

where: r = Rate of the L3 face advance = 5, m/week

$$d = 5(t_2 - t_1).$$

Along the L3 tail gate—in advance of the face:

$$A = d^{+0\cdot21} + 4\cdot75\,(1 + e^{-0\cdot22d})$$

$$= 0\cdot91(t_2 - t_1)^{+0\cdot21} + 1\cdot06[1 + e^{-1.1(t2-t1)}] \qquad (8)$$

Along the L3 tail gate—behind the face:

$$A = d^{+0\cdot61} + 3\cdot40\,(1 + e^{-0\cdot30d})$$

$$= 0\cdot37(t_2 - t_1)^{+0\cdot61} + 0\cdot76[1 + e^{-1.5(t2-t1)}] \qquad (9)$$

Along the F41 main gate: $A = d^{+0\cdot24} + 0\cdot45\,(1 + e^{-3\cdot42d})$

$$= 2\cdot01\,(t_2 - t_1)^{+0\cdot24} + 0\cdot10\,[1 + e^{-68.4\,(t2-t1)}] \qquad (10)$$

where: r = Rate of the F41 face advance = 20, m/week

$$d = 20(t_2 - t_1).$$

Along the F41 tail gate: $A = d^{+0\cdot32} + 0\cdot10\,(1 + e^{-0\cdot96d})$

$$= 2\cdot61\,(t_2 - t_1)^{+0\cdot32} + 0\cdot003\,[1 + e^{-19.2\,(t2-t1)}] \qquad (11)$$

Equations (5)–(11) can be used to predict the magnitude of roadway closure that may occur at any given distance from the faceline and any time interval, if the roadway conditions are similar to those investigated. A similar approach can be made to predict roadway closure due to roof lowering, hence giving an overall prediction of the roadway closure due to each component of roadway boundary. However, the above noted expressions do not provide a close fit to the floor lift near the facelines of the conventionally advancing faces. It is believed that the elastic beam analysis presented in a previous paper for publication is more applicable to the newly opened ground near the facelines. Where more uniform strata condition exist both above and below the opening and retreat faces or advanced headings encountered, only two or three sets of concentrated measurements along the roadways at regular monthly intervals are needed to obtain curves similar to that illustrated in Figs. 8 and 9. If the roadways were driven in partially overworked ground or where a non-uniform strata condition exists, more than two or three sets of measurements would then be necessary to achieve a representative average of the actual *in situ* behavior.

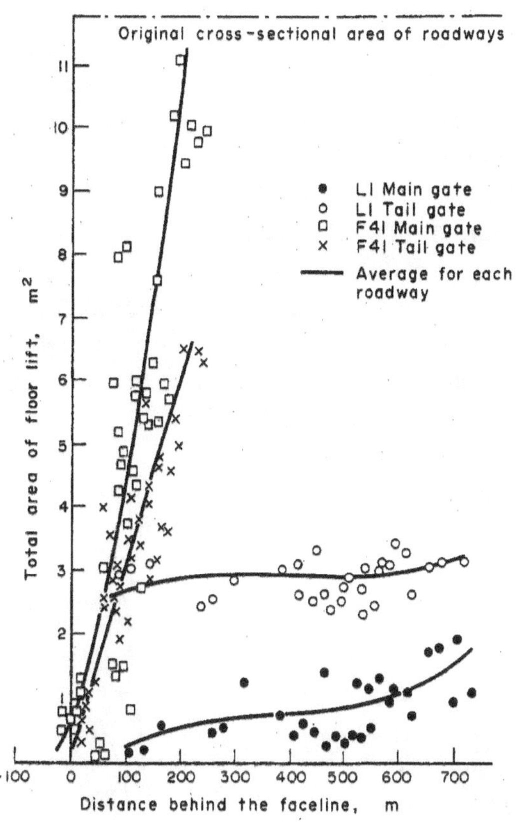

Fig. 8. Floor behavior along the L1 and F41 roadways, South Wales, U.K.

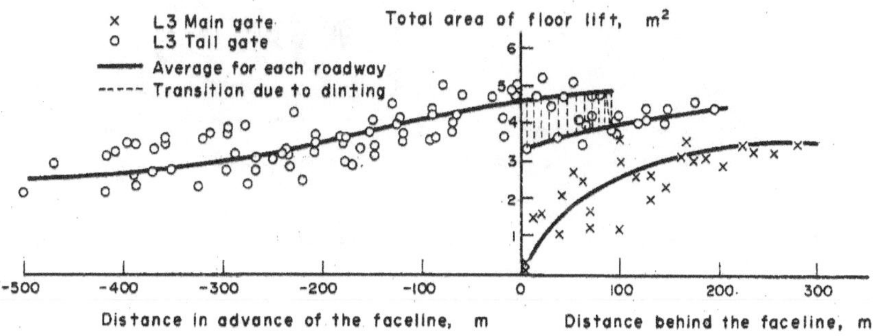

Fig. 9. Behaviour of the L3 Roadway floor, Britannia Colliery, S. Wales, U.K.

3.5. *Contribution of floor lift to the total roadway closure*

Measurements of cross-sectional area along the roadways revealed that the total rate of closure along the L1 main and tail gates was 282·5 and 295·0 cm²/week, respectively. Hence, the contribution of floor lift to the total closure along the L1 main and tail gates are 40 and 68 per cent, respectively (Fig. 10). The rate of total closure along the L3 main gate was 538·0 cm²/week, an increase of 48 per cent on the standard L1 main gate. The contribution of floor to this total closure was 71 per cent. The rate of total closure along the L3 tail gate, within the abutment zones was 650·0 cm²/week. The contribution of floor within these zones was 53 per cent. Outside the abutment zones, the rate of total closure was 150·0 cm²/week and the floor contributed 76 per cent to the total closure (Fig. 10). The rate of total closure along the F41 main and tail gates was 7256 and 7973 cm²/week, respectively. Therefore, the contribution made by floor was 72 and 70 per cent, respectively (Fig. 10).

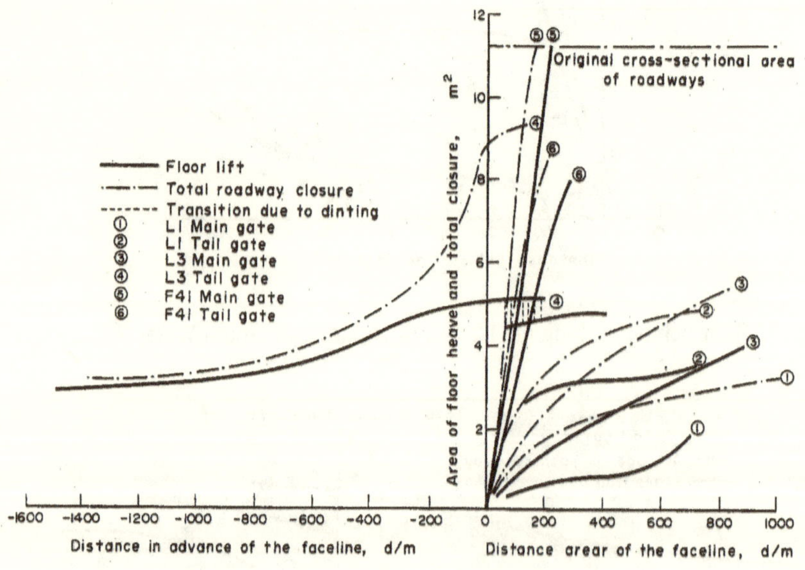

Fig. 10. Comparison of the floor lift and total closure along the roadways investigated.

It can be seen that the contribution of floor to the total closure does not change linearly with the total roadway closure, but its magnitude is influenced mainly by the position of the face relative to a point of the floor along the roadway. For instance, along the L3 tail gate, higher percentage of floor contribution was observed outside the abutment zones than that within the zones. This reveals that other boundaries of the roadway, especially the roof is more sensitive to the abutment pressure than the floor. Therefore, contribution of roof and sides is expected to increase at a larger rate than that of the floor as illustrated in Fig. 10.

4. MEANS OF REDUCING FLOOR LIFT

It is appreciated that under the tremendous strata pressure around longwall faces it is not economically possible to construct the access roadways so that they would stay intact over long periods. Therefore, certain deformation, especially that due to floor, should be accounted for in reducing roadway closure. Hence, the purpose is not to eliminate the floor lift, but to minimize it to acceptable limits. This can be achieved by two means discussed below:

4.1. *Floor reinforcement*

This method would increase the inherent strength of weak floor thereby improving the floor stability along roadways. Bolting, dowelling and injection are the present techniques used in strata reinforcement.

Bolting is suitable for hard and competent rock layers, where the bonding between the layers is weak, in order to hold the strata together. Floor beds of coal mines are normally of a medium to soft nature and do not efficiently respond to bolting.

Dowelling is a modified version of bolting, where in addition to the bolt (generally wooden) the column is filled with resin to hold the bolts to the rock. The advantage of dowelling over bolting is that the resin may tend to penetrate into the fissures within the floor and provide filling for the bored and fractured floor. To obtain maximum benefit from this type of reinforcement, the resin should have the ability to form a good bonding between the wooden bolts and the host rock. Moreover, the dowelling operation should be done as soon as the floor of the opening is exposed, and before any major movement and/or fracture develops. Alternatively, all existing fractured floor should be strengthened as soon as possible, by the injection of adhesive materials such as resin and cement.

Araldite resin reinforcement with a volumetric ratio of resin:hardener:plasticizers of 100:5:1 was conducted on the lower four feet underclay samples of the L1 and L3 faces under laboratory conditions. The results of uniaxial compression tests on the cured resin thus formed have shown an averaged value of 23·8 MN/m^2 as compared with 18·5 MN/m^2 of the sample. The reinforced Underclay samples formed the following empirical equation:

$$V_t \cdot \sigma_t = V_s \cdot \sigma_s + V_r \cdot \sigma_r \tag{12}$$

where V = Volume, cm^3

σ = Uniaxial compressive strength, MN/m^2

t = Total reinforced material

s = Rock sample

r = Resin.

However, it is important to know what strength and amount of reinforcement is needed to a floor under *in situ* conditions. For this reason, the most critical condition in the stability of floor and support was considered. This condition would occur when the floor under a support is failing and there exists little or no shear stress on the vertical plane A_0 extended from the support/base periphery as shown in Fig. 11.

The vertical stress on an element of the floor at depth D beneath the support (σ_{vs}) can be expressed as:

$$\sigma_{vs} = q + y.\, D \tag{13}$$

The corresponding horizontal stress (σ_{hs}) using Terzaghi's expression (10) is:

$$\begin{aligned}\sigma_{hs} &= \sigma_{vs} \,.\, \tan^2 (45 - \phi/2) - 2C \tan (45 - \phi/2)\\ &= (q - yD) \tan^2 (45 - \phi/2) - 2C \tan (45 - \phi/2)\end{aligned} \tag{14}$$

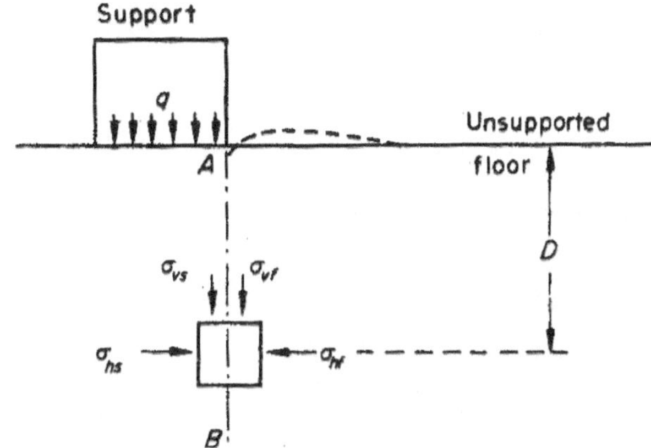

Fig. 11. Interaction of stresses in floor of roadways due to the strata and support pressures.

where: q = Load intensity of strata on floor beneath the support which is higher than the strength of soft floors

γ = Specific gravity of the floor rock

C = Cohesion of the floor rock, and

ϕ = Internal friction angle of the rock.

If the floor, along the roadway, is reinforced uniformly with a resin of uniaxial compressive strength (σ_r), then on the unsupported side of the reinforced floor, the vertical stress (σ_{vf}) at depth D can be expressed as:

$$\sigma_{vf} = \sigma_r + \gamma . D \tag{15}$$

The corresponding horizontal stress (σ_{hf}) is:

$$\sigma_{hf} = \sigma_{vf} . \tan^2 (45 - \phi/2) + 2C \tan (45 - \phi /2)$$
$$= (\sigma_{r'} + \gamma . D) \tan^2 (45 - \phi /2) + 2C \tan (45 - \phi /2) \tag{16}$$

For the floor to remain reasonably stable, the induced horizontal stresses on both sides within the floor should be equal (i.e. $\sigma_{hs} = \sigma_{hf}$). Then from equations (14) and (16):

$$(q + \gamma . D) \tan^2 (45 - \phi /2) - 2C \tan (45 - \phi / 2)$$
$$= (\sigma_r + \gamma . D) \tan^2 (45 - \phi /2) + 2C \tan (45 - \phi /2)$$

Therefore, $\sigma_r = q - 4C / \tan(45 - \phi/2)$ (17)

Substituting equation (17) into equation (12) yields:

$$V_r = V_t . \sigma_t - V_s . \sigma_s) \tan(45 - \phi/2) / [q . \tan(45 - \phi/2) - 4C] \quad (18)$$

For equilibrium to be reached throughout the roadway floor, it is necessary that the total strength of the reinforced floor (σ_t) be equal or greater than the strata pressure (q).

Considering: $\sigma_t = q$, then equation (18) can be expressed as:

$$V_r = V_s(\sigma_s - q) \tan(45 - \phi/2) / 4C \quad (19)$$

Equation (19) provides an approximate guidance on how much resin is expected to be used to reinforce a floor of known physical characteristics and dimensions under certain strata pressure.

Note: The negative result from equation (19) means that there is a volumetric deficiency in resin reinforcement per every unit volume of the floor rock to be treated. In that case, the actual volume of the resin to be used is equal to the total volume of floor rock to be reinforced multiplied by the negative value of V_r. If the result of equation (19) is a positive value, then there is a surplus strength in the floor to maintain its equilibrium by such positive amount. In the latter case, there will be no need for floor reinforcement unless the strata pressure (q) is increased so that a negative result is obtained.

4.2. *Mining or other excavation methods*

The principal methods may be categorized as follows:

(a) Use of inverted arches––which increase the total bearing of arch on floor and reduce the side closure or distortion of the arch legs wherever the side pressure along roadways are large. Splayed leg arches may also be used to counter the sides closure, but they have adverse effect on the function of stilts, if they are used, and are less effective than inverted arches. However, the use of inverted arches on soft floor may not be successful in minimizing floor lift and arch penetration into floor. They may present difficulty in dinting and floor repair work, especially when mechanized dinting is practiced. Similar drawbacks in dinting operating exist if the use of floor bolting, dowelling and injection fail to reduce the floor heave, but to a lesser extent.

(b) Provisions to use square steel base plates with rounded corner and edges—their dimensions can be matched to the bearing capacity of particular floor material and the strata pressure. However, base plates with bearing areas in the region of 0·1 m^2 and 0·6 cm thickness are expected to be satisfactory under average conditions. The plates should be made so that they can be bolted to the arch bases upon setting along the roadways to prevent slippage. This method is also feasible in conjunction with the splayed leg arch and its penetration into floor. Wooden chocks conventionally used beneath the arch bases tend to slip away from the base where excessive lateral movement in the floor is encountered. They are also susceptible to deterioration, especially in wet floor.

(c) Complete removal of the weak floor bed—generally, roadways having soft floor require two to three dintings and/or back rippings during the producing life of their corresponding faces. The total thickness of dinting in such cases may well exceed 1·5 to 2·0 m of soft floor during the initial 6 to 8 months of the opening. If the roof above the seam is competent, it may be advantageous to leave the roof bed intact during the roadway drivage and switch from ripping to dinting the soft floor. Therefore, the incompetent floor, which would be dinted at a later stage, is removed in one operation during the roadway drivage with the advantage that the competent roof is not broken and the normal ripping operation is eliminated. Moreover, since the soft floor is easier to remove than the hard roof material, the system lends itself to mechanized dinting. Therefore, the overall efficiency of the roadway drivage would increase while achieving better stability. In order to counter side pressure along the deepend roadway, splayed leg arches may be used. To reduce or possibly eliminate flow of soft rock from the sides into the roadway, it becomes necessary to place lagging board superimposed with wiremesh along the outer side of the arches. It would be preferable if this measure is taken to cover at least the thickness of the soft dinted seam floor positioned along the roadway sides.

The initial cost of the above noted support material and extra operation is believed to be less than the outlay for normal back ripping and/or dinting operations. Moreover, the harder floor of a deepened roadway is more adaptable to reinforcement than a soft and weak floor, if it becomes necessary.

(d) Elimination of either the main or the tail stables — this measure would reduce the total time and manpower used per face advance. Faster face advance results in a shorter life of face which subsequently leads to a shorter life of gate roadways. This measure does not necessarily stop floor lift, but indirectly reduces its time dependency.

(e) Elimination of advanced headings wherever abnormally soft floor is encountered—this will reduce the total life of the roadways and reduce the floor heave along roadways at given distances behind the faceline. In general, the average rate of floor lift along roadways with soft floor and advanced heading may be expected to be 33 per cent higher than those formed at or behind the face with similar strata conditions. This comparison was observed along the F41 roadways and is more or less in accordance with the findings of other investigators [8].

(f) To reduce asymmetrical floor lift along roadways—consideration may be given to provide effective yield facilities of solid coal ribsides. One method of achieving this is by cutting a slot of 1–2 m deep in the solid rib along the roadways and pack it with wood chocks and concrete blocks. This would transfer the peak side stress deeper into the coal rib and away from the immediate vicinity of the roadways. It would also equalize the strata pressure on both sides of the roadway thereby providing a more even stress distribution and so better stability along roadways. This method was tried successfully in an advanced heading at Pye Hill Colliery, U.K. [11].

(g) Provision for minimum utilization of water on floor and/or utilization of chemicals and organic additives which would increase surface tension of the Underclay floor thereby making it water repellent.

5. CONCLUSIONS

Analysis of results has shown that:

(a) Roadways driven along the strike and parallel to the main cleat of the coal seam would have a more stable floor than those driven along the full dip and perpendicular to the main cleat.

(b) Pillars left in the overworked areas tend to transfer the strata pressure towards the floor of underlying roadways at angles between 50 to 66 degrees to vertical projected downwards around the periphery of the pillar bases. The floor lift vertically below these pillars is usually less than the average along the roadways.

(c) Dinting medium strength underclay tends to increase the rate of floor lift and sides closure along roadways by up to 33 and 86 per cent respectively. Extensive dinting of soft underclay floor would increase the rate of floor lift and sides closure by 90 and 27 per cent, respectively.

(d) Linear extrapolation of results suggest that for every 1% increase in wetness of underclay floor, the rate of floor lift and sides closure increases by 0·7 and 0·5 per cent, respectively.

(e) Effect of advancing longwall face on floor lift and sides closure is within 150 to 220 m behind the face along its roadways. This effect for a retreating longwall face would be 75 m behind the faceline and 350 m in advance of the face.

(f) In general, contribution of medium strength underclay floor to the roadway closure in a typical coal measure strata and dry condition is 40%, and in a 60% wet condition, is 68%.

Contribution of soft and weak underclay floor to total closure in an otherwise typical coal measure strata and dry condition is 70%. A dry coal floor along roadways contributes up to 71% to the total roadway closure.

(g) Floor reinforcement by injection appear promising and lends itself to theoretical as well as practical analysis. There is much scope to investigate untilazation of strata and in particular floor injection by resin and/or cement under simulated pressures equivalent to the prevailing *in situ* pressures at each test location.

REFERENCES

1. Hobbs D. W. Scale model studies of strata movement around mine roadways. Apparatus, Technique and some preliminary results. *Int. J. Rock Mech. & Min Sci.* **3**, 110–127 (1966).
2. Hobbs D. W. Scale model studies of strata movement around mine roadways. Part 6: Circular roadways. NCB-MRE Report No. 2326, Feb. (1968).
3. Hobbs D. W. Scale model studies of strata movement around mine roadways. Part 7: Quality of roadway profile, support spacing and pack material. NCB-MRE Report No. 2336, June (1968).
4. Shepherd R. Conclusions from recent investigations of strata behaviour on mechanised coalfaces. NCB-MRE Report No. 2278 (1965).
5. Hobbs D. W. Scale model studies of strata movement around mine roadways. Part 8: Ribside support. NCB-MRE Report 2343, Oct. (1968).
6. Lawrence D. Scale model studies of strata movement around mine roadways—VII. Effect of horizontal and vertical pressure. *Int. J. Rock Mech. & Min. Sci.* **10,** 173–178 (1973).
7. Johnson G. The closure of gate roads related to the method of goaf control in the Low Main Seam at Bilsthrope Colliery. NCB-MRE Report No. 2309, Sept. (1967).
8. Thomas L. J. The effects of adjacent seams and methods of working on roadway closure in the Main Bright Seam at Hucknall Colliery. NCB-MRE Report No. 2330, June (1968).
9. Jones C. and Smith D. S. G. The mid-ordinate method of measuring mine roadway cross-sectional area. NCB Technical memorandum No. 202, January (1966).
10. Terzaghi K. and Peck R. B. *Soil Mechanics in Engineering Practice,* 2nd edn. John Wiley, New York (1967).
11. Shepherd R. and Kellet W. H. Destressing the ribside of an advanced heading at Pye Hill Colliery. NCB, MRE Report No. 2334, July (1968).

3.3.10 THE PERFORMANCE OF LONGWALL GATE ROADS WITH SOFT FLOORS

By: Andy A. Afrouz, F. P. Hassani, and M. J. Scoble

This work stems from field studies of convergence within gate roads leading to three longwall coal faces with different strata and mining conditions. The influence of production on the stability of these roads is illustrated. Empirical time-dependent equations are developed to express and predict the behaviour of the roadways. The particular contribution of the roadway floor and sides to the total closure of the gate roads has also been investigated. The mode of deformation and failure, observed by scanning electron microscopy within the immediate vicinity of underground gate roads, is also described. The application of this analysis is in the planning of production panels in soft tabular deposits, the dimensioning of gate roads, and the stability and performance prediction of such roadways and their supports.

Key words: soft mine floors, longwall mining, gate road stability, coal roads, soft underclay floors.

Ce travail découle d'études sur le terrain de la convergence de routes d'accès vers trois faces filantes de houille ayant des strates et des conditions d'exploitation différentes. L'influence de la production sur la stabilité de ces routes est illustrée. Des équations empiriques incluant les effets de temps sont développées pour représenter et prédire le comportement des voies routières. En particulier, la contribution de la fondation et des côtes de routes à la fermeture complète des voies d'accès a été étudiée. L'on décrit également le mode de déformation et de rupture dans le voisinage immédiat des routes d'accès souterraines stel qu'observé par microscope à balayage électronique.

Cette analyse s'applique à la planification des panneaux de production dans les dépots de houille molle en strates, au dimensionnement des voies d'accès et à la stabilité et la prédiction du comportement de telles voies et de leurs soutènements.

Mots clés: planchers mous de mines, faces filantes d'exploitation minière, stabilité de voies d'accès, routes sur houille, planchers d'argile molle.

[Traduit par la revue]

1. Introduction

Underground roadway stability is fundamental to the efficient operation of high-productivity longwall coal mining. The need in longwall coal mining to increase face production rates with minimum expense, to prolong the life of roadways with minimal maintenance, to consider multiple roadway usage for producing panels, and to develop advanced heading and retreat mining techniques has motivated research into gate roadway stability. This has included *in situ behavioral* studies (Shepherd 1965; Johnson 1967; Thomas 1968; Goetz 1982; Afrouz 1975*a*, *b;* Cain and Aston 1963) as well as laboratory and theoretical analysis (Whittaker and Singh 1979; Shearey and Dunham, 1978; Whittaker and Bonsell 1981; Jukes *et al.* 1983; Brown and Bray 1982; Ladanyi 1980). The results of the latter investigations, however, have been difficult to validate without accompanying data on the actual *in situ* performance of steel-arched gate roads.

This paper is an analysis of field and laboratory data relating to the performance over a 1.5 year period of gate roads with soft floors serving three longwall faces (F_1, F_2, F_3) located in the South Wales Coalfield, United Kingdom. The specifications of the gate roads are given in Table 1 and Figs. 1 and 2. The reevaluation and analysis of these data are aimed at improving prediction of longwall panel deformation for future Canadian mine design. Roadway closure related to soft floors is a problem encountered in the carboniferous coal measures of Europe and the Sydney Coalfield, Nova Scotia (Aston and Cain 1984).

Table 1. Physical character of the gate roads and faces investigated

	Longwall faces					
	F1		F2		F3	
Roadway investigation	Main	Tail	Main	Tail	Main	Tail
State of moisture along roadways	99% dry	60% wet	98% dry	98% dry	85% dry	85% dry
Size of arches (m): Main gate	4.5 × 3.5		4.5 × 3.5		Original 4.5 × 3.5 + 0.7 m stilts, then dinted 2.5 m and double skined with 4.5 × 3.0 m arches	
Tail gate	4.5 × 3.5		Originally 4.5 × 3.5, double skined at 15 m in advance of the face with 2.8 × 2.5 arches		Originally 4.5 × 3.5 + 0.7 m stilts, then dinted 2.5 m and double skined with 4.5 × 3.0 m arches	
Spacing of arches (m)	1		1		1	
Other support material used:						
Wood blocks under arches (cm)	15 × 15 × 10		15 × 15 × 10		30 × 30 × 15	
Concrete blocks along face side (cm)	50 × 25 × 25		50 × 25 × 25		50 × 25 × 25	
Road side packs (cm)	130 H × 50 W		130 H × 50 W		130 H × 50 W	
Barrier pillar (m)	25 – 30		25 – 30		25 – 30	
Dinting along main gate (m)	none		First dint, 0.81 coal floor at 13 behind the face, then dint 0.86 underclay at 50 behind face		First dint, 1.0 underclay at 10 behind the face then dint 1.5 at 100 behind the face	
Dinting along tail gate (m)	None		0.86 of underclay at the face line		1.5 of underclay at 180 behind the face	
Position of main gate relative to face line	Conventional		Conventional		23 m advanced	
Position of tail gate relative to face line	Conventional		Fully advanced		Conventional	

Depth below surface (m)	740	770	785
Average thickness of the seam (m)	2.8	2.3	2.4
Height of cut (m)	1.6	1.6	1.6
Total length of face (m)	118	192	192
Dip of seam	1:10	1:12	1:161
Average inclination of roadways	1:54	1:10 to 1:15	1:40
Direction of roadways relative to the dip of seam (deg)	87	172	90
System of face support	202 t, 6 legs Dowthy chocks		
Number of chocks along face	104	175	116
System of waste control	Total caving, except at the face ends where 3 m long stone and concrete packs are used		
Method of coal breaking along face	Unidirectional shearing		
System of stable control	3.5 m long H bars on dowty duke hydraulic props		
Spacing of stable supports	0.95 m centre to centre		
Method of coal breaking in stables	Conventional shotfiring		
Average rate of advance (m/week)	10	5	20
Type of coal	Steam bituminous		
Average uniaxial strength of coal (MPa)	5	5	6
Avreage thickness of underclay (m)	0.0	0.9	0.6 + 2 m of shale and underclay
Average uniaxial strength of underclay (MPa)	18	18	10
Type of roof rocks	S.St. & M.St.	S.St. & M. St.	St.St., Clift & Clod
Average uniaxial strength of roof (MPa)	66	59	83
State of bonding between coal and roof	Good	Moderate	Poor
State of bonding between coal and underclay	Good	Good	Moderate
Type of sidewall rock	Coal, S.St. & M.St.	Coal, S.St. & M.St.	Coal, Clift & S.St.

Working above	Five 9 m pillars 24 m above	Three 9 m pillars 25 m above	None
Working below	None	None	None

Notes: S.St. = sandstone; M.St. = mudstone; Clod = mudstone with lumps of clay; Clift = sandstone with other thin partings.

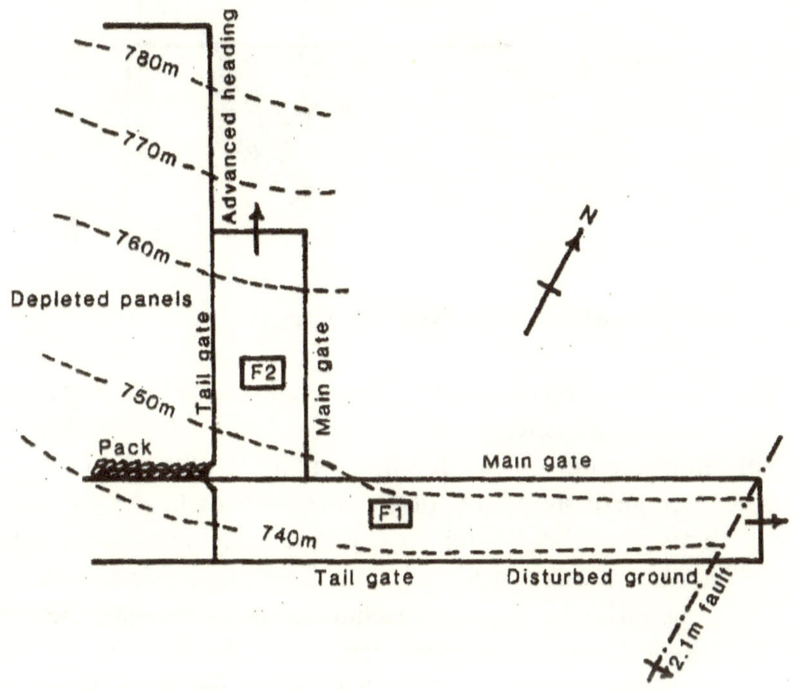

FIG. 1. Position of the F1 and F2 panels relative to coal seam dip.

2. Instrumentation

The method of roof, floor, and side closure measurement employed was a modification of the mid-ordinate system (Jones and Smith 1966). Permanent monitoring stations were established at 10 m intervals along the gate roads. In addition, arch leg and stilt penetration into floors was derived from the apparent floor lift readings. Similarly, where dinting operations were performed, dimensions were measured and corrections made to the measured values of the affected stations. The scheme provided the following data:

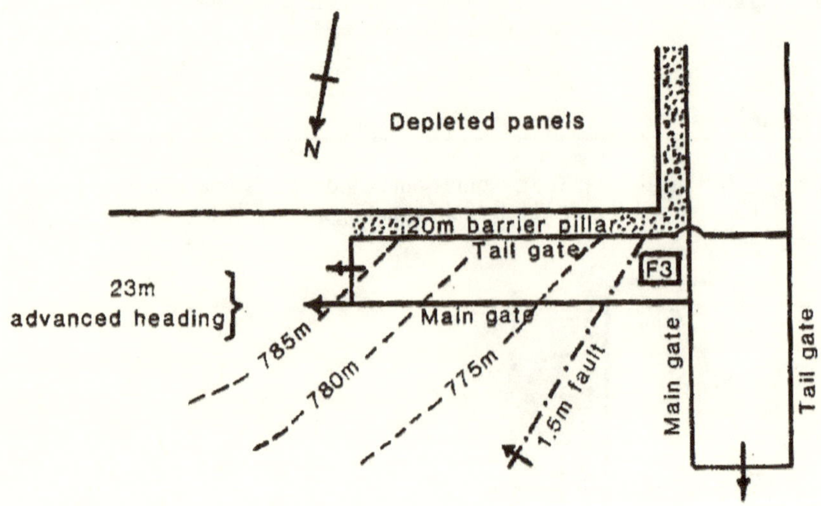

FIG. 2. Position of the F3 panel relative to coal seam dip.

(a) Roof lowering or closure – measured vertically upward from the mid-ordinate.
(b) Side closure – measured horizontally on the mid-ordinate. Average of shortening of the horizontal distances measured between the same arch legs and the unsupported wall of the roadways was considered to be the average side closure.
(c) Penetration or floor yield characteristics beneath the arch legs – derived from measurements of the distance between the mid-ordinate of the arch legs and the floor. Wherever distortion of the arch leg occurred, the curved length of the arch legs was measured and consequently deducted from the original reading. The same procedure was adopted where yielding arches or stilts were encountered.
(d) Floor lift across the roadways – the original total cross-sectional area between the mid-ordinate (as datum), floor, and sides of the roadway was computed as A_0. Successive cross-sectional areas obtained at time intervals of $t_1, ..., t_n$ were then deducted from A_0 to give the actual sectional area of floor lift $A_1, ..., A_n$, respectively.

3. Micro-structural deformation

To correlate the roadway behavior with stratigraphy and and facilitate a behavioral classification of roadway driven on soft floor, Scanning Electron Microscopy (SEM) studies were conducted. These also observed fracture and rupture characteristics in the longwall coal, the roof and floor rocks. They were classified as brittle, ductile or a combination of both. Fractures were observed to be:

(a) Inter-crystalline––where failure was observed to form by separation of crystal aggregates along grain boundaries;
(b) Trans-crystalline––where the fracture passed through or across crystals or grains.

Parallel brittle fractures were observed in the sandstone roof rock, exposing its layered structure (Fig. 3). These fractures tended to join together as large cleavage fractures (C), with the steps between the different fracture planes giving rise to characteristic "river-line" markings (→) on the fracture surface (Fig. 3). These "river-lines" tended to run together, so giving the local direction of mean crack propagation. They were also seen to run along and across the main cleats or other preexisting planes of weakness. Ductile fracture, caused by shear stress, was observed as glossy surfaces with plastic deformation and curved fibers in the argillaceous units, for example, the soft underclay of the F3 gate roads (Fig. 4).

A combination of brittle and ductile fracturing was observed especially in coal. Figure 5*a* shows a fibrous fusain layer (B), sandwiched between two layers of clarain (A and C). Magnification reveals the fracture character of the fusain (B) to be ductile and that of the clarain (C) to be brittle, creating an overall behaviour that is elastoplastic (Fig. 5*b*).

A similar character was observed where layers of soft (P) and hard (B) underclay were encountered along the F1 and F2 gate roads (Fig. 6).

FIG. 3. SEM of a cleavage fracture (C) in sandstone roof, showing "river line" markings (→) on the fracture surface. (Direction of scan = 45° to the bedding; × 1100.)

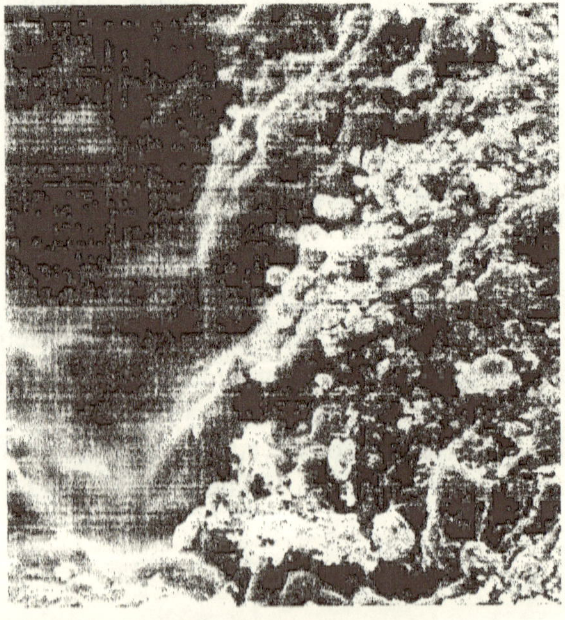

FIG. 4. SEM of a ductile fracture in soft underclay of F3 gate road. The white grains are calcitic and carbonic crystals. (Direction of scan = 45° to the bedding; × 1100.)

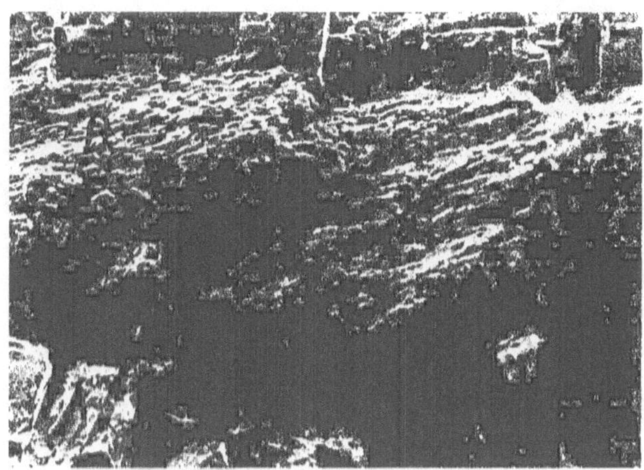

FIG. 5*a*. SEM coal showing clarain (A, C) and furain (B) layers with brittle and ductile fracture surfaces, respectively. (Direction of scan = parallel to the bedding; × 20.)

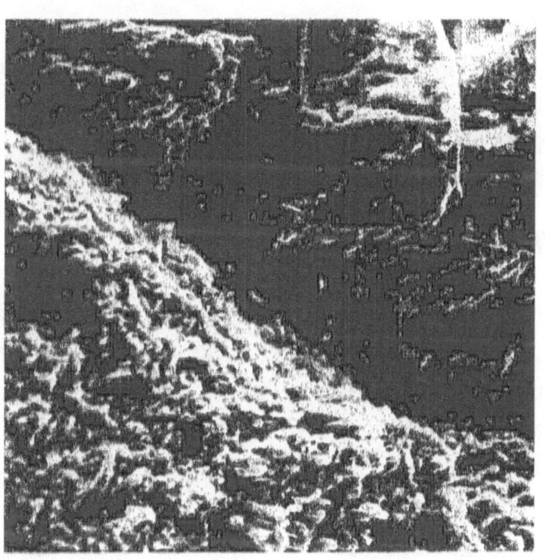

FIG. 5*b*. SEM of Fig. 5*a* magnified further. Clarain (B) has a brittle and fusain (C) has a ductile appearance. (Direction of scan = parallel to the bedding; × 500.)

FIG. 6. SEM of soft (P) and hard (B) underclay layers encountered along F1 and F2 gate roads, exhibiting ductile and brittle fracture, respectively. (Direction of scan = 45° to the bedding; × 550.)

4. Gate road deformation characteristics

4.1. Floor lift across the gate roads

The floor lift features observed were not symmetrical or uniform across roadways owing to several factors. *In situ* measurements showed the floor lift along the F1 main gate to be 33% higher at the centre than at the sides of the roadway cross section (Fig. 7*a*). The relatively hard underclay floor promoted semi-elastic behaviour and fractured predominantly in the central part of the cross section. A more asymmetrical floor lift was observed across the F2 main gate, which was 38% higher along the infinite pillar or solid rib side than along the gob side of the roadway (Fig. 7*b*). This floor behaviour appeared to arise from a higher pressure gradient along the rib side. The F2 main gate experienced a pattern of floor lift similar to that of the F1 main gate, although averaging 45% more floor lift between the centre and sides of the cross section (Fig. 7*c*). This is indicative of the reduced floor stability frequently observed along mine roadways driven perpendicular to the main cleat of a coal seam. The floor lift across the F1 tail gate was more uniform (Fig. 7*d*). This was due to excess water, which promoted plastic uplift of the saturated underclay.

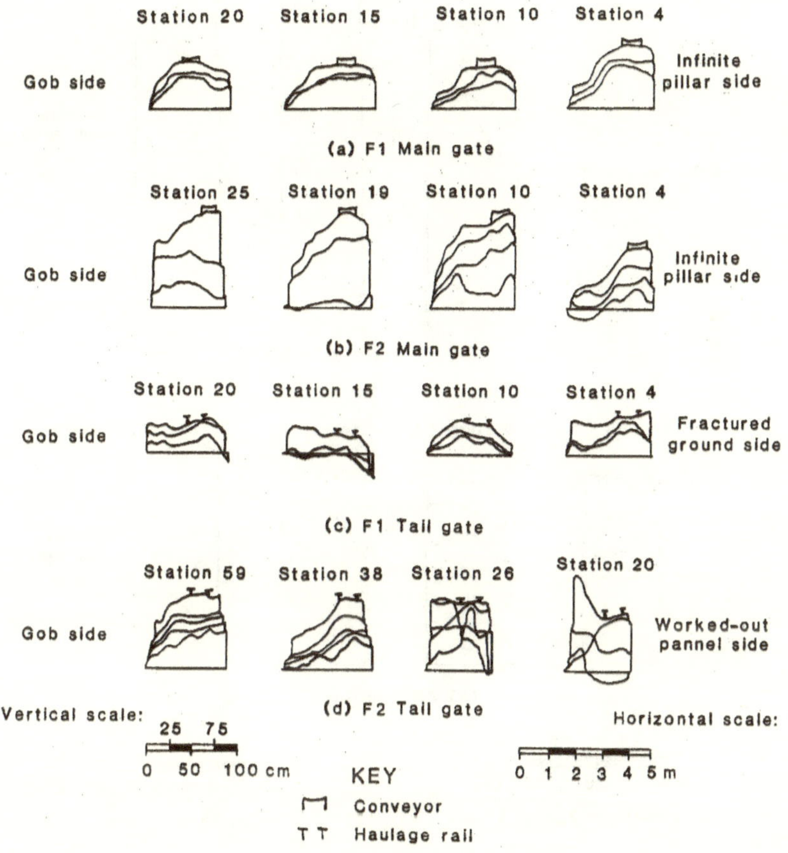

FIG. 7. Comparison of floor lift across various points in the F1 and F2 gate roads.

In the junction area of the F1 and F2 panels, the floor lift was most pronounced along the F1 main gate, because of higher abutment pressures and more frequent dinting. The F3 main gate floor lift was uniform and plastic (Fig. 8*a*). Although the F3 main gate was relatively dry, extensive dinting to keep the roadway open masked the characteristic floor lift that would otherwise be apparent. A floor lift similar to that of the F3 main gate, but of lower magnitude, was observed across the F3 tail gate (Fig. 8*b*).

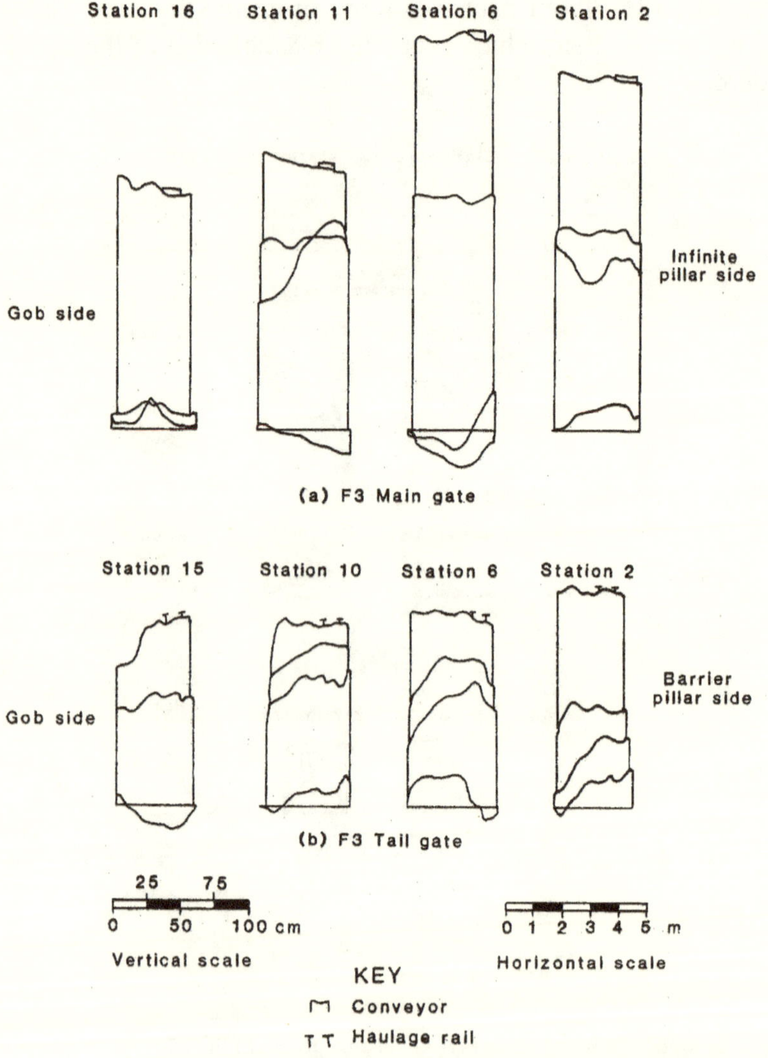

FIG. 8. Comparison of floor lift across various points in the F3 gate roads.

4.2 Floor lift along the gate roads

The F1 tail gate floor lift, with an average face advance of 10 m/week, was observed to be 2 – 3 times that of the F1 main gate with the same advance rate (Fig. 9). This arose from (a) proximity to disturbed ground and a washout and (b) water saturation of the underclay floor. The presence of 10 and 28 m wide coal pillars at 21 and 33 m respectively, in an overworked area above the F1 gate roads, resulted in 10–30% less floor lift in the regions immediately beneath the pillars. Floor lift around the periphery of the pillars, projected at 45 and 66° to the vertical respectively, was increased 30–40% from the average along the F1 roadways. The average rate of floor lift along the F1 main and tail gates was 113 and 203 cm^2/week, respectively.

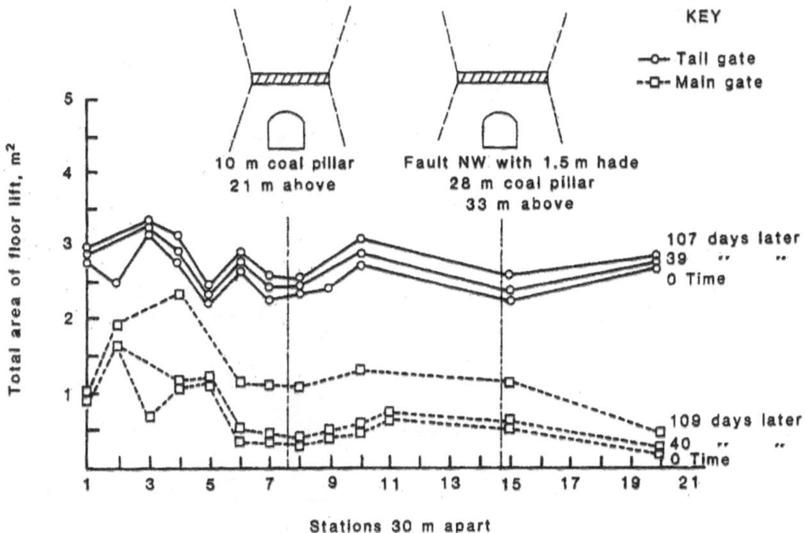

FIG. 9. Comparison of floor lift between the main and tail gates of the F1 panel.

The rate of floor lift along the F2 main gate, with an average face advance of 5 m/week, was observed to decrease with time and distance from the face line (Fig. 10).

The initial average rate of decrease was 384 cm^2/week, which gradually reduced by 75%, at 170 m from the face line. This roadway, was driven on full dip and perpendicular to the main cleat of the seam, experienced 54% more floor lift than the F1 main gate, which had the same underclay floor but was driven on the seam strike, parallel to the main coal cleat. This rate is illustrated in Fig. 11 and is characterized as follows:

(a) Beyond 180 m from the face, the average rate of floor lift decreased below 232 cm^2/week.
(b) Between 180 m to the rear and 160 m in advance of the face line, the average rate of floor lift was 245 cm^2/week, increasing with proximity to the face.
(c) At 25 m below the overlying 9 m wide pillars, the average rate of floor lift was 139 cm^2/week, which is 60% less than that experienced along the remainder of the roadway. The natural destressing effect of the overlying pillars appeared to account for a decrease in stress transfer to the underlying roadway floor.
(d) Outside the destressed zone, projected downward at 50° to the vertical from the perimeter of the pillars, the rate of floor lift was observed to triple to an average of 406 cm^2/week. This effect was also clearly apparent along the F1 roadways and F2 tail gate.

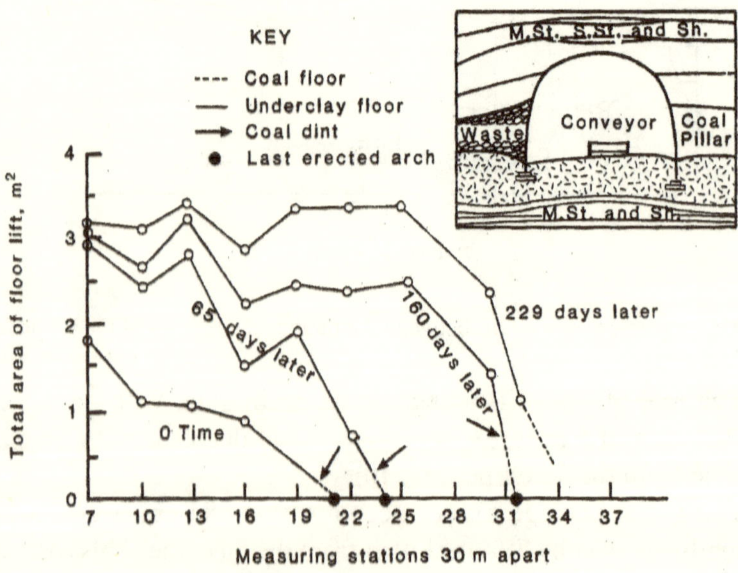

FIG. 10. Rate of floor lift along the F2 main gate, relative to distance from the face line.

Roadways driven in dry conditions along full dip and perpendicular to the main cleat of the coal seam, when used for a second time in order to serve a semi-retreat face with dinting, were observed to experience an average 31% more floor lift than when driven parallel to the strike.

The average rate of floor lift along the F3 main and tail gates, with an average face advance of 20 m/week, was 5225 and 3484 cm^2/week respectively (Fig. 12).

This contrastingly high rate of floor lift arose from the softness of the floor and extensive dinting. The rate of floor lift was 33% higher along the F3 main gate than along the tail gate because (a) the F3 main gate was driven in virgin ground, whereas the F3 tail gate was bounded by a 20 m wide pillar and worked-out panels in destressed, and (b) the effect of dinting along the F3 main gate and its 23 m of advanced heading was to decrease the frictional resistance of the floor rock to heave as well as to expose more floor area.

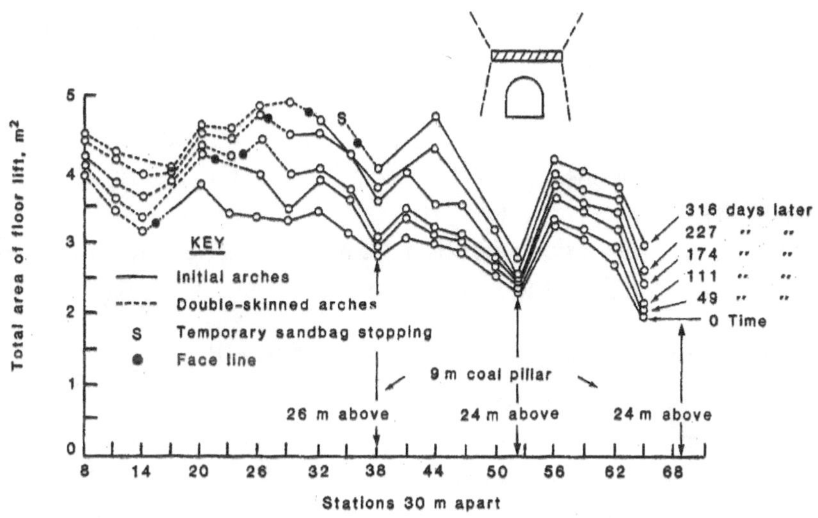

FIG. 11. Comparison of floor lift along the F2 tail gate, relative to distance from the face line.

The adverse effect of the rate of face advance on soft floor along the roadways was to increase the rate of floor lift. This was due to increased momentum in the strata loading on the soft floor. The effect of the presence of an underclay floor in dry conditions was increased floor lift—about 12 times that of the F1 main gate previously discussed. Dinting operations in such conditions gave rise to a floor lift increase of 19 times that of the F1 main gate.

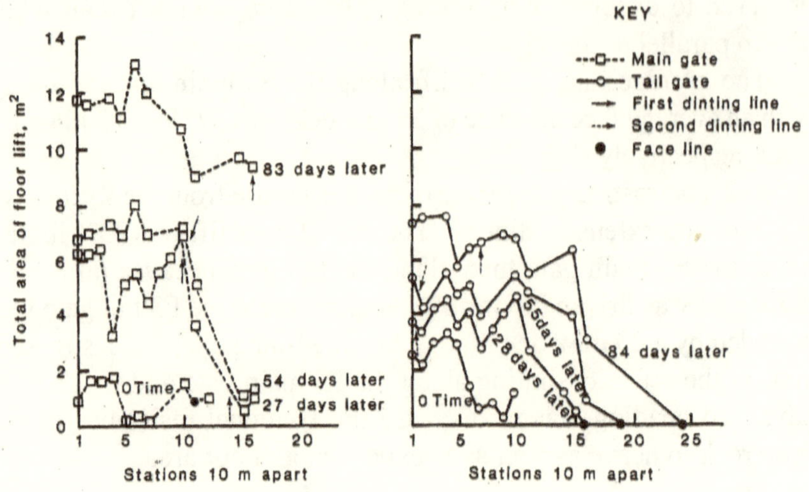

FIG. 12. Comparison of rate of floor lift along F3 gate roads, relative to the distance from the face line.

Dinting, although providing required space along the converging roadway, was not observed to furnish any long-term solution to floor instability.

The total area of floor lift observed along the gate roads is plotted against distance from the face line in Figs. 13 and 14.

These illustrate three distinct phases of floor behavior:

(a) A gradually increasing floor lift rate along the advanced headings and roadways serving the retreating face, that is, within the front and back abutment pressure zones. This was within the range of 400 m in front to 50 m to the rear of the face line.

(b) A decreasing, although significant, floor lift rate at 50 – 300 m behind the face line, that is, within the back abutment pressure zone.

(c) A general retarding floor lift due to time-dependent strata pressure and floor deformation. This creep phenomenon predominated along the full length of the roadways and was distinguishable outside the abutment pressure zones of the respective faces. It remained active until the roadway was either completely closed, or repaired and kept open for further use.

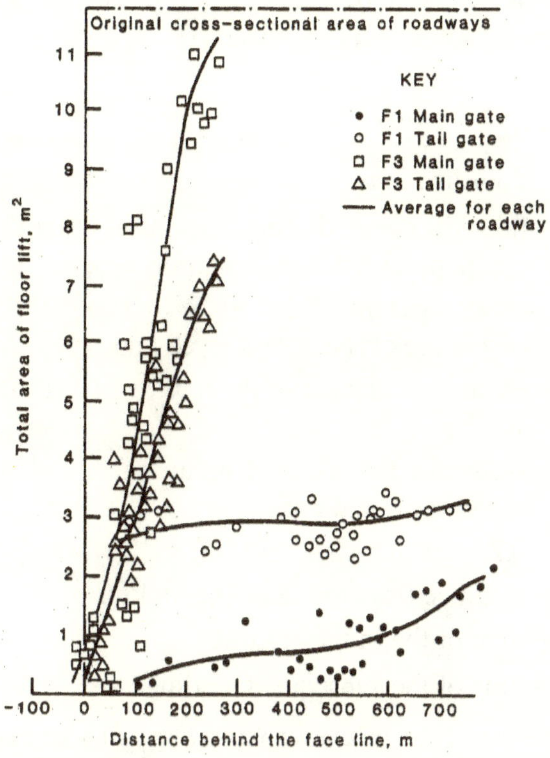

FIG. 13. Comparison of floor lift along the F1 and F3 gate roads, relative to distance from the face line.

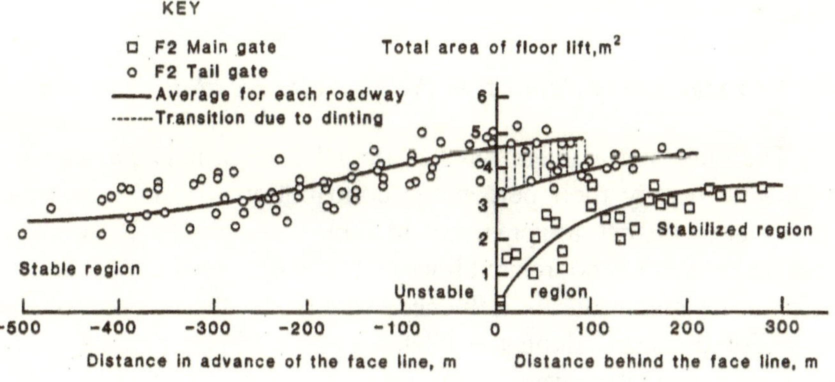

FIG. 14. Comparison of floor lift along the F2 gate roads, relative to distance from the face line.

Superimposing the above-noted three phases of floor lift, the following general equation is proposed [1]:

$$C_f = a + D^b + c \exp[-d \cdot D]$$
$$= a + [r(t_2 - t_1)]^b + c \exp[-d \cdot r(t_2 - t_1)]$$

where: C_f = floor lift, in m^3;

a = floor lift due to dinting and causes other than effect of creep, pillar, pack, and support type, in m^2;

b = creep constant = $0.1 \times 10^{-2}W$ to $0.6 \times 10^{-2}W$, dependent on the rock type, stress level, and time;

c = plasticity constant of the rock = $1.1 \times 10^{-2}W$ to $+0.5 \times 10^{-2}W$;

d = factor for the effect of pillar, pack, and support base dimensions and type
= $0.2 \times 10^{-3}W$ to $3.4 \times 10^{-3}W$;

D = distance from the face line, in km;

t_1 = time of the initial reading;

t_2 = time of the subsequent reading;

r = rate of the face advance = $D/(t_2 - t_1)$, in km/year;

W = load of the fractured roof rock on the support, measured by load cell placed under the arch legs, in kN.

Average values of the parameters and constants introduced above, utilizing a least squares method with an error of ± 16% are noted in Table 2.

4.3 Contribution of floor lift to the total closure of the gate roads

The observed average rates of floor lift (C_f) and total closure (C_t), and thereby their percentage contribution to each gate road deformation history, are presented in Table 3 and Fig. 15. It is evident that the floor lift is not related linearly to the observed total roadway closure. Its magnitude is influenced mainly by the position of the face. For instance, along the F2 tail gate, a higher percentage of floor lift contribution was observed outside the abutment zones. This indicates that other boundaries of the roadway, especially the roof, are more sensitive to abutment pressure than is the floor.

TABLE 2. Averaged parameters and constants for [1]

Location	r (m/week)	a					$b \times 10^{-2}W$					$c \times 10^{-2}W$					$d \times 10^{-3}W$				
		C_f	A_p	S	C_s	C_t	C_f	A_p	S	C_s	C_t	C_f	A_p	S	C_s	C_t	C_f	A_p	S	C_s	C_t
F1 main gate	10	0.3	0.5	—	0.4	0.6	0.3	0.1	—	0.2	0.3	0.2	0.4	—	0.3	0.4	0.2	0.4	—	0.3	0.4
F1 tail gate	10	2.3	2.5	—	1.5	2.5	0.5	0.2	—	0.3	0.4	0.3	1.0	—	0.4	1.2	0.3	0.6	—	1.3	0.7
F2 main gate	5	0.2	0.4	—	0.4	0.5	0.4	0.1	—	0.2	0.2	0.2	0.4	—	0.2	0.4	0.2	0.5	—	0.3	0.4
F2 tail gate (behind the face)	5	3.4	0.6	—	1.2	0.7	0.6	0.2	—	0.3	0.4	0.3	0.7	—	0.4	0.8	0.3	0.8	—	0.4	0.9
F2 tail gate (advanced heading)	5	4.8	0.5	—	0.7	0.7	0.2	0.3	—	0.4	0.3	0.3	0.6	—	0.5	0.7	0.4	0.5	—	0.4	0.6
F3 main gate	20	0.5	7.5	1.5	1.4	1.5	0.3	0.1	0.2	0.3	0.3	0.5	0.5	−1.1	0.4	1.5	3.4	0.3	0.4	0.9	0.7
F3 tail gate	20	0.2	6.1	1.5	1.5	1.5	0.3	0.1	0.2	0.4	0.3	0.4	0.4	−1.1	0.4	1.5	1.0	0.2	0.5	1.3	0.8

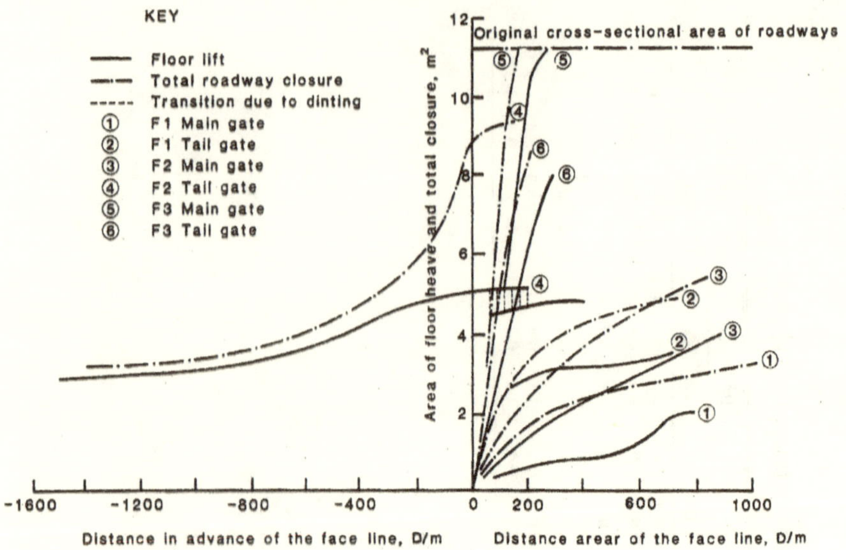

FIG. 15. Comparison of total roadway closure with floor lift for all gate roads.

4.4 Penetration of arch legs and stilts into floors

Along the F1 main and tail gates, the average penetration rate was 0.32 and 0.5 cm/week, respectively. The relatively dry conditions along the F1 main gate, driven on the strike and parallel to the main cleat of the coal seam, resulted in stable conditions with low penetration of the arch legs into floor. The effect of a 60% water accumulation along the floor of the F1 tail gate and the proximity to fractured and water-bearing strata resulted in the observed 36% greater floor penetration along the F1 tail gate than along the F1 main gate.

The effect of setting arches originally on the coal floor, then dinting the coal about 10 m behind the face along the F2 main gate, driven perpendicular to the main cleat of the coal seam in a dry-condition, was an observed 48% increase in the rate of arch penetration along the F2 main gate over that of the F1 main gate.

Along the F2 tail gate driven on full dip, conditions differed significantly from the above roadways, since this roadway was driven originally as the main gate of the previous district. Thus the majority of floor deformation and arch penetration had previously occurred. In general, the average rate of arch leg penetration along this roadway was 10% lower than that of the F2 main gate. The effect of dinting at the face line along the F2 tail gate, in addition to the above-noted features, caused a 43% greater floor penetration along the F2 tail gate than along the F1 main gate.

The support penetration into the softer underclay floor along the F3 main and tail gates averaged 5.85 and 5.08 cm/week, respectively. In addition, the arch penetration into stilts was at a rate of 2.85 and 1.40 cm/week, respectively. This yielded a total penetration and shortening of arch legs along the F3 main and tail gates of 8.70 and 6.48 cm/week, respectively. The effect of dinting was to nonlinearly increase the rate of arch penetration because of a reduction in the frictional resistance between the arch leg and floor material. Penetration measurements along the F3 roadways indicated a mean 33% increase due to dinting, in addition to the above-noted support penetration along the F3 main gate (i.e., on overall penetration rate of 11.5 cm/week).

Penetration of arch legs (A_p) and stilts (*s*) into floors is shown with respect to distance from the F3 face line in Fig. 16. The data indicate a higher support penetration into the floor along the F3 main gate than along the F3 main tail gate (Fig. 16*a*). The behaviour can be expressed empirically in the form of [1] for both arch leg (A_p) and stilt (*s*) penetration into the floor. None of the stilts utilized along the F3 gate roads traveled their full distance by the time the face had passed (Fig. 16*b*).

4.5 Side closure

The average rate of side closure along the F1 main and tail gates was 0.15 and 0.20 cm/week, respectively. The average rate of side closure along the F2 main, F2 tail, F3 main, and F3 tail gates was 0.28, 0.41, 0.64 and 0.53 cm/week, respectively.

Considering the F1 main gate as a standard, the effect of a 60% increase in wetness was to increase the closure rate by 33%. Roadways driven in a 98% dry condition, but along the full dip and perpendicular to the main cleat of the coal seam with the adoption of dinting, experienced an increase of 86% side closure. The average side closure rate along such a roadway, used for a second time and serving a semi-retreat face, would thus be expected to be 2.7 times the standard, that is, side closure would be almost double that experienced during production from the first face. Therefore, the effect of the face on side closure does not appear to diminish with the number of times a face passes through a given station along the roadway; in other words, the rate of side closure is directly proportional to the local strata pressure acting horizontally on roadways. Soft underclay floors appeared to reduce the resistance of arch legs to side pressure, thereby causing an increase of 40% in the rate of closure, compared with the F1 main gate standard. If such a roadway is extensively dinted, the rate of closure can be expected to increase a further 27%, giving an overall rate of side closure equal to 67% that of the standard. This shows the adverse effect of dinting on side closure: larger side walls are created, which then experience increased loading and closure rates.

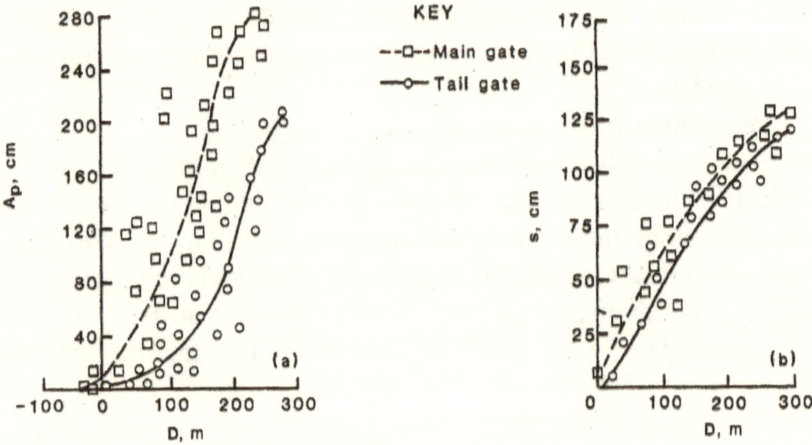

FIG. 16. Comparison of arch leg penetration into floor (*a*) and stilt (*b*) along the F3 gate roads.

The effect of face location on total side closure along all roadways was such that the rate of closure tended to diminish with distance from a face line (Fig. 17). Side closure appeared to have stabilized at 150 and 250 m behind the F1 face along the main and tail gates, respectively (Fig. 17*a*), and 220 m behind the F2 face along its main gate (Fig. 17*b*). The closure along the F2 tail gate appeared to gradually accelerate at about 400 m in advance of the face line. After the face passed, closure gradually decreased to stabilize at about 75 m behind the face. Since some arches along the F2 gate were already distorted owing to the effect of the previously worked face and the effect of time, much of their initial resistance appeared to have been overcome by the time the F2 face approached. The resistance of these distorted arches to the pressure induced by the approaching F2 face appeared to be much less than that exhibited by the new arches along other roadways in the same seam (Fig. 17). Hence, the resultant front area of influence of the F2 face appeared to be much larger than its rear. The limited data available were not adequate to locate the extent of influence of the F3 face on its roadways. It appears, from extrapolation of Fig. 17*c*, that the region lies at about 300 m behind the face.

In order to reduce side closure in roadways, splayed leg arches may be introduced, but their use has an adverse effect on the function of stilts. Where stilts are adopted, in order to provide yield for prevention of arch distortion, they would appear to be better used with straight legs, with matching of the yield properties of the stilt to that of the arch, floor, and the strata pressure. The behaviour of the sides of the gate roads, Fig. 17, was comparable to that for floor lift. An empirical equation was similarly fitted to characterize the side closure. The average evaluated parameters are noted with the ± 16% error in Table 2. The average rate of side closure (C_s) and its percentage contribution to the total closure of each of the gate roads appears in Table 3.

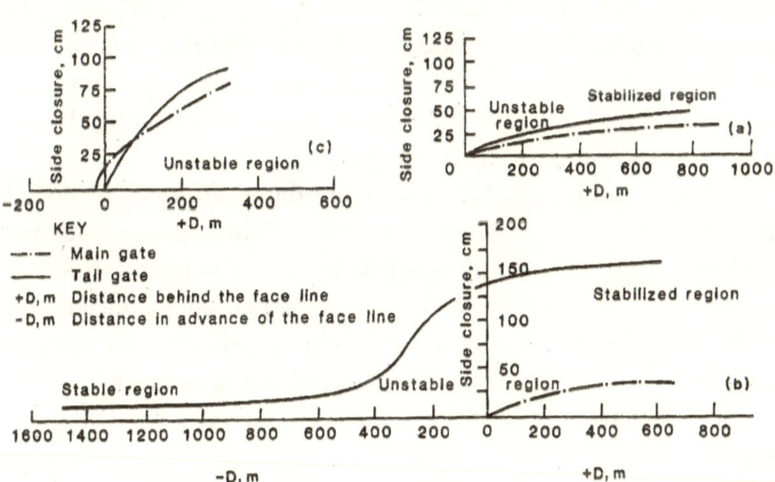

FIG. 17. Comparison of side closure of all gate roads investigated: (*a*) F1 roadways; (*b*) F2 roadways; (*c*) F3 roadways.

TABLE 3. Average rate of floor lift (C_f) and side closure (C_s), and their percentage contribution to the total roadway closure (C_t)

Location	C_t (cm²/week)	C_f (cm²/week)	$100C_f/C_t$	C_s (cm²/week)	$100C_s/C_t$
F1 main gate	705	282	40	30	11
F1 tail gate	454	295	65	40	14
F2 main gate	805	338	42	56	17
F2 tail gate	864	400	46	82	21
F3 main gate	7256	5224	72	128	3
F3 tail gate	7973	5581	70	106	2

5. Conclusions

A solid pillar left on one gate road side with a softer pack on the other side causes asymmetrical closure. Pillars left in overworked areas tend to transfer the strata pressure towards the floor of underlying roadways at an angle of between 50 and 66° to the vertical, projected downwards around the periphery of the pillar bases. The closure experienced in roadways below pillars due to such destressing was generally less than the average observed along the gate roads. Roadside pack rigidity exhibited an appreciable effect on observed floor lift. When packed very tightly, indications were that they tend to decrease side closure and increase soft floor lift. Gate roads driven along the seam strike and parallel to its main cleat appear to exhibit greater stability than those driven along the seam dip and perpendicular to its main cleat.

Dinting medium strength underclay increased rates of floor lift and side closure along gate roads by up to 33 and 86%, respectively.

The presence of water along a gate road resulted in increased arch leg penetration rate into floors. Linear extrapolation of data suggested that for every 1% increase in wetness of gate roads the rate of closure increases by 0.5 – 0.7%, up to the point of strata saturation.

The effect of an advancing longwall face on gate road closure appeared to be limited to within 150 – 250 m behind the face. This effect for a semi-retreating longwall face was from 75 m behind the face line and up to 400 m in advance of the face. Within the above-noted limits, the effect of face advance rate on the soft floor was to increase the rate of floor lift along roadways.

Gate roads used for a second time, to serve a semi-retreating face, exhibited an overall rate of arch leg penetration into floors 43% lower than on initial usage.

The contribution of soft and weak underclay floor to the total closure of a typical dry coal measures strata sequence was observed to be 70%. A dry coal floor along gate roads appeared to contribute up to 71% of total roadway closure. Total gate road closure can be empirically formulated as:

$$C = a + [r\,(t_2 - t_1)]^b + C \exp\,[-d \cdot r\,(t_2 - t_1)]$$

The overall behaviour of the ground was elastoplastic when in a dry condition and plastic when moist. However, when soft floors were encountered, then the ground behaviour was plastic. A design chart for coal gate roads, relating ground conditions to percentage roadway closure, is presented in Fig. 18.

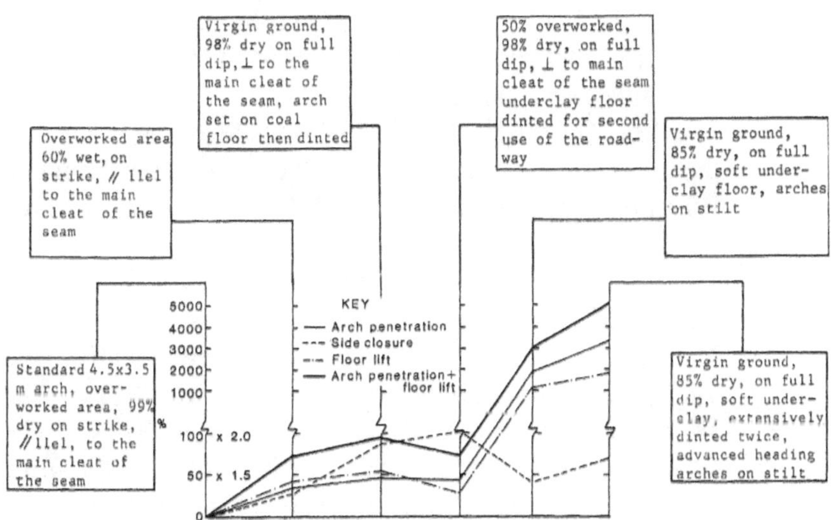

FIG. 18. Design considerations for gate roads.

References

1 - AFROUZ, A. 1975*a*. Yield and bearing capacity of coal mine floor. International Journal of Rock Mechanics and Mining Sciences & Geomechanics Abstracts, **12**: 241 – 253.

—— 1975*b*. Floor behaviour along longwall roadways. International Journal of Rock Mechanics and Mining Sciences & Geomechanics Abstracts, **12**: 229 – 240.

2 - ASTON, T., and CAIN, P. 1984. Strata control investigations in the Sydney Coalfield. Canada Center for Mineral and Energy Technology, Coal Research Laboratories, division report ERP/CRL 84-3(IR), p. 24.

3 - BROWN, E. T., and BRAY, J. W. 1982. Rock support interaction calculations for pressure shafts and tunnels. International Society for Rock Mechanics Symposium, Aachen, pp. 555 – 565.

4 - CAIN, P., and ASTON, T. 1983. Strata control investigations in the Sydney Coalfield. Canada Center for Mineral and Energy Technology report ERP/CRL 83-12(IR), p. 17.

5 - GOETZ, W. 1982. Planning roadways from the point of strata mechanics. Glueckauf, **118**: 13 – 23.

6 - JOHNSON, G. 1967. The closure of gate roads related to the method of goaf control in the Low Main Seam at Bilsthrope Colliery. National Coal Board – Mining Research Establishment Report 2309.

7 - JONES, C., and SMITH, D. S. G. 1966. The mid-ordinate method of measuring mine roadway cross-sectional area. National Coal Board – Mining Research Establishment Technical Memorandum 202.

8 - JUKES, S., HASSANI, F. P., and WHITTAKER, B. N. 1983. Characteristics of steel arch support systems for mine roadways (Part I). Mining Sciences & Technology, **1**: 43 – 56.

9 - LADANYI, B. 1980. Direct determination of ground pressure on tunnel lining in a non-linear viscous-elastic rock. Underground Rock Engineering, 13th Canadian Rock Mechanics Symposium, pp. 126 – 132.

10 - SHEAREY, P. R., and DUNHAM, L. K. 1978. An approximate analysis of floor heave occurring in roadways behind

advancing longwall faces. International Journal of Rock Mechanics and Mining Sciences & Geomechanics Abstracts, **15**: 277 – 288.

11 - SHEPHERD, R. 1965. Conclusions from recent investigations of strata behaviour on mechanized coalfaces. National Coal Board – Mining Research Establishment Report 2278.

12 - THOMAS, L. J. 1968. The effects of adjacent seams and method of working on roadway closure in the Main Bright Seam at Hucknall Colliery. National Coal Board – Mining Research Establishment Report 2330.

13 - WHITTAKER, B. N., and BONSALL, C. J. 1981. Design aspects relating to the stability of coal mining tunnels. *In* Stability in underground mining. Pergamon Press, London, United Kingdom, pp. 519 – 533.

14 - WHITTAKER, B. N., and SINGH, R. N. 1979. Evaluation of design requirements and performance of gate roadways. Mining Engineer (London), **138**: 535 – 553.

3.3.11 METHODS TO REDUCE FLOOR HEAVE AND SIDES CLOSURE ALONG THE ARCHED GATE ROADS

By: Andy A. Afrouz

Abstract

Back analysis of experimental results from the laboratory and longwall gate roads have been utilized to elaborate on the causes of floor heave and sides closure along the roadways. Techniques to control and reduce the floor heave and sides closure along mine gate roads with soft floor is discussed. Appropriate, practical and theoretical suggestion is given. Application of some of the techniques, especially those utilizing resin or grout injection into the boreholes, are in the mechanized, high productive and continuous mining necessitating continuous ground reinforcement or support system.

Introduction

Closure of gate roads with soft floor is a problem encountered in the carboniferous coal measures of the North America and Europe. Parameters influencing the stability and performance of the gate roads are numerous. They are often directly or indirectly inter-related. The most important of these are:

(a) Depth of cover and varying strata pressure.
(b) Dimensions and shape of the gate roads.
(c) Method of the gate road support and its size.
(d) Direction of the roadway advancement relative to the dip of the seam.
(e) Rate of advance.

(f) Geological factors, such as stratigraphy, discontinuities, cementation between the discontinuities, washouts and existence of the tectonic actions.
(g) Type of the roadside packs, its dimensions and behaviour as a support.
(h) Barrier pillar, its dimensions and behaviour.
(i) Panel width or the face length.
(j) Extracted height of the seam.
(k) Proximity of any mining activity to the gate roads.
(l) Method of the roadway development and the production technique.
(m) Rock/support interaction properties.
(n) Non-uniform stress distribution in the support system.
(o) Bearing capacity of the roadway floor.
(p) Type and quality of the support or reinforcement utilized along the roadway and behind the face.
(q) Support intervals along the roadways.
(r) Influence of water on the floor material.
(s) Physical, mechanical and time-dependent behaviour of the rocks surrounding the roadway.
(t) Influence of any ripping or dinting on the neighboring supports, i.e. pillar, pack and steel arches.
(u) Position of the roadways relative to the faceline and its direction of advance (or retreat) e.g. advanced heading, behind the face line with ripping lip, or in-line with the face.

Assessment of all the above noted parameters is prerequisite to a sound prediction, planning and the ground control along the gate roads. There has been several attempts on the basis of theoretical analysis and on the assumptions of circular roadways to evaluate the above noted parameters [1–3]. However, these results, although useful, do not close fit the in situ performance of gate roads with arched cross-section.

This paper is a result of laboratory and field data relating to the performance of arched gate roads with soft floors serving longwall faces in the Sydney Coalfield, Nova Scotia, Canada, and the South Wales Coalfield, U.K.

Causes of floor heave and sides closure along the gate roads

Inherent weakness of the rock - The floor beneath the coal seams is usually underclay. It consists of clay minerals with abundant randomly oriented stigmarian rootlets, carbonaceous layers and sometimes pyritic nodules [4]. The underclay thus formed has low uniaxial compressive strength values in the range of 10–18 MPa [5,6]. This soft rock is usually slickensided which promote flow of the floor under relatively low strata pressure [7].

Effect of water - There are 5 main groups of clay minerals present in the underclay floor of the coal seams. These are: kaolinite, montmorillonite, illite, chlorate and vermiculite. These minerals, especially the first two, have great water absorption properties, of up to 3 times their dry volume, with resulting swelling, cracking and disintegration of the rock mass.

Ground pressure - This is one of the main factors causing deformation and consequently floor heave and sides closure. This phenomenon is mostly encountered in the relatively deep gate roads, with more than 700 m cover depth. Roadways located under remnant pillars also experience substantial floor lift due to presence of zones of high ground pressure projected from the perimeter of the pillars downwards with an angle of fracture expressed as follows [5,6]:

$$\beta = k[(\pi/4) + (\phi/2)] \tag{1}$$

where: β = average fracture angle of the ground to horizontal. (= 24–40°; see Fig. 1);

ϕ = average angle of internal friction of the composite strata to horizontal

(= 10–30°); k = constant relating the laboratory uniaxial test results of the intact rock samples to the in situ measurements on the rock mass (= 0.4–0.8).

Fig. 1. Fracture characteristics of coal (left) and underclay (right) samples showing influence of the bedding. Dimensions of the samples: length 100 mm, diameter 50 mm.

Size and shape of the gate roads - Wider roadways are more prone to floor lift. On the other hand, increase in the height of a roadway, relative to its width results increase in the sides closure. This is more predominant in the deep workings [5]. Mining operations necessitate a minimum cross-sectional area for the gate roads, especially for the ventilation and materials handling purposes. Therefore, knowledge of the rock types encountered in the floor and sides of the gate roads—cover depth—magnitude, direction and ratio of the major to minor strata pressures on the gate roads can be utilized to determine the ratio of width to height of the gate roads in the planning stage.

Effects on the type and dimensions of the supports, packs and pillars along the gate roads - These can be classified into one or a combination of the following:

(1) Penetration of the arch legs into very soft floor (Fig. 2a)—in this case the floor exhibits no resistance to the load exerted on the arched support.
(2) Floor heave in the soft underclay (Fig. 2b)—which increase with moisture content of the rock mass. For asymmetrical ground loading, the floor heave is higher towards the more solid rib pillar or barrier pillar than the pack side.
(3) Buckling or elasto-plastic type of floor lift (Fig. 2c)—this occurs in the medium strength rocks under relatively dry mining conditions. It is generally accompanied with cracks or fractures.

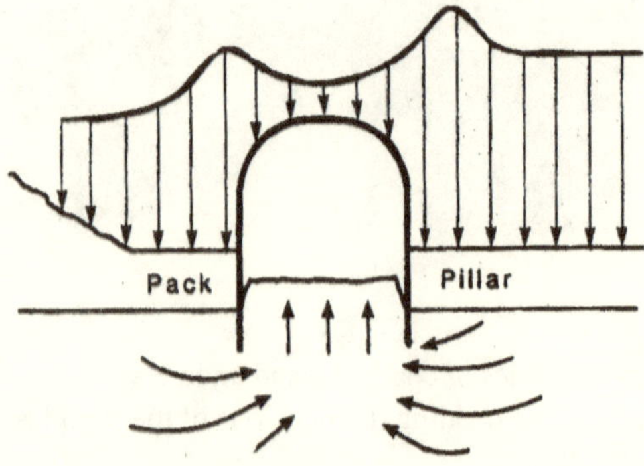

(a) Penetration of each leg into very soft floor.

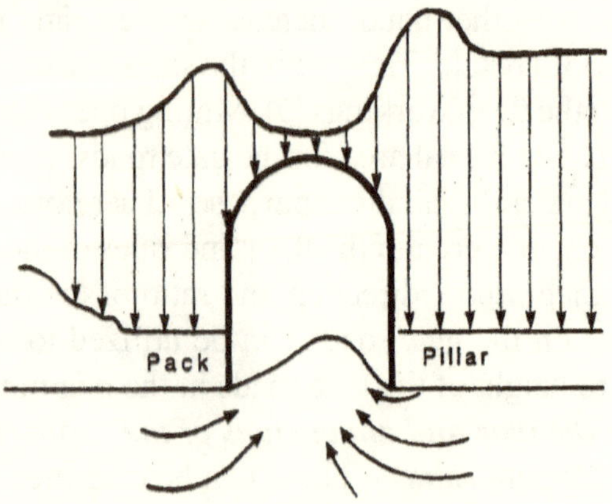

(b) Soft floor lift, mainly nearer to the pillar side.

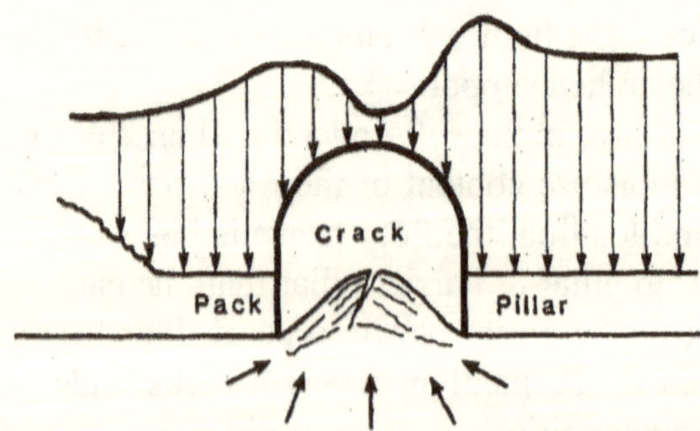

(c) Buckling floor lift in the medium strength rocks

Fig. 2. Effects of the support, rib pillar and the gob-side pack on the floor along the gate roads.

Figure 2d. An outburst from the floor of a gate road (20 meters from the lip looking towards the production face) in Ireland Mine, North Derbyshire, East Midland, U.K.

There are also two main causes of sides closure due to the effects of the ground pressure on the supports, pillars and packs. These are:

(a) Buckling of the arch leg (Fig. 3a)—this occurs generally in the more solid pillar side than the weaker pack side.
(b) Convergence of the arch legs (Fig. 3b)—this occurs usually where the horizontal ground pressure is high, there is no or little difference between the magnitude of the horizontal ground pressure exerted to the sides of the same arches or the strengths of the pillar and the pack are designed to be close to each other.

(a) Buckling of the arch leg in the pillar side of a weak roadside pack

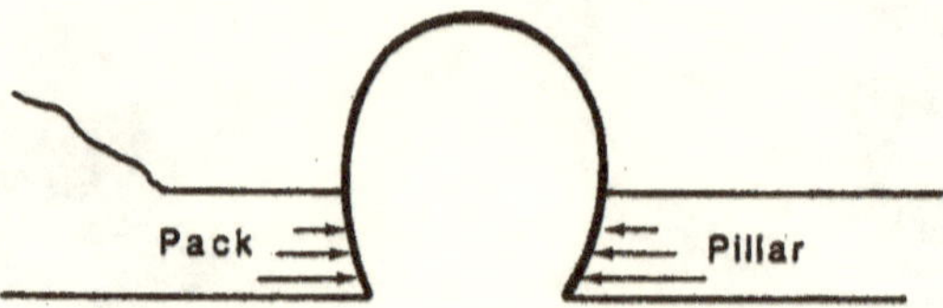

(b) Convergence of the arch legs along the hard packed gate roads having stronger floor

Fig. 3. Main types of sides closure along gate roads.

Methods of reducing floor heave and sides closure

Under the tremendous ground pressure around longwall faces. It is not economically possible to construct and support the gate roads so that they stay intact over their utilization period. Therefore, certain amount of deformation should be planned for when designing shape, dimensions and type of the support requirements. Hence, the purpose is not to eliminate the floor lift and sides closure completely, but to minimize them to acceptable limits. This can be achieved by the following techniques:

Provisions for minimum utilization of water on the soft floors

In situ measurements on coal and underclay floor [8] indicated that introduction of water to the dry floor along the gate roads increased their floor heave by an average of 18–30%, respectively. The effect was even greater where the floor was water saturated or when clay bands were present in the underclay. Where reduction of water on the floor is not possible (e.g. seepage of the groundwater to the gate road), spraying chemicals or organic additives which will increase surface tension of the floor material can be helpful. This will make the floor water repellent.

Installation of steel base plates

To reduce penetration of the arch legs into the soft floor. The steel base plates should have rounded corners and edges, secured under the arch legs. Dimensions of the base plates can be matched to the bearing capacity of the particular material and the prevailing strata load. Load/deformation relationship of the floor beneath the base plates can be expressed as follows [5,6]:

$$\delta = K_1 \cdot P^n = l \cdot P / A \cdot E \qquad (2)$$

where: δ = deformation or settlement of the floor, mm;

P = strata vertical load applied on the base plate, MN;

l = thickness of the floor layer affected by the strata loading, m;

A = area of contact between the base plate and the floor, m^2;

E = average Young's modulus of the floor material, G Pa.

K_1 and n are proportionality constants varying between zero to one; depending upon: the area (A) and perimeter (p) of the base plates, strength and Young's modulus of the floor material.

The ultimate bearing capacity of the floor beneath the base plate (q) is given empirically [6]:

$$q = -K + m\,(p/A)^n \qquad (3)$$

where: K = compressive stress in the floor directly under the base plate
(= –2.6 to –2.8, MPa; the negative sign indicates the compression);

m = proportionality constant, dependent on the perimeter of the base plates
(= –0.2 to +0.2, MPa).

Base plates with bearing area of A = 0.1 m^2 and 5 mm thickness are satisfactory under the ground pressure of up to 25 MPa, if the gate road is used once over a planned life of 1.5–2 years [8]. To prevent slippage, the plates should be provided with bolting facilities to bottom of the arch legs. The wooden chocks conventionally utilized beneath the arch legs tend to deteriorate or deform after which they will not be functional.

Elimination of advanced headings wherever abnormally soft floor is encountered

This will reduce the period for which the gate road is needed to stay open, thereby reducing the total floor lift along the roadway. In general, floor lift along the gate roads with advanced heading and soft floor are expected to be 33% higher than those without the advanced heading, under similar ground conditions [6].

Elimination of either main or tail stables

This measure reduces the total time and manpower utilized per face advance. Higher rate of advance results in shorter life requirement for the production panel, which subsequently leads to shorter life requirement for the gate roads. This measure does not stop the floor lift, but indirectly reduces its total heave.

Reduce asymmetrical floor lift along gate roads

Consideration may be given to provide yield facilities for the pillar side of the roadways to be equal to that of the pack side. One method of achieving this is by cutting a slot of 0.5–1 m deep and 5–20 cm high in the solid rib along the roadways before the arches are installed. Then pack the slots with wood chocks or concrete blocks. This action will transfer the peak stress deeper into the coal rib and away from the immediate vicinity of the gate road. In doing so, it will yield and equalize the ground pressure on both sides of the gate roads, thereby providing a more even stress distribution across the roadways. To facilitate the even stress distribution is site specific. It can only be done by trial and error, by changing the dimensions of the slot cut, type and quality of packing the slot to provide the same yield characteristics as the of the opposite side of the gate road.

Facilitate symmetrical yielding on both sides of the gate roads

This can also be achieved by designing the pack, to be place in the gob side of the roadway, so that it will provide a similar overall yielding (or stiffness) properties to the opposite side. Various commercially available monolithic and non-monolithic pack materials such as: Shotcrete, Aqualit, Aquapak, Astrapak, Tekpak, Anpak, Anfl and Antail can be used.

Complete removal of the soft floor

Generally, gate roads with soft floor require at least one and upto 3 times dinting, or back ripping, to keep the road open during the production life of their corresponding faces. The total thickness of dinting in such cases may well exceed 1.5–2.0 m of soft floor during the initial 6–8 months of the production period. Two situations may arise, requiring different technique to be used. These are:

(1) If the gate road has competent roof and soft floor, it is advantageous to leave the roof undisturbed and switch from ripping to dinting the soft floor down to a more stable bed rock. It is suggested to plan to remove the soft floor in one operation, during the roadway drivage. This technique will eliminate the conventional and hazardous ripping operation and lend itself to mechanized dinting of soft floor. Therefore, the overall efficiency of the roadway development, maintenance and safety will increase. To counter sides closure along the deepened gate roads. splayed leg arches can be used with lagging board or concrete blocks superimposed with wire mesh at the interface between the arches and the rock. The initial cost of the extra wire mesh is less than the outlay for conventional dinting or ripping operations. Moreover, the harder floor of a deepened gate road is more adaptable to reinforcement than a soft and weak floor, if it becomes necessary.

(2) If competency of the floor is the same or higher than that of the roof along the gate roads, it is advisable to resort to the ripping operation and only remove the obstructing bulges due to the floor heave.

Utilization of inverted arches

This technique is applicable where no dinting is anticipated through the life of the gate road. It will increase the total bearing of arch on the floor and reduce the sides closure or distortion of the arch legs. Splayed leg arches may also be used to counter the sides closure. They are less expensive than the inverted arches; but they have adverse effect on the function of stilts, if they are used, and are less effective then the inverted arches.

Floor reinforcement

Floor reinforcement will increase the inherent strength of the floor material, improving its stability. Bolting, dowelling and injection are the main techniques utilized.

Rock bolts and especially cable bolts are suitable for hard and relatively competent floor layers having weak bonding between the layers or joints. So the bolts knit the strata together. Soft or weak floors, especially in the coal mines do not efficiently respond to the steel bolting. Dowelling is modified version of bolting, where the bolts are wooden (or plastic) and the column is filled with cement grout or resin to hold the wooden bolts to the rock. To obtain maximum benefit from this type of reinforcement, the dowel should be installed as soon as the floor of the gate road is exposed and before any major floor movement or fracture is developed. The advantage of the wood dowelling over the conventional steel bolting is less cost and more compatibility to mechanized or continuous dinting, if it becomes necessary. This technique can also be utilized to support roof and walls on a continuous or mechanized basis.

Alternatively, all major existing fractures or open joints in the soft floor can be reinforced, as soon as possible, by the injection of cement, resin or other adhesive materials into boreholes. The injection pressure can be 20–60 kPa. This technique is more costly than the dowelling, but provides a more uniform reinforcement throughout the floor. It is adoptable to continuous mining systems and for roof or wall reinforcement. Araldite resin reinforcement with a volumetric ratio of Resin/Hardener/Plasticizers of 100:5:1 was applied to the underclay samples of 100 mm length and 50 mm diameter, under laboratory conditions, and cured in a well ventilated room temperature of 19–21°C for 24 hours. Comparison of the averaged unconfined compressive strength test results have shown:

σ_s = U.C.S. of the untreated underclay sample = 10.0–18.5, MPa
σ_t = U.C.S of the reinforced underclay sample = 12.8–20.1, MPa
σr = U.C.S. of the resin sample cured for 24 hours = 75.0, MPa

The following empirical expression holds within ± 17% error for the resin injected underclay samples:

$$V_t \cdot \sigma_t = B(V_s \cdot \sigma_s + V_r \cdot \sigma_r) \tag{4}$$

where: V_t = total volume of the reinforced samples, m³ (= $V_s - V_b + V_r \approx V_s$);
V_s = volume of the untreated underclay sample, m³;
V_b = volume of the borehole for the resin injection, m³;
V_r = volume of the resin used for reinforcement, m³; (= 0.1 V_s);
B = constant related to the efficiency of bonding between the rock and the resin (= 0.5–1).

Determination of the strength and amount of the reinforcement needed to the soft floor under the in situ conditions can be expressed as follows:

Considering the most critical condition in the weak floor/support stability, as shown in Fig. 4.

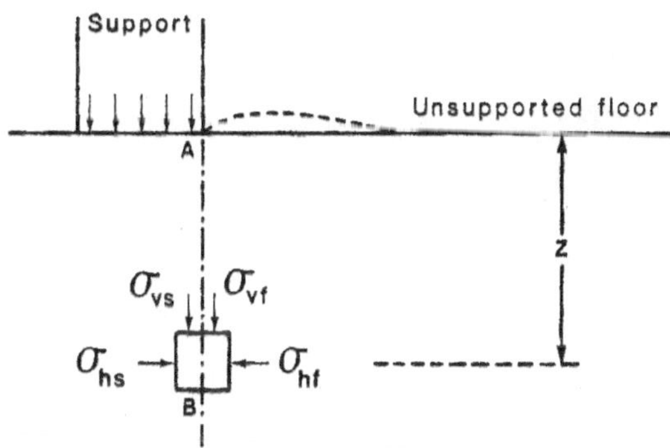

Fig. 4. Schematic of the support base and soft floor interaction at a depth *z* below the support perimeter.

This condition occur when the floor under the support is failing and there is no considerable shear on the vertical plane *AB* extended from the support base perimeter to the floor. The vertical stress of an element of the floor at depth *z* beneath the support (σ_{vs}) can be expressed as:

$$\sigma_{vs} = Q + \gamma \cdot z \tag{5}$$

where Q = load intensity of the strata on the floor beneath the support, which is higher than the bearing capacity of the soft floor, MPa;

γ = weight per unit volume of the floor material, MN/m³.

The corresponding horizontal stress (σ_{hs}) using Terzaghi's [9] expression is:

$$\begin{aligned}\sigma_{hs} &= \sigma_{vs} \tan^2(45 - 0.5\phi) - 2S_0 \tan(45 - 0.5\phi) \\ &= (Q - \gamma \cdot z) \tan^2(45 - 0.5\phi) - 2S_0 \tan(45 - 0.5\phi)\end{aligned} \tag{6}$$

where: ϕ = average internal friction angle of the floor rock, degree;

S_0 = Cohesion of the floor rock, MPa.

If the floor along the gate roads is reinforced uniformly, with an injection material of unconfined compressive strength (σ_r), then on the unsupported side of the reinforced floor, the vertical stress (σ_{vf}) at a depth *z* using eqn. (4) can be given as:

$$\begin{aligned}\sigma_{vf} &= \sigma_t - \gamma \cdot z \\ &= [B(V_s \cdot \sigma_s + V_r \cdot \sigma_r)/V_t] - \gamma \cdot z\end{aligned} \tag{7}$$

The corresponding horizontal stress (σ_{hf}) is:

$$\begin{aligned}\sigma_{hf} &= \sigma_{vf} \tan^2(45 - 0.5\phi) + 2S_0 \tan(45 - 0.5\phi) \\ &= (\sigma_f - \gamma \cdot z) \tan^2(45 - 0.5\phi) + 2S_0 \tan(45 - 0.5\phi)\end{aligned} \tag{8}$$

For the floor to remain stable, the induced horizontal stresses on both sides within the floor should be equal, i.e.:

$$\sigma_{hs} = \sigma hf \tag{9}$$

Substituting equations (6) and (8) into eqn. (9) gives:

$$(Q - \gamma \cdot z) \tan^2 (45 - 0.5\phi) - 2S_0 \tan(45 - 0.5\phi) = (\sigma_t + \gamma \cdot z) \tan^2(45 - 0.5\phi) + 2S_0 \tan(45 - 0.5\phi)$$

Therefore, the minimum required strength of the resin reinforced floor (σ_t) is:

$$\sigma_t = Q - [4S_0/\tan(45 - 0.5\phi)] \quad (10)$$

Substituting eqn. (4) into eqn. (10) for the required U.C.S. of the resin (σ_r):

$$\sigma_r = \{V_s \cdot \{Q - [4S_0/\tan(45 - 0.5\phi]\} - B \cdot V_s \cdot \sigma_s\}/V_r \quad (11)$$

Comparison of the required strength of the resin for the rock reinforcement along the gate roads under various conditions is given in Table 1 and illustrated in Fig. 5.

Compressive strength (σ_r) and elastic moduli (E_r) for some commercially available resins and polymers is given in Table 2. These can be prepared to have different mechanical properties, to be cuttable by rock cutting machines in the field. They can be cost effective where high production and continuous reinforcement or support system is sought.

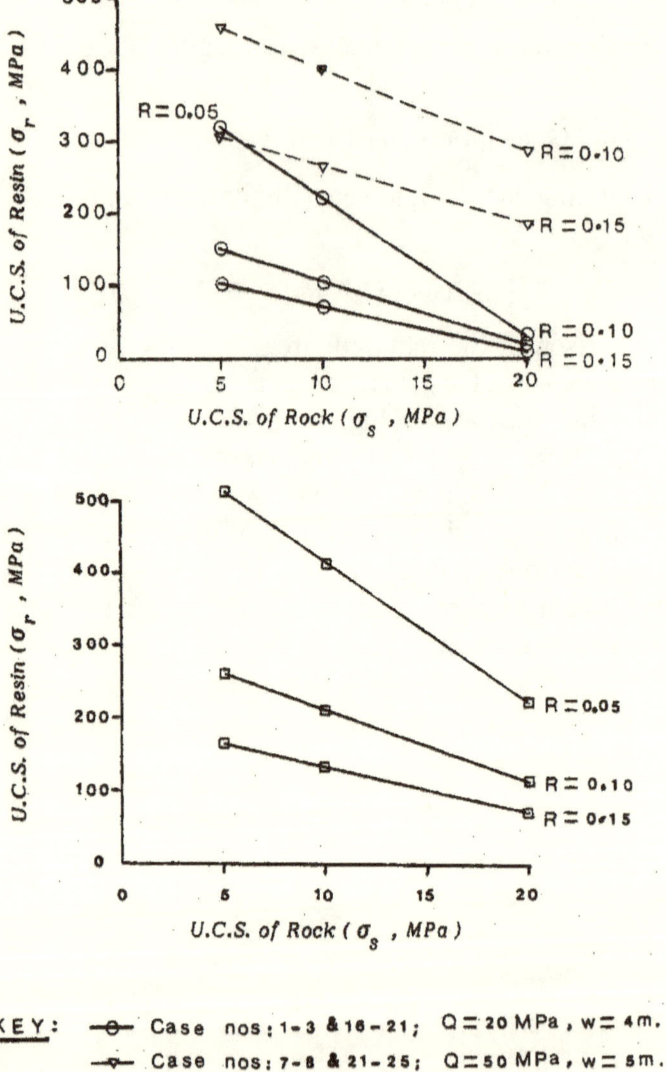

Fig. 5. Charts for selecting the U.C.S. of resin reinforcement under various ground conditions.

TABLE 1 The required resin strength for the floor reinforcement along the gate roads under various conditions utilizing equation (11)

Case No.	Strata pressure on the floor, Q (MPa)	Volume of rock to be reinforced, V_s (m^3)	Width, w (m)	Volume of reinforcement, V_r (m^3)	$V_r/V_s = R/3$	Bonding eff. of resin, B	U.C.S. of rock mass, σ_s (MPa)	Internal angle of rock, ϕ (°)	Cohesion of rock, S_0 (MPa)	Required U.C.S. of resin, σ_r (MPa)	Remarks
1	20	500×4×0.5	4	150	0.15	0.7	5	10	0.1	106.8	favourable
2	20	500×4×0.5	4	100	0.10	0.7	5	10	0.1	160.2	
3	20	500×4×0.5	4	50	0.05	0.7	5	10	0.1	320.5	not economically feasible
4	30	500×5×1.0	5	375	0.15	0.8	10	20	0.2	139.0	favourable
5	30	500×5×1.0	5	250	0.10	0.8	10	20	0.2	208.6	
6	30	500×5×1.0	5	125	0.05	0.8	10	20	0.2	417.1	not economically feasible
7	50	1000×5×1	5	750	0.15	0.9	20	30	0.4	194.9	
8	50	1000×5×1	5	500	0.10	0.9	20	30	0.4	292.3	
9	50	1000×5×1	5	250	0.05	0.9	20	30	0.4	584.6	not economically feasible
10	30	500×5×1.0	5	375	0.15	0.8	5	10	0.1	170.2	
11	30	500×5×1.0	5	375	0.15	0.8	20	30	0.4	74.9	favourable
12	30	500×5×1.0	5	250	0.10	0.8	5	10	0.1	255.2	
13	30	500×5×1.0	5	250	0.10	0.8	20	30	0.4	112.3	favourable
14	30	500×5×1.0	5	125	0.05	0.8	5	10	0.1	510.5	not economically feasible
15	30	500×5×0.1	5	125	0.05	0.8	20	30	0.4	224.6	
16	20	500×4×0.5	4	150	0.15	0.8	10	20	0.2	72.4	favourable
17	20	500×4×0.5	4	150	0.15	0.8	20	30	0.4	8.2	favourable
18	20	500×4×0.5	4	100	0.10	0.8	10	20	0.2	108.6	favourable
19	20	500×4×0.5	4	100	0.10	0.8	20	30	0.4	12.3	favourable
20	20	500×4×0.5	4	50	0.05	0.8	10	20	0.2	217.1	
21	20	500×4×0.5	4	50	0.05	0.8	20	30	0.4	24.6	favourable
22	50	1000×5×1	4	750	0.15	0.8	5	10	0.1	303.5	not economically feasible
23	50	1000×5×1	4	750	0.15	0.8	10	20	0.2	272.4	
24	50	1000×5×1	4	500	0.10	0.8	5	10	0.1	455.2	not economically feasible
25	50	1000×5×1	4	750	0.10	0.8	10	20	0.2	408.6	not economically feasible

TABLE 2. Compressive strength (σ_r) and elastic moduli (E_r) for some commercial polymers at 19–21°C

Type of polymer	σ_r (MPa)	E_r (GPa)
PVC	30	2.1
Hard rubber	40	2.8
Polymer-cement	45	3.2
Epoxy Resin	52	3.2
Polyester	93	1.9
CR 39	160	1.7
Makolon	270	2.6
Plexyglass	380	3.5
Prespec	380	3.3
Cataline 61–893	470	4.2
Araldite MY753	490	4.5
Castolite	510	4.8
Araldite CT 200	570	4.8

For the equilibrium to be reached throughout the floor along the gate roads, it is necessary that the total strength of the reinforced floor (σ_t) be greater than the strata pressure (Q) expressed as:

$$Q = \sigma_t (FS) \tag{12}$$

where: (FS) = the minimum required factor of safety for the floor (= 1.05–1.5).

Substitution of eqn. (12) into eqn. (11) provides the required volume of the resin as follows:

$$V_r = \{V_s\{\sigma_t(FS) - [4S_0/\tan(45 - 0.5\phi)]\} - B \cdot V_s \cdot \sigma_s\}/\sigma_r \tag{13}$$

Equations (11) and (13) provide an approximate guidance on the quality and quantity of resin to be used to reinforce a floor of known physical and mechanical characteristics and dimensions, under known ground pressure. In practice, if the result of calculation from the eqn. (13) is a negative value, it indicates the actual volume of the resin to be utilized, for a given volume of the floor to be reinforced. If the result of eqn. (13) is positive, then there is a surplus strength in the floor to maintain its equilibrium without any reinforcement. In the latter case, there is no need for the floor reinforcement, unless the ground conditions is changed so that a negative result is obtained.

Practical example

Floor of a gate road at depth of 650 m. Consists of 90 cm thick underclay, on average. Following averaged data is available:

Weight per unit volume of the underclay (γ) = 23, MN/m^3

Unconfined Compressive Strength of the underclay (σ_s) = 15, MPa

Constant relating the U.C.S. of underclay to that of the in situ rock mass (k) = 0.5

Internal friction angle of the underclay to horizontal (ϕ) = 25°

Cohesion of the underclay (S_0) = 0.3, MPa

Strata load intensity on the floor beneath the support (Q) = 36, MPa

Efficiency of bonding between the underclay floor and the resin reinforcement (B) = 0.9

Determine:

(a) The expected fracture angle of the ground to horizontal (β).
(b) The vertical stress in the untreated floor at a depth z = 0.9 m beneath the support (σ_{vs}).
(c) The minimum required strength of the resin reinforced floor (σ_t).
(d) The minimum factor of safety (FS).
(e) The required unconfined compressive strength of the resin, if the underclay floor is to be reinforced with a 10% volumetric ratio of the underclay: resin.

Solution

(a) Utilizing eqn. (1): $\beta = k[(\pi/4) + (\phi/2)] = 0.5\,[0.785 + 12.5] = 28.8°$
(b) From eqn. (5): $\sigma_{vs} = Q - \gamma \cdot z = 36 - (23 \times 0.9) = 15.3$, MPa

Since the U.C.S. of the underclay (σ_s) = 15 MPa, the floor will fail under the strata pressure and it requires reinforcement.

(c) Using eqn. (10): $\sigma_t = Q - [4S_0/\tan(45 - 0.5\phi)] = 36 - [4 \times 0.3/\tan(45 - 12.5] = 34.1$, MPa
(d) From eqn. (12): $(FS) = Q/\sigma_t = 36/34.1 = 1.06$ --->1.1
(e) Given $V_r = 0.1V_s$ and using eqn. (11):

$$\sigma_r = \{V_s\{Q - [4S_0/\tan(45 - 0.5\phi]\} - 0.9V_s \cdot \sigma_s\}/0.1V_s$$
$$= (36 - 1.9 - 13.5)/0.1 = 146, \text{ MPa.}$$

Conclusions

(1) Water increases the coal and underclay floor heave by 18–30%, respectively.
(2) Solid pillars left on one side of the gate roads contributes towards the asymmetrical floor heave, if the yielding characteristics of the gob-side packs is not similar to that of the pillar.
(3) Pillars left in the overworked area tend to transfer the strata pressure towards the floor of the underlying roadways at an angle between 50° and 66° to the horizontal.
(4) Gob-side packs have no appreciable effect on the floor heave unless they are packed very tightly and their compressive strength surpass that of the floor.
(5) Installation of steel base plates increases stability of the arched support and reduces penetration of the support legs into soft floor.
(6) Floor reinforcement by injection is promising and lends itself to the theoretical and in situ analysis. This technique can also be utilize to reinforce the roof or sides. It is feasible in mechanized, high productive and continuous mining systems necessitating continuous ground reinforcement or support. The injection can be done either as per-development, or per-production reinforcement, or as a post-production system. The reinforced sections are cuttable by rock cutting machines, if it becomes necessary.

References

1. Sheorey, P.R. and Dunham, L.K., An approximate analysis of floor heave occurring in roadways behind advancing longwall faces. Int. J. Rock Mech. Min. Sci., 15 (1978): 227–288.
2. Whittaker, B.N. and Bonsall, C.J., Design aspects relating to the stability of coal mining tunnels. In: Stability in Underground Mining. Pergamon, Oxford (1981), Chapter 24, pp. 519–533.
3. Goetz, W., Planning roadways from the point of strata mechanics. Glückauf, 118 (1982): 13–23.

4. Afrouz, A., The Lower Four Feet floor rocks of South Wales. Colliery Guardian, 224 (12) (1976): 656–660.
5. Afrouz, A., Floor behaviour along longwall roadways. Int. J. Rock Mech. Min. Sci., 12 (1975): 229–240.
6. Afrouz, A., Yield and bearing capacity of coal mine floor. Int. J. Rock Mech. Min. Sci., 12 (1975): 241–253.
7. Afrouz, A. and Harvey, J., Rheology of rocks within soft to medium strength range. Int. J. Rock Mech. Min. Sci., 11 (1974): 281–290.
8. Afrouz, A., Mechanical Behaviour of Mine Floor. Ph.D. Thesis, Univ. Wales, Cardiff (1973).
9. Terzaghi, K. and Peck, R.B., Soil Mechanics in Engineering Practice. J. Wiley, New York, N.Y. (1967) 2nd ed.

3.4 SURFACE EXCAVATIONS STABILITY AND REINFORCEMENT

3.4.1 COMPUTER DESIGN OF OPEN PIT SLOPE AND DESIGN OF REINFORCEMENT FOR PLANAR FAILURE

By: Andy A. Afrouz and X. Fu

ABSTRACT: This paper describes a computer program for design of open pit slopes under variable degrees of water saturation, and its reinforcement for planar failure conditions. Relationships between the static and dynamic slope factor of safety, the potential failure angle and the optimum angle of the reinforcing elements to the horizontal for conditions before and after the installation of reinforcements are analyzed. Solution to the design of passive (or untightened) and active (or tightened) reinforcement for static and dynamic conditions of the open pit slopes are addressed. Based on the slope conditions and tensile strengths of the reinforcing materials, design specifications such as: diameter, length, number, and spacings between the reinforcing elements are given as design charts.

1. INTRODUCTION

Computer design of optimum pit slope under the static and dynamic conditions with selection capabilities for passive (or untightened) and active (or tightened) reinforcement for all open pit conditions are desirable. The potential utilization of a design to provide the above noted capabilities are to provide accurate design with speed and simplicity.

Many open pit design computer programs exist that apply to specific planar failure conditions. Most do not take into account in their programs all of the in-situ ground conditions and mining operations encountered in developing an open pit.

This investigation considered the open pit slope specifications, under both static and dynamic conditions. The specifications considered slope height(H), angle of slope to horizontal(β), percent ground water saturation (s%), and weight per unit volume of the rock (γ). A Fortran 77 program determined if the factor of safety (FS) under different conditions satisfied a desired safety factor, or if reinforcement was required to achieve the desired safety factor. The program also determined the type of reinforcement; that is: static-passive, static-active, dynamic-passive, or dynamic-active, and the optimum design specifications as potential failure angle (α), dimensions of the reinforcing element, the resisting force (Fa), optimum slope angle (β), angle of placement of the reinforcing element to horizontal (Δ), number (N) and spacing (S) of the reinforcing elements, and the optimum factor of safety under different designs. The program tabulates and plots the output data as a chart that may be used for design.

2. THEORETICAL CONSIDERATIONS

Utilizing the analytical approach of Ucar et al. (1987) with respect to Figure 1,

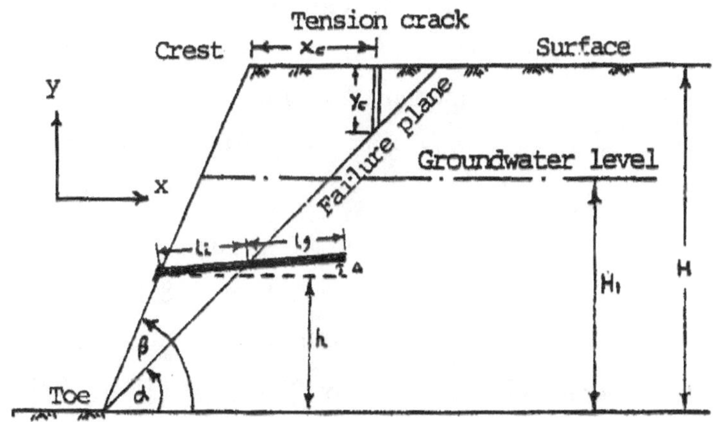

Figure 1. Schematic of a slope with planar failure.

The following design equations are programed:

2.1 Slope without reinforcement under static condition

$$D_f = W[(1 + K_v) \operatorname{Sin}\alpha + K_h \operatorname{Cos}\alpha] \qquad (1)$$

$$R_f = S_o H / \sin\alpha + [W(1 + Kv)\cos\alpha - U - W.\,K_h \sin\alpha]\tan\phi \quad (2)$$

$$FS = \{[S_0 H \sin\beta / [\psi.\ \sin(\beta-\alpha)] + \tan\phi[(1 + K_v)\cos\alpha - \psi_1 \sec\alpha / \psi - Kv \sin\alpha]\} / [(1+K_v)\sin\alpha + K_v \cos\alpha] \quad (3)$$

where: D_f = driving force,

R_f = resisting force,

FS = the safety factor,

W = weight of the sliding block, kN/m = $\psi.\ \sin(\beta-\alpha) / (\sin\alpha \sin\beta)$,

U = total force acting normal to the failure plane,

= $H.\ H_1\ \gamma_w\ (\cot\alpha - \cot\beta)\ \sec\alpha /2$, kN/m,

S_0 = shear strength of rock, MPa,

H = total height of the slope, m,

H_1 = water level, m,

β = inclination of the pit wall to horizontal, degrees,

α = inclination of failure plane to horizontal, degrees,

ϕ = angle of internal friction for the rock, degrees,

K_V = vertical component of seismic factor,

K_h = horizontal component of seismic factor,

ψ = total weight factor = $q.\ H + [\gamma (H - H_1)/2] + (\gamma_{sat} H_1 /2)$ (4)

ψ_1 = $\gamma_w\ .H_1 /2$

Y_c = tension crack depth, from the pit surface, m.

= $H[1-(\cot\beta \cot\alpha)^{0.5}]$[Hoek and Bray, 1981] (5)

X_c = tension crack location, from the crest, m.

= $H[1- (\cot\beta\ .\ \cot\alpha)^{0.5} - \cot\beta]$[Hoek and Bray, 1981] (6)

2.2 Passive reinforcement under static condition

$$Fa = \lambda_3 [(FS)_{passive} - (FS)] / f(\Delta) \quad (7)$$

where: Fa = anchor force along its length, MPa,

λ_3 = $w[(1 + K_v)\sin\alpha + K_v \cos\alpha]$ (8)

$F(\Delta) = \cos(\alpha-\Delta) + \cos(\alpha-\Delta)\tan\phi$ (9)

Δ = $\alpha - \phi$, the bolt angle to horizontal (10)

2.3 Active reinforcement under static condition:

$$Fa = \lambda_3 \,[(FA)active - (FS)] /[(FS)active + \tan\phi \,.\, \tan(\alpha\text{-}\Delta) = \tan\phi /(FS)active \quad (11)$$

2.4 The required number of reinforcing elements:

$$N = \psi[\mathrm{Sin}(\beta\text{-}\alpha)/(\mathrm{Sin}\alpha.\mathrm{Sin}\beta)]/\sigma_t \quad (12)$$

where: σ_t = tensile strength of the reinforcing element, MPa,

2.5 Distance between the reinforcing head and the failure surface:

$$1_i = (h/\mathrm{Sin}\beta)\,[\mathrm{Sin}(\beta - \alpha)/\mathrm{Sin}(\alpha - \Delta)] \quad (13)$$

2.6 Grouting Length

$$(lg)min = \ln\{[8P/(\pi\Theta.D.k)\tan(45+\phi/2)]+1\}D.K/4 \quad (14)$$

$$(lg)max = 2P\tan(45+\phi/2)/(\pi.\Theta.D) \quad (15)$$

where: P = working tensile load, MN,
D = diameter of borehole, m,
$\Theta = 2S_0 \tan(45° + \phi/2)$, degrees.

2.7 The static and dynamic safety factors:

$$(FS)dynamic = [(FS)static/(-\tan\epsilon.\tan\phi)/(1+0.88\tan\epsilon) \quad (16)$$

where: $\tan\epsilon = K_h/(1+K_v)$ (17)

3 - COMPUTER PROGRAMING

3.1 - Computer flow chart

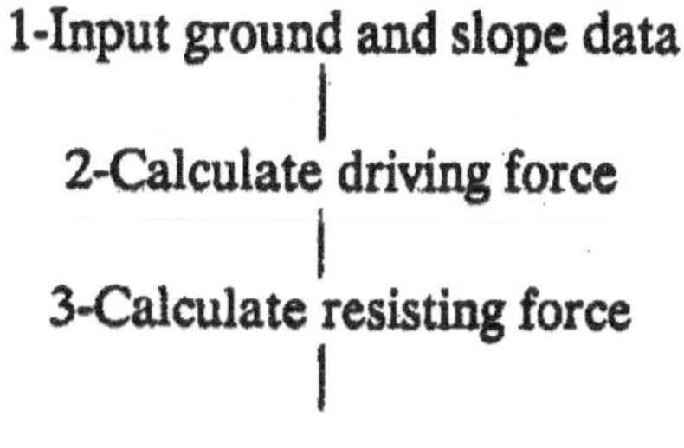

1-Input ground and slope data

2-Calculate driving force

3-Calculate resisting force

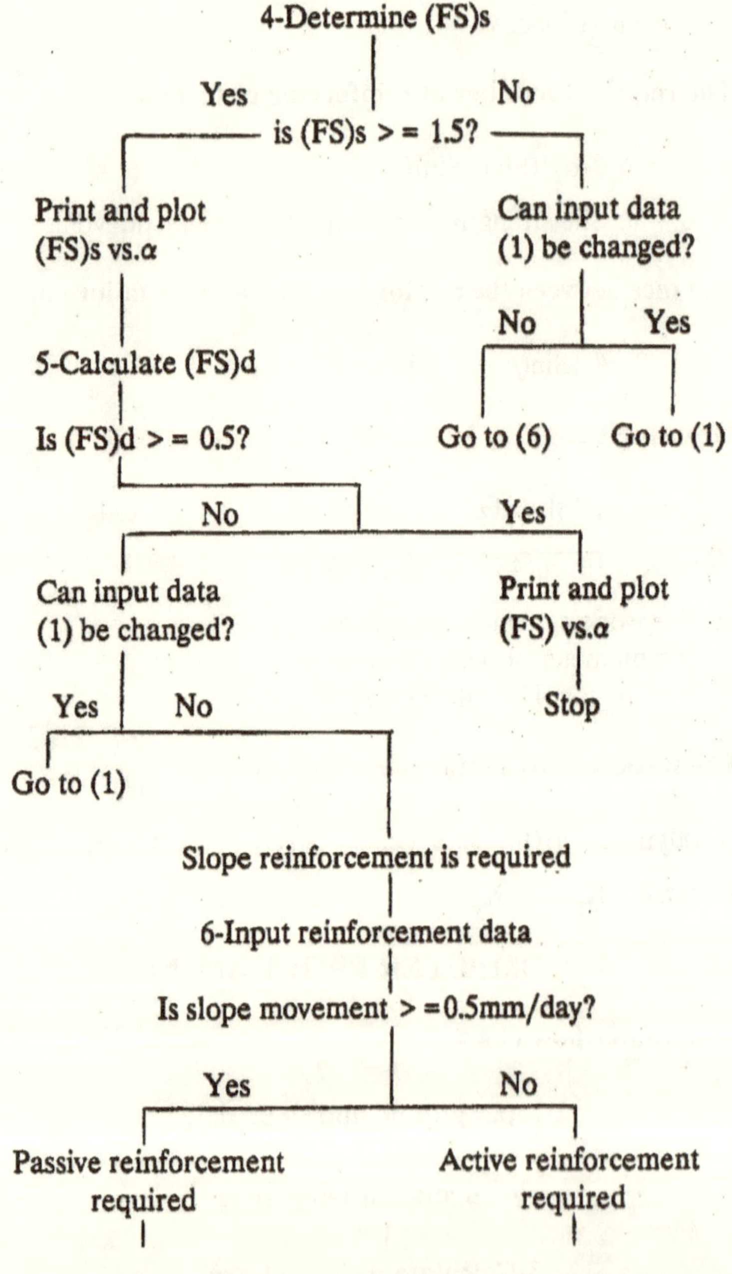
4-Determine (FS)s
Yes
No
is (FS)s > = 1.5?
Print and plot
(FS)s vs.α
Can input data
(1) be changed?
No
Yes
5-Calculate (FS)d
Is (FS)d > = 0.5?
Go to (6)
Go to (1)
No
Yes
Can input data
(1) be changed?
Print and plot
(FS) vs.α
Yes
No
Stop
Go to (1)
Slope reinforcement is required
6-Input reinforcement data
Is slope movement > =0.5mm/day?
Yes
No
Passive reinforcement
required
Active reinforcement
required

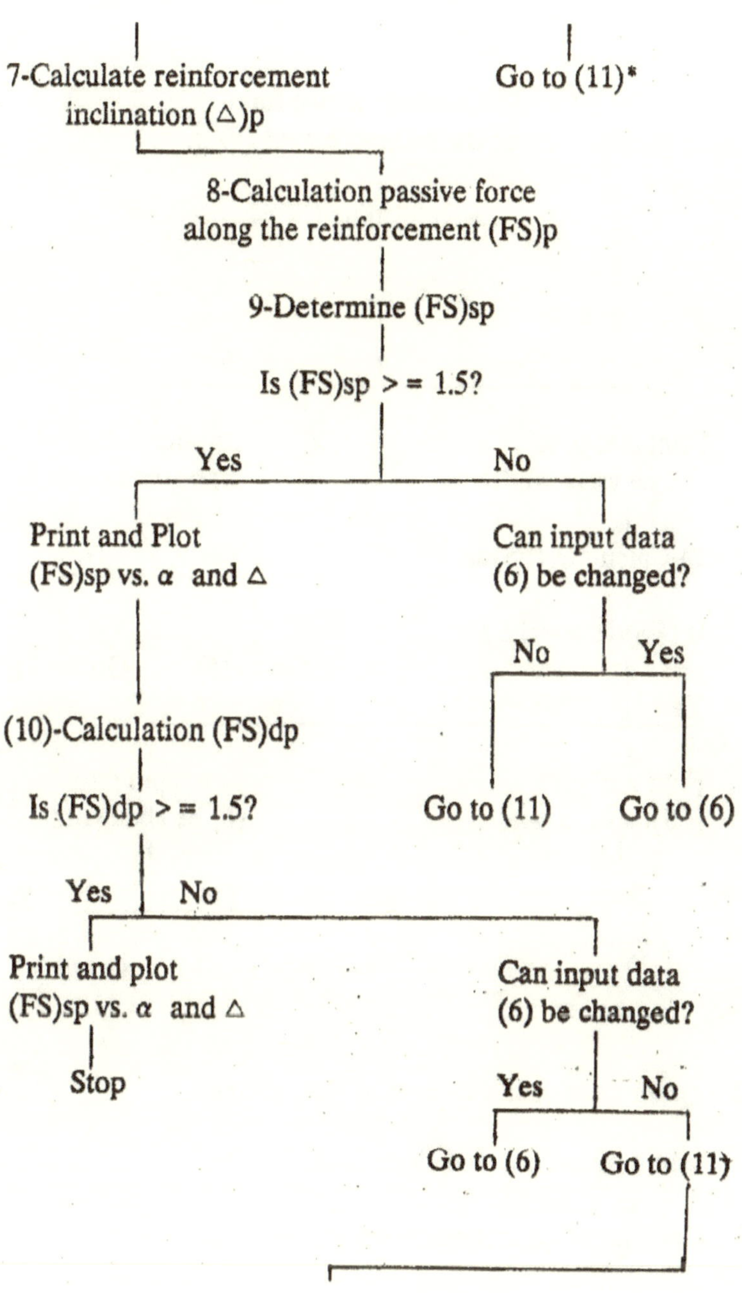
7-Calculate reinforcement
inclination (△)p
Go to (11)*
8-Calculation passive force
along the reinforcement (FS)p
9-Determine (FS)sp
Is (FS)sp > = 1.5?
Yes
No
Print and Plot
(FS)sp vs. α and △
Can input data
(6) be changed?
No
Yes
(10)-Calculation (FS)dp
Go to (11)
Go to (6)
Is (FS)dp > = 1.5?
Yes
No
Print and plot
(FS)sp vs. α and △
Stop
Can input data
(6) be changed?
Yes
No
Go to (6)
Go to (11)

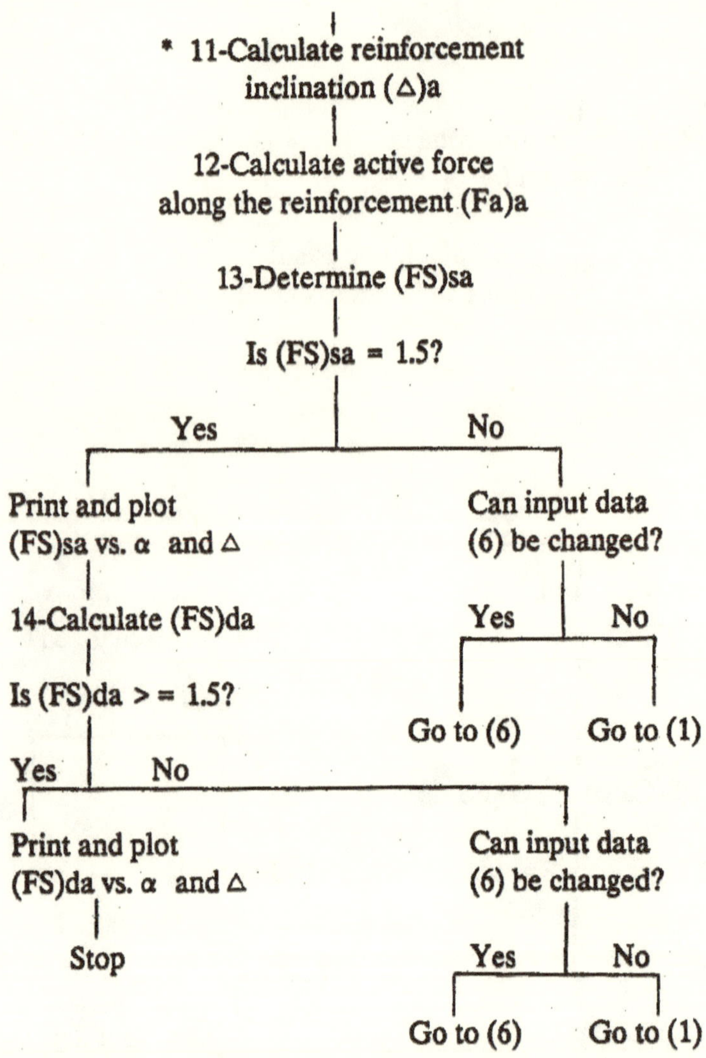

3.2 Explanation of the computer flow chart:

3.2.1 Imputing ground and slope data:

3.2.2 Calculation of driving force from equation (1).

3.2.3 Calculation of resisting force from equation (2).

3.2.4. Determination of (FS)s, equation(3).

3.2.5 Determination of (FS)d, equation (16).

Table 1 - Ground data input

Case Number

Input	1	2	3	4	5
H, m	10.	40.	70.	100.	150.
H_1, m	0.	5.	20.	50.	100.
β, degr.	35.	45.	55.	65.	75.
ϕ, degr.	20.	30.	40.	45.	50.
Kh	0.1	0.2	0.25	0.3	0.35
Kv	0.05	0.1	0.15	0.2	0.3
So, kPa;	200.	250.	300.	350.	400
n, %	0.05	0.1	0.15	0.2	0.25
q, N/ m.sq.	220.	250.	300.	350.	380.
Gs	2.0	2.5	3.0	3.5	4.0
s, %	0.05	0.15	0.25	0.35	0.40
γ, kN/m	10.	10.	10.	10.	10.

3.2.6 Input reinforcement data: $S_{r.}$ ϕ_{r}.

3.2.7 Calculate reinforcement inclination (Δ)p, equation (10).

3.2.8 Calculate passive force along the reinforcement (Fa)p, equation (7).

3.2.9 Determine (FS)sp, equations (3) and (7).

3.2.10 Calculate (FS)dp, equations (11) and (16).

3.2.11 Calculate reinforcement inclination (Δ)a, equation (11).

3.2.12 Calculate active force along the reinforcement (Fa)a, equation (11)

3.2.13 Determine (FS)sa, equations (3) and (11).

3.2.14 Calculate (FS)da, equation (16).

4 ANALYSIS OF RESULTS

4.1 Validation of the program with an in situ example:

4.1.1 Input from field measurements:

$H = 30$, m; $H_1 = 20$, m; $\beta = 76°$; $\phi = 30°$; $K_h = 0.2$; $K_v = 0.1$; $S_O = 295$, kPa;

$n = 10.7\%$; $q = 300$ kN /m; $G_s = 2.65$; s=32%; $w = 10$, kN/m.

4.1.2 Computer output:

(A) Conditions before the slope reinforcement: Considering Figure 1 for $\alpha = 45°$, and utilizing equations (3) and (16) gives:

minimum $(FS)_{static} = 1.15$; minimum $(FS)_{dynamic} = 1.05$, $X_c = 7.51$, m and $Y_c = 15.00$,m.

If $(FS)_{static} = 1.5$ is to be reached, the reinforcement will be required. Graph of the safety factor in static and dynamic conditions versus the angle of potential failure plane to horizontal (α) is given in Figure 2.

(B) Design of reinforcement under passive conditions: Substituting the shear strength properties of the rock into equations (3), (10), and (14) respectively, gives:

$a_{min} = 50°$; $\Delta_{min} = 22°$; $(FS)_{min.\ static} = 0.84$ and $(FS)_{min.\ dynamic} = 0.75$

Graph of the safety factor for passive reinforcement in static and dynamic conditions versus angle of reinforcement (Δ^O) and the angle of potential failure plane (α) to horizontal is given in Figure 3. Substituting the field data into equations (7) and (8) gives the minimum withstanding force of the reinforcement under passive condition to be

$(F_a)_{min.\ passive} = 6957$, kN/m length. Graph of $(F_a)_{passive}$ versus angle of the reinforcing element to horizontal (Δ^O) is given in Figures 4 and 5.

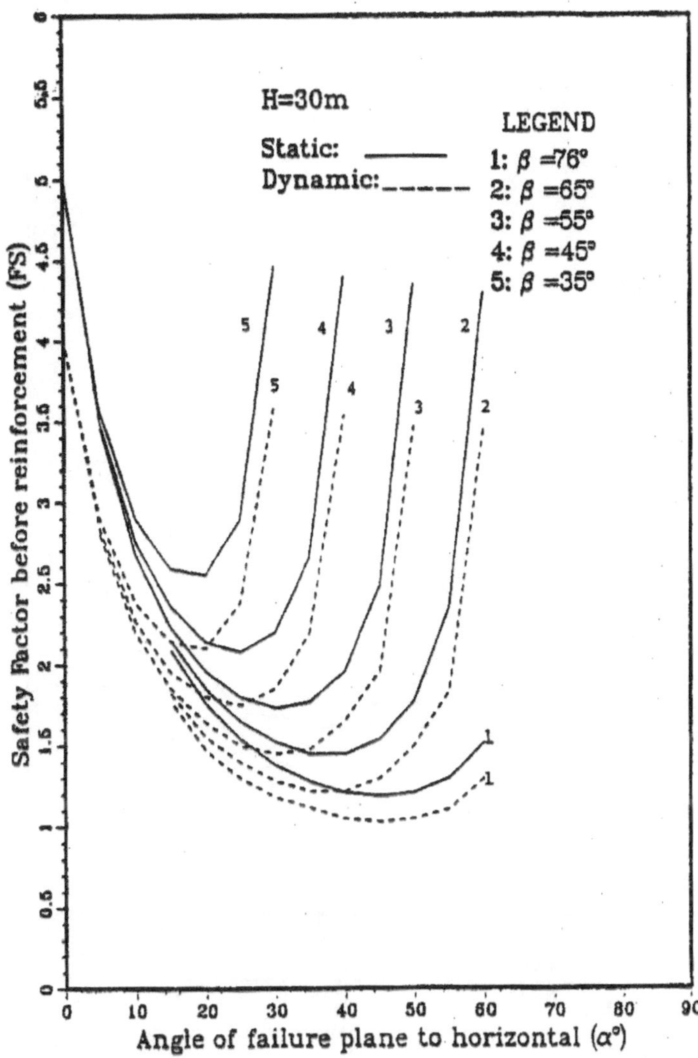

Figure 2 - Static and dynamic safety factors before reinforcement versus potential angle of failure plane to horizontal (α°)under the field conditions.

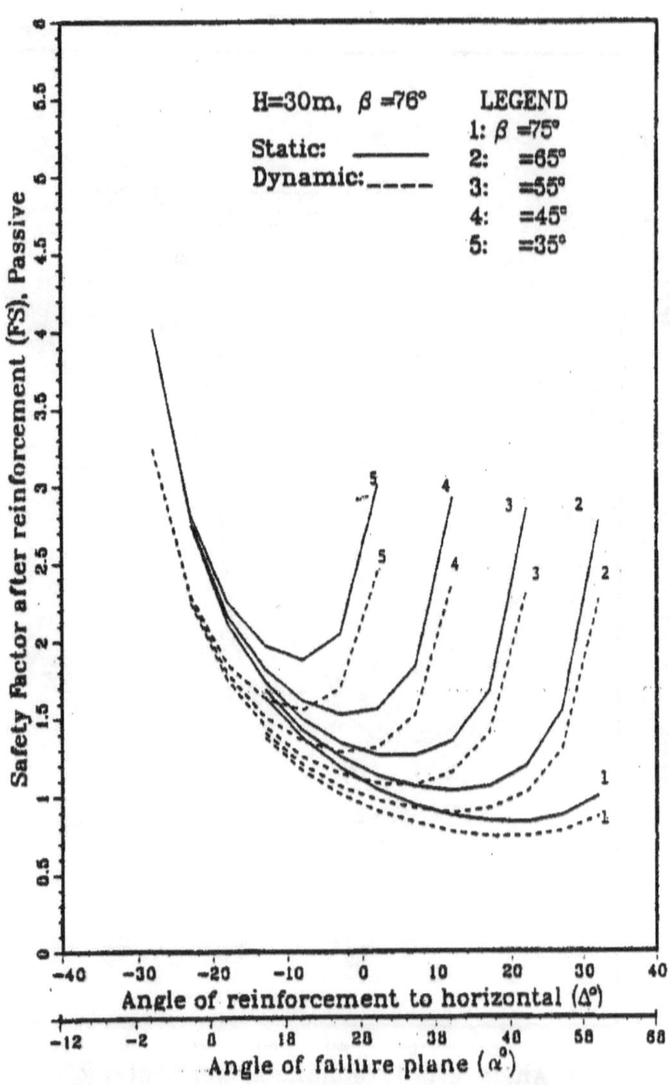

Figure 3 - Static and dynamic safety factors with passive reinforcement versus angle of the potential failure plane (α°) and angle of the reinforcement to horizontal (Δ°) under the field conditions.

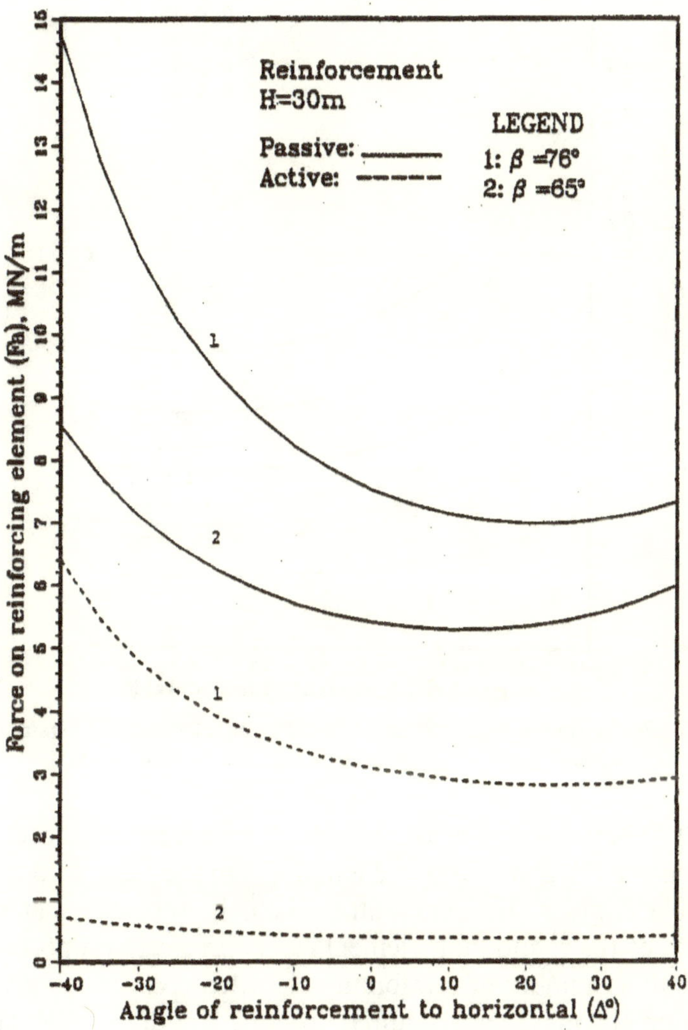

Figure 4 - Passive and active force on the reinforcing element (F_a) versus angle of the reinforcement to horizontal (Δ^0) under the field conditions.

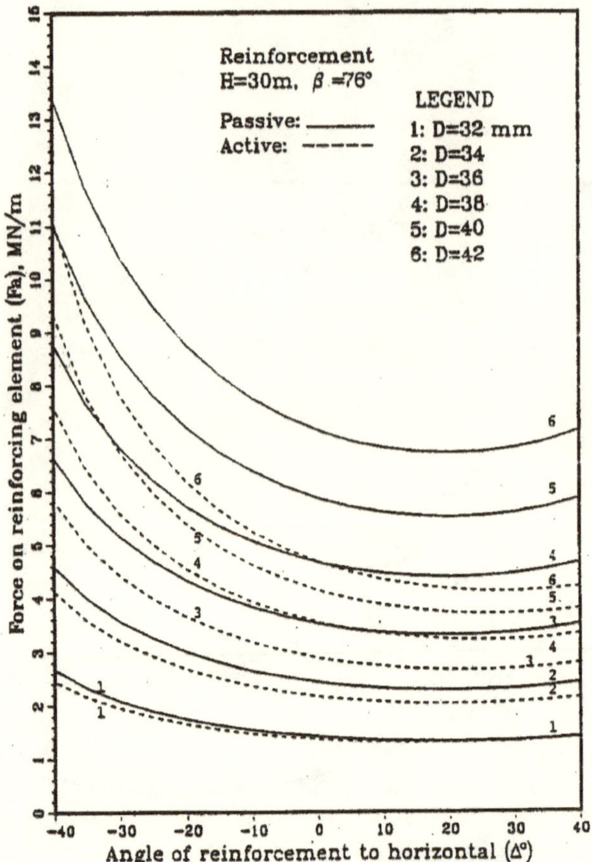

Figure 5 - Passive and active force on the reinforcing element (F_a) versus angle of reinforcement to horizontal (Δ^0) for various diameter (D).

(C) Determination of the length of reinforcement head to the failure plane (l_i), and its grouted length (l_g) in the passive conditions:

Table 2 gives the input/output data. It indicates that tensile strength of the reinforcing element (σ_t) is inversely proportional to the required number of reinforcing elements (N). It also gives the minimum and maximum required grouting length. By calculation

$$l_i = 19.25,\text{m for } h = 20,\text{m}.$$

Table 2 - Reinforcement data input in passive condition.

σt	P	D	Dh	N	(lg)min	(lg)max
MPa	MN	m m	m m		m	m
250	0.25	16	32	10	4.06	13.2
500	0.50	20	34	5	5.40	24.9
750	0.75	24	36	3	6.33	35.2
1000	1.00	28	38	2	7.08	44.5
1200	1.25	32	40	2	7.74	52.8
1500	1.50	36	42	2	8.34	60.4

(D) Design of reinforcement under active conditions: Table 3 gives the input data for active reinforcement.

Table 3 - Reinforcement data input in active conditions.

σt	P	D	Dh	N	(lg)min	(lg)max	(li)
Pa	MN	m m	m m		m	m	m
250	0.25	16	32	13	3.30	8.43	11.4
500	0.50	20	34	6	4.54	15.88	
750	0.75	24	36	4	5.39	22.49	
1000	1.00	28	38	3	6.08	28.41	
1250	1.25	32	40	3	6.68	33.74	
1500	1.50	36	42	2	7.23	38.56	

The data output pertinent to the active reinforcement is as follows:

From equ. (10): $^{\Delta}$minimum = 24°.

From equ. (3): (FS)min, static = 1.19.

From equ. (16): (FS)min, dynamic = 1.03.

From equ. (11): (Fa)min = 2816.54, kN/m.

Figure 6 shows variation of the static and dynamic safety factor under reinforcement with α and β.

(E) Determination of (l_i) and (l_g) under active reinforcement conditions:

Utilizing equations (13) and (14) for various tensile strength of the reinforcing element and the field conditions, gives number of the reinforcing elements (N), minimum and maximum grouting length (l_g). For h = 20m, l_i = 11.28,m.

4.2 Comparisons of the results with changing slope angle (β°), under the field conditions:

Figures 2, 3 and 6 show variation of the safety factor (FS) with α and Δ for slope height (H) = 30, m; slope angles (β) from 35 to 76 degrees and other factors remaining constant.

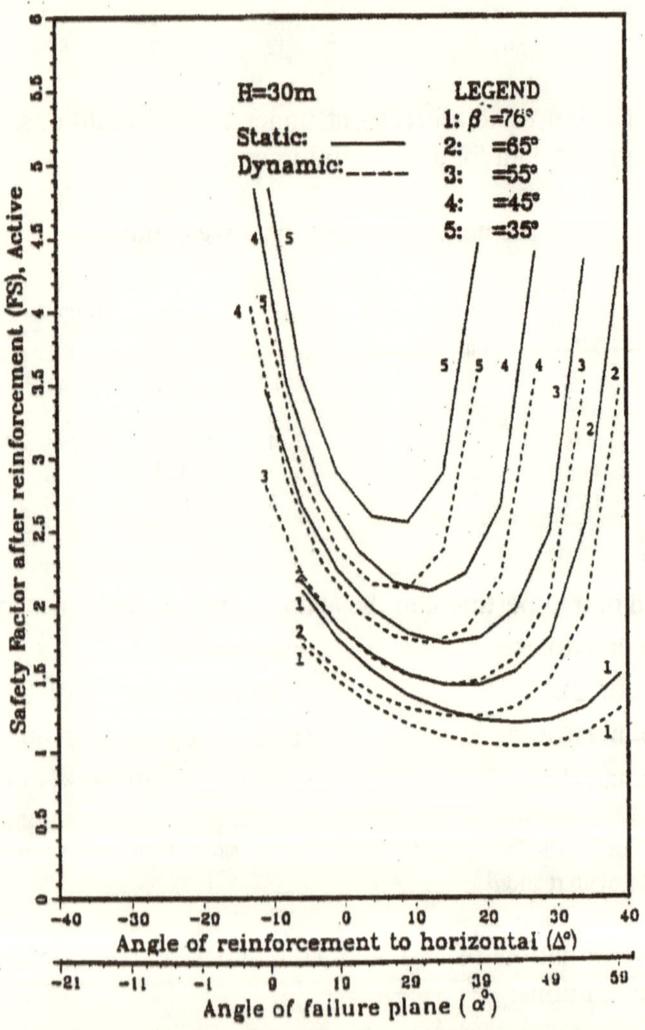

Figure 6 - Variation of static and dynamic safety factor under active reinforcement with α, β, and Δ, under the field conditions.

The charts indicate:

(a) $FS_{minimum} \propto 1/\beta \propto 1/\alpha$;

(b) $H \propto \alpha \propto \beta$;

(c) For (FS) = 1.5 and $\beta < 55°$, the $(FS)_{min} \geq 1.5$. This case requires no slope reinforcement ; and

(d) For $\beta \geq 55°$, $(FS)_{min} < 1.5$. In this case, the slope must be reinforced using passive or active techniques (Figures 3 and 6).

Figures 4 and 5 show variation of force on the reinforcing elements with Δ° for H = 30m; β of 65° and 76°; and D varying from 32 to 42, mm. The charts indicate:

(i) For the same slope angle (β), active Fa < passive Fa.
(ii) For the same reinforcement angle (Δ), active Fa > passive Fa.
(iii) Fa $\propto 1/\beta$. This suggestion has an important practical significance in the design of the slope reinforcement system.

4.3 - Comparisons of the results with input from different ground conditions: Figures 7 to 9 give a design chart for the slope safety factor (FS) under static and dynamic conditions, before reinforcement, with passive reinforcement and with active reinforcement. The above noted design is valid for:

(i) Slope angles (β) = 35 to 75 degrees.
(ii) Angle of slope failure (α) = 0 to 90 degrees.
(iii) Angle of reinforcement (Δ) = -5 to 40 degrees.
(iv) Slope height (H) = 30 to 150, m.

When H ≤ 40, m and $\beta \leq 45°$, the FS satisfies with the requirement of FS > 1.5.

When H ≥ 70, m and $\beta \geq 55°$, the FS < 1.5, prompting utilization of the passive or active reinforcement.

Figures 5, 10, and 11 provide selection of the diameter of the reinforcing element by knowing the force (Fa) in the passive or active conditions and angle of reinforcement to horizontal (Δ°). It can be seen that the optimum design is Δ = 5 to 15° up-hole. It is realized that up-hole drilling, placement of reinforcement and grouting (if required) is more time consuming and expensive than drilling the down-hole or even nearly horizontal angles. The latter case, however, exert more bending and shear force on the reinforcing elements where it has the weakest strength. Figure 11 indicates a non-linear limit where the force on the reinforcement element in the passive and active conditions are equal. These factors have design and placement significance.

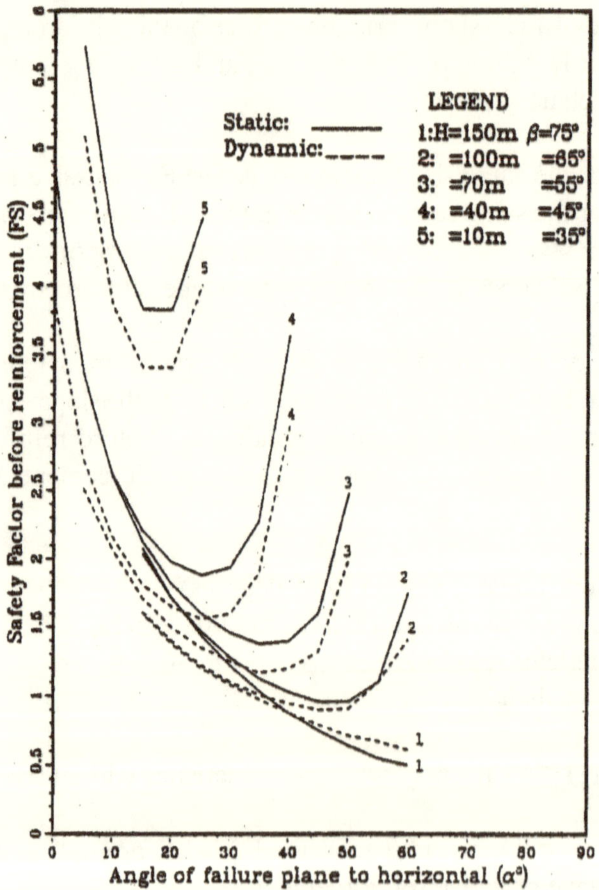

Figure 7 - Design chart to determine the static and dynamic safety factor of a slope before reinforcement in an average ground condition and various slope geometry.

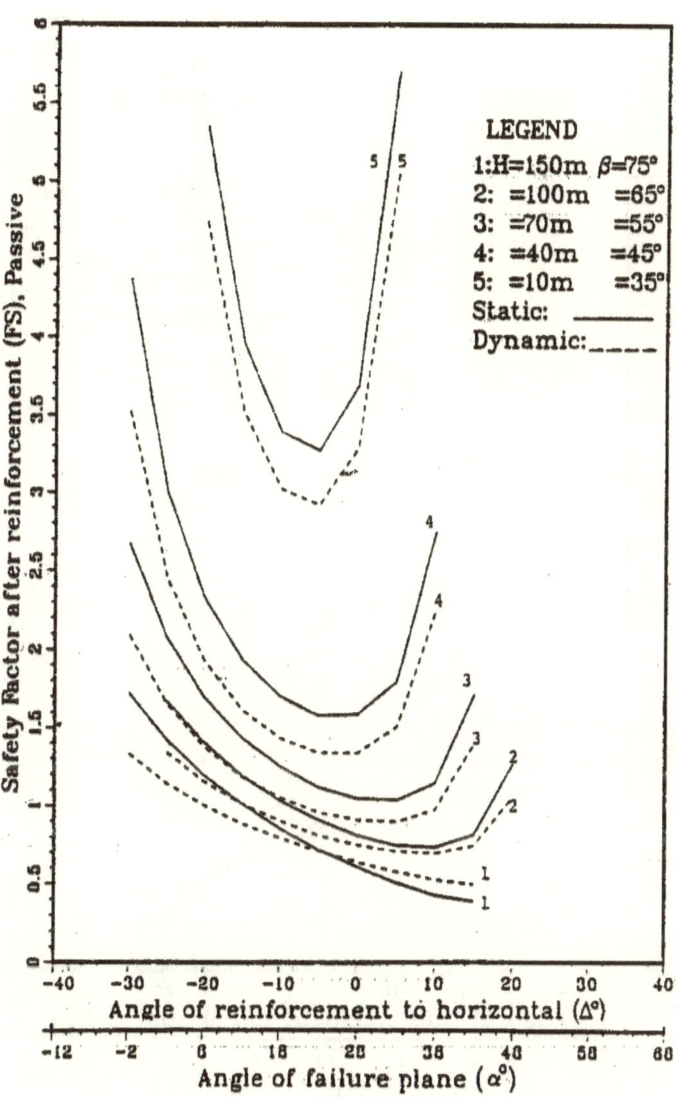

Figure 8 - Design chart to determine the static and dynamic safety factor of a slope with passive reinforcement, and its angle for average ground conditions and various slope geometry.

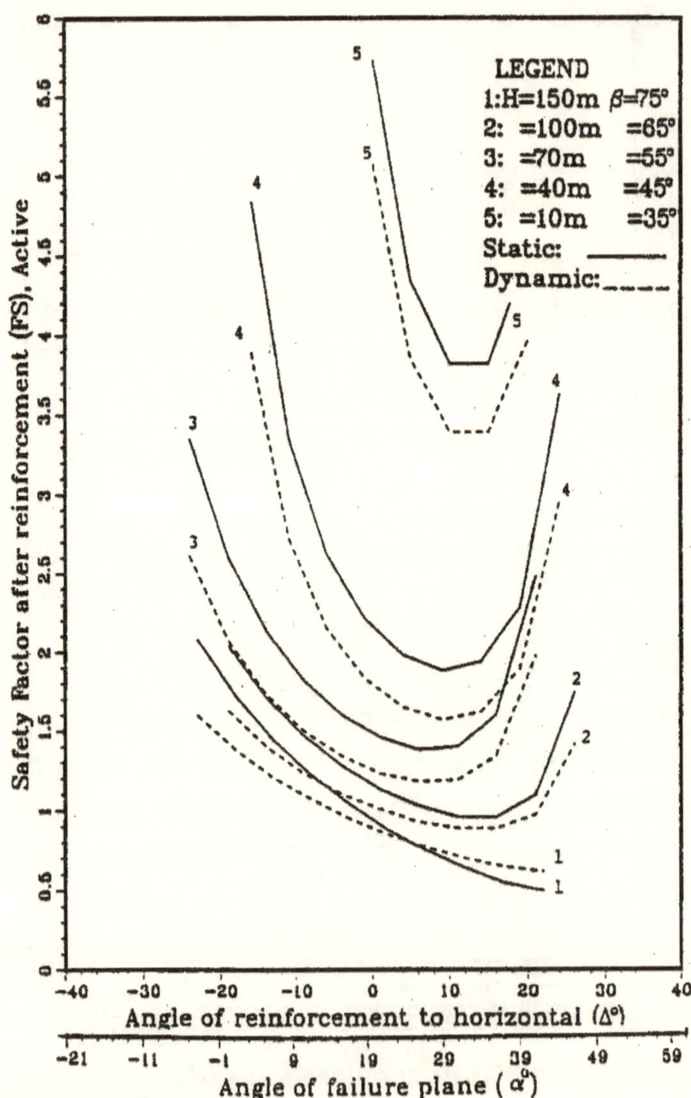

Figure 9 - Design chart to determine the static and dynamic safety factor of a slope with active reinforcement, and its angle for average ground conditions and various slope geometry.

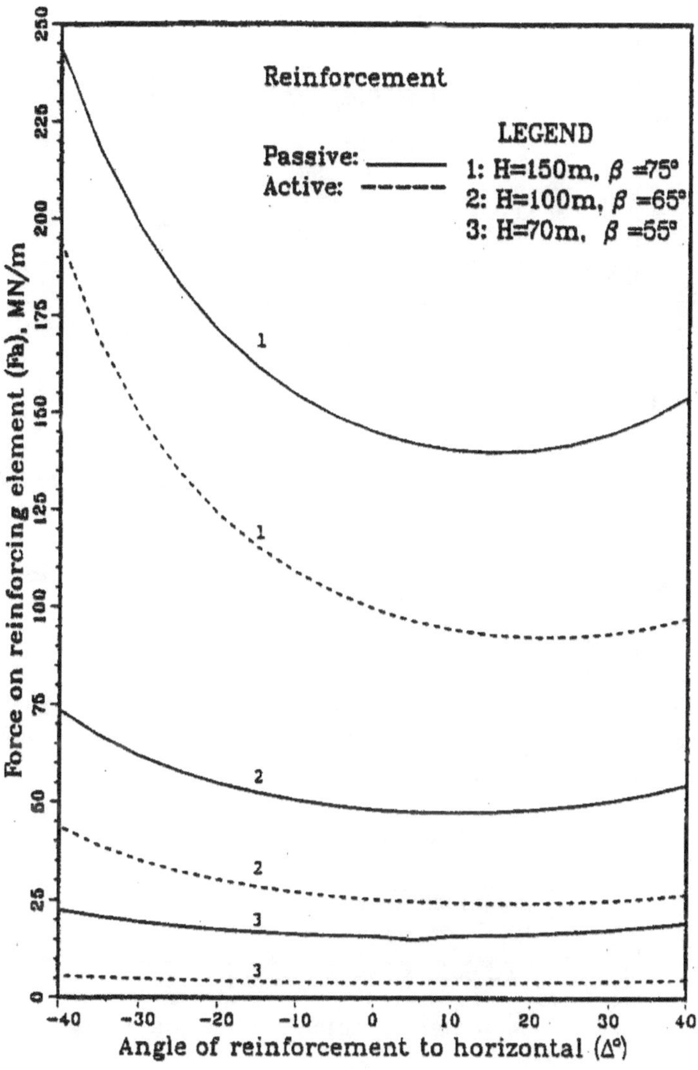

Figure 10 - Determination of force on the slope reinforcement (Fa) under passive band active conditions with varying slope geometry and reinforcement angle (Δ).

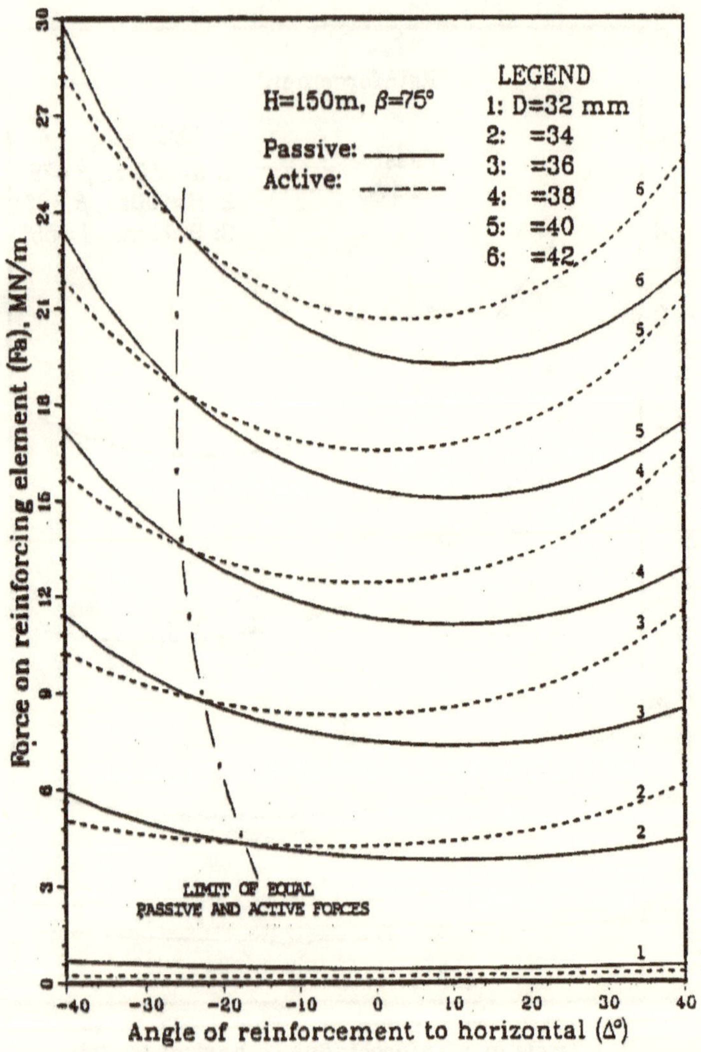

Figure 11 - Determination of Passive and Active Force on the Slope Reinforcement (Fa) Knowing Angle of Reinforcement to Horizontal (Δ) for Various Diameter (D).

5. CONCLUSION

Based on the analysis of the results:

(A) The new computer program to the design of open pit slopes and the reinforcement of slopes under planar failure condition is practical, quick and accurate. It considers for static and dynamic conditions as well as passive and active reinforcement techniques.

(B) Slope inclination and stability factors can be optimized using the design charts provided.

(C) When the slope factor of safety in a set of conditions is less than the requirement, the program will indicate: type of reinforcement, the optimum tensile strength, the number required, the angle of placement of reinforcement to the horizontal, the length, the spacing, and the grout length.

(D) The best angle of reinforcement (Δ) is 5 to 15° up hole from the horizontal.

(E) There is a limit where force (Fa) on the reinforcing element in passive and active conditions are equal.

REFERENCES

Hoek, E. & J.W. Bray 1981. Rock Slope Engineering, 3rd Ed. Institute of Mining and Metallurgy. London: 150-195.

Ucar, R., Hassani, P.F., & A. Afrouz 1987. Design of Anchoring in Rock With Planar Failure. International Journal of Surface Mining. Balkema Publishing. 199-208.

3.4.2 STABILITY ANALYSIS OF STEEP ROCK SLOPES

By: ANDY A. AFROUZ, F.P. HASSANI and R. UCAR

A simplified method of predicting the shape and location of the potential failure surface in steep rock slopes is suggested. The 1988 updated Hoek and Brown failure criterion is used with Priest and Brown *m* and *s* values and Bieniawski's jointed rock mass classification system. This method is demonstrated by an application to the design of slope reinforcement by rock anchoring and by a case study. The validity of the method is assessed by reference to the variational calculus approach, using the linear Mohr-Coulomb failure criterion.

Key words: benches, bench design, bench anchoring, quarrying, slope stability, ground control, rock mass rating.

Une method simplifiée pour la prediction de la forme et de la localisation de la surface de rupture potentielle dans les pentes rocheuses abruptes est suggérée. Le critère de rupture 1988 révisé de Hoek et Brown est utilisé avec les valeurs *m* et *s* de Priest et Brown et le système de classification de masse rocheuse fracturée de Bieniawski. L'on fait une démonstration de la méthode au moyen d'une application au calcul du comfortement d'un talus par boulonnage, et d'une étude de cas. La validité de cette méthode est évaluée en se référant à l'approche de calcul variationnel, faisant usage du critère de ruptura de Mohr-Coulomb.

Mots clés: banquettes, calcul de banquette, boulonnage de banquette, exploitation de carrière, stabilité de talus, contrôle de terrain, évaluation de masse rocheuse.

[Traduit par la revue]

Introduction

Steep rock slopes in this study are considered to have an angle of 80–90° with the horizontal. They are utilized predominantly where (*a*) the rock strength ranges from medium to competent, that is, $\sigma_c > 60$ MPa, (*b*) the rock mass has medium to low fracture and joint intensity of less than one per metre interval oriented parallel or perpendicular to the slope face, (*c*) the rock is either massive or the bedding planes are horizontal and more than 2 m thick with medium to good cementation, (*d*) the slope height is less than 60 m, and (*e*) the groundwater table is below the toe.

In practice, the curvilinear failure surface in the slope of a fractured rock mass can be expressed in parametric form by means of the following radius vector:

$$R(t) = \Omega(t)\, i + \epsilon\,(t).\, j \qquad [1]$$

where: R(t) = radius vector of the failure surface, in m;

$\Omega(t) = x(t)$ = abscissa of the points on the failure surface as a function of parameter t, in m;

i = unit vector in the x direction;

$\epsilon(t) = y(t)$ = ordinate of the points on the failure surface as a function of parameter t, in m; and

j = unit vector in the y direction.

In the case of massive non-fractured rocks, the failure surface can also be divided into two failure planes with two sliding blocks. The corresponding slope factor of safety is determined by taking the total normal and tangential forces acting on each of the two sliding blocks, as suggested by Gudehus (1972), Kranz (1972), Stocker *et al.* (1979), and Huang (1983).

Various methods have been adopted to determine the slope safety factor for an assumed failure surface condition. These methods range from simplified assumptions made by Fellenius (1936) and Bishop (1955) to more rigorous and accurate approaches utilizing differential equations by Kötter (1903), Morgenstern and Price (1965), Spencer (1967), and Janbu (1973). Determination of the shape of a slope failure surface as a logarithmic spiral was made by Baker and Garber (1977) and Ucar (1988), using variational calculus and applying the linear Mohr–Coulomb failure criterion.

The objective of this case study was to reduce the inaccuracy associated with predicting a failure surface location and to simplify calculations for the design of steep rock slopes. This was achieved by determining the actual shape and extent of the slope failure surface in a non-homogeneous rock mass utilizing differential equations, and then applying the nonlinear rock mass failure criterion suggested by Hoek and Brown (1988), together with Bieniawski's (1976) rock mass rating.

Determination of the failure surface shape

Consider a vertical slope of height H with a possible curved failure surface of AB as shown in Fig. 1.

In cases where the direction of the minor effective principal stress (σ_3) is at an angle α to the horizontal, then, from Fig. 1,

$$\tan \eta = \tan(\alpha + \epsilon) = dy/dx = f'(x) \qquad [2]$$

where: η = angle of the tangent to the failure surface with the horizontal, in deg, and

α = angle between the failure surface and the direction of the minor effective principal stress, in deg.

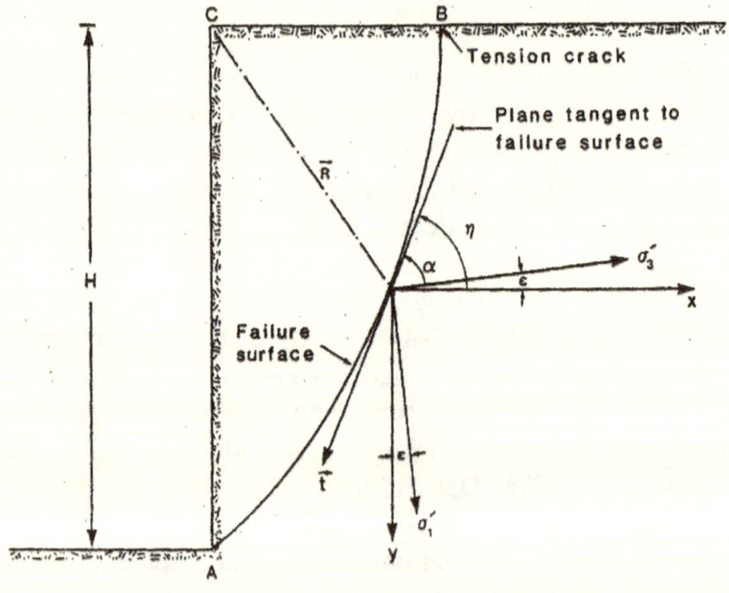

FIG. 1. Rock slope cross section showing direction of principal stresses acting on the failure surface.

In cases where the major effective principal stress acts in the vertical direction –– an accurate assumption for most vertical and near-vertical slope situations (Goodman 1980) –– the minor effective principal stress will act horizontally and $\epsilon = 0$. Therefore, the tangent to the failure surface (α), relative to the horizontal, can be expressed as follows (Ucar 1986):

$$\tan \alpha = dy/dx = (\partial\sigma_1/\partial\sigma_3)^{0.5}$$

$$= \{1 + [m / 2[(m.\ \sigma3/\sigma c) + s]^{0.5}\}^{0.5}$$

$$= \{1 + [m / 2(\sigma 1 - \sigma 3) / \sigma c]\}^{0.5} \qquad [3]$$

where: σ_1 = $\sigma 3 + \sigma c\ [(m.\sigma 3/\sigma c) + s]^{0.5}$, as suggested by Hoek and Brown (1988);

m = constant that depends on the properties of the rock mass and controls the curvature of the line describing the relationship between the major and minor principal stresses at failure = Rm_i, according to Hoek and Brown (1980) and Priest and Brown (1983);

m_i = value of m for the intact rock;

R for disturbed rock masses = exp[(RMR – 2100)/14] (according to Hoek and Brown 1988) = 0.007– 0.03 for poor-quality rock masses;

R for undisturbed or interlocking rock masses = exp[(RMR – 100)/28]

according to Hoek and Brown 1988) = 7–25 for intact rock samples (where $m = m_i$);

RMR = Bieniawski's rock mass rating (1976);

s = constant for a given rock type that controls the curve location between the principal stresses = $(\sigma_{c\,mass}/\sigma_c)^2$, according to Hoek and Brown (1988) and Priest and Brown (1983);

NOTE: s- value for disturbed rock masses = exp[(RMR – 100) /6] (according to Hoek and Brown 1988) = 0 to 3×10^{-6} for poor-quality rock samples;

s- value for undisturbed or interlocking rock masses = exp[(RMR – 100)/6] (according to Hoek and Brown 1988) = 1, for intact rock samples (where $m = m_i$);

$\sigma_{c\,mass}$ = uniaxial compressive strength of the rock mass, in kPa; and

σ_c = uniaxial compressive strength of the intact rock, in kPa.

Applying the Hoek and Brown failure criterion to [3] gives:

$$dy/dx = \{1 + [m / [m^2 + (4m\,\sigma_1/\sigma_c) + 4s]^{0.5} - m]\}^{0.5} \quad [4]$$

Equation [4] is the general equation for the tangent to the failure curve (Fig. 1).

In general, the following expression holds at shallow depths:

$$\sigma_1 = \sigma_v = \gamma y \quad [5]$$

where: σ_v = major vertical stress, in kPa;

γ = average weight per unit volume of the rock, in kN/m^3; and

y = depth below the surface to any designated point on the failure curve, in m.

Substituting σ_1 from [5] into [4] and squaring both sides yields:

$$(dy/dx)^2 = 1 + [m / [m^2 + (4m\,\gamma y/\sigma_c + 4s]^{0.5} - m] \quad [6]$$

Solution of differential equation to obtain x and y – Let:

$$m^2 + 4s = \psi \quad [7a]$$

$$4m\gamma/\sigma_c = \psi_1 \quad [7b]$$

Substituting [7] into [6] gives:

$$(dy/dx)^2 = 1 + \{m/[(\psi + \psi_1 y)^{0.5} - m]\} \text{ or} \quad [8a]$$

$$y = (m^2/\psi_1)\{(dy/dx)^2/[(dy/dx)^2 - 1]\}^2 - \psi/\psi_1 \quad [8b]$$

$$= (m^2/\psi_1)\,\{1/\{1 - [1/(dy/dx)^2]\}\}^2 - \psi/\psi_1$$

Let: $1 - [1/(dy/dx)^2] = 1/(t)^{0.5}$ [9]

Substituting [9] into [8] gives: $y = \epsilon(t) = (m^2/\psi_1)t - \psi/\psi_1$

$$=(\sigma_c/4m\gamma)[m^2t-(m^2+4s)] \qquad [10]$$

Therefore: $dy/dt=\epsilon'(t)=m^2/\psi_1=m.\sigma_c/4\gamma$ [11*a*]

$$\text{or } dy/dx = \theta(t) = y' = -[t^{0.5}/(t^{0.5}-1)]^{0.5} \qquad [11b]$$

$$\text{or } dx=dy/y'=\epsilon'(t)dt/\theta(t) \qquad [11c]$$

Equations [11] are a concise form of the general equation for the tangent to the failure curve given as [4] and shown in Fig. 1. The negative sign in [11*b*] results from the coordinate system adopted in Fig. 1.

Integrating [11] as a function of parameter t gives:

$$x=\int[\epsilon'(t)/\theta(t)]dt]+x_c \qquad [12]$$

where: x_c = integration constant.

Substituting [11] into [12] gives: $x=-(m^2/\psi_1)\int[(t^{0.5}-1)/t^{0.5}]^{0.5}dt+x_c$ [13]

Integrating [13] gives: $x = (m^2/2\psi_1)\{\ln[(t^{0.25}+(t^{0.5}-1)^{0.5}$

$$-(t-t^{0.5})^{0.5})-1]\}+x_c \qquad [14]$$

Substituting [7] into [14] gives:

$$x = \Omega(t) = (m.\sigma_c/8\gamma)[\ln(t^{0.25}+(t^{0.5}-1)-(t-t^{0.5})^{0.5}(2t^{0.5}-1)]+x_c \qquad [15]-$$

Substituting [7] into [10] gives:

$$y=\epsilon(t)=(\sigma_c/4m\gamma)[m^2t-(m^2+4s)] \qquad [16]$$

Equations [15] and [16] represent the position vector of the surface failure expressed in parametric form with the radius vector given in [1].

***Boundary conditions to determine* x_c, x_2, t_1, *and* t_2:**

Let failure of the vertical slope occur through its toe, as shown in Fig. 1, with the following boundary conditions:

(a) when $y = H$ = height of the slope, $x = x_1 = 0$ and $t = t_1$;
(b) when $y = 0$, $x = x_2$ and $t = t_2$.

Substituting the boundary conditions *a* into [15] gives:

$x_c = -(m.\sigma_c/8\gamma)[(t_1 - {}_t^{0.5})^{0.5}(2\,{}_t^{0.5} - 1) - \ln(t^{0.25} + [(t^{0.5} - 1)]$ [17]

and $t_1 = (4/m)(\gamma.\, H/\sigma_c + s/m) + 1$ [18]

Substituting the boundary conditions *b* into [14] and [16] gives

$x_2 = (m.\sigma_c/8\gamma)\{\ln[t^{0.25} + [(t^{0.5} - 1)^{0.5}]^{0.5} - [t_2 - t_2^{0.5}]^{0.5}[2\,t_2^{0.5} - 1)]\} + x_c$ [19]

and $t_2 = [(4s/m^2) + 1]$ [20]

Determination of the failure weight

The failure weight (W) is defined as the weight of the failed rock bound between the slope face and the failure surface. It is considered to act in the direction of gravity and the effective principal stress.

According to Fig. 1 and [5] and [10], the failure weight (W) can be expressed as

$W = \gamma\ \int$integral between x1 and x_2 of ydx

$= \gamma\ \int$integral between t1 and t2 of $\epsilon(t)\ (dx/dt)\ dt$ [21]

Substituting [11] and [16] into [21] gives:

$W = (\sigma_c^2/16\,\gamma)\ \int$integral between t1 and t2 of $[m^2.\, t - (m^2 - 4s)]$

$\times\ [(t^{0.5} - 1)/t^{0.5}]^{0.5}\ dt$

$= (\sigma_c^2/16)\{m^2/2[(t^{0.5} - 1)/t^{0.5}]^{0.5} \times [t^2 - (t^{1.5}/6) - (5t/24) - (5\,t^{0.5}/16)]\} -$

$(5m^2/32)\ln\{1/[t^{0.25} - (t^{0.5} - 1)^{0.5}] - (m^2 + 4s)\{[(t^{0.5} - 1)/t^{0.5}][t - (t)/2] -$

$0.5\ln\{1/[t^{0.25} - (t^{0.5} - 1)]\}\}$ [22]

Determination of the failure surface length

The vector tangent to the failure surface (Fig. 1) in parametric form can be expressed as follows:

$dR(t)/dt = R'(t) = \Omega'(t) + \epsilon'(t)$ [23]

and the unit tangent vector will become as follows:

$R'(t)\,/\,|R'(t)| = (\Omega'(t)\,.\,i + \epsilon'(t)\,j)\,/\,[(\Omega'(t)^2 + \epsilon'(t)^2)^{0.5}$ [24]

Therefore: $\cos\alpha = \Omega'(t) / [\Omega'(t)^2 + \epsilon'(t)^2]^{0.5}$ [25]

and $\sin\alpha = \epsilon'(t) / [\Omega'(t)^2 + \epsilon'(t)^2]^{0.5}$ [26]

where: $R'(t) = [\Omega'(t)^2 + \epsilon'(t)^2]^{0.5} = d\,L_f / dt$ [27]

L_f = AB = curved length of the failure surface, in m (see Fig. 1).

Integrating [27] with respect to dt gives:

$L_f = \int$ integral between t1 and t2 of $[\Omega'(t)^2 + \epsilon'(t)^2]^{0.5} dt$ [28]

Substituting [11] into [28] gives:

$L_f = (m.\sigma_c / 4\gamma) \int$ integral between t1 and t2 of $[(2\, t^{0.5} - 1)/ t^{0.5}]^{0.5}\, dt$

$= (m.\sigma_c / 0.04\gamma)\, \{(4\, t^{0.5} - 1)\, (4t - 2\, t^{0.5})^{0.5} - \ln [(2\, t^{0.5})^{0.5} + (2\, t^{0.5} - 1)]\}$ [29]

Determination of the stresses acting on the failure surface

Shear stress determination

Considering the failure criterion developed by Hoek and Brown (1980), the shear stress acting on the failure surface (τ_f) can be expressed as follows (Ucar 1986):

$$\tau_f = [m.\sigma_c(1 - \sin\beta)] / 8 \tan\beta \quad [30]$$

where: β = the inclination of the Mohr's failure envelope to the horizontal, in deg. $= 2\alpha - \pi/2$ [31]

Substituting [31] into [30] gives:

$$\tau_f = - m.\sigma_c(1 + \cos 2\alpha)/8 \cot 2\alpha = - m.\sigma_c(\tan 2\alpha + \sin 2\alpha) / 8 \quad [32]$$

Normal stress determination

The normal stress acting on the failure surface (σ_n) can be expressed as follows (Ucar 1986):

$$\sigma_n = (m.\sigma_c / 8)[1/(2 \sin^2\beta) + \sin\beta] - \sigma_c(3m/16 + s/m)$$

$$= (m.\sigma_c / 8)[1/(2 \cos^2 2\alpha) - \cos 2\alpha] - \sigma_c(3m/16 + s/m) \quad [33]$$

From Mohr's circle, the equilibrium equations between σ_n and τ_f are as follows:

$$[\sigma_n - (\sigma_1 + \sigma_3)/2]^2 + \tau_f^{\,2} = [(\sigma_1 - \sigma_3)/2]^2 \quad [34a]$$

or

$$\sigma_n - \sigma_3 = (\sigma_1 - \sigma_3)/(1 + \partial\sigma_1/\partial\sigma_3) \quad [34b]$$

or

$$\tau_f = [(\sigma_1 - \sigma_3)/(1 + \partial\sigma_1/\partial\sigma_3)](\partial\sigma_1/\partial\sigma_3)^{0.5} \quad [34c]$$

or

$$\tau_f/(\sigma_n - \sigma_3) = (\partial\sigma_1/\partial\sigma_3)^{0.5} \quad [34d]$$

From Fig. 1: $(\partial\sigma_1/\partial\sigma_3)^{0.5} = \tan\alpha$ [35]

Substituting [35] into [34] and rearranging gives:

$$\sigma_n = (\tau_f/\tan\alpha) + \sigma_3 \quad [36]$$

Determination of the minor principal stress

Substituting [30], [31], and [33] into [36] in terms of the minor principal stress (σ_3) gives:

$$\sigma_3 = \{m.\sigma_c[(1 - 2\sin\beta)^2 + 2\sin\beta]/16\sin^2\beta\} - \sigma_c(0.19m + s/m)$$
$$= \{m.\sigma_c[(1 + 2\cos 2\alpha)^2 - 2\cos 2\alpha]/16\cos^2 2\alpha\} - \sigma_c(0.19m + s/m) \quad [37]$$

Determination of the anchor force

Consider the ground pressure exerted on the anchor (P_a) to be equivalent to the average stress along the surface of failure, as shown in Fig. 2. This provides a good approximation to use in solving complex problems under field conditions.

With respect to the Fig. 2, the following expressions can be derived:

$$\lim_{\Delta s \to 0} \Delta T_a/\Delta s = dT_a/ds \quad [38]$$

$$\lim_{\Delta s \to 0} \Delta N_a/\Delta s = dN_a/ds \quad [39]$$

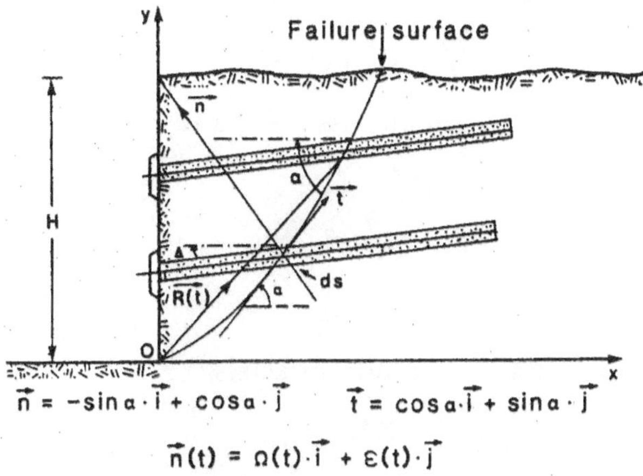

FIG. 2. Coordinate system and forces on the failure surface and anchors.

$$dT_a = P_a \cos(\alpha - \Delta) ds = P_a(\cos\Delta \cos\alpha \, ds + \sin\Delta \sin\alpha \, ds) \quad [40a]$$

$$T_a = P_a[\cos\Delta \int \cos\alpha \, (ds / dt)\, dt + \sin\Delta \int \sin\alpha \, (ds/dt)\, dt] \quad [40b]$$

$$dN_a = P_a \sin(\alpha - \Delta) ds$$
$$= P_a(\cos\Delta \sin\alpha \, ds - \sin\Delta \cos\alpha \, ds) \quad [41a]$$

$$N_a = P_a[-\sin\Delta \int \cos\alpha (ds/dt)\, dt + \cos\Delta \int \sin\alpha \, (ds/dt)\, dt] \quad [41b]$$

where: dT_a = differential tangential shear force on the surface of failure, kPa;

dN_a = differential normal component of force acting on the surface of failure, kPa;

α = angle of the failure surface to the horizontal, in deg;

Δ = angle of the anchor to the horizontal, in deg (positive when above and negative when below the horizontal);

ds = differential length of the curvilinear failure surface, in m.

For the coordinate system adopted, the tangent to the failure curve is positive; then equ.[11] becomes:

$$dy / dx = \tan\alpha = [t^{0.5} / (t^{0.5} - 1)]^{0.5} \quad [42]$$

Also, [16] can be transformed as follows:

$$\epsilon(t)=y=(\sigma c/4m\gamma)[m^2t-(m^2+4s)]-H \qquad [43]$$

Trigonometric transformations from [42] gives:

$$\sin\alpha = [t^{0.5} / (2\ t^{0.5} -1)]^{\ 0.5} \qquad [44a]$$

$$\cos\alpha = [(t^{0.5} -1) / (2\ t^{0.5} -1)]^{0.5} \qquad [44b]$$

$$\tan\alpha = t^{0.5} / (t^{0.5} -1)]^{0.5} \qquad [44c]$$

The vector tangent to the failure surface is:

$$R\,'(t) = (m.\sigma c / 4)\{[i(t^{0.5} -1) / t^{0.5}] - j\} \qquad [45]$$

Differentiating [13] and [43] with respect to the parametric function *t* gives, respectively,

$$dx / dt = (m.\sigma_c / 4\gamma)\,[(t^{0.5} -1) / t^{0.5}]^{0.5} = \Omega\,'(t) \qquad [46]$$

$$dy/dt = m.\sigma_c / 4\gamma = \epsilon'(t) \qquad [47]$$

Substituting [46] and [47] into [45] gives:

$$R\,'(t) = \Omega\,'(t).\,i + \epsilon'(t)\,j \qquad [48]$$

where: from vector calculus,

$$I\,R\,'(t)\,I = ds / dt = (m.\sigma_c / 4\gamma)\ [2\ t^{0.5}]^{0.5} - 1) / t^{0.5}]^{0.5} \qquad [49]$$

Substituting [44] and [49] into [40] gives:

$$T_a = (P_a .m.\sigma_c / 4\gamma)\{\cos\Delta \int \text{integral between t1 and t2 for } (t^{0.5} -1) / t^{0.5} dt + \sin\Delta \int \text{between t1 and t2 } dt\}$$

$$= (P_a .m.\sigma_c / 4\gamma)\{(\cos\Delta /2)\{(t - t^{0.5})^{0.5}\ (2\ t^{0.5} - 1) - \ln[t^{0.25}+(t^{0.5} -1)]\} + \sin\Delta(t1 - t2)\} \qquad [50]$$

Substituting [44] and [49] into [41] gives:

$$N_a = (P_a .m.\sigma_c / 4\gamma)\{-\sin\Delta \int \text{between t1 and t2 of } (t^{0.5} -1) / t^{0.5} dt + \cos\Delta \int \text{between t1 and t2 of } dt\}$$

$$= (P_a .m.\sigma_c / 4\gamma)\ \{\cos\Delta(t1 - t2) - (\sin\Delta /2)\ [(t - t^{0.5})^{0.5}\ (2\ t^{0.5} - 1) - \ln[t^{0.25} + (t^{0.5} - 1)^{0.5}]\} \qquad [51]$$

Hence the average anchor tangential stress (τ_a) and normal stress (σ_a) that act on the failure surface of curvilinear length L_f can be expressed from [50] and [51], respectively, as follows:

$$\tau_a = T_a / L_f = (Pa..m.\sigma_c / 4\gamma L_f)\{[(\cos \Delta)/2] - \ln(t^{0.25} + + \sin \Delta(t_1 - t_2)\} \quad [52]$$

$$\sigma_a = N_a / L_f = (Pa.m.\sigma_c / 4\gamma L_f)\{\cos \Delta(t_1 - t_2) - [(\sin \Delta)/2] \times$$
$$[(t - t^{0.5})^{0.5} + (2\, t^{0.5} - 1)] - \ln[t^{0.25}(2\, t^{0.5} - 1)^{0.5}]\} \quad [53]$$

For a range of stress levels, [32] can be expressed in terms of a linear relationship between τ_f and σ_{nn} as an average value, as follows:

$$\tau_f = \tau_0 + \sigma_{nn} . \tan\beta \quad [54]$$

where: σ_{nn} = average normal stress at the failure surface, in kPa;

τ_0 = intercept strength at zero normal stress (cohesion of failed rock, at the failure surface), in kPa; and

τ = ϕ = average angle of internal friction, at the cohesion of failed rock, in deg.

The stress developed along the anchor can be expressed in terms of the passive and active safety factor by use of [52]–[54], as follows:

$$FS = \tau_f / \tau_{nt} \quad [55]$$

$$(FS)_{passive} = (\tau_0 + \sigma_{nn} \tan \phi + \tau_a + \sigma_a \tan \phi) / \tau_{nt}$$
$$= (\tau_f + \tau_a + \sigma_a \tan \phi) / \tau_{nt} \quad [56]$$

$$(FS)_{active} = (\tau_0 + \sigma_{nn} \tan \phi) / (\tau_{nt} - \tau_a) = \tau_f / (\tau_{nt} - \tau_a) \quad [57]$$

where: FS = factor of safety before reinforcement;

τ_{nt} = tangential stress acting on the failure surface, in kPa;

$(FS)_{passive}$ = factor of safety with passive reinforcement, which results when the anchor is not stressed upon placement; under this condition, the tangential anchor stress (τ_a) behaves like the resisting cohesion strength until slope displacement or dilation takes place; and

$(FS)_{active}$ = factor of safety with active reinforcement, which occurs when the anchor is prestressed immediately after placement; in this case a reduction of the active tangential stress by τ_a occurs in the denominator of [57].

Case study

A medium-quality shale deposit is situated 40 km west of Merida, Venezuela. It has a thickness of 20–80, m. Bedding and joint spacing of the rock is characteristically around 10–15 cm. The beddings have an average full dip of 30° away from the slope daylight and directed towards N50°E (see Fig. 3). It is planned to develop a quarry operation with bench height of $H = 20$ m and slope of 85°. Results of Geo-mechanics measurements are as follows: $\gamma = 25$ kN/m^3; $\sigma_c =$ 8000 kPa; = 0; $m_i = m$ for intact shale = 10; and RMR = rock mass rating (Bieniawski 1976) = 45.

Determination of the shape and specification of the potential bench failure surface

From Hoek and Brown (1988), for disturbed rock masses:

$$R = \exp[(\text{RMR} - 100)/14] = \exp[(45 - 100)/14] = \exp(-3.93) = 0.02$$

From Hoek and Brown (1980): $m = Rm_i = 0.02 \times 10 = 0.20$ and

$$s = \exp[(\text{RMR} - 100)/6] = \exp[(45 - 100)/6] = \exp(-9.17) = 10^{-4}$$

Utilizing [18], [17], [20], and [19] gives, respectively:

$t_1 = 2.26$; $x_c = 8.6811$, m; $t_2 = 1.01$; $x_2 = 8.6735$, m

Coordinates of the failure surface are given in Table 1 and its shape is shown in Fig. 4.

FIG. 3. A steep-sloped shale quarry 40 km west of Merida, Venezuela, showing bedding and weathering; Scale 1:200.

Utilizing [18], [17], [20], and [19] gives, respectively:

$$t_1 = 2.26; \quad x_c = 8.6811, \text{ m}; \quad t_2 = 1.01; \quad x_2 = 8.6735, \text{ m}$$

Coordinates of the failure surface are given in Table 1 and its shape is shown in Fig. 4.

TABLE 1. Coordinates of the slope failure surface

y	$(m)\ t$	x (m)
0	$1.0100 = t_2$	$8.6735 = x_2$
1	1.0725	8.5361
2	1.1350	8.3176
3	1.1975	8.0461
4	1.2600	7.7335
5	1.3225	7.3873
6	1.3850	7.0126
7	1.4475	6.6131
8	1.5100	6.1917
9	1.5725	5.7508
10	1.6350	5.2922
11	1.6975	4.8176
12	1.7600	4.3283
13	1.8225	3.8255
14	1.8850	3.3103
15	1.9475	2.7834
16	2.0100	2.2458
17	2.0725	1.6980
18	2.1350	1.1407
19	2.1975	0.5746
20	$2.2600 = t_1$	$0.0000 = x_1$

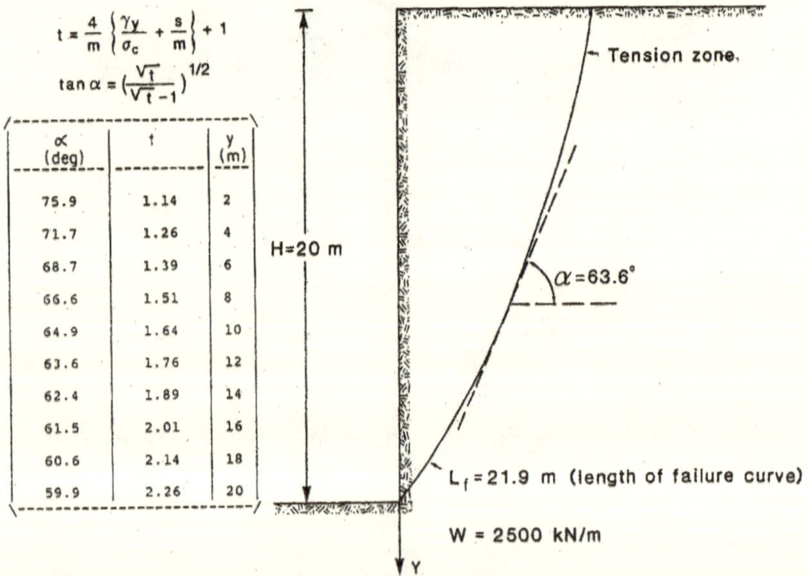

FIG. 4. Shape of the failure surface in a near-vertical slope.

The weight of the probable failure (W) can be determined using [22]: $W = 2500$, kN/m

The length of the failure curve (L_f) can be determined from [29]: $L_f = 21.9$, m

The failure surface curve shown in Fig. 4 can be divided into two sliding blocks. The upper block forms from the development of a tension crack, which reaches a depth equivalent to that where the minor principal stress (σ_3) approaches zero. Therefore, any of the equations [3], [37], or [44] for $\sigma_3 = 0$ gives:

$\alpha_{max} = \alpha$ at the surface ($y = 0$) = 85.9° (see Table 2)
α at the bottom of the tension crack = 73.22°
α_{min} at the toe of the slope = 59.9°
$\beta_{max} = 2\alpha_{max} - \pi/2$ = 81.8°
$\beta_{min} = 2\alpha_{min} - \pi/2$ = 29.8°

The maximum depth of the tension crack (y) and its horizontal distance from the slope toe can be determined from [16], [15], and [17], respectively, as follows:

$$y = (8000/4 \times 0.2 \times 25)\{0.2^2 \times 1.21 - [0.2^2 + (4 \times 10^{-4})]\} = 3.2, \text{ m}$$

$$x = [(0.2 \times 8000)/(8 \times 25)][\ln(t^{0.25} + (_t^{0.5} - 1)^{0.5} - (2\,_t^{0.5} - 1) + x_c] = 8.05, \text{ m}$$

Variation in the depth of the tension crack (y) and the normal stress acting on the potential failure surface (σ_{nn}) are given in Table 2 and the distribution of σ_{nn} is shown in Fig. 5.

Determination of the normal force (N) and the average normal stress (σ_{nn}) acting on the potential failure surface

Considering, as a first approximation, where the inter-slice force is neglected, then from equ. [58]:

$$N = \gamma \,.\, \text{Sum of } (y\, \Delta \,.\, \cos \alpha)$$

$$= \gamma \int \text{between t1 and t2 of } [\epsilon(t) .\, \cos \alpha\, (dx/dt)\, dt] \qquad [58]$$

TABLE 2. Variation in the depth of the tension crack (y) and the normal stress (***σnn***) with α.

α (deg)	y (m)	σ_{nn} (kPa)	$\sigma_{nn}/\gamma . H$
85.9	0.00	–3.96	0.008
76.6	1.92	0.00	0.000
75.9	2.00	1.01	0.002
71.7	4.00	11.72	0.023
68.7	6.00	27.77	0.055
66.6	8.00	46.30	0.093
64.9	10.00	68.07	0.136
63.6	12.00	90.48	0.181
62.4	14.00	117.16	0.234
61.5	16.00	142.04	0.284
60.6	18.00	172.25	0.344
59.9	20.00	200.00	0.400

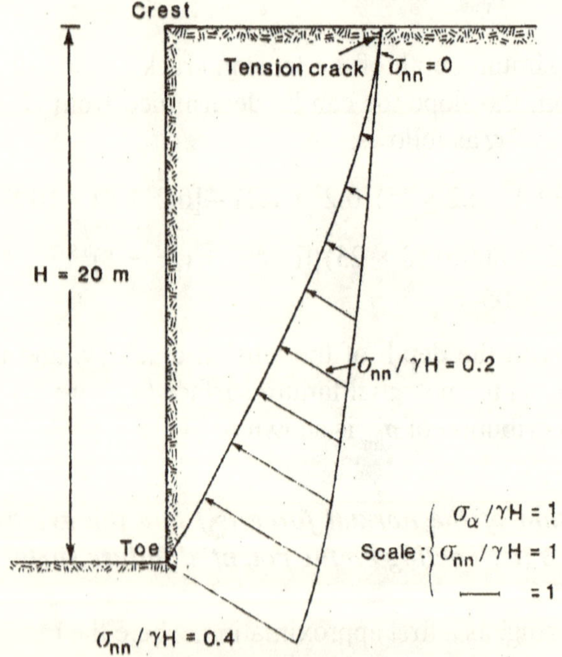

Fig. 5. Distribution of normal stress on the failure surface

where: Δx = width of each slice in the potential failure mass, m.

Substituting [10] and [44] into [58], after integration, gives:

$N = 1133.40$, kN/m and $\sigma_{nn} = N/L_f = 1133.4/21.9 = 51.75$, kPa

Determination of the average angles β and α

According to Ucar (1986):

$$\beta = \sin^{-1}\{(k/3)\, 2 \cos[1/3 \text{ arc} \cos(1\text{-}27/2k^3) + 240°] + 1\} = 42.2°$$

where: k = constant = $(8/m^2)\,(m\,.\sigma_{nn}/\sigma_c + s) + 3/2 = 1.78$ and

$$\alpha = [(\pi/2) + \beta]\,/2 = 66.10° \text{ (average value).}$$

Determination of the average shear strength (τ_f) acting on the potential failure surface

$$\tau_f = (m\sigma_c/8)[(1 - \sin\beta)/\tan\beta] = 72.41, \text{kN/m}^3$$

Determination of the tangential force (T) and the average disturbing shear stress (τ_{nt}) acting on the potential failure surface

Considering the inter-slice forces to be negligible, then:

$T = \gamma \sum y.\ \Delta x.\ \sin\alpha = \gamma \int$between t1 and t2 of $[\epsilon(t).\sin\alpha\ (dx/dt)\ dt]$ [59]

Substituting [10] and [44] into [59] and integrating gives:

$T = 2222.0$, kN/m (disturbing force)

$\tau_{nt} = T/L_f = 2222.0/21.9 = 101.46$, kPa

Determination of the slope factor of safety (FS)

From [55]: $FS = \tau_f/\tau_{nt} = 72.41/101.46 = 0.71$

This very low factor of safety indicates the requirement for ground reinforcement.

There is an alternative method for determining the slope factor of safety. It requires dividing the potential failure mass into two blocks, each having planar failure surfaces with angles of α_1 and α_2 to the horizontal (Huang 1983). The latter method gives a cubic equation for calculation of the safety factor. In the latter case the calculated safety factor (FS) was 0.65.

Comparison of the results obtained from these two methods shows a 9% difference. Each method offers advantages as well as inaccuracies due to the inherent assumptions. It is suggested that the average of the results of the two methods provides a more accurate FS values.

Determination of the amount of stress to be sustained by the anchor reinforcement* (P_a) *(when ground reinforcement is required)

Utilizing [54] gives: $\tau_0 = 72.41 - (51.75 \times \tan 42.2°) = 25.49$, kPa

Considering $\Delta = -10°$ (i.e., 10° below the horizontal), then [50] and [51] give respectively:

$T_a = 5.06P_{a,}$ kN/m; and $N_a = 21.19P_{a,}$ kN/m

Substituting the above findings into [52] and [53] gives, respectively,

$\tau_a = 0.23P_a$ kPa (as an average value)

$\sigma_a = 0.97P_a$ kPa (as an average value)

Utilizing [57] for an active and a more conservative case, with a $(FS)_{active} = 1.4$ and $\beta = \phi = 42.2°$, gives:

$(FS)_{active} = 1.4 = [72.41 + (0.97P_a \tan 2.2)] / (101.46 - 0.23P_a)$

Hence $P_a = 58$ in kPa, acting on the failure surface.

Validation of the simplified method

A more accurate method of predicting the potential slope failure surface involves the use of variational calculus, since it satisfies all of the equations of equilibrium in a search for the minimum slope safety factor (Ucar 1988). However, this method involves lengthy calculation and requires many factors to be defined and monitored in the field, adding to design costs. The simplified method suggested earlier utilized Bieniawski's rock mass rating in a nonhomogeneous media. This method has been compared with results obtained using the variational calculus method. A summary of the output from the two methods is given in Table 3 and compared graphically in Fig. 6. This shows that (*a*) predictions of the slope failure surface locations agree closely; slight differences up to a maximum of 12.3% occur near the tension crack region, and (*b*) predictions of the normal stress distribution (σ_{nn}) by the two methods agree very closely, with the maximum relative error of 10% occurring near the toe of the slope. Hence the assumption of $\sigma_v = \gamma/y$ for the near-vertical benches appears to have been a good approximation.

TABLE 3. Comparison of the simplified method, suggested here, with the Ucar variational calculus method

	Simplified method (this investigation)			Variational calculus method (Ucar 1988)		
y (m)	x (m)	σ_{nn} (kPa)	$\sigma_{nn}/\gamma.\ H$	x (m)	σ_{nn} (kPa)	$\sigma_{nn}/\gamma.\ H$
0	8.67	– 3.96	0.007	7.60	– 4.00	- 0.008
4	7.73	11.50	0.023	7.35	11.00	0.022
8	6.20	46.50	0.093	6.10	44.50	0.089
12	4.33	90.50	0.181	4.25	80.50	0.161
16	2.25	142.00	0.284	2.30	125.00	0.250
20	0.00	200.00	0.400	0.00	180.00	0.360

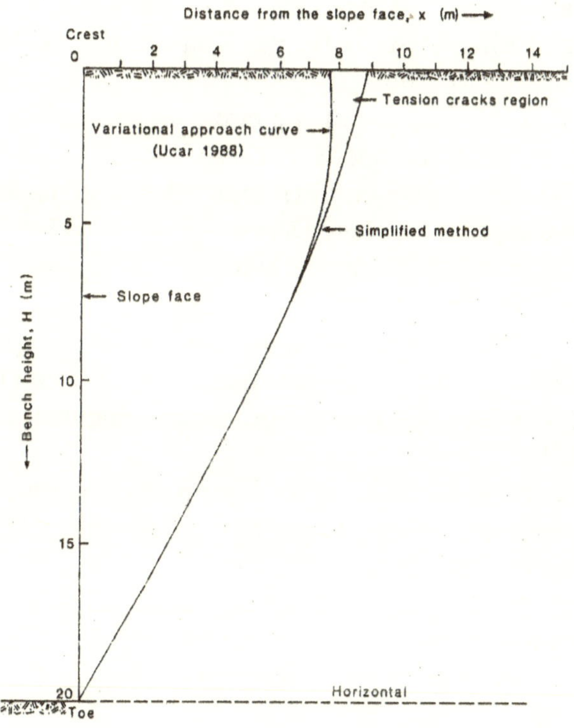

FIG. 6. Comparison of predicted slope failure surfaces obtained from the simplified and variational calculus methods.

Conclusions

The Hoek and Brown failure criterion, used in conjunction with Bieniawski's rock mass classification system, appears to be useful for the prediction of the potential shape, location, and extent of failure in steep slopes. This has been demonstrated by close agreement with analyses by the variational calculus method. This simplified technique presented also enables the prediction of the probable location and extent of tension cracks associated with failure relative to the crest of the slope.

References

BAKER, R., and GARBER, M. 1977. Variational approach to slope stability. Proceedings, 9th International Conference on Soil Mechanics and Foundation Engineering, pp. 9–12.

BIENIAWSKI, Z.T. 1976. Rock mass classification in rock engineering. Proceedings, Symposium on Exploration for Rock Engineering. A.A. Balkema, Rotterdam, The Netherlands, vol. 1, pp. 97–106.

BISHOP, A.W. 1955. The use of slip circle in the stability analysis of slopes. Géotechnique, **5**: 7-17.

FELLENIUS, W. 1936. Calculation of stability of earth dams. Proceedings, 2nd Congress on Large Dams, vol. 4, pp. 445–463.

GOODMAN, R.E. 1980. Rock mechanics. J. Wiley and Sons, New York, NY. pp. 99–100.

GUDEHUS, G. 1972. Lower and upper bounds for stability of earth-retaining structures. Proceedings, 5th European Conference on Soil Mechanics and Foundation Engineering, Madrid, vol. 1.

HOEK, E., and BROWN, E.T. 1980. Empirical strength criterion for rock masses. ASCE Journal of the Geotechnical Engineering Division, **106:** 1013–1035.

——— 1988. The Hoek-Brown failure criterion—a 1988 update. Rock Engineering for Underground Excavation, 15th Canadian Rock Mechanics Symposium, University of Toronto, pp. 31–38.

HUANG, Y.H. 1983. Stability analysis of earth slopes. Van Nostrand Reinhold Co., pp. 14–15. New York, NY.

JANBU, N. 1973. Slope stability computation. *In* Embankment-dam engineering, Casagrande volume. *Edited by* R.C. Hirschfeld and S.J. Poulos. J. Wiley and Sons Ltd., New York, NY. pp. 47–86.

KÖTTER, F. 1903. Bestimmung des Drucks an Gekrummten Geleiflächen, eine Aufgabe an der Lehre von Erddruck. Sitzungsberichte der Preussischen Akademie der Wissenschaften zu Berlin, pp. 229–233.

KRANZ, E. 1972. Ground anchors. *In* French code of practice. Éditeur Éyrolles, Paris, France, Recommendation TA. 72.

MORGENSTERN, N.R., and PRICE, V.E. 1965. The analysis of the stability of general slip surfaces. Géotechnique, **15**: 70–83.

PRIEST, S.D., and BROWN, T. 1983. Probabilistic stability analysis of variable rock slopes. Institution of Mining and Metallurgy, Transactions, Section A: Mining Industry, **92**: A1–A12.

SPENCER, E. 1967. A method of analysis of the stability of embankments assuming parallel interslice forces. Géotechnique, **17**: 11–26.

STOCKER, M.F., KORBER, G.W., GASSLER, G., and GUDEHUS, G. 1979. Soil nailing. Comptes Rendus, le Colloque International de Renforcement des Sols, Paris.

UCAR, R. 1986. Determination of shear failure envelopes in rock masses. ASCE Journal of Geotechnical Engineering, **112**; 303–315.

——— 1988. New design methods for ground anchoring. Ph.D. thesis, Department of Mining and Metallurgical Engineering, McGill University, Montréal, Quebec, Canada.

3.4.3 DESIGN OF ANCHORING IN ROCK WITH PLANAR FAILURE

By: R. Ucar, F.P. Hassani and Andy A. Afrouz

ABSTRACT: This paper reports on an analytical approach to the determination of a minimum safety factor for rock slopes subject to planar failure, prior to reinforcement. The safety factor determined by this approach may then be used in the design of active and passive anchor systems for slope support under static and dynamic conditions. It may also be utilized to evaluate the anchor force and inclination for optimum rock reinforcement. Practical design examples are given by taking into account parameters such as: groundwater pressure, surcharge, seismically, anchor dimensions and passive and active forces.

1. INTRODUCTION

Although the surface of a failure plane is three dimensional, it is generally approximated in two dimensions. In this case, most slope failures fall into the following categories:

(i) Sliding on a single, continuous plane. This is the simplest case and is represented by a mass resting on a persistent discontinuity.
(ii) Sliding on a set of intersecting planes. This is represented by a combination of planes such as soil - rock interface, composite surfaces of two or more discontinuous joint sets, or continuous intersecting planes.

Prior work exists on rock slope reinforcement under planar failure conditions (Hoek and Bray, 1977, Hanna, 1982, Attewell and Farmer, 1976, Brandle, 1979 and Sigmiller, 1975). Such reinforcement has been more successful in rock slopes with large blocks than in soft and fractured rock masses (Rabcewicz, 1955).

Reinforcement of planar slides in slopes was further studied when the maximum excess shear stress of the rock acting on the failure plane (τ_e) was considered to act along the failure plane whose critical angle was half the slope angle plus the angle of internal friction (Baoshen and Jinshow, 1983, Hanna, 1982, Hobst and Zajic, 1983, Dight, 1983 and Barron et al., 1971). However, the above approaches consider a slope safety factor of one under passive conditions and neglect the cohesion along the failure plane. This was clarified in the case of utilizing active support by Sigmiller (1982), but without incorporating the optimum inclination of anchors.

Some investigators simplified the failure plane into two planar wedge surfaces (Gudehus, 1972, Krauz, 1972, Goodman et al., 1982 and Jumikis, 1983). In this method the traction effect of the tension crack in the upper slope was considered to be linear. Other work indicated that maximum reinforcement may be achieved by horizontal anchors (Farmer, 1968). This condition is evident when the minor principal stress (σ_3) is horizontal.

This paper attempts to consider the complete range of design conditions and specifications for rock anchors in mine slope design. The Mohr-Coulomb failure criterion was considered in order to obtain the slope safety factor without any reinforcement. This is based on the following:

(a) Average cohesion (S_0) and friction angle (ϕ) for the slope occur on the failure surface after a relatively small displacement.
(b) Shear failure occurs on an inclined planar surface passing through the slope toe.
(c) Effects of seismic activity, pore water pressure and surcharge are included for cases (a) and (b).
(d) Sliding is conditional upon the following geometrical controls being satisfied:

(i) the strike of the failure surface is within ± 20° of the slope strike.
(ii) the failure plane must "daylight" in the slope face, i.e. $\beta > \alpha$
where: α = potential failure plane inclination to horizontal

β = slope face inclination to horizontal

(iii) the dip of the failure plane be greater than the internal friction angle of the rock (ϕ) i.e. $\alpha > \phi$.

Therefore, for sliding to take place: $\beta > \alpha > \phi$

2. ANALYSIS WITHOUT REINFORCEMENT

Fig. 1 shows the general sedimentary, igneous and metamorphic range of structural environments encountered in surface mine slope design.

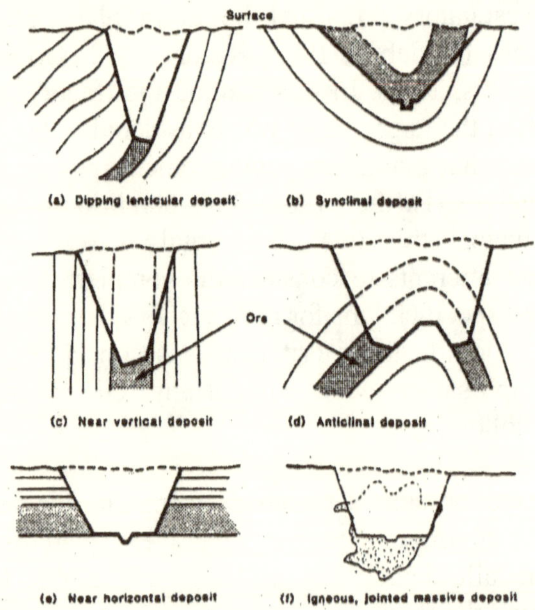

Fig. 1 - Generalized Surface Mine Environments

Considering a mine slope configuration where the groundwater table is above the pit bottom and a potential failure plane exists at angle (α) to the horizontal, less than the final pit slope (β). In this case, the minimum safety factor is considered to exist when the failure plane develops in the slope when no reinforcement is in place. Fig. 2 indicates a surcharge (q), blasting or seismic coefficient (K), erratic surface, the weight of the failure wedge and the water force acting through the centroid of the sliding mass without rotation.

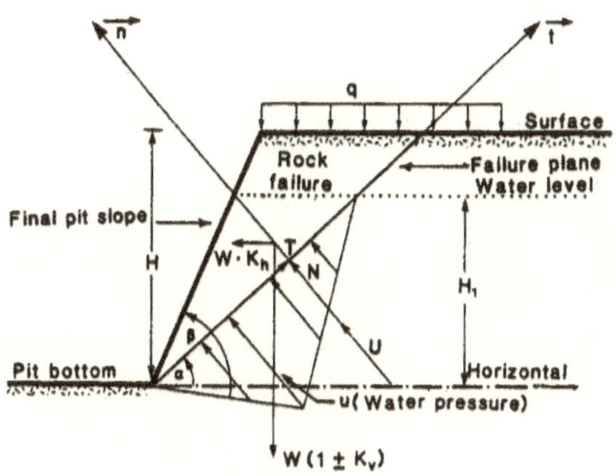

Fig. 2 - Planar Failure Slope Configuration

$$\text{(A) At } \Sigma F_n = 0 : N + U - W(1 \pm K_V) \cos\alpha + \tilde{W} K_h . \sin\alpha = 0 \quad (1)$$

where: F_n = Sum of the forces acting normal to the failure plane, kN/m

N = Reaction force acting normal to the failure plane, kN/m

U = Total neutral force acting normal to the failure plane
$= H_1^2 \gamma_w (\cot\alpha - \cot\beta) \sec\alpha / 2$, kN/m

W = Weight of the sliding block, kN/m

K_v = Vertical component of the seismic factor = $K_h/2$, in practice

K_h = Horizontal component of the seismic factor
= 0.10 to 0.33, for a = 0.1 g_2 to 0.33 g

a = Seismic acceleration, m/sec^2

g = Acceleration due to gravity, m/sec^2

α = Inclination of failure plane to horizontal, degrees

β = Inclination of the pit wall to horizontal, degrees

γ = Unit weight of rock, kN/m^3

γ_w = Unit weight of water, kN/m^3

H_1 = Height of the horizontal groundwater table above the final pit bottom, m

From equation (1): $N = W(1 \pm K_V) \cos\alpha - U - K_h \tilde{W} \sin\alpha$ (2)

The magnitude of the horizontal seismic force (F_h), in kN per metre of the anchor can be calculated as follows:

$$F_h = m \cdot a = W \cdot a / g = W \cdot K_h \qquad (3)$$

where: m = Failure mass, $kN.sec^2/m^2$

(B) At $\Sigma F_t = 0 : T - W(1 \pm K_v) \sin\alpha - \tilde{W} K_h . \cos\alpha = 0$, or

$$T = W(\pm K_v) \sin\alpha + \tilde{W} K_h . \cos\alpha \qquad (4)$$

where: ΣF_t = Sum of the forces acting along the failure plane, kN/m
T = Force along the failure plane, kN/m

According to its definition, the factor of safety (FS) = Resisting force / Disturbing force

= (Cohesional force + Frictional resistance force) / T

$$= \{(S_o . H / \sin\alpha) + [W(1 \pm K_v) \cos\alpha - U - W. K_h . \sin\alpha] \tan\phi\} / [W(1 \pm K_v) \sin\alpha + W . K_h . \cos\alpha] \qquad (5)$$

Considering surcharge (q) and water level at H_1, then the total weight of rock above the failure plane (W) can be calculated as follows:

$$W = q. H(\cot\alpha - \cot\beta) + (H - H_1)\{[H(\cot\alpha - \cot\beta) + H_1(\cot\alpha - \cot\beta)]/2\} + [H_1^2 . \gamma_{sat} (\cot\alpha - \cot\beta) / 2]$$

$$= (\cot\alpha - \cot\beta)\{q. H + [\gamma(H^2 - H_1^2)/2] + (H_1^2 . \gamma_{sat}/2)\} \qquad (6)$$

where: H = Total height of the slope, m

γsat = Unit weight of saturated rock, kN/m^3

$\cot\alpha - \cot\beta = \sin(\beta - \alpha) / \sin\alpha . \sin\beta$

Total weight factor (ψ), in kN/m = $q. H + [\gamma(H^2 - H1^2)/2] + (H1^2 . \gamma sat/2)$ (7)

Then equation (6) can be written as:

$$W = \psi[\sin(\beta - \alpha)/(\sin\alpha . \sin\beta)] \qquad (8)$$

The water weight factor $(\psi_{1)} = \gamma w \,.\, H1^2 / 2$ (9)

The neutral force due to water uplift (U)
$= \psi_1 \sec \alpha \,.\, \sin(\beta - \alpha) / \sin\alpha \,.\, \sin\beta$ (10)

Therefore, the safety factor (FS) in terms of ψ and ψ_1 from equations (5), (8), (9), and (10) can be expressed as follows:

$$FS = \{S_o.\ H + [\sin(\beta - \alpha)/\sin\beta]\ [\psi\ (1 \pm K_v)\ \cos\alpha - \psi 1\ .\ \sec\alpha - Kh\ \psi.\ \sin\alpha]\ \tan\phi\}/\{[\psi\ (1 \pm K_v)\ \sin\alpha + Kh\ \psi.\ \cos\alpha]\ \sin(\beta - \alpha)/\sin\beta\}$$

$$= \{[S_o.\ H.\ \sin\beta\ /\ \psi\ \sin(\beta - \alpha)] + \tan\phi\ [(1 \pm K_v)\ \cos\alpha - (\psi_1\ .\ \sec\alpha\ /\ \psi) - K_h\ .\ \sin\alpha]\}/\ [(1 \pm K_v)\ \sin\alpha + 'K_h\ .\ \cos\alpha] \quad (11)$$

Let: $1 \pm K_v = \Omega$ and $\psi_1/\psi = \Omega_1$ (12)

Substituting equ. (12) into equ. (11) in a dimensionless form yields:

$$FS = \{S_o.\ H.\ \sin\beta/\psi.\ \sin(\beta - \alpha)] + \tan\phi\ (\Omega\ .\ \cos\alpha - \Omega_1\ .\ \sec\alpha - K_h\ .\ \sin\alpha)\}/\ (\Omega\ .\ \sin\alpha + K_h\ .\ \cos\alpha) \quad (13)$$

It should be noted that for any constant value of S_o, ϕ, ψ, H, H_1, β, q, K_v and K_h, then the safety factor will be a minimum when $\partial FS/\partial\alpha = 0$. Thus, differentiating equation (11) with respect to (α) and equating to zero gives:

$$[S_o.\ H\ .\ \sin\beta\ /\psi.\ \sin^2(\beta - \alpha)]\ [K_h\ .\ \cos(\beta - 2\alpha) - \Omega\ .\sin(\beta - 2\alpha)] - \tan\phi\ [\Omega^2 + \Omega\ .\ \Omega_1\ (\tan^2\alpha - 1) + 2\ \Omega_1\ .\ K_h.\ \tan\alpha + K_h^2] = 0 \quad (14)$$

For a particular case when there is no surcharge (q = 0), on a fully drained slope, ($H_1 = 0$ and $K_h = K_v = 0$), then equations (6), (8) and (10) give:

$$[\sin^2(\beta - \alpha)/\sin(\beta - 2\alpha)] + (2\ S_0\ .\ \sin\beta/\gamma\ H)/\tan\phi = 0 \quad (15)$$

3. SLOPE ANALYSIS WITH ROCK ANCHORING

There are two methods of rock anchoring, passive and active.

3.1 - Passive anchoring results when the anchor is not pre-stressed upon placement. Under this condition, the tangential passive anchor force $(T_a)_{passive}$ behaves similar to the resisting cohesive force until slope displacement or dilation takes place. In this case, the normal and shear components of the support force both appear in the numerator. Therefore:

The tangential passive force, $(T_a)_{passive} = (F_a)_{passive} \cos(\alpha - \Delta)$, and (16a)

The normal passive force, $(N_a)_{passive} = (F_a)_{passive} \sin(\alpha - \Delta)$ (16b)

where: F_a = anchor force along its length (Fig. 3), kN/m

Δ = angle of anchor to horizontal (Fig. 3)

T_a = tangential anchor force, kN/m

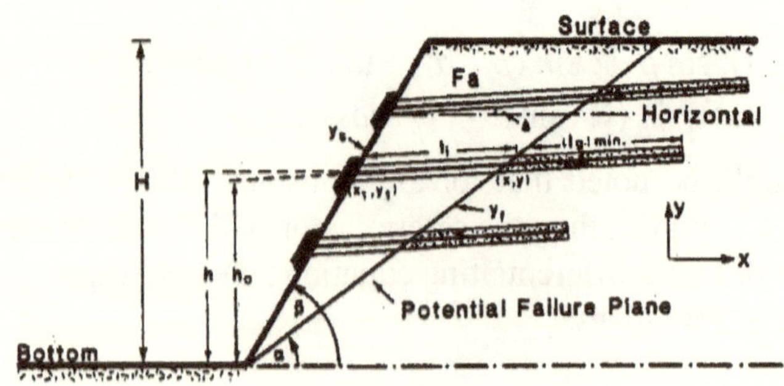

Fig. 3. Position of Anchor Relative to the Pit Slope.

From equations (5) and (16):

$$(FS)passive = \{[(S_o \cdot H / \sin\alpha) + F_a \cdot \cos(\alpha - \Delta)] + [W(1 \pm K_v)\cos\alpha - U - K_h \cdot W \cdot \sin\alpha + F_a \cdot \sin(\alpha - \Delta)]\tan\phi\} / [W(1 \pm K_v)\sin\alpha + K_h \cdot W \cdot \cos\alpha] \quad (17)$$

3.2- Active anchoring results when the anchor is pre-stressed immediately after placement. Therefore, a reduction of the active force by T_a, ($= F_a . \cos(\alpha - \Delta)$) occurs in the denominator, and the additional resisting force is:

$$(N_a)_{active} = (F_a)_{active} . \sin(\alpha - \Delta) \qquad (18)$$

where: N_a = normal active anchor force component, kN/m

From equations (5) and (18):

$$(FS)_{active} = \{(S_o . H / \sin\alpha) + W(1 \pm K_v)\cos\alpha - U - K_h . W . \sin\alpha + F_a . \sin(\alpha - \Delta)]\tan\phi\}/[W(1 \pm K_v)\sin\alpha + K_h . W . \cos\alpha] - F_a . \cos(\alpha - \Delta) \qquad (19)$$

4. OPTIMUM ANCHOR INCLINATION

The angle between the rock anchor and the horizontal (Δ) is considered as positive when above horizontal and negative when below horizontal. This is in accordance with the co-ordinate system shown in Fig. 3. Let:

$$Y_o = \{(S_o . H / Fa \sin\alpha) + [W(1 \pm K_v)\cos\alpha - U - W . K_h . \sin\alpha]\tan\phi\}/F_a \qquad (20a)$$

$$Y1 = Y_o . F_a = \text{Resistive force along the failure plane} \qquad (20b)$$

$$Y2 = W[(1 \pm K_v)\sin\alpha + K_h . \cos\alpha]/F_a \qquad (20c)$$

$$Y3 = Y2 . F_a = \text{Disturbing force along the failure plane}$$

$$= W[(1 \pm K_v)\sin\alpha + K_h . \cos\alpha] \qquad (20d)$$

Then, FS without the reinforcement = $Y_0 / Y_2 = Y_1 / Y_3$ (20e)

Considering both the passive and active cases, then the following analysis can be made:

4.1- For Passive Anchoring: Substituting equations (20) into equation (18) gives:

$$(FS)_{passive} = \{Y_1 + F_a[\cos(\alpha - \Delta) + \sin(\alpha - \Delta)\tan\phi]\}/ Y_3 \qquad (21)$$

and $F_a = [Y_3 . (FS)_{passive} - Y_1] / [\cos(\alpha - \Delta) + \sin(\alpha - \Delta)\tan\phi]$

$$= [Y_3 . (FS)_{passive} - Y_1] / f(\Delta) \qquad (22)$$

where: $f(\Delta) = \cos(\alpha - \Delta) + \sin(\alpha - \Delta)\tan\phi$ (23a).

$$Fa = Y3[(FS)_{passive} - (FS)]/f(\Delta) \qquad (23b)$$

Differentiating equ. (23) with respect to d f Δ and equating to zero results:

$$f'(\Delta) = d.\, f(\Delta) / d\Delta = 0 \tag{24}$$

$$\sin(\alpha - \Delta) - \cos(\alpha - \Delta) \tan\phi = 0 \tag{25a}$$

$$\tan(\alpha - \Delta) = \tan\phi \tag{25b}$$

$$\Delta = (\alpha - \phi) \tag{25c}$$

Equations (24), and (25) are similar to those obtained by Barron et al. (1971), except that they now consider the cohesion of the rock mass.

In passive anchoring, three conditions are assumed to prevail:

(a) when displacement of the rock mass occurs, then the necessary anchor force (F_a) to provide equilibrium must correspond to a safety factor of $1 \leqslant (FS)_{passive} \leqslant 2$
(b) the value of the cohesive strength (or bonding strength due to cohesion) decreases gradually to a residual value (S_r) and finally approaches zero when the slide occurs.
(c) the internal friction angle (ϕ) reduces to the residual value (ϕ_r).

Hence, equation (20) will become:

$$Y1 = [W[(1 \pm K_v) \cos\alpha - U - W.\, K_{h'}.\, \sin\alpha] \tan\phi_r \tag{26}$$

From equations (23), (26) and for $(FS)_{passive} = 1$:

$$F_a = F_e / f(\Delta) = \{W[(1 \pm K_v) \sin\alpha + W.\, K_{h'}.\, \cos\alpha] - [W[(1 \pm K_v) \cos\alpha] - U - W.\, K_{h'}.\, \sin\alpha] \tan\phi_r\} / [\cos(\alpha - \Delta) + \sin(\alpha - \Delta) \tan\phi_r] \tag{27}$$

where: F_e = The excess shear force acting downwards along the failure plane, kN/m

= Disturbing force (Y3) - Resistive force (Y1)

Equation (27) is similar to that of Barron et al. (1971) at SF = 1 and $S_o = 0$. However, the following differences exist between the two:

(a) the anchor force (F_a) in equation (27) is for the passive case, whereas in the case of Barron et al. (1971) it implies a prestress force, i.e. active anchor force.
(b) in this study, F_e (equation 27) is acting on a failure plane of inclination equivalent to a minimum safety factor. In Barron et al. (1971) the F_e is a maximum value acting on a failure plane of inclination $\alpha = (\beta + \phi)/2$.

4.2- For Active Anchoring: Substituting equations (20) into equation (19) gives:

$$(FS)active = [Y0 + \sin(\alpha - \Delta) \tan\phi] / [Y2 - \cos(\alpha - \Delta)] \quad (28)$$

For Y0, Y2, α and ϕ to be constant, then:

$$(FS)_{active} = f(\Delta) \quad (29a)$$

$$\text{and } f'(\Delta) = \delta F / \delta \Delta = (FS)_{active} / d\Delta = 0 \quad (29b)$$

Substituting equations (20) and (28) into equations (29) gives:

$$\delta (FS)active / \delta \Delta = \{[Y0 + \sin(\alpha - \Delta) \tan\phi] \sin(\alpha - \Delta) - \cos(\alpha - \Delta) \tan\phi\, [Y2 - \cos(\alpha - \Delta)]\} / [Y2 - \cos(\alpha - \Delta)]^2 = 0 \quad (30)$$

$$\text{Then: } \sec(\alpha - \Delta) = (Y3 / F_a)\, [1 - (FS). \tan(\alpha - \Delta) / \tan\phi] = Y2\, [1 - \mu. \tan(\alpha - \Delta)] \quad (31)$$

where: (FS) = Safety factor without reinforcement
$\mu = (FS) / \tan\phi$

$$\text{Hence: } \tan(\alpha - \Delta) = \{\mu. Y2^2 \pm [Y2^2(1 - \mu^2) - 1]^{0.5}\} / Y2^2 . \mu^2 - 1] \quad (32)$$

Generally, $(\alpha - \Delta) < \pi/2$, and for $(FS)_{active}$, $\tan(\alpha - \Delta)/\tan\phi < 1$, equation (32) can be presented as follows:

$$\tan(\alpha - \Delta) = [(FS).Y2^2 / \tan\phi] - \{Y2^2 [1 + (FS)^2 / \tan^2\phi - 1]^{0.5}\} / \{Y2^2 (FS)^2 / \tan^2\phi - 1]\} \quad (33)$$

To obtain the inclination angle for anchor (Δ) for an optimum anchor force (F_a) and a known active safety factor $(FS)_{active}$, substituting equations (20) into (28) gives:

$$(Y3 / F_a)\,[(FS)_{active} - (Y1 / Y3)] = (FS)_{active} \cdot \cos(\alpha - \Delta) + \sin(\alpha - \Delta)\tan\phi \qquad (34)$$

Therefore:

$$F_a = Y_3\,[(FS)_{active} - (FS)] / [\cos(\alpha - \Delta)\,(FS)_{active} + \sin(\alpha - \Delta)\tan\phi]$$
$$= Y_3 \,.\, \delta(FS) / f(\Delta)_{active} = Y_3 \,.\, \delta(FS) / [(FS)^2_{active} + \tan^2\phi]^{0.5} \qquad (35)$$

where: $\delta\ (FS) = (FS)_{active} - (FS)$

$Y2 = f(\Delta)_{active} / \delta(FS)$

$f(\Delta)_{active} = [(FS)^2_{active} + \tan^2\phi]^{0.5}$, for the optimum Δ- angle; and

$f(\Delta)_{active} = \cos(\alpha - \Delta)\,(FS)_{active} + \sin(\alpha - \Delta) \,.\, \tan\phi$, for all Δ - angles.

$$\text{Thus: } \tan(\alpha - \Delta) = \tan\phi / (FS)_{active} = \tan(\alpha - \Delta_1) \qquad (36)$$

where: $\Delta_1 = \Delta = (\alpha - \phi)$ for $(FS)_{active} = 1$

It is suggested that equation (36) be used rather than equation (33) for design purpose. In this case, the active safety factor $(FS)_{active}$ is selected according to Table 1 and substituted together with ϕ and α- values in equation (36) to obtain the optimum angle of anchor to the horizontal (Δ). This - value is then substituted in equation (35) to calculate the optimum anchor force (F_a).

Table 1 - Significance of Active Safety Factor $(FS)_{active}$ in Various Ground Conditions

Ground Condition	$(FS)_{active}$ of structure	
	Temporary	Permanent
Unsafe	1<	1.1<
Low confidence	1.0 – 1.2	1.1 – 1.3
Safe for cut and fill, pillar and most mining structures or excavations	1.2 – 1.4	1.3 – 1.6
Safe for dams	1.5 – 2.0	

5. DETERMINATION OF THE DISTANCE BETWEEN ANCHOR HEAD AND FAILURE SURFACE (l_i)

The above noted length (l_i) is dependent upon:

(a) boundary conditions of the slope (Fig. 3),
(b) location of the anchor in the slope and
(c) angle of the anchor to the horizontal (Δ).

Referring to Fig. 3, the following conditions exist at (x,y), $y_a = y_f$:

$y_{slope} = y_s = x \,.\, \tan\beta;\ y_{failure} = y_f = x \,.\, \tan\alpha;\ y_{anchor} = y_a = x \,.\, \tan\Delta + h_0$

and $\tan\Delta = (y\text{-}h)/(x\text{-}x_1)$

Therefore, $y_a = x \,.\, \tan\Delta + (h - x_1 \,.\, \tan\Delta) = x \,.\, \tan\Delta + h_0$ (37)

where: $h_0 = h - x_1 \,.\, \tan\Delta$ (38)

Thus, at (x,y): $x \,.\, \tan\alpha = x \,.\, \tan\Delta + h_0$

$$h_0 = x\,(\tan\alpha - \tan\Delta), \text{ and}$$
$$x = h_0 / (\tan\alpha - \tan\Delta)$$

From the geometry of Fig. 3:

$$l_i = (x - x_1)/\cos\Delta$$
$$= \sec\Delta\{[h_0/(\tan\alpha - \tan\Delta)] - x_1\} \quad (39)$$

At (x_1, y_1): $y_a = y_s$ and $x_1 \,.\, \tan\Delta + h_0 = x_1 \,.\, \tan\beta$

Therefore, $h_0 = x_1(\tan\beta - \tan\Delta) = h - x_1 \,.\, \tan\Delta$ (40)
Hence, $x_1 = h/\tan\beta$ (41)

Thus from equations (39) and (41):

$$l_i = \sec\Delta\left\{[(h - x_1 \,.\, \tan\Delta)/(\tan\alpha - \tan\Delta)] - (h/\tan\beta)\right\}$$
$$= (h/\sin\beta)\,[\sin(\beta - \alpha)/\sin(\alpha - \Delta)] \quad (42)$$

where, angle Δ is taken as positive above horizontal and negative below horizontal.

Sigmiller (13) suggested the following expression for determination of the length (l_i):

$$l_i = h\left\{(\cot\alpha - \cot\beta)\,[\sin\Delta.\,\mathrm{Cot}\,(\alpha - \Delta) + \cos\Delta]\right\} \tag{43}$$

Equations (42) and (43) give identical results but equation (42) provides a simpler approach. For practical purposes the minimum l_i- value has been indicated to lie from 4.5 to 7 meters (Barron et al. 1971 and Sigmiller 1982).

6. DETERMINATION OF MINIMUM GROUTING LENGTH (l_g)

The minimum length of grouting depends mainly on:

(a) the anchor force (P, in kN),
(b) the hole diameter (D, in m),
(c) the bonding strength along the grout-rock and grout-anchor interfaces, and
(d) the rock strength and frictional parameters.

Experience has shown that the bonding strength along the grout-anchor interface is significantly greater than the bonding strength between the rock and anchor. Hence, the larger the length of grouting the higher is the bonding strength of anchors.

Utilizing the Mohr-Coulomb failure criteria to calculate the minimum length of grouting $(l_g)_{min}$ will give:

$$(l_g)\mathrm{min} = \ln\{\{8P/2\pi.\,Sb.\,\tan[(45 + \phi b/2)]\,D^2.\,K\}\tan[45 + (\phi b/2)] + 1\}D.\,K/4 \tag{44}$$

where: ϕ_b = internal friction angle of the rock and grouting, in degree, to be evaluated in the field, or laboratory condition.

S_b = bonding cohesion between rock and grouting, in MPa, to be evaluated in the field and laboratory condition.

K = dimensionless constant relating to the shear and tensile strength of the anchor, to be evaluated in the field = 250 in soft fissured rocks.

In heavily fissured rock masses, the values of ϕ and S_o can be taken as that of the rock alone. In such case, it is conservative to utilize the following expression in equation (44):

$S_o = S_b$ = cohesion of the rock mass, MPa

$\phi = \phi_b$ = internal friction angle of the rock mass, degrees

$$Q = 2\, S_o \,.\, \tan [45+(\phi /2)] \tag{45}$$

Substituting equation (45) into equation (44) gives the following simplified general expression:

$$(l_g)_{min} = \ln\{8P/(\pi\, Q.\, D2.\, K) \tan [45 +(\phi\, b/2)] + 1\}D.\, K/4 \tag{46}$$

7. DETERMINATION OF MAXIMUM GROUTING LENGTH $(l_g)_{max}$

The worst condition exists when the ground is heavily fractured and the normal stress at the rock and grout interface (Q_n), approaches to zero. Then K approaches ∞ and in this case, the grouting length to promote ground stability is a maximum, $(l_g)_{max}$. In this case, let:

$$8P\,.\, \tan [45+(\phi /2)] /\pi\,.\, Q\,.\, D^2 = \mu \tag{47}$$

Substituting equation (47) into (46):

$$(l_g)_{min} = D\,.\, K.\, \ln [(\mu /K) + 1] /4 = \ln \{[(\mu /K) + 1]^{(k/\mu)}\}(\mu.\, D/4) \tag{48}$$

When limit of K ---> ∞ $[(\mu /K) + 1]^{(k/\mu)}$ ---> e, the equation (48) becomes:

$$(l_g)_{max} = \ln\,.\, e^{(D.\,.\mu/4)} = D.\, \mu /4 = 2P\,.\, \tan [45 + (\phi /2)] /\pi\,.\, D\,.\, Q \tag{49}$$

8. DETERMINATION OF THE ANCHOR LENGTH (l)

According to Fig. 3, the anchor length (l) can be determined by the following expression:

$$l = l_i + (l_g)_{min} + \Delta_l \tag{50}$$

where: Δ_1 = excess length of anchor which stays out of the hole and holds the reinforcement plate with screw and knots, in m.

In equation (50), l_i and $(l_g)_{min}$ can vary depending upon the rock type, state of fracture in the rock mass, position and slope of the failure plane, position of the anchor relative to the mine or any excavation slope and type of the grouting material, i.e. as $(l_g)_{min}$ and or $(l_g)_{max}$.

9. ESTIMATION OF DYNAMIC SAFETY FACTOR FOR PLANAR FAILURE SLOPES

From equation (11), the dynamic safety factor $(FS)_{dynamic}$ can be expressed as follows:

$$(FS)_{dynamic} = \{[S_o . H. \sin\beta / \psi .\sin(\beta - \alpha)] + [(1 \pm K_v) \cos\alpha - (\psi_1 \cdot_{\sec\alpha} / \psi) - K_h. \sin\alpha] \tan\phi\} / [(1 \pm K_v) \sin\alpha + K_h. \cos\alpha]$$

$$= [1 / (1 \pm K_v)] [1 + (\tan\epsilon. \cot\alpha)]\{\{[S_o . H. \sin\beta / \psi .\sin(\beta - \alpha)] + [\tan\phi . \cos\alpha - \psi_1 / \psi]\} / \sin\alpha\} \pm [\cos\alpha. \tan\phi. \tan\epsilon / \sin\alpha (1 + \tan\epsilon / \cot\alpha)] - [\tan\epsilon.\tan\phi / (1 + \tan\epsilon.\cot\alpha)] \quad (51)$$

where: Horizontal component of the seismic force = W. K_h, Vertical component of the seismic force in downward and upward directions = W $(1 \pm K_v)$, and $\tan\epsilon = K_h / (1 \pm K_v)$

Then: $$(FS)\ static = \{[S_o . H. \sin\beta / \psi. \sin(\beta - \alpha)] + [\cos\alpha - (\psi_1 / \psi)] \tan\phi\} / \sin\alpha \quad (52)$$

Substituting equation (52) into equation (51) gives:

$$(FS)_{dynamic} = [(FS)\ static / (1 \pm K_v)(1 + \tan\epsilon.\cot\alpha)] \pm [n. \tan\epsilon. \tan\phi / (\tan\alpha + \tan\epsilon) - (\tan\epsilon. \tan\phi) / (1 + \tan\epsilon. \cot\alpha)] \quad (53)$$

where: $n = [K_v / (1+K_v)\} / \tan\epsilon$

α = Failure angle to horizontal
= generally between $\alpha_1 = 35^0$ and $\alpha_2 = 65^0$

On average, tan α = Integral between α_1 and α_2 of [tan α. $d\,\alpha$ /(α_2 - α_1)] = 5/4 = 1.25
On average, cot α = Integral between α_1 and α_2 of [cot α. $d\,\alpha$ / ($\alpha_2 - \alpha_1$)] = 7/8 = 0.88

Then equation (53) can be presented as follows, when the vertical seismic force is downward:

$$(FS)_{dynamic} = \{[(FS)\ static/(1+K_v)] + (0.8\ n - 1)\tan\epsilon\ .\ \tan\phi\} / (1+ 0.88\tan\epsilon) \quad (54)$$

In the case of a vertical seismic force being upward, equation (54) can be expressed as:

$$(FS)_{dynamic} = \{[(FS)\ static/(1+K_v)] - (1 + 0.8\ n)\tan\epsilon\ .\ \tan\phi\} / (1+0.88\tan\epsilon) \quad (55)$$

For a case where $K_v = 0$, then equation (55) gives:

$$(FS)_{dynamic} = [(FS)\ static - \tan\epsilon\ .\ \tan\phi] / (1+ 0.88\tan\epsilon) \quad (56)$$

10. PRACTICAL EXAMPLE

Considering an example of a mine or an excavation slope with respect to the Figure 2 and the following input data:

$H = 30$ m; $H_1 = 20$ m; $\beta = 76^o$; $\phi = 30^o$; $K_h = \Omega_2 = 0.2$;
$K_v = 0.1$; $S_o = 295$ kPa ;
$\gamma = 24$ kN/m^3 = 2.4 Tonnes/m^3; n = porosity = 10.7%;
q = 300 kN/m^2;
G_s = rock specific gravity = 2.65;
s = Degree of saturation = 32%;
γ_w = Unit weight of water = 10 kN/m^3

Sequential slope reinforcement design will give:

10.1 - Ground conditions before reinforcement:

Void ratio of the ground = e = n/(1-n) = 0.12

$$\gamma = \gamma_w (G_s + s.e) / (1 + e) = 24, kN/m^3$$

$$\gamma_{sat} = \gamma_{water} (G_s + e)/(1 + e) = 10(2.65 + 0.12)/(1 + 0.12) = 25 kN/m^3$$

$$\psi = q \cdot H + (H^2 - H_1^2) / 2 + \gamma_{sat} \cdot H_1^2 / 2$$

$$= 300 \times 30 + 24(900 - 400)/2 + (25 \times 400) / 2 = 20{,}000, kN/m$$

$$\psi_1 = \gamma_{water} \cdot H_1^2 / 2 = 10 \times 20^2/2 = 2000, kN/m$$

$$\Omega_1 = \psi_1 / \psi = 2000 / 20{,}000 = 0.1$$

$$\Omega = (1 + K_v) = 1 + 0.1 = 1.1$$

Utilizing equation (14) gives: α = Angle of failure plane to horizontal = 45°

Utilizing equation (13) gives: minimum (FS) = 1.22

Variation of the safety factor before rock reinforcement with Δ-values is given in Table 2 and Fig. 4.

Table 2 - Variation of Safety Factor before reinforcement (F S) with angle of failure plane to the horizontal (α), from equation (13).

α^0	Minimum (F S)
35	1.30
40	1.24
45	1.22
50	1.25
55	1.35
60	1.58

Note: In this case, f (α) = f(45°), see Fig. 4.

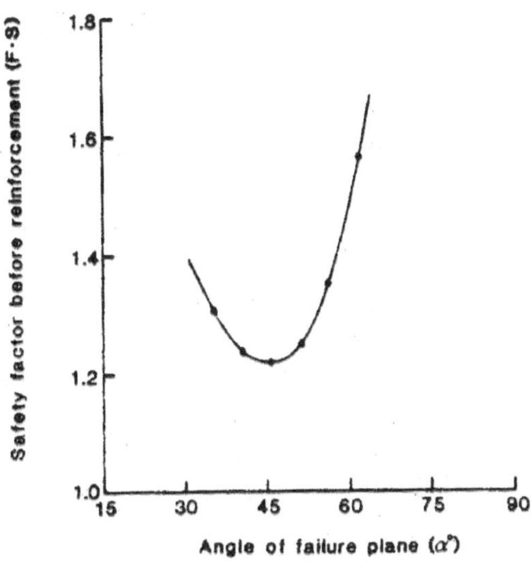

Fig. 4. Variation of Safety Factor before Reinforcement with Angle of Failure Plane to Horizontal, from Equation (13).

10.2 - Design of active anchor supports:

Considering a required $(FS)_{active}$ = 1.5 and utilizing equation (36) to calculate the anchor angle to the horizontal (Δ): $(\alpha - \Delta)$ = arc tan (0.385) = 21°; Δ = 24°

From equation (8): W = 20,000 x sin (76 – 45) /sin45 x sin76
= 15,013.44 kN /m

From equation (20): Y_3 = 15013.44(1.1 sin 45 + 0.2 cos 45)
= 13,801.00 kN /m

From equation (35): δ (FS) = 1.50 - 1.22 = 0.28, and
f(Δ) = f (24°)
= $[(FS)^2\ active + \tan^2\phi]^{0.5}$ = 1.61

Therefore: F_a = 13801 x 0.28/1.61 = 2400, kN/m

Variation of the active anchor force $(F_a)_{active}$ with Δ - values is given in Tables 3 and 4 and Fig. 5.

Table 3. - Variation of active anchor force $(F_a)_{active}$ with angle of anchor to the horizontal (Δ), from equation (36).

A^0	$(F_a)_{active}$, kN/m	f(A)/f(0)
40	2500	0.95
35	2450	0.93
30	2418	0.92
24	2400	0.91
20	2410	0.92
15	2434	0.93
10	2477	0.94
5	2542	0.97
0	2629	1.00
-5	2747	1.05
-10	2897	1.10
-15	3091	1.10
-20	3339	1.27

Note: Optimum $\tan(\alpha - \Delta) = \tan\phi / (F\ S)_{active}$

The optimum angle of the anchors to the horizontal (Δ) for a maximum Safety Factor corresponds to when Δ is positive. However, the optimum angle may not necessarily be the most practical installation angle. Field observation indicates that negative Δ or down-hole anchors are easier to place and grout.

For average field condition, it is recommended that be kept to zero and if this is not possible in certain cases then: $+5^0 \ngtr \Delta \ngtr -10^0$

Table 3 and Fig. 5 illustrate the significance of the above noted remark.

Considering for practical field purposes an anchor angle $\Delta = 0^0$. The anchor force then required is $F_a = 2629$ kN/m (see Table 3). At the same time the stress acting on the slope face is:

$F_a/(H/\sin\beta) = 2{,}629/20 \sin 76\ kN/m^2 = 127.54\ N/m^2$.

If an anchor working load of P = 500 kN, and a square pattern $(S_a \times S_a)$, ik. lateral spacing equal to row spacing, is in effect then:

$S_a = (500, kN/127.54, kN/m^2)^{1/2} = 2, m$

Table 4. - Variation of passive anchor force $(F_a)_{passive}$ with angle of anchor to the horizontal (Δ), from equation (23).

A	$(F_a)_{passive}$, kN/m	f(A) /f(0)
40	7449	1
35	7247	0.97
30	7108	0.95
25	7027	0.94
20	7000	0.93
15	7027	0.94
10	7108	0.95
5	7247	0.97
0	7450	1.00
-5	7723	1.03
-10	8083	1.00
-15	8545	1.15
-20	9138	1.23

Note: Optimum $\Delta = (\alpha - \phi)$

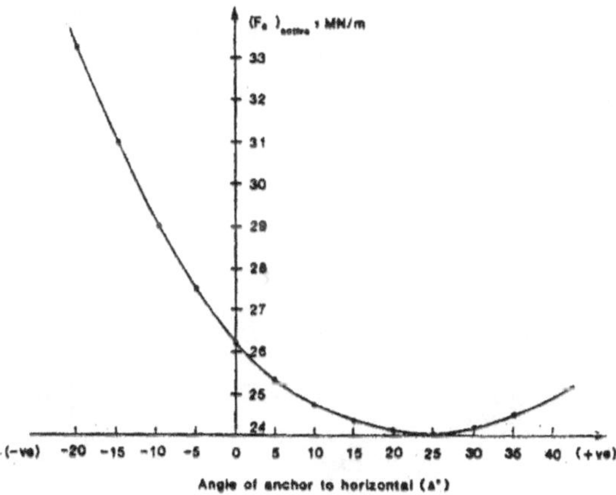

Fig. 5. Variation of Required Active Anchor Force $(F_a)_{active}$ with Angle of Anchor to Horizontal (Δ) using Equation (36).

Thus the design for the anchored slope is (2 m x 2 m) at 500, kN of tension per anchor.

To illustrate the importance of the variation of anchor angle (Δ), let the anchor inclination be $\Delta = -20^0$ (downward) with the remaining factors constant and an active safety factor of 1.5. Then from Table 3: F_a = 3339, kN/m. Under this condition the stress acting on the slope face is:

$F_a /(H /\sin \beta) = (3339/20) \sin 76 = 162$, kN/m^2

$S_a = (500, kN) /[(162\ kN/m^2)]^{1/2} = 1.75$

Hence, the anchor system design is: (1.75, m x 1.75, m) at 500 kN of tension per anchor.

The latter alternative implies an increment in the drill cost of 33%, plus the increased grouting, amount of cable anchor and the other accessories used in the anchor support. Therefore even though it is difficult to achieve the optimum anchor angle (above horizontal) in the field, due to installation conditions, it is possible and easy to drill horizontal hole (near optimum angle) as a practical field operation.

10.2.1 - Determination of the Distance (l_i):

According to Fig. 3, and equation (42), the value of l_i is given by the expression:

$l_i = (h /\sin \beta) \sin(\beta - \alpha) /\sin (\alpha - \Delta)$

For example, taking into account the 5th row, we have: h = 10.5 m, then:

l_i = (10.5/sin76)(sin31/sin45) = 7.88, m ---> 8, m

10.2.2 Design of Minimum Grouting Length $(l_g)_{min}$:

Considering equation (46) we have:

$(l_g)_{min} = \ln \{[8P /(_\pi Q. _D^2. K)] \tan [45 + (\phi /2)] + 1\}(DK /4)$

Now if the borehole diameter D = 0.1, m and K = 250, then for:
P = 0.5, MN (working tensile load/anchor); $\phi = 30^0$, and
$Q = 2S_o \tan(45 + \phi/ 2) = 2 \times 0.295 \tan 60 = 1.02$, MPa

Then: $(l_g)_{min}$ = ln {[(8x 0.5) / [3.14x 1.02x 0.1x 0.1x 250](tan 60 + 1)}(25 /4) = 3.90, m

10.2.3 - Determination of Maximum Grouting Length $(l_g)_{max}$:

Equation (49) gives: $(l_g)_{max} = \{[2P\ (_\pi\ Q.\ D^2.\ K)]\ \tan\ [45 + (\phi\ /2)\}$

$= [(2x\ 0.5)\ /\ (3.14x\ 0.1x\ 1.02]\ \tan 60 = 5.4$, m

If a safety factor of 2 for the tensile load (P) is considered, then: P = 1, MN. In this case, the new values of $(l_g)_{min}$ and $(l_g)_{max}$ will be:

$(l_g)_{min} = 6.27$, m and $(l_g)_{max} = 10.80$, m

Thus using the above noted approach, it is possible to save approximately 4 to 5 m/hole in the embedding length.

10.3 - Design of passive anchor supports:

Considering the input data given at the start of this example with a passive safety factor of $(FS)_{passive} = 1.5$. Also assuming that from in situ and laboratory test results, the statistically determined residual values for shear strength parameters are:

$S_r = 188.41$, kPa and $\phi_r = 28^0$. It is then possible to predict the most probable angle of the failure plane (α), minimum safety factor without reinforcement (FS) and the passive anchor force, $(F_a)_{passive}$.

From equation (14): $\alpha = 48°$. Utilizing equation (13), modified to take account for the residual shear strength values and data input: $\beta = 76°$; $\alpha = 48°$; $\Omega = 1.1$; $\Omega_2 = K_h = 0.2$; $\Omega_1 = 0.1$, = 20000, kN/m

Then: (FS) = 0.86, minimum value, when $\alpha = 48°$

From equations (22), (7) and (8):

W = 20,000 (sin28 /sin48 x sin76)
= 13,021.52 kN/m

Y_3 = 1,3021.52 (1.15 sin 48 + 0.2 cos 48)
= 12,387.18 kN/m

$(F_a)_{passive}$ = 12387.18 (1.5 – 0.86) /(cos48 + sin48 . tan28)
= 7,450, kN/m.

NOTE: The Δ- value is taken as zero instead of the optimum, ($\Delta = 20^{0)}$ for practical purposes.

Variation of the passive anchor force $(F_a)_{passive}$ with Δ- values is given in Table 4 and Fig.6.

Comparison of examples in the passive case with that of the above active case demonstrate that:

(a) The mobilized shear strength of the rock mass plus the available force for the passive anchor must be larger than that for the active case to stop any progressive failure.

(b) In the passive case the anchor will exhibit a tensile force when the rock mass has started to slide. In this condition, the mobilized sliding friction along the weakness plane is active and the sliding friction is less than that of static friction.

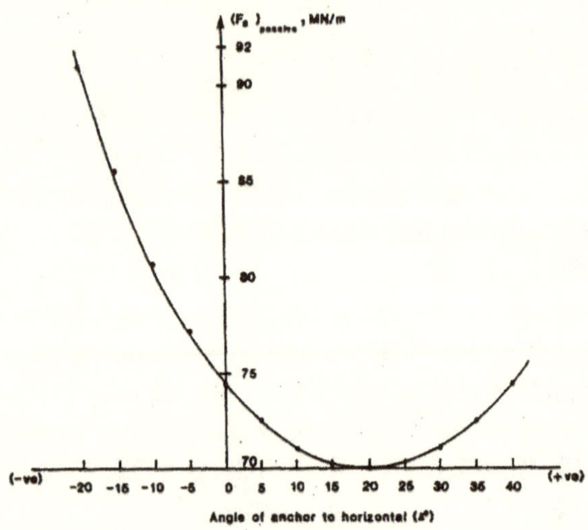

Fig. 6. Variation of Required Active Anchor Force $(F_a)_{passive}$ with Angle of Anchor to Horizontal (Δ) using Equation (23).

10.4- Determination of grouting length (l_g):

Considering the following borehole geometry and rock properties:

Case (a): D = 0.05 m; $\phi = 30^0$; Q = 4, MPa

Case (b): D = 0.10 m; $\phi = 35^0$; Q = 6, MPa

If the applied tensile force (P) is 1/4, 1/2 and 1.0 MN, then the minimum and maximum grouting length of the anchor, $(l_g)_{min}$ and $(l_g)_{max}$, can be determined. Utilizing equations (46) and (49) the grouting length can be determined, see Table 5.

The theoretical normalized stress distribution as a ratio of normal shear / uniaxial compressive strength (τ_α /Q) along the grouting section of the anchor, for the case of

$D = 0.1$ m; $\phi = 30^0$ and $(l_g)_{min} = 1$, is shown in Fig. 7.

The normalized stress is uniformly and near linearly distributed along the grouting length of the anchor, as is shown in Fig. 7. This is in accordance with the predictions of Coates et al. (1970) for soft rocks i.e. when the ratio of elastic moduli of the anchor material (E_a) to that of the rock (E_r) is greater than 10. In this case, the uniaxial compressive strength of the rock (Q) is less than 7 MPa. In the case of hard rocks, the above noted figure is non-linear and exponential.

Table 5 - Minimum and Maximum Grouting Length of the Anchor under various conditions.

	D = 0.050, m	$\phi = 30^0$	Q = 4, MPa
P, MN	0.25	0.50	1
(l_g), m	0.84	1.55	2.71
$(l_g)_{max}$, m	1.38	2.76	5.52
	D = 0.10, m	$\phi = 35^0$	Q = 6, MPa
P, MN	0.25	0.50	1
(l_g), m	0.49	0.94	1.76
$(l_g)_{max}$, m	0.5	1.0	2.00

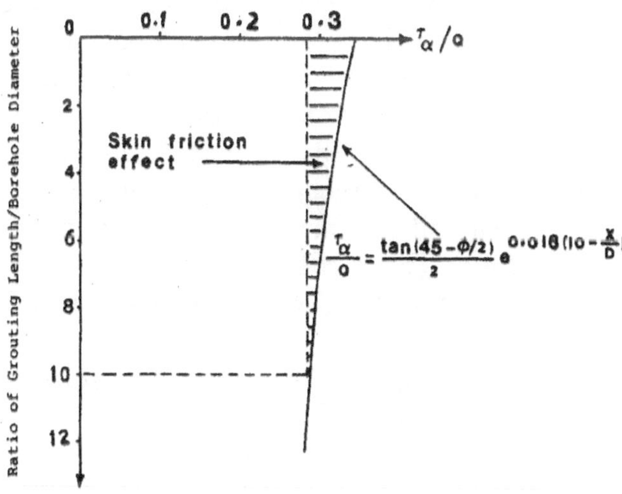

Fig. 7 - Shear stress distribution along an anchor for $(l_g)_{min} = 1$ m, $D = 0.1$ m and $\phi = 30^0$.

10.5- Determination of the dynamic safety factor of an anchor as a function of its static safety factor:

Considering the conditions given at the start of this section, the dynamic safety factor of an anchor can be determined utilizing equation (52), as follows:

Static safety factor = $(FS)_{static}$

$= [(0.4425x \sin 76 / \sin 31) + \tan 30 (\cos 45 - 0.1)] / \sin 45 = 1.675$

$\tan \epsilon = K_h / (1 + K_v) = 0.2 /(1 + 0.1) = 0.182$

From equation (55): Dynamic safety factor = $(FS)_{dynamic}$

$= \{(1.675 / 1.1) + [0.182 (0.8 / 2) - 1]\} /$
$[1 + (0.88x\ 0.182) - 1] = 1.219$

This value is very close to the quantity determined utilizing equation (12) in section 10.1.

11. CONCLUSION

1 - A new approach to anchor design has been proposed, based upon an initial prediction of the minimum safety factor for a planar slope failure. Parameters taken into account are: surface surcharge, seismic effect and groundwater pressure. The minimum value can then be used in a dimensionless analysis for determination of optimum anchor force.

2 - The calculation based on the minimum safety factor provides a conservative design to contend with uncertainties associated with the determination of shear strength parameters (i.e. S_o, S_r, ϕ and ϕ_r). This is considered to be especially difficult when the effects of weathering and groundwater are difficult to formulate accurately.

3 - For a more realistic prediction of the failure plane inclination (α) in passive anchoring, the residual shear strength parameter has been considered.

4 - Determination of the active anchor force by this approach has economical significance, since it provides the optimum anchor angle to the horizontal as a function of: internal friction angle, failure plane angle and active safety factor.

Since the optimum angle of anchoring (Δ) has been proven to be upward this creates difficulty in drilling, anchor placement and grouting. It is therefore suggested to angle anchors as close to horizontal as possible. As a result, significant reduction in drilling and grout placement costs can be realized with consequent savings in total reinforcement costs.

REFERENCES

Attewell, P.B. and Farmer, I.W., 1976, Principles of Engineering Geology. Chapman and Hall, p. 1045, London.

Baoshen, L. and Jinshow, H., 1983, Rock Bolting in Republic of China. Int. Symp. on Rock Bolting, Sweden, Aug. 28 - Sept. 2, p. 44-56.

Barron, K., Coates, D.F. and Gyenge, M., 1971, Artificial Support of Rock Slopes. Mining Research Centre, CANMET, Department of Energy, Ottawa, Research Report No. R228, p. 144.

Brandl, H., 1979, Design of High Flexible Retaining Structures in Steeply Inclined Slopes. Proc. 7th European Conference on Soil Mechanics and Foundation Engineering, Vol. 7, p. 157-166.

Coates, D.F. and Yu, Y.S., 1970, Three Dimensional Stress Distribution Around a Cylindrical Hole and Anchor. Proc. 2nd Congress of International Society for Rock Mechanics, Vol. 3, p. 175-180, Belgrade.

Dight, P.M., 1983, A Case Study of the Behaviour of a Rock Slope Reinforced with Fully Grouted Rock Bolts. Int. Symp. on Rock Bolting, Aug. 28 - Sept. 2, Sweden, p. 358-373.

Farmer, I.W., 1968, Engineering Properties of Rocks. E. and F.N. Spon Ltd., London.

Goodman, R.E., Gen-Hua, Shi and Boyle, W., 1982, Calculation of Support for Hard Jointed Rock Using the Key Block Principle. Proc. 23rd U.S. Symp. on Rock Mech. AIME/SME, p. 883-898.

Gudehus, G., 1972, Lower and Upper Bounds for Stability of Earth-Retaining Structures. Proc. 5th European Conference, SMFE, 1, Madrid.

Hanna, T.H., 1982, Foundations in Tension: in Ground Anchors, McGraw Hill Co., 573 p., U.K.

Hobst, L. and Zajic, J., 1983, Anchoring in Rock and Soil. 2nd Ed., Elsevier Scientific Publishing Co., 570 p.

Hoek, E. and Bray, J.W., 1977, Rock Slope Engineering. Institution of Mining and Metallurgy, 2nd Edition, London.

Jumikis, A.R., 1983, Rock Mechanics. 2nd Ed., Trans Tech Publications, p. 460-466.

Krauz, E., 1972, Bureau Securitas, Ground Anchors, French Code or Practice. Editions Eyrolles.

Rabcewicz, L., 1955, Bolted Support for Tunnels. Mine and Quarry Engineering, p. 113-116.

Seegmiller, B.L., 1975, Cable Bolts Stabilize Pit Slopes. World Mining, Vol. 28, No. 8, 36 p.

Seegmiller, B.L., 1982, Artificial Support of Rock Slopes. 3rd Int. Conf. on Stability in Surface Mining, Soc. of Mining Engineers, AIME, p. 249-288.

3.4.4 A DESIGN ANALYSIS FOR ANCHORING SOIL AND FRACTURED ROCK MASSES PRONE TO CIRCULAR FAILURE

By: R. Ucar, F.P. Hassani and Andy A. Afrouz

ABSTRACT: This analytical study considers anchoring of soil and highly fractured rock masses under circular failure conditions. A new formula is developed to calculate the total weight of the failing material, and its consequent total normal and tangential forces acting on the circular failure zone utilizing the Fellenius method. Also, the driving moment is calculated based on the most unstable conditions using Janbu's charts for critical toe circles. Comparison is made between the above noted method and that of Kötter's differential equation for calculation of the safety factor. An analytical solution is suggested to determine the optimum anchor inclination, related either to safety factor or minimum support requirement. Finally, a practical example is given to clarify the proposed method under active conditions.

1. INTRODUCTION

Failure of soil or highly fractured rock slopes occurs usually along a surface approaching a circular arc of finite length offering least resistance to sliding. A tendency to circular failure may be attributed to several contributory factors such as the soil/rock size and shape classification, slope height, cohesion and groundwater.

The relationship between the resultant stress (P) and the radius of the curvature (ρ) is given by Kötter (1903) as a differential equation. This method is useful to compare the slope safety factor before reinforcement with the traditional method of slices described below.

The traditional method to determine the anchor force in soil reinforcement is to divide the potential sliding section of the slope into a number of slices, each with different base inclination (α) and calculate the anchor force for each slice; the total anchor force being then the summation for all the slices (Hobst et al., 1983 and Hunt, 1986). In Hobst's method, the anchor is considered most effective when it is inclined to the plane normal to the slide at an angle (90 - ϕ'), (where ϕ' = the effective friction angle on the failure plane, in degrees). In some cases a logarithmic spiral has been considered for the failure plane (Lochor, 1969, and Ostermayer, 1977). This can derive a minimum safety factor for soil to obtain the potential slip surface utilizing variational calculus (Baker et al., 1977). Moreover, some investigators have suggested double wedge failure planes for analysis (Gudehus, 1972, Kranz and Stocker et al., 1979). The simplest of all methods to determine the slope safety factor under any shape of failure described above, is proposed by Fellenius (1936). However, the associated simplification neglects interslice forces acting parallel to the failure surface. To overcome the above drawback, Janbu proposed a method which satisfies both force and moment equilibrium and is applicable to failure surfaces of any shape (Janbu, 1954, Janbu, 1973). He assumed that the point at which the interslice forces act can be defined by a line of thrust. This method is based on analytical determination which can also be utilized graphically. Work towards the graphical determination of the location of the critical failure circle (Taylor, 1948, Cousin, 1978 and Hoek et al., 1981). In the simplified Bishop method (1955) the interslice shear forces (friction) was neglected. This simplified procedure, which set the vertical interslice forces to zero, gives approximately the same results as the rigorous procedure, which satisfies all the equilibrium conditions. This is the most popular method for effective stress analysis of slope stability.

A more accurate but time consuming method of slope stability analysis is that of Morgenstern et al. (1965). This method assumes an arbitrary mathematical function to describe the direction of the interslice forces given as follows:

$$\lambda f(x) = X(x)/E(x) \tag{1}$$

where: λ = a constant to be evaluated in solving the slope safety factor.

f(x) = function variation with respect to x.

$X(x)$ = shear force along each slice.

E(x) = normal force to each slice.

A similar method to that of Morgenstern et al. is developed by Spencer (1967). In the latter, two equations for tangential and normal forces on each slice are utilized to solve the factor of safety and a constant angle of inclination of the interslice forces to horizontal (θ). The result of the above noted equation is expressed as follows:

$$\tan \theta = X(x) / E(x) = X(x + \Delta x) / E(x + \Delta x) \qquad (2)$$

where: x = coordinate of the slice.
Δx = thickness of the slice.

Equation (2) is a special case of equation (1), where $\lambda . f(x) = \tan \theta$ = constant. However, determination of the anchor force from a pre-determined safety factor by a simple, accurate and more practical solution is much to be desired.

This investigation is concerned with developing an analytical solution to determine the weight of failing soil, its driving moment, the necessary anchor inclination and force for a circular failure condition utilizing the average anchor stress acting on the failure surface.

2. STABILITY ANALYSIS OF SOIL SLOPES

Presently available stability charts are utilized mainly as a guide or rough check during the preliminary assessment of slope stability. An accurate determination can be performed based on the critical sliding surface. Assumptions made are as follows:

(a) Homogeneous medium.
(b) The shear strength of the material (τ) is characterized by the Mohr-Coulomb failure criterion, expressed as follows:

$$\tau = S_0 + \sigma_n . \tan \phi \qquad (3)$$

where: S_o = average cohesion of the material in MPa.
= shear strength of the material at zero normal stress, i.e. $\sigma_n = 0$.

$\sigma_n = \sigma_\alpha$ = normal stress acting on the failure surface in MPa.

ϕ = average internal friction angle of the material in degrees.

In the equation (3), it may be necessary to consider:

(i) total stress analysis based on the undrained shear strength (τ), then ($S_o = S_u$) and ($\phi = \phi_u$) usually utilized to determine short term stability during or at the end of slope construction.

ii) effective stress analysis based on the drained shear strength (τ') commonly called effective shear strength parameters (S'_0) and (ϕ') and effective normal stress (σ'_n).

The major difference between a total stress analysis and effective stress analysis is that the former does not require a knowledge of the pore pressure, in contrast to effective stress analysis.

(c) Slope failure occurs on a circular surface passing through the slope toe. This was suggested by Taylor (1948) as being especially true for $\phi > 3$ and the slope angle $\beta > 10$. A number of methods are available to determine the slope failure following Janbu, Kötter, Morgstern and Spencer). All of these methods are based on the limit equilibrium concept, in turn based on the slip surface theory. Some other methods consider log-spiral failure surface (Baker et al. 1977, and Taylor, 1948), however, their results are similar to those obtained using a circular arc surface (Bishop, 1955, Janbu, 1954, Morgstern, 1965, Spencer, 1967 and Taylor, 1948). Common to all methods is a free body under gravitational force loaded by external forces (i.e. pore water and seismic), with a normal and shear stress distribution along the failure surface.

The equilibrium condition with regard to the slices of the total sliding mass can be expressed in a general form as follows:

$$(dQ_i / dx) + \gamma\,[F(x) - D(x)] + dS_i / dx - \sigma_a - \tau_a\,[dF(x)/dx] = 0$$

Therefore, $- S_i + E_i\,[dG(x)/dx] - (dE_i / dx)\,F(x) + (dE_i / dx).G(x) = 0$

Hence, $(dE_i / dx) - \tau_a + \sigma_a\,[dF(x)/dx] = 0$

Definition for the above noted symbols is given in Figure 1. The above expressions can be simplified according to the conditions noted in the section 1.

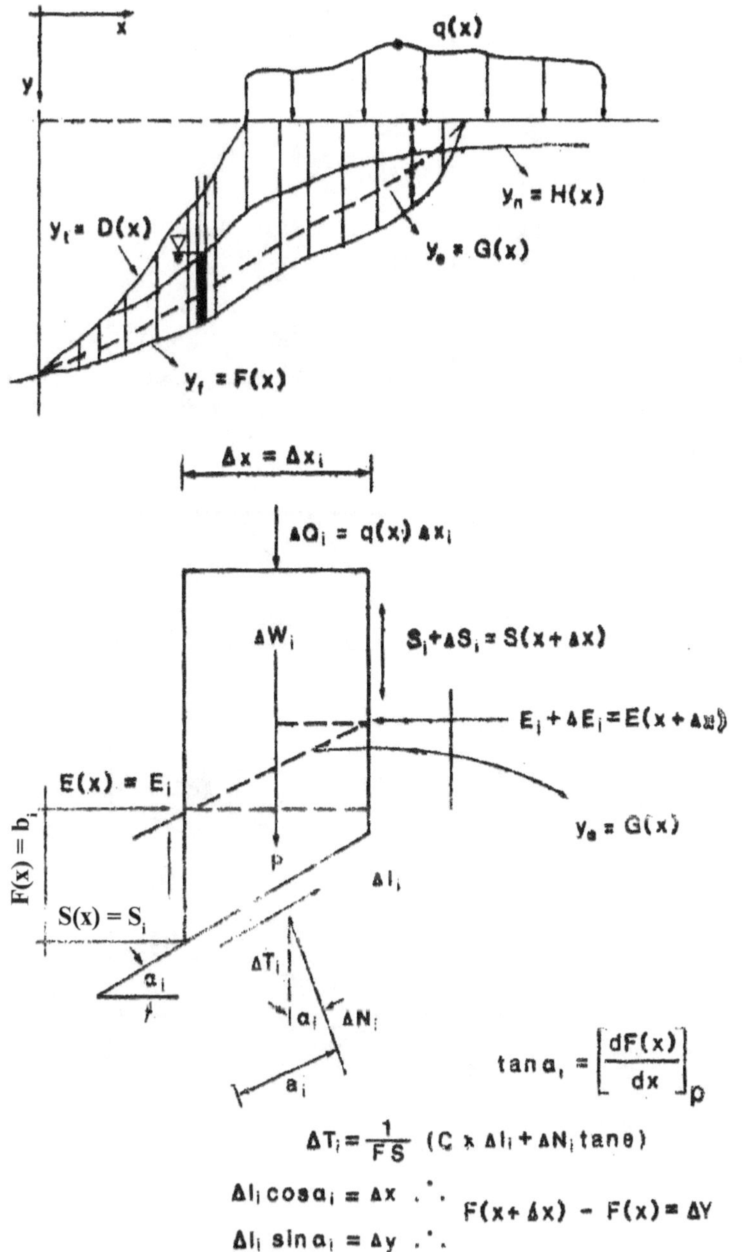

Fig. 1 - Graphical presentation of the forces acting on a potential failure slice.

Slope stability charts given by Janbu (1973) relate the minimum stability number N to the slope angle (β) in terms of a dimensionless factor ($\lambda . S_o . \phi$). For a special case where there is no tension crack and ground water pressure, the slope safety factor (FS) can be expressed as follows:

$$FS = N . S_o / \gamma . H = \gamma . \mathrm{H} . \tan\phi / \mathrm{S}_o$$

where: γ = unit weight of the sliding material in kN/m^3.
H = slope height in meters.

Slope stability charts to determine the location of the centre of the critical circle and the minimum slope safety factor are given in Figures 2 and 3.

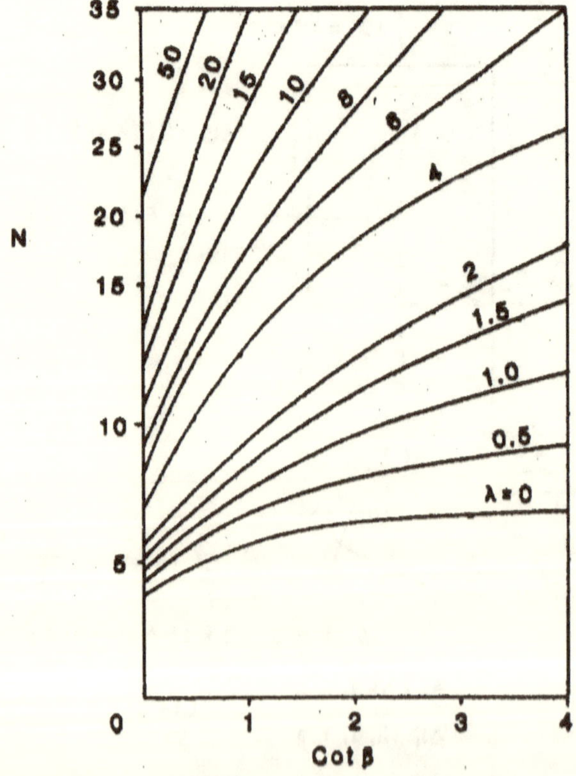

Fig. 2- Variation of the stability number (N) as a function of the slope angle (β) [after Janbu, (1954)].

3. AN ANALYTICAL APPROACH TO DETERMINE CIRCULAR SLOPE FAILURE PARAMETERS

The popularity of the method of slices (Fellenius, 1936) relates to its ability to accommodate complex and irregular geometries, variable soil and water pressure conditions. In the ordinary or Fellenius method (1936) the solution is simplified by assuming the forces on the two sides of a slice are parallel to the failure surface, at the bottom of the slice. Therefore, they do not affect the force normal to the failure surface. This produces small inaccuracy in calculations, especially where the forces are not parallel to the failure surface. In order to express a theoretical solution to the above noted approach, simple analytical solution by integration is suggested here under circular failure conditions as follows:

3.1. Determination of total failed weight of soil/rock (W), kN.

$W = \gamma \,.\, \int$ Integral between between Θ_1 and Θ_b of $[(R\sin\theta - a)\, R \sin\theta \,.\, d\theta] + \gamma \,.\int$ Integral between between Θb and $\Theta 2$ of $\{[R^2(\sin\theta + \cos\theta .\tan\beta) - b] \sin\theta .\, d\theta\}$

$$= (\gamma \,.\, R/2)\{R(\theta_2 - \theta_1) - R[(\sin 2\theta_2 - \sin 2\theta_1)/2] + 2[a \cos\theta_b - \cos\theta_1) + b(\cos\theta_2 - \cos\theta_b)] + R \,.\tan\beta(\sin^2\theta_2 - \sin^2\theta_b)\} \quad (3)$$

where: γ = average weight of unit volume of the failing soil or rock mass, kN/m^3.

Θ_1 = angle of the radius vector (R) at the top of the potential failure surface on the bench with the horizontal, degrees.

Θ_2 = angle of the radius vector (R) at the toe of the slope with the horizontal, degrees.

Θ_b = angle between the horizontal and the radius vector (R) which intercepts the vertical projection of the slope face crest on the potential failure surface, degrees (Fig. 4).

R = radius of the circular failure, m.

θ = angle between the radius vector (R) and the horizontal, degrees.

β = angle of slope face to the horizontal, degrees.

a = ordinate of the potential failure curve at the top of the bench, metre (Fig. 4).

b = ordinate intercept of the line that defines the slope faces (see Figure 5).

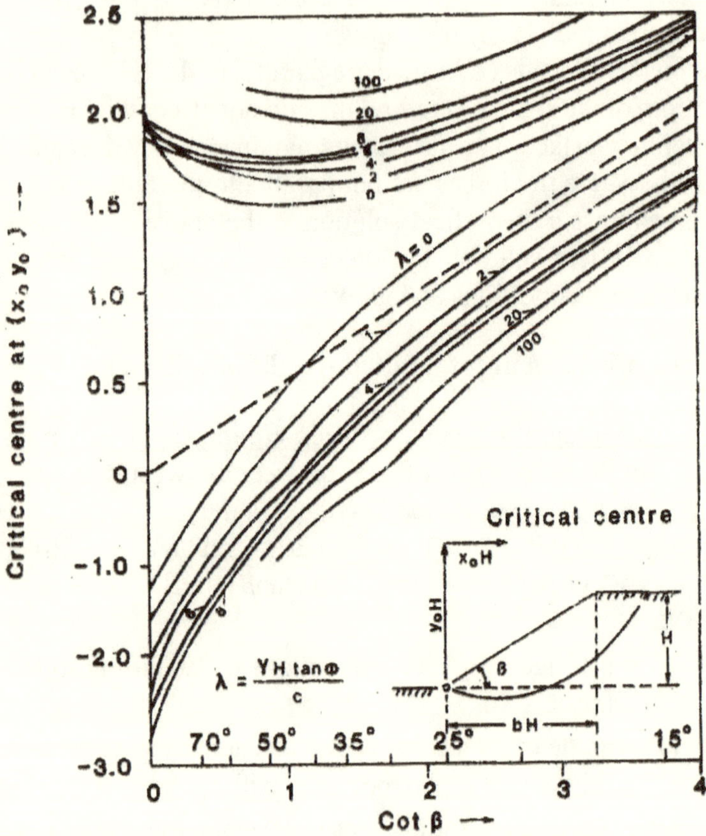

Fig. 3- Location of the critical circle centre for failure through the toe - after Janbu, (1954).

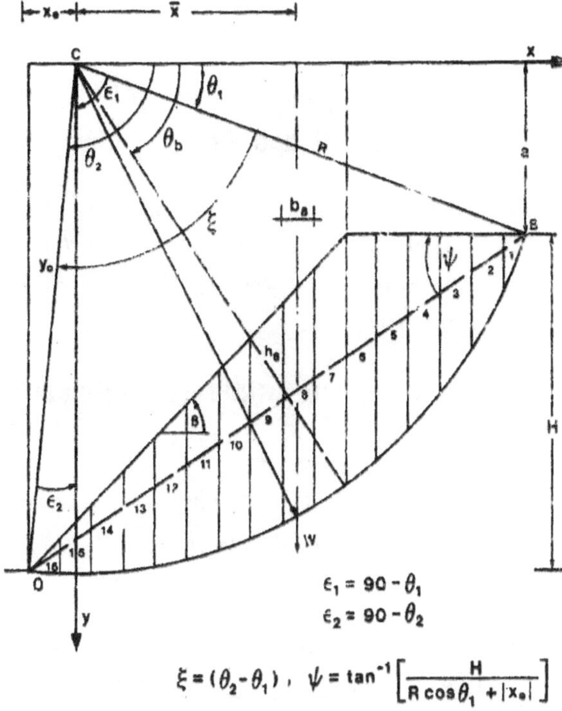

Fig. 4 - Method of dividing the potential failure mass into slices.

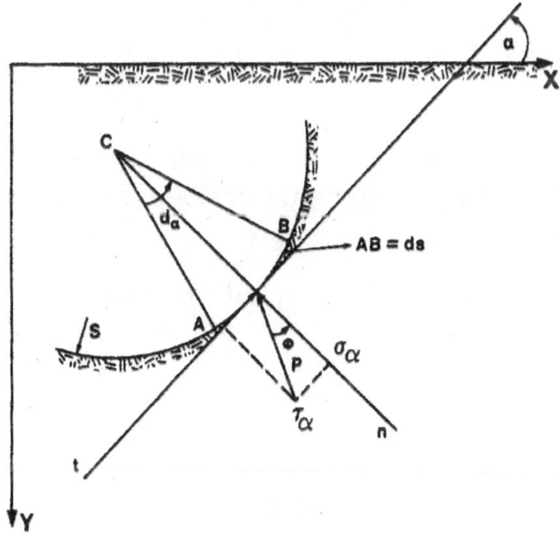

Fig. 5- Schematic of the reaction pressure (P) acting on the potential failure surface.

3.2. Determination of total normal force acting on the failure surface (N), kN/m.

$N = \gamma$.∫ Integral between θ_1 and θ_b of [R. $\sin^2\theta - a$) $R.\sin^2\theta.d\theta$] + γ . ∫ Integral between θ_b and $\theta_2$of {[R^2 ($\sin\theta + \cos\theta$. $\tan\beta$) - b] $\sin^2\theta.d\theta$}

$= (\gamma R/2)\{[2R \tan\beta(\sin^3\theta_2 - \sin^3\theta_b)/3] - a[(\theta_b - \theta_1) - (\sin 2\theta_b - \sin 2\theta_1)/2] - b[(\theta_2 - \theta_b) - (\sin 2\theta_2 - \sin 2\theta_b)/2] - 2R[\cos\theta_2(\sin^2\theta_2 + 2) - \cos\theta_1(\sin^2\theta_1 + 2)]/3\}$ (4)

3.3. Determination of total tangential force acting on the failure surface (T), kN/m.

$T = \gamma$. ∫ between θ_1 and θ_b of [(R. $\sin\theta$ - a) $\sin\theta$.$\cos\theta$. $d\theta$] + γ. ∫ *between* θ_b *and* θ_2 *of* {R^2 ($\sin\theta + \cos\theta$. $\tan\beta$) - b] $\sin\theta$. $\cos\theta$. $D\theta$}

$= (\gamma . R/2)\{[2R(\sin^3\theta_2 - \sin^3\theta_1)/3] - a(\sin^2\theta_b - \sin^2\theta_1) - b(\sin^2\theta_2 - \sin^2\theta_b) - [2R . \tan\beta(\cos^3\theta_2 - \cos^3\theta_b)/3]\}$ (5)

3.4. Determination of the driving moment (M_d) with respect to the critical centre (C):

$M_d = (\gamma.R^2/6)\{2R[(\sin^3\theta_2 - \sin^3\theta_1) - \tan\beta(\cos^3\theta_2 - \cos^3\theta_b)] - 3[a(\sin^2\theta_b - \sin^2\theta_1) - b(\cos^2\theta_2 - \cos^2\theta_b)]\}$ (6)

A more compact equation for the total weight (W) and driving moment (M_d) was suggested by Taylor (1948) for the circular failure surfaces. This is expressed as follows:

$W = \gamma\{R^2[(\zeta/2) - \sin(\zeta/2)\cos(\zeta/2)] + H^2(\cot\varphi - \cot\beta)/2\}$ (7)

$M_d = (\gamma . H^3/12)[1 - 2\cot^2\beta + 3\cot\beta . \cot\varphi - 3\cot\beta . \cot(\zeta/2) + 3\cot(\zeta/2) . \mathrm{Cot}\varphi]$ (8)

where: $\zeta = \theta_2 - \theta_1 = \tan^{-1}(H/R . \cos\theta_1 + X_o)$

H = height of the slope bench, m.

X_0 = the abscissa of the critical centre with reference to the coordinate of the slope toe, m.

It is to be noted that the accuracy of both methods is very similar, an analytical comparison is given in the practical application later.

3.5. Determination of the resisting moment (M_r):

In order to determine the slope safety factor in terms of moments, it is necessary to obtain the resisting moment of the soil or fractured rock mass (M_r). This is done by using the Kötter's differential equation (1903), for a homogeneous soil: as follows:

$$(dp/ds) - 2p \,.\, \tan\phi \,(d\alpha/ds) - \gamma \,.\, \sin(\alpha - \phi) = 0 \qquad (9)$$

where: dp = differential reaction pressure on the failure surface in kPa.

Ds = differential length of arc of circular failure surface in metre.

P = reaction pressure on the failure surface in kPa.

$d\alpha$ = differential angle at any segment on the potential failure surface, degree.

The differential equation of the shear strength along the circular failure surface τ_a is:

$$d\tau_\alpha / d\alpha = 2\tau_\alpha \,.\, \tan\phi - \gamma \,.\, R \,.\, \sin\phi \,.\, \sin(\alpha - \phi) \qquad (10)$$

Hence:

$$\tau_a = [m \,.\, f \,.\, \sin(\alpha - \phi)/(1 + f^2)] - [m \,.\, \cos(a - \phi)/(1 + f^2)] + (\Omega_1 / e^{f\alpha}) \qquad (11)$$

where: m = $-\gamma \,.\, R \,.\, \sin\phi$

f = $-2 \tan\phi$

Ω_1 = integration constant to be evaluated from the boundary conditions.

At= σ_α 0 and from the Mohr-Coulomb equation $\tau_\alpha = S_o$ at the top of the bench.

To obtain a more realistic result, it is necessary to account for the tensile stress or if possible to observe the presence of the tension cracks on the top of the slope. Under the above noted conditions:

$$\Omega_1 = \{S_a + [m/(1+f^2)]\cos(\epsilon_1 - \phi) \\ -[m \,.\, \sin(\epsilon_1 - \phi)/(1+f^2)]\}e^{f\alpha} \quad (12)$$

where: $\epsilon_1 = 90 - \theta_1$, (see Figure 6).

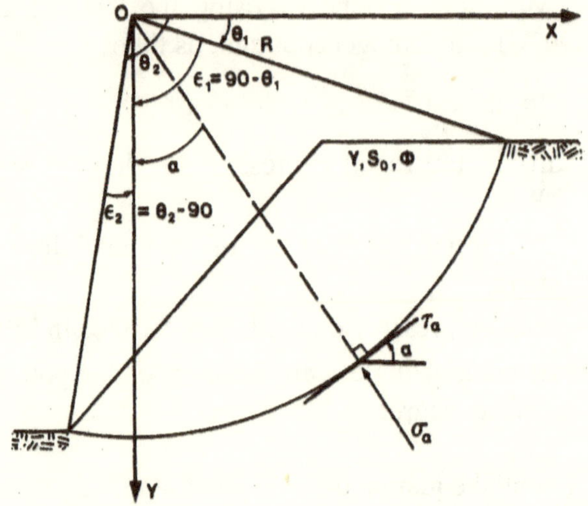

Fig. 6- Schematic of tangential and normal stresses acting on the potential failure surface [Kötter's differential equation, (1903)].

If the shear strength of soil or rock mass is defined by the Mohr-Coulomb failure criterion, then normal stress (σ_a) can be determined as follows:

$$(d\tau_\alpha/d\alpha) = (d\sigma_\alpha/d\alpha)\tan\phi \quad (13)$$

Substituting equation (13) into (10) gives:

$$(d\sigma_\alpha/d\alpha)\tan\phi = 2\tan\phi\,(S_o + \sigma_\alpha \,.\, \tan\phi) - \gamma \,.\, R \,.\, \sin\phi \,.\, \sin(\alpha - \phi) \quad (14)$$

The differential equation of the normal stress (σ_α) on the circular failure surface is:

$$\sigma_\alpha = (2S_o/f) + [B.\,f/(1+f^2)]\sin(\alpha-\phi) - [B/(1+f^2)]\cos(\alpha-\phi) + \Omega_2/e^{-f\alpha} \quad (15)$$

where: B $= - \gamma . R . \cos\phi$

e = base of naperian log =2.17

Ω_2 = the integration constant at $\sigma_\alpha = 0$ and at $\alpha = \epsilon_1$ on the top of the bench.

The Ω_2-value is determined as follows:

$$\Omega_2 = \{(2S_o / f)+[B . f/(1 + f^2)]\sin(\epsilon_1 - \phi) - \beta . \cos(\epsilon_1 - \phi) / (1 + f^2)\}e^{-f\epsilon_1} \quad (16)$$

where: $\epsilon_1 = 90 - \theta_1$

Finally, from the geometry of Figure 6, the resisting moment by Kötter (M_r) can be determined as follows:

$$M_r = \int \text{between } \epsilon 2 \text{ and } \epsilon 1 \text{ of } [\tau\alpha . R^2 . d\alpha]$$

$$= R^2 \int \text{between } \epsilon 2 \text{ and } \epsilon 1 \text{ of } \{[m . f. \sin(\alpha - \phi) / (1 + f^2)] - [m . \cos(\alpha - \phi) / (1 + f^2)] + (\Omega_1 / e^{-f\epsilon})\}d\alpha$$

$$= [R^3 . \gamma . \sin\phi / (1 + 4\tan^2\phi)] \{\sin(\epsilon_1 - \phi) + 2\tan\phi [\cos(\epsilon_2 - \phi) - \cos(\epsilon_1 - \phi)] - \sin(\epsilon_2 - \phi)\} + (R^2\Omega_1 / 2\tan\phi)(e^{-f\epsilon_1} - e^{-f\epsilon_2}) \quad (17)$$

where: $\epsilon_2 = \theta_2 - 90$, (see Figure 6)

3.6. Determination of the slope safety factor (FS):

3.6.1- Utilizing Fellenious and Bishop methods, the slope safety factor before reinforcement is given below in terms of moments:

$$FS = R(S_o L_f + N . \tan\phi) / M_d \quad (18)$$

where: L_f = length of the perimeter of failure surface in metre.

N = total normal force acting on the failure surface.

3.6.2- The passive slope safety factor after anchoring according to Figure 7 is is given below:

$$(F_a)_{passive} = S_o . + T_o = (N + Na)\tan\phi / T \quad (19)$$

where: $AB = \pi . R(\alpha_1 - \alpha_2) / 180°$

N_a = normal component of the anchor force acting on the potential failure surface, kN/m.

T_a = tangential component of the anchor force acting on the potential failure surface, kN/m.

3.6.3- The active slope safety factor after anchoring according to Figure 7 is given below:

(i) in terms of the shear force using Bishop's method:

$$(F_a)_{active} = S_o . AB + (N + N_\alpha)\tan\phi / (T - T_\alpha) \quad (20)$$

(ii) in terms of moments using Kötter equation:

$$(F_a)_{active} = [M_r + R(N_\alpha . \tan\phi)] / (M_d - R . T_\alpha) \quad (21)$$

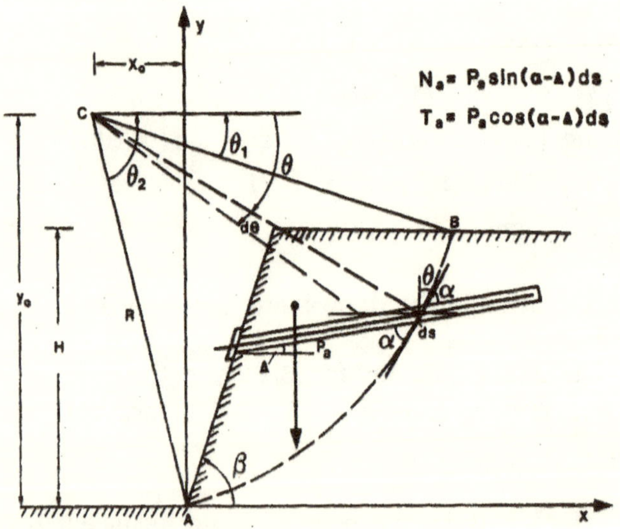

Fig. 7- Schematic showing anchor force.

4. DESIGN OF TOTAL ANCHORING FORCE (F_a)

Considering uniform stress acting on the failure surface (P_a) along the circular failure surface Figure 7, the following apply:

limit of Δ_a to zero for $(\Delta T_a / \Delta s) = dT_a / ds$

limit of Δ_a to zero for $(\Delta N_a / \Delta s) = dN_a / ds$

$$dT_a = P_a . \cos(\alpha - \Delta)ds$$

$$dN_a = P_a . \sin(\alpha - \Delta)ds \tag{22}$$

where: dT_a = differential tangential (shear) force along the failure surface.

dn_a = differential normal component of force acting on the circular failure.

P_a = average surface pressure acting on the potential surface failure (= F_a / AB)

ds = differential arc length of the circular failure surface (= R. dα)

Δ = inclination angle of the anchor to the horizontal.

The above equations can be transformed to give:

$T_a = R.\ P_a$ Integral between α_1 and α_2 of $[(\cos(\alpha - \Delta)\ d\,\alpha]$

$$= R.\ P_a\,[\sin(\alpha_1 - \Delta) - \sin(\alpha_2 - \Delta)] \tag{23}$$

$N_a = R.P_a$ Integral between α_1 and α_2 of $[\sin(\alpha_2 - \Delta) - \cos(\alpha - \Delta)]$

$$= R.P_a\,[\cos(\alpha_2 - \Delta) - \cos(\alpha_1 - \Delta)] \tag{24}$$

Under passive conditions, considering equations (19), (23), and (24):

$$\begin{aligned}(FS)_{passive} &= (S_o . AB + N . \tan\phi + N_a \tan\phi + T_a) / T \\ &= \{S_o .AB + N . \tan\phi + R . P_a \{\tan\phi [\cos(\alpha_2 - \Delta) - \cos(\alpha_1 - \Delta)] + \sin(\alpha_1 - \Delta) - \sin(\alpha_2 - \Delta)\}\}/T \\ &= \{S_o . AB + N . \tan\phi + (R.F_a / AB) \{\tan\phi [\cos(\alpha_2 - \Delta) - \cos(\alpha_1 - \Delta)] + \sin(\alpha_1 - \Delta) - \sin(\alpha_2 - \Delta)\}\}/T\end{aligned}$$

Therefore:

$$F_a = AB/R\ \{[T(FS)_{passive} - S_o . AB - N_a . \mathrm{Tan}\ \phi] / \{\tan\phi [\cos(\alpha_2 - \Delta) - \cos(\alpha_1 - \Delta)] + \sin(\alpha_1 - \Delta) - \sin(\alpha - \Delta)\}\} \tag{25}$$

In equation (25), can be expressed as:

$f(\Delta)_{passive} = \tan\phi\,[\cos(\alpha - \Delta) - \cos(\alpha_1 - \Delta)] + \sin(\alpha_1 - \Delta) - \sin(\alpha_2 - \Delta)$ (26)

The optimum angle of the anchor to the horizontal (Δ) is obtained by taking the derivative of equation (25) and equating to zero, as follows:

$f'(\Delta) = 0 = \tan\phi\,[\sin(\alpha - \Delta) - \sin(\alpha_1 - \Delta)] - \cos(\alpha_1 - \Delta) + \cos(\alpha_2 - \Delta)$

Then, $\Delta_{passive} = (\alpha_2 + \alpha_1 - 2\phi)/2$ (27)

Substituting equation (27) into equation (25) gives the minimum anchor force (F_a). Under active conditions, utilizing equations (20), (23) and (24) gives:

$$(FS)_{active} = \{S_o\,.\,AB + N\,.\,\tan\phi + R\,.\,P_a\,\{\tan\phi\,[\cos(\alpha_2 - \Delta) - \cos(\alpha_1 - \Delta)]\}\}/\,\{T - R.P_a\,[\sin(\alpha_1 - \Delta) - \sin(\alpha_2 - \Delta)]\}$$

$$= \{S_o\,.\,AB + N\,.\,\tan\phi + (R\,.\,F_a\,/\,AB\,\{\tan\phi\,[\cos(\alpha_2 - \Delta) - \cos(\alpha_1 - \Delta)]\}\}/\,\{T - R.\,F_a\,/\,AB) + [\sin(\alpha_1 - \Delta) - \sin(\alpha_2 - \Delta)]\}$$

Therefore:

$$F_a = \{[T\,(FS)_{active} - S_o\,.\,B + N\,.\,\tan\phi]\,/R.P_a\,\{\tan\phi\,[\cos(\alpha_2 - \Delta) - \cos(\alpha_1 - \Delta)] + (FS)_{active}\,[\sin(\alpha_1 - \Delta) - \sin(\alpha_2 - \Delta)]\} \quad (28)$$

In equation (28), let:

$$f(\Delta)_{active} = \tan\phi\,[\cos(\alpha_2 - \Delta) - \cos(\alpha_1 - \Delta)] + (FS)_{active}\,[\sin(\alpha_1 - \Delta) - \sin(\alpha_2 - \Delta)]$$

The optimum angle of the anchor to the horizontal (Δ) can be determined by taking the derivative of equation (29) and equating it to zero, as follows:

$$f'(\Delta)_{active} = 0 = \tan\phi\,[\sin(\alpha_2 - \Delta) - \sin(\alpha_1 - \Delta)] + (FS)_{active}\,[\cos(\alpha_2 - \Delta) - \cos(\alpha_1 - \Delta)] \quad (29)$$

Then:

$$[\sin(\alpha_1 - \Delta) - \sin(\alpha_2 - \Delta)]\,/\,[\cos(\alpha_2 - \Delta) - \cos(\alpha_1 - \Delta)] = (FS)_{active}/\tan\phi \quad (30)$$

Substituting Δ- value from the equation (30) into equation (28) gives the minimum anchor force (F_a).

In a special case when $(FS)_{active} = 1$, then: $\Delta_{active} = \Delta_{passive}$

Finally the average stress acting on the slope face (P_{sf}) can be given as follows:

$$P_{Sf} = P_a . \sin\beta / H \tag{31}$$

5. DESIGN OF ANCHOR DIAMETER (D_a) AND SPACING (S)

For field conditions: $\sigma_{ws}(FS)_{load} = \sigma_b$ (32)

where: σ_{ws} = working stress of anchor, kPa.

Σ_b = breaking strength of anchor, kPa.

(FS) = safety factor of anchor material under load = between 1.5 to 2

Working load per anchor (Q) is expressed as:

$$Q = \sigma_{ws} . \pi . D_a^2 / 4, kN \tag{33}$$

where: D_a = diameter of the anchor, meter.

Therefore from equations (32) and (33): $D_a = [4Q(FS)_{load}/\pi . \sigma_b]^{0.5}$ (34)

The projected vertical spacing between the anchors (S_v) on a slope of β degrees to the horizontal (Figure 8) can be determined as:

$$S_v = S_o / \sin\beta \tag{35}$$

where: S_v = vertical spacing between the anchors, in meter.

The lateral spacing between anchors (S_l) can then be expressed as:

$$S_l = Q / S_v . P_{sf} \tag{36}$$

6. DESIGN OF THE ANCHOR LENGTH (l_a)

The length of anchor depends upon: (a) location of the anchor in the slope, (b) slope geometry, (c) geometry of the circular failure surface and (d) anchor angle to the horizontal (Δ). Referring to Figure 8, the circular failure surface can be represented as follows:

$$(x - X_o)^2 + (y - Y_o)^2 = R^2 \quad (37)$$

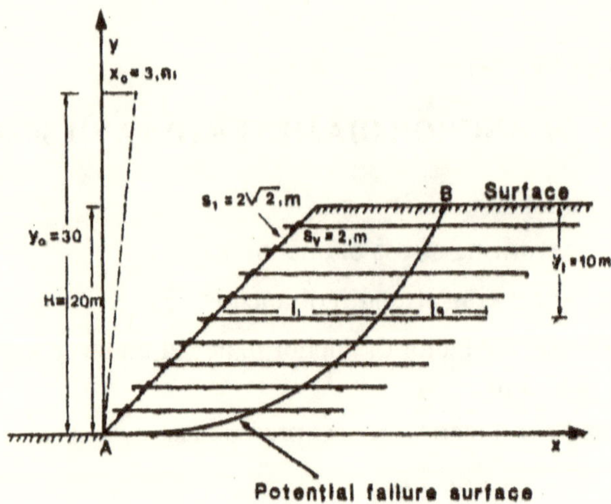

Fig. 8- Schematic diagram of anchoring pattern.

The straight line defining the anchor is expressed as follows:

$$y_{anchor} = d - x \,.\tan\Delta \quad (38)$$

where: $d = y_1 - x_1 \tan\Delta$

$y = y_{anchor}$ at the intersection of the failure surface and anchor (Figure 6).

$x = X_o - \tan\Delta(d - Y_o) + [X_o - \tan\Delta\,(d - Y_o)^2 - K\sec^2\Delta]^{0.5}\} / \sec^2\Delta$

$K = (d - Y_o)^2 - (R^2 - X_o^{\;2})$

(x_1, y_1) = coordinates of the anchor at the slope day light with respect to the toe as origin (Figure 7).

The length of the anchor between the slope wall and circular failure surface (l_i) can be given as follows: $l_i = (x - x_1)\sec\Delta$ (39)
The length of the anchor grouting (l_g) can be expressed as follows:

$$l_g = (FS)_{gr}\, Q / \pi \,.\, D_h \,.\, \tau_b \quad (40)$$

where: $(FS)_{gr}$ = factor of safety against anchor extraction at the grout rock interface

= 1.5 to 2.0.

D_h = anchor hole diameter, in meter.

T_b = cohesive strength of the anchor grout with the soil or fractured rock

= $S_o + \sigma_n \tan \phi$,, *in kPa*

σ_n = normal stress to the major axis of the anchor, in kPa (Figure 8).

Note: Since the cohesive strength of the grout is usually greater than that of the soil or the fractured rock, the smaller S_o and ϕ-values for soil was considered.

Hence, anchor length (l_a) at any location, along the slope, in meters, from equations (39) and (40) is expressed as: $l_a = l_i + l_g$ (41)

$$\text{or } l_a = \{\sec \Delta\{\{X_o - \tan \Delta(d - Y_a) + [X_o - \tan \Delta(d - Y_o)^2 - K.\sec^2 \Delta]^{0.5}\}/\sec^2 \Delta\} - x_1\} + [(FS)_{gr} Q/\pi . D_h . T_b] \quad (42)$$

7. PRACTICAL APPLICATION

Input data, for short-term stability immediately after excavation is given as follows:

$\phi = \phi_u = 10$; $S_o = S_u = 50$ kPa; $\gamma = 18$ kN/m^3; H = 20 m; $\beta = 45°$ (i.e. 1:1:1);

$(FS)_{active} = 1.4$; Q = 150 kN; $\sigma_b = 600$ MPa and $S_v = 2$ m.

7.1. Determination of factor of safety (FS) for unsupported slope, applying various methods:

Method (a) - utilizing Janbu's charts [1954 and 1973] with reference to Figures 2 and 3, then: $\lambda . \phi . S_o = \gamma . H. \tan \phi_u / S_u = 18 \text{ x } 20 \text{ x } \tan 10°/50 = 1.27$

N = minimum stability number from the geometry of Figure 3 (for = 1.27 and $\cot\beta = \cot 45° = 1$) = 8.2

Therefore: $FS = N . S_u / \gamma . H = 8.2 \text{ x } 50 / 18 \text{ x } 20 = 1.139$

Method (b) - coordinates of the critical circle is found from Figure 3 to be as follows:

$X_o = 0.15\ H = 3m$; $Y_o = 1.5\ H = 30m$, and

R = radius of the failure arc = 30.15m.

Considering Figure 4, the following can be obtained:

$\theta_1 = 19.37° = 1.338 rad$; $\theta_2 = 95.71° = 1.670 rad$

$\theta_b = 56.00° = 0.977 rad$; $\zeta = \theta_2 - \theta_1 = 76.34°$

$a = 10.00, m$; $b = 27.00, m$

$AB = R\ .\ \zeta = 30.15 \times 76.34\ (\pi/180) = 40.17, m$

From equations (3), (4), (5) and (6):

W = weight of the failure circle = 5008.10, kN/m

N = normal force on the failure surface = 4215.60, kN/m

T = tangential force on the failure surface = 2364.60, kN/m

M_d = driving moment = 70808.80, kN m /m

Therefore, applying the Mohr-Coulomb criterion:

$FS = AB.\ S_u + N\ .\ \tan\phi_u\ /\ T$

$= (40.17 \times 50) + (4215.60 \times \tan 10°)\ /2364.60 = 1.16$

Method (c) - dividing the failure cross section area into vertical slices (by Bishop's method) as shown in Figure 4 and table 1, then:

$\sum W_i = 5160.60, kN/m$

$\sum N_i = 4580.28, kN/m$

$\sum T_i = 2426.20, kN/m$

The above values are close to those obtained by method (b). Therefore the safety factor in terms of moments is expressed as follows:

$FS = R\ (AB \times S_u + N\ .\ \tan\phi_u)/M_d$

$= 30.15(40.17 \times 50 + 4215.60 \times 0.176)/70808.80 = 1.171$

Method (d) - utilizing equation (17) and considering Figure 4 gives:

$M_r = 83231.3$, *kN m /m for* $\alpha = \epsilon_1 = 70.63°$; $\epsilon_2 = -5.71°$

$\Omega_1 = -11.10$, kN/m^2; and $\sigma_\alpha = 0$

Therefore: $M_r / M_d = 83231.3 / 70808.80 = 1.175$

Method (e) - considering Bishop's simplified method of slicing (1955), Table 1 and Figure 4 gives: $\sum N_i = N = 4580.28$, *kN/m*

$W = \gamma\{R^2[(\zeta/2) - \sin(\zeta/2)\cos(\zeta/2)] + H^2(\cot\psi - \cot\beta)/2\}$

$= 5000.76$, *kN/m*

$Md = \gamma \cdot H^3 [1 - 2\cot^2\beta + 3\cot\beta \cdot \cot\psi - 3\cot\beta \cdot \cot(\zeta/2) +$

$3\cot(\zeta/2)\cot\psi]/12 = 706800$, *kN m/m*

Therefore: $FS = M_r / M_d = 832313.3 / 706800 = 1.178$

Comparing the above noted methods (a to e) it can be seen that the FS- value changes from 1.139 to 1.178 (i.e. 1.14 to 1.18 or 3.5% variation). Therefore, the minimum factor of safety is 1.14.

7.2. Determination of the required anchor force (F_a):

Utilizing equation (29) and considering $(FS)_{active} = 1.4$; $\tan \phi_u = \tan 10$;

$\alpha_1 = (90 - \theta_1) = 70.63$ and $\alpha_2 = (90 - \theta_2) = -5.71$; gives:

$\Delta_{active} = +25.28$ (positive means above horizontal).

Horizontal or downward drilling, anchoring and grouting is easier and more efficient in practice. Therefore, in this particular case, it is recommended $\Delta = 0$.

Assuming $\Delta = 0$, using equations (23) and (24) with regard to Figure 6, the tangential anchor force (T_a) and the normal anchor force (N_a) are:

$T_a = 30.15 P_a (\sin 70.63° + \sin 5.71°) = 31.44, P_a$

$N_a = 30.15 P_a [\cos(-5.71°) - \cos 70.63°] = 20.00, P_a$

From equations (20) and (21) and Table 1:

N = 4580.28 kN/m; T = 2462.20 kN/m and $(FS)_{active} = 1.4$;

the stress on anchor (P_a) re 12.21 kN/m^2 and 11.10 kN/m^2 respectively.

Table 1 - Solution for the Simplified Bishop's Method (after Bishop, 1955)

Slice	h (m)	ΔW_i (kN/m)	x_i (m) (1)	y_i (m) (2)	$\alpha_i = \tan^{-1}\left(\frac{x_i}{y_i}\right)$ (3)	$C_i \times \Delta x_i$ (kN/m) (4)	m_α for FS=1.30 (5)	m_α for FS=1.17 (6)	$\Delta W_i \tan\phi_i$ (kN/m) (7)	$\Delta W_i \sin\alpha_i$ (kN/m) (8)	$[(4)+(7)]/m_\alpha$ FS=1.3	$[(4)+(7)]/m_\alpha$ FS=1.17
1	2.00	64.80	28.0	12.0	66.80	90	0.52	0.53	11.42	59.60	195.0	191.4
2	5.90	221.24	26.0	15.9	58.55	100	0.64	0.65	39.01	181.20	217.2	213.9
3	8.75	315.00	24.0	18.8	52.00	100	0.72	0.73	55.54	248.20	216.0	213.1
4	11.0	396.00	22.0	21.0	46.33	100	0.79	0.80	69.82	286.40	215.0	212.3
5	13.0	468.00	20.0	23.0	41.00	100	0.84	0.85	82.52	307.00	217.3	214.7
6	14.5	522.00	18.0	24.5	36.40	100	0.88	0.89	92.04	309.00	218.2	215.8
7	14.8	532.80	16.0	25.8	31.80	100	0.92	0.93	93.95	280.08	210.8	208.5
8	14.0	504.00	14.0	27.0	27.40	100	0.95	0.96	88.87	231.90	198.8	196.7
9	12.8	460.80	12.0	27.8	23.34	100	0.97	0.98	81.25	182.60	186.8	184.9
10	11.5	414.00	10.0	28.5	19.33	100	0.99	0.99	73.00	137.00	174.7	174.7
11	10.2	367.20	8.0	29.3	15.30	100	1.00	1.00	64.74	96.90	164.7	164.7
12	8.75	315.00	6.0	29.5	11.50	100	1.01	1.01	55.54	62.80	154.0	154.0
13	7.00	252.00	4.0	30.0	7.59	100	1.01	1.01	44.43	33.30	143.0	143.0
14	5.10	183.63	2.0	30.2	3.78	100	1.01	1.01	32.37	12.21	131.1	131.1
15	3.15	113.40	0.0	30.3	0.00	100	1.00	1.00	19.20	0.00	119.2	119.2
16	1.10	39.60	-2.0	30.2	-3.78	100	0.98	0.98	6.98	-0.26	109.2	109.2
$\sum$		5160.06								2426.2	2871.0	2847.2

$$W = \sum \Delta W_i \quad ; \quad m_\alpha = \left(\cos\alpha_i + \frac{\sin\alpha_i \cdot \tan\phi_i}{FS}\right) \quad ; \quad T = \sum \Delta W_i \sin\alpha_i$$

$$N = \sum \Delta N_i \quad ; \quad \Delta N_i \approx \left\{ \frac{\Delta W_i - \frac{C_i \cdot \Delta x_i \cdot \tan\alpha_i}{FS}}{m_\alpha} \right\}$$

Variation of the anchor stress (P_a) with anchor inclination to the horizontal (Δ) is shown in Figure 9.

Therefore, using equation (28), with ϵ_1 = 70.63 = 1.23 rad, the total anchor force (F_a), considering the greater value of P_a obtained before, is as follows:

$$F_a = 12.21 \text{ x } 30.15 \text{ x } 1.23 = 452.80, \text{ kN/m}$$

Thus, the average stress acting on the slope face, from equation (29), is as follows:

$$P_{sf} = 452.80 \times /20 \times 2 = 16.00, \ kPa$$

7.3. Determination of the required anchor diameter, spacing and length:

From equation (34): $D_a = (4 \times 150 \times 1.4/3.14 \times 600{,}000)^{0.5}$

$$= 0.0253, \ m = 25.3, \ mm$$

Utilizing equation (35) and (36) respectively, for $S_v = 2$, m then:

$$S_1 = 2 / \sin 45° = 2.8, \ m$$

$$S_2 = 150/2.8 \times 16.00 = 3.35, \ m$$

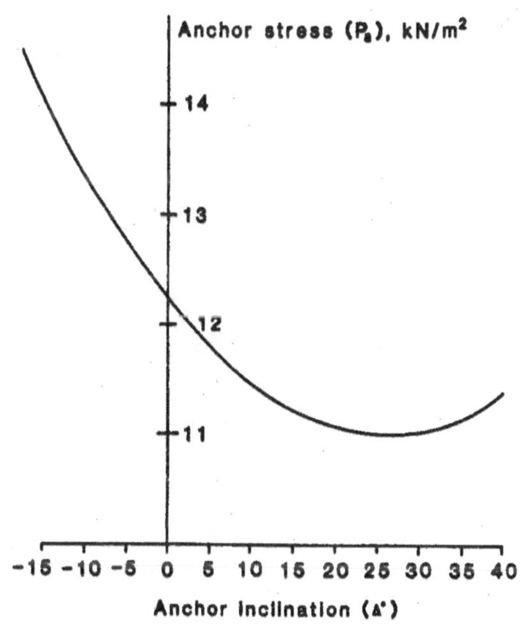

Fig. 9- Variation of the anchor stress (P_a) acting on the potential failure surface with the anchor inclination (Δ).

Therefore the anchoring pattern can be square or staggered at 2.8 by 3.35 meter. The anchor length (l_a) for the middle row of coordinate P(10,10) from equation (41) is as follows: $l_a = l_i + l_g$

From equation (39): $l_i = 15.60$, m.

$l_g = (FS)_{gr}\ Q / \pi \, . \, D_h \, . \, \tau_b$

$\tau_b = S_u + \sigma_n \, . \tan \phi$

$\sigma_n = (-\sin \Delta . \cos \Delta) \begin{bmatrix} \sigma_h & 0 \\ 0 & \sigma_v \end{bmatrix} \begin{bmatrix} -\sin \Delta \\ \cos \Delta \end{bmatrix}$

$= \sigma_v \, . \cos^2 \Delta + \sigma_h \, . \sin^2 \Delta$

where: σ_v = major vertical stress = γ . γt, (see Figure 8)

γt = depth of the anchor from the surface = 5, m.

σ_h = major horizontal stress = $K_{o.}\ \sigma_v$

K_o = earth pressure at rest = $v / (1 - v) = 2/3$

$v = 0.4$

$l_g = 1.5 \times 150 / 3.\ 14 \times 0.1(50 + 180 \tan 10°) = 8.76$, m

Therefore: $l_a = 15.60 + 8.76 = 24.36$ m; for the middle of the row (see Figure 8).

In this case, due to the large anchor length, it is advisable to practice cable anchoring as compared with anchor rod (or bar).

8. CONCLUSIONS

In order to provide a conservative anchor design, the minimum factor of safety obtained by the Janbu and Bishop methods of slices has been utilized. This consideration was made due to: relative ease and simplicity of the approach as well as accuracy and adaptability of the calculations to low computerization costs. A new relationship has been proposed to calculate the weight of circular failure mass and its driving moment by differential equations. This has provided an analytical approach to that of the Taylor (1948), with the same precision. The optimum anchor inclination in the method developed can be determined analytically by simple equations without resort to the more tedious method of slices, for both passive and active conditions. It has been proved theoretically that the optimum anchor inclination is upwards, depending largely on the internal friction angle of the mass. This will reduce promotion of the resultant driving moment by the anchor. If practical difficulties in upward drilling, cabling and grouting are severe then it is suggested that the anchors be placed horizontally or near horizontal.

REFERENCES

Baker, R. and Garber, M., 1977, Variational approach to slope stability, Proceedings of the 9th International Conference on Soil Mechanics and Foundation Engineering, p. 9-12.

Bishop, A. W., 1955, The use of slip circle in the stability analysis of slopes, Geotechnique, 5, p. 7-17.

Cousin, B. F., 1978, Stability charts for simple earth slopes, Proceedings of ASCE, J. Geotech. Engr. Division, Vol. 104, No. GT2, February, p. 267-279.

Fellenius, W., 1936, Calculation of the stability of earth dams, Proceedings of the 2nd Congress on large dams, 4, p. 445-463.

Goodman, R. E., 1980, Introduction to Rock Mechanics, J. Wiley and Sons, p. 338-349.

Gudehus, G., 1972, Lower and upper bounds for stability of earth-retaining structures, Proceedings of the 5th European Conference SMFE, 1, Madrid.

Hobst, L. and Zajic, J., 1983, Anchoring in Rock and Soil, 2nd ed., Vol. 3, Elsevier Scientific Publishing, Co., 570p.

Hoek, E. and Bray, J.W., 1981, Rock Slope Stability, 3rd ed, The Inst. of Mining and Met., p228-242, London.

Huang, Y. H., 1983, Stability Analysis of Earth Slopes, Van Nostrand Reinhold Co., p. 14-15.

Hunt, R. E., 1986, Geotechnical Engineering Analysis and Evaluation, McGraw-Hill Co. 729p.

Janbu, N., 1954, Stability Analysis of Slopes with Dimensionless Parameters, Ph.D. dissertation, Harvard University, 81p, Cambridge, Massachusetts.

Janbu, N., 1973, Slope stability computation, Embankment-Dam engineering Casagrande volume, edited by R. C. Hirschfeld and S. J. Poulos, J. Wiley and Sons Ltd., p47-86, New York.

Kötter, F., 1903, Bestimmug des Drucks an gokrummten Géléitfleǐchen, ueine Aufgabe ans der Lebre von Erddruck,

Sitzungsberichte der Königlich Preussischen Akademic der Wissenscherften, p229-233, Berlin.

Kranz, E., 1972, Bureau Securitas, Ground anchors, French code of practice, Editions Eyrolles, Recommendation TA. 72.

Locher, H. G., 1969, Anchored retaining walls and cut-off walls, Losinger and Co., p. 1-23, Berne.

Morgrsten, N. R. and Price, V. E., 1965, The analysis of the stability of general slip surfaces, Geotechnique, 15, p.70-03.

Ostermayer, H., 1977, Practice on the detail design application of anchorages, A review of diaphragm walls, Institution of Civil Engineering, p. 55-61, London

Spencer, E., 1967, A method of analysis of the stability of the embarkments assuming parallel interslice forces, Geotechnique, 17, p. 11-26.

Stocker, M. F., Korber, G. W., Gassler, G., Gudehas, G., 1979, Soil nailing, C. R. Coll. Int. Renforcement des sols, Paris.

Taylor, D. W., 1948, Fundamentals of soil mechanics, J. Wiley and Sons, Inc., p. 455-461, London.

3.4.5 PARAMETRIC STUDY OF VARIABLE SLOPE DESIGN FOR A MULTIPLE LAYERED OPEN PIT

By: X. Fu, M. Sengupta and Andy A. Afrouz

ABSTRACT: A quantitative assessment of the stability of an open pit slope is clearly important when a judgment is needed about whether the slope is stable or not, and decisions are to be made as a consequence. This study describes a deterministic stability analysis together with a sensitivity analysis of variable slopes of an open pit with different layers subjected to planar failure. This approach is a combination of circular failure method and planar failure method, in which the strength parameters are based on Mohr - Coulomb criterion or derived from the rock mass classification (RMR) system and applied to the Hoek and Brown rock failure criterion. A computer program has been developed to analyze the stability of slopes. The results of the analysis are presented as plots of factor of safety (FS) with potential failure angle α_i, in which the stability of the slopes can be determined. It is expected that the optimum parameters FS = f(α_i, β_i, γ_i, H_i, Hw_i, C_i and ϕ_i) could be used in the design of a multiply layered open pit slope. In this study, the influences of the parameters (α_i, β_i, γ_i, H_i, Hw_i, C_i and ϕ_i) on the stability of the slopes are discussed and some remedial measures are suggested.

Note: H_i = the height of the slope; Hw_i = the height of water level;
β_i = the angle of slope surface; α_i = the potential failure angle,
γ_i = the angle of ith layer; q = surcharge.

1. INTRODUCTION

The slope failure problem is often related to surface mining. Evaluation of stability is a necessary consideration prior to any construction involving existing slopes of an open pit mines. The objective of a slope stability analysis is to identify the most likely mechanism of failure and to determine the associated minimum factor of safety. Most stability analysis are based on a two dimensional characterization. Generally, the circular method or the planar method is used to analyze slope stability. In the circular method, the limit equilibrium method is recommended, such as Bishop (1955), Morgenstern and Price (1965), Spencer (1967), Sarma (1973), Janbu (1973), Fredlund et al (1980), and others, which are used in the soil slope or heavily jointed rock mass slope. In the planar method, the limit analysis method is used, such as Ucar et al (1987), Afrouz et al (1991). For a medium or large size open pit, it is possible to have several different layers and several slope surface angles. Each of the layers can have different physical and mechanical properties of rock mass, such as the cohesion and the internal angle of friction, and the geometry of each layer is different. So a more accurate method is needed to deal with the problem of open pit slopes with different layers.

In this study, the deterministic stability and sensitivity analysis of slopes, with the predominant failure mechanism of being planar in nature, are carried out. The analysis is based on the planar failure model and circular failure model in which the rock mass of each layer in the slope is divided into a number of potential failure surfaces of different inclinations (failure angle α_i (i=1, 2, 3...n) from 0° to 90°). By the analysis of the geometry (H_i, β_i, γ_i) and the ground water level (Hw_i) of each layer, the forces acting on sides and bases of the layers (N_i, T_i) can be calculated. Two strength calculation methods are chosen in this work: Mohr - Coulomb criterion (if the strength parameters C_i and ϕ_i are known) and the Hoek and Brown non-linear criterion (if the C_i and ϕ_i are not known). The relationship of factor of safety with potential failure angle α_i is developed: $FS = f(\alpha_i, \beta_i, \gamma_i, H_i, Hw_i, C_i \text{ and } \phi_i)$.

Finally, as an example of the proposed computer program and deterministic procedure, stability analysis on open pit slopes with five different layers is carried out and the influences of the parameters on the stability are discussed. The report also gives some remedial measures for stabilizing the open pit slopes.

2. ANALYSIS METHOD

2.1 *Geometry of open pit slopes with different layers:*

Fig. 1 illustrates an open pit slope composed of different layers. For clarity, only five layers of rock mass are shown; however, the method analysis is applicable to any number of layers.

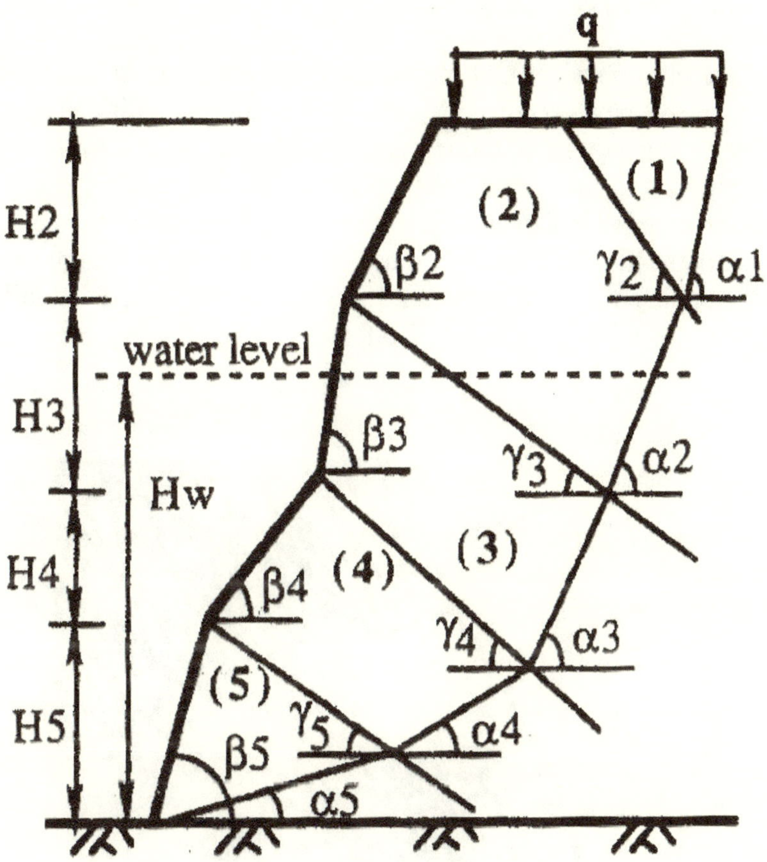

Figure 1. Example of open pit slope with different layers.

The cohesion C_i and the internal friction angle ϕ_i of each layer (i =1,2...n) are assumed to be known, or can be derived from the rock mass rating (RMR) system and applied to the Hoek and Brown rock failure criterion. The rock mass, for which the factor of safety is to be determined, is contained between the known slope surface,

$Yb(x)_i$, and the potential failure surface $Y(x)_i$. The surface $Y(x)_i$ is subjected to a distributed normal stress $N(x)_i$. Both functions $Y(x)_i$ and $N(x)_i$ are unknown and are assumed to be continuous but not necessarily smooth at their transition from one layer to another. It is assumed, however that y = dy /dx and N = dN(x) /dx are continuous within each layer.

2.2 The method of analysis:

Since the physical and mechanical properties of each layer are different, we have to consider all possible failure conditions. In this problem, the potential failure angle α_i should be at any possible inclination and location in each layer of the slope, i.e. it must be kinetically possible for the rock to move. Fig. 2 illustrates the possible failure paths in each layer. As we see, the more possible failure angles divided in each layer, the greater the certainty that a modeled failure path has the minimum factor of safety.

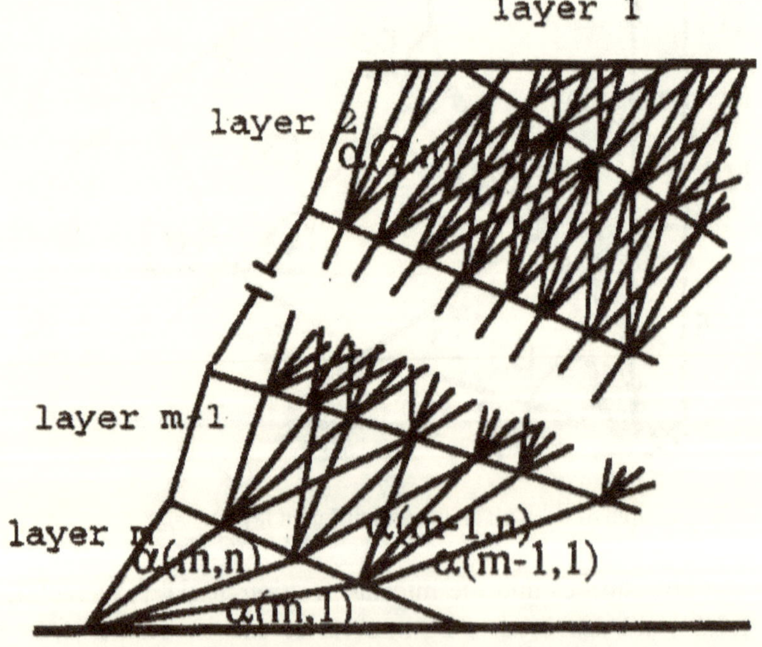

Figure 2. The possible failure angles for each layer.

Supposing the slope to have m layers (counting from the top of the pit slope), the mth layer (at the toe of the slopes) is divided into n possible failure angles, and the m-1th layer is divided into n^2 failure angles, and so on. The 2nd layer divided into n^{m-1} angles and the 1st layer is divided into n^m possible failure angles. According to the analysis method, the sum of the potential failure angles dividing the slopes is $n^m!$ (= $n^m + n^{m-1} + \ldots + n^2 + n$). Each failure surface of every layer has two formulae of factor of safety (FS) versus failure angle (α_i):

(a) The factor of safety of ith layer - This considers the stability of the ith layer.

By the calculation below, the weakest layer could be represented. More attention should be paid on the weak layer. Therefore,

$$(FS)_i = f(\alpha_i, \beta_i, \gamma_i, H_i, Hw_i, C_i \text{ and } \phi_i) = \tau_i / T_i \qquad (1)$$

Where: τ_i = the shear strength of the layer,

T_i= the shear forces required for the stability of the layer.

(b) The total factor of safety of the whole system - This considers the stability of the whole slope. By the calculation below, we can tell if the factor of safety of the whole slope satisfies the requirements, needing supports or not. Therefore,

$$(FS)_t = f(\alpha_i, \beta_i, \gamma_i, H_i, Hw_i, C_i \text{ and } \phi_i) = \text{Sum of } \tau_i / \text{Sum of } T_i \qquad (2)$$

Thus, by the $n^m!$ times calculation, the minimum $(FS)_i$ and minimum $(FS)_t$ can be obtained, and the potential failure surface (α_i) on each layer can be located.

3. FORMULATION

3.1 *The forces acting on the sliding rock mass:*

Recognition of the appropriate case is made by checking the water level, layer geometry and the relevant water force components (U_i, $P.w_i$, $P.w_{i+1}$) acting on the base and the other sides of the layer. Considering the equilibrium of the ith layer of the potential sliding mass shown in Fig. 3.

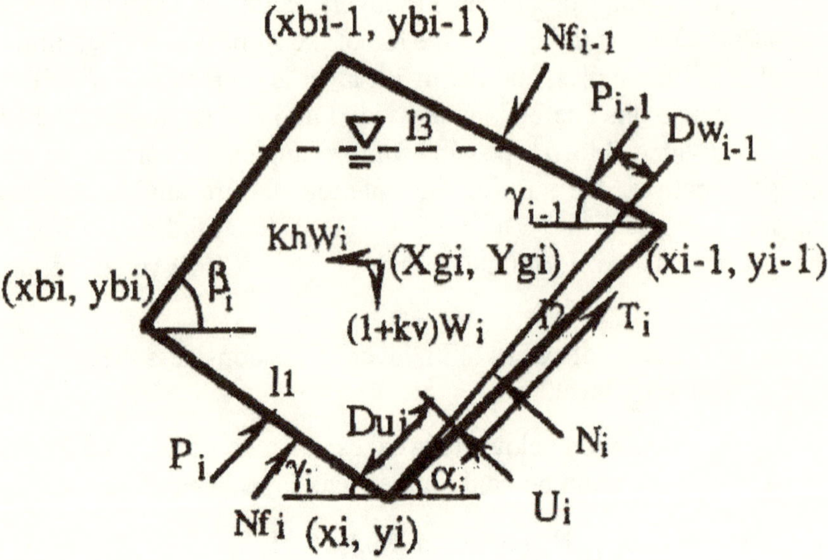

Figure 3. The forces acting on the ith layer.

Where: W_i = the weight of the sliding mass,

k_h = the horizontal seismic component,

k_v = the vertical seismic component,

U_i = uplift force by water on the potential failure side,

Du_i = the distance of Ui to moment center,

Pw_i = the water force acting on the layer side,

Dw_i = the distance of moment of Pwi,

Xg_i, Yg_i = the geometry center of the sliding mass,

l_1, l_2 and l_3 = the side length of the sliding mass.

The sliding mass is divided into m layers. Just before failure the body of the mass must be in equilibrium. It is assumed that all the layers are of thickness sufficiently small to assume that the normal forces N_i, N. f_i, and N. f_{i-1} act at the mid points of the sides of each layer.

From the vertical, horizontal and moment equilibrium of the layer (see Fig. 3), the following equations may be obtained:

For $\subset X = 0$:

$$T_i \cos \alpha_i - N_i \sin\alpha_i + Nf_i \sin\gamma_{i+1} - N.\ f_{i-1} \sin\gamma_i$$

$$= KhW_i + U_i \sin\alpha_i - Pwi_{i+1}\sin\gamma_{i+1} + Pw_{i-1}\sin\gamma_i$$

For $\subset Y = 0$:

$$T_i\sin\alpha_i + N_i\cos\alpha_i + N.\ f_i \cos\gamma_{i+1} - N.\ f_{i-1}\cos\gamma_i$$

$$= (1+Kv)W_i - U_i\cos\alpha_i \_ P.wi_{i+1}\cos\gamma_{i+1} + Pw_i\cos\gamma_i$$

For $\subset M_o = 0$:

$$N_iL2_i - N.f_iL1_i = 2\{[-(1+kv)(Xg_i - X_{i+1}) - Kh(Yg_i - Y_{i+1})]W_i - Pw_i\ Dw_i +$$

$$Pw_{i+1}(Ybw_i - Y_{i+1}) /3 / \sin\gamma_{i+1} - U_iDu_i\} - Nf_{i-1}L3_i$$

The right-hand side of the above equations are known, letting them equal to CGB. The unknowns, T_i, N_i and Nf_i should be:

$$N.\ f_i = \{\{[(1+k\ v)\cos\alpha_i - k\ h\sin\alpha_i]\ W_i - U_i + (P.\ w_i + N.\ F_{i-1})\cos(\alpha_i+\gamma_i)\ P.\ w_{i+1}.\cos(\alpha_i+\gamma_{i+1})\}\ L2_iCGB\}/[L1_i + L2_i\cos(\alpha_i+\gamma_{i+1})]$$

$$N_i = [(1+kv)\cos\alpha_i - Kh\sin\alpha_i]\ W_i + (P.\ w_i + N.\ f_{i-1})\cos(\alpha_i+\gamma_i) - U_i - (P.\ w_{i+1} + N.\ f_i)\cos(\alpha_i+\gamma_{i+1}) \quad (3)$$

$$T_i = [(1+kv)\sin\alpha_i + Kh\cos\alpha_i]\ W_i + (P.\ w_i + N.\ f_{i-1})\sin(\alpha_i+\gamma_i) - (Pw_{i+1} + N.\ f_i)\sin(\alpha_i+\gamma_{i+1}) \quad (4)$$

4. COMPUTER SOLUTION

A computer program has been written in FORTRAN 77. The program is not dependent on the development system and can be run on any computer with a FORTRAN 77 compiler. Figure 4 shows the computer flow chart.

The Flow Chart of Computer Program

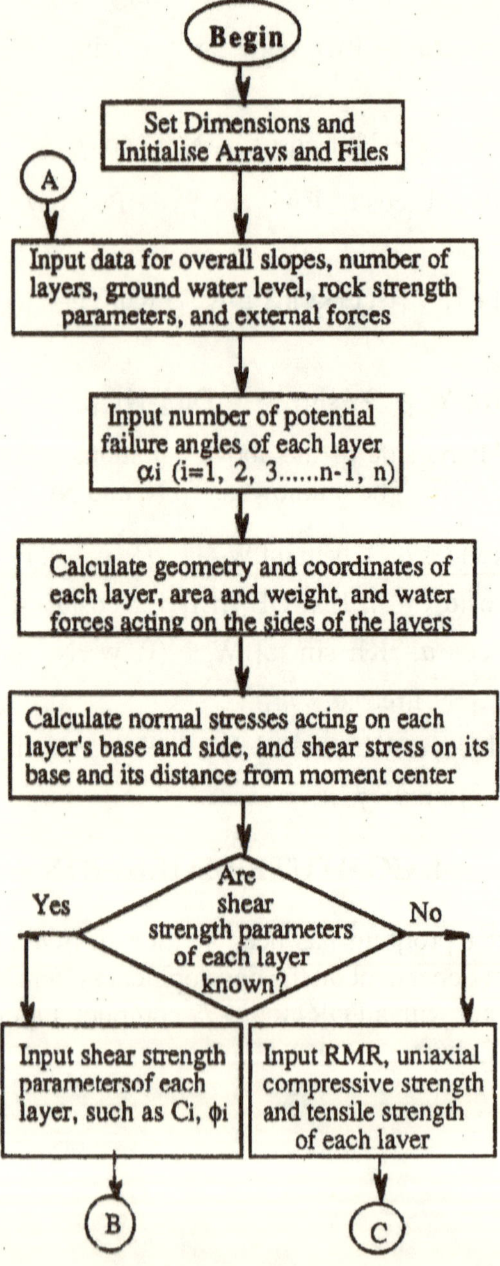

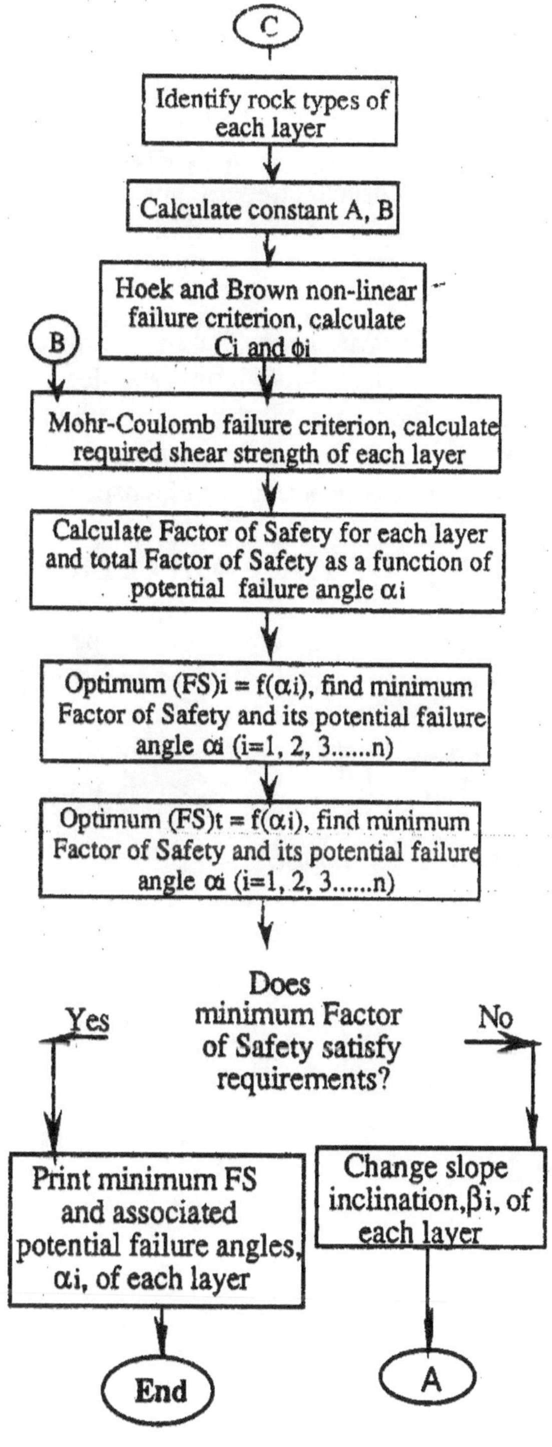
C
Identify rock types of each layer
Calculate constant A, B
Hoek and Brown non-linear failure criterion, calculate Ci and φi
B
Mohr-Coulomb failure criterion, calculate required shear strength of each layer
Calculate Factor of Safety for each layer and total Factor of Safety as a function of potential failure angle αi
Optimum (FS)i = f(αi), find minimum Factor of Safety and its potential failure angle αi (i=1, 2, 3......n)
Optimum (FS)t = f(αi), find minimum Factor of Safety and its potential failure angle αi (i=1, 2, 3......n)
Does minimum Factor of Safety satisfy requirements?
Yes
No
Print minimum FS and associated potential failure angles, αi, of each layer
Change slope inclination,βi, of each layer
End
A

It is noted that the geometrical coordinates of layers (for the most critical one), the coordinates of intersection points of layer sides and phreatic surface, water unit weight, rock mass unit weight and external forces associated with their acting directions (for each layer) are input as fixed input data or generated by the input data. Two different methods of shear strength data input are incorporated in the program, with keyboard selection of the input mode. These two modes are:

(1) Input of the known shear strength parameters (C_i, ϕ_i) of layers base and sides through the data file.
(2) Calculating the required shear strength parameters (C_i, ϕ_i) from keyboard input data for rock types, RMR rating, uniaxial compressive strength and tensile strength.

5. APPLICATION OF THE COMPUTER PROGRAM

Considering an example of an open pit slope with five layers, the paper gives a complete application of the computer program including the input data, the calculation results and its analysis. The effect of parameters on the stability of the open pit slope, such as the parameters α_i, β_i, γ_i, H_i, Hw_i, C_i and ϕ_i, etc. is also discussed.

5.1 *Input data for overall slopes:*

Table 1. Input data for overall slopes

Number of layers	k_h	K_v	q, kN/m^2	γw, kN/m$_3$
5	0.2	0.1	300	10

5.2 *Input data of each layer:*

Table 2. Input data for each layer

layer	1	2	3	4	5
H, m	x	30	25	21	32
Hw, m	x	22	25	21	32
γ, degrees	x	25	24	23	24
β, degrees	x	75	70	65	60
γ_s, kN/m^3	22	24	27	26	21
γr, kN/m^3	21	23	25	24	20

Note: The total height of the slopes is 108 m, the total height of water level is 100 m;

γ_S = the density of saturated rocks; γr = the rock density above the water table.

Table 3. The strength parameters

layer	1	2	3	4	5
C, kPa	900	1000	1200	1100	800
ϕ, degrees	25	30	32	29	26

5.3 *Calculation results:*

(1) The factor of safety of each layer as a function of α - Fig. 5 show the relationship of the factor of safety of each layer with the potential failure angle α (given in Table 3). The minimum factor of safety (0.92) occurs at potential failure angle α = 35° in layer 5, which means layer 5 is weaker than other layers, so more attention should be paid to layer 5. Even though the FS <1.0 in layer 5, for the equilibrium of the whole slope the minimum total factor of safety may not be less than 1.0. Therefore this does not mean the slope is unstable.

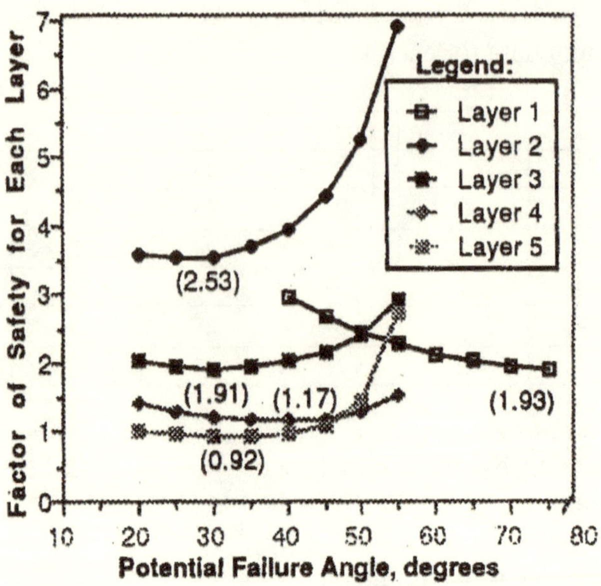

Figure 5. Variation of the factor of safety of each layer versus potential failure angle (α).

(2) The total factor of safety $(FS)_t$ of the whole slope as a function of α - Fig. 6 illustrates the relationship of the factor of safety for each layer as a function of potential failure angle α. With the increasing of α, the total factor of safety of layer 2 to layer 5 is decreasing to their minimum and then slowly increasing, but the factor of safety of layer 1 is decreasing. (The tendency is the same as the analysis of Hoek and Bray (1981), in which they concluded that, when plane failure occurs, slopes have a vertical tension crack on the top of the slope). The minimum total factor of safety is 1.63, and its relative potential failure angle $\alpha = 75°$ (layer 1), 40° (layer 2), 35° (layer 3), 35° (layer 4) and 25° (layer 5), which means that the slope tries to slide down along the failure surface, shown in Fig. 7.

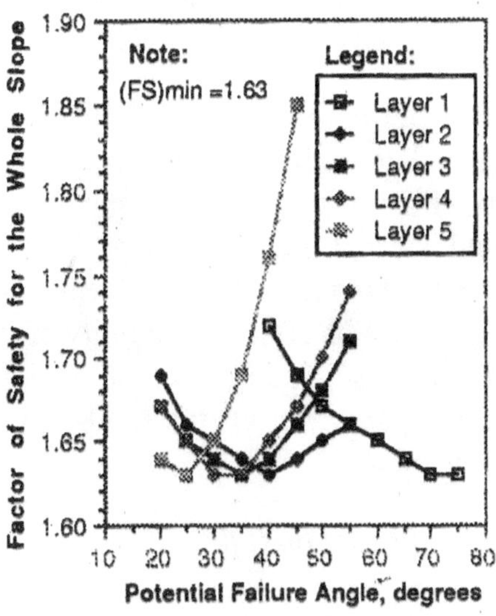

Figure 6. Variation of total factor of safety of each layer versus α.

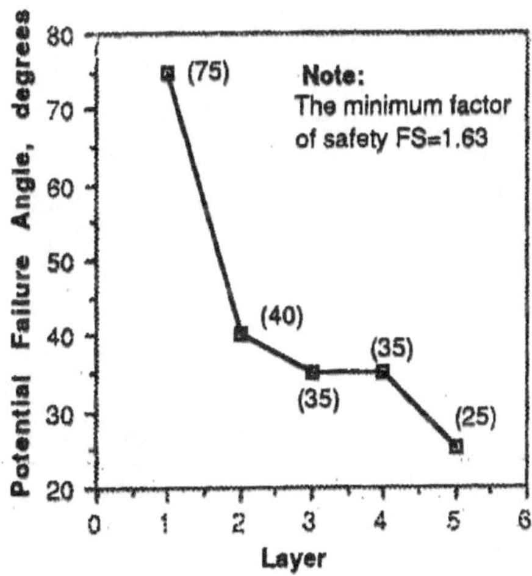

Figure 7. The potential failure angle in each layer.

Fig. 7 illustrates the potential failure angle α in each layer. It shows that, for the same factor of safety, the highest potential failure angle (75°) occurs in layer 1, and it gets reduced in layer 2 to layer 5 (25°). Therefore, in the design of open pit slopes, we should consider the potential failure angles of the layers respectively.

6. INFLUENCE OF PARAMETERS ON THE STABILITY OF THE SLOPES

The minimum factor of safety and its potential failure angle are not fixed. They depend upon all the parameters, such as the slope face angle β, the layer inclination γ, the slope height H, ground water level Hw, cohesion C and internal friction angle ϕ. Some parameters have strong influence, but others have little influence. This section will discuss the parameters which effect the stability of the open pit slope.

Comparisons of each section is based on the input data given in Table 1 to 3. When a parameter changes, other parameters remain the same as the basic data.

6.1 *The inclination of the slopes, β:*

Fig. 8 shows that, if the other parameters are assumed as constant, the factor of safety decreases as the slope angle increases. In the design of an open pit slope, if no other steps are taken to stabilize the slope, the slope angles should be decreased. However, in order to reduce to a minimum the amount of waste rock which has to be excavated in recovering an ore body, the ultimate slopes of the mine are generally cut to the steepest possible angle. Since the economic benefits gained in this way can be negated by a major slope failure, evaluating the stability of the ultimate slopes is an important part of open pit mine planning.

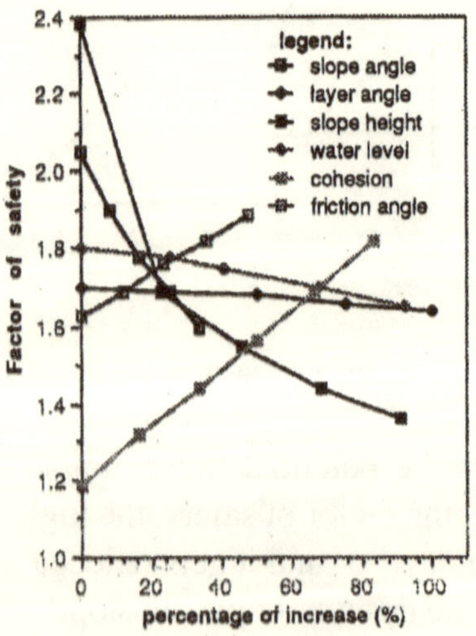

Figure 8. The influences of parameters on the Factor of Safety.

6.2 *The inclination of rock mass layers, γ:*

If the layer angle, γ, increases, the minimum factor of safety, FS, decreases. Large variation in layer inclination had only a minor effect on slope stability.

6.3 *The height of the slope, H:*

The relationship of factor of safety with the slope height H. The increasing of slope height causes a serious reduction of the factor of safety. With the excavation of the open pit, the higher the slope height is, the more unstable the slope will be at a given slope angle.

6.4 Height of the ground water level, Hw:

In this analysis, assuming the slope height of 108 meters, if the Hw = 0 (the slope is assumed to be completely drained), the factor of safety FS= 1.80. If Hw = 100 meters,

FS = 1.65. Even though varying ground water has a minor effect on factor of safety compared to slope angle and slope height, if the ground water flow is considered, the factor of safety will be reduced again. Therefore, it becomes clear that the presence of ground water in a slope can have a very important influence upon its stability and that drainage of this ground water is one of the most effective means of improving the stability of the slope.

6.5 *The cohesion of the rock mass* (C):

Thc cohesion of rock mass alters with the rock types, the water content, the joints and weathering of rock mass. The cohesion of some rock masses can be as high as 50 MPa, such as igneous rocks (granite, basalt), but the cohesion of some other rock masses can be as low as 400 kPa, such as soft sedimentary rock (coal, shale or jointed rocks). The cohesion of most of the rock masses is usually between 400 - 1500 kPa (Hoek, 1981). The cohesion has a linear influence on the factor of safety according to the Fig. 9. The factor of safety increases with increasing cohesion. In this example, when cohesion C = 400 - 800 kPa, the factor of safety FS = 1.19; But when C = 900 - 1300 kPa, FS = 1.82, increasing 53%.

6.6 *The internal friction angle, φ :*

The friction angle f is another strength parameter which has an important influence on the slope stability. The relationship of factor of safety with friction angle φ is linear over the range selected. With the increasing of the friction angle φ, the factor of safety FS increases. The average friction angle φ = 25.4°, FS = 1.63. While the average friction angle φ = 37.4°, FS = 1.89. Generally, the friction angle of rocks φ = 25 - 45°, which means that friction angle can be reduced 80%. According to the discussion, the factor of safety will be decreased 27%. So the friction angle has a significant effect on the stability of the slope.

7. CONCLUSIONS

A computer solution for stability analysis of open pit slopes with different layers written in FORTRAN 77 is proposed, in which a new approach, the combination of the circular failure method and the planar failure method, based upon an initial prediction of minimum factor of safety for a planar slope failure in each layer, has been developed to perform the stability analysis. Mohr - Coulomb failure criterion is used to define strength if the strength parameters of rock mass are known. Since estimating the required strength parameters of rock mass by field and laboratory tests is usually an expensive process, the Hoek and Brown empirical strength criterion is used as an alternative means of estimating the rock mass strength parameters from the simple laboratory tests.

The sensitivity and stability analysis based on the application of the computer program have suggested the follow points:

(1) The factor of safety for each layer $(FS)_i$ alters with the potential failure angle α_i. The minimum $(FS)_i$ indicates the ith layer is relatively weak compared with the other layers, but it does not mean the whole slope unstable.

(2) The total factor of safety for overall slopes $(FS)_t$ depends upon the strength parameters C_i and ϕ_i, the layer angle γ_i the slope angle β_i, the slope height H_i and the water level Hw_i. The potential failure angle α_i is different with each layer, and

the minimum total factor of safety indicates the stability of the whole slope.

(3) This project shows degree of influence of the various parameters in Stability of a slope and its factor of safety that: (a) Increases with increase in the strength parameters, C_i and $\phi_{i;}$ (b) Decreases with increase in the slope angle to horizontal $\beta_{i,}$ slope height H_i, water level Hw_i, and the layer angle γ_i.

REFERENCES

1. Afrouz, A. and Fu, X. (1991). "Computer design of open pit slope and design of reinforcement." International journal of Surface Mining and Reclamation, Vol. 5, No. 4, 177-183.
2. Bishop, A.W. (1955). "The use of the slip circle in the stability analysis of slopes, Geotechnique, Vol. 5, 7-17.
3. Fredlund, D. G. (1984). "Analytical methods for slope stability analysis", Stat-of-the Art. Proc. 4th Int. Symp., Landslides, Toronto, Canada, Vol. 1, 229-250.
4. Hoek, E. (1981). "Analysis of slope stability in very heavily jointed or weathered rock mass", Third International Conference on Stability in Surface Mining, C.O. Brawner, Editor, New York, 375-406.
5. Hoek, E and Bray J. (1981). Rock Slope Engineering, Revised third edition, The Institution of Mining and Metallurgy, London.
6. Janbu, N. (1973). "Slope stability computations", Embankment-dam Engineering, edited by Casagrande, V., Hirschfeld R.C. and Poulos S.S., John Wiley and Sons, New York, 47-86.
7. Leshchinsky, D. (1990). "Slope stability analysis: Generalized approach", Journal of Geotechnical Engineering, Vol. 116, No. 5, 851-867.
8. Morgenstern, N. R. and Peice, V.E. (1965). "The analysis the stability of general slip surfaces." Geotechnique, Vol. 15, No. 1, 79-93.
9. Nash, D. (1989). "A comparative review of limit equilibrium method of stability analysis", Slope Stability, Edited by Anderson M.G. and Richards, K.S., John willy and Sons, New York, 11-75.

10. Sarma, S.K. (1973). "Stability analysis of embankments and slopes", Geotechnique, Vol. 23, No. 3, 423-433.
11. Singh, R.N. and Gahrooee, D.R. (1990). "Deterministic stability and sensitivity analyses of slopes in jointed rock masses", Mining Science and Technology, Vol. 10, 265-286.
12. Spencer, E. (1967). "A method of analysis of the stability of embankments assuming parallel inter-slice forces", Geotechnique, Vol. 17, 11-26.
13. Ucar, R., Hassani F.P. and Afrouz A. (1987). "Design of anchoring in rock with planar failure", International Journal of Surface Mining, No. 12, 199-208.

3.4.6 SLOPE REINFORCEMENT BY ROCK ANCHORING

A. A. Afrouz, F.P. Hassani, R. Ucar and H.S. Mitri
Department of Mining and Metallurgical Engineering
McGill University, Montreal, Canada

ABSTRACT

Design of anchors as a means of slope reinforcement in soil and rock has been previously achieved by a variety of techniques. Some of these methods either do not cater for all the parameters and variations which exist in the ground, or are non-systematically lengthy and do not lend themselves to computerization. In this investigation, a theoretical approach is adopted, which is based on in situ observation, in order to calculate the factor of safety for anchor reinforcement. Based on this, the anchor specifications such as: optimum angle of inclination, length of the anchor, length of the grouting, anchor diameter, and spacing between the anchors is determined.

KEY WORKDS:
Rock Anchoring, Slope Stability, Slope Reinforcement, Ground Control.

1. INTRODUCTION

There are about fifteen modes of classical ground failure in soil and rock, which numerous combinations of the above. Two of the most familiar modes of slope failure are:

(a) Circular failure, most predominant in soil or highly fractured rock slopes - it occurs usually along a surface approaching a circular arc of finite length offering least resistance to sliding.

(b) Planar failure, mostly encountered in rock slopes with bedding planes, joints and other discontinuity planes. This type of failure can either occur on a single continuous plane represented by a mass sliding on a persistent discontinuity plane, or can occur on a set of intersecting planes.

An approach to determine the optimum anchor design and slope reinforcement under each of the above noted conditions varies one another. Prior work exists on soil and rock reinforcement by anchoring, utilizing variety of techniques [1-14]. Some of these methods either do not cater for all the parameters and variations which exist in the field, or are non-systematically lengthy and do not lend themselves to computerization.

In this investigation, a theoretical approach is adopted which is based on in situ observation in order to calculate the factor of safety for anchor reinforcement. Based upon the safety factor, the anchor specifications such as: optimum anchor inclination, length of the anchor, length of grouting, anchor diameter, and spacing between anchors can be determined.

2. SLOPE REINFORCEMENT ANALYSIS

Parameters considered were: total weight of unstable ground, its consequent total normal and tangential forces acting on the failure zone, its driving moment, ground water pressure, surcharge, seismic effect, anchor dimensions and inclination to horizontal, together with passive and active forces.

Circular and planar slope failures without reinforcement was compared with that utilizing anchor reinforcement.

Result of the above noted analysis are as follows:

2.1 Determination of the slope Factor of Safety (FS) prior to reinforcement:

(a) For circular failure conditions:

$$FS=R(S_O.L_f+\sum N.\tan\phi)/M_d \quad (1)$$

(b) For planar failure conditions:

$$FS = \{[S_O .H.\sin \beta/\psi .\sin (\tau-\alpha)]+\tan \phi\ \Omega.\cos\alpha - \Omega_1.\sec \alpha - K_h.\sin \alpha)\}/\Omega.\sin \alpha+K_h.\cos \alpha) \quad (2)$$

Variation of the safety factor before rock reinforcement with α - values is given in Table 1 and Fig. 1, using equation (2).

Table 1 - Variation of Safety Factor before reinforcement (FS) with angle of failure plane to the horizontal (α) - Ref. [15].

α^o	Minimum (FS)
35	1.30
40	1.24
45	1.22
50	1.25
55	1.35
60	1.58

Note: In this case, f(α) = f(45°), see Fig. 1.

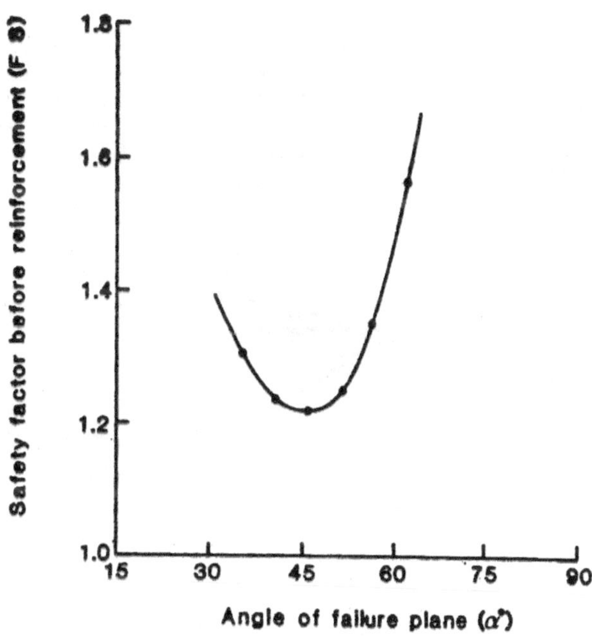

Fig. 1 - Variation of Safety Factor before Reinforcement with Angle of failure Plane to Horizontal, Ref. [15].

2.2 Determination of the slope factor of safety after reinforcement:

(I) For the passive anchoring:

(a) In circular failure conditions (Fig. 2):

$$(FS)_{passive} = S_O \,.\, AB + T_a + (N + N_a) \tan \phi / T \tag{3}$$

(b) In planar failure conditions (Fig. 3):

$$(FS)_{passive} = \{[S_O \,.H/\sin \alpha) + F_a.\cos(\alpha - \Delta)] + [W(1 \pm K_v) \cos \alpha - U - k_h \,.\, W.\sin \alpha + F_a.\sin(\alpha - \Delta)] \tan \phi\} / [W(1 \pm K_v) .\sin \alpha + W.k_h.\cos \alpha] \tag{4}$$

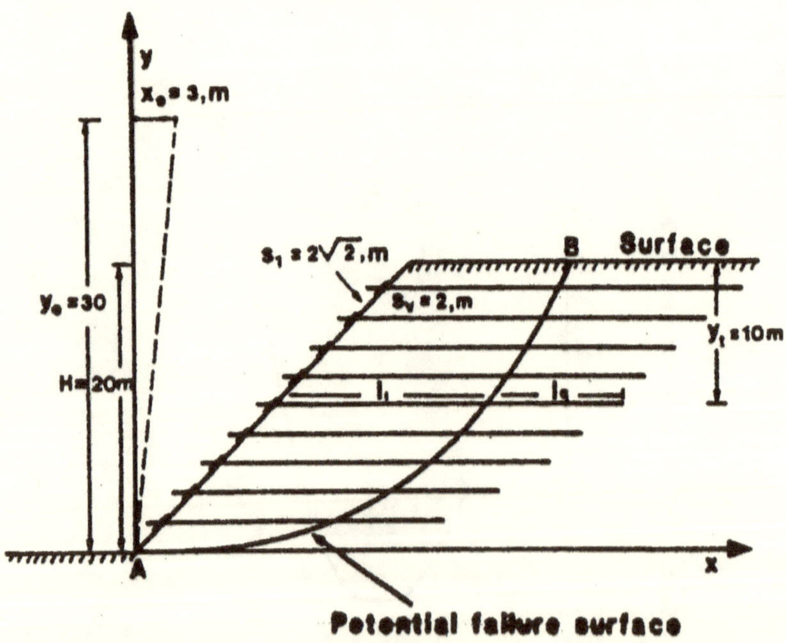

Fig. 2 - Position of anchor relative to the pit slope, in circular failure conditions [16].

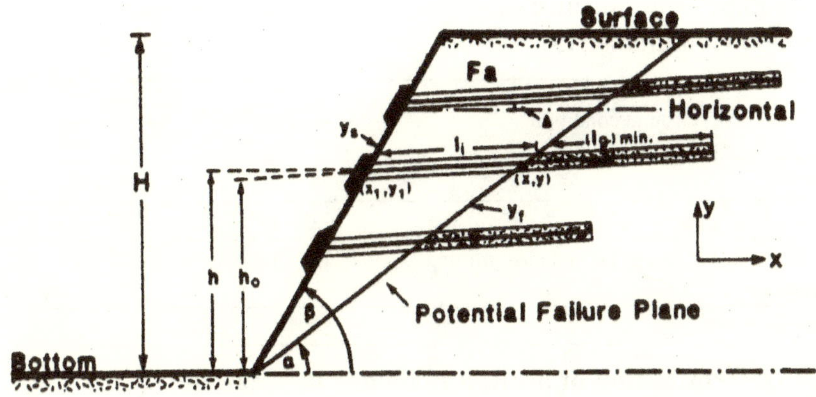

Fig. 3 - Position of anchor relative to the Pit slope, in planar failure conditions [15].

(II) For the active anchoring:

(a) In circular failure conditions:

$$(FS)_{active} = [M_r + R(N_a.\ \tan\phi)]\ /(M_d - R.\ T_a) \tag{5}$$

(b) In planar failure conditions:

$$(FS)_{active} = \{[S_O\ .H\ /\sin\alpha) + [W(1 \pm K_v)\cos\alpha - U - K_h.\ W.\sin\alpha + F_a.\sin(\alpha - \Delta)]\tan\phi\}\ /\ [W(1 \pm K_v)\sin\alpha + W.\ K_h\cos\alpha - F_a.\ \cos(\alpha - \Delta)] \tag{6}$$

Table 2 indicates significance of active safety factor $(FS)_{active}$ in various ground conditions.

The factor of safety under dynamic conditions $(FS)_{dynamic}$ can be determined from the following expression:

$$(FS)_{dynamic} = [(FS)_{static} - \tan\epsilon\ .\ \tan\phi]\ /(1 + 0.88\tan\epsilon) \tag{7}$$

Table 2- Significance of Active Safety Factor $(FS)_{active}$ in Various Ground Conditions [15].

Ground condition	$(FS)_{active}$ of structure	
	Temporary	Permanent
Unsafe	1<	1.1 <
Low confidence	1.0 – 1.2	1.1 – 1.3
Safe for cut and fill, pillar and most mining structure	1.2 – 1.4	1.3 – 1.6
Safe for dams	1.5 - 2.0	

2.3 Determination of optimum anchor inclination:

(I) For passive anchoring:

(a) In circular failure conditions (see Figure 2):

$$\Delta_{passive} = (\alpha_1 + \alpha_2 - 2\phi)/2 \quad (8)$$

(b) In planar failure conditions (see Figure 3):

$$\Delta_{passive} = \alpha - \phi \quad (9)$$

Variation of passive anchor force $(F_a)_{passive}$ with anchor inclination to horizontal (Δ), using equation (23) of reference [15] is given in Table 3 and illustrated in Figure 4.

Table 3 - Variation of Passive Anchor Force $(F_a)_{passive}$ with Angle of Anchor to Horizontal (Δ) from Equation (23) of Ref. [15]

Δ^{O}	$(F_a)_{passive}$, kN/m	f(Δ) / f(0)
40	7449	≈1.00
35	7247	0.97
30	7108	0.95
25	7027	0.94
20	7000	0.93
15	7027	0.94
10	7108	0.95
5	7247	0.97
0	7450	1.00
-5	7723	1.03
-10	8083	1.00
-15	8545	1.15
-20	9138	1.23

Note: Optimum $\Delta = (\alpha - \phi)$

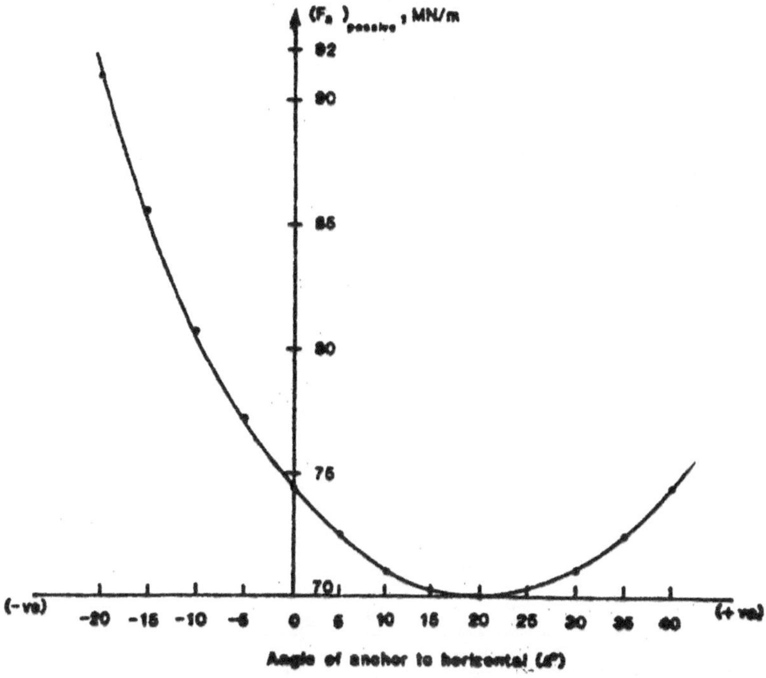

Fig. 4 - Variation of Required Active Anchor Force $(F_a)_{passive}$ with Angle of Anchor to Horizontal (Δ), using Equation (23) of Ref. [15].

(II) For active anchoring:

(a) In circular failure conditions (see Figure 2):

$$[\sin(\alpha_1 - \Delta) - \sin(\alpha_2 - \Delta)] / \cos(\alpha_2 - \Delta) - \cos(\alpha_1 - \Delta) = (FS)_{active} / \tan\phi \quad (10)$$

(b) In planar failure conditions (see Figure 3):

$$\tan(\alpha + \Delta)_{active} = \tan\phi / (FS)_{active} \quad (11)$$

Variation of active anchor force $(F_a)_{active}$ with anchor inclination to horizontal (Δ), using equation (36) of reference [15] is given in Table 4 and illustrated in Figure 5.

Table 4 - Variation of Active Anchor Force $(F_a)_{active}$ with Angle of Anchor to the Horizontal (Δ), from Equation (36) of Ref. [15].

Δ^o	$(F_a)_{active}$, kN/m	f(Δ) / f(0)
40	2500	0.95
35	2450	0.93
30	2418	0.92
24	2400	0.91
20	2410	0.92
15	2434	0.93
10	2477	0.94
5	2542	0.97
0	2629	1.00
-5	2747	1.05
-10	2897	1.10
-15	3091	1.10
-20	3339	1.27

Note: Optimum $\tan(\alpha - \Delta) = \tan\phi / (FS)_{active}$

2.4 Determination of anchor length and length of grouting:

(a) For circular failure conditions (see Figure 2):

$$l_a = l_i + l_g \tag{12a}$$

$$l_i = (x - x_1) \sec \Delta \tag{12b}$$

$$l_g = (FS)_{gr}. Q/\pi.D_h/\tau_b \tag{12c}$$

(b) For planar failure conditions (see Figure 3):

$$l_a = l_i + l_g \tag{13a}$$

$$l_i = h\{(\cot \alpha - \cot \beta)\ [\sin \Delta . \cot(\alpha - \Delta) + \cos \Delta]\} \tag{13b}$$

$$(l_g)_{min} = \ln \{(8P/\pi. Q. D_h^2 . K) \tan [45 + (\phi/2)] + 1\}\ D_h. K/4 \tag{13c}$$

$$(l_g)_{max} = 2P.t\ an[45 + (\phi/2)] / \pi .D.Q \tag{13d}$$

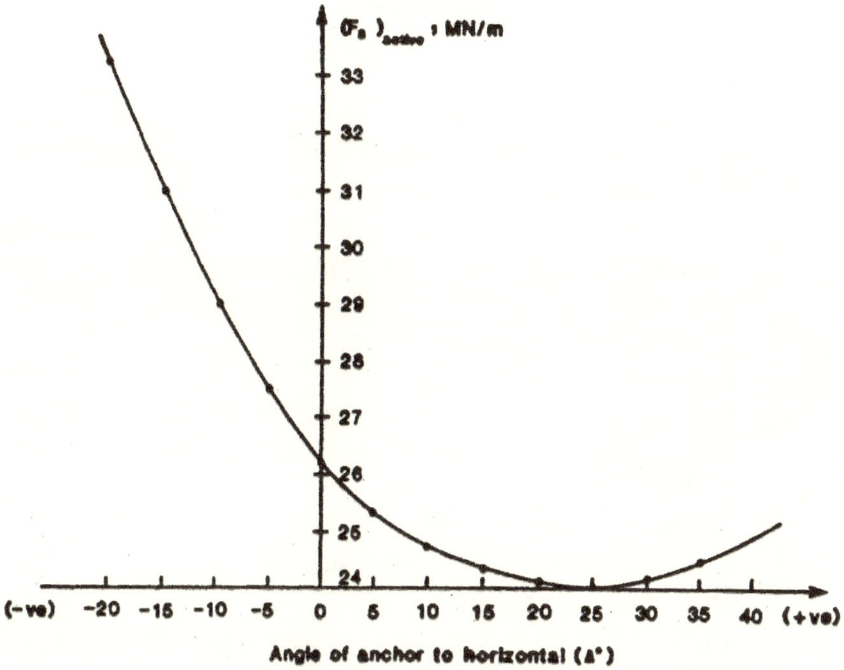

Fig. 5 - Variation of Required Active Anchor Force $(F_a)_{active}$ with Angle of Anchor to Horizontal (Δ) using Equation (36) of Ref. [15].

2.5 Determination of the anchor diameter and spacing:

$$D_a = [4Q\ (FS)_{load} / (\pi \cdot \sigma_b)]^{0.5} \tag{14}$$

$$S_h = Q / (S_1 \cdot P_{sf}) \tag{15}$$

$$S_v = S_1 . \sin\beta \tag{16}$$

3. CONCLUSION

The proposed method provides a new approach to predict the optimum factors of safety upon which design of the anchor under circular, planar, static, dynamic, passive and active slope reinforcement conditions is facilitated. The optimum anchor inclination (Δ) has been proved to be upwards, which creates difficulty in drilling, anchor placement and grouting. It is therefore suggested to angle the anchor to horizontal as closely as possible.

REFERENCES

1 - Kotter, F., Bestimmug des Drucks and gokrummten Geleitfleichen, ueinie Aufgabe ans der Lebre von Erddruck, Sitzungsberichte der Koniglich Preussischen Akademic der Wissenscherften, p. 229-233, Berlin (1903).

2 - Fellenius, W., Calculation of the stability of earth dams, Proceedings of the 2nd. Congress on large dams, 4, p. 445-463, (1936).

3 - Locher, H.G., Anchored retaining walls and cut-off walls, Losinger and Co., p. 1-23, Berne (1969).

4 - Janbu, N., Slope stability computation, Embankment dam engineering Casagrande volume, Edited by R.C. Hirschfeld and S.J. Poulos, J. Wiley and Sons Ltd., p. 47-86, New York (1973).

5 - Hoek, E. and Bray, J.W., Rock slope stability, 3rd, Ed., The Inst. of Mining and Met., p. 228-242, London (1981).

6 - Baker, R. and Garber, M., Variational approach to slope stability, Proc. 9th. Int. Conf. on Soil Mechanics and Foundation Engineering, p. 9-12, (1977).

7 - Bishop, A.W., The use of slip circle in the stability analysis of slopes, Geotechnique, 5, p. 7-17, (1955).

8 - Gudehus, G., Lower and upper bounds for stability of earth-retaining structures, Proc. 5th. European Conf. SMFE, 1, Madrid (1972).

9 - Morgestern, M.R. and Price, V.E., The analysis of the stability of general skip surfaces, Geotechnique, 15, p. 70-83, (1965).

10 - Seegmiller, B.L., Artificial support of rock slopes, 3rd. Int. Conf. on stability in surface mining, Soc. of Mining Engineers, AIME, p. 249-288, U.S. (1982).

11 - Dight, P.M., A case study of the behaviour of rock slope reinforced with fully grouted rock bolts, Int. Symp. on Rock Bolting, Aug. 28-Sept. 2, p. 358-373, Sweeden (1983).

12 - Hanna, T.J., Foundations in Tension: in Ground Anchors, Mc Graw Hill Co., 573 p., U.K. (1982).

13 - Hobst, L. and Zajic, J., Anchoring in rock and soil, 2nd. Ed., Elsevier Scientific Publishing Co., 570 p., (1983).

14 - Spencer, E., A method of analysis of the stability of embankments assuming parallel interslice forces, Geotechnique, 17, p. 11-26, (1967).

15 - Ucar, R., Hassani, F.P. and Afrouz, A., Design of anchoring in rock with planar failure, International Journal of Surface Mining, Balkema Publishers, p. 199-208, Netherlands, 1, Dec. (1987).

16 - Ucar, R., Hassani, F.P., and Afrouz, A., A design analysis for anchoring soil and fractured rock masses prone to circular failure, International Journal of Surface Mining, Balkema Publishers, p. 267-276, Netherlands, 1, Dec. (1987).

5. LIST OF SYMBOLS

AB = curved length of A to B = π. R $(\alpha_1 - \alpha_2)$ /180°

D_a = diameter of the anchor, m

D_h = diameter of the anchor hole, m

F_a = anchor force along its length, kN/m

FS = slope factor of safety before reinforcement

$(FS)_{active}$ = slope factor of safety after active reinforcement

$(FS)_{passive}$ = slope factor of safety after passive reinforcement

$(FS)_{dynamic}$ = slope factor of safety under dynamic conditions

$(FS)_{static}$ = slope factor of safety under static and planar failure conditions

$(FS)_{gr}$ = factor of safety of the grout-rock interface

$(FS)_{load}$ = safety factor of anchor under load

H = bench height, m

H_1 = height of the horizontal ground water table above the final pit bottom, m

K = dimensionless constant relating to the shear and tensile strength of the anchor

K_h = horizontal component of the seismic factor

K_v = vertical component of the seismic factor

l_a = anchor length, m

l_g = length of anchor grouting, m

l_i = distance between the anchor head and failure surface, m

L_f = curved length of the perimeter of failure surface, m

Md = driving moment

M_r = resisting moment

N = total normal force acting on the failure surface, kN/m

N_a = normal force acting on the anchor, kN/m

P_{sf} = average stress acting on the slope face, kPa

q = surcharge, kPa

Q = working load per anchor, kN

R = radius of the circular failure, m

S_O = average cohesion of the failure mass, MPa

S_h = horizontal spacing between anchors, m

S_v = vertical spacing between anchors, m

S_1 = projected vertical spacing between the anchors on the slope of degrees to horizontal, m

T = tangential force acting on the failure surface, kN/m

T_a = tangential force acting on the anchor, kN/m

U = the neutral force due to water uplift, kN/m

W = weight of rock mass above the failure surface, kN

(x, y) = co-ordinate of the centre of failure circle from the x-y axis to pass from the slope toe, m

α_1 = minimum angle of potential failure surface, degree

α_2 = maximum angle of potential failure surface, degree

α = angle of potential failure surface in horizontal degree

β = angle of slope face to horizontal, degree

γ = unit weight of rock, kN/m^3

γ_{sat} = unit weight of saturated rock, kN/m^3

γ_w = unit weight of water, kN/m^3

Δ = angle of anchor to horizontal, degree

ϵ = centre angle = $\tan^{-1}[K_h/(1 \pm K_v)]$

σ_b = breaking strength of anchor, kPa = cohesive strength of the anchor grout

τ_b = cohesive strength of the anchor grout with the soil or fractured rock mass, kPa

ϕ = average internal friction angle of the failure mass, degree

ψ = total weight factor, kN/m = g . H + [γ(H² - H²₁) /2] + (γ_sat . H²1) / 2)

ψ_1 = water weight factor, kN/m = γ_W . H²1) / 2

$\Omega = 1 \pm K_v$

$\Omega_1 = \psi_1 / \psi$

3.5 STEEL ARCH SUPPORTS FOR TUNNELS

3.5.1 DESIGN OF SEMI-CIRCULAR SUPPORT SYSTEM IN MINE ROADWAYS

Design de supports d'acier semi-circulaires dans les galeries de mines

Entwurf eines Ausbausystems aus halbrundem Stahl in der Strecke

By: F. P. Hassani and Andy A. Afrouz
Department of Mining and Metallurgical Engineering,
McGill University, Montreal, Quebec, Canada

ABSTRACT: The results of experimental and theoretical investigations into the behaviour of the Semi-Circular steel support under fractured roof loading is presented.

RESUME: Les résultats d'études expérimentales et théoriques relatives au comportement de supports d'acier semi-circulaires avec des conditions du chargement d'un toit fracturé sont présentés.

Zusammenfassung: Die Ergebnisse einer experimentalen und theoretischen Erforschung in das Verhalten von halbrundem Stahlausbau unter gebrochener Deckenbelastung sind vorgestellt.

1. INTRODUCTION

Steel sets are utilized commonly as a passive support in the main access roadways and gate-roads in coal mines. They serve to:

(a) Control the strata around the roadways,
(b) Reduce closure of the opening,
(c) Increase safety along the roadways.

Although considerable capital expenditure is involved in the support and associated repairs along roadways in coal mines, there is no standard rule in the coal mining industry for the determination of the support specifications. The design of support systems has tended to be based on experience and availability of support material.

This paper stems from part of an overall research program into rock support interaction in soft and hard rock mining excavations. It is devoted to determining the size and collapse load of semi-circular steel H-section as a roadway support by utilizing Castigiliano's theorem (Timoshenko, 1968) which experimentally has proved to give a good approximation to physical modeling (Jukes et al., 1983).

2. THEORETICAL ANALYSIS

Considering a semi-circular steel support and a point load acting on its crown downwards (Fig. 1), the load carrying capacity of the support can be expressed as follows:

$$\delta_y = M_p / Z_p = R(p - h)/M_s - f \qquad (1)$$

where: δ_y = Yield stress of the support (kPa)

M_p = Plastic moment R(p - h)

Z_p = Plastic modulus (m^3)

M_s = Section modulus (m^3)

R = Radius of the support material (m)

f = Shape factor = Z_p / M_S (between 1 and 1.6)

p = Vertical reaction of the support legs, (kN)

h = Horizontal component of the load acting on the support at its legs (kN)

The bending moment at the crown of the support (M_C) can be expressed as:

$$M_C = p.\ R - h.\ R \qquad (2)$$

The bending moment at any point along the support can be expressed as:

$$M = p.\ x - h.\ y \qquad (3)$$

According to Jukes et al, (1983):

$$p = P/2 \; ; \; h = P/\pi \; ; \; x = R(1 - \cos\theta) \; ; \; y = R\sin\theta \tag{4}$$

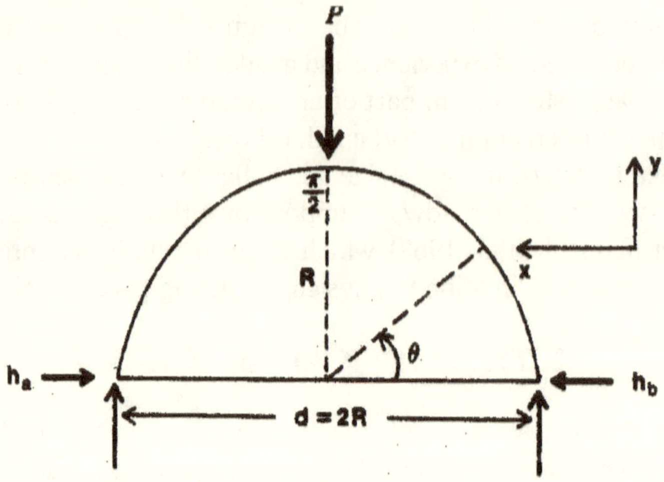

Fig. 1 - Schematic of a semi-circular support, representing forces and notations.

It is known that a plastic hinge will form at the crown of the support under the load, then at this point utilizing equations (2) and (4):

$$M_C = M_p = (P_C.R/2) - (P_C.R/\pi) = P_C.R(1/2 - 1/\pi) \tag{5}$$

where: P_C = Critical load to cause yield of the support, kN

Hence the yield stress at the crown of the support using equations (1) and (5) is:

$$\delta_y \text{ at the crown} = [P_C.R(1/2-1/\pi)] / M_S.\, f = [P_C.R(\pi - 2)] / 2\pi\, M_S.\, f$$

$$= 0.18\, P_C.\, R / M_S.\, f$$

$$P_C = 2\pi\, \delta_y.\, M_S.\, f / R(\pi - 2) = 5.56\, \delta_y\, .\, M_S.\, f / R \tag{6}$$

The yield stress at any other point along the support, from equations (1), (3) and (4) is:

$$\delta_y \text{ at any point} = [(P_C.R(1 - \cos\Theta)/2) - (P_C.R \sin\Theta/\pi)] / M_S.\, f$$

$$= P_C.\, R(\pi - \pi\cos\Theta - 2\sin\Theta)/2\pi\, M_S.\, f$$

$$P_{C}. = 2\pi\, \delta_y.\, M_S.\, f / R(\pi - \pi\cos\Theta - 2\sin\Theta) \tag{7}$$

One of the most likely plastic yield point along the semi-circular steel support is at $\Theta = 45^{o}$. Therefore, values of δ_y and P_C at this point, using equation (7) is:

$$\delta_y \text{ at } \Theta \text{ of } 45^{o} = -0.08\, P_C.R\, /M_S.\, f$$

$$P_C \text{ at } \Theta \text{ of } 45^{o} = -12.82\, \delta_y.\, M_S.\, f/R \qquad (8)$$

The negative sign in the above expressions indicate compressional failure. All of the above noted expressions are applicable for the supports erected along horizontal or near horizontal level roadways. For roadways dipping with an angle of β^{o} to the horizontal, the P_C-values should be divided by Sin β. In this case, the support material may not fail under the load equivalent to the vertical collapse load, but it will be tilted, resulting in the reduction of support efficiency.

3. DESIGN PROCESS

To select a support system, five main steps should be considered. These are as follows:

3.1- The effect of section modulus (M_S) and shape factor (f) on the yield stress of the support (δ_y) and its collapse load (P_C) is seen from the equations (6) and (7). The values of M_S and f are directly proportional to P_C and indirectly proportional to δ_y. The value M_S is determined from the following expression:

$$M_S = A.\, z/3 \times 10^{-9}\, (m^3) \qquad (9)$$

where: A = Cross-sectional area of the H-section support ($mm^{2)}$
z = Thickness of the H-section support (m.m) as shown in Fig. 2.

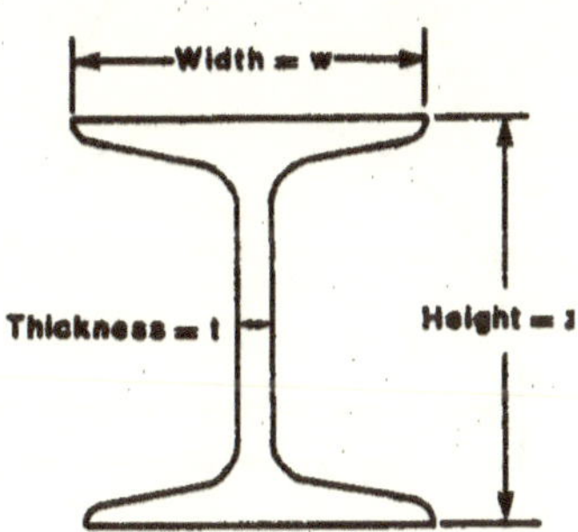

Figure 2. Dimensional notations of H-section supports used in Table 1.

The M_S-values for various dimensions of the H-section supports is given in Table 1. The f-values depend on the cross-sectional shape of the support. Since the steel supports in mines have generally H-shape, f-value is a constant and commonly accepted to be 1.15 (Jukes et al., 1983). This value is conservative in comparison with the value of 1.55 also found experimentally (Sadler et al., 1965).

3.2- The support should have sufficient load capacity to hold the envelope of fractured strata caused by the strata pressure redistribution. This fractured zone is usually in the shape of a dome, cone or sinosoidal. Fig. 3 shows 4main loading conditions described as follows:

3.2A- Fractured envelope extending up to half the diameter of the support (d /2 = R) is shown as curve A. This is most commonly observed where either the roof strata is strong or the roadway is machine-cut, well formed or a good system of packing is employed along the roadway. In this case, the load of fractured roof rock on the support (W_A) and its corresponding pressure (P_A) can be determined from:

$$W_A = \gamma.\ d.\ R/2 = 9.81\ \gamma.\ R^2 = 24.5\ R^2\ (kN/m)$$

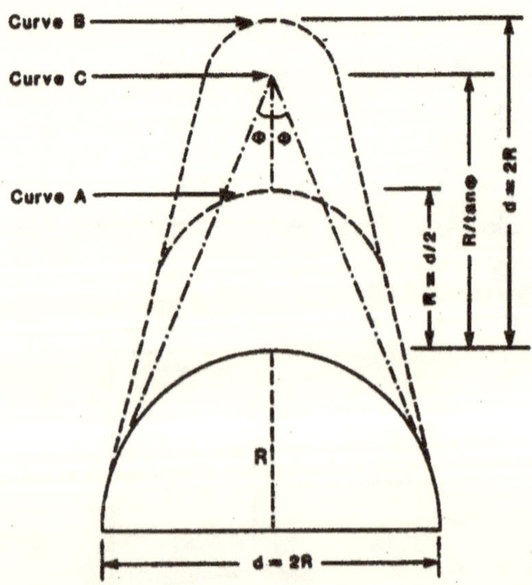

Figure 3. Three main conditions of fractured roof rock above the roadway supports.

Table 1 - Specifications of steel H section supports and fractured rock load (W)

Item No.	Support Radius (R), m.	Sectional dimensions, mm. (Fig.4)	Section Modulus (M_S), m^3X10^{-6} (eqn. 9)	Mass per unit length of the support, kg/m	Load of the fractured roof rock on the support (W), kN /0.9 m			
					At height R (eqn.10)	At height 2R (eqn. 11)	At height R/tan φ (eqn.12)	Terzaghi's formula (eqn.14)
1 2 3 4	1.52 1.67 1.82 2.42	z = 76 w = 76 t = 8.9	45.0	14.67	51.06 61.63 73.20 129.43	102.02 123.27 146.41 258.85	65.89 79.54 94.47 167.02	91.60 100.63 109.67 145.83
5 6 7 8	1.52 1.67 1.82 2.42	z = 89 w = 89 t = 9.5	67.4	19.35	51.06 61.63 73.20 129.43	102.02 123.27 146.41 258.85	65.89 79.54 94.47 167.02	91.60 100.63 109.67 145.83
9 10 11 12	1.52 1.67 1.82 2.42	z = 102 w = 102 t = 9.5	97.0	23.06	51.06 61.63 73.20 129.43	102.02 123.27 146.41 258.85	65.89 79.54 94.47 167.02	91.60 100.63 109.67 145.83
13 14 15 16	1.52 1.67 1.82 2.42	z = 114 w = 114 t = 9.5	132.0	26.78	51.06 61.63 73.20 129.43	102.02 123.27 146.41 258.85	65.89 79.54 94.47 167.02	91.60 100.63 109.67 145.83
17 18 19 20	1.52 1.67 1.82 2.42	z = 127 w = 114 t = 10.2	158.0	29.76	51.06 61.63 73.20 129.43	102.02 123.27 146.41 258.85	65.89 79.54 94.47 167.02	91.60 100.63 109.67 145.83

However, most roadways driven in coal measure rocks utilize steel supports at 0.8 metre intervals. Therefore, the above equation in terms of kN/0.9 m can be expressed as follows:

$$W_A = 24.5\ R^2 \times 0.9 = 22.1\ R^2\ (kN/0.9m)$$

$$P_A = 9.81\ \gamma.\ R^2 = 24.5\ R^2\ (kN/m^2) \qquad (10)$$

where: γ = Average density of the fractured roof rock, excluding the coal.

= 2.50 (tonnes /$m^{3)}$

Variation of the load (W_A) and pressure (P_A) per support diameter (d) is shown as curve A in Fig. 4.

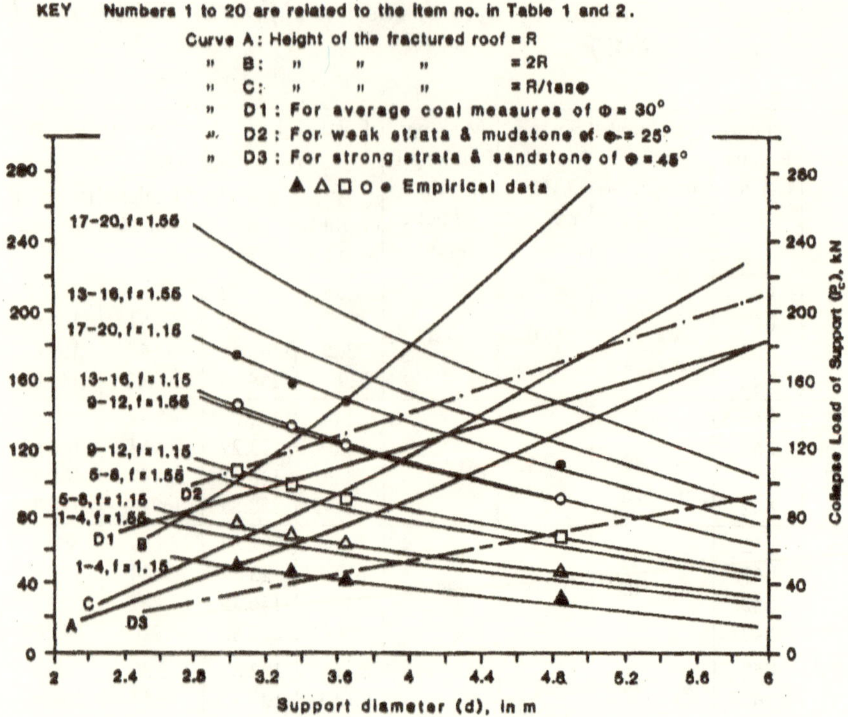

Figure 4. Load capacity of semi-circular H-section steel supports at the crown.

3.2B- Fractured roof envelope extending up to a distance equal to the diameter of the support (d = 2R). This condition prevails when either the roof strata is weak, or intense pressure persists around the roadway. In this case, the load of fractured roof rock on the support (W_B) and its corresponding pressure (P_B) can be determined from:

$$W_B = 9.81\ \gamma.\ d.\ R = 2 \times 9.81\ \gamma.\ R^2 = 49.0\ R^2\ (kN/m)$$

$$= 0.9 \times 49.0\ R^2 = 44.2\ R^2\ (kN/0.9\ m)$$

$$P_B = 2 \times 9.81 \gamma . R^2 = 49.0 R^2 (kPa) \qquad (11)$$

Variation of the load (W_B) and pressure (P_B) per support diameter (d) is shown as curve B in Fig. 4.

3.2C- Fractured roof envelope bond by the average angle of internal friction of the roof rocks to the vertical (ϕ) shown in as curve C. This condition exists when highly stratified strata is over stressed. In this case the load of fractured roof rock on the support (W_C) and its corresponding pressure (P_C) is determined from the following expression:

$$W_C = 9.81\gamma\ (R/\tan\phi)\ 2R/2 = 9.81\ \gamma.R^2/\tan,\ (kN/m)$$

For characteristic coal measure rocks $\phi = 30^o$, therefore,

$$W_C = 9.81\ \gamma.\ R^2/\tan\phi = 42.5\ R^2 (kN/m) = 38.3\ R^2\ (kN/0.9m)$$

$$P_C = 42.5\ R^2\ (kPa) \qquad (12)$$

Variation of the load (W_C) and pressure (P_C) per support diameter (d) is shown as curve C in Fig. 4.

3.2D- Adaption of Terzaghi's formula for the vertical load on the beam shaped supports (Terzaghi, 1943). The formula can be expressed in the following form:

$$\begin{aligned} W_D &= 9.81\gamma[2a + 2b\tan(45 - \phi/2)]/2\tan\phi \\ &= 24.5\ [a + b\tan(45 - \phi/2)]/\tan\phi,\ (kN/m) \\ &= 22.1[a + b\tan(45 - \phi/2)]/\tan\phi,\ (kN/0.9m) \end{aligned} \qquad (13)$$

where:a = Half width of the support, (m)
b = Height of the support = a, (mm)

For an average angle of internal friction, $\phi = 30^o$, expression (13) can be written as:

$$W_D = 42.46(a + 0.577b),\ (kN/m)$$

$$= 38.21(a + 0.577b),\ (kN/0.9m) = 60.26a,\ (kN/0.9m)$$

$$P_D = 42.46(a + 0.577b) = 66.96a,\ (kPa) \qquad (14)$$

Variation of W_D and P_D per support diameter (d) for typical friction angles (ϕ) is shown as curves D_1, D_2 and D_3 in Fig. 4.

3.3- Calculation of the collapse load (P_C) - The theoretical value of P_C is given by equations (6-8). Empirically the P_C-value can be derived for the section modulus (M_S) of 132 x 10^3 mm^3 as follows (2).

$$P_C = 45.10/w \qquad (15)$$

where: P_C = Collapse load on the crown, (tonnes)

w = Width of the semi-circular support = 2R = d, (meter)

Equation (15) can be expressed for f = 1.15 in terms of kN and section modulus (M_S) in the following form:

$P_C = (45.10 \times 9.806/1.15 \times 260 \times 10^3 \times 132 \times 10^{-6} \times 2)\ (\delta_y.\ M_S.\ f/R)$

$= 5.6\ \delta_y.\ M_S.\ f/R$, (kN) (16)

The P_C-values for prime quality grade 43A of H-section structural steel of yield stress (δ_y) = 260 x 10^3 kPa, with various dimensions, section modulus (M_S) and shape factors of f = 1.15 and 1.55 is given in Table 2 and Figure 4.

Table 2 - Empirical and theoretical collapse load of steel supports (P_C) relevant to the specifications noted in Table 1

Item No. (from Table 1)	Collapse load (P_C) at the crown, in kN			Collapse load (P_C), in kN at ϕ = 45 °, (eqn. 8)	
	Empirically (eqn. 16)	Theoretically (eqn. 6)			
		f = 1.15	f = 1.55	f = 1.15	f = 1.55
1	49.53	48.67	65.74	113.39	152.76
2	45.08	44.39	59.83	103.20	139.04
3	41.36	40.73	54.90	94.70	127.58
4	31.11	26.96	41.29	71.22	95.95
5	74.19	73.05	98.44	169.83	228.80
6	67.52	66.49	89.60	154.57	208.25
7	61.96	61.01	82.21	141.83	191.03
8	46.60	45.88	61.83	106.67	143.71
9	106.77	105.13	141.67	244.41	392.29
10	97.18	95.69	128.94	222.46	299.71
11	89.17	87.81	118.32	204.12	275.01
12	67.06	66.03	88.98	153.51	206.82
13	145.29	143.07	192.79	332.60	448.10
14	132.24	130.22	175.47	302.73	407.85
15	121.35	119.49	161.01	277.78	374.24
16	91.26	89.86	121.09	208.91	281.45
17	173.91	171.25	230.76	398.12	536.37
18	156.37	155.87	210.03	362.36	488.19
19	145.24	143.02	192.72	332.49	447.95
20	109.23	107.56	144.94	250.06	336.89

3.4 - Effect of strutting on the collapse load - The H-section steel supports have about 3 times lower load capacity in the direction of its minor axis, in comparison with its major axis (Lewis, 1970). In order to improve the strength of the support in the direction of its minor axis, two steps may be considered:

3.4.1- Closer spacing between the supports, and

3.4.2- Utilization of suitable strutting, thereby to strengthen the support system as a whole.

The stability of struts under loads is related to slenderness ratio 1/K (Jukes et al., 1983). where: 1 = Effective length of the strut, (m)

K = The least radius of gyration of the strut section, (m)

For a solid circular strut (rod): K = r /2
where: r = Radius of strut section (cross section of the rod).

For a thin-walled tabular strut: $K = [r/(r_o + r_j)]^{1/2}\ \Omega\ r_o/2^{1/2}$
where: r_o = Outer radius of the tube section, (mm)
r_j = Inner radius of the tube section, (mm)

Laboratory tests on scaled model steel supports (Jukes et al., 1983) have shown that the collapse load of supports at the crown is dependent on the slenderness ratio (1/K) and the number of struts (N), when strutting is adopted. The relevant empirical formulae projected for the full size supports in terms of P_{CS} and kN, yields the following expressions:

(a) In general: $P_{CS} = 124.10 - 0.17\ 1/K$, (kN) (17)

(b) For weak struts of 1/K = 96: $P_{CS} = 84.95 + 2.94N$, (kN) (18)

(c) For strong struts of 1/K = 48: $P_{CS} = 84.07 + 4.71N$, (kN) (19)

The difference between the constants in the equations (18) and (19) is very small. It can be averaged to (84.95+84.07)/2 = 84.51, kN. This is the value of collapse load (P_C) when there is no strutting on the support system (N=0).

Hence, the equations (17), (18), and (19) can be expressed in the following forms, respectively:

In general: $P_{CS} = 1.47P_C - 0.17\ 1/K$, (kN) (20)

For weak struts of 1/K = 96: $P_{CS} = P_C + 2.94N$, (kN) (21)

For strong struts of 1/K = 48: $P_{CS} = P_C + 4.71N$, (kN) (22)

Giving actual values of 1/K, N and P_C (from Table 2) the P_{CS}-values are determined from the equations (20), (21) and (22) and shown graphically in Figs. 5 to 9.

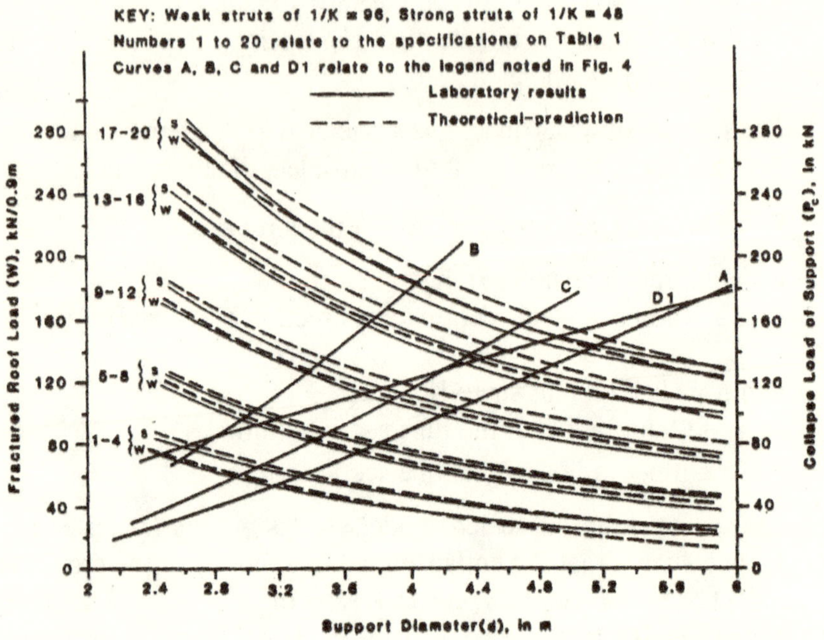

Figure 5 Design chart for the empirically and theoretically determined collapse load of supports at the crown, with f = 1.15, using weak and strong struts.

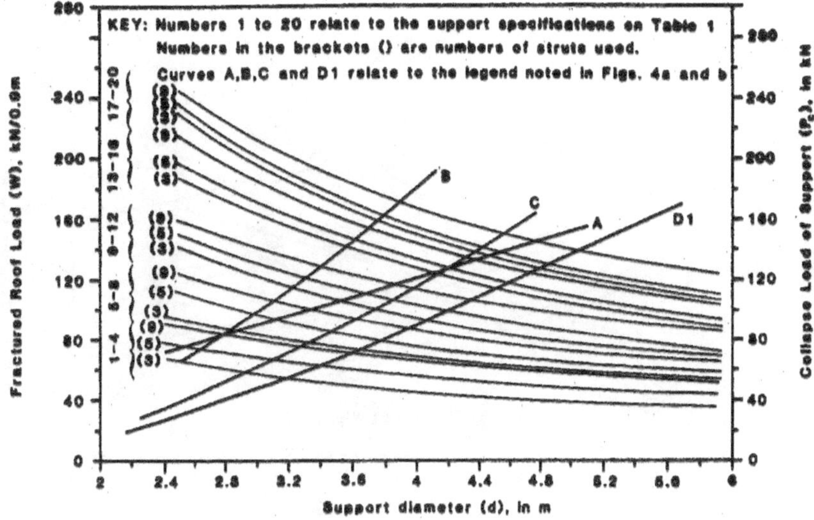

Figure 6. Design chart for the collapse load of supports at the crown with f = 1.15 and varying number of weak struts of 1/k = 96.

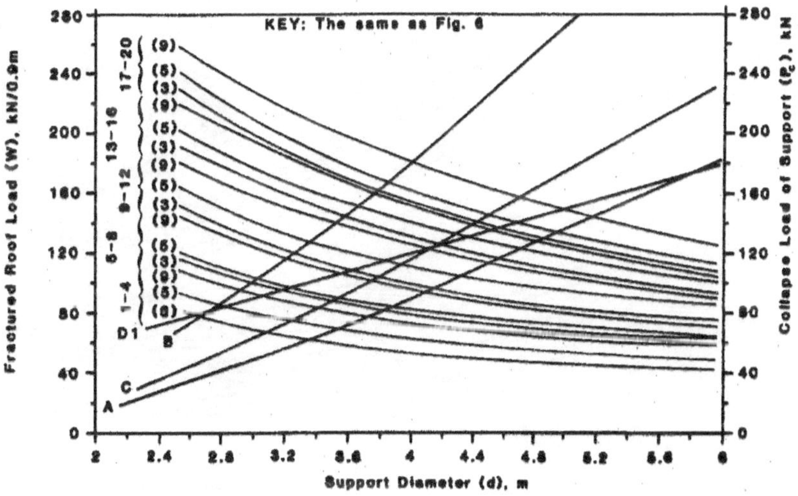

Figure 7. Design chart for the collapse load of supports at the crown with f = 1.15 and varying number of strong struts of 1/k = 48.

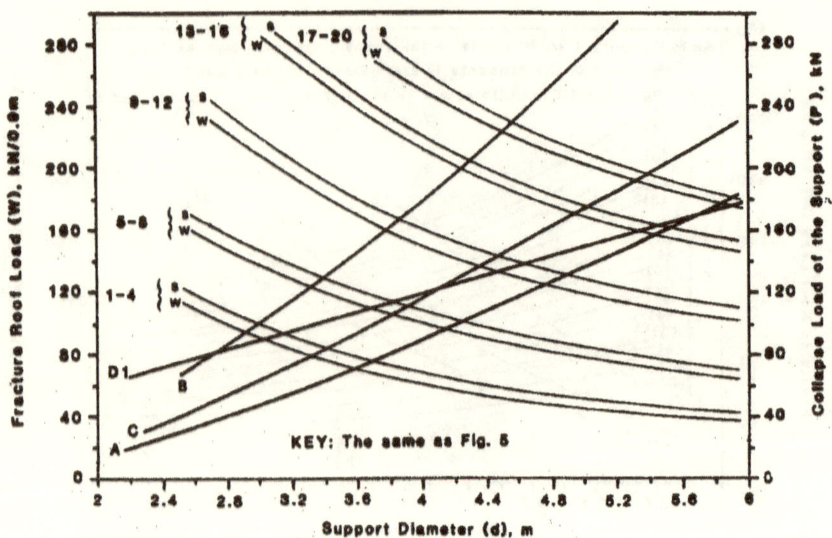

Figure 8. Design chart for the collapse load of supports at the crown with f = 1.55, using strong and weak struts.

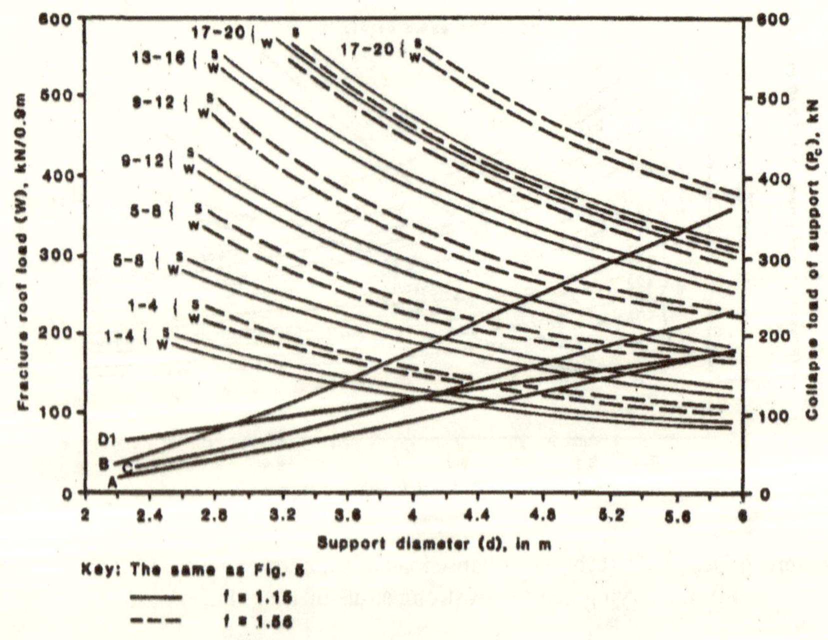

Figure 9. Design chart for the collapse load of supports at the shoulders, with f = 1.15, using weak and strong struts.

3.5 - The support should accommodate the initial roof lowering caused by compression of the gate-side pack. This can be done by measuring the roadway convergence onto the packs, and then accounting for an equal amount of convergence to take place in the support by provision of wooden foot blocks, wood chocks, stilts or selecting appropriate support height.

4. CONCLUSIONS

In-situ observations, empirical and theoretical expressions utilized in this investigation suggest the followings:

1 - The weakest point of the semi-circular steel support lies on its crown.

2 - Curves A to D in Fig. 4 provide a prediction of the fractured roof load (W) on a semi-circular steel support system for various rock types, strata conditions and method of analysis.

3 - The collapse load of semi-circular steel supports (P_C), without on the radius (R), sectional modulus (M_S) and shape factor (f) of the support expressed in the following forms:

a) At the crown of the supports:

Theoretical $P_C = 11.11\ \delta_y.\ M_S.\ f/R$, (kN)

Empirical $P_C = 11{,}65\ \delta_y.\ M_S.\ f/R$, (kN)

b) At any other point along the supports:

$P_C = 12.56\ \delta_y.\ M_S.\ f/R(\pi - \pi\cos\theta - 2\sin\theta)$, (kN)

c) For $\theta = 45^{o}$ the above equation can be expressed as:

$P_C = 25.63\ \delta_y.\ M_S.\ f/R$

4 - Figure 4 provides design charts for the semi-circular, H-section, steel supports, without strutting, along mine roadways, by considering crown and shoulders of the supports, respectively. Figures 5 to 9 provide design charts for the above noted supports with strutting.

5 - High quality and well placed struts can increase the load bearing of steel supports by up to 39%, dependent upon the

slenderness ratio (1/K) and number of struts (N) utilized, according to the following expressions:

a) In general: $P_{CS} = 1.47\ P_C - 0.17\ 1/K$, (kN)
b) For weak struts of 1/K = 96: $P_{CS} = P_C + 2.94\ N$, (kN)
c) For strong struts of 1/K = 48: $P_{CS} = P_C + 4.71N$, (kN)

6 - The above noted conclusions are valid for semi-circular, H-section, steel supports. Work is on hand to extend the investigation towards various type of steel arched supports under different loading conditions.

REFERENCES

Jukes, S.G., Hassani, F.P. and Whittaker, B.N., 1983/84. Characteristics of Steel Arch Support Systems for Mine Roadways, Parts I and II, Mining Science and Technology, Elsevier, 43-58.

Lewis, S., 1970. Systems of Roadway Supports, Conf. on Strata Control, in Mine Roadways, University of Nottingham, 33-49.

Sadler, G.W., Campbell, S.G., 1965. Approximate Equations for the Yield and Collapse of H-Section Roadway Arches, N.C.B. Laboratory Testing Branch, Report No. 652.

Terzaghi, K., 1943. Theoretical Soil Mechanics, J. Wiley, pp. 194-197.

Timoshenko, S., 1968, Elements of the strength of materials, 5th Ed., Van Nostrand Reinhold.

3.5.2 DESIGN OF ROCK SUPPORTS IN UNDERGROUND TUNNELS

By: F.P. Hasani, H.S. Mitri, A. A. Afrouz and J. Du

ABSTRACT

A design package for steel arch support systems frequently employed in underground semi-circular tunnels is developed. The package is presented in a graphical form as a set of nomograms giving: (1) the ultimate strength of the support for different support diameters, section sizes and loading distributions and (2) the roof loading for a variety of strata conditions commonly encountered in mine environments. A nonlinear finite element model which accounts for both material and geometric nonlinearities are developed to predict the ultimate strength of the support, whereas a set of roof loading formulae are proposed to estimate the fractured strata load.

INTRODUCTION

The basic requirement of any underground support system is that it controls the movements of the surrounding strata during the period of operation of the tunnel in such a way that the various functions of the tunnel are not impaired. Such functions include: (1) maintaining an opening of sufficient dimensions and (2) holding any loose material in place to ensure the safety of the tunnel. Cambered girders were first used in mine roadways as steel supports but they were eventually replaced by steel arch supports; see Fig. 1. The use of steel arches has resulted in a dramatic reduction of the number of fatalities occurring in mine roadways (Carver, 1970). Due to their excellent structural performance, the steel arches are now regarded as the most suitable forms of mine tunnel support.

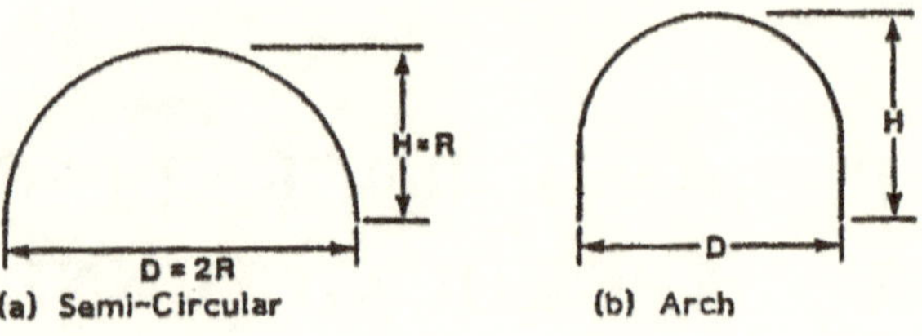

Fig. 1. Common Shapes of Steel Supports in Underground Tunnels

In order to achieve a successful design of a steel support system, information about the roof load and the structural behaviour of the support should be made available. The amount of roof load is determined from the size and the shape of the fractured roof envelope. The latter depends on the rock type, the opening size and the strata condition. The behaviour of steel supports has been the subject of many investigations carried at over the past two decades. Several methods have been pursued. These included full scale laboratory tests, full scale underground tests, physical scale model tests and simple theoretical models based on Castigliano's theorem. Full scale laboratory tests have been carried out by Brown et al. (1965) and Paul et al. (1974). Underground full size testing was conducted, whereby the loading was achieved by either allowing the strata load to develop naturally (Cunliffe and Johnston, 1957) or by jacking the support into the rock and thus simulating its loading (Round and Lewis, 1981). The use of physical scale models is wide-spread (Jukes et al., 1983). These models have the advantage of low cost as well as being an exact replica of the full size support. Jukes et al. (1983) proposed two empirically based formulae for the prediction of collapse load of semi-circular and arch supports. They also derived an expression for the collapse load of a semi-circular support using a simple linear elastic analysis based on Castigiliano's theorem. Despite the approximate nature of the analysis, comparison with test results showed good agreement. No attempt was made to extend the analysis to other shapes.

Computer-based, finite element models have received considerably less attention than other modeling techniques to study the behaviour of steel supports. In this paper, a nonlinear finite element model which accounts for both plasticity and large deformations is presented and used to predict the ultimate strength of steel supports. While applicable to any shape, the model has been applied only to the semi-circular arch with an H-shape cross section since it is one of the most popular types currently used in mine roadways. The finite element model is then employed to produce a series of design curves giving collapse load values for different section sizes, support diameters and loading distributions. A variety of roof loading formulae are presented and others are proposed in this paper. The latter are plotted and superimposed on the support design curves to generate a set of nomograms for design use.

FINITE ELEMENT ANALYSIS

Many studies of nonlinear behaviour of steel, planar frames have been undertaken over the past twenty years or so; see for example Morris and Fenves (1970). The present analysis is aimed to examine the strength characteristics of steel arches to the point of collapse and hence the incorporation of both material and geometric nonlinearities deemed necessary.

The beam-column element having six degrees of freedom is utilized in the analysis; see Fig. 2. The element formulation is based on large deformation theory and does not account for shearing deformations. Assuming a linear interpolating function for the axial displacement and a cubic function for the lateral displacement, it is possible to formulate the element strain mat rices (Bathe, 1982) corresponding to the linear and nonlinear components of Green's strain (Fung, 1977) at any point within the element. The tangent stiffness matrix of the element is 6 x 6 and is evaluated by means of numerical integration which is performed both in the element length and element depth directions. As can be seen from Fig. 3, twenty one integration points are used; three Gaussian stations along the element and seven Newton Cotes points across the depth at each Gaussian location. This scheme of sampling points permits an accurate representation of plastic zone progression within the element. The initial stress matrix is evaluated by three Gaussian points along the element. All weight factors and local co-ordinates of both Gauss quadrature rule and Newton Cotes are given by Bathe (1982). The initial stress matrix is added to the tangent stiffness matrix to obtain the total stiffness matrix of the element.

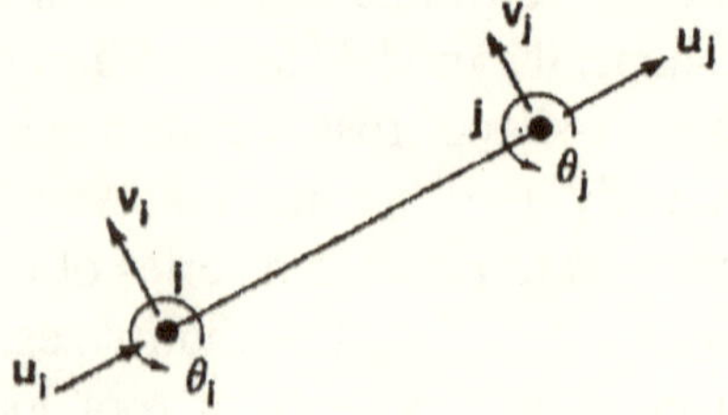

Fig. 2. Beam-Column Element.

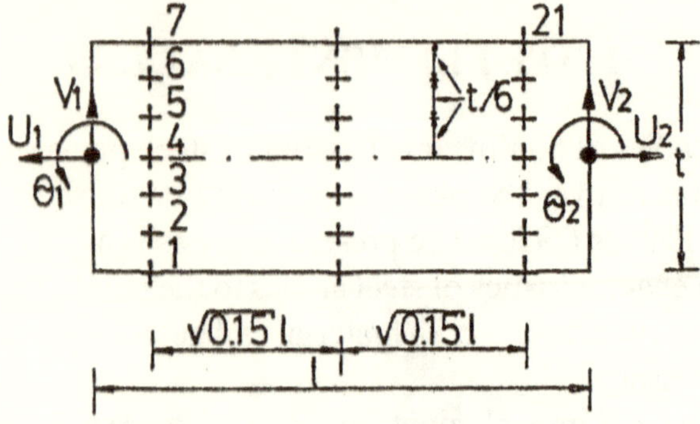

Fig. 3. Numerical Integration Points.

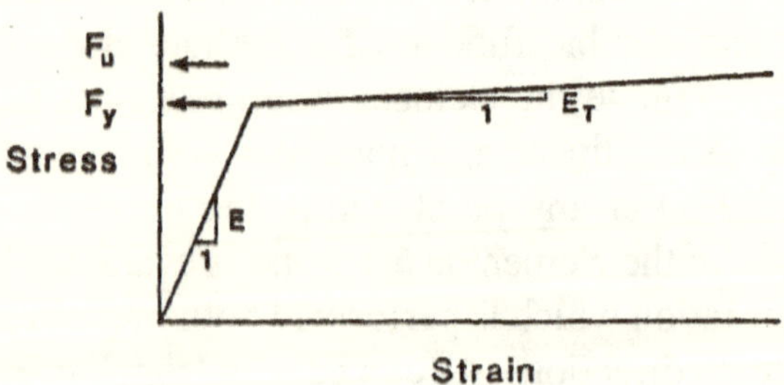

Fig. 4. Material Model.

An elastic-plastic material model with strain hardening is assumed. The stress-strain relation is shown in Fig. 4. The material is described by the Young's modulus of elasticity E, the yield stress F_y, the tangent modulus E_T and the ultimate stress F_U. No further strain hardening is allowed beyond the point of ultimate stress.

Because of the combined material and geometric nonlinearities, the strain-displacement relation as well as the stress-strain relation are written in incremental form for the calculation of strain and stress increments respectively. The algorithm employed in the incremental stress analysis follows the general procedure described by Owen and Hinton (1980).

The global incremental equations of equilibrium are established using the principle of virtual incremental displacements (Cichon, 1984). The resulting system of simultaneous equations is solved by Gauss elimination. To achieve a better economy of the solution, the skyline of the global stiffness matrix is first determined so that Gauss reduction is applied only to the elements below it (Bathe, 1982). The nonlinear analysis is carried out incrementally and iteratively by means of Newton Raphson method whereby the tangent stiffness matrix is updated after every iteration and so are the displacements, strains and stresses. Having reached an equilibrium configuration at the end of a load increment, another load increment, equal to or less than 20% of initial yield load, is applied. This is added to the residual load from the previous increment to form the total current load increment. The application of load increments ceases to continue when the peak load is reached. This is detected when there is sufficient plasticity in the structure to render the global tangent stiffness matrix non-positive definite. Within each load increment, the iteration is continued until the convergence criterion is satisfied. The latter is based on computing the increment in internal energy during the current iteration and comparing it with the initial work done by the out-of-balance forces on the initial displacements (Bathe and Cimento, 1980). When the ratio of current to initial energy increment is small enough (typically 0.001), the iteration is terminated.

MODEL SIMULATION

The finite element analysis just described is to be used to predict the strength characteristics of steel supports. In order to verify the model, a comparison with test results deemed necessary. Jukes et al. (1983) tested a number of physical scale models of semi-circular and arch types under concentrated, centrally applied loads. The properties of such models are listed in Table 1. Due to symmetry of both geometry and loading only half of the support need be considered for the finite element simulation. A total of 18 elements for the semi-circle and 23 elements for the arch have been used to discretize the support. The nodal distribution and the boundary conditions of both support models are indicated in Fig. 5. The H-section is idealized by three rectangular regions representing the web and the two flanges of the original section. The average thickness of the flange is used as the thickness of the equivalent rectangle; see Fig. 6. The numerical integration scheme described earlier can accurately evaluate the stiffness properties of this idealized section.

Table 1: Physical Scale Model Test Data by Jukes et al.

Model	Shape	D (mm)	H (mm)
1	Semicircle	355.83	177.91
2	Semicircle	305.00	152.50
3	Arch	305.00	254.16
4	Semircircle	406.66	203.33
5	Arch	355.83	254.16
6	Arch	406.66	300.00

$F_y = 0.670$ kN/mm^2, E = 210 kN/mm^2, Z = 76.4mm^3

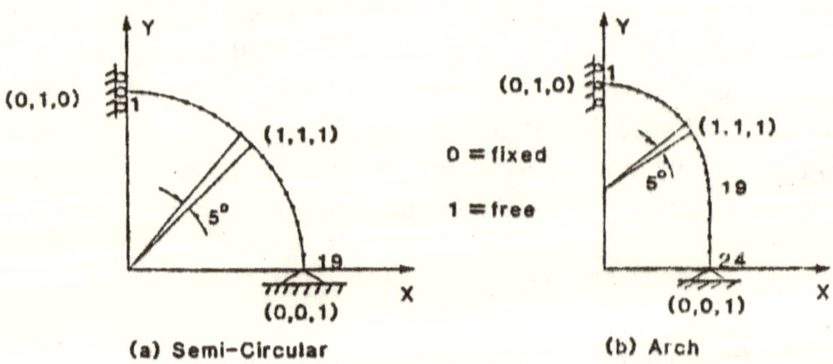

Fig. 5. Descretized Finite Element Models of Steel Supports.

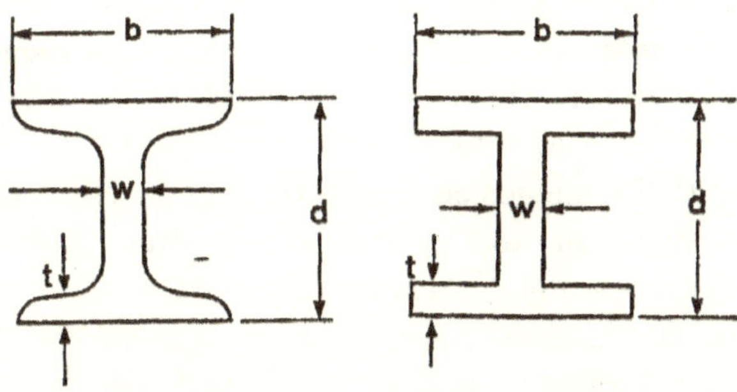

(a) Actual H- Section (b) Idealized H- Section

Fig. 6. Actual and Idealized Geometry of an H-Section

A comparison between experiments and finite element models has been made. This is presented in Table 2. As can be seen, excellent correlation between experimental and finite element results exists. The mean value of computed to experimental collapse load ratio is 1.015 with a coefficient of variation of 0.031.

Table 2: Comparison Between Finite Element and Experimental Results

Model	Pc (kN) (computed)	Pe (kN) (Exp.)	Pc/Pe
1	1879.32	1765.45	1.06
2	2243.4	2325.75	0.9645
3	1623.5	1615.45	1.004
4	1652.62	1643.93	1.005
5	1513.56	1495.45	1.012
6	1266.46	1210.22	1.046

Note: Mean (Pc/Pe) = 1.015 Cov = 0.031

DESIGN PROCEDURE

The finite element model is now extended for use in design. While the concept of design nomograms is not new, a problem is often encountered, which is the determination of strata load distribution. It is generally agreed that the assumption of a concentrated load at the crown leads to overly conservative design whereas the assumption of a uniformly distributed load is not safe. An acceptable distribution would thus be one in which the load distribution varies from zero at the support base to a maximum at the crown. Whittaker and Hodgkinson (1971) suggested an elliptic load envelope with heights above the crown equal to D when the roof strata is weak and D/2 when it is very strong. No explicit formula was reported in the paper for the roof load. A convenient load envelope which is easy to integrate and yet reflects a realistic condition is the sinusoidal curve shown in Fig. 7a. The load per unit distance of the projected length of the support crown is $p = \lambda \gamma \sin \theta$, where λ depends on the type of strata. Typically, two values are considered; $\lambda = D/2$ for a strong strata and $\lambda = D$ for a weak one. It can be shown that after performing the integrations, the total load $P = 0.785\ \lambda \gamma D$, where γ = weight per unit volume of the material.

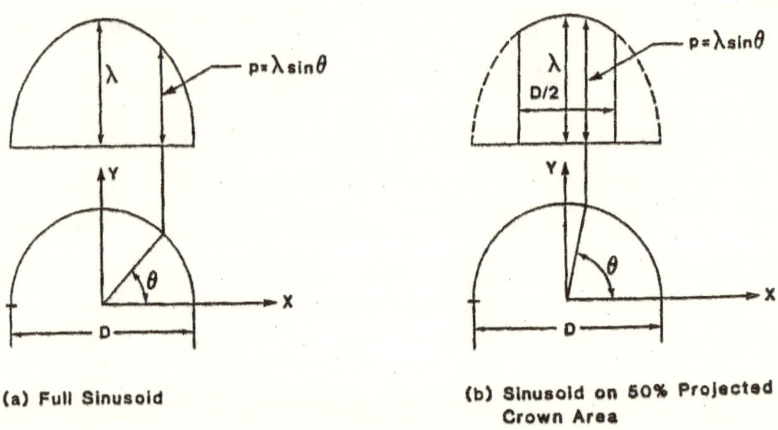

Fig. 7. Suggested Load Distributions

Substituting for λ, the following expressions are obtained:

For a strong strata, $P = 0.393\ \gamma D^2$ (1)

For a weak strata, $P = 0.785\ \gamma D^2$ (2)

Another important consideration is the area of crown subjected to roof loads. It is not unusual that due to the large deformations of the rock media above the crown, most of the load is taken by a portion of the support crown and not its full length. Fig. 7b shows a sinosoidal load distribution applied to only 50% of the projected crown area. Naturally, this loading condition is more severe since the loads are lumped around the vicinity of the crown. Hence, when a sinosoidal load applied over a distance D/2 of the support crown for strong and weak strata respectively:

$$P = 0.239\,\gamma\,D^2, \qquad \lambda = D/2 \tag{3}$$

$$\text{and } P = 0.478\,\gamma\,D^2, \qquad \lambda = D \tag{4}$$

The finite element model is then subjected to the four loading conditions described above. While doing so, the diameter is varied from 2 to 7 meters every 0.5 m and the section size is kept constant. The section size is then altered and the process is repeated. This procedure is implemented and a series of support strength curves are produced. These are plotted in Fig. 8. On the other hand, roof loading formulae are plotted on the same graphs thus generating the required design nomograms. The properties of H-sections used in the simulation are listed in Table 3. These sizes are commonly employed in mine roadways. It should be pointed out that other loading distributions, as well as support geometries and section sizes are possible to include using the finite element analysis described in this paper. More elaborate design charts are currently being implemented. The nomograms shown in Fig. 8, however demonstrate the potential usefulness of the adopted modeling technique.

Table 3: Properties of H-Sections Used as Steel Supports

$F_y = 0.268$ kN/mm, $E = 200$kN/mm^2

Section	b (mm)	t (mm)	d (mm)	w (mm)	z (mm^3)
H1	76	8.4	76	8.9	42,200
H2	89	9.9	89	9.5	65,600
H3	102	10.3	102	9.5	90,600
H4	114	9.9	127	10.2	148,400
H5	127	13.2	127	10.4	217,087

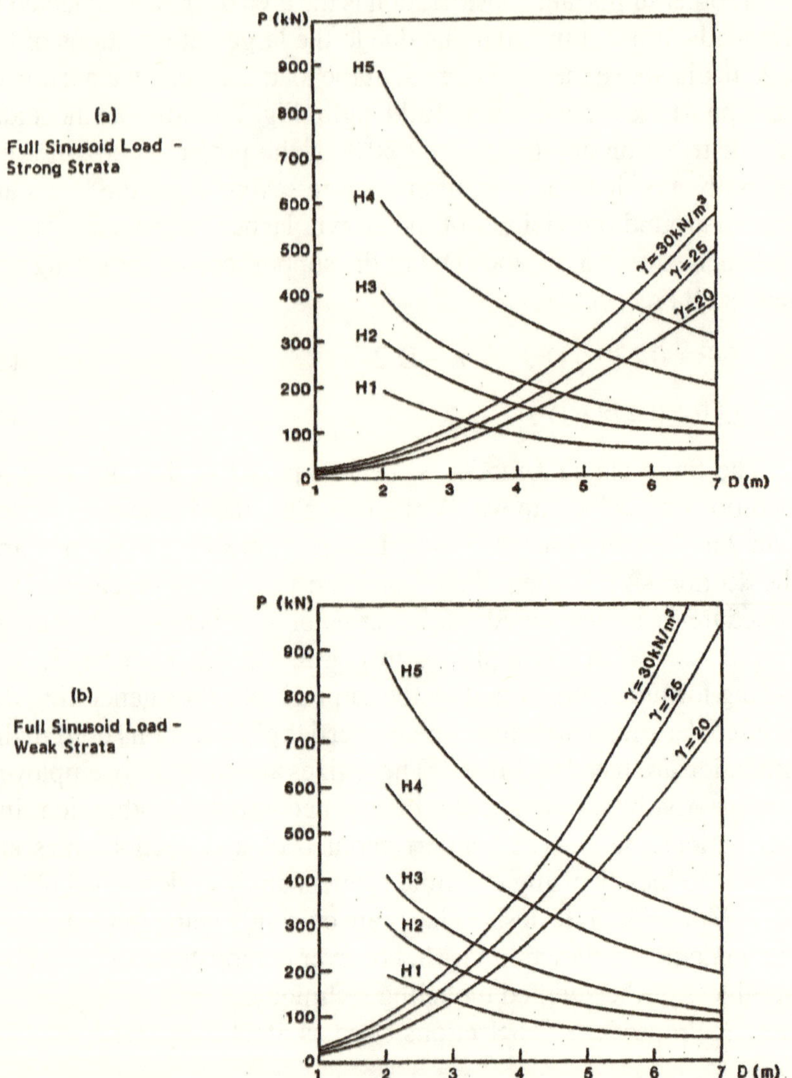

Fig. 8. Design Nomograms for Semi-Circular Steel Supports

CONCLUSION

A nonlinear finite element analysis which accounts for both material and geometric non-linearity in planar steel supports of underground semi-circular and arch tunnels has been presented. The finite element simulation is shown to give very good correlation with test results from previous work and is then extended for use in design. By combining the model simulation and proposed loading formulae for the fractured strata envelope, it has been possible to generate a set of design nomograms relating the semi-circular support section, diameter, strength and anticipated roof load. While applied only to the semi-circular support, it is believed that the modeling technique presented can be extended to other support shapes and loading conditions which may be encountered in practice.

REFERENCES

- Carver, J., 1970. Safety Problems in Roadways. Symposium on Strata Control in Roadways, University of Nottingham.
- Brown, A., Campbell, S.G., 1965. A comparison of Arches of Different Construction under Single Point Loading. National Coal Board Central Engineering Establishment, Report No. LT 651.
- Paul, S.L., Siess, C.P. and Gaylord, E.H., 1974. Large Scale Tests of Tunnel Supports. R.E.T.C. proc. 2(C.H. 91), pp. 1395-1417.
- Cunliffe, H., Johnston, A.G., 1957. Roadway Supports with Special Reference to Yielding Arches. Transactions of the Institute of Mining Engineers, Vol. 117, pp. 804-818.
- Round, C., Lewis, S. Royton Drift: New Mine, New Techniques. The Mining Engineer, Vol. 141, pp. 17-24.
- Jukes, S.G., Hassani, F.P. and Whittaker, B.N., 1983. Characteristics of Steel Arch Support Systems for Mine Roadways, Part I, Modeling Theory, Instrumentation and Preliminary Results. Mining Science and Technology, Vol. 1, pp. 43-58.
- Morris, G.A. and Fenves, S.J., 1970. Elastic Plastic Analysis of Frameworks. Journal of the Structural Division, Vol. 96, No. ST5, pp. 931-946.
- Bathe, K.J., 1982. Finite Element Procedures in Engineering Analysis. Prentice Hall Inc., New Jersey.
- Fung, Y.C., 1977. A First Course in Continuum Mechanics. 2nd Ed., Prentice Hall, Inc., New Jersey.
- Owen, D.R.J. and Hinton, E., 1980. Finite Element in Plasticity - Theory and Practice. Pineridge Press Ltd., Swansea, U.K.

- Cichon, C., 1984. Large Displacements In-Plane Analysis of Elastic-Plastic Frames. Computers & Structures, Vol. 19, No. 5/6, pp. 737-745.
- Bathe, K.J. and Cimento, A.P., 1980. Some Practical Procedures for the Solution of Nonlinear Finite Element Equations. Computer Methods in Applied Mechanics and Engineering, Vol. 22, pp. 59-85.
- Whittaker, B.N. and Hodgkinson, D.R., 1971. The Influence of Size on Gate Roadway Stability. The Mining Engineer, January, pp. 203-214.

3.5.3 AN INVESTIGATION INTO DESIGN OF STEEL ARCH SUPPORT SYSTEM IN MINE ROADWAYS

By: F.P. Hassani, A.A. Afrouz and H.S. Mitri

ABSTRACT

This paper presents a review of some theoretical and experimental studies relating to two types of steel arches used as mine roadway supports. Such types are the vertical legs and the splay legs arches. In order to determine the collapse load of the supports, relevant parameters including support dimensions, steel properties, type and number of strutting, strata conditions and fractured roof load envelope have to be evaluated. Appropriate design charts are then presented.

1 - INTRODUCTION

The demand for higher productivity of coal mines is directly reflected in an increasing dependence upon roadways and their stability. It requires that roadways be constructed and supported under more per-established circumstances. While the technology to drive perfect roadways and to properly support them does exist; this can only be achieved at a high cost not compatible with the present mining budgetary conditions in Canada.

Support systems of coal mine roadways are predominantly made from steel H-sections. These have excellent supporting characteristics in a variety of underground and mining conditions, in comparison with other types of support systems, e.g. wood sets, pre-fabricated concrete linings, and straight steel sections such as portal frames, pitched roof frames, etc.

Physical scale models of steel supports were fabricated using 1/12 scale and tested in the laboratory under one point load acting at the crown (1). Also in situ roof loading measurements were done previously using load cells installed between the support and the hosting rock (2). The results of the latter study indicated the development of a non-uniform loading pattern on the steel support. The magnitude of the load and consequent deformations of the support were found to be greater at the crown than at the sides, as shown in Fig. 1. In situ arch deformations and roadway closure measurements indicated the development of steel arch failure at three locations. These locations are one point at the crown and two points at the shoulders (3-4). There have been also numerous investigations into the failure criteria in relation to the design of steel arches (5-7). However, most of published materials were linked to specific mine locations or mining conditions.

In this paper, it is considered that (a) the majority of the strata load around a roadway, acting upon the support is due to the fractured roof load. (b) The roof load acts at the crown of the steel arch as a point load. This implies a conservative loading condition for the assessment of arch collapse load, Table 1 and Fig. 2).

2 - THEORETICAL CONSIDERATIONS

2.1 -Forming General Equations:

Having been subjected to the strata load, the steel arch undergoes deformations. In the following, elastic analysis of the support structure and its collapse mode is considered. However, when the deformation is to be predicted, a complete elasto-plastic analysis would be more appropriate. The latter case has been presented previously by the authors (8).

As noted earlier, the loading pattern considered is that of a single point load (P) acting at the crown as illustrated in Fig. 3. The support is assumed to be of uniform cross section. The reactions to the load (P) are taken by the arch legs which are assumed to be pinned to the floor.

The load carrying capacity of steel arch depends on the value of the bending moment (M) which the members can sustain. This can be expressed as:

$$\sigma_y = M/Z = M/M_S \,.\, C \,.\, f \qquad (1)$$

where: σ_y = Yield stress of the steel arch, kPa

= $260 \times 0.6 \times 10^3$, = 156×10^3, kPa for steel arches used underground

M = Bending moment of the arch, kN. m

Z = Modulus of the arch, M_S . C . f, m^3

M_S = Section modulus of the arch, m^3

= A . z/3, $m^3 \times 10^{-9}$, (see Fig. 2)

C = Proportionality constant

= 20, by scaled laboratory determination (1).

f = Shape factor = Z/M_S . C

= Between 1 to 1.3 = 1.15 (9).

Considering symmetry of both loading and geometry, the vertical reaction of the arch legs can be expressed as: $p_a = p_b = P/2$ (2)

The bending moment (M), yield stress (σ_y) and collapse load (P_C) of the vertical leg and splay leg arches at the crown can be related as follows:

A - For the vertical leg arch (Fig. 3a):

$$M = p_a . R - h_a (R+L) = (P.R/2) - h.R (1+n) \quad (3)$$

where: n = L/R

$h_a = h_b$ = horizontal component of the reactions at the support back, kN.

= P/3, according to Jukes et al. (1).

Thus, M = (P.R/2) + P . R (1+n)/3

$$= P.R [0.50 + 0.33 (1+n)] \quad (4)$$

Substituting equation (4) into equation (1) gives:

$$\sigma_y = P . R [0.50 + 0.33 (1+n)] / C. M_S. f \quad (5)$$

Therefore:

$$P_C = C.\sigma_y.M_S.f/d(0.16n+0.41) \quad (6)$$

Variation of the collapse load in vertical leg arches with factors noted in equation (6) is given in Table 2 and shown in Fig. 4.

B - For splay leg arches (Fig. 3b): $M = p_a . R - h_a (R+L)$ (7)

where: $R' = R + (L/\tan \theta_o) = R [1+(n/\tan \theta_o)]$

θ_o = angle of splay leg to horizontal = 45 to 85°

= 83°, for coal mines under study.

Substituting equation (2) into equation (7):

$$M = P . R [0.50 + (n/2 \tan \theta_{o)} + 0.33 (1+n)] \quad (8)$$

Substituting equation (8) into equation (1) gives:

$$\sigma_y = P . R [0.50 + (n/2 \tan \theta_o) + 0.33 (1+n)]/C. M_S. f \quad (9)$$

Hence: $P_C = C. \sigma_y. M_S. f / d [n(0.16 + 0.25/\tan \theta_{o)} + 0.41]$ (10)

Variation of the collapse load in splay leg arches with factors noted in equation (10) is given in Table 2 and shown in Fig. 5. A comparison between vertical and splay leg arches at n = 0.5 is also shown in Fig. 6 which indicates slightly lower load capacity for the splay leg arches of $\theta_o = 83^o$.

2.2 - Effect of Strutting on the Collapse Load:

The H-section steel supports have about 3 times lower load capacity in the direction of its minor axis, in comparison with its major axis (4). In order to improve the support strength in the direction of minor axis two steps may be considered:

(I) Reduce spacing between supports, and

(II) Uutilize suitable strutting system, thereby strengthen the support system as a whole.

The stability of struts under compressive longitudinal stresses is related to slenderness ratio 1/K.

where: 1 = Effective length of the strut, m

K = The least radius of gyration of the strut section, m.

(a) For a solid circular strut (rod): $K = r/2$
where: r = Radius of strut section (cross-section of the rod).

(b) For a thin walled cylindrical strut: $K = [r_O{}^3/(r_O + r_i)]^{0.5}$
$= r_O/2^{0.5}$
where: r_O = Outer radius of the tube section, mm
r_i = Inner radius of the tube section, mm

Laboratory tests on scaled model steel supports (1) have shown collapse load dependency of the supports at the crown with slenderness ratio (1/K) and number of struts (N). The relevant empirical formulae projected for full size supports yield the following expressions:

In general: $P_{CS} = 1.47 P_C - 0.17\, 1/K$ (11)

where: P_{CS} = Collapse load of supports with struts, kN
P_C = Collapse load of supports without struts, kN

For weak struts of 1/K = 96: $P_{CS} = P_C + 2.94\ N$ (12)

For strong struts of 1/K = 48: $P_{CS} = P_C + 4.71\ N$ (13)

Given the actual values of 1/K, N and P_C (from Table 2) the P_{CS}-values can be determined from equations (11), (12) and (13). The P_{CS} - values for n = 0.5 are displayed in Figs. 7 to 10.

3 - ESTIMATION OF ROOF LOAD

This can be determined employing one of the following expressions, depending upon the depth of roadway and state of fractured roof load:

A - At depths greater than 300 metres with strong roof, or where the roadway is machine-cut, well formed and good system of packing is employed. In this case, the fractured roof envelope extends near elliptically upto a height equivalent to half of the support width (d/2 = R), therefore:

$P = \gamma\,.\,d^2/4 = 0.25\,\gamma.d^2$ (14)

where: P = Fractured roof load per metre length of the roadway, kN/m
d = 2R = Diameter of the roadway, m

γ = Average weight per unit volume of the fractured roof rocks, kN/m^3.

For coal measure rocks, $\gamma = 24.53$, kN/m^3

For $\gamma = 24.53\ kN/m^3$, equation (14) can be expressed as follows:

$$P = 6.13\ d^2 \tag{15}$$

Variation of the roof load (P) with support diameter (d) is given in Table 1 and shown as curve A in the design charts Figs. 4 to 10.

B - At depths greater than 300 m with weak roof, or where intense pressure exists around the roadway. In this case, the fractured roof envelope extends near elliptically upto a height equivalent to the support diameter (d = 2R), (9);

$$P = \gamma\ .\ d^2/2 = 0.5\ \gamma\ .\ d^2 \tag{16}$$

For $\gamma = 24.53\ kN/m^3$, equation (16) can be expressed as:

$$P = 12.26d^2 \tag{17}$$

Variation of the roof load (P) with support diameter in this case, is given in Table 1 and shown as curve B in the Figs. 4 to 10.

C - At shallow depths of less than 200 m, and in broken stratified roof rocks with fractured roof envelope bond by a linear internal friction angle of the roof rock (10);

$$P = \gamma\ .\ d^2/4 \tan \phi \tag{18}$$

For $\gamma = 24.53\ kN/m^3$ and $\phi = 30^{\circ}$, equation (18) can be expressed as follows:

$$P = 10.62\ d^2 \tag{19}$$

Variation of P with d, in this case, given in Table 1 and shown as curve C in Figs. 4 to 10.

D - At shallow depths in unconsolidated, heavily fractured and weathered strata, where the roof fractures extend up to the surface (i.e. chimney type of subsidence):

$$P = \gamma . A . H = \pi . d . \gamma . H/2 \tag{20}$$

where: A = The roof area to be supported per metre length of the roadway
= $\pi d/2$, m^2
H = Average depth below surface = Cover depth, m.

For $\gamma = 24.53$ kN/m^3 and H = 150 m, equation (20) can be written as follows:

$$P = 5776.8\, d \tag{21}$$

This is a very high load which may be encountered where no mining artificial support could withstand economically. In this case pillars should be utilized as a means of support.

The above noted fractured roof load can act on an arched support in three main conditions shown in Fig. 11. In the case of sinosoidal loading, the cases A, B and C described above are shown in Fig. 12.

A comparison of the steel arched roadway closure obtained from various theoretical and empirical expressions, and measured data from the Canadian and British Mines is shown in Fig. 13.

4 - NUMERICAL MODELLING OF ARCH SYSTEMS

Despite its application to a pair of commonly used support systems, the above described analysis is valid when the load is assumed to be concentrated and is applied at the crown. A nonlinear finite element study is currently underway and is aimed to account for the reduction in plastic moment and prediction of arch deformation. The study would eventually lead to a more thorough understanding of the structural behaviour of the support system. It will also permit the inspection of a wide range of parameters such as distributed and nonuniform loads and other geometric configurations of the arch.

5 CONCLUSIONS

In situ observation, empirical and theoretical expressions utilized in this investigation suggest that the weakest point on the steel arch support is at its crown. In a condition of symmetry, the majority of the roof load is likely to act on the arch crown. Therefore, from a design view point, it is appropriately conservative to make the arch specifications according to the collapse load at the crown.

High quality and well placed struts may result in an increase the load bearing capacity of steel supports (P_{CS}). Such an increase will depend mainly on the slenderness ratio (1/k) and number of struts (N).

The overall closure of gate roads, in medium to deep mines, supported by specific steel arches is affected by load of the fractured roof (P) rather than the cover load (q) and width of the roadways (or equivalent arch diameter d).

Roadway closure equations noted in references (11-13) are mainly useful in tunnels having full lining or driven in competent ground. Reference (14) is mainly of use in Coal measures having strong floor and relatively stable conditions. Reference (15) is applicable in soft rocks with soft floor and unstable ground.

Table 1- Specifications of steel H-section supports and fractured roof load (P)

item No.	support dia., m (d)	sectional dimension, mm (Fig. 2)	section modulus Ms m**3 *10E-6	weight of support, kg/m	load of fractured roof rock on the support (P), kN/m		
					Height		
					d /2 equ. 15	d equ. 17	d*tan ϕ equ.20
1	2	z=76	45.0	14.67	24.50	49.0	42.5
2	4	w=76			98.1	196.2	169.9
3	6	t=8.9			220.7	441.4	382.3
4	8				392.3	784.6	679.7
5	2	z=89	67.4	19.35	24.5	49.0	42.5
6	4	w=89			98.1	196.2	169.9
7	6	t=9.5			220.7	441.4	382.3
8	8				392.3	784.6	679.7
9	2	z=102	97.0	23.06	24.5	49.0	42.5
10	4	w=102			98.1	196.2	169.9
11	6	t=9.5			220.7	441.4	382.3
12	8				392.3	784.6	679.7
13	2	z=114	132.0	26.78	24.5	49.0	42.5
14	4	w=114			98.1	169.2	169.9
15	6	t=9.5			220.7	441.4	382.3
16	8				392.3	784.6	679.7
17	2	z=127	158.0	29.76	24.5	49.0	42.5
18	4	w=114			98.1	169.2	169.9
19	6	t=10.2			220.7	441.4	382.3
20	8				392.3	784.6	679.7

Table 2 - Empirical and Theoretical Collapse Load (P_C) at the Crown of Steel Arches for f = 1.15 Relevant to Specifications given in Table 1.

Item No.	Vertical leg					Splay leg				
	Empirical		Theoretical			Empirical		Theoretical		
	n=0.0	n=0.5	n=0.0 (t2v)	n=0.5 (t3v)	n=1.0 (t4v)	n=0.0	n=0.5	n=0.0 (t2s)	n=0.5 (t3s)	n=1.0 (t4s)
1	197.5	165.9	196.9	164.8	141.6	197.5	159.5	196.9	158.3	132.3
2	99.5	83.1	98.5	82.4	70.8	99.5	80.4	98.5	79.2	66.2
3	67.2	55.6	65.6	54.9	47.2	67.2	53.5	65.6	52.8	44.1
4	50.1	41.9	49.2	41.2	35.4	50.1	40.7	49.2	39.6	33.1
5	296.3	248.2	294.9	246.8	212.1	296.3	237.9	294.9	237.1	198.2
6	148.2	123.9	147.5	123.4	106.1	148.2	119.6	147.5	118.5	99.1
7	99.5	83.5	98.3	82.2	70.7	99.5	80.5	98.3	79.0	66.1
8	74.5	62.7	73.7	61.7	53.0	74.5	60.3	73.7	59.3	49.6
9	425.4	356.7	424.4	355.1	305.3	425.2	342.5	424.4	341.2	285.3
10	213.2	178.6	212.2	177.6	152.7	213.2	171.6	212.2	170.6	142.6
11	142.5	119.5	141.5	118.3	101.8	142.5	114.8	141.5	113.7	95.1
12	106.8	89.4	106.1	88.8	76.3	106.8	86.9	106.1	85.3	71.3
13	578.7	484.3	577.6	483.3	415.5	578.7	465.7	577.6	464.3	388.2
14	289.9	242.5	288.8	241.6	207.7	289.9	234.5	288.8	232.1	194.1
15	193.8	161.8	192.5	161.1	138.5	193.8	155.9	192.5	154.8	129.4
16	145.3	121.3	144.4	120.8	103.9	145.3	117.9	144.4	116.1	97.1
17	692.1	579.6	691.4	578.5	497.3	692.1	557.1	691.4	555.8	464.7
18	346.5	290.5	345.7	289.2	248.6	346.5	279.5	345.7	277.9	232.3
19	230.9	193.2	230.5	192.8	165.8	230.9	186.7	230.5	185.3	154.9
20	173.4	145.2	172.8	144.6	124.3	173.4	140.6	172.8	139.0	116.1

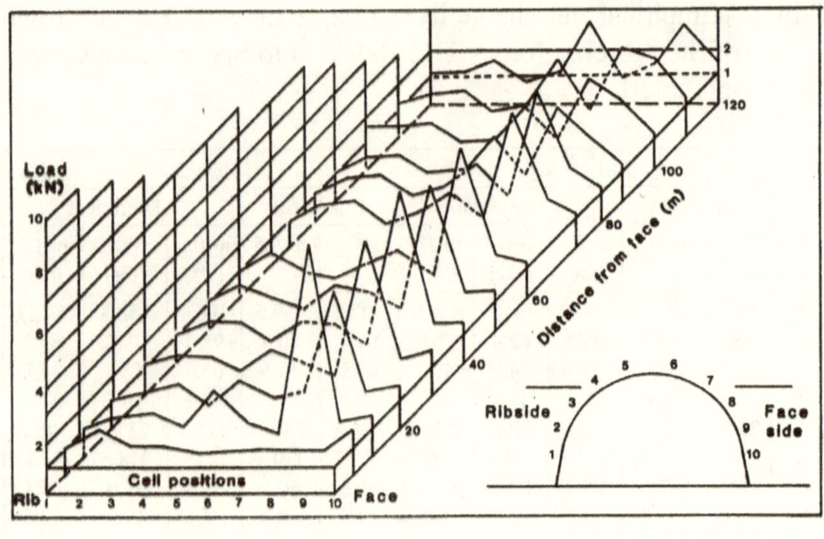

(b) Tail gate

Fig. 1 - Typical fractured roof load distribution versus distance from the faceline along mine gate roads (2).

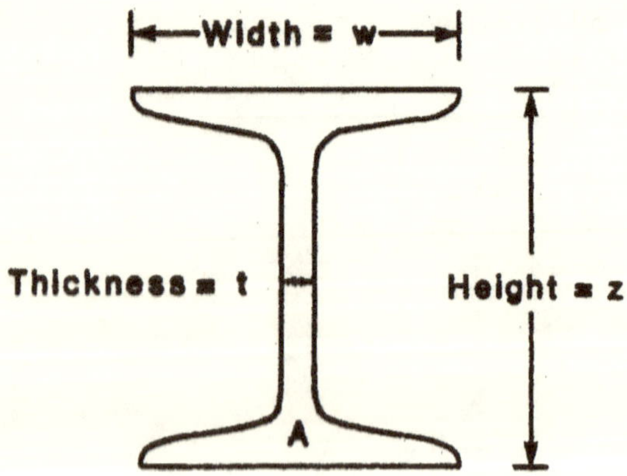

Fig. 2 - dimensional notations of H-section steel supports utilized in Table 1.

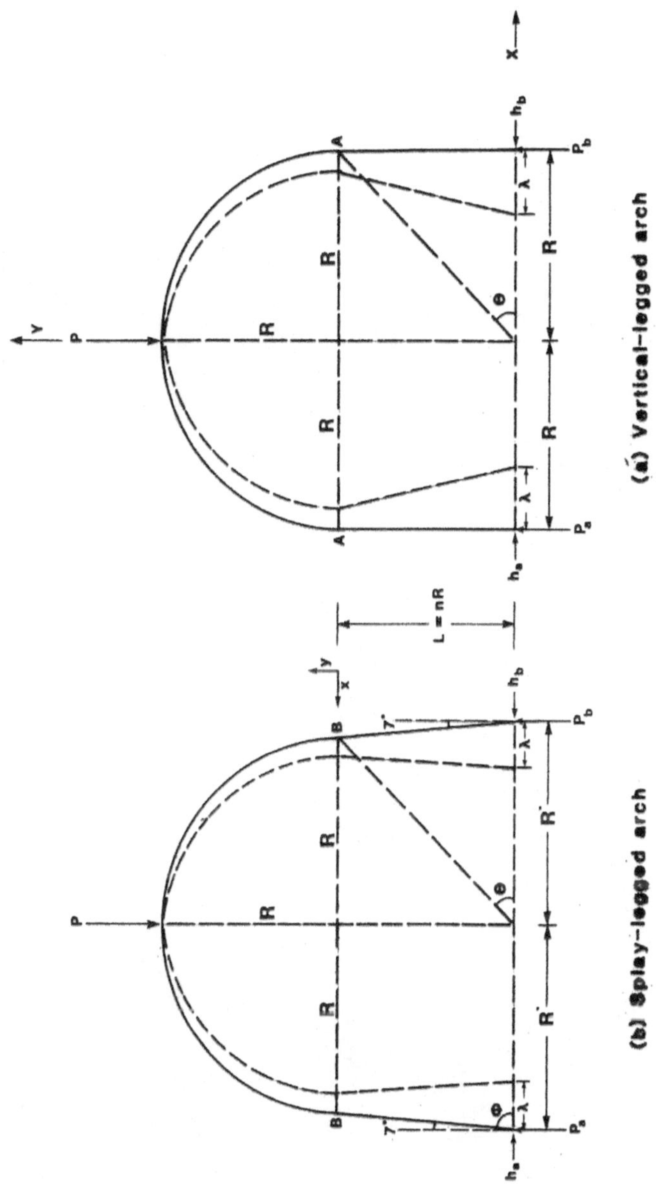

Key: R = Radius of the arch; L = Straight part of the arch legs; P = Load on the crown; = Horizontal deformation of the arch leg; p = Vertical reaction of the arch leg; h = Horizontal reaction of the arch leg.

Fig. 3 - Schematic diagram representing major forces acting on a steel arch support and notation used for determination of collapse load.

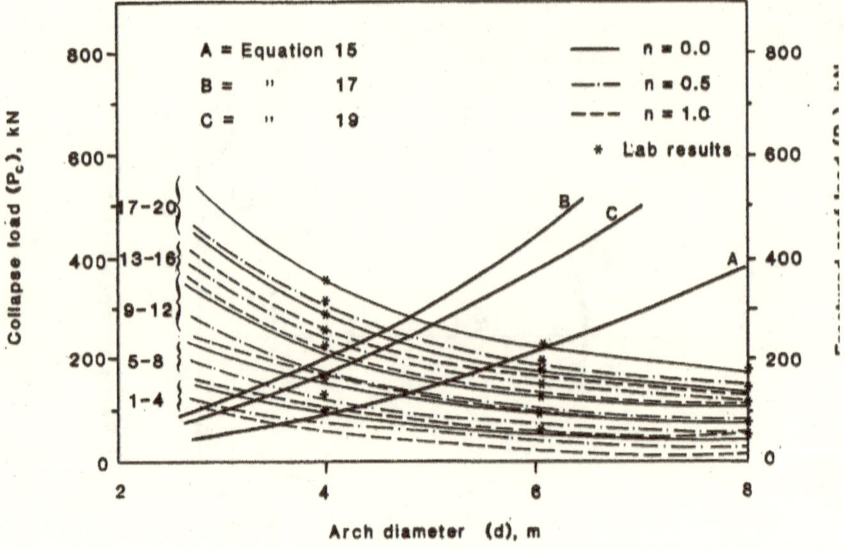

Note: Numbers 1 to 20 denote arch specifications given in Table 1.

Fig. 4 - Design chart for collapse load of single vertical leg steel arch support.

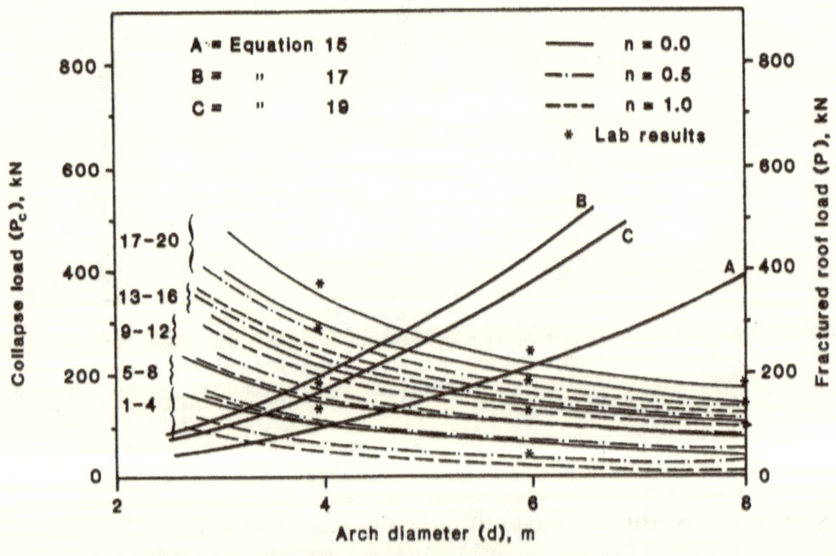

Note: Numbers 1 to 20 denote arch specifications given in Table 1.

Fig. 5 - Design chart for collapse load of single splay leg steel arch support.

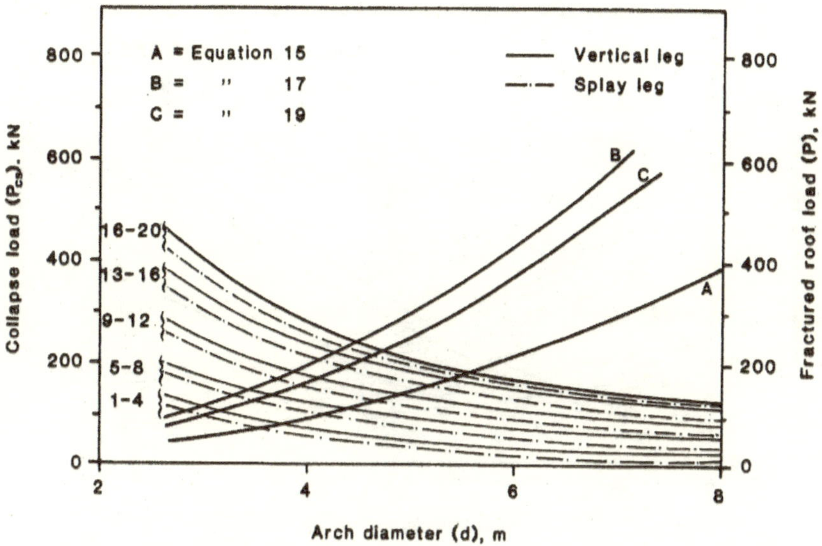

Note: Numbers 1 to 20 denote arch specifications given in Table 1.

Fig. 6 - Comparison between vertical and splay leg arches with n = L/R = 0.5.

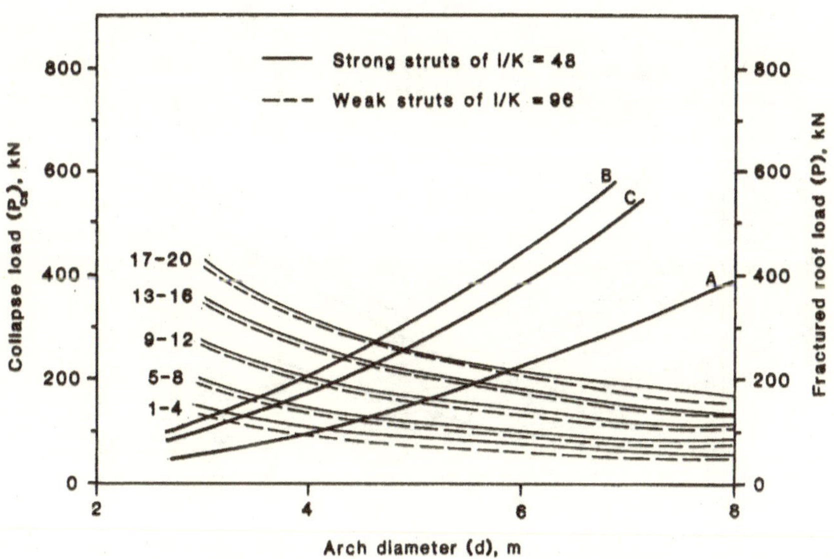

Note: Numbers 1 to 20 denote arch specifications given in Table 1.

Fig. 7 - Design chart for collapse load of vertical leg steel arches with n = 0.5 and N = 5 struts.

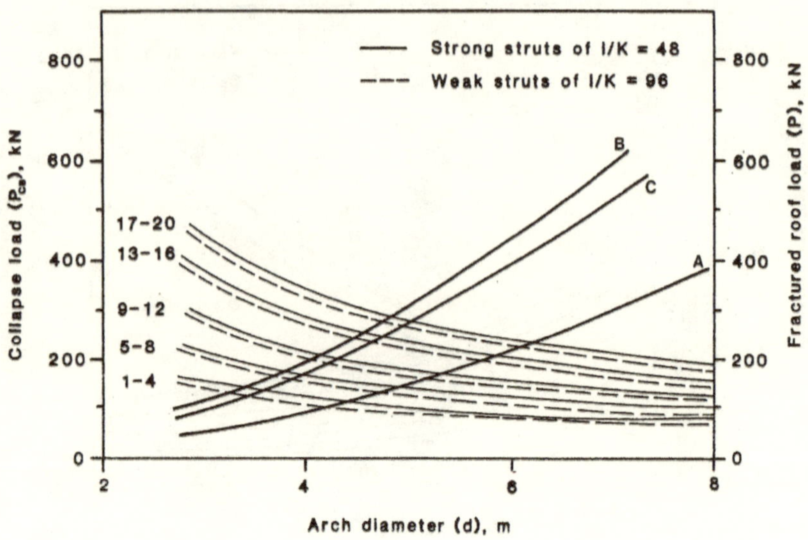

Note: Numbers 1 to 20 denote arch specifications given in Table 1.

Fig. 8 - Design chart for collapse load of vertical leg steel arches with n = 0.5 and N = 9 struts.

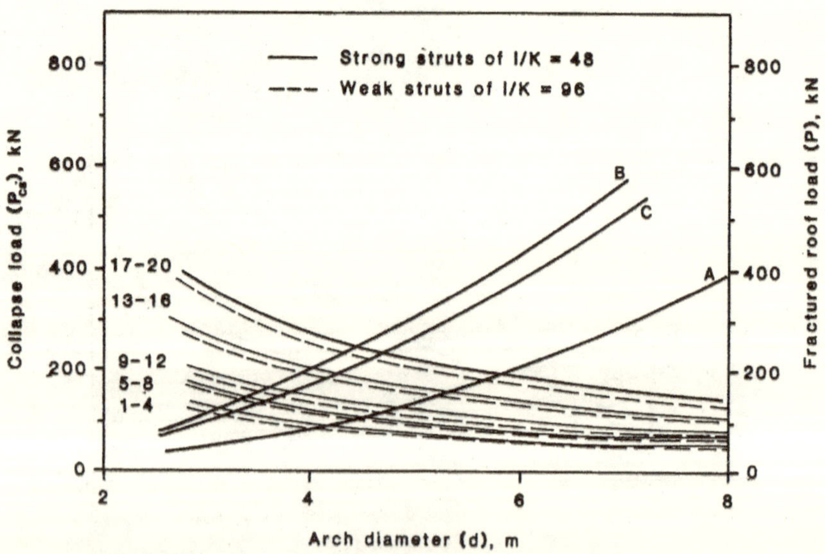

Note: Numbers 1 to 20 denote arch specifications given in Table 1.

Fig. 9 - Design chart for collapse load of splay leg steel arches with n = 0.5 and N = 5 struts.

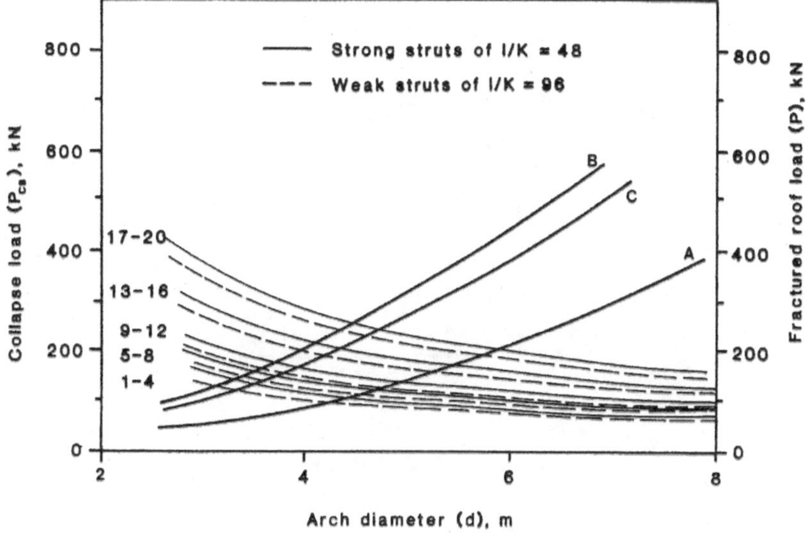

Note: Numbers 1 to 20 denote arch specifications given in Table 1.

Fig. 10 - Design chart for collapse load of splay leg steel arches with n = 0.5 and N = 9 struts.

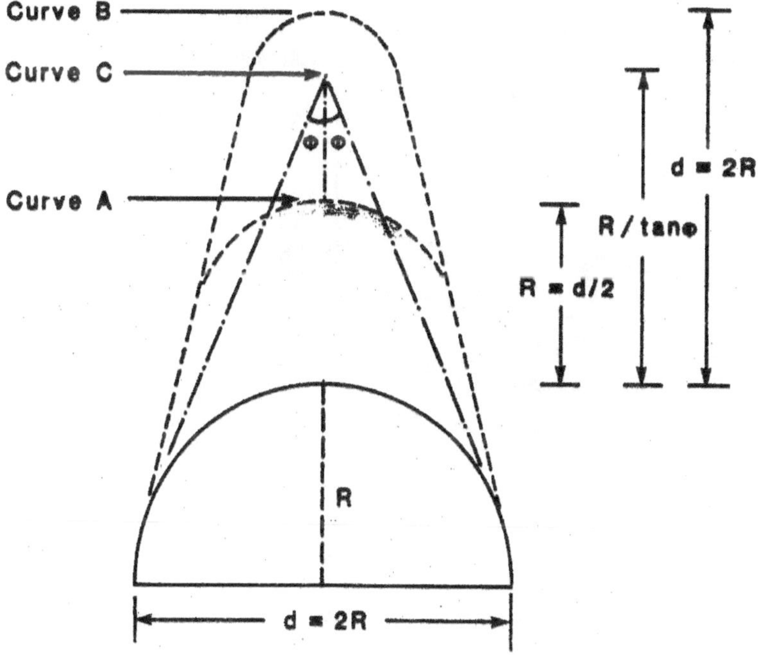

Fig. 11 - Three conditions of fractured roof rock on the gate road support.

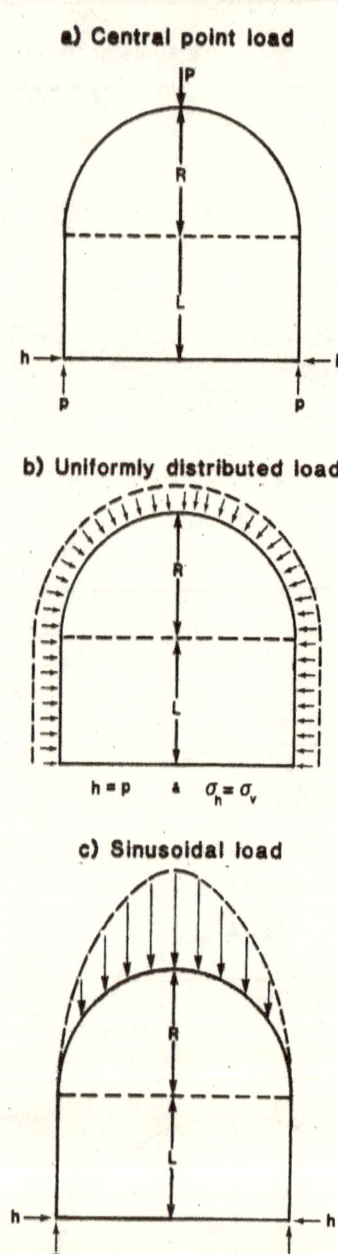

Fig. 12 - Various types of fractured roof load on the steel arched supports.

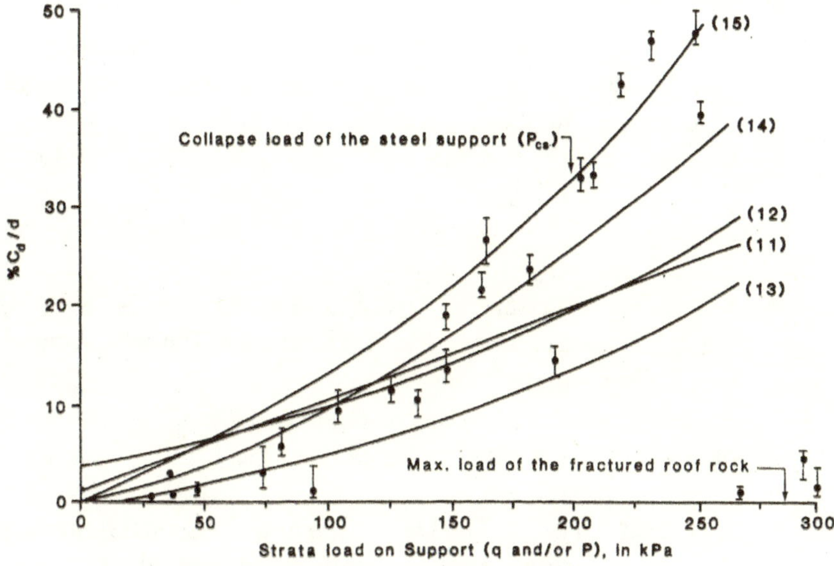

KEY: Numbers 13 to 15 relate to the references.

Fig. 13 - Comparison between gate road closure obtained from various theoretical and empirical expressions, and measured data from the Canadian and British Mines.

REFERENCES

1. Jukes, S.G., Hassani, F.P. and Whittaker, B.N., (1983), Characteristics of Steel Arch Support Systems for Mine Roadways, Mining Science and Technology, Elsevier Science Publishers B.V., 1, pp. 43-58.
2. Miller, R.R., (1981), A Study of Design and Structural Stability Aspects of Mine Roadway Support Systems, Ph.D. Thesis, University of Nottingham, U.K.
3. Afrouz, A., (1975), Floor Behaviour Along Longwall Roadways, Int. J. Rock mech. Min. Sci, Pergamon Press, Vol. 12, pp. 229-240.
4. Lewis, S., (9170), Systems of Roadway Supports, Conf on Strata Control in Mine Roadways, University of Nottingham, pp. 33-49.

5. Hassani, F.P. and Afrouz, A., (1987), Design of Semi-Circular Support System in Mine Roadway, accepted for presentation and publication in the International Congress on Rock Mechanics, Montreal, Aug. 30-Sept. 3, p. 8.
6. Choquet, P., (1986), Design of Steel Arch Supports for Gate Roadways, C.I.M., Vol. 79, No. 891, pp. 88-96.
7. Ryder, G.H., (1969), Strength of Materials, 3rd Ed., Macmillan Press, London.
8. Mitri, H.S., Hassani, F.P., Du, J. and Afrouz, A., (1987), Design of Rock Supports in Underground Tunnels, Annual CSCE Conference, Montreal, pp. 15.
9. Whittaker, B.N. and Hodgkinson, D.R., (1971), The Influence of Size on Gate Roadway Stability, the Mining Engineer, January, pp. 205-206.
10. Airey, E.M., (1974), The Derivation and Numerical Solutions Relating to Stresses around Mining Roadways, Ph.D. Thesis, Dept. of Maths, University of Surrey, U.K.
11. Bray, J.W. (1967), A History of Jointed and Fractured Rock, Part 2 - Theory of Limiting Equilibrium, Vol. 4, Rock Mech. and Engineering Geology, pp. 197-216.
12. Ladanyi, B., (1974), Use of Long-Term Stress Concept in the Determination of Ground Pressure on Tunnel Linings, Vol. 11B, Proceedings of the 3rd Rock Mech. Congress, Denver, pp. 1150-56.
13. Kaiser, P.K., (1980), Effect of Stress History on the Deformation Behaviour of Underground Openings, Proceedings of 13th Canadian Symposium, Toronto, pp. 133-140.
14. Wilson, A.H., (1980), Stability of Underground Openings in Soft Rocks of Coal Measures, Ph.D. Thesis, University of Nottingham, U.K.
15. Afrouz, A., Hassani, F.P. and Scoble, M.J., (1987), The Performance of Longwall Gate Roads with Soft Floors, Canadian Geotechnical Journal, Under review, pp. 15.

3.6 BACK-FILL DESIGNS FOR EXCAVATIONS

3.6.1 INTEGRATED DESIGN AND INSTALLATION OF STOPE DEWATERING SYSTEM FOR BACK-FILL OPERATION

By: Andy A. Afrouz, Ph.D.(U.K.), M.Sc.(U.S.A.), P. Eng.(Ontario, Canada and Australia)

SYNOPSIS

Success of back-fill operation in mined stopes depends largely on the efficiency of the back-fill dewatering system. Low rate of drainage will cause build up of water pressure behind the bulkhead. This condition reduces the designed stability of the bulkhead, decreases the strength of the back-fill, slows the rate of curing and hardening if cement or any other binders such as fly-ash or anhydrite is used in the fill. High rate of drainage requires large void ratio and high permeability in the back-fill. This situation causes the binding material and fines to flow out of the stope reducing the designed strength of the fill. This paper integrates efficient and popular design and installation techniques for dewatering system in stopes to be back-filled. Optimum drainage system can be achieved by: (a) utilization of high density fill, (b) optimizing the placed permeability of the fill at the lower elevation in the stope, especially near the bulkhead.

INTRODUCTION

Water from back-filling and flushing operations contributes to:

(a) Formation of mud and its deposition in box-holes and cross-cuts, or
(b) Increase in the hydrostatic pressure and instability of the bulkhead, or
(c) Combination of (a) and (b).

Therefore, drainage characteristics of the fill and the stope dewatering system is critical to success of the back-fill operation.

Design and installation of stope dewatering system is part of the stope preparation, prior to back-filling. It comprises the following three site specific activities or stages:

Stage A - Installation of stope and back-fill dewatering system.
Stage B - Installation of monitoring system.
Stage C - Installation of the bulkhead.

This article discusses Stage A of the above noted activities in a manner that it can be used in most mining and back-filling conditions requiring minor site specific modifications to suit the type and dimensions of a stope; accumulation and quality of groundwater; ground conditions around the stope; constituent, size distribution and water content of the fill. These factors are affected by each other and by the mining method. Preparation of the stope for the stage A can be classified as follows:

(i) Stope preparation and dewatering in non-cyclic production (or open stopes) - in which the preparation and installation activity is delayed until the stope depletion. It is then commenced as one continuous operation to completely back-fill the stope.

(ii) Stope preparation in cyclic production - which requires an initial or basic preparation followed by cyclic adjustments and up-keeping.

MECHANISM OF BACKFILL DEWATERING

As far as the drainage problem is concerned, the best back-fill is a dry fill. However, to have a good cementation between the fill constituents it becomes necessary to use a binder in water saturated state (or over saturated state in the case of hydraulically pumped fill). The drainage water can be minimized by increasing the placed density of the hydraulic fill to a certain extent. The efficiency of dewatering system can be optimized by increasing the placed permeability of the fill at the lower elevation in the stope, particularly in areas near the bulkhead. These measures will also increase the placed strength of the fill, permitting earlier hardening and readiness to working conditions.

During transportation of the hydraulic back-fill (as slurry) from the preparation site to the stope, the solid particles are kept in suspension by the velocity of the carrier fluid (i.e. water and fine particles). On deposition in the stope, the velocity is reduced permitting the particles to settle out. The back-fill reaches a fully saturated state where the particles are in contact with each other and the voids between them are filled with the carrier fluid. The fluid continues to drain between the particles until a stage is reached where air enters the particle matrix. This condition is referred to as partial saturation. A typical profile of the boundary between fully and partially saturated back-fill in a stope shortly after placement is shown in Figure 1.

At the partially saturated state, the back-fill drainage proceeds with a relatively slow rate. This process is accompanied with clogging or colmatage, as permeability of the matrix is increasingly restricted.

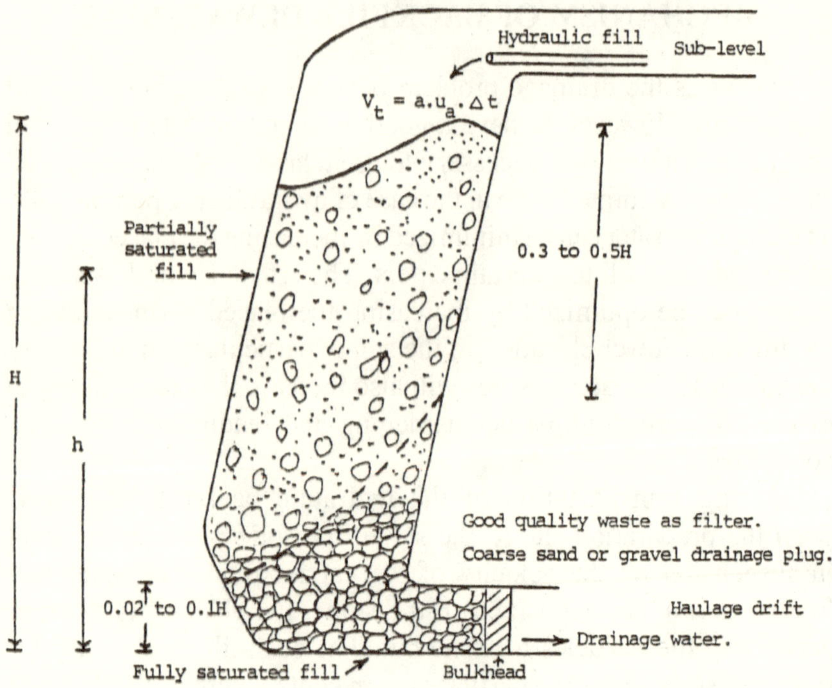

Figure 1 - Schematic of a stope with hydraulic back-fill and typical profile for various saturation states.

DESIGN OF THE DEWATERING SYSTEM

1 - Permeability of hydraulic back-fill (k) is given as follows [Thomas, 1978 and Piper, 1989]:

$$k = Q \bullet L \bullet \mu / h \bullet A \bullet \rho^2 \bullet g \qquad (1)$$

where: k = permeability of back-fill = 2 X 10^{-11} to 94 X 10^{-11}, m^2.

Q = mass rate of water flow through back-fill, kg • s^{-1}

L = length of the stope, m

μ = dynamic viscosity of the water flow, Pa • s

h = water head in the back-filled stope, m

A = cross sectional area of the stope normal to the flow direction, m^2

ρ = density of the fill water, kg • m^{-3}

g = acceleration due to gravity = 9.81, m • s^{-2}.

After time "t" from placement of the back-fill in the stope, permeability of the fill will decrease due to deposition of the solids in the pores and gradual consolidation of the fill in the stope. The decrease in permeability of the fill (k_d) is given as follows [Senyur, 1989]:

$$k_d = k - k_O \bullet e^{-n \bullet t} \tag{2}$$

where: k_O = constant pertinent to the final permeability of the fill after time (t, seconds) from the fill placement = 1 X 10^{-11} to 90 X 10^{-11}, m^2

e = base of the log = 2 • 17

n = constant = 0.03 to 0.15

2 - Total volume of hydraulic fill entering the stope via a back-fill pipe or hole (V_t) is determined as follows:

$$V_t = a \bullet u_a \bullet \Delta t \tag{3}$$

where: a = cross sectional area of the fill placement pipe or the hole, m^2

u_a = outgoing velocity of the hydraulic fill from the pipe, $m \bullet s^{-1}$

Δ t= time interval, s.

3 - Volume of solids in the hydraulic fill entering the stope (V_S) is given by:

$$V_S = V_t \bullet v_S \tag{4}$$

where: v_S – volume fraction of solids in the hydraulic fill, m^3

= volume of solids in unit volume of the hydraulic fill, m^3.

4 - Volume of water in the hydraulic fill entering the stope (V_W)

$= V_t - V_S$

$$= A \bullet u_a \bullet \Delta t(1 - v_S) = V_t(1 - v_S) \tag{5}$$

5 - Percolation rate (u_f) = rate of water flow through back-fill in the stope

$$= (q/A)(dh/dH) \tag{6}$$

where: q = volumetric rate of fluid flow through back-fill in the stope, $m^3 \cdot s^{-1}$

dh = change in the water head in the stope, m. (see Figure 1).

dH = change in the height of back-fill in the stope, m. (see Figure 1).

6 - Decrease in the percolation rate (u_d) after a time (t) can be expressed as follows [Senyur, 1989]:

$$u_d = u_f - u_O \cdot e^{-m \cdot t} \qquad (7)$$

where: u_O = constant = 0.6 X 10^{-4} to 78 X 10^{-4}, $m \cdot s^{-1}$

m = constant = 0.05 to 0.15

t = time elapsed from the start of the flow, s.

7 - The water drained out of stopes is directed, by sloped open channels (or ditches) and gravity flow away from the bulkhead and production areas into a sump.

$$\text{Average depth of the open channels } (d) = (A/2)^{0.5} \qquad (8)$$

where: A = cross sectional area of the ditch, m^2.

Slope of ditch along the water flow (s) = 1:500 to 1:1500

= 0.7 X 10^{-3} to 2 X 10^{-3} degree to horizontal (9)

Velocity of water flow along the ditch (v) = q /A

= 0 • 5 to 2 • 5, $m \cdot s^{-1}$ (10)

Side slope of the ditch (ss) = 1:0 to 1:1.5

= 90 to 34 degrees to horizontal, depending on the material (11)

8 - Size of the drainage sump is chiefly dependent on quantity of inflow (q) from one or number of stopes. It should be large enough to contain the inflow while pumps are stopped for services, repair or other reasons. Minimum sump capacity should contain 8 hours maximum inflow without inundating the pump or the surrounding. Capacities of 24 hours or more are not unusual.

9 - Water pumps are used to empty the sump and direct the water to the surface or for recirculation purposes. Various types

of pumps and the drive energy can be selected. Electrical or compressed air driven centrifugal pumps are most popular. Pumping by compressed air is noisier and less efficient than electric pumps. However, in mines which are either gassy or have high concentration of airborne explosive dust (such as sulfide and coal), use of compressed air pump is warranted. Depending on the purity of water in the sump and vertical head of water to be delivered, the pump can be single stage or multistage. For mine service, the head for single-stage pumps is limited to about 150 meters. In practice the maximum head per stage is considered to about 100 meters for contaminated water.

Following relationships exist for the pump characteristics:

a) Capacity of the pump $\propto$ RPM of the pump motor,
b) Capacity of the pump $\propto$ 1/total head of delivery,
c) Total head of delivery $\propto$ $(RPM)^2$,
d) HP of the pump motor $\propto$ $(RPM)^3$,
e) Required HP for the pump $\propto$ total head of delivery,
f) Required HP for the pump $\propto$ quantity of inflow to handle (q),
g) Mechanical efficiency of the pump (η) = 100X theoretical power to deliver water,
h) Theoretical power to deliver water = $q \cdot H_d \cdot \gamma$, watts/m. Delivery. (12)

where: γ = specific gravity of the water = 1000 to 1100 kg/m^3

i) Total head delivered by pump $(H_d) = h_s + h_d + (v_d^2/2g) - v_S^2/2g) + d$ (13)

where: h_S = suction head measured by a Gage on the suction side of the pump, m.
= pipe entrance head less + $v_S^2/2g$ + suction pipe friction loss + suction lift

h_d = discharge head measured by a Gage at foot of the column pipe, m
= static head at pump + discharge pipe friction loss to Gage point.

v_d = velocity of water in discharge pipe at gage point, m • s^{-1}

v_S = velocity of water in suction pipe at gage point, m • s^{-1}

d = vertical distance between the suction and discharge Gage centers, m.

j) Minimum required power for the pump = $q \cdot H_d \cdot \gamma / k \cdot \eta$ (14)

where: k = conversion factor to Watt = 0.102

1.W = 1/735 • 5 metric HP(= 1/745 • 7 imperial HP).

10 - Pipes or borehole in rock are sometimes used for directing water from an upper level to a lower level or a sump. Internal diameter of the required piping or the borehole can be determined from Hazen-Williams chart, given in Figure 2. Flow velocity (v) can be determined as follows [Taggart, 1956]:

$$v = 1 \cdot 32C \cdot R^{0.63} \cdot s^{0.5} \quad (15)$$

where: R = hydraulic radius or hydraulic mean depth, m.

= cross sectional area of pipe /wetted perimeter of pipe

= approximately average depth of water in the pipe / 4.

s = slope of hydraulic gradient, 1:x meter, or degrees to horizontal.

C = Chezy constant; varies between 80 to 150 dependent upon pipe material, Diameter and new or old pipe is classified in Table 1.

= $[41.6 + (0.0028 /s) + (1.81/n)] /\{1+[41.6 + (0.0028/s) + (n/R^{0.5})]\}$, according to Kutter's formula. (16)

= $158/[1+(N/R^{0.5})]$, according to Bazin's formula (17)

n = coefficient of pipe roughness in Kutter's formula, given in Table 2.

N = coefficient of pipe roughness in Bazin's formula, given Table 2.

TABLE 1 - Range of Chezy Constant (C) for Various Pipe Material

Type of Pipe	C		
Tuberculated pipe, and old pipe	80	-	110
Old cast iron pipe, and ordinary rough iron pipe	90	-	110
New riveted steel pipe and vitrified pipe	100	-	120
Smooth wood-lined pipe, and cement lined pipe	110	-	130
New cast iron pipe	120	-	140
Smooth clean masonry pipe, lead, brass, and tin pipe	130	-	140
Smooth anti-acid polymer pipe	130	-	150

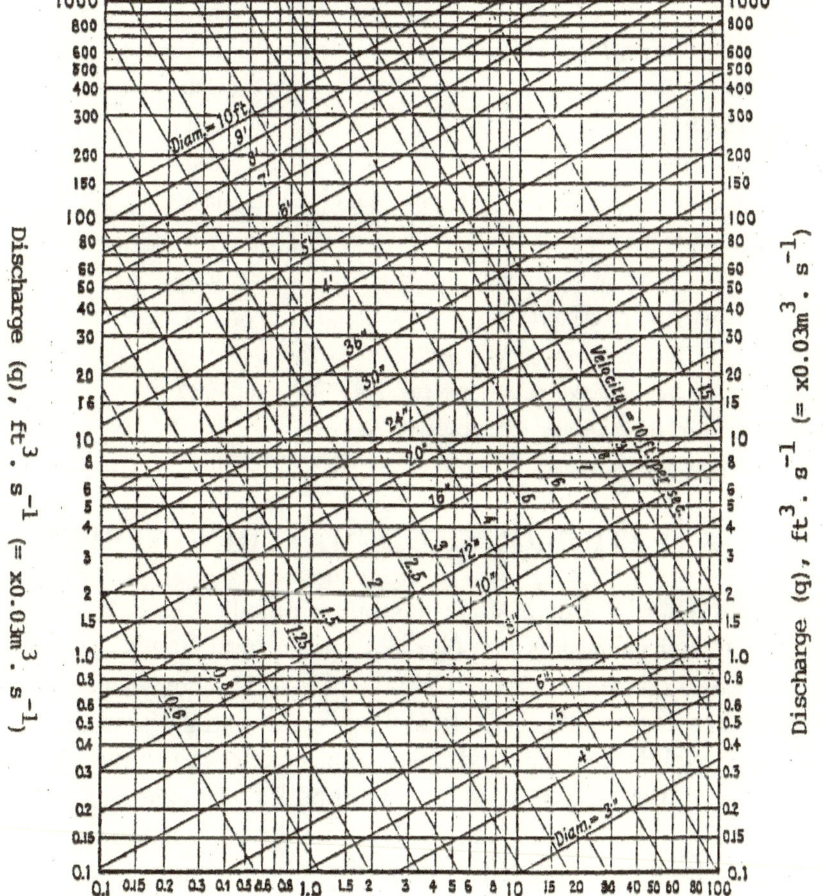

Friction loss of head, ft. per 1000 ft. (= 0.3m. per 303m.)

Figure 2 - Design of dewatering pipes, utilizing Hazen-Williams formula [Taggart, 1956].

NOTE: (a) Figure 2 was constructed for C = 100; (b) For any other value of C, Multiply the flow velocity (v) and discharge (q) by C/100; (c) to select the pipe diameterfor a discharge (q), multiply (q) by 100/C, giving q' and use this value to find the required pipe diameter from Figure 2; (d) For channels, multiply the hydraulic radius ® by 4 to get the equivalent diameter.

TABLE 2 - Experimental Values of Roughness in Kutter and Bazin Formulas [Taggart, 1956]

Type of Channel/pipe	Kutter, n	Bazin, N
Channels of uniform section		
Well-planned timber, evenly laid	0.009	
Neat cement; best pipe	0.010	0.11
Cement, one-third sand; smooth pipe	0.011	
Unplaned timber; ordinary pipe	0.012	
Ashlar; brick work; new pipe	0.013	0.29
Ordinary brick work and sewers; foul pipe	0.015	
Rubble masonry; rough concrete	0.017	0.83
Channels of non-uniform section		
Canals in firm gravel, section nearly uniform	0.02	1.54
Earth canals and rivers free from large stones and weeds	0.025	2.35
Canals and rivers in bad order	0.03-0.04	3.20

INSTALLATION OF THE DEWATERING SYSTEM

Once a horizon or a stope has been mined, slotted 75 to 150 mm. diameter galvanized or polythene agricultural pipe is laid along the footwall of the stope as a drainage pipe. It is recommended that hessian or a coarse cotton sleeve be wrapped around the pipes as filter to stop the solids entering the drainage pipe [Ashby and Hunter, 1982].

Drain towers of 3 to 3.5m height are built on brick and concrete foundations at either end of the excavation close to each drain raise, or at the filling limit for each section being filled in large excavations. A successful example of this arrangement in Sweden is illustrated in Figures 3 and 4. The drain tower consists of tightly rolled lengths of 150 mm sides square wire mesh and are wrapped with 4 layers of hessian to act as filter.

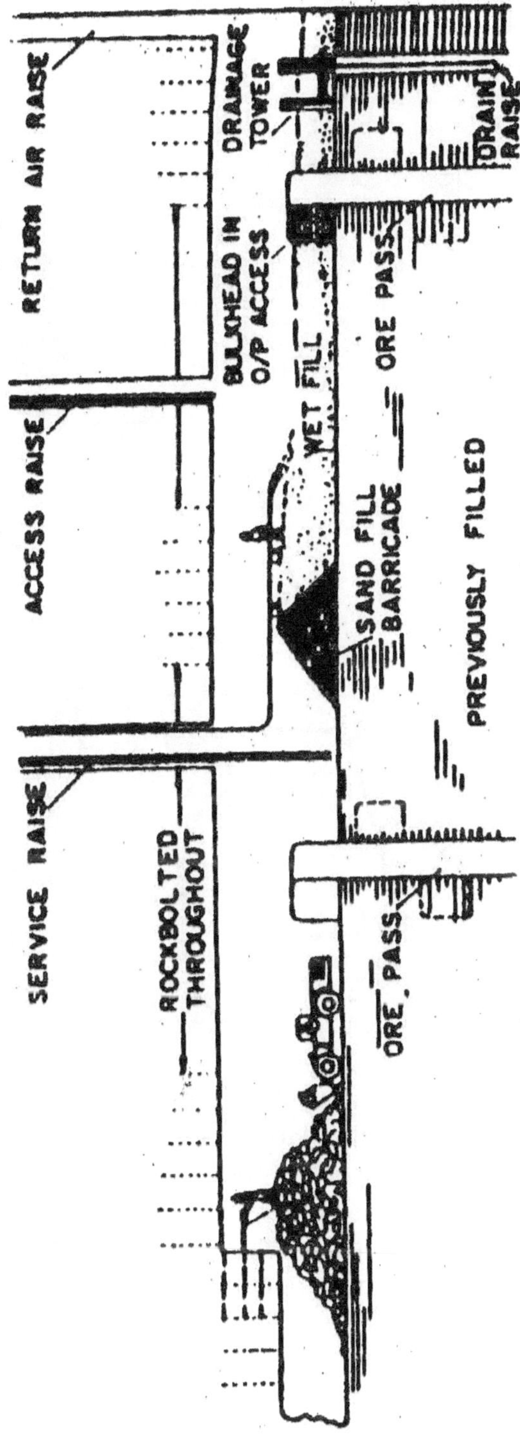

Figure 3 - Typical layout of a back-filling stope arrangement [Neindrof, 1983].

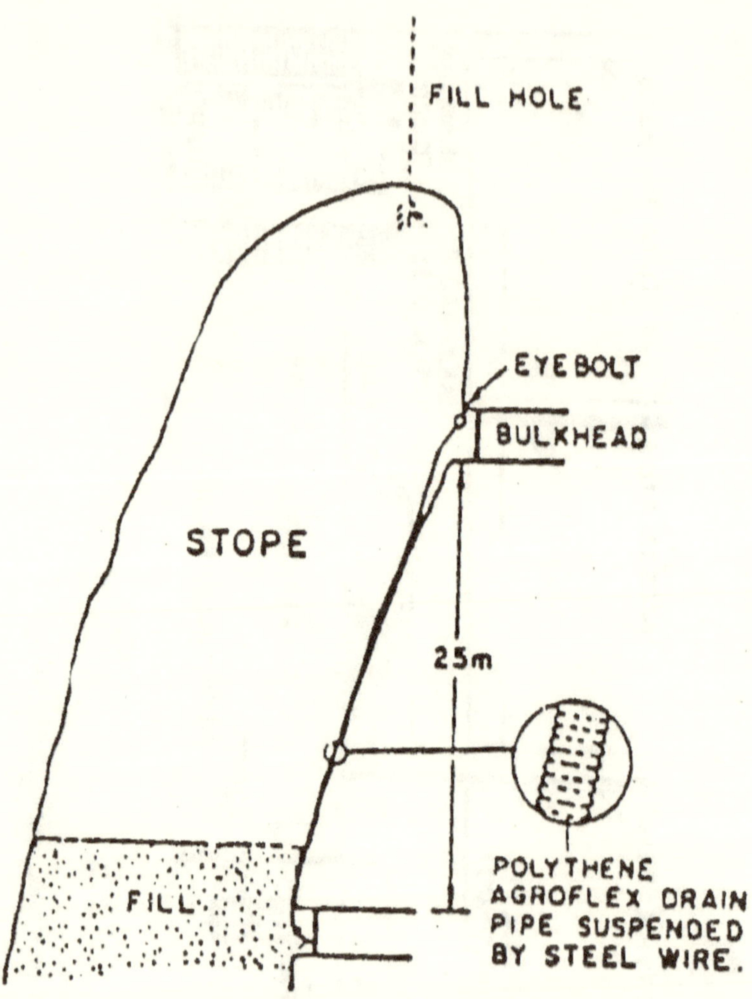

Figure 4 - Level to level stope drainage system [Neindrof, 1983].

Drain raises are kept open by bolted pre-cut crib timbers. The raise is then wrapped with chicken wire, hessian and polythene. Fill water is allowed to escape only through the drain raises via the drainage towers and connecting 100mm diameter X 5 to 6m length drain pipes. The water is directed to sumps located on the level below. Before each new lift, drain pipes from the drain towers are blanked-off in the drain raise.

Suction pipe should be short and with fewer bends as possible. Bends and elbows increase the resistance to flow, they should be as large a radius as possible. Suction pipes should dip directly into the water. For acid water, anti-acid polymer pipes or cement lined pipes are used. In the latter case, a mixture of 1 part finely sieved Portland cement and 2 parts sharp silicon sand is applied by a long handle trowel: This forms a layer of 1 to 1.5 cm thick; after which both ends of the pipe are closed with damp canvas to keep out the air until cement has set. If wet cement is exposed to air it will crack. Good cement linings should last for years.

Discharge piping should be well supported and braced to reduce vibration and effect of water hammer transmitted to the pump. Diameter of the pipes should be determined to minimize friction head. Increased power requirement due to frictional resistance may cost more than the larger piping.

Prior to filling all stopes and ore passes, accesses and crosscuts are closed off with 0.4 to 0.9 m thick brick bulkheads. The bulkheads are constructed on concrete pads and secured at the walls 2.5 to 3.5 m from the stope edge. A typical drainage system and bulkhead layout is shown in Figure 5.

Generally, the fill will drain sufficiently in 3 days after placement to act as the working floor for up-hole drilling to commence for the next 3 m lift.

In the case of open stoping, a 1.5 m X 0.9 m access hole is installed in the bulkhead (Figure 3). This access allows cement grouting of the space between the brick work and the rock inside the bulkhead. Later, it can be used for installation of percolation pipes and for stope inspection before the fill level reaches that horizon. Rubber flaps are usually installed over the access hole for ventilation control. Bulkheads are also grouted and sealed from outside of the stope. Sometimes it becomes necessary to shotcrete cracked ground in the vicinity of bulkheads to prevent fill leakage and solids loss from the stope. When the fill in the stope approaches the level of the access hole, it is bricked off.

The peculation water is either pumped into the sub-level drainage system or allowed to gravitate via drain holes to the main level drainage system.

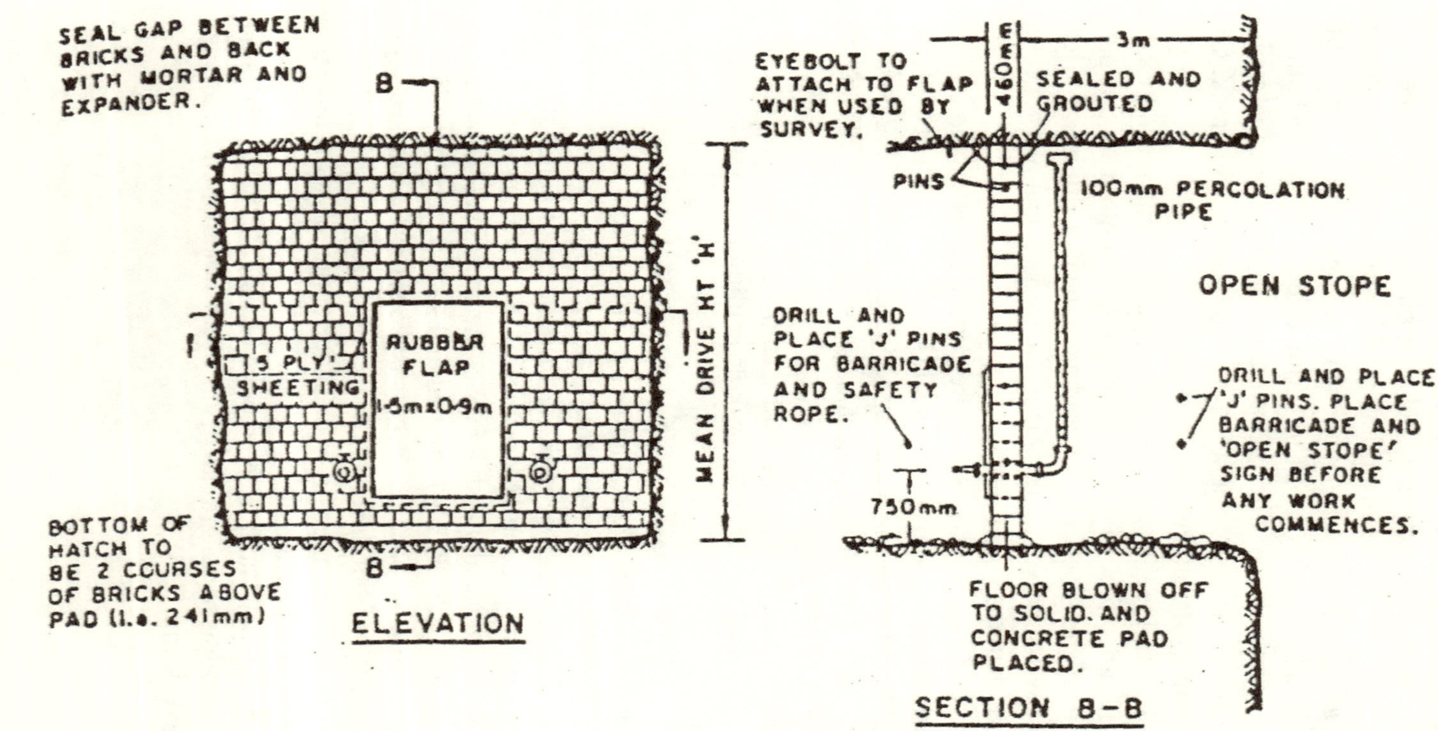

Figure 5 - Typical drainage and bulkhead layout [Neindrof, 1983].

CONCLUSION

Initial permeability of a fill will decrease after placement. It is time dependent and affected by deposition of solids in the pores and gradual consolidation of the fill in the stope.

Percolation rate of water also decreases with time. This should be taken into account when designing: the sump, suction and discharge pipes, and pump capacities to discharge water away from the workings or for recirculation with known delivery head and flow velocity.

Success of a back-fill dewatering system is affected by correct implementation and installation of the system design, and practicality of the design.

REFERENCES

Ashby, I.R., and Hunter, G.W., 1982, Filling operation at Mount Isa, Underground Operations Conference, Queens-town, Australia.

Piper, P.S., 1989, Guidelines for alleviating classified tailing placement problems, Presented in Back-fill Seminars Exploitation Technology of COMRO, September 7-12, South Africa, 17 p.

Senyur, E.G., 1989, The time effect on flow through mine back-fill materials, International Symposium on Mine Back-fill, Montreal Canada, 9 p.

Taggart, A.F., 1956, Handbook of Mineral Dressing, 6th Edition, J. Wiley and Sons, USA, pp. 20-26.

Thomas, E.G., 1978, Fill permeability and its significant in mine fill practice, Mining with back-fill 12th Canadian Rock Mechanics Symposium, CIM Special Volume 19, Ontario, pp. 139-145.

3.6.2 DESIGN AND CONSTRUCTION OF BULKHEADS FOR UNDERGROUND BACK-FILL SYSTEMS

By: F.P. Hassani and Andy A. Afrouz

Abstract—*The types of bulkhead that contain back-fill in stopes under various pressures and mining conditions are outlined. Conventionally adopted theoretical and empirical techniques were integrated in the design and construction of each type of bulkhead utilized underground. Methods to monitor the performance of bulkheads are discussed.*

Introduction

A bulkhead is defined as a structure that contains back-fill material in an excavated stope. Bulkheads are installed at the draw-points of stopes or on remnants of sub-levels.

Bulkheads to retain dry fill do not require any special consideration unless water flows into or onto the fill. Rock-fill containing an appreciable amount of fines, when saturated, could generate mud flow into adjacent workings. Bulkheads, to retain hydraulic fill, require particular features such as drainage system, solids retaining devices (or filtration), and water-level and flow-monitoring arrangements.

This paper relates existing information concerning the design and construction of bulkheads for each of the above noted sets of conditions with the objective of facilitating an integrated approach to design.

Estimation of maximum pressure on bulkheads

Consider a bulkhead placed some distance from the bottom of a stope, as shown in Fig. 1.

The total maximum pressure imposed on a bulkhead

$$(P) = \sigma_h + \sigma_v - F_r = u + P_a - F_r, \text{kPa} \tag{1}$$

where: σ_h = horizontal fill pressure, kPa

σ_v = vertical pressure due to the fill and water weight, kPa = $\gamma \cdot h$

γ = average weight per unit volume of the fill, kN /m^3

h = height of the back-fill, m

F_r = frictional resistance of the fill material reducing the lateral fill pressure in the bulkhead, kPa = 0.1P to 0.4P (Smith and Mitchell, 1982)

u = hydrostatic water pressure in the fill, kPa

P_a = active fill pressure acting horizontally on bulkhead, kPa

ϕ = angle of internal friction of the fill, degrees

$$\text{and } \sigma_v \cdot \tan^2 [45 - (\phi / 2)] = \gamma \cdot h \cdot \tan^2[45 - (\phi / 2)] \tag{2}$$

Five cases are reviewed with respect to the imposed pressure (P) on a bulkhead. These cases (modified from Nantel, 1981) are delayed where maximum pressure exertion

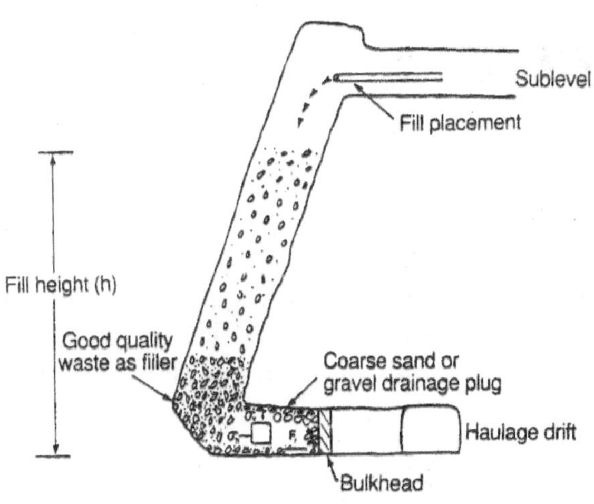

Fig. 1 — Schematic of a stope being filled with bulkhead in place.

Five cases are reviewed with respect to the imposed pressure (P) on a bulkhead. These cases (modified from Nantel, 1981) are delayed where maximum pressure exertion in a relatively short time will prevail:

Case A — For stope filled with standing water:

This situation occurs when the stope is left empty but becomes water logged. Therefore,

$$P = u + P_a - F_r = \gamma_w \cdot h + 0 - 0 = 9.8 \text{ x } 30 = 294, \text{ kPa } (= 42.7 \text{ psi})$$

where: γ_w = weight per unit volume of water = 9.8, kN/m^3
h = height of water in the stope = e.g. 30, m

Case B — For stope with saturated fill and no drainage:

This situation occurs when the drainage system — e.g. mousetraps, pipes, permeability of the fill or pump — becomes inoperative. Therefore,

$$P = u + P_a - F_r = \gamma_{sf} \cdot h \cdot \tan^2 [45 - (\phi_S / 2)] - F_r$$

$$= \{21 \text{ x } 30 \tan^2 [45 - (\phi_S / 2)]\} - F_r$$

$$= 210 - F_r, \text{ kPa } (= 30.5 - F_r, \text{ psi}) \quad (3)$$

where: γ_{sf} = weight per unit volume of saturated fill = 21, kN/m^3
ϕ_s = internal friction angle of saturated fill = 30°

Case C — For a stope with 50% saturated fill and a good drainage system:

This condition prevails during good filling practice. Ideally, this situation exists when the fill grain size is 2 to 500 mm, with 0 to 2-mm sieve sizes removed. This involves considerable expense to prepare. Poor quality waste such as clay, chlorite, mica or sericite schist (which would readily decompose and produce fines) should be avoided (Smith and Mitchell, 1982). In this case:

$$P = u + P_a - F_r$$

$$= (Q \cdot l_d \cdot \gamma_w / R_p \cdot A) + \{\gamma_{hsf} \cdot h \tan^2 [45 - (\phi_S / 2)]\} \quad (4)$$

where: Q = maximum quantity of drainage water, t /hr or m^3/hr
$= (W_{ds} / t) \{[(1 - \gamma_p) / \gamma_p] - [n(1 - S_w) / P_s (1 - n)]\}$ (5)
l_d = length of drainage plug = 3, m
γ_w = weight per unit volume of water = 9.8, kN/m^3
R_p = percolation rate of drainage plug material = 10, m/hr

A = cross sectional area of drainage drift = 3.8 x 4.5 = 17.1, m^2
W_{ds} = weight of dry solids in the fill per day = 3600, t/day
t = duration of fill placement = 18, hr/day
γ_p = pulp density as a decimal = 0.65
S_w = water saturation as a decimal = 0.5
n = porosity of settled fill = 0.45
ρ_s = specific gravity of solids in the fill relative to the fill water = 2.5
γ_{hsf} = weight per unit volume of 50% saturated fill = 15, kN/m^3

Therefore: $P = \{(3600/18)\ \{[(1 - 0.65)/0.65] - [0.45(1 - 0.5) / 2.5(1 - 0.45)]\}\ 3 \times 9.8 / 10 \times 17.1\} + \{15 \times 30 \tan^2[45 - (30/2)]\} - F_r$

$= (74.8 \times 3 \times 9.8/171) + 150 - F_r$
$= 162.9 - F_r$, kPa (= 23.6 – F_r, psi)

Case D — For a fully drained stope containing 15% water saturated fill:

This condition exists after completion of the fill placement. In this case:

$$P = u + P_a - F_r = O + \gamma_f \cdot h \tan^2[45 - (\phi_p / 2)] - F_r$$

where: γ_f = weight per unit volume of 15% by weight water saturated fill = 21, kN/m^3
ϕ_p = internal friction angle of the 15% saturated fill = 37°

Therefore: $P = O + 20 \times 30 \tan^2(45 - (37/2)] - Fr$
$= 149.2 - F_r$, kPa (21.6 – F_r, psi)

Case E — For stope containing liquefied fill:

This condition involves a reduction to zero of the effective horizontal stress within the mass of granular fill. This situation exists if energy is imparted to the mass by occurrences such as rock-bursting, blasting, earthquakes, etc. Under such conditions, the fill behaves as a liquid with a density nearly double that of water. The result is a sudden pressure increase on the bulkhead equivalent to the hydrostatic head of a fill slurry.

Therefore: $P = u + P_a - F_r = \sigma_v + O - O$
$= \gamma_{lf} \cdot h = 19.6 \times 30 = 588$, kPa (= 85.3, psi)
where: γ_{lf}= weight per unit volume of the liquefied fill material
$= 2\gamma_w = 2 \times 9.8 = 19.6$, kN/m^3

Results of the conditions A to E are summarized in Table 1.

Table 1 — Maximum pressure on bulkheads under most delayed back-fill open stoping conditions.

Case no.	Material in the stope	Max. pressure on bulkhead (kPa) Hydrostatic pressure (u)	Active fill pressure (P_a)	Total pressure (P)
1	Water	294	0	294
2	Saturated fill, no drainage	59	151 – F_r	210 - F_r
3	50% saturated fill, good drainage	12.9	150 – F_r	162.9 - F_r
4	15% saturated fill, fully drained	0	149.2 – F_r	149.2 - F_r
5	Liquefied fill	588	0	588

Notes:
1) F_r = frictional resistance of the fill reducing the lateral fill pressure on the bulkhead = 0.1P to 0.4P.
2) Under the above noted conditions, timber bulkheads with maximum allowable total pressure (P) of 90 to 99 kPa are recommended for cases 3 and 4. However, if the fill is to be placed in a cyclic condition (e.g. cut-and-fill mining), the above noted values will be reduced by 20% to 50%.

Determination of specifications for various types of bulkhead

1- Pressure-treated, free-draining timber bulkhead:

This bulkhead type is used mostly at stope draw-points and for fill containment. It can resist fill pressures of up to 207, kPa (1, kPa = 0.145, psi).

A wooden bulkhead is relatively easy and quick to install. The wood is easily available in most North American mining areas, it has a higher ratio of deflection to load than concrete or steel, and it is less costly. However, wood is combustible in high heat areas, and it can deteriorate rapidly in water or humid conditions.

The types of timber used can be maple, oak or aspen, or, among the soft woods, fir, pine or spruce. Mechanical properties of these are given in Table 2. Bulkheads for back-fill containment are generally made from hardwood.

Table 2 — Mechanical properties of commonly used bulkhead timber (modified from USDA, 1974).

Properties/types	Specific gravity (γ, kN/m^3)	Modulus of rupture (E, MPa)	Compressive strength (σ_c parallel to the grain, MPa)	Shear Strength (τ parallel to the grain, MPa)
Maple, green wood	4.3-4.5	39.5-40.5	17.0-17.5	7.3 – 7.7
Maple, 12% moisture	6.0-6.5	10.80-109.5	53.5-54.5	15.7 -16.0
Red oak, green wood	5.0-5.4	50.5-51.5	21.2-21.5	6.3 – 6.5
Red oak, 12% moisture	6.6-7.1	105.1-106.4	59.5-60.2	14.1 – 14.4
White oak, green wood	5.5-6.0	49.3-50.0	22.6-22.9	8.2 – 8.5
White oak, 12% moisture	7.0-7.4	126.5-127.0	61.3-61.5	18.1 – 18.4
Aspen, green wood	3.5-3.8	35.1-35.3	14.4-14.6	4.0 -4.2
Aspen, 12% moisture	3.6-4.0	62.5-62.9	36.4-36.7	7.3 – 7.5
Fir, green wood	3.2-3.5	33.7-34.0	16.3-16.6	4.1 – 4.3
Fir, 12% moisture	4.2-4.5	73.6-73.9	44.7-44.9	7.4 – 7.6
Pine, green wood	3.3-3.5	32.3-32.6	16.7-16.9	4.6 – 4.9
Pine, 12% moisture	5.7-6.2	99.5-101.0	51.8-52.1	10.3 – 10.5
Spruce, green wood	3.1-3.5	32.2-32.6	14.9-15.1	4.2 – 4.5
Spruce, 12% moisture	3.8-4.2	70.1-70.5	40.5-40.8	8.1 – 8.4

Preservative treatment can increase the service life of wooden bulkheads. The wood is heated slightly to decrease its moisture content to 19% or less prior to treatment. This provides good penetration of the preservative solution into the wood. It is then soaked in a mixture of:

- creosote, 431 to 574 parts per m^3 (12 to 16 parts per ft^3), and
- chromated copper arsenide, 21 to 54 parts per m^3 (0.6 to 1.5 parts per ft^3).

This mixture preserves the wood for 20 to 50 years, dependent on the site conditions. Timber members in bulkheads generally behave as uniformly loaded beams. However, the ingredients produce toxic gas, especially in mine fire situations. Dimensions of the timbers are determined by the allowable working stresses in shear, bending or deflection. For bulkhead construction, deflection should not exceed 1/240 of the timber length (or span). In short, heavily loaded beams, the horizontal shear is likely to control the beam size. In long beams, deflection may control the size determination.

Design specifications (modified from Smith and Mitchell, 1982) are as follows:

- **End bearing area of timber bulkhead (A_b)**

$$= 0.5\ FS\ .\ P\ .\ W_w\ .\ h_d\ /\ \sigma_c,\ cm^2 \qquad (6)$$

where: FS = factor of safety for the bulkhead = 3 to 5
P = max. pressure on the bulkhead, kPa
W_w = load width of the timber, cm = $t_w + s$
t_w = timber width, cm
s = spacing between timbers = 1 to 1.5, cm
h_d = height of the drift, where the bulkhead is to be erected, cm
σ_c = compressive strength of the timber perpendicular to its grain, kPa

- **Bearing length of the timber (lb)** = A_b / T_w, cm (7)

- **Spacing between the center of the timber bolt (S_b) is determined as follows (see Fig. 2):**

For rock bolt diameter > 19 mm: S_b = 40, cm
For rock bolt diameter = 16 to 19 mm: S_b = 35, cm
For rock bolts diameter < 16 mm: S_b = 30, cm

Pins can be placed behind these timbers to increase the support capacity. Pinning without rock bolting is not recommended due to lack of resistance to timber rotation.

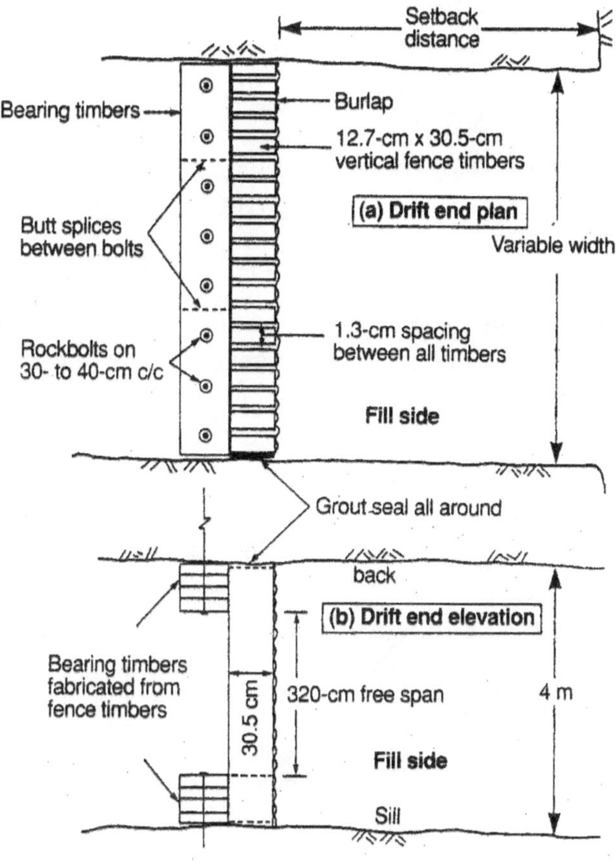

Fig. 2 — A timber bulkhead design (Smith and Mitchell, 1982).

For rock bolt diameter > 19 mm: $S_b = 40$, cm

For rock bolt diameter = 16 to 19 mm: $S_b = 35$, cm

For rock bolts diameter < 16 mm: $S_b = 30$, cm

Pins can be placed behind these timbers to increase the support capacity. Pinning without rock bolting is not recommended due to lack of resistance to timber rotation.

- **Maximum shear stress across laminated timber beams between bolts (τ_a)**

$$= 0.5\ P\ .\ h_d\ .\ S_b\ /\ t_w\ .\ L^3,\ \text{kPa} \tag{8}$$

where: L = length of timber, cm

- **Maximum longitudinal shear stress along the timber** (τ_l)

$$= 0.75\ P\ .\ W_w\ (h_d - E_i)\ /\ t_w\ .\ L,\ \text{kPa} \tag{9}$$

where: E_i = end bearing influence of the timber, cm = 65 to 85, cm

- **Maximum fiber stress in timber due to bending** (σ_t)

$$= \tau_l\ (h_d - E_i)\ /\ L = 0.75\ P\ .\ W_w\ (h_d - E_i)^2\ /\ t_w\ .\ L^2,\ \text{kPa} \tag{10}$$

Where the dimensions of excavation are larger than 4.3 m, external bracing will be required to reduce bending of timber under the fill pressure. Steel I-beams are better than timber for this purpose. The beams should be notched into the excavation walls at mid-height of the opening with sufficient bearing areas to carry 50% of the design loading (P), as shown in Fig. 3. Depth of the bearing notches is site specific. It depends mainly on the ground stability.

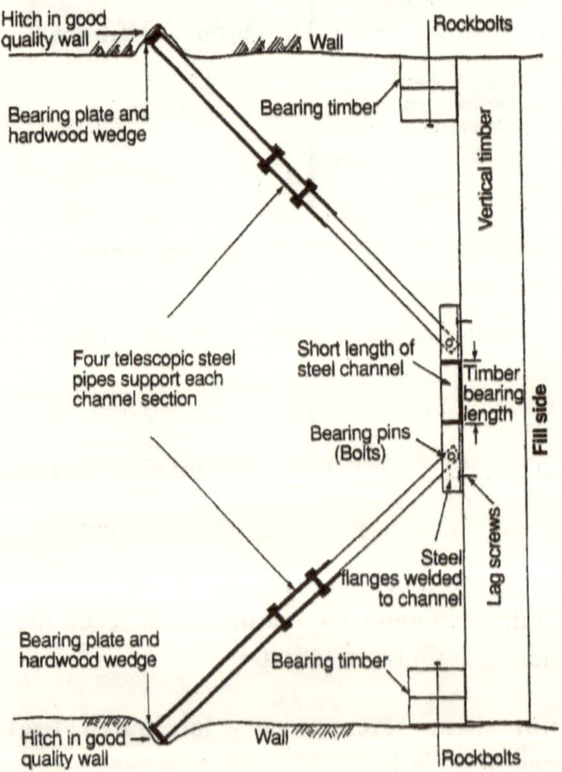

Fig. 3 — Plan of a removable, metal, external bracing system (Smith and Mitchell, 1982).

When the bulkhead is to be constructed near to the draw-point, the design pressure (P) used should be 0.5γ . h. In cases where strong arching is provided by good quality rock fill in draw-point cones above the drift opening, the actual P value may be less than 0.5γ . h (Smith and Mitchell, 1982).

Removal and reuse of timber bracing is time-consuming and results in damage to the timbers. Telescopic, light metal pipes are better braces and more efficient. Such pipes are 1 to 1.5 m long and readily removable (Fig. 3). A horizontal beam on the timber bulkhead serves as a contractor between the telescopic braces and the timber bulkhead. It is screwed and bolted as required. Variations in the design of timber bulkhead and its bracing and steel bearing plate systems are shown in Figs. 4 to 7.

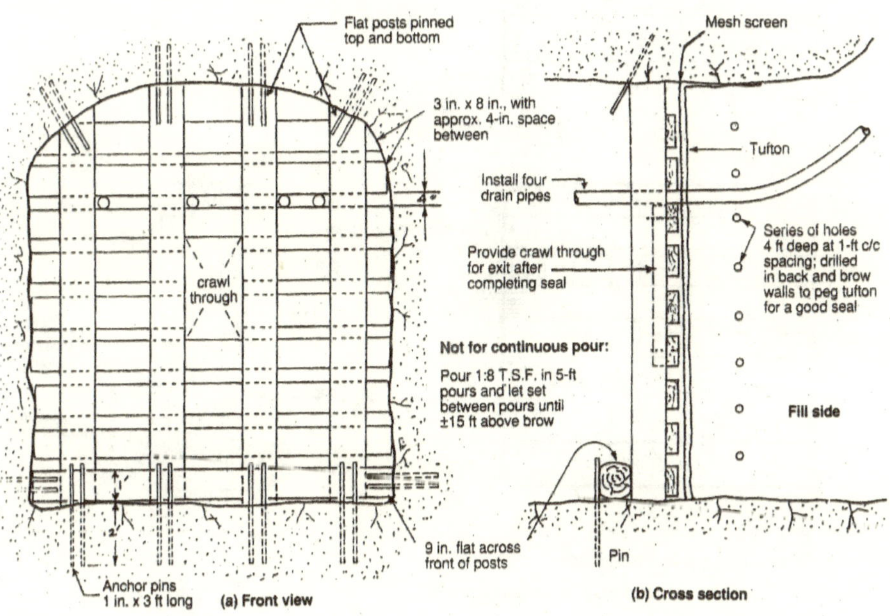

Fig. 4 — Fill bulkhead for blasthole stope in Falcon bridge mines, Canada (Courtesy of Falcon bridge, Ltd.).

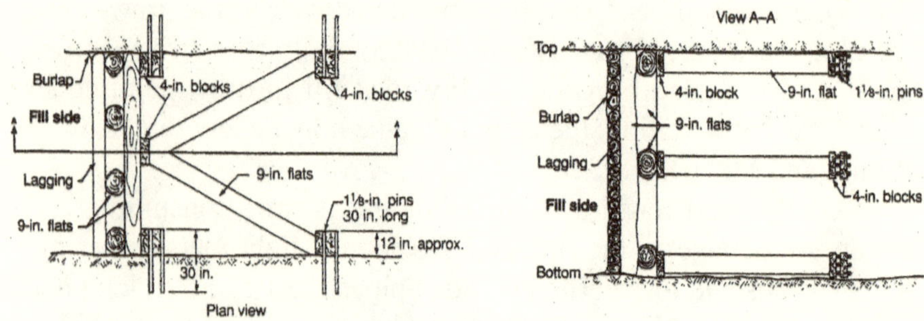

Fig. 5 — Bulkhead bracing system in Falcon bridge nickel mines, Canada (Courtesy of Falcon bridge, Ltd.).

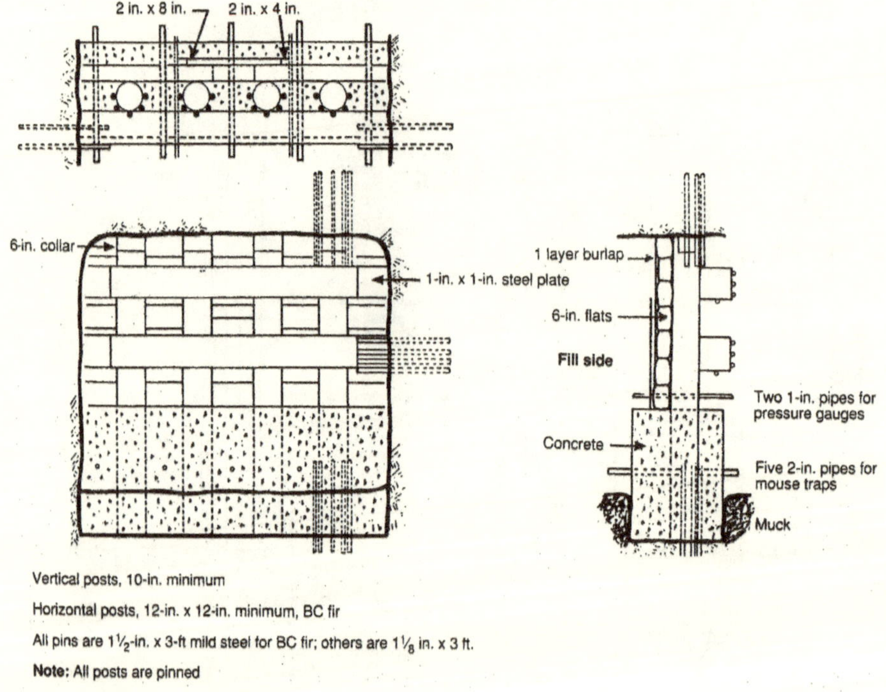

Note: All posts are pinned; All pins are 1½-in. × 3-ft mild steel for BC fir; others are 1⅛ in. x 3 ft.; Vertical posts to be 10-in. Minimum; Horizontal posts to be 12-in. x12-in. Minimum; BC fir.

Fig. 6 — Bulkhead steel bearing plates in Dome mines, Canada (Courtesy of Dome Mines, Ltd.).

2- Brick and cement, concrete and reinforced concrete bulkheads:

These bulkheads are designed to resist back-fill pressures of 275, kPa and higher. They are more expensive than the timber bulkheads and require a 28 day curing time before withstanding the fill pressure. However, they offer a useful alternative when high performance characteristics are needed. They are even capable of withstanding liquefied fill pressure. The following conditions may occur at the bulkhead:

1 - If mousetrap drains behind a concrete bulkhead perform adequately, there will be very little pressure behind the bulkhead.

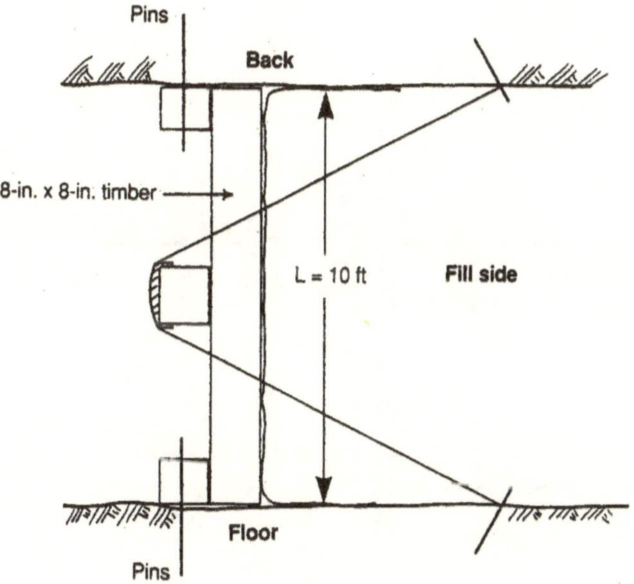

Fig. 7 — Use of steel cables as bracing for wooden bulkheads in the Horne Mine, Canada (Courtesy of Horne Mine, Ltd.).

2 - If drains do not perform adequately, or become blocked, the hydrostatic pressure can build up to the full fill height because the drainage water has no alternative scape route.

3 - If the fill is water saturated, it becomes subject to potential blast liquefaction. The worst condition must be considered

in the design of impervious concrete bulkheads. In this case, monitoring (e.g. piezometer installations) should be carried out to establish a control on the hydrostatic pressure and to limit (or discontinue) the placement rate of the water-saturated fill according to the piezometric levels.

- **The design pressure on the bulkhead (P) can be estimated in the worst condition (c) and according to Fig. 8 as follows:**

$$P = FS \cdot \gamma \cdot h, \text{ kPa} \tag{11}$$

where: FS = factor of safety for the bulkhead = 3 to 5

γ = weight per unit volume of saturated fill = 20 to 30, kN/m^3

h = height of the fill, m

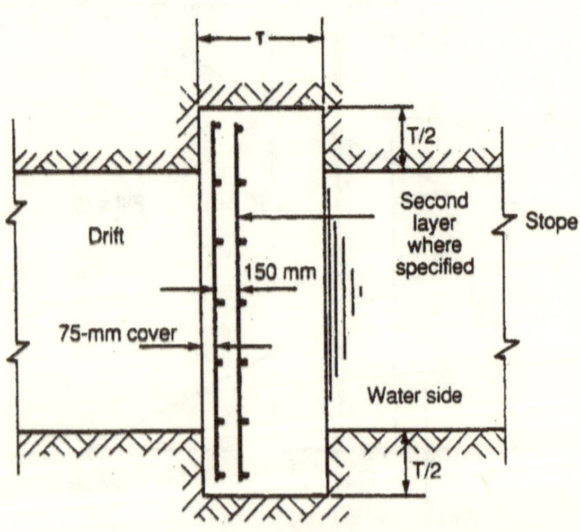

Fig. 8 — Cross section of a typical concrete bulkhead and its specifications (Neindrof, 1983).

- **Applied or allowable shear stress (τ_a)** $= 6\,\sigma_c^{0.5}$, kPa (12)

where: σ_c = compressive strength of concrete = 3×10^3 to 4×10^3, kPa

- **The critical section for shear is at 0.5 L to 0.75 L from the face of the support,** where L = length of the bulkhead clear span, m.
- **Design shear per meter width of the bulkhead (τ_d)** = u . A . $W_f/p = \gamma_{sf} . h . A . W_f/p$, kPa (13)

where: u = water pressure of the saturated fill on the bulkhead, kPa
A = cross sectional area of the drift to erect the bulkhead = 7 to 14, m^2
W_f = load factor for ultimate strength in bulkhead design = 0.4 to 0.6, m^{-1}
p = perimeter of the drift
γ_{sf} = weight per unit volume of saturated fill = 21, kN/m^{-3}
h = height of the back-fill, m

- **Required thickness of the bulkhead (T)**

= effective depth from compression face to centroid of tension reinforcement

$= \tau_d . C_R / b . \tau_a$, m (14)

where: C_R = capacity reduction factor = 1.1 to 1.3, m^2

b = width of the beam section, m

- **Thickness of the concrete bulkhead to overcome the shearing resistance of the concrete (Ts)** = P . A / $2h_d . \tau_S$, m (15)

where: h_d = height of the drift to erect the bulkhead = 2 to 4, m
τ_s = shear resistance of the concrete bulkhead = 6×10^3 to 8×10^3, kPa

Note 1: **If concrete beams of width (x)and length (y) are used, then:**

- **Maximum bending moment of the concrete beams with fixed ends (M_b)**

$= P . x . y^2 / 12$, kN . m (16)

where: x = width of the beam = 0.5 to 1.5, m
y = length of the beam = width of the drift = 3 to 5, m

- **Maximum tensile strength of the concrete beams (σ_t)**

  $= n \, . \, \sigma_{y,}$ kN (17)

  where: n = number of concrete beams installed per meter of the bulkhead height, which is perpendicular to the beam installation; and σ_y = yield strength of each beam, kN

- **Thickness of the concrete beam to overcome the compression half of the applied moment, from the neutral axis (T)** $= (M_b / 2\,\tau_s)^{0.5}$, m (18)

Note 2: **If the required thickness of the concrete is too large, the following measure can be taken:**

(a) **Use of steel reinforcement in the concrete.** This will increase the shear and tensile strength of the beams (i.e. τ_x and τ_t, respectively);

(b) **Notch the entire perimeter of the drift at the bulkhead location.** This will promote a smaller-magnitude slab bending, instead of beam bending (M_b). This will require heavy reinforcing steel with grout injection into concrete at the slots.

There are three methods used in mines or other excavations for anchoring reinforced concrete bulkheads to the mine openings. These are:

(A) Cutting a hitch in the rock to a depth of T/2. This is the most common approach in mines. If the rock strength is known, the required depth of the hitch (d) can be determined as follows: $d = \tau_t / \sigma_b$, m (19)

where: τ_t = shear strength of rock at the support, kPa. m

σ_b = bearing capacity of the rock or concrete, whichever is smaller

$= \sigma_c / 3 \sim 1000$ to 1300, kPa

σ_c = uniaxial compressive strength of rock or concrete, whichever is smaller, kPa

(B) Pinning the concrete bulkhead to the rock. Re-bar size pins are chosen. Using the working stress method:

- **The required total cross-sectional area of the steel pin (a_t)**

$$= P/\sigma_y, m^2 \qquad (20)$$

where: σ_y = yield strength of the steel pin, kPa

- **Required cross-sectional area of individual steel pin (a)**

$$= a_t / n, m^2 \qquad (21)$$

where: n = number of steel pins to be utilized

- **Required embedment length of the pin into the rock (l_v)**

$$= 3 F_1 . FS / d_p . \sigma_{cp}, m \qquad (22)$$

where: F_1 = lateral force per pin, kN

FS = factor of safety for reinforced bulkhead = 1.5 to 3
d_p = pin diameter, m
σ_{cp} = uniaxial compressive strength of the pin, kPa

- **Required embedment length of the pin into concrete (l_c)**

$$= f_1 . FS / d_p . \sigma_b, m \qquad (23)$$

where: σ_b = bearing capacity of concrete = σ_{cp} / 3, kPa

(C) Using a concrete plug or possibly a cement-fill plug. This method has been successfully used in South Africa for water containment pressure exceeding 145 kPa (Piper, 1989). The general design incorporates a concrete plug with the thickness greater than the longest span of the bulkhead, thereby eliminating the need for reinforcement to resist bending moment.

In the selection of any of the above noted bulkhead reinforcement methods, one should consider the economical comparisons for the material and installation costs.

3- Arched concrete-block bulkheads:

These have been used successfully in Australia and Ireland (Ashby and Hunter, 1982; Neindrof, 1983). Concrete blocks are relatively easy to make and construction is fast. Usually a pad is poured at the bulkhead location, and sometimes hitches are made in the drift walls. The arch design shown in Fig. 9 has a higher shear resistance than common concrete blocks. Dimensions of the concrete blocks are 40 x 20 x 20 cm^3 with three holes. Two-hole blocks are also suitable but would not provide as much concrete area. These blocks should be mortared in place with the holes oriented horizontally to provide drainage ports. A 20-cm arch depth is sufficient for drifts up to 4.3 m wide. For openings wider than 4.3 m, a double row of blocks or buttress walls is needed.

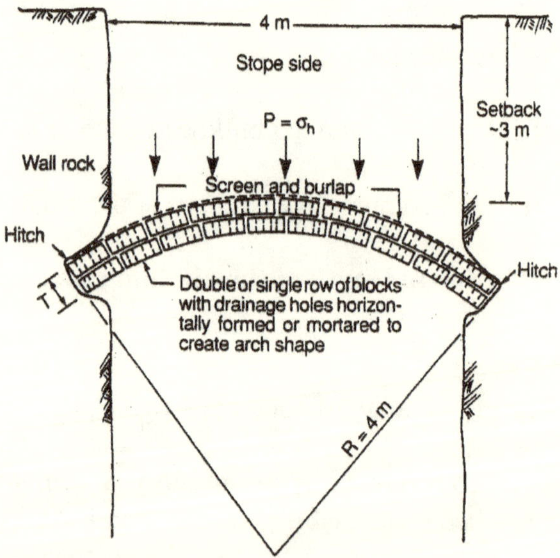

Fig. 9 — Plan of an arched concrete-block bulkhead (Neindrof, 1983).

The concrete blocks are mortared in place to form the arch. Filter screens are to be placed on the stope side and sealed with mortar on all edges. The screens are wire mesh covered with burlap.

Thickness of arched bulkhead (T) = FS . R . P / σ_c, m (24)

where: FS = required factor of safety
R = radius of the bulkhead arch, m
P = pressure on the bulkhead, kPa
σ_c = Compressive strength of concrete block = 3000 to 4000, kPa

This type of bulkhead construction allows placement of the bulkhead filter screen, drift plug and gravel as the height of the arch is raised. Tight filling between the bulkhead and the waste rock drainage blanket is essential to the stability of bulkhead. Close supervision of bulkhead construction is required to ensure that a correct arch is formed.

Concrete-block bulkheads to contain cemented fill have been constructed in Ireland using tapered solid blocks, sometimes without mortar. These blocks, when placed side by side, automatically produced an arch shape (Smith and Mitchell, 1982). Spaces between the blocks provide sufficient drainage for seepage through the fill. However, drainage pipes are also used to connect to decant pipes.

Bulkhead pressure measurement

One or more dial gauges placed along the mid-span of the bulkhead beam can measure bulkhead deflection with time. Figure 10 illustrates a dial gauge installation on a timber bulkhead. The measuring device consists of a pole with fixed location points and a dial gauge with ± 0.025 mm accuracy (Nantel, 1982). The measuring device can be moved onto various bulkheads.

Pressure on the bulkhead (P) = $384\ E\ .\ I\ .\ \Delta\ /\ 5 \times 141^4$, kPa (25)

where: E = average modulus of deflection for timber, kPa
= 7×10^6 to 9×10^6 kPa, for spruce
I = moment of inertia of the beam, m^4
Δ = mid-span deflection in the timber, m
L = free span or length of the beam = 3 to 4, m

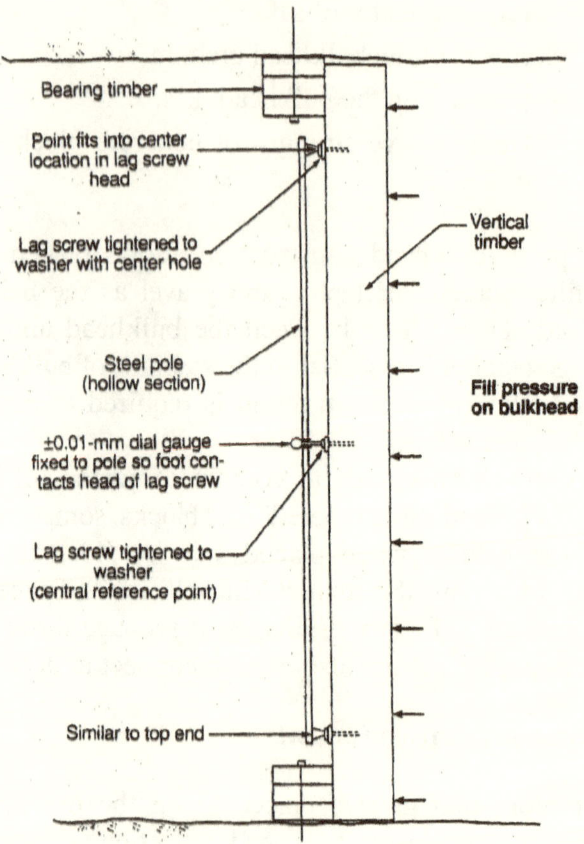

Fig. 10 — Arrangement for deflection monitoring of timber bulkheads (Nantel, 1981).

At designed pressure of P = 96.5, kPa (=14 psi) on a spruce timber bulkhead, for example, the timber deflection (Δ) would be 7 to 8 mm.

An alternative technique is to replace the movable measurement pole and dial gauge with a fixed wire tightly stretched between the upper and lower leg screw locations.

On initial installation, the allowable deflection (7 or 8 mm) can be set as a spacing between the wire and the central reference point. This arrangement can be connected to an automatic alarm circuit to sound when the allowable deflection is exceeded.

Total pressure cells are commercially available to be utilized for bulkhead pressure measurements. They are more pertinent to the brick and concrete bulkheads. However, they are more expensive than the above mentioned instruments.

In addition to these measurements, bulkheads should be inspected by a supervising engineer during and after construction to ensure that the design specifications and performance prediction have been met. Inspection should also be continued during stope back-filling, making sure adverse stress, deformation, tilting, rotation or cracking in the bulkhead does not occur. Any unusual or abnormal behavior of the bulkhead system should be recorded and remedied.

Conclusion

It is not possible to recommend one general bulkhead design and installation formulation to satisfy the variety of back-fill material constituents and conditions occurring under static and dynamic loading and pillar recovery situations. Each of these conditions requires site-specific design and installation. However, the practical methods presented in this article can be used as a basis for assessment of the suitability for each type of bulkhead and category of conditions with minor on-site modifications and adjustments.

References

Ashby, I.R., and Hunter, G.W., 1982, "Filling operations at Mount Isa," Underground Operators Conference, Queenstown, Australia.

Nantel, J., 1981, "Design and installation procedures of timber bulkheads for fill containment," Canadian National Committee on Underground Bulkheads and Dams, p. 31.

Neindrof, L.B., 1983, "Fill operating practices at Mount Isa Mines," *Proceedings of the International symposium on Mining with Back-fill*, Luleá, Sweden, June 7-9, pp. 179-187.

Piper, P.S., 1989, "Guidelines for alleviating classified tailing placement problems," Presented at Back-fill Seminars, Exploitation Technology of COMRO, Sept. 7-12, South Africa, p. 17.

Smith, J.D., and Mitchell, R.J., 1982, "Design and control of large hydraulic backfill pours," *CIM Bulletin*, Vol. 75, No. 838, Canada, February, pp. 102-111.

US Department of Agriculture, 1974, *Wood Handbook*, Forest Product Laboratory, Agr. Handbook, No. 72, pp. 4.41-4.55.

3.6.3 PLACEMENT OF BACK-FILL

By Andy A Afrouz, PhD, MSc.

SYNOPSIS

This paper condenses part of an ongoing investigation to integrate various techniques for the design and implementation of back-fill placement in mined stopes. It consists of the practical methods for cyclic and delayed placement of a variety of back-fill constituents such as rock-fill, hydraulic fill, cemented fill, paste fill, and ice fill under different underground mining conditions. Discussion is also made on mechanics of placed back-fill, solids losses, arching effects, fill consolidation or shrinkage, liquefaction, placement problems and suggestions to solve the problems.

1. INTRODUCTION

Placement of waste rock in mines as back-fill is a process with the following merits:

(i) Provides a better environmental control on the waste rock and mill tailing.
(ii) Fills the excavated areas, promoting better support and ground control.
(iii) Facilitates pillar recovery between adjacent stopes.
(iv) Reduces ventilation short circuiting between adjacent workings.
(v) Reduces cost of waste transportation to the mine surface dumps and tailing ponds, and associated up-keeping and monitoring of these facilities.

Ideally, placement techniques should facilitate:

(i) Least segregation of fill constituents during placement.
(ii) Sufficient mechanical compaction of the placed fill.
(iii) Least arching effect after placement and dewatering.

(iv) Minimum liquefaction potential.
(v) Complete filling of the mined out spaces, reducing surface subsidence.
(vi) Cost effectiveness in comparison with alternative methods.

In general, the average cost of dumping the underground waste rock on the surface, depending on the site specific factors, varies between 1.30 to 2.50 $/m^3$ of waste to be handled, excluding the environmental monitoring and reclamation costs. Alternatively, the total cost of back-filling and associated environmental monitoring operations varies between 1.20 to 2.60 $/m^3$ of the fill. Hence, it is economically advantageous, with simpler and less monitoring and reclamation requirements, to place mine wastes in the excavated area and not on the surface dumps.

This investigation integrates and discusses various fill placement techniques, placed fill mechanisms and solutions to potential problems in situ.

2. PLACEMENT TECHNIQUES

The method of fill placement depends on a variety of factors, the most important being ground conditions, mining methods, constituent and state of the material to be back-filled in the mined excavation. Placement techniques can be categorized into cyclic placement and delayed placement, each of which comprise numerous methods described herein. The material can be vibrated in place if excess or quick compaction resulting stronger fill is desired. The vibration can be continuous or intermittent using either prop type concrete vibrators or a system of submerged pipes extending through the length of the fill. This practice can be utilised in all mining methods. It is not labor intensive and is economical. The vibration not only causes compaction of the fill, but also causes the water in hydraulic fill to decant or drain off faster. As the solids sink and become more compact, the water rises to the fill surface.

The main placement techniques are as follows:

2.1 Cyclic Placement of Back-fill

In cyclic placement of back-fill the fill is placed as part of the mine production cycle, eg cut and fill operation, staggering or timber supported systems, longwall or short-wall mining. The placement can be done in overhand-cut-and-fill, underhand-cut-and-fill, behind or beside the excavated area.

2.1.1 Fill Placement in Overhand-cut-and-fill

In which the production proceeds upward and the fill is placed in the floor of the excavated area or on top of the previous fill, eg in conventional cut and fill mining. This type of fill placement is utilized where stable roof conditions prevail. It can be sub-divided as follows:

2.1.1.1 Placement in Overhand-cut-and-fill

This is the simplest and least expensive method. However, it does not develop high strength necessary for pillar recovery, if needed. In this technique the waste rock is in any transportable size range. It can be poured into the excavation from an access at a higher elevation to the stope of adequate height, by a truck or conveyor (**Figure 1**). The fill is then leveled in the stopes as desired. It is also possible to rock-fill the stope with two different fill placements, eg rock and then mill tailings. The latter is to fill the voids in the rock-fill by means of gravity. The geometry of such a technique is shown in **Figure 1**.

2.1.1.2 Cemented Rock-fill

This provides a higher strength back-fill, useful to stabilize weak hanging wall or for pillar recovery. The unconfined compressive strength of cemented rock-fill ranges 1 to 2 MPa, depending on the ratio of rock, cement, and the size distribution of the rock aggregate. In this technique, waste rock and cement grout are poured into the stope alternately, so that they provide a mixture of cemented rock to the desired stope height. The arrangement is similar to that in **Figure 1**. Uniformity of the mixture is very important in the overall strength development of the back-fill. The cement grout consists of sand, cement, water and small amounts of additives (or retarders) to reduce the grout hardening in the pipeline during pumping. The volumetric ratio of the cemented rock-fill constituents is as follows:

- Sand: 10
- Cement: 0.2 to 2.1
- Water: 10 to 15
- Retarder: 0 to 0.1

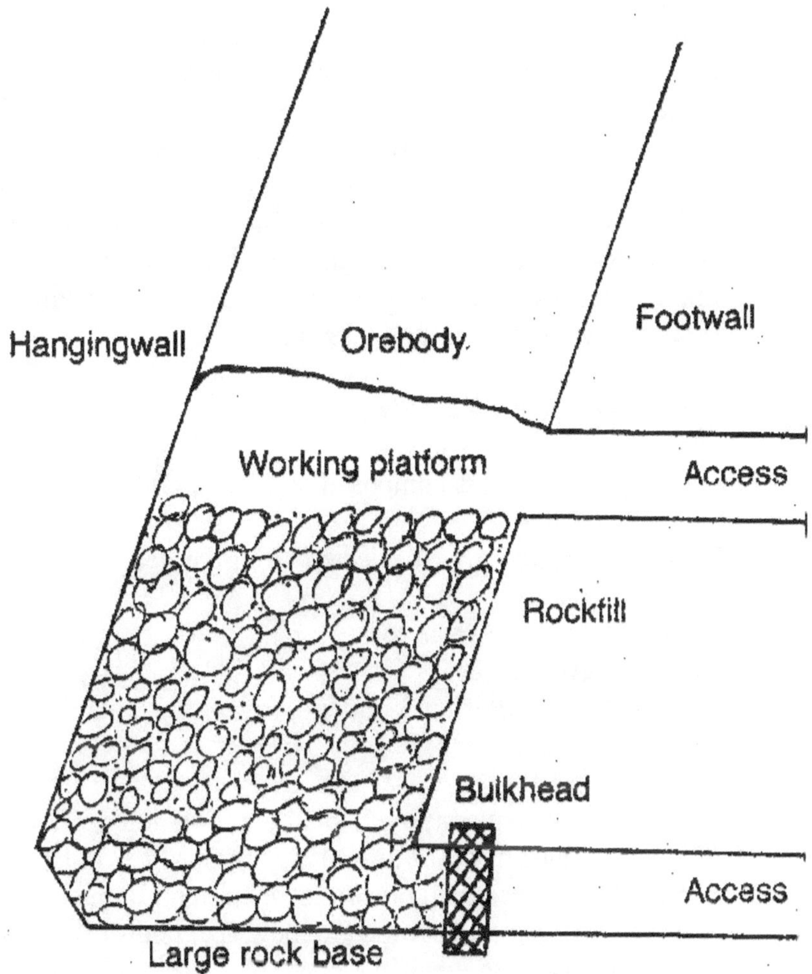

Figure 1 Schematic of a Rock-fill Operation in Over-cut-and-fill Mining Method

It is possible to substitute mill tailing for the sand and water constituents of the cement grout. In this case, the volumetric ratio of tailing is cement ranging from 4:1 to 30:1 with slurry density ranging from 65 to 80% by weight solids.

Higher strength cemented rock-fill can be achieved by either increasing the cement content, decreasing the water content, or utilizing the 'Prepacked Method'. The first two measures will cause occasional hardening of the cement grout in the transportation pipeline but the prepacked method overcomes this problem. It has shown less shrinkage than the common cemented fill. However, it is relatively more expensive and it is described as follows [Stout, 1975]:

(i) Grout pipes are installed at strategic locations to the bottom of the stope, as shown in **Figure 2**. The stope is then filled with coarse waste rock in 6 to 100 mm size range, up to a per-determined height.
(ii) Grout is prepared consisting of sand, cement, water and retarder as described previously. The grout is then pumped to the bottom of the grout lines. The mortar fills all the voids between the course rock-fill and produces a compact concrete which becomes stronger with time to an unconfined compressive strength of up to 3 MPa.

2.1.1.3 Placement of Cemented Aggregate Fill

Cemented aggregate fill is composed of a weight ratio of:

(a) 20 to 30% coarse gravel rejected from the heavy medium separation plant.
(b) 30 to 45% crushed waste rock.
(c) 5 to 10% furnace slag.
(d) 2 to 4% ordinary Portland cement.
(e) 20 to 30% water.

Unconfined compressive strength of the above noted mixture cured for 28 days ranges from 0.7 to 1 MPa [Thomas, 1973]. Higher strength is possible by increasing the cement content. Items (a), (b) and (c) are mixed together and dumped in the stope. Items (d) and (e) are mixed and pumped onto the above noted mixture, alternatively.

Cemented aggregate fill segregates when placed into the stopes. The segregation involves formation of a cemented hydraulic fill beach or lens against exposures in the aggregate slope. The bulk of the remaining void is also filled with cement grout.

Dry aggregate and cemented hydraulic fill placed in stopes at 1:1 by weight ratio, showed large loose zones of uncemented aggregate which had been washed out by the hydraulic fill against the stope walls [Grice, 1990]. Dry aggregate has a repose angle of 35° to 40° to the horizontal. The shape of the cone formed at the top of a filled stope does not tightly fill or provide any crown support. Moderate introduction of water or cement grout to the above noted cone tends to reduce the angle of repose by 4° to 10° and permits a tighter fill to the back of the stope.

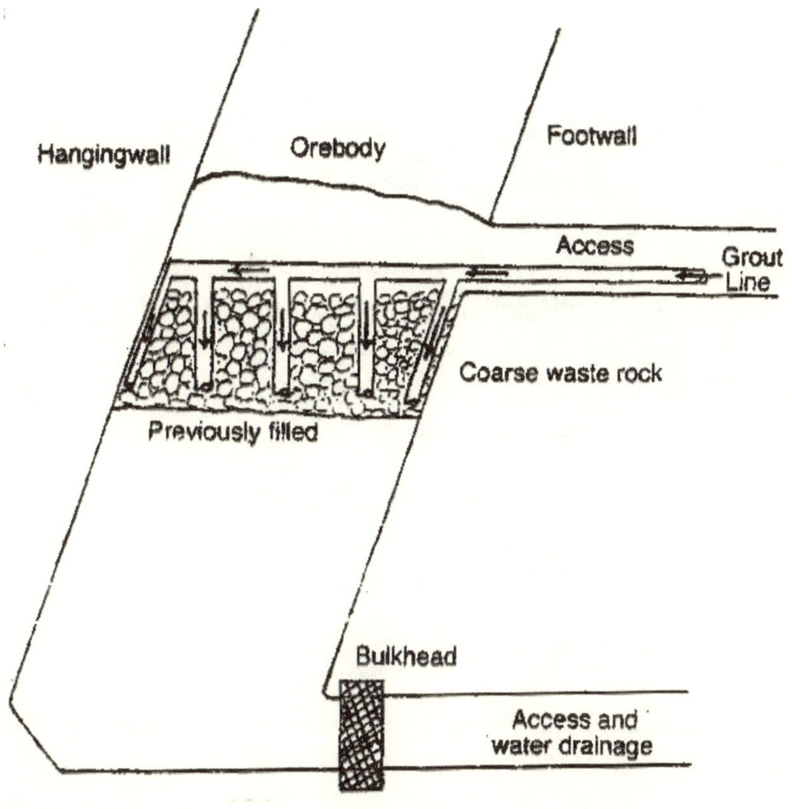

Figure 2 Prepacked Method of High Strength Cemented Rock-fill Placement [Stout, 1975]

2.1.1.4 Drift and Cemented Fill

Where ground conditions cannot withstand large width of stopes, the ore may be mined by drifts driven in the stopes. In this case, the width of the drift is equivalent to ⅓ to ½ width of a large stope and is usually 3 m to 5 m, as shown in **Figure 3**. The ore is extracted in drift 1 and the excavation is then filled by cemented rock-fill or cemented aggregate before the next drift is developed beside the previous one (eg drifts 2 and 3). When the full width of the orebody is exploited and filled, the next drift will be developed on top of the previous raw (eg drifts 4, 5 and 6). The access is provided by ramps as shown in **Figure 3**. The above noted method has been tried in the Swedish Garpenberg Mine with success [Krauland, 1990].

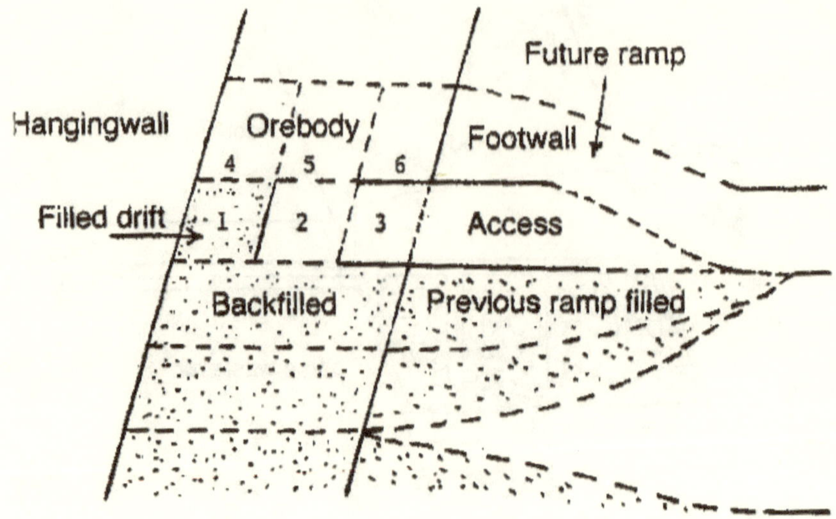

Figure 3 Drift and Fill Technique in Overhand-cut-and-Fill Operation

2.1.1.5 Placement of Cemented Tailing

Two different strengths of tailing/cement mixture can be used. The procedure is as follows:

(i) Bulkhead is erected at the bottom access, near the stopes lowest point, as shown in **Figure 4**.
(ii) Gravel (or large rocks) is placed at the stope bottom. It is dropped from the top access by truck or conveyor.
(iii) Stope is filled 15 cm below the required floor grade by pumping (or gravity flow) a mixture of mill tailing, cement and water with the following volumetric ratio:

Mill tailing = 25 to 35; Ordinary Portland cement = 0.8 to 1.2; and Water = 7 to 12. This base pour is left for 2 to 4 hours to be water drained. This fill is soft with low cement content. Its unconfined compressive strength is 0.2 to 0.4 MPa after 28 days of curing.

(iv) A harder fill with higher cement content is poured on top of the previous cemented tailing to form a 15 cm thick working platform. The volumetric ratio of the harder fill is as follows:

Mill tailing = 6 to 16; Ordinary Portland cement = 1 to 1.2; Water = 2 to 6. This harder fill is left for one to two days to be drained and cured before mining commences. Unconfined compressive strength of the harder fill after 28 days of curing is 0.6 to 1.5 MPa. When slushing, a shallow bed of muck is left on the working platform to protect the freshly placed floor until sufficient time has passed for the fill to cure and harden properly.

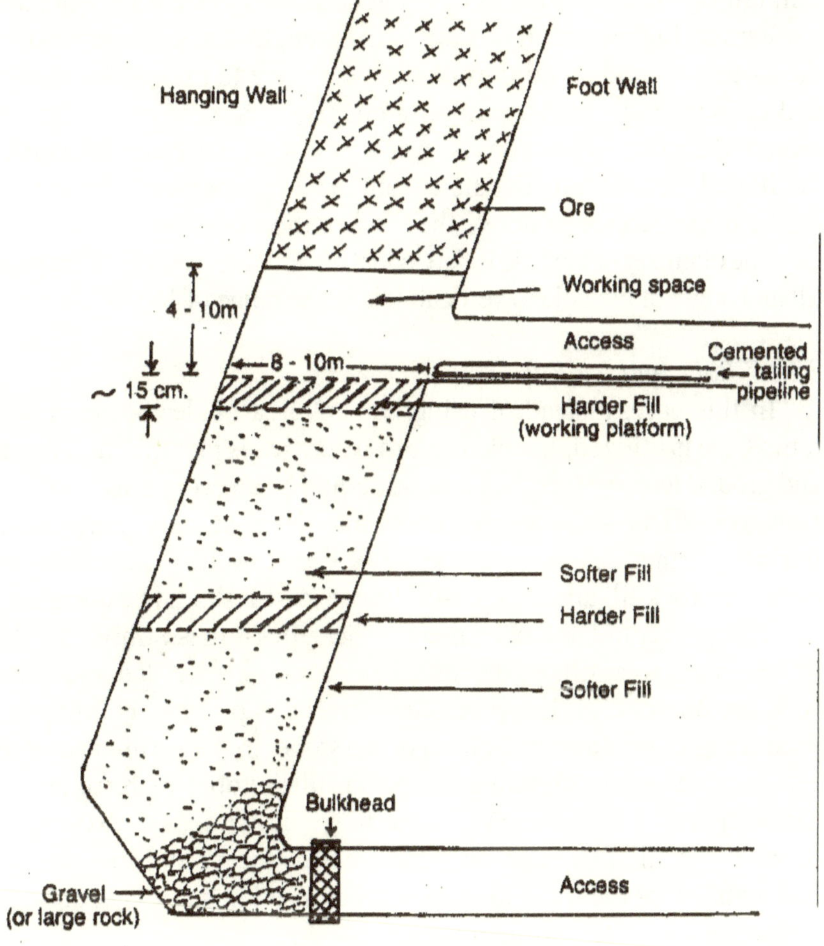

Figure 4 Schematic of an Over-cut-and-fill Placement Using Two Different Strengths of Fill Material

2.1.1.6 Placement of Cemented Paste Fill

In this case the mill tailing are water drained by a tail spinner or high speed cyclone. This will produce a paste of fine mill rejects with low water content. The paste can be pumped or dumped into the stope as it is, or cement added to it for enhanced strength. The volumetric ratio of the paste fill shows cement from 10:1 to 40:1 producing unconfined compressive strength of 0.9 to 2 MPa after 28 days of curing. The placement procedures are the same as cemented mill tailing. The advantages of using cemented paste fill are quicker hardening, higher and more uniform strength and tight support of the stope roof and hanging wall. However, due to excess dewatering and consequently higher pumping energy, it is more expensive. To reduce the pumping expenses, gravity flow of the paste fill can be facilitated through a drilled hole from the fill preparation site located on the mine surface or above the level of the stope roof.

The diameter of the hole is dependent on the water content of the paste fill and angle of the hole to vertical. It is in the range of 150 to 350 mm.

2.1.1.7 Rock and Ice Fill

In this case, instead of using cement as a binder between rock lumps, ice is utilized. In cold regions where ice is plentiful and the air and ground temperatures are low, this could be an economical method. However, air or stope refrigeration may be necessary, particularly during warmer seasons. In the latter case, overall costs between cemented rock-fill and ice rock-fill or just ice-fill should be compared.

The procedure for placement of ice rock-fill is similar to that of cemented rock-fill, as discussed in section 2.1.1.2. Ice and waste rock are dumped in the stope alternatively to provide as uniform a mixture as possible at the bottom of the stope. In the warmer seasons, refrigerated air circulation around the filled stopes or refrigeration pipes, spaced 2 to 5 m apart in the fill, will act as heat exchangers to keep the fill frozen at all times, hence increasing the strength and stability of the fill. This method was used in some Swedish and Canadian mines to a limited extent.

2.1.2 Fill Placement in Underhand-cut-and-fill

In this case, the production advances downwards and the fill is placed in the roof of the excavation. The fill used for underhand-cut-and-fill constitute over 4% cement by weight to provide the higher strength required to keep a stable working roof in the stope. The main advantages in utilizing high cement back-fill are:

(i) Better control in heavy ground and rock-burst conditions.
(ii) Higher pillar recovery due to stronger fill material, higher standing height and longer time before failure.
(iii)Quicker availability of stope to resume production, due to higher bearing capacity of the fill.

Some of the disadvantages for using high cement back-fill are:

(a) Higher cost, since 80 to 85% of total back-fill cost is due to the cost of cement.
(b) Skin burning from contact with alkaline in cement water poses hazard to the fill placement crew. These can be serious enough to require hospitalization. To reduce the above noted effects, the crew should be required to use skin protection cream, such as Kerodex, if contact with cement water is anticipated.

Placement techniques in the underhand-cut-and-fill operation are as follows:

2.1.2.1 Placement of Cemented Tailing

The fill is made with the following volumetric constituents: Mill tailing or sand = 6 to 20; Ordinary Portland cement = 1.5 to 2; and Water = 2 to 7. This mixture gives an unconfined compressive strength of 3 to 4 MPa after 28 days of curing.

There are two main techniques to place fill in the roof of the stope and these are:

(i) Flatted stringer technique in which the following steps can be taken:
    a) Stope crew clean out each entry panel, and place horizontal stringers which consist of laminated beams of 3.7 m long x 30 cm wide x 7.5 cm thick planks. In the main extraction crosscut, sometimes 30.5 x 30.5 cm wide, flange steel beams are used instead of wooden stringers. Two or more rows of horizontal stringers, 1.8 m apart, will be needed for each

stage of back-fill placement. The horizontal stringers are secured to the walls of the stope and are connected to each other by 2 m centre to centre vertical stringers, via joint plates, as shown in **Figure 5**. Cross poles of 15 to 20 cm diameter (or 20 x 20 cm square timbers) with appropriate length are then placed on top of the horizontal stringers on 1 m centers and nailed to the stringers. Then a 10 x 10 cm wire mesh screen is rolled out on top of the cross poles, and nailed or wire tied to the cross poles to keep the whole assembly from floating or moving during the fill placement. The whole assembly can be tightened and secured within one day.

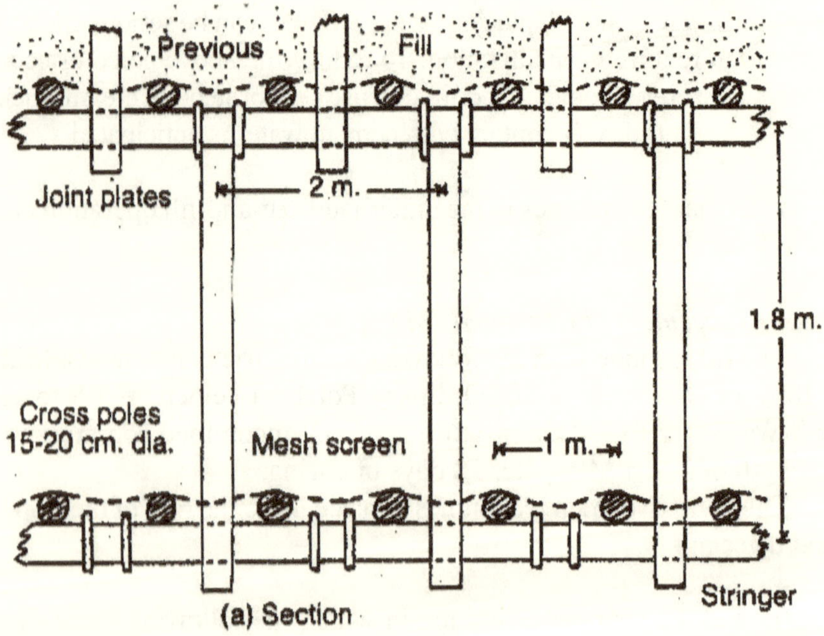

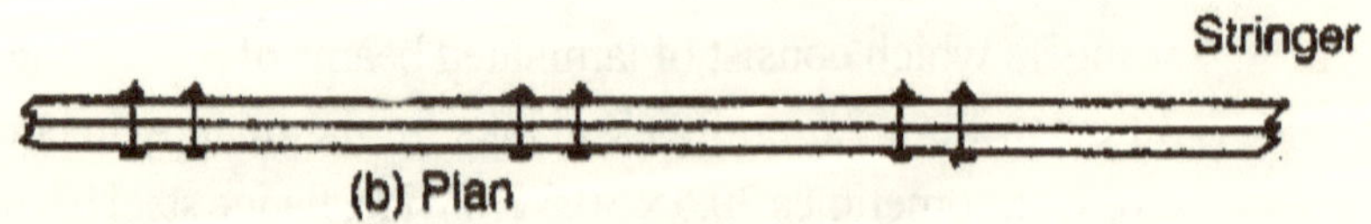

Figure 5 Detail of Stringer Technique Fill Placement for Underhand-cut-and-fill Mining [McCreedy and Hall, 1966]

b) Filter walls are constructed by placing a 25 x 25 cm centre post in the front of the panel and blocking it to the drift post in the extraction crosscut.

c) The ground seals are made from standard lagging, cut on one end to the contour of the rib. Each piece should be angled from a side post to the ground so that the placed fill pressure tends to force the lagging against the ground.

d) Burlap is to be run in a continuous line from one ground seal around the filter-wall to the other ground seal. Each piece of timber and pipe which penetrates the wall must be individually wrapped with burlap to reduce the possibility of major leak occurrence. The burlap is stapled over the wire and the sides should be constructed in the same way. If wire is not readily available, solid lagging can be used. Lagging on the front filter-wall should be staggered from one side of the centre post to the other, to allow the pore crews to drain water more continuously. Staggering of the lagging is not necessary on the sidewalls.

e) One or two workers spray a 0.3 m strip of ground adjacent to the ground seal with an undercoating emulsion and a strip of burlap stuck on it. The strip should be overlapped onto the wooden ground seal and stapled.

f) Small dams, to collect drainage water, should be built in the extraction crosscut on either side of each panel or stope. They are made of three staging 3.7 m x 30 cm x 7.5 cm boards, with a polyethylene sheet stapled to them. It is advisable to drop burlap over the plastic to prevent tearing. A small diaphragm pump should be set near each dam to handle the decant water and slimes out of the working area to a slime dump. Sometimes a stope panel has to be used temporarily as an intermediate slime dump. The difference between slime dumps and conventional water dumps is the use of a nylon mesh fabric, called 'fabrene', instead of burlap, for the slime. It retains the smaller particles better and is more durable. The water and slime is then pumped from the intermediate slime dump to the main slime dump. The recommended specification of a large slime-pump for a 182 m vertical lift on an incline in the Magma Mine, Arizona is as follows [Hall, 1978]:

- Pump horsepower = 100 hp
- Flow rate = 380 liter/min.

One person attends to the pump when it is in operation.

(ii) Scissors mat or dome technique, in which 12.5 cm flattened timbers are scissored at the mid span of the stope by a pair of bolted metal plates, is shown in **Figure 6**.

The ends of these timbers rest on clip angles, rock bolted to the side walls. The scissor timbers are held in position by a longitudinal key-block timber. A 20 cm round log is placed longitudinally from one set of scissor members to the next and the framework is covered with spaced 2.5 cm planks. The slice is then filled with the above specified cemented mix by pumping. When under-cut-and-fill mining is resumed under a scissor mat or dome, posts are not required for support. On a wide span, angle braces are run from key-block timbers to clip angles bolted to the wall. In addition, the elimination of vertical posts removes an obstruction to the scraping operation, and is a particular advantage in the final clean-up of the completed slice.

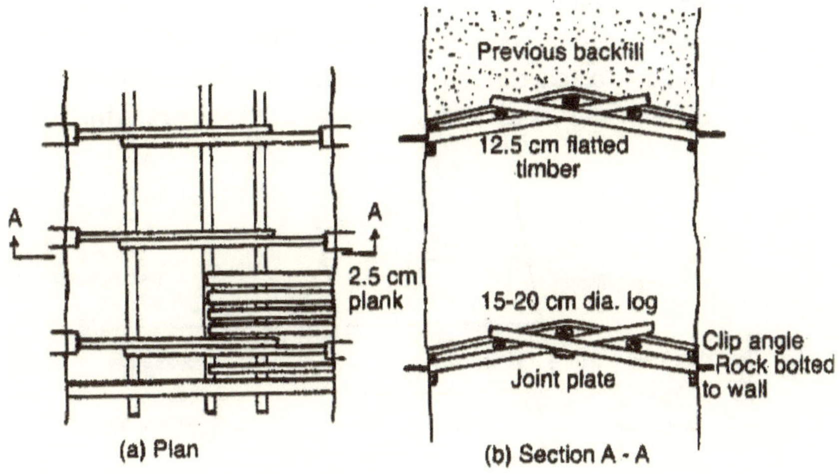

Figure 6 Detail of Scissors Mat in Underhand-cut-and-fill Stope [McCreedy and Hall, 1966]

After completion of the stingers or scissors mat, and the drainage system, the back-fill boss initiates the pouring of the fill by telephoning the fill preparation supervisor and specifying the mill tailing; cement:water ratio. It then takes 25 to 50 minutes for the grout to arrive by gravity. Pouring is to be moved from one panel to another to bring the grout up as evenly as possible in each panel. If a leak develops, grout is to be diverted to another panel until the leak is fixed.

A fill placement crew usually consists of the superintendent and three workers. One worker is usually sent to watch the slime dump; one is assigned to watch the stope pumps, and a third watches each panel for leaks. Crews are scheduled to give continuous 24 hour coverage to avoid having to interrupt an ongoing fill placement at the shift change.

2.1.2.2 Drift and Cemented Fill Technique

This technique is utilized where the stope dimensions are large and ground conditions are poor. Therefore, mining and back-filling is done in ⅓ to ½ of the stope width in sequences which is reverse of that given in section 2.1.1.4 and **Figure 3**. However, the ratio of tailing:cement:water for the fill should be increased to 6:1:2 to 6:2:7 to provide unconfirmed compressive strength of 3 to 4 MPa. The operation can be carried out either with a uniform strength fill (**Figure 7 a**), or with two different strength fills (**Figure 7 b**).

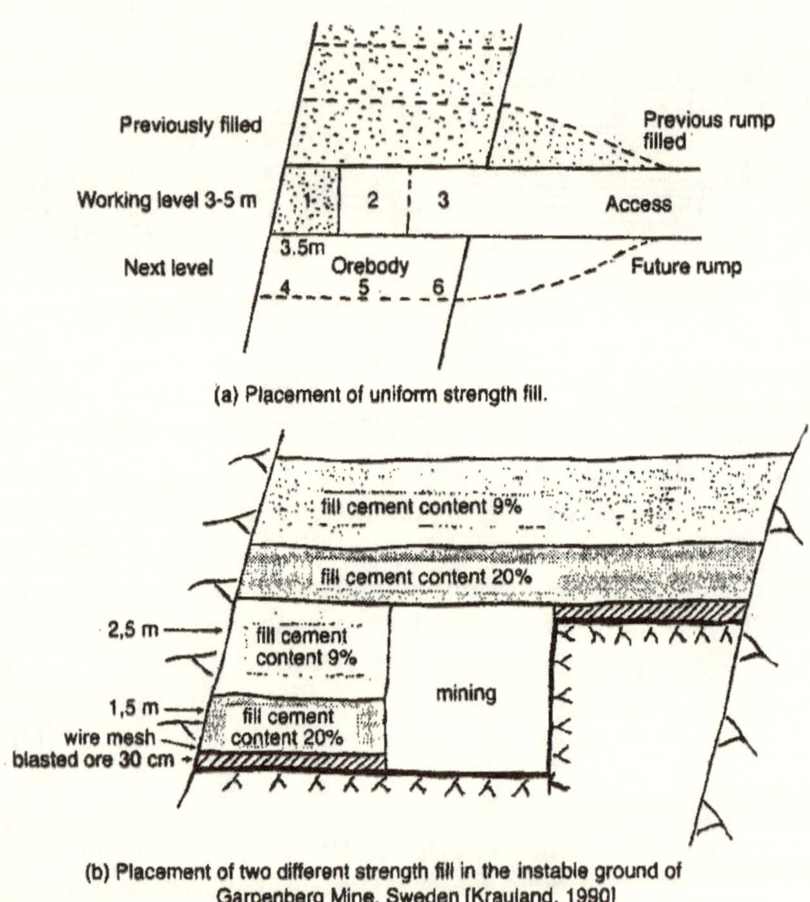

Figure 7 Drift and Fill Placement Technique in Underhand-cut-and-fill Mining.

2.1.2.3 Placement of Cemented Paste-fill

This type of fill provides higher compressive strength to 5 MPa. Due to the low viscosity of the fill, it is more expensive than cemented tailing. However, it requires a less elaborate, smaller and inexpensive drainage system, and the placed fill hardens quicker with low shrinkage, allowing better roof stability and a higher production rate. Placement assembly is as noted in section 2.1.2.1. A paste-fill pump and a cement grout pump are used to place the material through a large diameter pipe to the roof of the excavated stope for tight filling.

2.1.3 Placement of Back-fill Behind or Beside the Excavated Area

This technique is also called stowing or packing. It is practiced mainly to back-fill horizontal to vertical seam excavations such as exists in longwall, short-wall, room and pillar and longhole coal and potash mining. Modern stowing techniques are as follows:

2.1.3.1 Gravity Stowing

In which the fill placement is done by the force of gravity. This technique is economically feasible in workings steeper than 42° horizontal, which is an average internal friction angle of the fill material. The stowing material is filled by a dump truck or dozer operation and held in place by a stowing screen, wooden or metal dam.

2.1.3.2 Conveyor Stowing

This is also called mechanical stowing, in which the fill is transported by a conventional conveyor to the excavation. It is then transferred to a jet or high speed conveyor at a speed of 10 m/s and is thrown to the back of the production face, behind the support system. The jet conveyor is pulled to the faceline slowly as the stowing progresses and the production advances. A wire screen keeps the stowing away from the faceline. Since two conveyor arrangements require a height of more than 1.5 m, this technique is feasible in thick and flat seams [Biron and Arioglu, 1983].

2.1.3.3 Pneumatic Stowing

In this case, the stowing material is transported in pipes by compressed air, and thrown to the back of the face. The set up is shown in **Figure 8**. It requires smaller installation than the conveyor technique and is suitable in low seam heights. However, the mine should have an ample supply of compressed air, as the air consumption by one stowing machine is approximately equivalent to a medium sized compressor at the surface.

Capacity of the stowing machine varies with the size of the equipment, and the rock to be filled, availability of air pressure and working conditions. It is generally in the range of 70 to 150 m^3/hour at a pressure of 5 to 7 kg/cm^2. The pipes on tailing roadways are lined inside with rubber or basalt to minimize the wear. The working life of the pipes are as follows:

- Straight, lined pipes for up to 500 000 t of the fill.
- Elbows at 90 for up to 6000 t of the fill.
- Manganese steel pipes at the face for up to 300 000 t of the fill.

In the case of stoppages, the material is cut first and the air is kept blowing to eliminate settling in the pipes.

2.1.3.4 Hydraulic Stowing

In which the waste rock is mixed with water and cement, if required. It is the most advanced filling system in mining. The material is pumped into the excavation via pipeline. The assembly is similar to that noted for the cemented rock- fill in section 2.1.1.2. It requires a preparation plant at the mine surface or near the excavation to be stowed, fill pump, pipelines, sump and ditches (or channels) and water pump to direct excess water to the preparation plant for recycling (**Figure 9**). The stowing material must be within the size range of 0.1 to -80 mm, so that it is small enough to be pump-able and yet large enough not to stay in suspension in water. The best material is mine tailing, smelter slug suddenly cooled in water, crushed rock aggregates and beach or river sand from which the slime has already flushed away by the river.

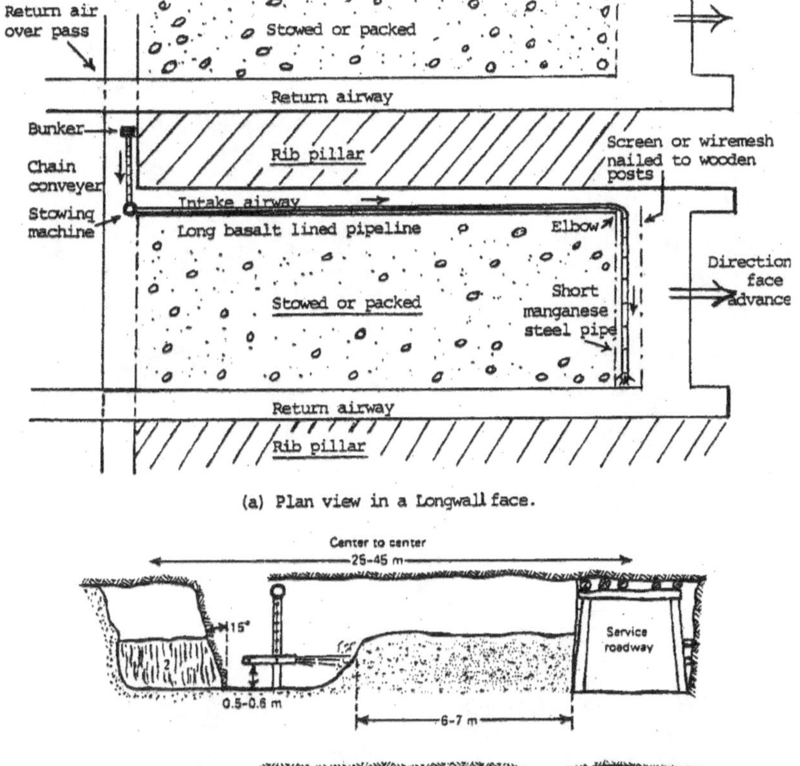

Figure 8 Schematic of Two Pneumatic Stowing Techniques.

Compressive strength of the hydraulic stowing, with sand to cement ratio of 5:1 after water is drained off is 0.5 to 2.5 MPa.

2.1.3.5 Consolidated Stowing

In this technique the fill material is consolidated by chemical or mineral additives called accelerators, such as sulphides of copper, iron, aluminum or even sulfur. These additives oxidize in the fill, raising the temperature to 60 - 70°C and cementing the material. If cement is used as a fill constituent, the additives will accelerate the rate of hardening. The amount of additives is generally 2 to 6% by weight in the fill. Unconfined compressive strength of cemented consolidated stowing, at a cement content of 10 to 20% by weight, ranges from 1.5 to 6.5 MPa after 48 hours of curing.

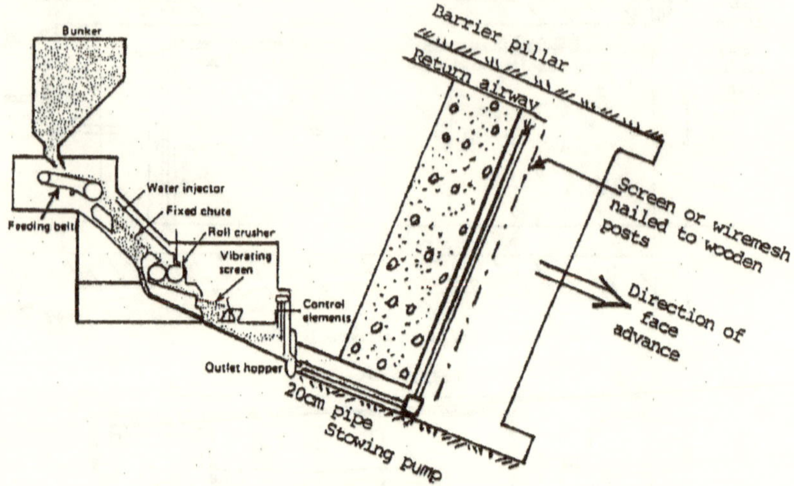

Figure 9 Schematic of a Hydraulic Stowing Assembly in Longwall Mining.

2.2 Delayed Placement of Back-fill

Fill is placed after depletion of the stope production. All open stope mining use delayed back-fill, e.g. shrinkage, block cut and fill, blasthole stoping, vertical crater retreat, sub-level stoping and top slicing. The following are fill placement techniques for some of the above noted mining methods:

2.2.1 Placement of Cemented Fill in Shrinkage Mining

*P*illar recovery with the shrinkage method of mining can be made feasible if the footwall of the shrinkage stope is rock bolted, screened if necessary, and filled with cemented hydraulic fill. A typical layout is shown in **Figure 10**. The mining block consists of 6 to 9 m wide stopes, separated by 5 to 6 m width pillars. Timbered raises, driven in the pillar are used initially for access to the stopes and finally as chutes for recovery of the pillars by underhand-cut-and-fill mining.

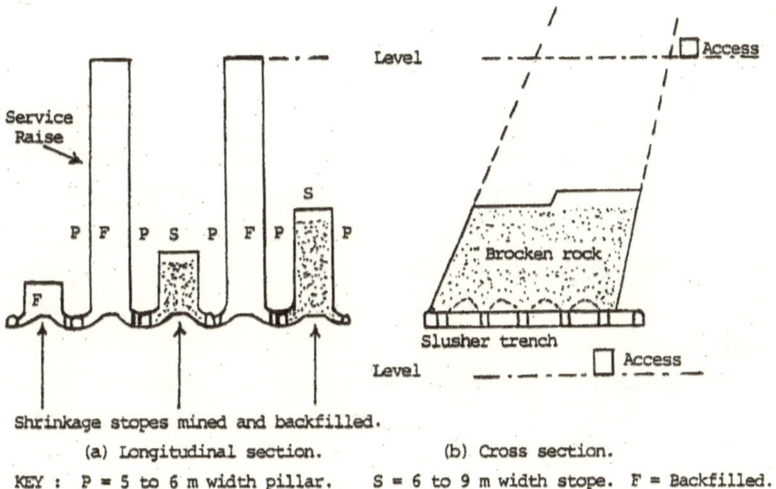

Figure 10 Placement of Cemented Fill in the Shrinkage Stopes for Pillar Recovery [Modified from McCreedy and Hall, 1966].

The sequence of the operation is as follows:

(i) Coned box-holes are excavated at the bottom of each shrinkage stope.

(ii) A mixture of cemented fill is poured to form a floor coned to the box-holes.
Volumetric constituents of the cemented fill at this stage are: Classified tailing or sand = 9 to 11 parts; Ordinary Portland cement = 1 part; and Water = 4 to 6 parts.

(iii) Shrinkage mining is then commenced. The broken ore is scraped in a slusher trench to a short pocket storage raise.

(iv) When mining and removal of the broken ore is completed, the stope is filled in one step with cemented fill having the following volumetric ratio:
Mill tailing or sand = 23 to 27 parts; Ordinary Portland cement = 1 part; and Water = 6 to 8 parts.

(v) After completion and hardening of the placed fill, the intervening shrinkage stopes are mined and filled as noted above.

(vi) Underhand-cut-and-fill mining of the pillars follows the completion of shrinkage stoping using the same technique, noted in section 2.1.2.

2.2.2 Placement of Rock and Cemented Fill in Blasthole Mining -

Sub-level drifts are driven in ore on the hanging wall at 15 to 20 m vertical intervals. Where the ore is narrow, the drifts are widened to the footwall to permit down-dip drilling.

The blasthole stopes have to be filled immediately after being depleted. A series of 1.5 m diameter raises are bored from surface down to an access 90 to 150 m from the top of the stope. From this point 3 to 4 raises are bored in a fan pattern to service the strike length of the stope. A reinforced concrete collar, measuring 3 x 4.2 m inside and 5.2 m high is to be constructed over the surface raise, with 1.2 cm thick steel wear plates used to protect the concrete from abrasion. To form a grizzly at this point, 2 to 3 I-beams are to be placed on the surface raise.

If the fill consists of only cemented tailing (or sand) without rock, the diameter of raises can be reduced 5-10 cm and lined by pipe. If rock-fill is utilized, the stope is filled with rock to a height of 9 m first, the stope walls ore is then filled with cement, having the following: Mill tailing or sand = 5 to 10 parts; Ordinary Portland cement = 1 part; and Water = 2 to 4 parts.

The above procedure is adopted to consolidate the rock-fill. The height of rock-fill is increased to another 9 m and the above sequence repeated until the stope is filled. In general, the weight ratio of rock-fill to cement mix to give unconfined compressive strength of 1.7 to 5 MPa is as follows: Rock-fill = 3 to 5, t; Cement mix = 1, t.

Once the fill has drained, mining can proceed as before in an adjacent slice.

2.2.3 Placement of Cemented Fill in Block Cut and Fill Mining -

This technique is modified from a combination of low cost blasthole stoping with overhand-cut and-fill mining. The method is illustrated in **Figure 11**. After the necessary development is completed, a 6 m high undercut is mined at the sill elevation. A timber slusher trench is then constructed in the undercut and the balance of the opening is filled with cemented fill to within approximately 1.2 m of the stope back. Box-holes are developed in the fill from the slusher trench. When mining is completed in the block, it is filled with the following volumetric mixture and the cycle is repeated in the next block above: Mill tailing or sand = 25 to 35 parts; Ordinary Portland cement = 1 part; and Water = 7 to 12 parts.

Success of the above note placement techniques are dependent upon:

(i) Sufficient void and permeability in the rock fill, ie the larger the rock size the better is the grout penetration to consolidate the back-fill.
(ii) Quality control in the rock-fill and grout placement and their segregation properties. The coarser material tends to gravitate to the walls of the stope while the finer material remains in the centre.
(iii)Cost of the fill material and placement. This item varies widely from mine to mine. The cost of waste rock-placement and mill tailing (or sand) placement in Thompson Mine, Manitoba, Canada, is given in **Table 1**.

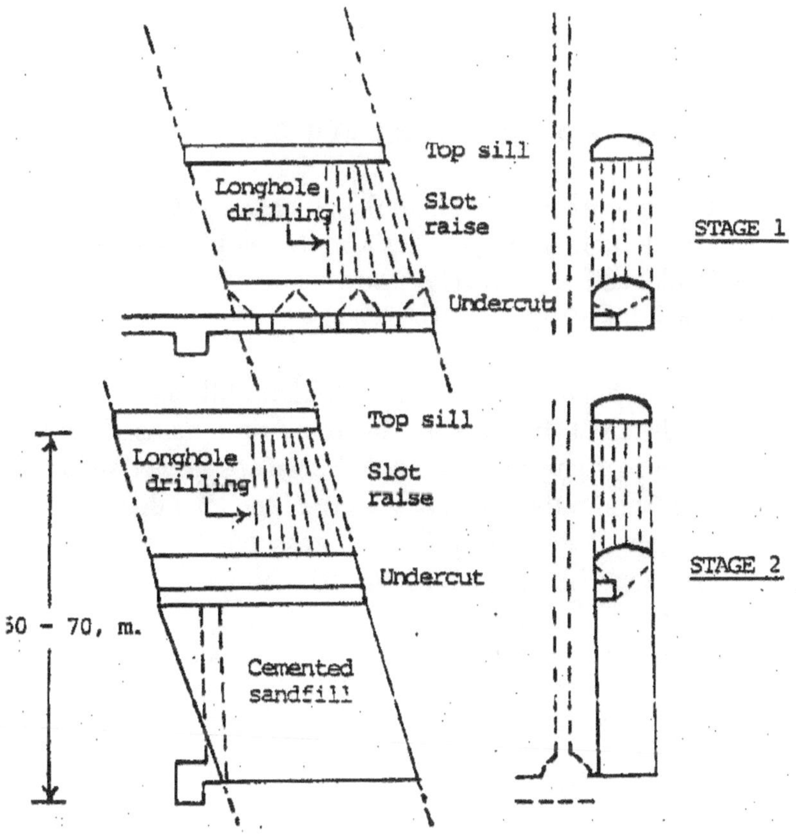

Figure 11 Schematic of a Block Cut and Fill Mining and Cemented Fill Placement [McCreedy and Hall, 1966]

Table 1 Cost of Waste Rock vs Mill Tailing (or sand) from Thompson Mine, Manitoba, Canada [Kerr, 1978]

Case A - Waste Rock	**Fill Cost Per Imperial Ton of Ore (cents)**
Fill Pass Development to 1500 Level at $65 per foot	20
Rock Handling on Surface	15
Ripping in Winter	15
Summer Cost per Imperial Ton of Ore	35
Winter Cost per Imperial Ton of Ore	50
Case B - Mill Sand Place Hydraulically	**Fill Cost Per Imperial Ton of Ore (cents)**
Cycloning and Filtering at Thompson Mill	12
Trucking to Birchtree	14
Sand Handling at Birchtree	15
Sand-placing Labor	15
Fill Preparation - Concrete Bulkheads and Drainage	15
TOTAL COST	71

3. MECHANICS OF PLACED BACKFILL

Large stresses may build up in the pillars and walls of the stope before the back-fill in the stope is sufficiently compacted and hardened to provide adequate support. The following cases may occur in mines utilizing back-fill:

(i) The mechanical properties of pillars, fill, and stope walls are different. In this case, the more rigid material will take a greater share of the load and develops higher stress, eg a more rigid pillar will carry more of the load than a yielding fill, thereby developing higher stress than the back-fill.

(ii) The stope is filled with high density or modulus material, having mechanical properties near that of the stope walls and pillars. This causes a greater portion of the load to be carried by the fill, resulting in a reduction of stress on the supporting pillars. Therefore, the stress zone around the stope will be smaller.

The most desirable mechanical properties of a fill material would be to exhibit stress- strain characteristics similar to those of rocks surrounding the fill in a stope. Concrete is the nearest to this ideal material. However, it is extremely expensive to fill the mine stopes with concrete, and it does not help to dispose the waste rock underground. An alternative is to use cemented rock fill or cemented tailing. Unconfined compressive strength (σc) of 90 different cemented wastes worldwide, after 28 days of curing, fall within 0.1 to 10 MPa [Swan, 1985].

α_p = specific surface of the aggregate particles, assuming spherical

= $6k[N]_2/[N]_3$ (1)

k = aggregate shape normalizing factor = 1.7 to 1.9.

$[N]_i$= integral between u_o to u_m of ui . Pu du (2)

P_u = a function describing the cumulative percentage of aggregate particles passing for a particle diameter u = 40 to 60%.

u_o and u_m are lower and upper limits of particle diameters, respectively. These are chosen as: $u_o = u_{10}$, and $u_m = u_{96}$, thereby avoiding uncertainty in the extreme values of the particle size distribution.

To convert the volumetric content of cement (cw) the following expression can be used:

$$C_W = C_V \cdot Pc/ Pbd \quad (3)$$

where: Pc = cement density = 30.9, kN/m^3.

Pbd = back-fill bulk density, $kN/m^3 = (1-p)Ps$ (4)

p = back-fill porosity = $95\ C_\upsilon/\sigma_c$ (5)

σ_c = unconfined compressive strength of back-fill, kPa.

Ps = density of solids in the back-fill, kN/m^3.

Critical standing height of back-fill (Hc) can be determined as follows [modified from Terzaghi and Peck, 1948]: $H_C = 0.1\ S_O \cdot N \cdot \phi / \gamma$, m (6)

where: S_O = Cohesive strength of the fill in placed, kPa.

γ = weight per unit volume of the fill in placed, kN/m^3.

ϕ = internal friction angle of the fill in placed, degrees.

$N\phi = \tan^2 [45 + (\phi /2)]$ (7)

4. SOLIDS LOSSES

Solids losses occur when the back-fill particles are suspended in the water drain through the bulkhead, after placement. The losses depend on:

(i) Size distribution of the particles reaching the bulkhead. In general, solids losses α 1/particle size.
(ii) Type of back-fill and its placement method.
(iii)Rate of water drainage from the fill α solids loss.
(iv) Placement density of the fill α 1/solids losses.
(v) Porosity, permeability and cracks present in the bulkhead, or back-fill containment.
(vi) Use of flocculent in the fill. It accelerates the natural settlement rate of the fill particles before they reach the bulkhead. However, control of the optimum flocculent dosage is difficult accompanied with added cost.

5. ARCHING

When fill is pored in a stope, much of its weight is transferred by friction to the stope walls. This causes the fill to arch across the stope, as shown in **Figure 12**. In the case of underhand back-filling (**Figure 12a**) arching appears in the form of a depression situated approximately in the middle of the back-fill floor. Whereas, in the case of overhand back-filling (**Figure 12b**) the fill mat or bulkhead, placed as the working roof, does not have to carry the full weight of the fill above after it is drained and set. This is due to the formation of a natural arch when the fill becomes self-supporting. In this case only a small portion of the weight of the overlying fill below the natural arch is carried by the fill mat and bulkhead.

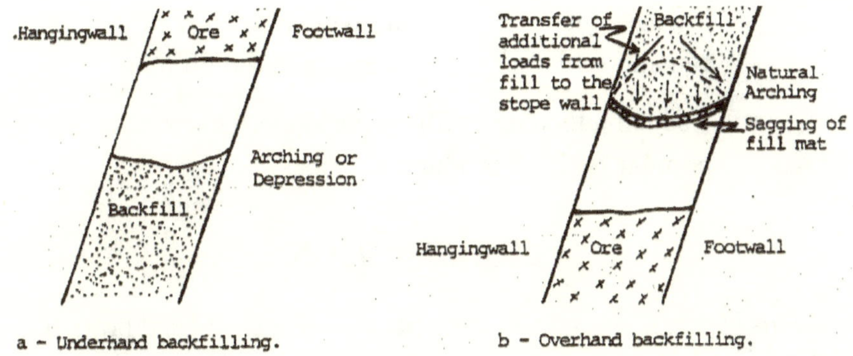

Figure 12 Arching Effect after Placement of the Back-fill.

6. FILL CONSOLIDATION OR SHRINKAGE

This occurs when large quantities of water are drained from the hydraulic fill after placement. This process is most apparent in stopes dipping less than 15° to horizontal [Ontario MHSTP, 1983]. Shrinkage will develop a gap or enlarge the existing gap between the back-fill and the hanging wall. It is undesirable due to causing reduction in the supporting capability to the hanging wall and the stope walls. Shrinkage can be reduced or alleviated by one or a combination of the following measures:

(i) Eliminating excess water through increasing the fill slurry density.

(ii) Increasing the water drainage rate during the fill placement. This action will minimize the quantity of the water left in the fill after placement, which in turn minimizes the fill shrinkage. The drainage can be increased by:

(a) Using large sizes of well-classified-back-fill to provide larger permeability and reducing to 10 micron size fractions. This action will also cause higher velocity of water flow which will wash away the cement from the fill mix, if cemented fill is used. Therefore, the fill permeability should be optimized.

(b) Increasing the drainage area or points, especially adjacent to the hanging wall, while decreasing the water flow rate from individual drainage point. This technique will increase the total stope drainage at a given time, while causing settlement of the cement in the fill mix.

(c) Reducing the rate of filling the stope. This method is not popular for high production stopes.

7. LIQUEFACTION

This is a state when solid mass behaves like liquid and flows like water. Liquefaction is a serious hazard to uncemented fill in the stope. Conditions that may result in liquefaction of the fill are as follows:

(i) In fully saturated back-fill where all voids are filled with water.
(ii) In cohesion-less uncemented fill, especially when the bulkhead fails, removing the constraints on the fill, or promoting flow of the fill.
(iii) Vibration due to nearby blasting, rock-burst, or earthquake.

To reduce or prevent liquefaction, the following steps can be taken:

(i) Desliming (classifying) the tailing so that fines are removed and the percolation rate of the fill is improved.
(ii) Adding cement or other consolidating material to the back-fill. Once the fill sets, about 1 to 3 days after placement, chances of liquefaction diminishes.
(iii) Use of paste fill or fill with high pulp density. These actions reduce quantity of water to be removed for quick hardening.
(iv) Ensuring good drainage during the fill placement so that the fill does not remain water logged or fully saturated. An efficient technique to facilitate good drainage is to install a drain tower made of fine mesh wire, PVC or nylon pipes at strategic points in the stope.
(v) Designing stronger bulkheads where possibility of liquefaction exists.
(vi) Utilizing a plug or cemented fill near bulkhead, when uncemented fill is to be placed in the stope.

8. PROBLEMS IN BACKFILL PLACEMENT

The main problems can be classified by one or a combination of the following:

(i) Excessive viscosity of the back-fill due to low slurry density, high flocculent in the slurry, and small particle sizes. These will cause water from the pipeline, loss of solids through the bulkhead and higher pressure on the bulkhead increasing possibility of its failure.

(ii) Rupturing of the bulkhead during or immediately after the fill placement due to inadequate containment for the hydraulic head, insufficient safety factor in designing the bulkhead under dynamic conditions, inherent weakness in construction of the bulkhead, and external damage to the bulkhead due to mining operations.

(iii) Uniformity of the fill constituents and quality control during placement due to inadequate design and procedure for the fill preparation and transportation, segregation of fill during placement, solids losses after placement, infrequent monitoring of the fill material during placement, and inadequate supervision.

(iv) Gaps between back-fill and stope hanging wall or sill pillars (stope roof) caused by incomplete filling of the stope, insufficient constraint of the bulkhead to prevent bulging, large surface pool remaining after placement due to low fill density, and self-weight consolidation of the fill after placement.

9. SOLUTION TO THE PROBLEMS

Identification of the placement problems and causes for the specific site outlined above will indicate where the shortcomings are and modification or correction requirements. This aims at reviewing fill material, mining conditions, fill placement procedure and quality control. Naturally, the solutions should be site specific and their adoption to various sites should consider appropriate modification. The following recommendations are helpful:

(i) Increase the slurry density as economically and practically as possible. A placement density of 1.7 to 1.8 t m^{-3} is satisfactory; higher densities are better.
(ii) Keep the fine particles (-10 microns) in the fill below 4% of total fill weight.
(iii) Monitor the fill quality and consistency during placement on a regular basis and take account of segregation of the fill constituents in the stopes.
(iv) Fill the stopes as tight as possible.
(v) Facilitate adequate water drainage.
(vi) Account for the compaction or shrinkage of the fill after placement, and fill the stope again with high density material after the shrinkage.

REFERENCES

1. BIRON, C and ARIOGLU, E (1983). Design of Supports in Mines, J Wiley and Sons, pp 215-223.
2. GRICE, A G (1990). Fill Research at Mount Isa Mines Ltd, International Symposium on Back-fill, Montreal, Canada, 8 pps.
3. HALL, B (1978). Magma's Sand-fill System as Employed at the Magma Mine, Superior, Arizona, Mining with Back-fill - State of the art review, pp 75-83.
4. KERR, R W (1978). The Fill System for 108 Blasthole Stope at Inco's Birchtree Mine, Mining with Back-fill - State of the art review, pp 16-20.
5. KRAWLAND, N (1990). Developments in Sweden of the Rock Mechanics in Cut and Mining. International Symposium on Back-fill, Montreal, Canada, 10 pps.
6. MCCREEDY, J and HALL, R J (1966). Cemented Sand-fill at Inco. CIM Bulletin, Canada, July, pp 888-892.
7. Ontario Mining, Health and Safety Training Program (1983). Mine Backfill, Module 23, 44 pps.
8. STOUT, K S (1975). Soil and Rock Mechanics Illustrated. Bul. 96, Montana Bureau of Mines and Geology, Butte, Montana, USA, pp 135-140.
9. SWAN, G (1985). A New Approach to Cemented Back-fill Design. CLM Bulletin, Canada.
10. TERZAGHI, D and PECK, R B (1948). Soil Mechanics in Engineering Practice. J Wiley and Sons, Inc, New York.
11. THOMAS, E G (1973). A Review of Cementing Agents for Hydraulic Fill. Jubilee Symposium on Mine filling, Australian Inst. of Mining and Metallurgy, Mount Isa, pp 65-75.

Appendix

DETERMINATION OF VARIOUS SUPPORT REQUIREMENTS USING ROCK MASS QUALITY (Q)1.16

The support recommendations listed in Tables 1.17 to 1.20 were derived from the description of permanent roof support given in the numerous case records. The methods of estimating permanent wall support and temporary support that are summarized in this appendix are unlikely to give as reliable an estimate of support as that for permanent roof support. However, in the feasibility and planning stages, estimates of permanent wall support and temporary support also play a part in the cost predictions, so some form of support estimation is required. In the excavation stage of a project the estimates of roof support can continue to serve as a useful guide to actual practice. However, at this stage the less reliable wall support estimate should be critically reviewed. The temporary support will be largely in the competent hands of the engineer in charge at the mine.

1. PERMANENT WALL SUPPORT

An approximate rule of thumb for estimating wall support in medium rock conditions is to use 1.5 times the roof bolt spacing (= approx. half the support pressure) and ⅔ times the thickness of roof shotcrete. However in difficult rock conditions the wall (and invert) support may need to be similar to that of the roof arch. Conversely, in very favorable conditions there may be no need for any general wall support. Exceptions to these general assumptions may be encountered in the case of high walls. Special support might be required to stabilize deep-seated wedges.

An empirical method of modifying the roof support estimates is to multiply the rock mass quality, Q, by a factor which ranges in value from one to five. The resulting wall factor, Q_w, is used in place of Q for determining wall support from Figure 1.6 and Tables 1.17 to 1.20.

Range of Q	Wall factor Q_w
Q > 10	5.0 Q
0.1 < Q < 10	2.5 Q
Q < 0.1	1.0 Q

The equivalent dimension axis of Figure 1.6 is evaluated in terms of the total excavation height for the case of wall support (Height/ Excavation Support Ratio (ESR) ≠ wall height/ESR). The worked examples given in this appendix illustrate the above method.

2. TEMPORARY SUPPORT (FOR FEASIBILITY AND PLANNING ONLY)

The method of modifying the estimates to permanent support to take care of temporary support is to select a support category (box numbers 1 to 38, Figure 1.6) closer to the "no support" diagonal given in Figure 1.6. It has been found from trial and error that the following modifications to Q and ESR give reasonable estimates:[1.16]

1. Increase ESR to 1.5 x ESR
2. Increase Q to 5Q (for roof arch)
3. Increase Q_w to $5Q_w$ (for walls)

These factors are applied equally to both roof and walls such that any differences in the permanent roof and wall support will also be in operation for temporary support. The worked examples given in this appendix illustrate the method.

3. RECOMMENDED BOLT AND ANCHOR LENGTHS

Bolt and anchor lengths for permanent support depend on the dimensions of the excavations. Lengths used in the roof arch are usually related to the span, while lengths used in the walls are usually related to the height of the excavations. The ratio of bolt length to span tends to reduce as the span increases.[1.18] Accordingly, the following recommendations are given as a simple rule of thumb, to be modified as *in situ* conditions demand:[1.16]

For roof: bolts $L = 2 + 0.15\ B/ESR$
anchors $L = 0.40\ B/ESR$

For walls: bolts $L = 2 + 0.15\ H/ESR$
anchors $L = 0.35\ H/ESR$

where: L = length, m
B = span, m

H = excavation height, m

ESR = excavation support ratio

(Bolt lengths used as temporary support will usually be only loosely dependent on excavation dimensions. Lengths of between 1.5 and 3.0 m seem to be used in many types of excavations.)

4. WORKED EXAMPLES

Two hypothetical examples are given[1.16] to illustrate the various stages of the method outlined in this appendix. It is assumed that estimations of permanent and temporary support are required for a machine hall of 20-m span, and a tail-race tunnel of 9-m span, both to be excavated in the same phyllitic rock mass. It is assumed that the estimates are required for the planning stage of a project. At this stage the following geotechnical information has been produced: surface mapping and bore core analysis; rock stress estimates; and rock compression tests.

4.1 Rock Mass Classification

Joint set 1 Strongly developed foliation likely to act as fully developed joint set

Smooth, planar: $J_r = 1.0$

Chloride coatings: $J_a = 4.0$

Ca. 15 joints/m

Joint set 2 Smooth, undulating: $J_r = 2$

Slightly altered walls: $J_a = 2$

Ca. 5 joints/m

$J_v = 15 + 5 = 20$, RQD = 50 (from Equation 1.3)

$J_n = 4$

Most unfavorable $J_r/J_a = 1/4$

Minor water inflows: $J_w = 1.0$
Unconfined compression strength of phyllite: $\sigma_c = 400$ kg/cm^2
Major principal stress: $\sigma_1 = 30$ kg/cm^2
Minor principal stress: $\sigma_3 = 10$ kg/cm^2

The following are the virgin stress levels:

$(\sigma_1/\sigma_3) = 3$

$\sigma_c/\sigma_1 = 13.3$ (medium stress), SRF = 1.0

Q = × ¼ × = 3.1 (poor) (from Equation 1.4)

4.2 Estimates for 20-m Span Machine Hall

I. For permanent support
Type of excavation: machine hall B = 20 m, and H = 30 m (ESR = 1.0) B/ESR = 20 and H/ESR = 30
A. Roof
Q = 3.1: Category 23 (from Figure 1.6)
Table 9: B(tg) 1.4 m + S(mr) 15 cm (Notes II, IV, VII)

B. Walls

Q_w = 3.1 × 2.5: Category 20 (from Figure 1.6)

Table 1.18: B(tg) 1.7 m + S(mr) 10 cm (Notes II, IV)

C. Mean length of bolts and anchors:

1. For Roof
   a. Bolts = 5.0 m
   b. Anchors = 8.0 m
2. For Walls
   a. Bolts = 6.5 m
   b. Anchors = 10.5 m

II. For temporary support:

B/1.5 × ESR = 13.3, H/1.5 ×ESR = 20

A. For Roof

"Q" = 3.1 × 5: Category 14 (from Figure 1.6)

Table 1.17: B(utg) 1.6 m + clm (Notes I, III)

B. For Walls

"Q" = (3.1 × 2.5) × 5: Category 14 (from Figure 1.6)

Table 1.16: B(utg) 2.0 m (Notes I, III)

4.3 Estimates for 9-m Span Tail-race Tunnel

I. For permanent support:

Type of excavation: tail-race tunnel B = 9 m, H = 9 m

(ESR = 1.6), B/ESR = H/ESR = 5.6

A. Roof : Q = 3.1: Category 21 (from Figure 1.6)

Table 1.18: B(utg) 1.0 m + S 2-3 cm (Note I)

B. Walls: Q_w = 3.1 × 2.5: Category 17 (from Figure 1.6)

Table 1.18: B(utg) 1.4 m (Note I)

C. Mean length of bolts: 1. Roof 2.9 m

2. Walls 2.9 m

II. For temporary support:

A. Roof: "Q" = 3.1 × 5: Category 0 (no support)

B. Walls "Q_w" = (3.1 × 2.5) × 5: Category 0 (no support)

Terminology

Active reinforcement — Reinforcing element which is prestressed or artificially tensioned in the rock mass when installed.

Admissible load of a cable bolt — Determined by the upper bound of its bearing capacity, computed or ascertained during tests with subtraction of a safety margin.

Cable bolt — Reinforcing element, made of steel wires, strands or ropes which are layed to strand or rope configuration and installed in the rock mass as passive or active ground reinforcement. It is secured inside the rock mass with resin or cement grouting.

Cement grout — A pumpable thin slurry consisting primarily of a mixture of cement, sand, and water injected into rock formation through boreholes.

Ground control — Arresting or minimizing deformation and closure of the strata around an excavation.

Ground reinforcement — Installation of any type of element, such as cable bolt, rock bolt, doweling, resin, and cement grouting, inside the rock mass to increase its inherent stability. It forms part of the rock mass.

Ground stabilization — Combined application of ground reinforcement and ground support to prevent failure of the rock mass.

Ground support — Installation of any type of engineering structure around or inside an excavation, such as steel sets, wooden cribs, timbers, concrete blocks, or lining, which will increase its stability. This type of support is external to the rock mass.

Grouting — Injecting cement or chemical into the rock mass, usually through a borehole drilled into the rock to be reinforced.

Grout injector — A machine that mixes the dry ingredients for a grout with water or other liquid chemicals and injects it, under pressure, into a grout hole.

Passive reinforcement — Reinforcing element which is not prestressed or tensioned artificially in the rock, when installed. It is sometimes called rock dowel.

Permanent cable bolt — Has a service life of more than 2 years, with higher safety demands than the temporary cable bolt.

Pillar — A column or area of coal or ore left to support the overlaying strata or hangingwall in mines.

Pre-reinforcement — Installation of reinforcement in a rockmass before excavation commences.

Raveling — Progressive release of small loose blocks from a rock surface.

Rock bolt — Reinforcing element, made of solid or tube formed steel, installed in the rock mass as passive or active ground reinforcement. It is secured inside the rock mass with or without resin or cement grouting.

Rock mass — *In situ* rock, composed of various pieces the dimensions of which are limited by discontinuities.

Rock mass failure — Loosening or fallout of rock from the rock mass into the surrounding excavation.

Safety factor of cable bolt — Ratio of the limit load or limit deformation load of the cable bolt, to its admissible or working load.

Stope — An excavation from which ore has been excavated in a series of steps.

Temporary cable bolt — Has a service life of less than 2 years.

Testing load of cable — Short-term loading to which the test cable is subjected in order to check the quality of its manufacture and its maximum load.

Uprooting or bearing capacity of a cable bolt — That load under which the resistance of any functional part of the cable bolt system (i.e., ground, cable bolt, grout, and slab or plate, etc.) fails and the cable bolt ceases to function.

Working cable bolt — Fulfills a static function in the overall reinforcement of the ground.

Working load of cable bolt — The force which the cable should be capable of transmitting continuously throughout its service life.

Symbols

Definition	Symbol	Unit
Length	l	m
Width	b	m
Thickness	d	m
Radius	r	m
Diameter	d	m
Height or depth	h	m
Time	t	sec or min
Area	A	m^2
Volume	V	m^3
Mass	M	kg
Weight	W	kN
Density	γ	kN/m^3
Force	F	kN
Moment	Nm	kN·m
Work	W	kN·m
Pressure	P	kPa
Stress	σ	kPa
Strain	ϵ	μ cm/cm
Gravitational acceleration	g	m/sec^2
Energy	E	KW· h

Efficiency	η	%
Poisson's ratio	ν	—
Modulus of elasticity	E	MPa
Modulus of deformation	D	MPa
Shear modulus	G	MPa
Bulk modulus	K	MPa
Compressive strength	σ_c	MPa
Tensile strength	σ_t	MPa
Shear strength	τ	MPa
Cohesion	S_0	MPa
Internal friction angle	ϕ	degree
Coefficient of friction	μ	—
Void Ratio	e	—
Porosity	n	—
Pore pressure	u	kPa
Proportionality symbol	$\propto$	

Prefixes

Prefix	**Symbol**	**Unit**
Meter	m	1
Decimeter	d	10^{-1} m
Centimeter	cm	10^{-2} m
Millimeter	moment	10^{-3}
Micro	μ	10^{-6}
Mono	n	10^{-9}
Pico	p	10^{-12}
Femto	f	10^{-15}
Atto	a	10^{-18}
Kilo	k	10^{3}
Mega	M	10^{6}
Giga	G	10^{9}
Tera	T	10^{12}
Gram	g	
Pascal	Pa	N/m^2
Newton	N	$kg \cdot m/sec^2$
Day	d	
Hour	h	
Minute	min	
Second	sec	
Tonne	T	10^{3} kg

Conversion Factors

Length	1 m	= 39.37 in	1 in.	= 2.54 cm
		= 3.28 ft	1 ft.	= 0.30 m
		= 1.09 yd	1 yd	= 0.91 m
	1 km	= 0.62 mile	1 mile	= 1.61 km
Area	1 m^2	= 10.76 ft^2	1 ft^2	= 929.03 cm^2
		= 1.20 yd^2	1 yd^2	= 0.84 m^2
		= 0.01 acre	1 $mile^2$	= 2.589 km^2
	1 km^2	= 247.11 acre	1 acre	= 4046.86 m^2
		= 0.39 $mile^2$	1 hectare	= 400 acres
				= 10000 m^2
Volume	1 liter	= 61.02 in^3	1 ft^3	= 28.32 liters
		= 0.22 U.K. gallon	1 U.K. gallon	= 4.55 liters
		= 0.26 U.S. gallon	1 U.S. gallon	= 3.79 liters
	1 m^3	= 1.32 yd^3	1 yd^3	= 0.76 m^3
Mass	1 kg	= 2.20 lb	1 lb	= 453.49 g
		= 35.27 oz	1 oz	= 28.35 g
		= 0.98 U.K. ton	1 U.K. ton	= 1016.5 kg
		= 1.10 U.S. ton	1 U.S. ton	= 907.19 kg
Density	1 kg/m^3	= 0.06 lb/ft^3	1 lb/ft^3	= 16.02 kg/m^3
		= 0.98 lb/in^3	1 lb/in^3	= 27679.9 kg/m^3
Force	1 N	= 1 kg·m/sec^2	1 kg·f	= 9.81 N
		= 0.22 lb·f	1 lb·f	= 4.45 N

Pressure or Stress	1 Pa	= 1 N/m²	1 bar	= 100 kPa
		= 0.01 bar	1 at	= 98.07 kPa
		= 0.01 kb /cm² (at)	1 psi	= 6.89 kPa
	1 kPa	= 0.145 psi		= 0.07 kg·f/cm²
		= 760 mm Hg	1 mmHg	= 1.32 Pa
Energy	1 Joule	= 1 N·M	1 Wh	= 3600 J
		= 0.24 Cal	1 Cal	= 4.19 J
		= 1 W/sec	1Btu	= 1055.06 J
Moment	1.36Nm	= 0.14 kg·f·m	1 ft·lb·f	= 1.36
			1 lb·f·ft	= 1.36 Nm
Power	1 W	= 1 J/sec		
Velocity	1 m/sec	= 3.28 ft/sec	1 ft/sec	= 0.31 m/sec
		= 0.04 mile/min	1 mile/min	= 88 ft/sec
		= 2.24 mile/h	1 mile/h	= 1.61 km/h
Frequency	1cm/sec	= 1 Hz		

Index

D

E

F

G

I

O

P

Q

R

T

www.ingramcontent.com/pod-product-compliance
Lightning Source LLC
Chambersburg PA
CBHW020717180526
45163CB00001B/8

* 9 7 8 1 5 0 3 5 2 9 4 8 9 *